KT-460-545

# Chemistry in Context

# Chemistry in Context

**Graham C. Hill** MA (Cantab)

Deputy Headmaster, Dr Challoner's Grammar School, Amersham, formerly Head of Science, Bristol Grammar School

**John S. Holman** MA (Cantab)

Head of Science, Watford Grammar School

Nelson

To Elizabeth and Wendy

Thomas Nelson and Sons Ltd
Nelson House Mayfield Road
Walton-on-Thames Surrey KT12 5PL
PO Box 18123 Nairobi Kenya
Watson Estate Block A 13 Floor
Watson Road Causeway Bay Hong Kong
116-D JTC Factory Building
Lorong 3 Geylang Square Singapore 14

Thomas Nelson Australia Pty Ltd
19–39 Jeffcott Street West Melbourne Victoria 3003

Nelson Canada Ltd
81 Curlew Drive Don Mills Ontario M3A 2R1

Thomas Nelson (Nigeria) Ltd
8 Ilupeju Bypass PMB 21303 Ikeja Lagos

© G. C. Hill, J. S. Holman 1978
First Published 1978
Reprinted 1979 twice, 1981
ISBN 0 17 448061 X
NCN 2594 21 4

All Rights Reserved. No part of this
publication may be reproduced, stored in a
retrieval system, or transmitted, in any form
or by any means, electronic, mechanical,
photocopying, recording or otherwise,
without the prior permission of Thomas
Nelson and Sons Limited.

Art direction by Kevin Mangan
Diagrams by Colin Rattray and Hedgehog Design
Layout by Phil Kay
Phototypeset by Tradespools Ltd.,
Frome, Somerset
Printed in Great Britain by
Butler & Tanner Ltd.,
Frome & London

# Preface

This book is about chemistry and chemists and the contribution they make towards science, industry and society. It is designed primarily for students studying chemistry to Advanced level. In planning the book, we have been considerably influenced by the Nuffield 'A' level chemistry project (with which one of the authors was associated) and by the newer and updated syllabuses recently introduced by the examining boards. We have not written specifically for any one syllabus but hope that the book will be helpful to students studying for all 'A' level examinations. Between us we have worked in various capacities with several examination boards including the University of Cambridge Local Examinations Syndicate, the Joint Matriculation Board, the University of London Examinations Board, the Associated Examining Board and the Southern Universities Joint Board.

Man's interest in chemistry is essentially a practical one. He is interested in the way materials in the universe behave and react under different conditions and the ways in which he can use these materials for his own purposes. Complementary to this practical approach is a logical scientific method of working, the discovery of laws to summarise information and the development of ideas and theories to interpret and explain observations.

With these important points in mind, we have tried to ensure that *Chemistry in Context* fulfils certain important criteria. In a single volume, we have presented chemistry as a unified and integrated subject using physical principles as a basis for the inorganic and organic sections which follow. Our aim has been to show chemistry in its wider context as a relevant and developing science which makes an essential contribution towards society, industry and civilization. We have tried to show, both in the text and by carefully chosen photographs, how chemical knowledge and chemical industry have influenced human life and society.

The text is designed for use with the present modern syllabuses incorporating the best of traditional courses and the best of recent innovations. It draws on experimental evidence in establishing laws and theories. Throughout the text, we have relied heavily on experimental results and data in developing key ideas and concepts. The book is intended to complement a course of practical work and consequently space is not devoted to instructions or suggestions for experiments.

We have also tried to ensure that *Chemistry in Context* will be suitable for use by A-level students of all abilities, not just the most able and we have borne in mind that many students starting an A-level course have little experience of using advanced texts. Consequently, each chapter is divided into fairly short sections and concludes with a summary of the more important facts and ideas. In addition, there are frequent questions in the text which will encourage the student to read more carefully and critically. It is important that an effort is made to answer these questions, as they often form an integral part of the text.

The study questions at the end of each chapter vary in style and may be used in tests, in class discussion, in revision or for homework. Many of these questions are designed to test understanding and application rather than factual recall.

Considerable care has been taken over the layout of each page so as to present the subject clearly and attractively using figures, tables, photographs, charts, flow diagrams and graphs to reinforce the text.

The first three chapters begin by revising and extending key ideas concerning the mole, redox and periodicity. The next ten chapters (4–13) develop the major themes of structure, bonding, energetics, periodicity and competition which can then be used throughout the following chapters of inorganic chemistry (14–19). Here, emphasis is placed on patterns and on the use of the periodic table and the activity series to summarise information. Four chapters (20–23) on equilibria and reaction rates complete the development of principles and illustrate the importance of these factors in industrial processes. The book concludes with nine chapters (24–32) of organic chemistry. This is developed systematically in terms of functional groups, and the patterns of behaviour are related to structure, kinetics and energetics. Simple ideas of mechanism are introduced in the treatment of organic chemistry.

The order in which topics are presented is not intended as an ideal learning or teaching sequence. It is simply a convenient classification of topics within a book. Different teachers will wish to cover the material in different ways according to the interest, ability and aptitude of their students.

At present, there is no universal agreement regarding nomenclature, units and abbreviations. Throughout the text, we have been guided by the Royal Society publication (*Symbols, Signs and Abbreviations Recommended for British Scientific Publications*, London 1969) and have used the International System of Units (SI) with very few exceptions. Consequently, we have expressed volumes in cubic decimetres ($dm^3$) rather than litres and concentrations are given as moles per cubic decimetre rather than moles per litre. Similarly, we have abandoned the Ångstrom unit and used the SI alternative, the nanometre (nm), in describing atomic radii, bond lengths and unit cell lengths.

Units of pressure are completely altered by SI recommendations. In general, we have used the SI unit of pressure which is a newton per square metre ($N\ m^{-2}$), sometimes called a pascal (Pa), although atmospheres (atm) have been used in making comparisons with atmospheric pressure and in expressing the units for $K_p$. In a similar fashion, we have used degrees centigrade (°C) to express temperature when this has seemed more suitable or more appropriate than kelvins (K).

The SI unit for the quantity referred to by international agreement as 'amount of substance' is 'mol'. Unfortunately, the word 'molecule' is often abbreviated to 'mol.' and to avoid ambiguity we have used the unit mole rather than mol for amount of substance. Thus, the units of concentrations are given as mole $dm^{-3}$ not mol $dm^{-3}$, and those for molar enthalpies are written as kJ $mole^{-1}$ not kJ $mol^{-1}$.

Modern nomenclature has been used throughout the book, following recommendations in the report *Chemical Nomenclature, Symbols and Terminology for use in School Science* published by the Association for Science Education. In general, organic chemicals are named systematically following I.U.P.A.C. Rules, though trivial names of common substances, such as ethanoic acid and phenylamine, are sometimes bracketed after the systematic name. In the case of inorganic compounds, our only major departure from ASE recommendations has been to omit the oxidation numbers of elements with invariable oxidation numbers. Hence, NaCl is named sodium chloride rather than sodium(I) chloride and $CaSO_4$ is named calcium sulphate(VI) not calcium(II) sulphate(VI).

Many friends and colleagues have influenced us in planning and writing *Chemistry in Context*. Countless students at different schools have influenced our styles of teaching and the way in which topics are introduced and developed in the book.

Our sincere thanks must go to Professor Malcolm Frazer who read the entire manuscript and suggested many improvements in both content and style. At a later stage, we were fortunate in the efficient handling of the manuscript by our publishers, and the competent and sympathetic organization of the entire operation by Miss Elizabeth Johnston.

Finally, we must thank our wives, Elizabeth and Wendy for their patience, help and encouragement at all times.

Graham Hill
John Holman
February 1978

# Contents

# Atoms, Atomic Masses and Moles 1

## 1.1 Atoms

The first chemist to use the name '**atom**' for the smallest particle of an element was John Dalton (1766–1844). Dalton used the idea of atoms to explain how elements could react together to form **molecules** which he called 'compound atoms'.

*Far left*. John Dalton (1766–1844), the quiet genius. Dalton was born in the remote village of Eaglesfield in Cumberland, the son of a handloom weaver. For most of his life, Dalton lived and worked in Manchester at what was then the Presbyterian College.

*Left*. Dalton's symbols for the elements. In 1803, Dalton published his Atomic Theory. He suggested that all matter was composed of small particles which he called 'atoms'. Later, Dalton went on to suggest symbols for the atoms of different elements and these are shown in the photograph. What do we now call 'Azote'? Which substances in Dalton's list are not elements, but compounds?

Do not confuse the terms atom and molecule. *An atom is the smallest part of an element which can ever exist, whereas a molecule is the smallest part of an element or compound which can exist alone under ordinary conditions.*

Thus, chlorine gas consists of molecules of $Cl_2$ under ordinary conditions, but at very high temperatures these molecules will split up to form chlorine atoms, Cl.

- ○ Name an element (other than chlorine) whose molecules consist of two atoms under normal conditions.
- ○ Name an element whose molecules consist of one atom under normal conditions.
- ○ Is it possible to have an atom of a compound?

Most atoms have a radius of about $10^{-10}$m (0.1nm). At one time, the unit used in measuring atomic distances was the Ångström (1Å = $10^{-10}$m = 0.1nm). Nowadays, the unit most frequently chosen is the nanometre (nm). 1 nanometre is equivalent to 10 Ångström units.

Atoms, of course, are far too small to be seen even with the most powerful light microscope. However, chemists have discovered that it is possible to pick out single atoms of certain elements. In 1958, scientists in the USSR reported that they could pick out atoms of barium (which have a diameter of about 0.4nm) using an electron microscope with a magnification of 2 000 000. In 1967, Japanese scientists provided evidence for their identification of particles as small as 0.1nm.

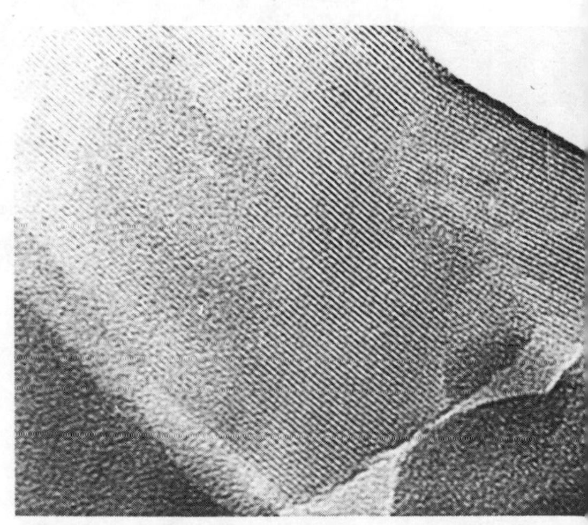

An electronmicrograph of copper phthalocyanine, a complex copper compound. The parallel dark lines indicate rows of copper atoms which have been magnified 583 330 times.

Aston's mass spectrometer. A vaporized sample of the element was bombarded by electrons in the large glass bulb on the left. The ions produced were then accelerated by an electric field towards the magnetic field on the right (produced by the many coils of the electromagnet).

## 1.2 Comparing the masses of atoms

Atoms are too small to be weighed. However, Dalton had suggested early in the nineteenth century that it should be possible *to compare* the masses of atoms even though the atoms themselves could not be weighed. During the nineteenth century various methods were devised for obtaining approximate values for the relative masses of different atoms.

It was not until Aston invented the mass spectrometer in 1919 that chemists had a reliable and accurate method of comparing the relative masses of atoms. At one time, the relative masses of atoms were known as atomic weights, but nowadays we refer to them as **relative atomic masses**.

The basic idea of a mass spectrometer can be demonstrated using the apparatus in figure 1.1. Wooden balls of a *different* size, but with *identical* iron cores, roll down a sloping plane. At the bottom of the slope, a powerful magnet attracts the iron cores and the moving balls are deflected.

○ Does the magnet attract each ball the same?
○ Which size of ball will be deflected the most? Why?

As the balls have identical iron cores, they are all attracted equally by the magnet, but the smaller balls are lighter and are therefore deflected the most. The balls collect in different compartments depending on their mass and all balls of the same mass will collect in the same compartment. Using this simple apparatus, it is possible to separate the different-sized balls according to their mass and to find the relative numbers of each present.

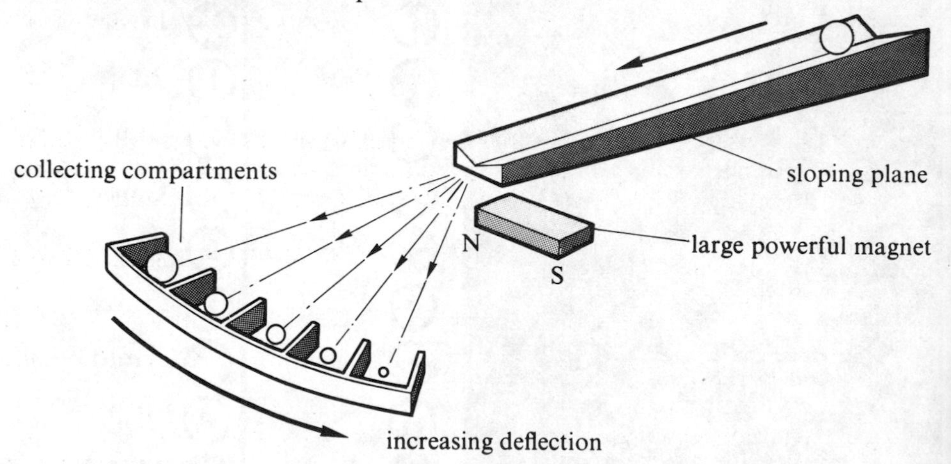


**Figure 1.1** A simple model to illustrate the working of a mass spectrometer.

In a similar fashion, a real mass spectrometer separates atoms according to their mass and shows the relative amounts of the different atoms present. Before the atoms can be deflected and separated by a magnetic and/or, an electric field they must be converted to their positively-charged ions.

Figure 1.2 shows a simple mass spectrometer. There are five main stages.

1 The sample of element is vaporized.
2 Positive ions are obtained from the vapour.
3 The positive ions are accelerated by an electric field.
4 The positive ions are deflected by a magnetic field.
5 The ions are detected and a record is made.

A stream of the vaporized element enters the ionization chamber. Gases, liquids and volatile solids can be injected directly into the instrument just before the ionization chamber. Less volatile solids must be pre-heated and vaporized. Positive ions are formed by bombarding the neutral atoms in the vapour with a stream of high-energy electrons. One (or occasionally two) electrons are knocked out of the atoms leaving positive ions. Thus, for an atom X we have:

$$e^- + X \rightarrow X^+ + e^- + e^-$$

(high-energy electron) + (atom) → (positive ion) + (electron knocked out of X) + (high-energy electron retreating)

or occasionally $e^- + X \rightarrow X^{2+} + \underbrace{e^- + e^-}_{\text{2 electrons knocked out of X}} + e^-$

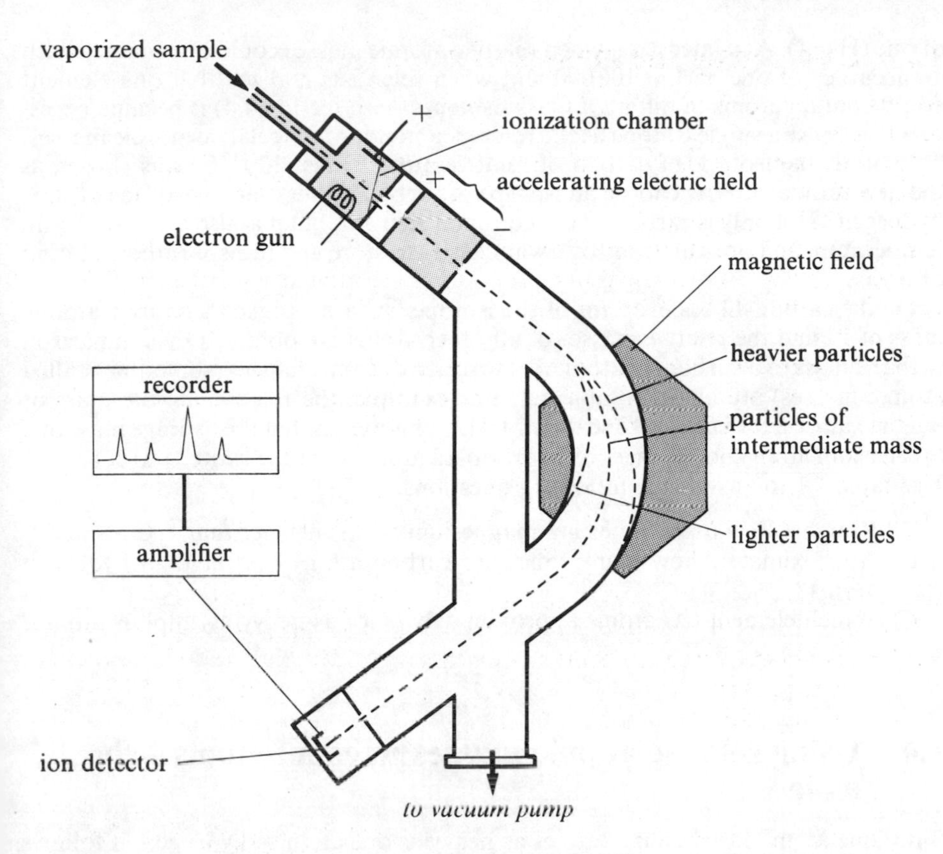


Figure 1.2 A diagram of a mass spectrometer.

These positive ions, such as $X^+$ and $X^{2+}$, now pass through holes in parallel plates to which an electric field is applied. The electric field accelerates the ions which then enter the magnetic field where they are deflected according to their mass and their charge.

For one particular size of the accelerating electric field and the deflecting magnetic field, only those ions of one particular mass/charge ratio will hit the ion detector at the end of the apparatus. Ions of smaller mass/charge ratio will be deflected too much, ions of greater mass/charge ratio will be deflected too little. The ion detector is usually linked through an amplifier to a recorder. As the strength of the magnetic field is slowly increased, ions of increasing mass will be detected and a mass spectrum similar to that shown in figure 1.3 is traced out by the recorder. The mass of a particular ion can be calculated from the values of the magnetic and electric fields and the relative heights of the peaks in the mass spectrum give a measure of the relative proportions of the different ions present*. In practice, a reference peak using a known substance is first obtained on the mass spectrum and the relative masses of other particles can then be obtained by comparison with this. Look closely at figure 1.3.

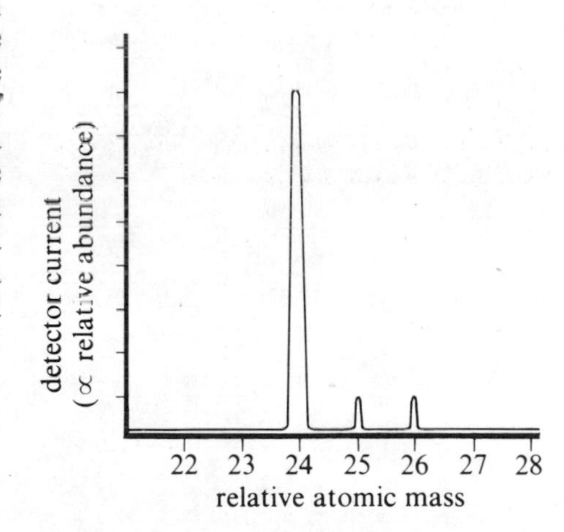


Figure 1.3 A mass spectrometer trace for naturally occurring magnesium.

- ○ How many different ions are detected in the mass spectrum of naturally-occurring magnesium?
- ○ What are the relative masses of these different ions?
- ○ What are the relative proportions of these different ions?

The interpretation of mass spectrometer traces and the calculation of relative atomic masses is taken up again in sections 4.5 and 24.4.

## 1.3 Relative atomic masses: the relative atomic mass scale

Chemists use a **relative atomic mass scale** to compare the masses of different atoms. This scale is sometimes called the **relative atomic weight scale**. At first the element hydrogen was chosen as the standard against which the masses of other atoms were compared. Hydrogen was chosen initially because chemists realized that it had the smallest atoms which could conveniently be assigned a relative atomic mass

* Strictly speaking, it is the areas under the peaks and not the peak heights which give the relative abundancies.

of one (H = 1). At a later date, when relative atomic masses could be obtained with an accuracy of one part in 100 000 and when scientists realized that one element could contain atoms of different mass (**isotopes**—see section 4.4) it became necessary to choose a single isotope as the reference standard for relative atomic masses.

In 1961, the isotope of carbon of relative atomic mass 12 ($^{12}C$) was chosen as the new standard. Why choose an isotope of carbon rather than an isotope of, say, hydrogen? Not only is carbon a very common element, but it is also a solid at room temperature and pressure and somewhat easier to store and transport than a liquid or a gas.

**Table 1.1** The relative atomic masses of some elements.

| Element | Symbol | Relative atomic mass |
|---|---|---|
| Carbon-12 | $^{12}C$ | 12.000 |
| Carbon | C | 12.011 |
| Chlorine | Cl | 35.453 |
| Copper | Cu | 63.540 |
| Hydrogen | H | 1.008 |
| Iron | Fe | 55.847 |
| Magnesium | Mg | 24.312 |
| Sulphur | S | 32.064 |

On the carbon-12 scale, atoms of the isotope $^{12}C$ are assigned a relative atomic mass of 12 and the relative masses of all other atoms are obtained by comparison with the mass of a carbon-12 atom. A few relative atomic masses (sometimes called atomic masses) are listed in table 1.1. For example, the relative atomic mass of magnesium, on the carbon-12 scale is 24.312. This means that the average mass of a magnesium atom and the mass of a carbon-12 atom are in the ratio 24.312:12.000. Use table 1.1 to answer the following questions.

- ○ Roughly how many times are magnesium atoms heavier than $^{12}C$ atoms?
- ○ Approximately how many times are carbon atoms heavier than hydrogen atoms?
- ○ Which element has atoms approximately twice as heavy as sulphur atoms?

## 1.4 Using relative atomic masses to count atoms—the mole

Since one atom of carbon is 12 times as heavy as one atom of hydrogen, it follows that 12 grams of carbon and one gram of hydrogen contain the same number of atoms. In the same way, since an atom of sulphur is 32 times as heavy as an atom of hydrogen, it follows that 32 grams of sulphur will also contain the same number of atoms as one gram of hydrogen. In fact, *the relative atomic mass in grams of all elements will contain the same number of atoms*. Experiments show that this number is $6.023 \times 10^{23}$; it is called the **Avogadro constant**. At one time, this number was referred to as the Avogadro number, but the term Avogadro constant (rather than number) is now used since its value is constant for all elements.

The Avogadro constant, represented by the symbol $L$, is now defined as *the number of atoms in exactly 12 grams of* $^{12}C$.

Since 12g of carbon contains $6.0 \times 10^{23}$ atoms,

$$1\text{g of carbon contains } \frac{6.0 \times 10^{23}}{12} \text{ atoms, and}$$

$$5\text{g of carbon contains } 5 \times \frac{6.0 \times 10^{23}}{12} \text{ atoms.}$$

In this way, it is easy for chemists to count the number of atoms in a sample of an element by weighing. Chemists are not the only people who 'count by weighing'.

Chemists are not the only people who 'count by weighing'.

Bank clerks use the same idea when they count money (silver or bronze) by weighing it. Since 100 pennies weigh 356 grams, it is quicker to take one hundred 1p coins by weighing out 356 grams of them than by counting.

- ○ What is the mass of $6 \times 10^{23}$ atoms of copper?
- ○ What is the mass of $6 \times 10^{24}$ atoms of magnesium?
- ○ How many atoms are there in 6g of carbon?
- ○ How many atoms are there in 20g of hydrogen?
- ○ How heavy is one atom of carbon?

Since the atomic mass in grams of all elements contains $6.0 \times 10^{23}$ atoms, chemists refer to $6.0 \times 10^{23}$ atoms of an element as one **mole** of atoms and they use the term mole to mean *the relative atomic mass in grams.* In fact, the mole is defined as the amount of substance which contains the same number of particles (atoms, ions or molecules) as there are atoms in exactly 12 grams of $^{12}C$. Thus, one mole of any substance contains $6 \times 10^{23}$ particles.

In equations, the symbol for an element can be used to represent one mole of the element as well as one particle of the element. Thus, C represents one mole of carbon atoms (12g of carbon) and one-tenth C represents one-tenth of a mole of carbon atoms (1.2g of carbon). O represents one mole of oxygen atoms (16g of oxygen) but $O_2$ represents one mole of oxygen molecules (32g of oxygen).

Notice how important it is to specify precisely which particles you mean in discussing the number of moles of different substances. For example, the statement 'one mole of oxygen' is ambiguous. It could mean one mole of oxygen atoms (O), i.e. 16g; or it could mean one mole of oxygen molecules ($O_2$), i.e. 32g. Because of such ambiguity, it is important to state the formula of the substance involved.

In considering giant structures such as sodium chloride, $Na^+Cl^-$, it is again important to quote the formula in discussing amounts of substance. Thus, one mole of sodium chloride, $Na^+Cl^-$ is the formula weight in grams, i.e. 58.5g. It contains one mole of sodium ions, $Na^+$(23.0g) and one mole of chloride ions, $Cl^-$(35.5g). On the other hand, one mole of magnesium chloride, $Mg^{2+}(Cl^-)_2$ (95.3g) contains one mole of $Mg^{2+}$ ions (24.3g) and two moles of $Cl^-$ ions ($2 \times 35.5$g).

- ○ What is the mass of one mole of hydrogen molecules, $H_2$?
- ○ What is the mass of one mole of sulphur dioxide, $SO_2$?
- ○ What is the mass of 0.5 moles of $SO_2$?
- ○ How many moles of $SO_2$ are there in 16g of $SO_2$?
- ○ How many molecules of $SO_2$ are there in 16g of $SO_2$?

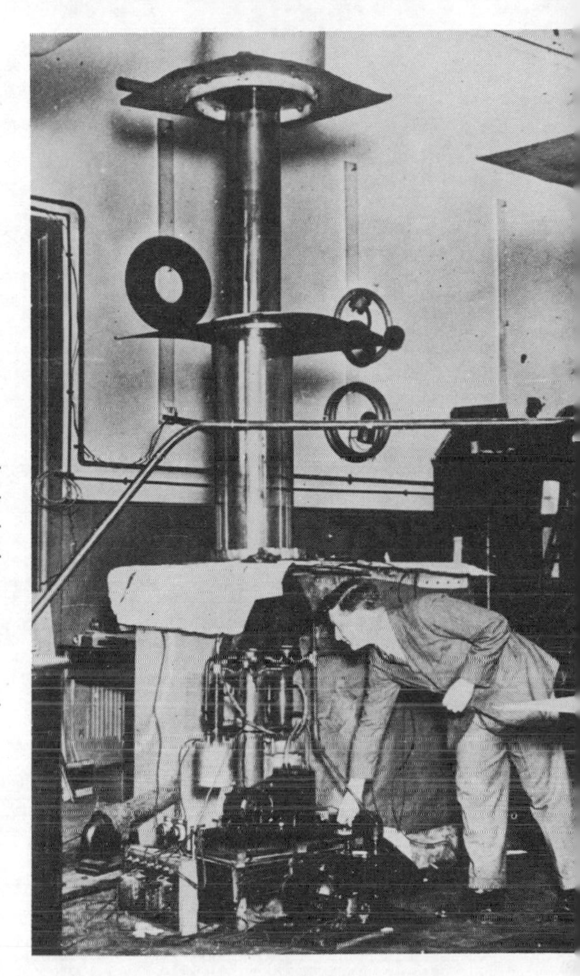

Sir John Cockcroft beside the apparatus with which he and Dr. E. T. Walton first 'split the atom' in 1932. Protons were accelerated down the large central tube towards a lithium target. The bombarding protons split some of the lithium atoms into helium.

From this discussion you will see that it is more sensible for chemists to consider amounts which contain the same number of particles (atoms, ions, molecules or formula units) rather than amounts which contain the same number of grams. Thus, in making iron(II) sulphide from iron and sulphur ($Fe + S \rightarrow FeS$), it is necessary to weigh out 56 grams of Fe and 32 grams of S (and not 44g of Fe and 44g of S) to make one mole of FeS (88g).

## 1.5 Using relative atomic masses to find the formulae of compounds

*The formula of a substance shows the relative number of atoms of each element present in the compound.* The masses of different elements present in a sample of the compound can be used together with atomic masses to find its formula.

The following example shows how the formula of a compound can be obtained after its composition has been determined by experiment.

When ethene is analysed it is found to contain 85.72% carbon and 14.28% hydrogen.

Ratio of the masses of carbon: hydrogen in ethene
$= 85.72:14.28$

$\Rightarrow$ Ratio of the moles of C: H in ethene

$$= \frac{85.72}{\text{mass of 1 mole of C}} : \frac{14.28}{\text{mass of 1 mole of H}}$$

$$= \frac{85.72}{12} : \frac{14.28}{1}$$

$$= 7.14:14.28$$

$$= 1:2$$

$\Rightarrow$ Ratio of atoms of C: H in ethene
$= 1:2$

This suggests that the formula of ethene is $CH_2$.

This formula for ethene shows only the simplest ratio of carbon atoms to hydrogen atoms. The actual formula showing the correct number of carbon atoms and hydrogen atoms in one molecule of ethene could be $CH_2$, $C_2H_4$, $C_3H_6$, $C_4H_8$, etc since all these formulae give $CH_2$ as the simplest ratio of atoms.

The simplest formula for a substance, such as $CH_2$ for ethene, is called the **empirical formula**. This shows the simplest whole number ratio for the atoms of different elements in the compound. In the case of molecular compounds such as ethene, it is generally better to use the **molecular formula**. This shows the actual number of atoms of the different elements in one molecule of the compound. The molecular formula for ethene is not $CH_2$ but $C_2H_4$.

In substances composed of giant lattices, whether these are giant covalent structures, such as silicon(IV) oxide, $SiO_2$, or giant ionic structures, such as sodium chloride, $Na^+Cl^-$, it is of course meaningless to talk of molecules and molecular formulae. In this case, the formulae we use are empirical formulae. Experimental methods of determining the structures and formulae of compounds are described in sections 7.5, 9.2, 9.3, 10.10 and 24.4.

## 1.6 Solutions and moles

When two solutions react together there is usually a simple ratio between the number of moles of solute in one solution which react with a given number of moles of solute in the other solution. For this reason, it is convenient to represent the concentration of solute in a solution in terms of the number of moles of it in a particular volume of solution. Thus, chemists measure the concentration of solutions in terms of **molarity**.

**A molar solution (1.0M)** *contains one mole of solute in one cubic decimetre* (litre) of solution.* Its molarity is one. For example, a 1.0M solution of HCl(aq) contains 36.5g of HCl in $1dm^3$ of solution. A 2.0M solution of HCl(aq) (molarity = 2.0) contains 73g ($2 \times 36.5$g) in $1dm^3$ of solution. Notice that a molar solution contains one mole of solute in $1dm^3$ of *solution*, *not* in $1dm^3$ of *solvent*.

- ○ How many moles of $H_2SO_4$ are there in 9.8g?
- ○ What is the molarity of the sulphuric(VI) acid produced when 9.8g of $H_2SO_4$ are dissolved in $1dm^3$ of solution?
- ○ How many moles of NaCl are there in $5dm^3$ of a 1.0M solution?
- ○ How many moles of NaCl are there in $200cm^3$ of 2.0M solution?

* The cubic decimetre ($dm^3$) is now the agreed international name for the unit of volume also known as the litre. In this book, we shall use cubic decimetres (not litres) and cubic centimetres, $cm^3$ (not millilitres, ml).

## Summary

1 An atom is the smallest part of an element which can ever exist.
2 A molecule is the smallest part of an element or a compound which can exist alone under ordinary conditions.
3 Most atoms have a radius between 0.07 and 0.20 nanometres (nm). 1 nanometre ≡ $10^{-9}$m ≡ 10Å
4 Chemists use the relative atomic mass scale to compare the masses of different atoms. Atoms of the isotope $^{12}C$ are assigned a relative atomic mass of 12 and the relative masses of all other atoms are obtained by comparison with the mass of a carbon-12 atom.
5 The relative atomic mass in grams of any element contains $6 \times 10^{23}$ atoms.
6 The relative molecular mass in grams of any compound contains $6 \times 10^{23}$ molecules.
7 The Avogadro constant ($6.0 \times 10^{23}$) is defined as the number of atoms in exactly 12g of $^{12}C$.
8 The mole is the amount of substance which contains the same number of particles (atoms, ions, molecules or formula units) as there are atoms in exactly 12g of $^{12}C$.
9 An empirical formula shows the simplest whole number ratio for the atoms of different elements in a compound.
10 A molecular formula shows the actual number of atoms of the different elements in one molecule of a compound.
11 A molar solution contains one mole of solute in 1dm³ of solution.

## Study questions

1 What is the mass of
(a) $6 \times 10^{23}$ atoms of O,
(b) $6 \times 10^{23}$ atoms of P,
(c) $6 \times 10^{23}$ molecules of $O_2$,
(d) $6 \times 10^{23}$ molecules of $P_4$,
(e) one mole of carbon dioxide, $CO_2$,
(f) two moles of silver, Ag,
(g) 0.2 moles of sulphur dioxide, $SO_2$,
(h) NaOH in 2dm³ of 1.5M solution?

2 How many moles of
(a) $Cl_2$ are there in 7.1g of chlorine,
(b) $CaCO_3$ are there in 10.0g of calcium carbonate,
(c) Ag are there in 10.8g of silver,
(d) $NH_3$ are there in 3.4g of ammonia,
(e) S are there in 32g of sulphur,
(f) $S_8$ are there in 32g of sulphur,
(g) NaOH are there in 1dm³ of 3.0M solution,
(h) NaOH are there in 20cm³ of 0.1M solution?

3 How many atoms are there in
(a) two moles of iron, Fe,
(b) 0.1 moles of sulphur, S,
(c) 18g of water, $H_2O$,
(d) 0.44g of carbon dioxide, $CO_2$?

4 Read section 1.2 again.
(a) Why must a mass spectrometer be evacuated to a very low pressure before being used?
(b) How would the accelerating field differ in its effect on $X^+$ and $X^{2+}$?
(c) How would the deflecting magnetic field differ in its effect on $X^+$ and $X^{2+}$?
(d) Atomic masses are not really masses. What are they?
(e) What units do atomic masses have?

5 15.3g of element X (X = 27) will combine with 13.6g of oxygen to form an oxide.
(a) Express this in moles.
(b) How many moles of oxygen, O, will combine with one mole of X?
(c) What is the simplest formula for the oxide?

6 5.34g of a salt of formula $M_2SO_4$ (where M is a metal) were dissolved in water. The sulphate(VI) ion was precipitated by adding excess barium chloride solution when 4.66g of barium sulphate(VI) ($BaSO_4$) were obtained.
(a) How many moles of sulphate(VI) ion were precipitated as barium sulphate(VI)?
(b) How many moles of $M_2SO_4$ were in the solution?
(c) What is the formula mass of $M_2SO_4$?
(d) What is the atomic mass of M?
(e) Use a table of relative atomic masses to identify M.

7 2.4g of a compound of carbon, hydrogen and oxygen gave, on combustion, 3.52g of $CO_2$ and 1.44g of $H_2O$. The relative molecular mass of the compound was found to be 60.
(a) What are the masses of carbon, hydrogen and oxygen in 2.4g of the compound?
(b) What are the empirical and molecular formulae of the compound?

8 $25cm^3$ of a solution of NaOH required $28cm^3$ of M $H_2SO_4$ to neutralize it.
(a) Write the equation for the reaction.
(b) How many moles of $H_2SO_4$ were needed?
(c) How many moles of NaOH were thus neutralized?
(d) How many moles of NaOH are there in $25cm^3$ of solution?
(e) What is the molarity of the NaOH?

9 10g of an impure iron(II) salt were dissolved in water and made up to $200cm^3$ of solution. $20cm^3$ of this solution, acidified with dilute sulphuric(VI) acid required $25cm^3$ of 0.04M $KMnO_4$(aq) before a faint pink colour appeared.
(a) Write a balanced ionic equation (or half equations) for the reaction of acidified manganate(VII) (permanganate) ions with iron(II) ions.
(b) How many moles of iron(II) ions react with one mole of $MnO_4^-$ ions?
(c) How many moles of $Fe^{2+}$ react with $25cm^3$ of 0.04M $KMnO_4$(aq)?
(d) How many grams of $Fe^{2+}$ are there in the $200cm^3$ of original solution? (Fe = 56)
(e) Calculate the % by mass of iron in the impure iron(II) salt.

10 Figure 1.4(*a*) shows the mass spectrum of HCl. The peak at mass 36 corresponds to the molecular ion, $(H^{35}Cl)^+$.
(a) What particle is responsible for the prominent peak at mass 38?
(b) What particles are responsible for the two lower peaks?
(c) How do you explain the relative heights of the peaks at mass 36 and 38?
Figure 1.4(*b*) shows the mass spectrum of methane. The peak at mass 16 corresponds to the molecular ion $(CH_4)^+$.
(d) How do you explain the peaks of relative mass 1, 2, 12, 13, 14, 15, and 17?

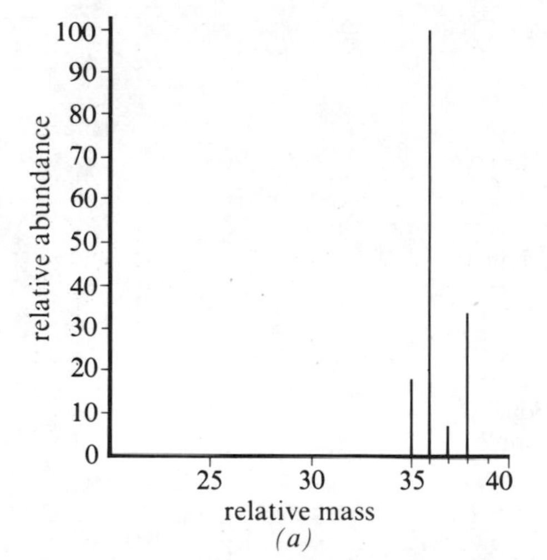


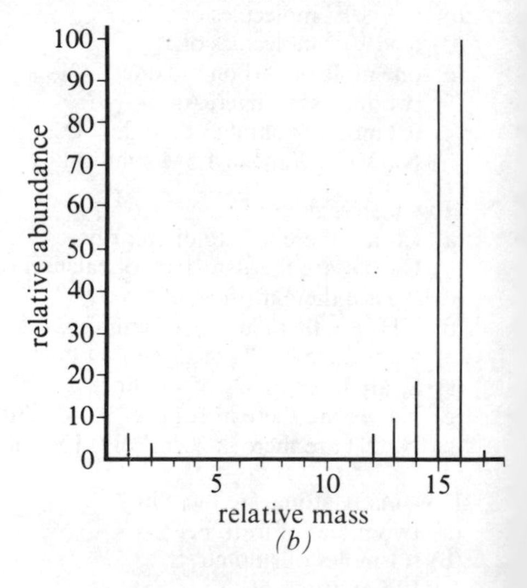


**Figure 1.4** Mass spectra for (*a*) hydrogen chloride and (*b*) methane.

# Redox 2

## 2.1 Introduction

The term '**redox**' is used by chemists as an abbreviation for the processes of **red**uction and **ox**idation which occur simultaneously. Redox reactions include such diverse processes as burning, rusting and respiration. Originally, chemists had a very limited view of redox using it to account for only the reactions of oxygen and hydrogen. Nowadays, our ideas of redox have been extended to include all electron transfer processes. An important feature of several electron-transfer redox reactions is that the energy of the chemical reaction may be released in the form of electrical energy and harnessed to provide electricity. This is what happens in the dry cell of a small torch, in the Mallory-cell of a hearing-aid and in the battery of a motor car. Commercial cells and batteries are discussed in some detail in section 13.7.

*Above*. Antoine Lavoisier (1743–1794), the father of modern chemistry. Lavoisier, the son of a rich Parisian lawyer, was the first chemist to explain the redox reactions which occur during burning.

*Above*. Lavoisier's apparatus for preparing oxygen. Red mercury(II) oxide was heated in the retort on the left. The oxygen evolved was collected in the bell jar above mercury.

*Left*. In 1775, Lavoisier was appointed to a post at the French government munitions factory. Here he carried out experiments in combustion and respiration. This print shows Lavoisier collecting the exhaled air from a seated volunteer. Madame Lavoisier, who acted as her husband's secretary and assistant, is on the right making notes.

## 2.2 Redox processes in terms of electron transfer

When metals react with oxygen they form oxides.
For example

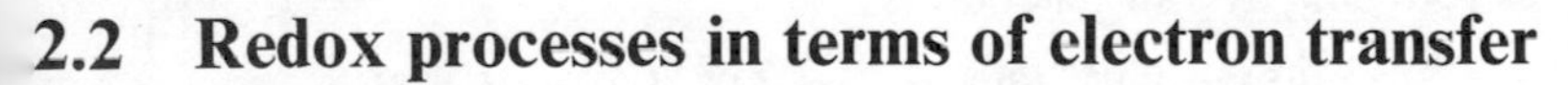

$$2Mg + O_2 \longrightarrow 2Mg^{2+}O^{2-}$$

$$4Na + O_2 \longrightarrow 2(Na^+)_2O^{2-}$$

The metal is oxidized and the oxygen is reduced. During this process the metal atoms lose electrons to form positive ions and oxygen gains electrons to form negative oxide ions, $O^{2-}$. The oxygen takes the electrons given up by the metal.

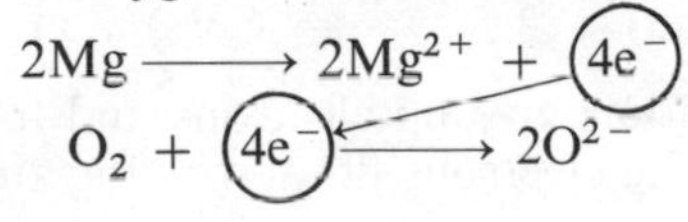

$$2Mg \longrightarrow 2Mg^{2+} + 4e^-$$

$$O_2 + 4e^- \longrightarrow 2O^{2-}$$

This cartoon (published in the eighteenth century) pokes fun at Joseph Priestley (1733–1804), the nonconformist English clergyman who carried out experiments on combustion. Priestley's discovery of oxygen provided Lavoisier with the vital information he needed for his redox theory of combustion.

Electron transfer reactions such as this are called **redox reactions**. The separate equations showing which substance gains electrons and which loses electrons are known as **half-equations**.

In the half equations above Mg loses electrons and is oxidized to $Mg^{2+}$; $O_2$ gains electrons and is reduced to $O^{2-}$.

Thus, **oxidation** is defined as *loss of electrons* and **reduction** is the *gain of electrons*. **Oxidizing agents**, such as oxygen, are defined as substances which *accept electrons*; **reducing agents** are substances which *donate electrons*.

Generally, when a metal reacts it loses electrons and forms its ions. Thus, metal atoms are oxidized and act as reducing agents in their reactions. Notice that the oxidized substance (magnesium in the above example) acts as the reducing agent and the reduced substance (oxygen in the above example) acts as the oxidizing agent.

When redox is viewed in terms of electron transfer it is easy to see why oxidation and reduction always take place together. One substance cannot lose electrons and be oxidized unless another substance gains electrons and is reduced.

Do the following processes involve oxidation, reduction, both oxidation and reduction or none of these?

- ○ $2H^+ + 2e^- \rightarrow H_2$
- ○ $Cu^+ \rightarrow Cu^{2+} + e^-$
- ○ $Mg \rightarrow Mg^{2+} + 2e^-$
- ○ $Ag^+ + Cl^- \rightarrow AgCl$
- ○ $NH_3 + H^+ \rightarrow NH_4^+$
- ○ $2Cu^+ \rightarrow Cu^{2+} + Cu$

## 2.3 Electron transfer in redox reactions

When powdered zinc is added to copper(II) sulphate solution an exothermic reaction occurs. Copper ions are reduced to a deposit of red-brown copper and zinc goes into solution as zinc ions.

$$Zn(s) + Cu^{2+}(aq) \longrightarrow Zn^{2+}(aq) + Cu(s)$$

The overall reaction can be separated into two simpler processes involving electron transfer.

$$Zn(s) \longrightarrow Zn^{2+}(aq) + \boxed{2e^-}, \text{ and}$$
$$Cu^{2+}(aq) + \boxed{2e^-} \longrightarrow Cu(s)$$

The apparatus in figure 2.1 shows how these two half-reactions can be separated to demonstrate that electron transfer is occurring.

As soon as the circuit is complete, the bulb lights to show that electrons are flowing through the wire. An ammeter used in place of the bulb will measure the electric current.

Zinc dissolves from the zinc rod which loses weight, whilst copper is deposited on the copper rod. The overall reaction is

$$Zn(s) + Cu^{2+}(aq) \longrightarrow Zn^{2+}(aq) + Cu(s)$$

This is identical to the reaction which occurs when zinc is added to copper(II) sulphate solution, but in this experiment the energy of the reaction is liberated as electrical energy (electron transfer) whereas on direct mixing the energy is liberated as heat.

Using the apparatus in figure 2.1 the overall process has been separated into two distinct half-reactions.

$$Zn(s) \longrightarrow Zn^{2+}(aq) + 2e^- \quad \text{at the zinc rod, and}$$
$$Cu^{2+}(aq) + 2e^- \longrightarrow Cu(s) \quad \text{at the copper rod.}$$

At the zinc rod, zinc atoms give up electrons and form zinc ions which go into solution as $Zn^{2+}(aq)$. The electrons flow from the zinc rod through the externa circuit including the bulb to the copper rod where they combine with $Cu^{2+}$ ions to form copper atoms.

- ○ Will the zinc rod be positive or negative?
- ○ What is the function of the salt bridge?
- ○ What happens when the salt bridge is removed?

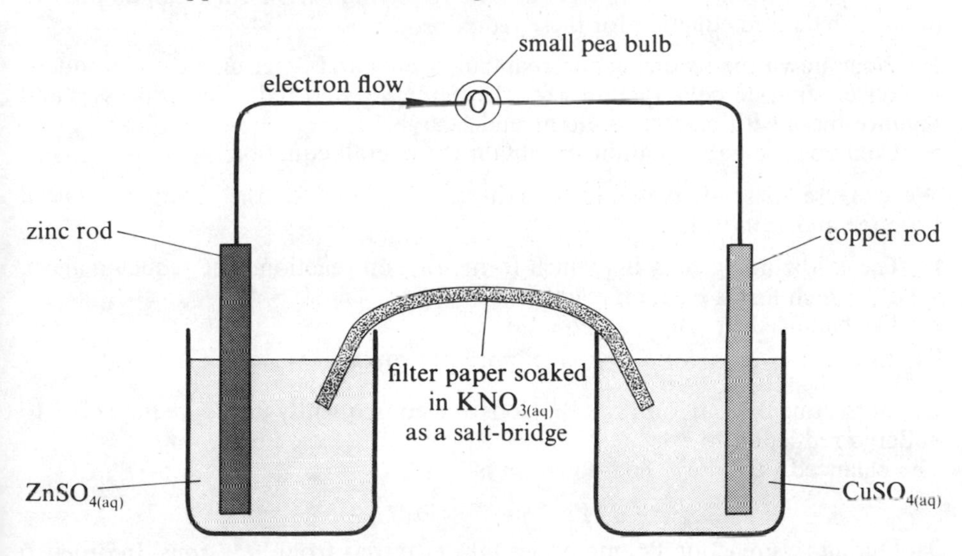


**Figure 2.1** Electron transfer between zinc and copper(II) sulphate solution.

When the circuit in figure 2.1 is complete, the zinc rod dissolves away and the concentration of $Zn^{2+}$(aq) in the left-hand beaker rises. If this nett increase in positive charge is not 'neutralized', the reaction will soon stop because the excess positive charge in the solution will repel any more $Zn^{2+}$ ions from coming into the solution. The positive charge on the $Zn^{2+}$ ions can be reduced by nitrate ions, $NO_3^-$, moving out of the salt bridge or by excess zinc ions moving into the salt bridge.

In the right-hand beaker, copper ions are converted to copper atoms leaving an excess of sulphate(VI) ions, $SO_4^{2-}$, in solution. The excess negative charge in the right-hand half-cell is reduced by $SO_4^{2-}$ ions moving into the salt bridge or potassium ions, $K^+$, diffusing out of the salt bridge into the solution.

The salt bridge serves two main functions:

(a) It completes the circuit by allowing ions carrying charge to move from one half-cell to the other. When the salt bridge is removed, the current ceases since charge (whether it is ions or electrons) can no longer flow around the circuit.
(b) It provides cations and anions which replace those consumed at the electrodes or which balance the charges on any ions formed from the electrodes. The movement of charge around the circuit is shown in figure 2.2.

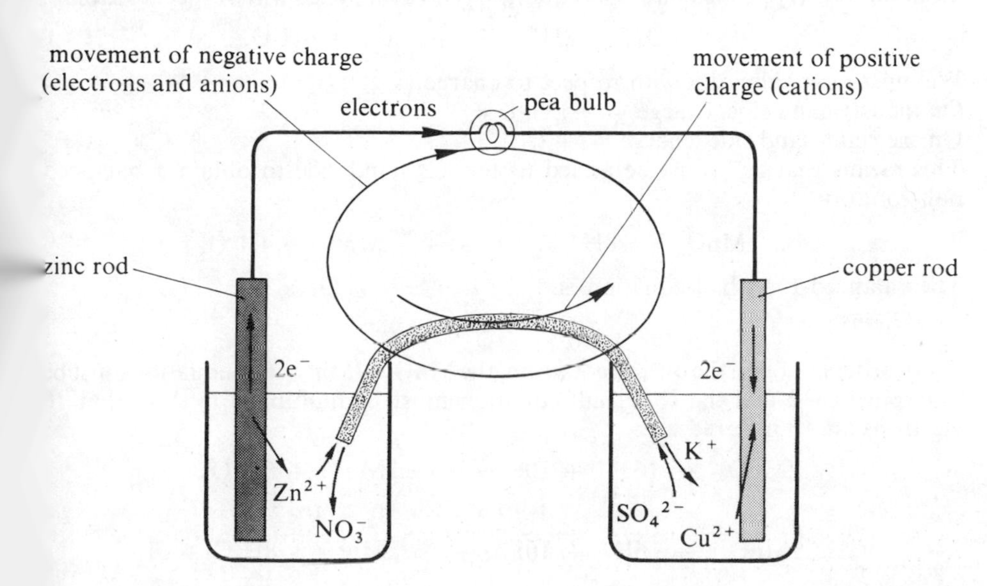


**Figure 2.2** Movement of charge around a circuit.

## 2.4 Balancing redox reactions

Many different species can act as oxidizing agents (e.g. $MnO_2$, $Cl_2$, $H_2O_2$, $Fe^{3+}$), there is a similar large number of reducing agents (e.g. $I^-$, $Fe^{2+}$, Zn). Thus a large number of possible redox reactions can result from various combinations of oxidizing agents and reducing agents. The following simple rules can be used to obtain a balanced equation for these redox processes.

1 Write down the oxidizing and reducing agents and determine their products.
2 Write separate half-equations for the oxidation and reduction processes and balance these with respect to atoms and charge.
3 Combine the half-equations to obtain the overall equation.

We can use these simple rules to write an equation for the reaction between bromine and iron(II) ions.

1 The oxidizing agent is $Br_2$ which forms $Br^-$ on reaction. The reducing agent is $Fe^{2+}$ which forms $Fe^{3+}$ on reaction.
2 The balanced $Br_2/Br^-$ half-equation is

$$Br_2 + 2e^- \longrightarrow 2Br^-.$$

Of course, one $Br_2$ can form two $Br^-$ ions and consequently the $Br_2$ requires $2e^-$ to undergo reduction to $2Br^-$.
The balanced $Fe^{2+}/Fe^{3+}$ half-equation is

$$Fe^{2+} \longrightarrow Fe^{3+} + e^-.$$

3 During this reaction $Br_2$ molecules take electrons from $Fe^{2+}$ ions. In order to obtain a balanced overall equation, it is necessary to double the $Fe^{2+}/Fe^{3+}$ equation so that $2Fe^{2+}$ gives up two electrons which are taken by one $Br_2$.

$$2Fe^{2+} \longrightarrow 2Fe^{3+} + 2e^-$$
$$Br_2 + 2e^- \longrightarrow 2Br^-$$

Adding these we get $2Fe^{2+} + Br_2 \rightarrow 2Fe^{3+} + 2Br^-$ which is the overall equation.

As a second example consider the reaction between manganate(VII), $MnO_4^-$, and iodide ($I^-$) in acid solution.

1 The oxidizing agent is $MnO_4^-$. In acid solution this is reduced to $Mn^{2+}$ and $H_2O$. Many other oxyanions including $Cr_2O_7^{2-}$, $BrO_3^-$, $IO_3^-$, $ClO^-$ and $ClO_3^-$ act as oxidizing agents in acid solution. Oxygen in the oxyanions combines with $H^+$ ions during the reaction to form water, and atoms such as Cr or Br in the oxyanion are reduced to stable ions.
2 The unbalanced $MnO_4^-/Mn^{2+}$ half-equation is therefore

$$MnO_4^- + H^+ \longrightarrow Mn^{2+} + H_2O.$$

Since the $MnO_4^-$ ion contains four oxygen atoms it can produce four $H_2O$ molecules. This means that eight $H^+$ ions are required to balance the hydrogen atoms in four $H_2O$ molecules. The equation is now balanced with respect to atoms:

$$MnO_4^- + 8H^+ \longrightarrow Mn^{2+} + 4H_2O.$$

We must now balance it with respect to charge.
On the left hand side, charge $= -1 +8 = +7$.
On the right hand side, charge $= +2$.
This means that $5e^-$ must be added to the left hand side to obtain a balanced half-equation:

$$MnO_4^- + 8H^+ + 5e^- \longrightarrow Mn^{2+} + 4H_2O.$$

The balanced $I^-/I_2$ half-equation is

$$2I^- \longrightarrow I_2 + 2e^-.$$

3 In order to obtain the overall equation, the $MnO_4^-/Mn^{2+}$ half-equation must be multiplied by 2 and the $I^-/I_2$ half-equation must be multiplied by 5 so that 10 electrons are transferred.

$$2MnO_4^- + 16H^+ + 10e^- \longrightarrow 2Mn^{2+} + 8H_2O$$
$$10I^- \longrightarrow 5I_2 + 10e^-$$
$$\overline{2MnO_4^- + 16H^+ + 10I^- \longrightarrow 2Mn^{2+} + 8H_2O + 5I_2}$$

An unusual redox reaction!

Try to write balanced half-equations and an overall equation for the following:

- ○ The reaction between $I_2$ and $S_2O_3^{2-}$. $S_2O_3^{2-}$ is oxidized to $S_4O_6^{2-}$ during the reaction.
- ○ The reaction between $H_2O_2$ and $I^-$ in acid solution.

Tutankhamun's gold funerary mask; gold is such an unreactive metal that it remains unoxidized and untarnished after centuries.

## 2.5 Important types of redox reactions

1 THE REACTION OF METALS WITH NON-METALS

In this case, metals give up electrons to form positive ions and non-metals ($O_2$, $Cl_2$, S) take these electrons to form negative ions ($O^{2-}$, $Cl^-$, $S^{2-}$). For example:

$$Fe + S \longrightarrow Fe^{2+}S^{2-} \begin{cases} Fe \longrightarrow Fe^{2+} + 2e^- \\ S + 2e^- \longrightarrow S^{2-} \end{cases}$$

$$2Fe + 3Cl_2 \longrightarrow 2FeCl_3 \begin{cases} 2Fe \longrightarrow 2Fe^{3+} + 6e^- \\ 3Cl_2 + 6e^- \longrightarrow 6Cl^- \end{cases}$$

The metal is oxidized and the non-metal is reduced. Metals higher in the activity (electrochemical) series can lose electrons more easily than the less reactive metals lower down in the activity series. Thus, moving down the activity series metals become weaker reducing agents. The reactions of metals and the activity series are discussed more fully in section 13.2 and in chapter 19.

2 THE REACTION OF METALS WITH WATER

Metals at the top of the activity series (K, Na, Ca and Mg) are sufficiently reactive to form hydrogen with water.

$$Ca + 2H_2O \rightarrow Ca^{2+}(OH^-)_2 + H_2$$

The next few metals in the activity series (e.g. Al, Zn and Fe) do not react noticeably with water, but they will react with steam to form hydrogen.

In these reactions the metal atoms are oxidized to form positive ions. The electrons which they release are accepted by water molecules which are reduced to hydroxide ions ($OH^-$) and hydrogen ($H_2$).

$$Ca \longrightarrow Ca^{2+} + 2e^-$$
$$2H_2O + 2e^- \longrightarrow 2OH^- + H_2$$

3 THE REACTION OF METALS WITH ACIDS

In this case, the metal atoms lose electrons which are taken by aqueous $H^+$ ions in the acid to form $H_2$. For example:

$$Zn(s) + 2H^+(aq) \longrightarrow Zn^{2+}(aq) + H_2(g)$$

The half-equations are:

$$Zn \longrightarrow Zn^{2+} + 2e^-$$
$$2H^+ + 2e^- \longrightarrow H_2$$

The reactivity of metals with acids also depends on the ease with which the metal loses electrons to form its aqueous ions. Metals which are keener to form ions than hydrogen will therefore react with acids releasing electrons which then react with $H^+$ ions.

Metals below hydrogen in the activity series, such as copper and silver, are less keen to form ions than hydrogen and so do not react with acids to form hydrogen.

4 REACTIONS AT THE ELECTRODES DURING ELECTROLYSIS

During electrolysis cations are attracted to the negatively charged cathode where they gain electrons and are reduced. For example during the electrolysis of molten lead(II) bromide:

**at the cathode (−)**

$$Pb^{2+} + 2e^- \longrightarrow Pb$$

Electrical gear and batteries under the bonnet of an experimental Daf car powered by fuel cells.

At the same time, $Br^-$ anions are attracted to the positively charged anode and oxidized by the loss of electrons.

**at the anode (+)**

$$2Br^- \longrightarrow Br_2 + 2e^-$$

Some important industrial processes involving electrolysis are discussed in sections 14.8, 15.6, 16.3 and 19.4.

## 2.6 Oxidation number

Many redox reactions involve a *complete* transfer of electrons from one substance to another. These redox processes usually have ions as either the reactants or the products or both.

However, there are some reactions which are regarded as redox processes in spite of the fact that they involve only molecules and do not have a *complete* transfer of electrons from one substance to another.

For example: the reactions

$$2H_2 + O_2 \longrightarrow 2H_2O \text{ and}$$
$$C + O_2 \longrightarrow CO_2$$

clearly involve redox, yet there is no complete transfer of electrons from one substance to another.

In order to overcome this problem, the concept of **oxidation number** (or oxidation state) was introduced. This provided a similar, but alternative definition of redox to that involving electron transfer whereby atoms were assigned a number to describe their relative state of oxidation or reduction. Using these oxidation numbers it is possible to decide whether redox has occurred in processes involving either transfer or re-sharing of electrons.

For example:

The oxidation numbers of the atoms in uncombined elements are given an oxidation number of zero. Thus, the oxidation number of Mg is 0, as is the oxidation number of chlorine atoms in $Cl_2$. For simple ions, the oxidation number is simply the charge on the ion. Thus the oxidation numbers of $Cl^-$, $Fe^{2+}$ and $Fe^{3+}$ are $-1$, $+2$ and $+3$ respectively.

For compounds and complex ions, the oxidation numbers of the atoms or ions within them are obtained by considering the compounds and complex ions to be *wholly ionic* and then working out the charge associated with each atom or ion.

For example:

| | |
|---|---|
| $H_2O((H^+)_2O^{2-})$ | Ox. No. of H in $H_2O$ = $+1$ |
| | Ox. No. of O in $H_2O$ = $-2$ |
| $HCl\ (H^+Cl^-)$ | Ox. No. of H in HCl = $+1$ |
| | Ox. No. of Cl in HCl = $-1$ |

In assigning oxidation numbers in this way it is necessary to assume that the electrons in each bond of the molecule or ion belong to the *more electronegative atom* (i.e. the atom with the greater attraction for electrons).

Since hydrogen is the least electronegative non-metal, the oxidation number of H in its compounds is usually $+1$.

For example:

| | |
|---|---|
| $PH_3\ (P^{3-}(H^+)_3)$ | Ox. No. of H in $PH_3$ = $+1$ |
| | Ox. No. of P in $PH_3$ = $-3$ |
| $H_2S\ ((H^+)_2S^{2-})$ | Ox. No. of H in $H_2S$ = $+1$ |
| | Ox. No. of S in $H_2S$ = $-2$ |

In metal hydrides, however, the metal is more electropositive than hydrogen and in this case the oxidation number of H is $-1$.

Early Japanese workmen smelting copper ore. What redox processes are involved in this?

For example:

$Na^+H^-$ Ox. No. of H in NaH $= -1$

Ox. No. of Na in NaH $= +1$

Excluding fluorine, oxygen is the most electronegative element. This means that the oxidation number of oxygen in its compounds is usually $-2$.

$Na_2O$ $((Na^+)_2O^{2-})$ Ox. No. of O in $Na_2O$ $= -2$

Ox. No. of Na in $Na_2O$ $= +1$

$CO_2$ $(C^{4+}(O^{2-})_2)$ Ox. No. of O in $CO_2$ $= -2$

Ox. No. of C in $CO_2$ $= +4$

However, in $OF_2$ the oxidation number of oxygen is $+2$ and in peroxides the oxidation number of oxygen is $-1$.

$OF_2$ $(O^{2+}(F^-)_2)$ Ox. No. of O in $OF_2$ $= +2$

Ox. No. of F in $OF_2$ $= -1$

$Na_2O_2$ $((Na^+)_2(O_2)^{2-})$ Ox. No. of O in $Na_2O_2$ $= -1$

Ox. No. of Na in $Na_2O_2$ $= +1$

These points concerning oxidation numbers can be summarized in a few simple rules.

### RULES FOR ASSIGNING OXIDATION NUMBERS

**1** The oxidation number of atoms in uncombined elements is 0.
**2** In neutral molecules, the algebraic sum of the oxidation numbers is 0.
**3** In ions, the algebraic sum of the oxidation numbers equals the charge on the ion.
**4** In any substance, the more electronegative atom has the negative oxidation number, the less electronegative atom has the positive oxidation number.
**5** The oxidation number of hydrogen in all compounds, except metal hydrides, is $+1$.
**6** The oxidation number of oxygen in all compounds, except in peroxides and in $OF_2$, is $-2$.

○ What are the oxidation numbers of each element in the following?

$MgCl_2$; $SO_2$; $CO$; $NaOH$; $PCl_3$; $SO_4^{2-}$; $MnO_4^-$.

○ What are the oxidation numbers of sulphur in the following compounds?

$NaHSO_4$, $CS_2$, $SO_2Cl_2$, $Na_2S$, $S_2Cl_2$.

Some elements have as many as five or more possible oxidation states. The principal oxidation states of sulphur are shown in an oxidation number chart in figure 2.3.

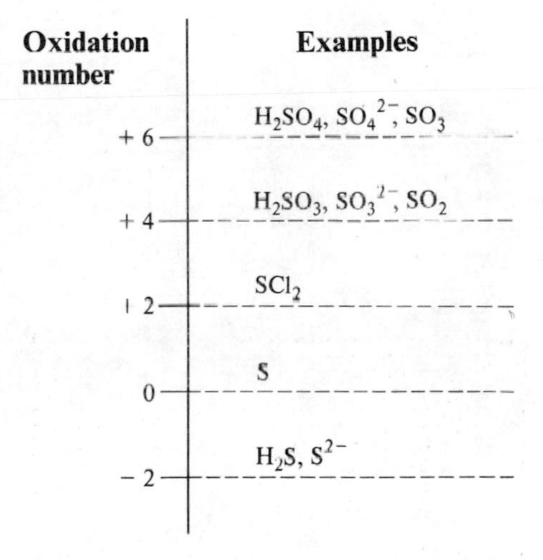


**Figure 2.3** An oxidation number chart for sulphur.

### OXIDATION NUMBERS AND NOMENCLATURE

The oxidation number concept is often important in naming compounds.

Simple molecular compounds containing only two elements, such as $CO_2$, $SiCl_4$ and $S_2Cl_2$, are usually named by reference to the numbers of their different atoms without stating oxidation numbers. Hence, $CO_2$—carbon *di*oxide, $SiCl_4$—silicon *tetra*chloride and $S_2Cl_2$—*di*sulphur *di*chloride.

The systematic names of more complex molecular compounds, such as $HNO_3$ and $H_2SO_4$, are obtained by reference to the oxidation numbers of any constituent elements which can have variable oxidation numbers.

For example:

| | *Recommended name* | *Common (trivial) name* |
|---|---|---|
| $H_2SO_3$ | sulphuric(IV) acid | sulphurous acid |
| $H_2SO_4$ | sulphuric(VI) acid | sulphuric acid |
| $HNO_2$ | nitric(III) acid | nitrous acid |
| $HNO_3$ | nitric(V) acid | nitric acid |

In the same way, the systematic names of giant ionic compounds and giant molecular compounds usually include the oxidation number of constituent elements with variable oxidation state:

For example:

| | |
|---|---|
| $FeSO_4$ | iron(II) sulphate(VI) |
| $FeSO_3$ | iron(II) sulphate(IV) |
| $FeCl_3$ | iron(III) chloride |
| $NaClO$ | sodium chlorate(I) |
| $Cu(NO_3)_2$ | copper(II) nitrate(V) |

Remember that Roman numerals are used in writing the oxidation number of an element within a compound to prevent any confusion with the real charge on an ion. Thus, CuO is named copper(II) oxide *not* copper(2) oxide and $MnO_4^-$ is named manganate(VII).

## 2.7 Explaining redox in terms of oxidation numbers

We are now in a position to consider an alternative definition of redox to that involving electron transfer, although the two definitions are quite closely related.

*An atom is said to be oxidized when its oxidation number increases and reduced when its oxidation number decreases.*

Consider the following reactions

(a)
$$2\overset{0}{Na} + \overset{0}{Cl_2} \longrightarrow 2\overset{+1}{Na}\overset{-1}{Cl}$$

The oxidation number of sodium has increased from 0 to +1: it has been oxidized. The oxidation number of chlorine has decreased from 0 to −1: it has been reduced.

(b)
$$\overset{+1}{H^+} + \overset{-2}{O}\overset{+1}{H^-} \longrightarrow \overset{+1}{H_2}\overset{-2}{O}$$

This ionic equation summarises the neutralization of an acid with an alkali. Notice that the oxidation number of each element remains unchanged and so the reaction does not involve redox.

(c) Another important class of ionic reactions which does not involve redox is precipitation. Here again the oxidation number of each element remains unaltered during the reaction.

$$\overset{+1}{Ag^+}(aq) + \overset{-1}{Cl^-}(aq) \longrightarrow \overset{+1}{Ag}\overset{-1}{Cl}(s)$$

$$\overset{+2}{Ba^{2+}}(aq) + \overset{+6}{S}\overset{-2}{O_4^{2-}}(aq) \longrightarrow \overset{+2}{Ba}\overset{+6}{S}\overset{-2}{O_4}(s)$$

(d)
$$\overset{+2}{C}\overset{-2}{O} + \tfrac{1}{2}\overset{0}{O_2} \longrightarrow \overset{+4}{C}\overset{-2}{O_2}$$

In this case, the oxidation number of carbon rises from +2 to +4: it has been oxidized. The oxidation number of the oxygen atom in CO remains unchanged but the elemental oxygen is reduced: its oxidation number falls from 0 to −2.

Which of the following reactions involve redox?

- O $Cl_2 + 2OH^- \rightarrow Cl^- + ClO^- + H_2O$
- O $Cu^{2+} + 2OH^- \rightarrow Cu(OH)_2$
- O $H_2O + SO_3 \rightarrow H_2SO_4$
- O $H^- + H_2O \rightarrow H_2 + OH^-$
- O $2CrO_4^{2-} + 2H^+ \rightarrow Cr_2O_7^{2-} + H_2O$

## 2.8 The advantages and disadvantages of the oxidation number concept

Using oxidation numbers it becomes clear whether or not redox is involved in a particular process. Reactions such as neutralization and precipitation are shown to be non-redox reactions even though they involve ions. This point highlights the importance of oxidation numbers as an electron book-keeping device that allows us to recognize redox processes.

The second advantage in using oxidation numbers is that they allow us to see exactly which part of a molecule or complex ion is reduced or oxidized. For example, the half-equation

$$MnO_4^- + 8H^+ + 5e^- \longrightarrow Mn^{2+} + 4H_2O$$

shows that $MnO_4^-$ and $H^+$ ions are reduced to $Mn^{2+}$ and $4H_2O$. But, which element or elements in $MnO_4^-$ and $H^+$ is reduced? Once oxidation numbers have been assigned to the atoms in the half-equation,

$$\overset{+7\ -2}{MnO_4^-} + \overset{+1}{8H^+} + 5e^- \longrightarrow \overset{+2}{Mn^{2+}} + \overset{+1\ -2}{4H_2O}$$

it is possible to regard manganese as the reduced element since its oxidation number changes from +7 to +2.

The main disadvantage of the oxidation number concept is that it can cause a misunderstanding concerning the structure of molecular substances. It is important to realize that no physical or structural significance can be attached to oxidation numbers. The assignment of +4 as the oxidation state of the carbon atom in $CO_2$ was quite arbitrary, and it must *not* be supposed that there is a charge of +4 on the carbon atom.

In a few cases ambiguities can arise with oxidation numbers. For example, the rules for assigning oxidation numbers suggest that each sulphur atom in the thiosulphate(VI) ion, $S_2O_3^{2-}$, has an oxidation number of +2. However, the structure of the $S_2O_3^{2-}$ ion shows that the two sulphur atoms within it are quite dissimilar. One sulphur atom is at the centre of a tetrahedron bonded to the other four atoms (one S and three O atoms) similar to the S atom in the $SO_4^{2-}$ ion. With this in mind, we might reasonably assign an oxidation number of +6 to the central S atom in $S_2O_3^{2-}$ (similar to the central S atom in $SO_4^{2-}$) and an oxidation number of −2 to each of the surrounding atoms, including the second sulphur atom.

Two further problems with oxidation numbers concern their use with organic compounds.

○ What is the oxidation number of carbon in

(i) $CH_4$; (ii) $C_2H_6$; (iii) $C_3H_8$?

The bonding in carbon and its oxidation state in each of these compounds is essentially the same, yet the carbon atoms have different oxidation numbers. The other problem (which $C_3H_8$ highlights) is that in some compounds, atoms can have oxidation numbers which are not whole numbers.

In spite of these disadvantages the concept of oxidation number is still very useful.

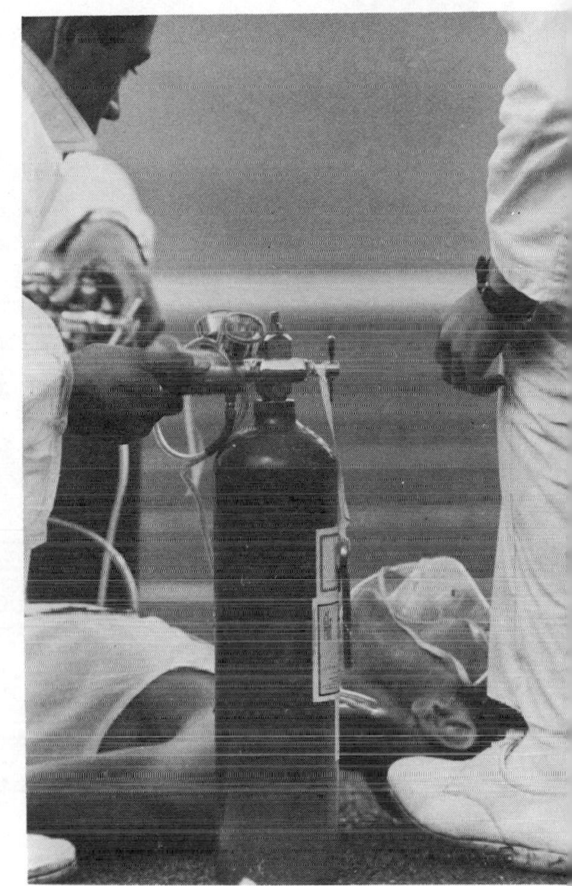

An exhausted competitor is given oxygen. Oxygen is essential for respiration, an important redox reaction.

## Summary

1 There are two related definitions of oxidation, one involving electron transfer, the other involving the concept of oxidation number.

2 In terms of electron transfer, oxidation is defined as a loss of electrons, and reduction is defined as a gain of electrons.

3 Electron-transfer reactions are called redox reactions.

4 The energy evolved by many electron-transfer reactions can be released in the form of electrical energy and harnessed to provide electricity.

5 An oxidation number is a number assigned to an atom or ion to describe its relative state of oxidation or reduction. Using these oxidation numbers it is possible to decide whether redox has occurred.

6 In terms of oxidation numbers, oxidation is defined as an increase in oxidation number and reduction is defined as a decrease in oxidation number.

## Study Questions

1 What are the oxidation numbers of
(a) chlorine in HCl, HClO, $ClO_3^-$, $PCl_3$, $Na_3AlCl_6$, $POCl_3$;
(b) nitrogen in $N_2O$, NO, $NO_2$, $NO_3^-$, $N_2H_4$, HCN?

2 (a) Write the formulae for substances containing sulphur in which it shows the following oxidation states: −2, −1, 0, +1, +2, +4, +6.

(b) The following equations summarize redox reactions involving sulphur compounds. Deduce the oxidation number of all the atoms and ions in these equations and hence determine precisely which species is oxidized and which is reduced.

(i) $2MnO_4^- + 6H^+ + 5SO_3^{2-} \longrightarrow 2Mn^{2+} + 3H_2O + 5SO_4^{2-}$

(ii) $2NaI + 3H_2SO_4 \longrightarrow 2NaHSO_4 + 2H_2O + I_2 + SO_2$

(iii) $S_2O_3^{2-} + 2H^+ \longrightarrow S + SO_2 + H_2O$

(iv) $SO_3^{2-} + H_2O + 2Ce^{4+} \longrightarrow SO_4^{2-} + 2H^+ + 2Ce^{3+}$

(v) $2S_2O_3^{2-} + I_2 \longrightarrow S_4O_6^{2-} + 2I^-$

3 Which of the following may be regarded as redox reactions? Explain your answers.

(a) $Cu^{2+} + 4NH_3 \longrightarrow Cu(NH_3)_4^{2+}$

(b) $Cl_2 + 2OH^- \longrightarrow Cl^- + ClO^- + H_2O$

(c) $Ca^{2+} + 2F^- \longrightarrow CaF_2$

(d) $Ca + F_2 \longrightarrow CaF_2$

(e) $2CCl_4 + CrO_4^{2-} \longrightarrow 2COCl_2 + CrO_2Cl_2 + 2Cl^-$

(f) $NH_3 + H^+ \longrightarrow NH_4^+$

4 Write redox half-equations for the following reactions:
(a) When copper is added to concentrated nitric(v) acid the solution becomes pale blue and brown fumes of nitrogen dioxide are produced.
(b) When potassium iodide is added to acidified hydrogen peroxide a brown colour appears.
(c) Sodium sulphate(IV) (sodium sulphite) reduces an acidified solution of orange dichromate(VI) ions, $Cr_2O_7^{2-}$, to a green solution containing $Cr^{3+}$ ions.
(d) Manganese(IV) oxide will oxidize concentrated hydrochloric acid to chlorine.
(e) When zinc is added to silver nitrate solution crystals of silver form on the zinc surface.

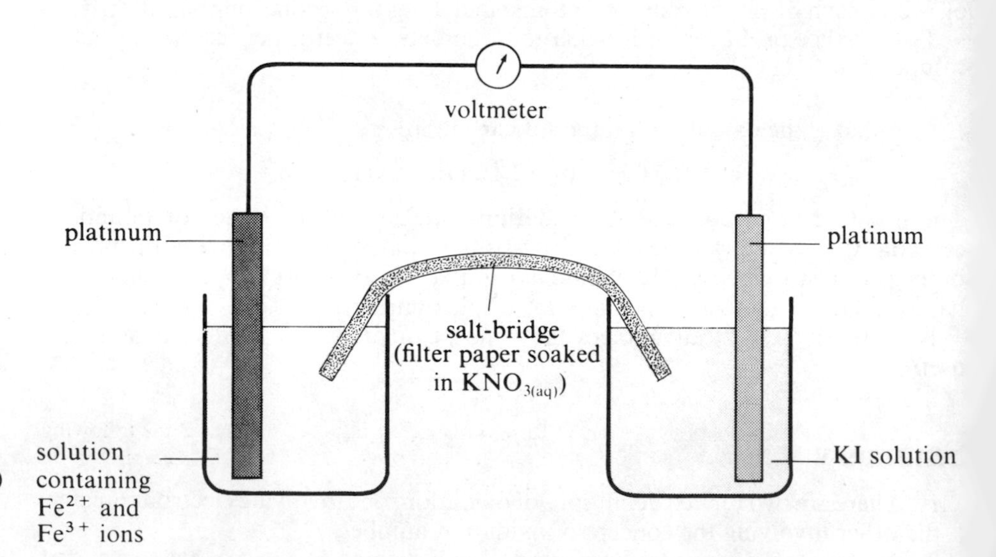


**Figure 2.4** Electron transfer between KI(aq) and a solution containing $Fe^{2+}$ and $Fe^{3+}$ ions.

5 Look at the apparatus in figure 2.4.
When the circuit is complete a brown colour appears around the platinum in the right hand beaker.
(a) Write a half equation to summarize the reactions at each of the platinum terminals.
(b) In which direction do electrons flow?
(c) Explain the function of the salt bridge, stating which ions are moving into it and out of it in each beaker.
(d) Do you think the voltage of the cell will increase, decrease or remain the same if the concentration of KI in the right hand beaker is increased? Explain your answer.

6 (a) What is the oxidation number of chromium in each of the substances A–F in the reaction scheme in figure 2.5?
(b) Which steps in the scheme involve redox?
(c) Write equations or half equations to describe each step in the scheme.

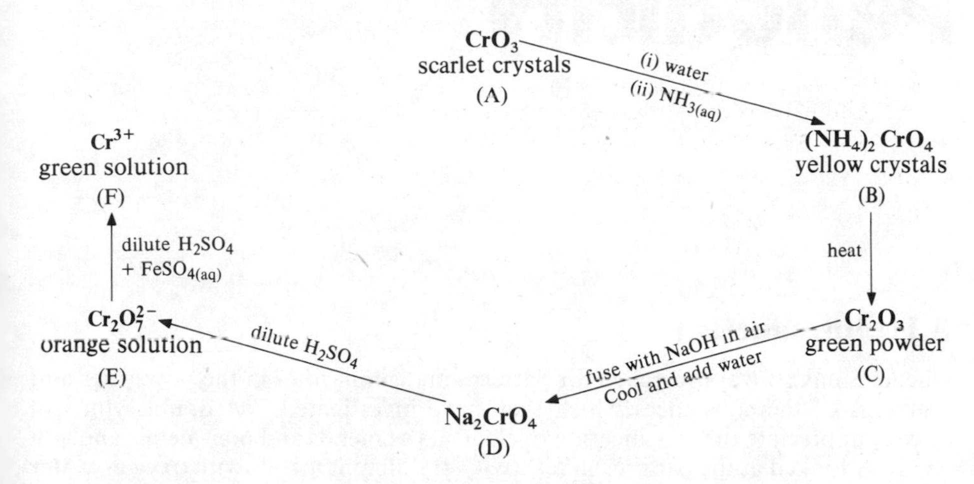


**Figure 2.5** A reaction scheme involving compounds of chromium.

7 Certain features of the element vanadium, V, are presented in figure 2.6. Consider the diagram carefully.
(a) What is the oxidation number of vanadium in the compounds A–I?
(b) What can you deduce about the oxidizing power of the halogens towards vanadium?
(c) How does the action of chlorine on vanadium compare with the action of hydrogen chloride?

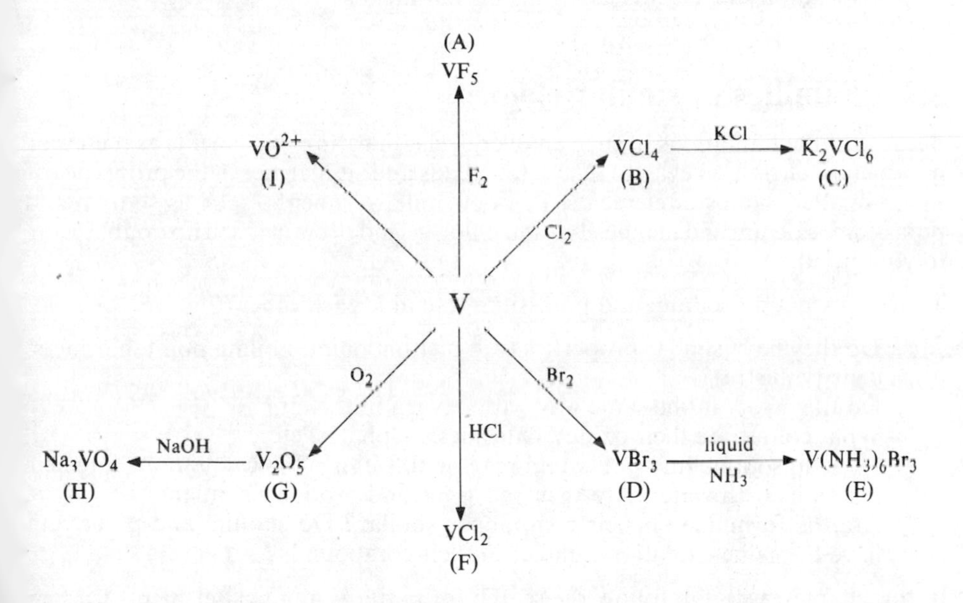


**Figure 2.6** Products from vanadium.

8 What do you understand by the terms ‘oxidation’ and ‘reduction’? In each of the following reactions say what (if anything) has been reduced and what has been oxidized, and write electron transfer equations to explain your answers:

(a) $2FeCl_2 + Cl_2 \longrightarrow 2FeCl_3$

(b) $CuO + H_2 \longrightarrow Cu + H_2O$

(c) $3Cu + 8HNO_3 \longrightarrow 3Cu(NO_3)_2 + 4H_2O + 2NO$

(d) $2Na + H_2 \longrightarrow 2NaH$

(e) $AgNO_3 + NaCl \longrightarrow AgCl + NaNO_3$

0.50g of hydrated iron(II) sulphate(VI) ($FeSO_4 . 7H_2O$), was dissolved in dilute sulphuric(VI) acid and titrated with 0.1 molar potassium manganate(VII) (potassium permanganate). What volume of the potassium manganate(VII) is required to complete the titration? What assumptions have you made in the calculations?

Fe = 55.8; S = 32; O = 16; H = 1.

# 3 Patterns and Periodicity

3.5 *Johann Wolfgang Döbereiner.*

Born 1780 at Hof; died 1849 at Jena. Döbereiner studied as a pharmacist at Münchberg, studied chemistry at Strasbourg, and in 1810 succeeded Göttling in the chair of chemistry and pharmacy at Jena. Döbereiner made the first observations on platinum as a catalyst and on the periodic system of the elements. He also contributed to technical questions, and was, for example, a keen protagonist of the erection of gas-works. Döbereiner and the poet Goethe were in close contact both personally and academically.

| Li<br>6.9 |
|---|
| Na<br>23.0 |
| K<br>39.1 |

**Figure 3.1** Döbereiner's triad of alkali metals.

| Cl<br>35.5 |
|---|
| Br<br>79.9 |
| I<br>126.9 |

**Figure 3.2** Döbereiner's triad of halogens.

## 3.1 Introduction

Chemists have always searched for patterns and similarities in the properties and reactions of the substances which they have investigated. No doubt, you will already appreciate the classification of elements as metals and non-metals and you will have looked at the patterns in reactivity of different metals with oxygen, water and acids. Unfortunately, there are some elements that cannot be classified easily and unambiguously as either a metal or a non-metal. Non-metals are usually volatile and non-conductors of electricity, but graphite (carbon) and silicon (see p. 33), usually classed as non-metals, have very high melting points and very high boiling points, and both conduct electricity. Mercury is untypical of metals in being liquid at room temperature. Can you think of other elements which do not fit neatly into the classification of elements as metals or non-metals?

## 3.2 Families of similar elements

Because of the limitations to such an overall classification of elements as metals or non-metals, chemists began to search for trends and similarities in the properties of much smaller groups of elements. Pairs of similar elements, such as sodium and potassium, calcium and magnesium and chlorine and bromine, will no doubt spring to your mind.

To what extent do sodium and potassium resemble each other?

- ○ Do they have similar properties (e.g. melting point, boiling point, hardness, density, lustre)?
- ○ Do they react in the same way with oxygen in the air?
- ○ What colour are their oxides, chlorides, sulphates, etc.?
- ○ How do sodium and potassium react with water? Write equations for their reactions with water. Are their reactions and products similar?
- ○ Are the formulae of their compounds similar? Do sodium and potassium have the same oxidation number in their compounds?

In this chapter, we shall follow the search for patterns amongst elements leading to the development of the periodic table and then describe in detail the physical and chemical periodicity of the elements in periods 2 and 3. In chapter 12, these periodic trends are explained in terms of atomic properties such as electronic structure, bonding and electronegativity.

Early in the nineteenth century the German chemist, Döbereiner had pointed out that many of the known elements could be arranged in groups of three similar elements. He called these families of three elements '**triads**'. Two of Döbereiner's triads were lithium, sodium and potassium (alkali metals—figure 3.1) and chlorine, bromine and iodine (halogens—figure 3.2). Döbereiner showed that when the three elements in each triad were written in order of atomic mass, the middle element had properties intermediate between those of the other two elements and what is more, the atomic mass of the middle element was very close to the average of the atomic masses of the other two elements. In the triad, chlorine, bromine and iodine, the atomic mass of bromine (79.9) is close to the average of the atomic masses of chlorine and iodine $((35.5 + 126.9)/2 = (162.4)/2 = 81.2)$.

○ How close to the atomic mass of sodium is the average of the atomic masses of lithium and potassium?

## 3.3 Newlands' Octaves

The relationship that Döbereiner had discovered encouraged other chemists to look for trends between the properties of elements and their atomic masses.

In 1866, John Newlands, an English chemist, read a paper to the Chemical Society in which he suggested that when the elements were arranged in order of their atomic masses (atomic weights), any one element had properties similar to those of the elements eight places in front of it and eight places behind it in the list.

Newlands called this a '**Law of Octaves**'. He claimed that the eighth element was a kind of repetition of the first like the eighth note of an octave in music.

Table 3.1 shows the first five of these 'octaves'. Although Newlands' 'octaves' showed similar elements in the same column, his ideas met with considerable scepticism.

**Table 3.1** Newlands' octaves

| | | | | | | |
|---|---|---|---|---|---|---|
| H | Li | Be | B | C | N | O |
| F | Na | Mg | Al | Si | P | S |
| Cl | K | Ca | Cr | Ti | Mn | Fe |
| Co and Ni | Cu | Zn | Y | In | As | Se |
| Br | Rb | Sr | Ce and La | Zr | Di and Mo | Ro and Ru |

○ Can you find three groups of similar elements in the columns of table 3.1?
○ Are Döbereiner's triads included in the columns of Newlands' octaves? Find one triad in one of the columns.
○ Why are the noble gases (He, Ne, Ar, etc.) missing from Newlands' octaves?
○ Why do you think chemists were sceptical about Newlands' octaves?

The periodic repetition of similar elements at regular intervals in Newlands' octaves led to the name **periodic table**. Newlands' classification was criticized for three important reasons.

**1** It assumed that all elements had been discovered. The discovery of one more element could throw out the whole idea of 'octaves'. This was a particularly potent criticism as four elements (thallium, indium, caesium and rubidium) had been discovered in the few years previous to Newlands' suggestions.
**2** In order to ensure repeating octaves Newlands had found it necessary to place two elements (e.g. Co and Ni, Ce and La) into only one space in certain cases.
**3** Newlands' classification grouped together some elements which were very dissimilar. For example, cobalt and nickel were placed in the same family as fluorine, chlorine and bromine; copper was placed in the same family as lithium, sodium, potassium and rubidium.

For these reasons, Newlands' ideas were rejected and even ridiculed by his fellow cientists. Indeed, one of his more sceptical critics enquired whether 'Mr Newlands ıad ever examined the elements according to the order of their initial letters'.

PROCEEDINGS OF SOCIETIES.

CHEMICAL SOCIETY.
*Thursday, March 1.*
*Professor* A. W. WILLIAMSON, *Ph.D., F.R.S., Vice-President, in the Chair.*

MR. JOHN A. R. NEWLANDS read a paper entitled "*The Law of Octaves, and the Causes of Numerical Relations among the Atomic Weights.*" The author claims the discovery of a law according to which the elements analogous in their properties exhibit peculiar relationships, similar to those subsisting in music between a note and its octave. Starting from the atomic weights on Cannizzaro's system, the author arranges the known elements in order of succession, beginning with the lowest atomic weight (hydrogen) and ending with thorium (= 231·5); placing, however, nickel and cobalt, platinum and iridium, cerium and lanthanum, &c., in positions of absolute equality or in the same line. The fifty-six elements so arranged are said to form the compass of eight octaves, and the author finds that chlorine, bromine, iodine, and fluorine are thus brought into the same line, or occupy corresponding places in his scale. Nitrogen and phosphorus, oxygen and sulphur, &c., are also considered as forming true octaves. The author's supposition will be exemplified in Table II., shown to the meeting, and here subjoined :—

*Table II.—Elements arranged in Octaves.*

| | No. | | No. | | No. | | No. | | No. | | No. | | No. | | No. |
|---|---|---|---|---|---|---|---|---|---|---|---|---|---|---|---|
| H | 1 | F | 8 | Cl | 15 | Co & Ni | 22 | Br | 29 | Pd | 36 | I | 42 | Pt & Ir | 50 |
| Li | 2 | Na | 9 | K | 16 | Cu | 23 | Rb | 30 | Ag | 37 | Cs | 44 | Os | 51 |
| G | 3 | Mg | 10 | Ca | 17 | Zn | 24 | Sr | 31 | Cd | 38 | Ba & V | 45 | Hg | 52 |
| Bo | 4 | Al | 11 | Cr | 19 | Y | 25 | Ce & La | 33 | U | 40 | Ta | 46 | Tl | 53 |
| C | 5 | Si | 12 | Ti | 18 | In | 26 | Zr | 32 | Sn | 39 | W | 47 | Pb | 54 |
| N | 6 | P | 13 | Mn | 20 | As | 27 | Di & Mo | 34 | Sb | 41 | Nb | 48 | Bi | 55 |
| O | 7 | S | 14 | Fe | 21 | Se | 28 | Ro & Ru | 35 | Te | 43 | Au | 49 | Th | 56 |

Dr. GLADSTONE made objection on the score of its having been assumed that no elements remain to be discovered. The last few years had brought forth thallium, indium, cæsium, and rubidium, and now the finding of one more would throw out the whole system. The speaker believed there was as close an analogy subsisting between the metals named in the last vertical column as in any of the elements standing on the same horizontal line.

Professor G. F. FOSTER humorously inquired of Mr. Newlands whether he had ever examined the elements according to the order of their initial letters? For he believed that any arrangement would present occasional coincidences, but he condemned one which placed so far apart manganese and chromium, or iron from nickel and cobalt.

Mr. NEWLANDS said that he had tried several other schemes before arriving at that now proposed. One founded upon the specific gravity of the elements had altogether failed, and no relation could be worked out of the atomic weights under any other system than that of Cannizzaro.

A note concerning Newlands' octaves from the magazine *Chemical News*, published by the Chemical Society in March 1866.

## 3.4 Patterns and periodicity

espite the criticism of Newlands' suggestions many scientists continued to search or a pattern relating the properties and relative atomic masses of the elements.

OTHAR MEYER'S CURVES

n 1869, the German chemist, Julius Lothar Meyer and the Russian chemist, mitri Mendeléev, separately published results which supported the ideas of eriodicity suggested a few years earlier by Newlands. Lothar Meyer plotted arious physical properties (melting point, boiling point, density) of the known ements against their relative atomic masses (atomic weights). Two of Lothar leyer's curves are shown in figure 3.3 and figure 3.4.

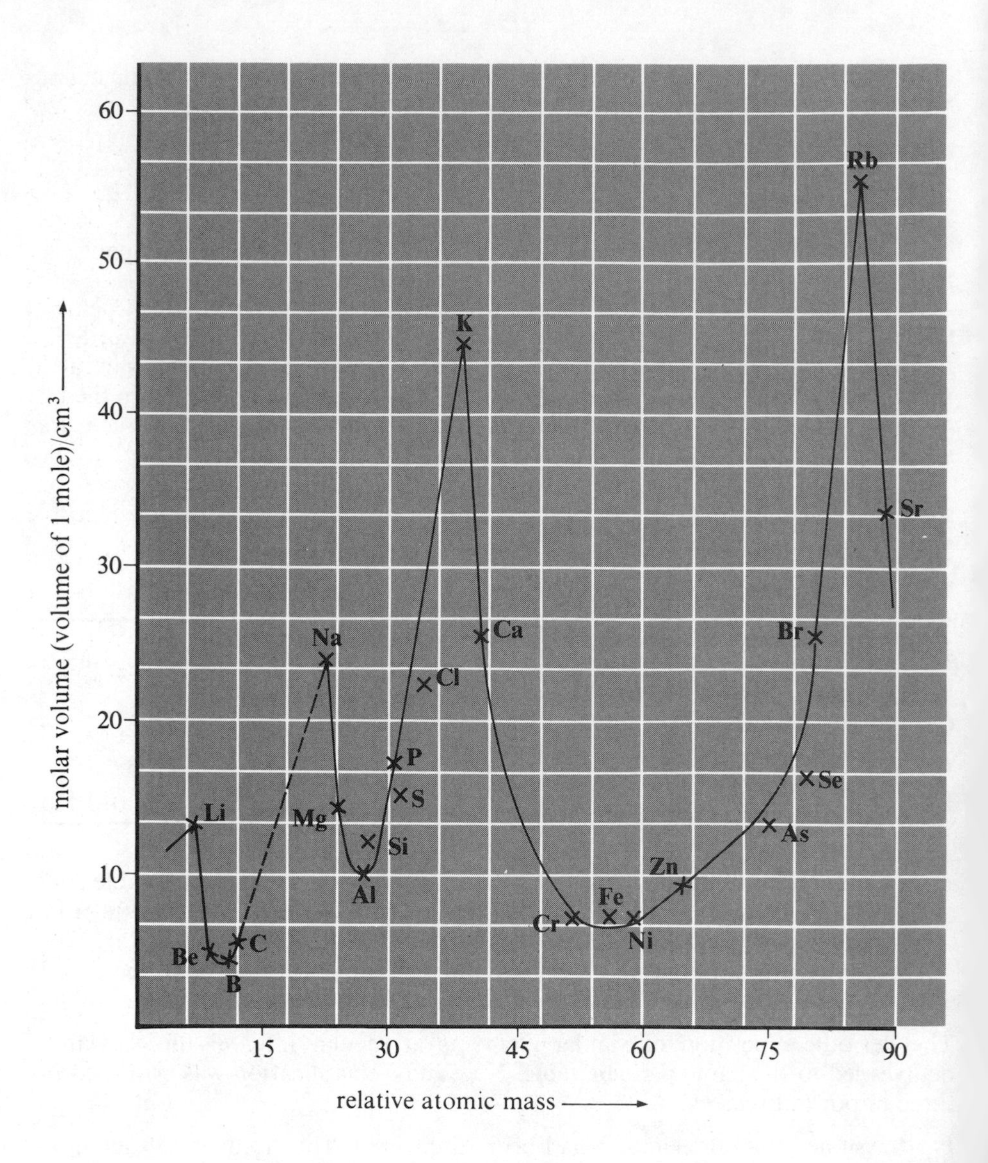


**Figure 3.3** Lothar Meyer's curve of molar volume against relative atomic mass.

The graph in figure 3.3 shows the periodic repetition obtained by plotting the **molar volume** (the volume occupied by one mole of the element in the solid state) against the relative atomic mass. Notice that elements in the same chemical family occur at similar points on the curves. For example, lithium, sodium and potassium (alkali metals) appear on the peaks, whilst fluorine, chlorine and bromine (halogens) occur just before the peaks.

The graph in figure 3.4 shows the periodic variation in melting point with relative atomic mass. As in figure 3.3, similar elements fall at similar points on the curves.

- ○ Where do the alkali metals appear on the curves in figure 3.4? Do they occupy similar positions?
- ○ Where do the halogens appear on the curves in figure 3.4? Do they occupy similar positions?
- ○ Which elements appear on the peaks in figure 3.4? Are these elements alike in their properties?

### MENDELÉEV'S PERIODIC TABLE

Although Lothar Meyer's curves showed a periodic repetition of properties with relative atomic mass (atomic weight), most of the credit for arranging the elements in a periodic table is given to Mendeléev. Mendeléev arranged the 60 or so elements known to him in order of increasing relative atomic mass and showed that elements with similar properties recurred at regular intervals. Figure 3.5 shows part of Mendeléev's periodic table. Elements with similar properties fall in the same vertical column. These vertical columns of similar elements are called **groups** and the horizontal rows of elements are called **periods**.

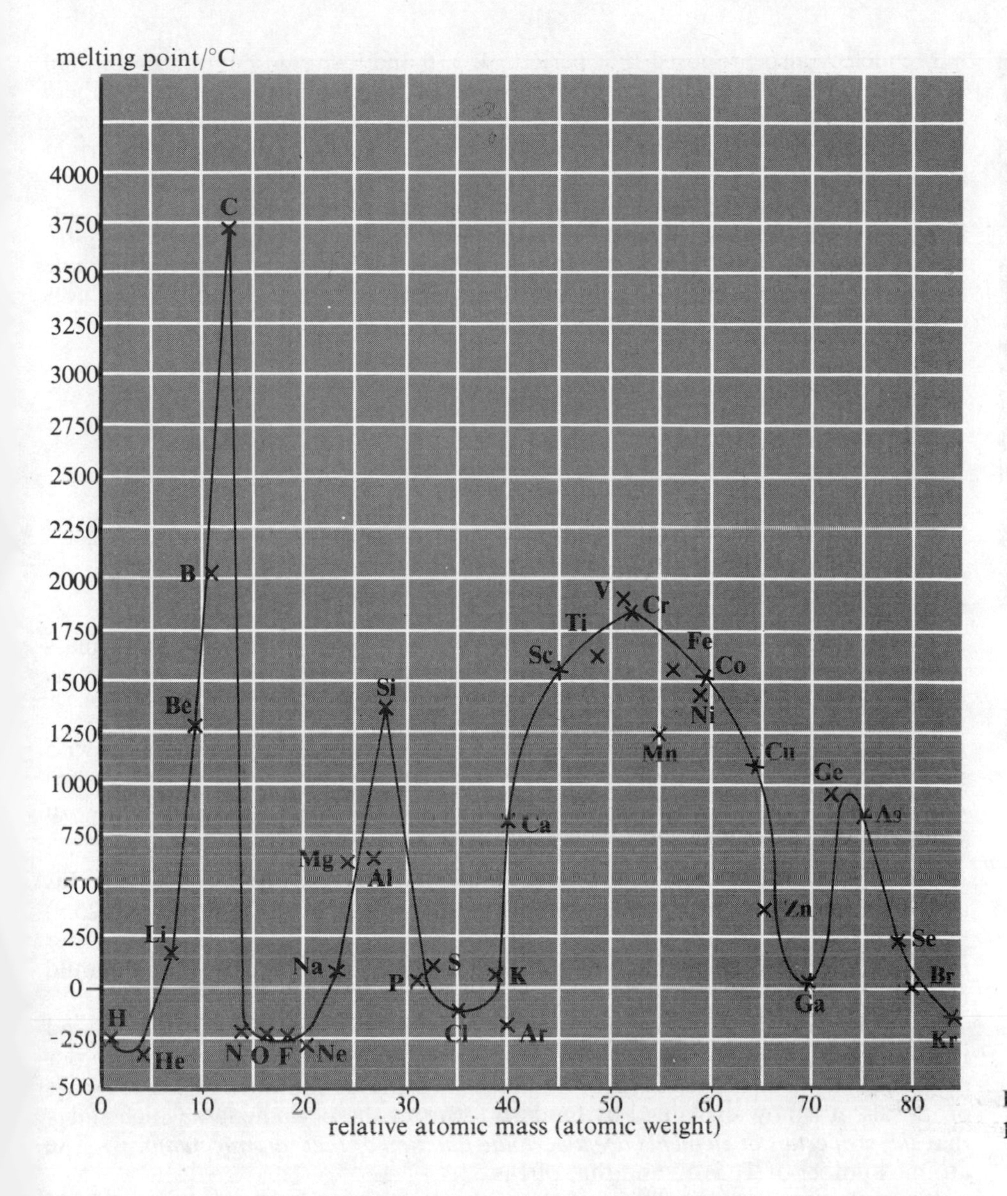


**Figure 3.4** Lothar Meyer's curve of melting point against relative atomic mass.

| | Group I | Group II | Group III | Group IV | Group V | Group VI | Group VII | Group VIII |
|---|---|---|---|---|---|---|---|---|
| **Period 1** | H | | | | | | | |
| **Period 2** | Li | Be | B | C | N | O | F | |
| **Period 3** | Na | Mg | Al | Si | P | S | Cl | |
| **Period 4** | K<br>Cu | Ca<br>Zn | *<br>* | Ti<br>* | V<br>As | Cr<br>Se | Mn<br>Br | Fe Co Ni |
| **Period 5** | Rb<br>Ag | Sr<br>Cd | Y<br>In | Zr<br>Sn | Nb<br>Sb | Mo<br>Te | *<br>I | Ru Rh Pd |

**Figure 3.5** Mendeléev's periodic table.

Why was Mendeléev's periodic table more successful than the one proposed by Newlands?

In the first place, Mendeléev left gaps in his table in order that similar elements would fall in the same vertical column. What is more, he suggested that, in due course, elements would be discovered to fill these gaps and he also predicted properties for some of the missing elements. There are four gaps (indicated by *) in the portion of Mendeléev's periodic table reproduced in figure 3.5. Which elements have since been discovered to fill these gaps? Within Mendeléev's own lifetime some of these missing elements were discovered and their properties coincided very closely with his predictions.

Dmitri Ivanovich Mendeléev (1839–1907). In 1869, Mendeléev published his periodic table, the forerunner of modern periodic tables.

Mendeléev also proposed that periods 4, 5, 6 and 7 should contain more than seven elements. In order to fit these longer periods into his pattern he divided them into halves and placed the first half of the elements in the top left hand corner of their space (e.g. K, Ca, etc. in period 4) and the second half in the bottom right hand corners (e.g. Cu, Zn, etc.).

The accuracy of Mendeléev's predictions quickly convinced scientists that his ideas were correct and his periodic table was accepted as a valuable overall summary of the properties of elements. Indeed, Mendeléev's concept of a periodic table has absorbed more and more new knowledge and the importance and usefulness of his original idea has been demonstrated many times.

## 3.5 Modern forms of the periodic table

Mendeléev's periodic table arranged the elements in order of relative atomic mass and his periodic law stated that *the properties of the elements are a periodic function of their relative atomic masses*. This law fulfilled two important functions.

**1** It summarised the properties of elements and classified them into groups with similar properties.
**2** It enabled predictions to be made about the properties of known and unknown elements and led to considerable research activity.

Although Mendeléev used the order of relative atomic masses as a basis for his periodic table, he wrote the elements tellurium (Te = 127.6) and iodine (I = 126.9) in reverse to the expected order. Mendeléev found that this was necessary in order that Te and I should fall in their correct vertical groups. Obviously, iodine should occupy the same group as chlorine and bromine.

Mendeléev's periodic table was proposed long before chemists could understand its fundamental relationship to electronic structure. Although the reverse order of Te and I was somewhat worrying and intriguing at the time of Mendeléev's proposals, it is now explained by modern forms of the periodic law which states that *the properties of elements are a periodic function of their atomic numbers*. (The atomic number of Te is 52 and that of I is 53.)

If the elements are numbered along each period from left to right starting at period 1, then period 2 etc., the number given to each element is called its **atomic number**. The real significance of atomic number is discussed in sections 4.2 and 4.4.

*In modern periodic tables, all the elements are in strict order of their atomic numbers.*

All modern forms of the periodic table are based on Mendeléev's original idea. Although there are various forms of the periodic table suitable for one purpose or another, the 'wide form' shown in figure 3.6 is probably the most useful. In this figure the atomic numbers are shown below the symbols for each element.

Those groups with elements in periods 2 and 3 are numbered from I to VII followed by Group O. Some of these groups have names besides numbers. The most common names used for particular groups are:

| Group number | Group name |
|---|---|
| I | alkali metals |
| II | alkaline-earth metals |
| VII | halogens |
| O | noble (inert) gases |

The most obvious difference between the periodic table in figure 3.6 and that proposed by Mendeléev is the removal of rows of 10 or more similar elements (the **transition elements**) from the simple groups. In period 3, for example, 10 transition elements (Sc, Ti, V, Cr, Mn, Fe, Co, Ni, Cu, Zn) have been taken out of the simple groupings suggested by Mendeléev and placed between Ca in Group II and Ga in Group III.

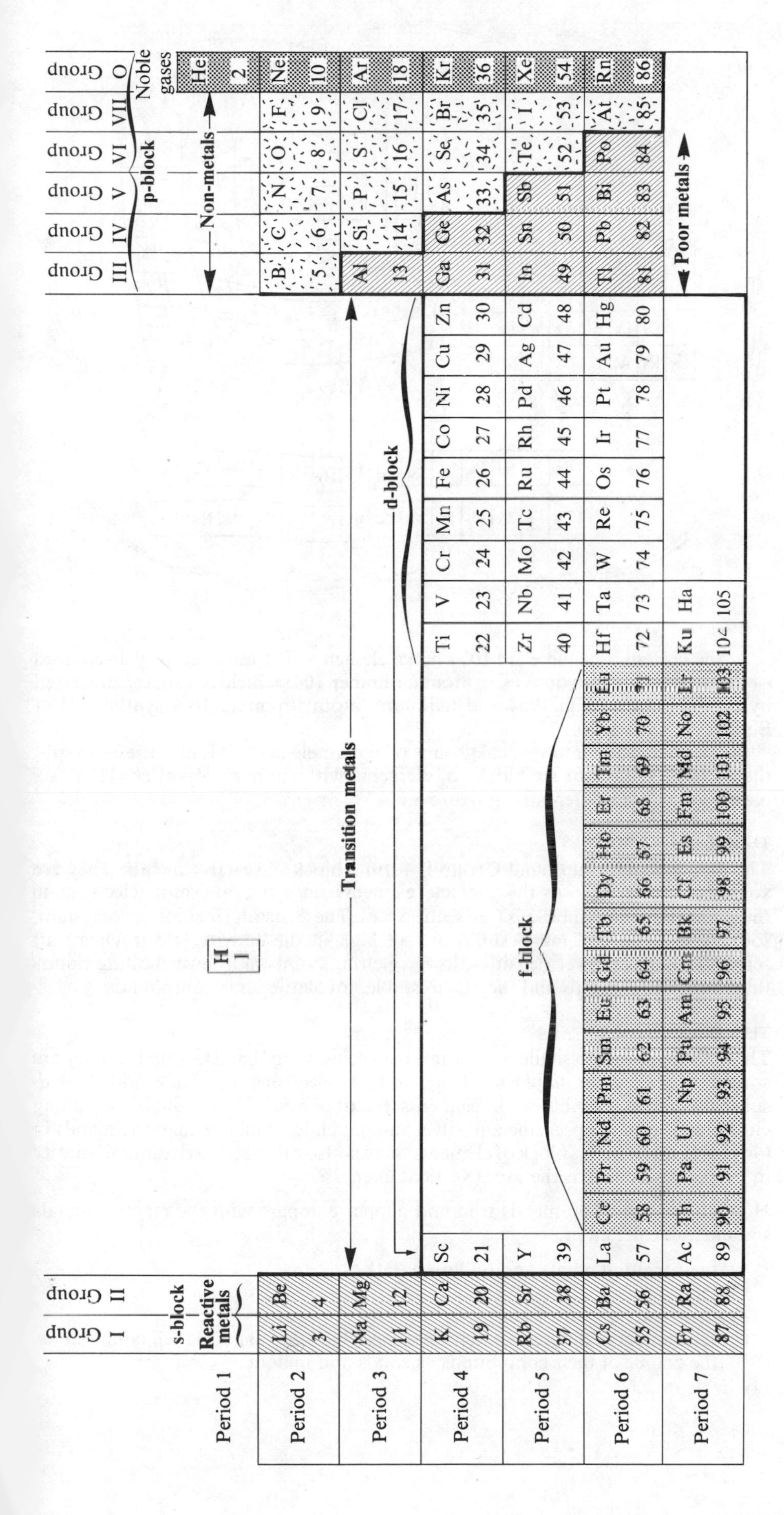


**Figure 3.6** The modern periodic table (wide-form).

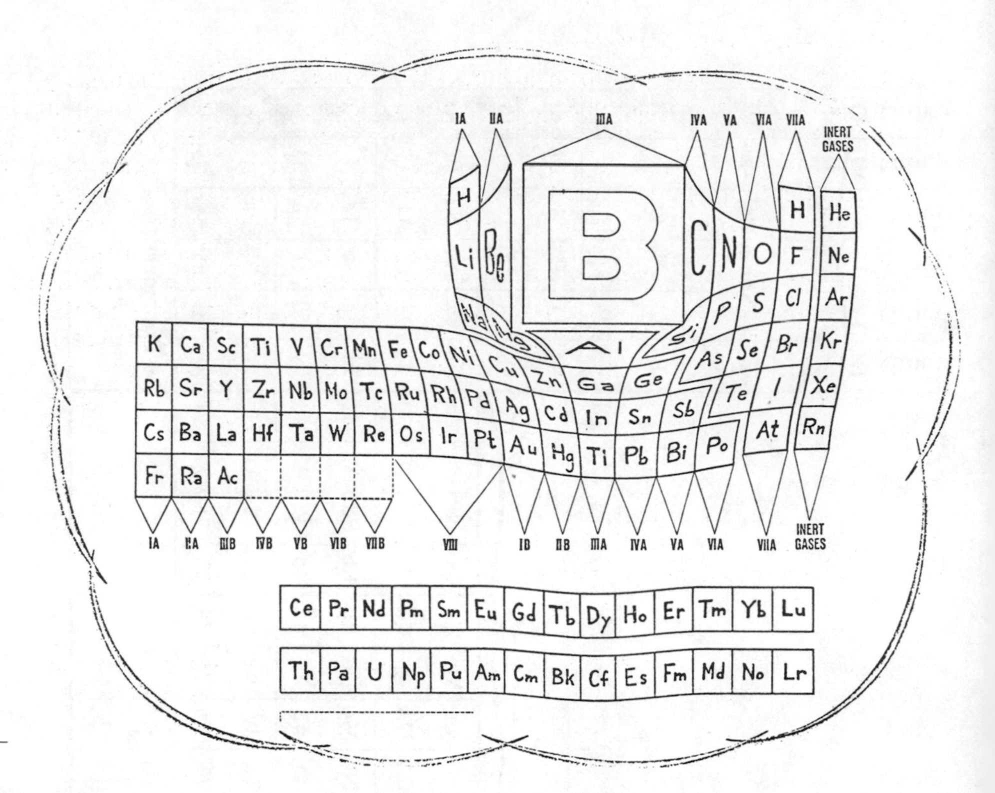


An unusual form of the periodic table – emphasizing the position of one element.

At the present time there are 105 known elements. The most recently discovered elements are kurchatovium (Ku, atomic number 104) which was first synthesized by Russian chemists in 1964 and hahnium (atomic number 105) synthesized at Berkeley, USA in 1970.

Besides a division into vertical groups of similar elements it is also useful to split the periodic table into six blocks of elements with similar properties. These six blocks are shaded differently in figure 3.6.

### THE REACTIVE METALS

The elements in Group I and Group II form a block of reactive metals. They are sometimes referred to as the **'*s*-block'** elements since the outermost electrons in these metals are in *s*-subshells (see section 5.6). These metals (including potassium, sodium, calcium and magnesium) are all high in the activity (electrochemical) series. They have lower densities, lower melting points and lower boiling points than most other metals and they form stable, involatile ionic compounds.

### THE TRANSITION METALS

These elements form a shallow rectangle between Group II and Group III. They are sometimes called the **'*d*-block'** elements since electrons are being added to *d*-subshells across this block of elements (section 5.6). These metals (including chromium, iron, copper, zinc and silver) are much less reactive than the metals in Groups I and II. In this block of elements there is also a marked horizontal similarity in properties as well as the usual vertical likenesses.

How do the transition metals iron and copper compare with the reactive metals sodium and calcium in:

- ○ their melting points and boiling points,
- ○ their density,
- ○ their reaction with water,
- ○ the number of different oxidation states which they show in their compounds
- ○ the colour of their compounds as solids and in aqueous solution?

### THE LANTHANIDES AND ACTINIDES

The lanthanides and actinides form a block of elements within the transition metals. Indeed, they are sometimes called the **inner transition elements**. Another name for these elements is the **'*f*-block'** elements, since electrons are being added to *f*-subshells across this block of elements. The lanthanides consist of the 14 elements from cerium (Ce) to lutetium (Lu) which come immediately after lanthanum (La) in the periodic table and which resemble each other greatly. In fact, the horizontal similarities across this block are so great that chemists experienced considerable difficulty in separating the lanthanides from one another. Lanthanum and the lanthanides are sometimes known as the rare earth elements. The actinides are the 14 elements from thorium (Th) to lawrencium (Lr) which follow actinium (Ac) in the periodic table. Only the first three elements in the actinide series (thorium (Th), protoactinium (Pa) and uranium (U)) are naturally occurring. All the elements beyond uranium have been synthesized by chemists since 1940.

### THE POOR METALS*

These elements (including tin, lead and bismuth) fall in a triangular block of the periodic table. They are usually low in the activity (electrochemical) series and they have some resemblances to non-metals.

### THE NON-METALS

These elements also form a triangular block in the periodic table. The elements in this block and the last one are sometimes called the **'*p*-block'** elements since the outermost electrons in these elements are going into *p*-subshells.

### THE NOBLE GASES

The atoms in these elements have outer *s*- and *p*-subshells of electrons which are completely filled. They are very unreactive and it was not until 1962 that the first noble gas compound was obtained. Because of their chemical unreactivity these elements were originally called 'inert gases'. Nowadays, several compounds of these elements (mainly oxides and fluorides of xenon and krypton) are known and the name 'inert' has been replaced by 'noble'.

*Below* The first crystals of xenon fluoride isolated at the Argonne National Laboratory, France.

## 3.6 Metals, non-metals and metalloids

Although the periodic table does not classify elements as metals and non-metals there is, however, a fairly obvious division between the two. '*Fairly obvious*' you need to note, not '*clear cut*'. Separating the elements unequivocally and completely into either metals or non-metals is rather like trying to separate all the shades of grey into either black or white. It cannot be done without ambiguity. The fairly obvious division between metals and non-metals is shown by a thick stepped line in figure 3.6. The 20 or so non-metals are packed into the top right-hand corner above the thick stepped line. Some of the elements next to the thick steps, such as germanium, arsenic and antimony, have similarities to both metals and non-metals and it is difficult to place these, with certainty, in one class or the other. Chemists sometimes use the name **metalloid** (or **semi-metal**) for these elements which are difficult to classify one way or the other.

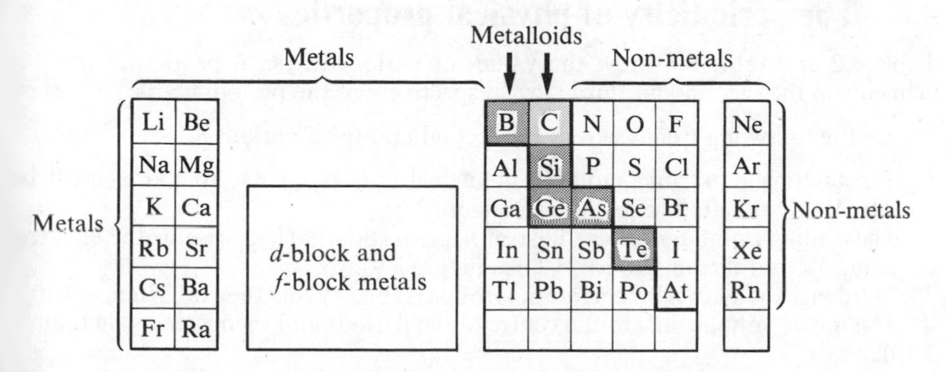


**Figure 3.7** Classifying the elements as metals, metalloids and non-metals on the basis of their electrical conductance.

Figure 3.7 shows a classification of elements as metals, metalloids and non-metals on the basis of their electrical conductance. In this classification:

* The term 'poor metals' is not widely used, but we believe it is a useful and apt description for these elements.

1 *Metals are good conductors of electricity* with atomic conductance* (atomic electrical conductivity) greater than $10^{-3}ohm^{-1}cm^{-4}$ and whose conductivity slowly falls as the temperature rises.
2 *Metalloids are poor conductors of electricity* with atomic conductance usually less than $10^{-3}$ but greater than $10^{-5}ohm^{-1}cm^{-4}$. The conductivity of a metalloid increases as the temperature rises and is also considerably affected by the presence of impurity.
3 *Non-metals are virtually non-conductors* (insulators). Their atomic conductance is usually less than $10^{-10}ohm^{-1}cm^{-4}$.

Notice in figure 3.7 that the cell for carbon has been shaded less heavily than those for the other metalloids. This is because carbon exists in two different solid forms called allotropes. One of these allotropes, graphite, is a poor conductor of electricity and would be classed as a metalloid; the other allotrope, diamond, is an insulator and therefore classed as a non-metal. In spite of problems such as this, the classification of elements into metals, metalloids and non-metals is both useful and convenient.

One other criterion which has been used to classify elements as metals, metalloids and non-metals is the acidic/basic nature of their oxides.

*A metal can be defined as an element whose oxide (with the group oxidation number) is basic. A metalloid is an element whose oxide is amphoteric and a non-metal is an element whose oxide is acidic.*

- ○ Draw an outline of the periodic table as in figure 3.7 (omitting the *d*- and *f*-block elements) and shade in the cells for those elements which you consider to be metalloids on the criterion of amphoteric oxides.
- ○ How does this classification of metals, metalloids and non-metals compare with that in terms of electrical conductivity?

## 3.7 Periodic properties

In the following sections of this chapter we shall look at the variation in properties of the elements across the periodic table. Some of these features and other properties of the elements will be taken up again and extended in chapter 12. In chapters 14, 16 and 17, we shall look at the vertical trends in properties down the periodic table.

In this chapter we shall pay particular attention to elements in the second and third periods, Li, Be, B, C, N, O, F, Ne, and Na, Mg, Al, Si, P, S, Cl, Ar respectively.

## 3.8 The periodicity of physical properties

Table 3.2 and table 3.3 show the values of various physical properties for the elements in the second and third rows respectively of the periodic table.

Notice the following trends across the second and third periods.

1 The melting point and molar heat of fusion ($\Delta H_{fus.}$) rise to the element in Group IV and then fall to low values (figure 3.8).
2 The boiling point and molar heat of vaporization ($\Delta H_{vap.}$) rise to the element in Group IV and then fall to low values (figure 3.9).
3 The density rises to the elements in Groups III and IV and then falls (figure 3.10).
4 The molar volume falls to the centre of the periods and then rises again (figure 3.10).
5 The electrical and thermal conductivity is relatively high for the metals, on the left of each period, lower for metalloids in the centre of the periods and almost negligible for non-metals on the right of each period.

* The atomic conductance (atomic electrical conductivity) is the conductivity of a block of the substance 1cm² in cross section but long enough to contain one mole of atoms of the element. It is a measure of the conductivity of one mole of atoms of the element. Notice that the units of atomic conductance (atomic electrical conductivity) are $ohm^{-1}cm^{-4}$. The electrical conductivity of a substance is usually defined as the reciprocal of the electrical resistance of a section of the substance one square centimetre in cross section and one centimetre long. Its units are therefore $ohm^{-1}$ $cm^{-1}$. This compares equal volumes of substances. More appropriately, the atomic conductance (atomic electrical conductivity) compares the conductivity of equal numbers (i.e. 1 mole) of atoms and its value is obtained by dividing the electrical conductivity of an element by its molar volume. Hence, the units of atomic conductance are $ohm^{-1}cm^{-4}$.

**Table 3.2** The values of various physical properties for the elements in the second row of the periodic table.

| | Li | Be | B | C (graphite) | C (diamond) | N | O | F | Ne |
|---|---|---|---|---|---|---|---|---|---|
| **Melting point /°C** | 180 | 1280 | 2030 | 3700 | 3550 | −210 | −219 | −220 | −250 |
| **Heat of fusion/kJ mole$^{-1}$** | 3.0 | 11.7 | 22.2 | — | — | 0.36 | 0.22 | 0.26 | 0.33 |
| **Boiling point /°C** | 1330 | 2480 | 3930 | sublimes | 4830 | −200 | −180 | −190 | −245 |
| **Heat of vaporization/kJ mole$^{-1}$** | 135 | 295 | 539 | 717 (sub) | — | 2.8 | 3.4 | 3.3 | 1.8 |
| **Density*/g cm$^3$ (at 25°C)** | 0.53 | 1.85 | 2.55 | 2.25 | 3.53 | 0.81 | 1.14 | 1.11 | 1.21 |
| **Molar volume/cm$^3$ mole$^{-1}$ (conditions as for density)** | 13.1 | 4.9 | 4.6 | 5.3 | 3.4 | 17.3 | 14.0 | 17.1 | 16.7 |
| **Atomic conductance × 1000 /ohm$^{-1}$cm$^{-4}$** | 8 | 51 | — | 0.14 | — | — | — | — | — |
| **Thermal conductivity/J cm$^{-1}$ s$^{-1}$ K$^{-1}$ (at 25°C)** | 0.71 | 1.6 | 0.01 | 0.24 | — | 0.00025 | 0.00025 | — | 0.00042 |
| **Type of element** | Metals: Li | Metals: Be | Metalloids: B | Metalloids: C graphite | Non-metals: C diamond | Non-metals: N | Non-metals: O | Non-metals: F | Non-metals: Ne |
| **Type of structure** | Giant metallic | Giant metallic | Giant molecular | Giant molecular | Giant molecular | Simple molecular: $N_2$ | Simple molecular: $O_2$ | Simple molecular: $F_2$ | Simple molecular: Ne |

* For those elements which are gaseous at 25°C, the density quoted is that of the liquid at its boiling point.

**Table 3.3** The values of various physical properties for the elements in the third row of the periodic table.

| | Na | Mg | Al | Si | P (white) | S (rhombic) | Cl | Ar |
|---|---|---|---|---|---|---|---|---|
| **Melting point /°C** | 98 | 650 | 660 | 1410 | 44 | 119 | −101 | −189 |
| **Heat of fusion/kJ mole$^{-1}$** | 2.60 | 8.95 | 10.75 | 46.4 | 0.63 | 1.41 | 3.20 | 1.18 |
| **Boiling point /°C** | 890 | 1120 | 2450 | 2680 | 280 | 445 | −34 | −186 |
| **Heat of vaporization/kJ mole$^{-1}$** | 89.0 | 128.7 | 293.7 | 376.7 | 12.4 | 9.6 | 10.2 | 6.5 |
| **Density* (at 25°C)/g cm$^{-3}$** | 0.97 | 1.74 | 2.70 | 2.33 | 1.82 | 2.07 | 1.57 | 1.40 |
| **Molar volume/cm$^3$ mole$^{-1}$ (conditions as for density)** | 23.7 | 14.6 | 10.0 | 12.1 | 16.9 | 15.6 | 22.8 | 28.5 |
| **Atomic conductance × 1000 /ohm$^{-1}$cm$^{-4}$** | 10 | 16 | 38 | 4 | $10^{-16}$ | $10^{-22}$ | — | — |
| **Thermal conductivity /J cm$^{-1}$ s$^{-1}$ K$^{-1}$ (at 25°C)** | 1.34 | 1.6 | 2.1 | 0.84 | — | 0.00029 | 0.00008 | 0.00017 |
| **Type of element** | Metals: Na | Metals: Mg | Metals: Al | Metalloid: Si | Non-metals: P | Non-metals: S | Non-metals: Cl | Non-metals: Ar |
| **Type of structure** | Giant metallic: Na | Giant metallic: Mg | Giant metallic: Al | Giant molecular: Si | Simple molecular: $P_4$ | Simple molecular: $S_8$ | Simple molecular: $Cl_2$ | Simple molecular: Ar |

* For those elements which are gaseous at 25°C, the density quoted is that of the liquid at its boiling point.

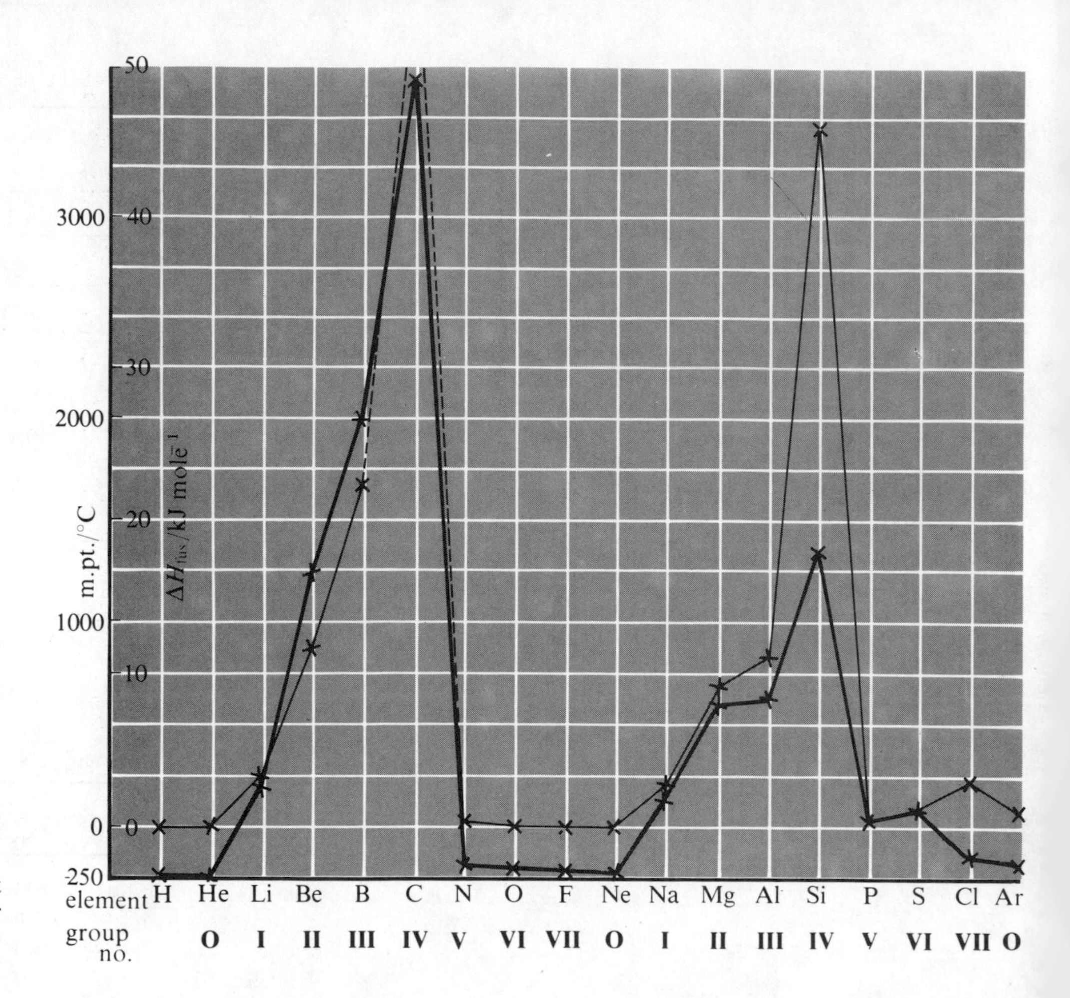


**Figure 3.8** Variation of melting point (thick line) and $\Delta H_{fus.}$ (thin line) for the elements hydrogen to argon.

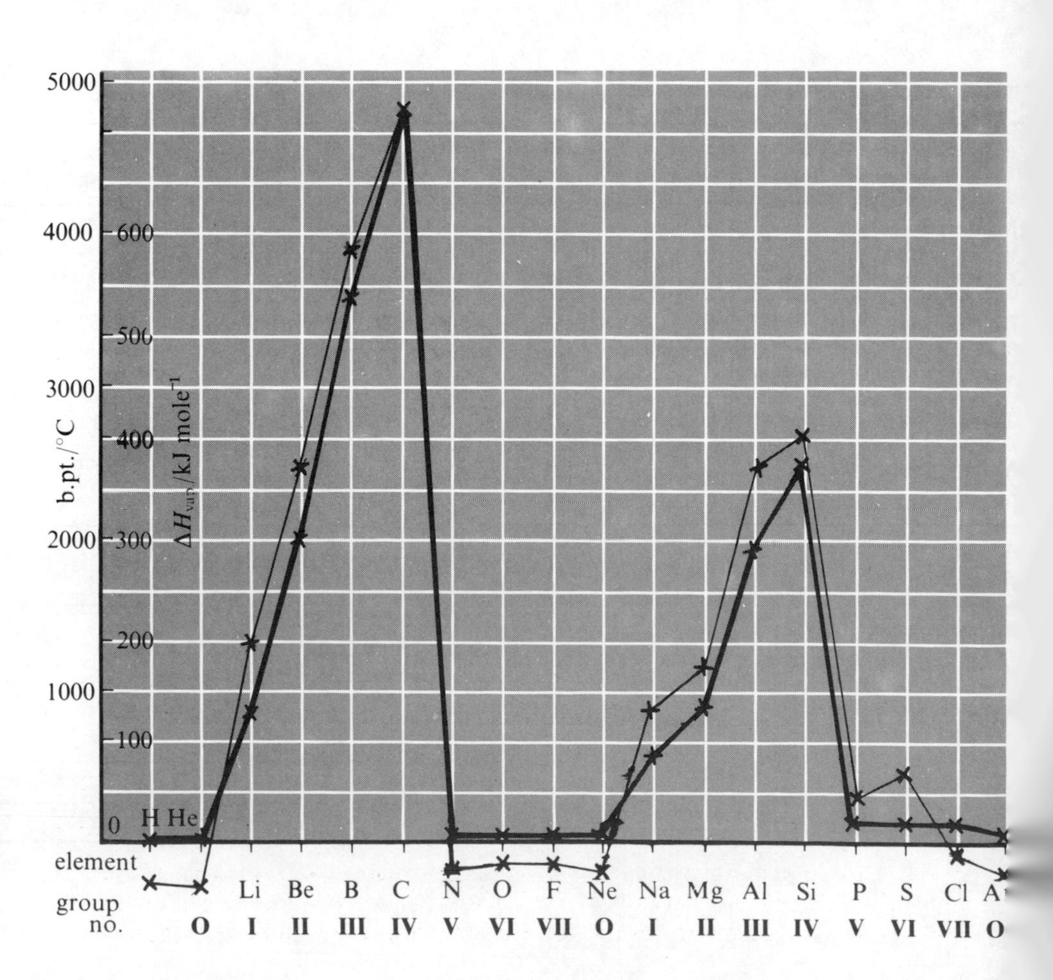


**Figure 3.9** Variation of boiling point (thin line) and $\Delta H_{vap.}$ (thick line) for the elements hydrogen to argon.

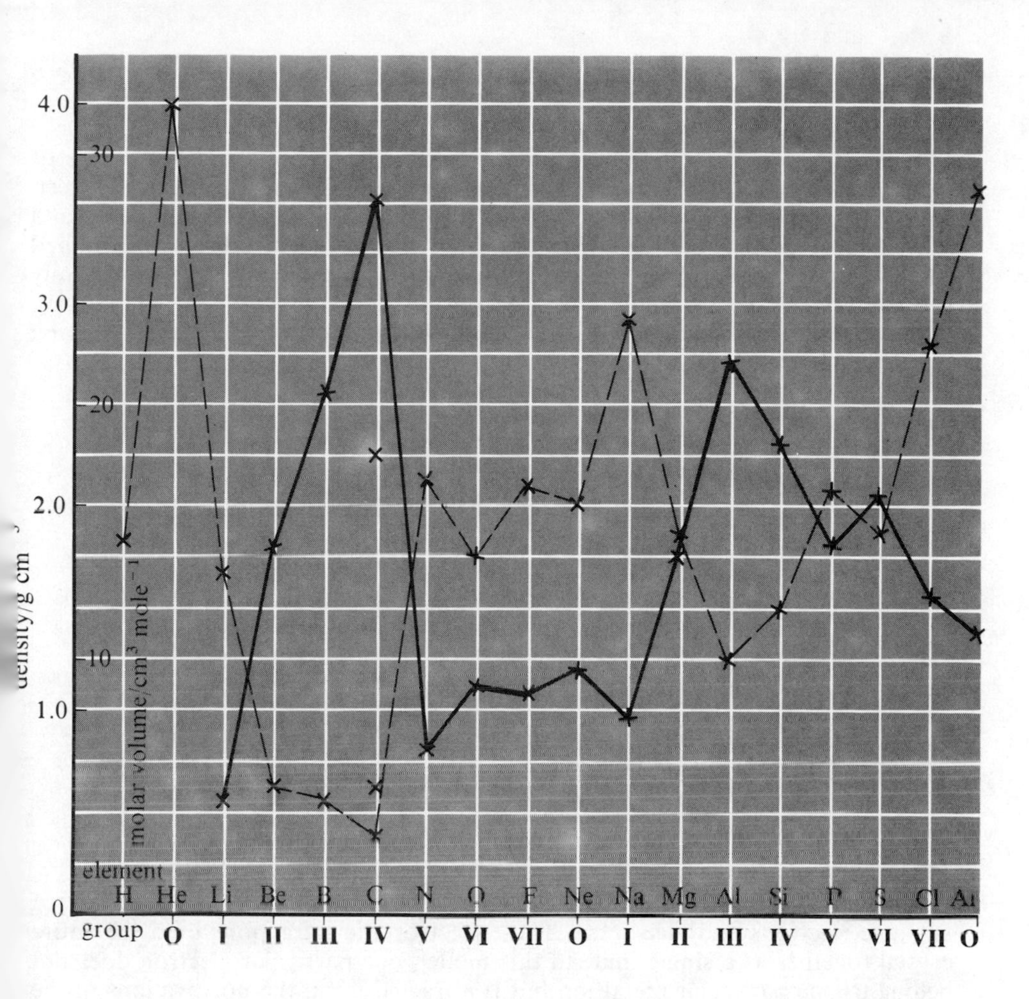


**Figure 3.10** Variation of density (thick line) and molar volume (dashed line) for the elements hydrogen to argon.

This periodicity of physical properties is, of course, related to a periodicity in the type of element and structure. Across a period, the elements change from metals through metalloids to non-metals. In period 3, sodium, the left hand element is a very reactive metal, whereas chlorine, next to the extreme right, is a very reactive non-metal. In between, the elements show a gradual transition from metals to non-metals. These periodic changes in the physical properties of the elements across the able are reflected in a periodic change in structure. The structure in the elements varies from metallic, through giant molecular in the metalloids to simple molecular tructures in the non-metals.

- ○ Which elements occur at or near the peaks on the graph in figure 3.9? What does this tell you about the strength of forces between their particles? In which groups of the periodic table are these elements found? What type of structure do these elements possess?
- ○ Which elements occur at or near the troughs on the molar volume graph in figure 3.10? What does this tell you about the average distances between their atoms?
- ○ Why is it that metals have a high electrical and thermal conductivity?
- ○ Why are the heat of fusion, heat of vaporization and molar volume values in tables 3.2 and 3.3 given for one mole of the element and not one gram?

'otice that elements on the left hand side of the periodic table exist as giant ructures (metallic or giant molecular), whereas those on the right consist of small olecules (simple molecular). The graphs in figures 3.8 and 3.9 reflect this change structure across periods 2 and 3 by the sharp drops from C to N and from Si to P. period 2, carbon is the last element with a giant structure and nitrogen is the rst element with simple molecules ($N_2$); in period 3, silicon is the last element with giant structure and phosphorus is the first element with simple molecules ($P_4$).

## 3.9 Relating the properties of elements to their structure

GIANT METALLIC STRUCTURES

Metals usually have high melting points, high boiling points and high heats of fusion and vaporization. These high values suggest that strong forces exist between the separate atoms in the metal. How does the structure of a metal explain these and other typical metal properties? The physical properties of a metal can be explained using a model in which the outer shell electrons of the metal move randomly throughout a lattice of regularly spaced positive ions (figure 3.11). The moving electrons are sometimes described as a 'sea' or 'cloud' of moving and fluctuating negative charge.

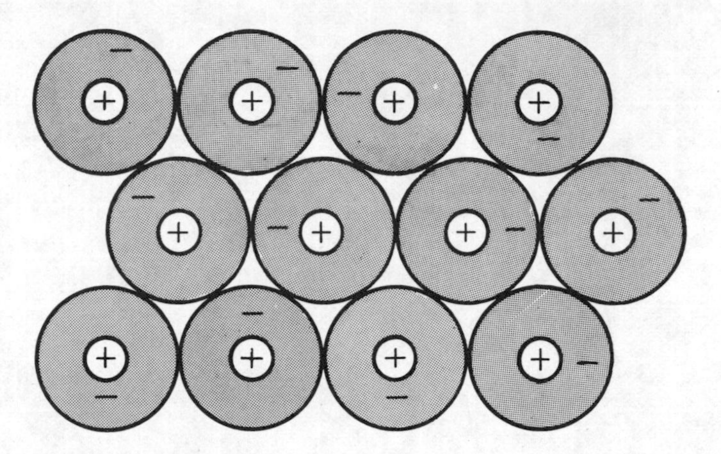

**Figure 3.11** A model of metallic structure.

In the metal lattice, each positively charged ion is attracted to the 'cloud' of negative electrons and vice versa. These electrostatic attractions bind the entire crystal together as a single unit. In this model, one particular electron does not belong to one particular metal ion but is attracted to all the positive ions in the lattice.

Since the mobile outer shell electrons are responsible for the bonding in metals, it is not surprising that moving from sodium (one outer shell electron) through magnesium (two outer shell electrons) to aluminium (three outer shell electrons) the bonding gets gradually stronger. Thus, the melting point, the boiling point, and the heats of fusion and vaporization rise from Na to Mg to Al. A similar trend is evident in the second period from Li→Be→B (see figures 3.8 and 3.9). The bonding in these elements is discussed more fully in sections 12.3 and 12.4.

The stronger bonding from Na → Mg → Al means that the atoms are pulled closer in Mg than Na and even closer in Al. This explains the increasing density (mass per unit volume) and the decreasing molar volume (volume of one mole o atoms) from Na → Mg → Al. Similar trends are observed in the third period from Li → Be → B (see figure 3.10). The mobile outer electrons in metals account fo their high electrical and thermal conductivity. Under normal conditions th electrons will move fairly randomly throughout the lattice of positive ions in th metal crystal. When the metal is connected across a difference of potential there i an overall movement of electrons (superimposed on their random motion) awa from the repelling negative electrode towards the attracting positive electrode Thus, metals have good electrical conductivity.

Why do you think the electrical conductivity rises from Na → Mg → Al? (Se table 3.3.)

The high thermal conductivity of metals can also be explained in terms of thei mobile electrons. Electrons in the regions of high temperature (i.e. electrons wit high kinetic energy) move rapidly and randomly towards the cooler regions of th metal, transferring their energy to other electrons throughout the metal.

- Why do you think the thermal conductivity rises from Na → Mg → Al an from Li → Be?
- Why does B have a lower thermal conductivity than Be? What type c structure does B have?
- Why do metals have a shiny appearance? Use the model of metals with mobi outer electrons to explain their lustre.

GIANT COVALENT STRUCTURES

The metalloids (boron, graphite and silicon) and the non-metal, diamond, have giant covalent structures. Figure 3.12 shows the arrangement of atoms in the structure of silicon and diamond. Each atom can be imagined to be situated at the centre of a regular tetrahedron strongly bound by covalent bonds to four other atoms. The covalent linking in these elements extends from one atom to the next through the whole lattice forming a three dimensional giant molecule (**macromolecule**). The strong covalent bonds hold each atom tightly in the crystal close to its neighbours and it is extremely difficult to break one atom away from the lattice. Thus, these elements have very high melting points and boiling points, very high heats of fusion and vaporization and they are very hard. Indeed, they have even higher values than metals for their melting points, boiling points and heats of fusion and vaporization (figures 3.8 and 3.9). The strong forces pull the atoms close and result in high densities and small molar volumes (figure 3.10). However, the electrons within the covalent bonds are held much more tightly in these elements than in metals and consequently their thermal and electrical conductivity is lower than metals. Refer to table 3.3 and compare the thermal and electrical conductivity of Si with the corresponding values for Na, Mg and Al.

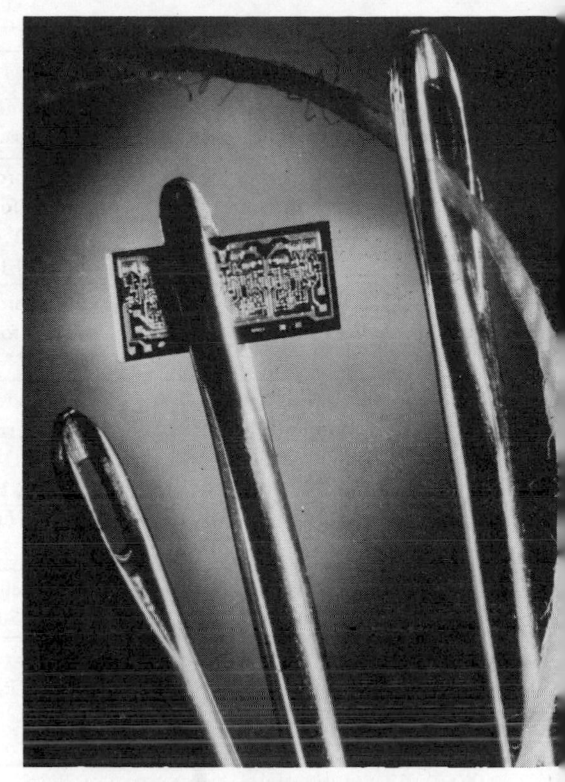

*Above* Through the eye of a needle! This photograph shows just how small it is possible to make 'silicon chips' on which extremely complex circuits can be made. 'Silicon chips' such as this are used extensively in transistors and other electrical equipment. The 'chip' shown contains 120 components and the 'rope' is ordinary sewing cotton.

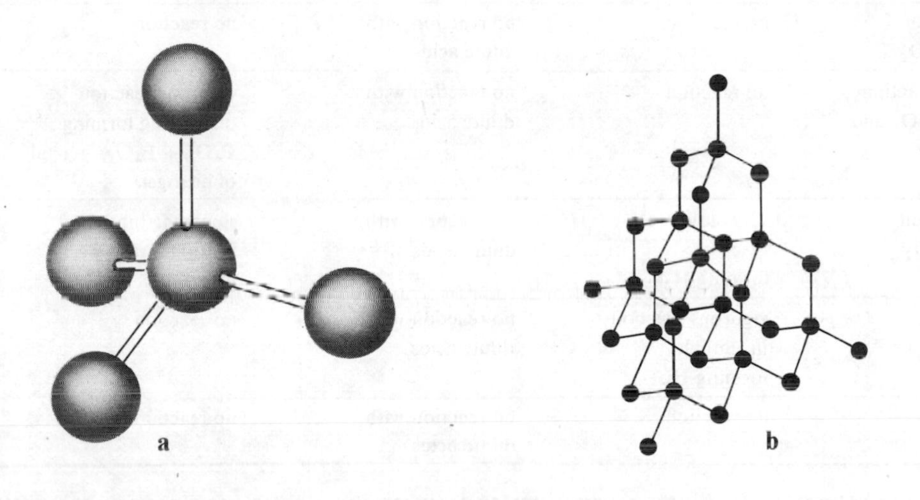


**Figure 3.12** Models showing the arrangement of atoms in the structure of silicon and diamond. (**a**) shows the tetrahedral arrangement of atoms around a central atom in silicon and diamond. (**b**) shows an extended section of the structure of silicon and diamond.

SIMPLE MOLECULAR STRUCTURES

All the non-metals (except diamond) in periods 2 and 3 form simple molecular structures. Each of these elements consists of separate, small molecules, i.e. $N_2$, $O_2$, $F_2$, Ne, $P_4$, $S_8$, $Cl_2$, Ar. There are strong covalent bonds between the atoms within these molecules (i.e. between the two O atoms in an $O_2$ molecule), but only very weak molecular (Van der Waals') forces (sections 8.3 and 8.4) between the separate molecules (i.e. between one $O_2$ molecule and other $O_2$ molecules). Consequently, the small distinct molecules can be separated easily and these non-metals have low melting points, low boiling points and low heats of fusion and vaporization. (See figures 3.8 and 3.9.) The weak Van der Waals' forces between such molecules as $N_2$, $O_2$, $P_4$ and $Cl_2$ mean that their crystals are not packed very tightly and so these non-metals have relatively low densities and high molar volumes (figure 3.10). There are no mobile electrons in the crystal structures of these elements and so they have very low conductivities.

## 3.10 The periodicity of chemical properties

Table 3.4 describes the reactions of elements in the third period with various reagents. Notice how the reactivities of the elements change across the period and also that the trends in reactivity vary with different reagents. For example, the reactivity of the elements with chlorine gradually falls across the period from Na to Ar. However, the reactivity with hydrogen falls at first to elements in the centre of the period and then rises for S and $Cl_2$. In spite of these differences, all the trends in table 3.4 can be related with considerable accuracy to the action of the elements as oxidizing or reducing agents.

**Table 3.4** Some reactions of the elements in the third period

| Element | Heat element in dry chlorine | Heat element in dry oxygen | Heat element in dry hydrogen | Add cold dilute acid (e.g. $H_2SO_4(aq)$) to the element | Add conc. $HNO_3$ as oxidising agent |
|---|---|---|---|---|---|
| Na | very vigorous reaction forming $Na^+Cl^-$ | very vigorous reaction forming $(Na^+)_2O^{2-}$ $+ (Na^+)_2(O_2^{2-})$ | very vigorous reaction forming $Na^+H^-$ | violent reaction $\rightarrow Na_2SO_4(aq)$ $+ H_2(g)$ | explosive reaction $\rightarrow Na^+$ salt $+ H_2O$ + oxides of nitrogen |
| Mg | vigorous reaction forming $Mg^{2+}(Cl^-)_2$ | very vigorous reaction forming $Mg^{2+}O^{2-}$ | vigorous reaction forming $Mg^{2+}(H^-)_2$ | very vigorous reaction $\rightarrow MgSO_4(aq)$ $+ H_2(g)$ | violent reaction $\rightarrow Mg^{2+}$ salt $+ H_2O$ + oxides of nitrogen |
| Al | vigorous reaction forming $Al_2Cl_6$ | vigorous reaction at first forming $(Al^{3+})_2(O^{2-})_3$, then the oxide layer prevents further attack | no reaction | vigorous reaction (after oxide layer is removed) | oxide layer on Al reduces vigour of reaction |
| Si | slow reaction forming $SiCl_4$ | slow reaction forming $SiO_2$ | no reaction | no reaction with dilute acids | no reaction |
| P | slow reaction forming $PCl_3$ $+ PCl_5$ | vigorous reaction forming $P_4O_6$ and $P_4O_{10}$ | no reaction | no reaction with dilute acids | vigorous reaction on heating forming $P_4O_6 + P_4O_{10}$ + oxides of nitrogen |
| S | slow reaction forming $SCl_2$ $+S_2Cl_2$ | slow reaction forming $SO_2$ | very slow reaction forming $H_2S$ | no reaction with dilute acids | slow reaction on heating forming a little $SO_2$ |
| Cl | no reaction | no reaction | vigorous reaction in sunlight forming HCl | no reaction with dilute acids | no reaction |
| Ar | no reaction | no reaction | no reaction | no reaction with dilute acids | no reaction |

The metals, (sodium, magnesium and aluminium) are strong reducing agents. They readily give up electrons to form their corresponding ions.

$$Mg \longrightarrow Mg^{2+} + 2e^-$$

Of course, these three metals never exist freely in nature, but only as $Na^+$, $Mg^{2+}$ or $Al^{3+}$ ions in their compounds. Thus, these three metals react vigorously with non-metals and acids which are oxidizing agents.

$$Cl_2 + 2e^- \longrightarrow 2Cl^-$$
$$O_2 + 4e^- \longrightarrow 2O^{2-}$$
$$H_2 + 2e^- \longrightarrow 2H^-$$
$$2H^+ + 2e^- \longrightarrow H_2$$

Silicon in Group IV is a very weak reducing agent. It will react slowly with $Cl_2$ and $O_2$ which are strong oxidizing agents, but not with $H_2$ and acids which are weaker oxidizing agents.

$$Si + 2Cl_2 \longrightarrow SiCl_4$$
$$Si + O_2 \longrightarrow SiO_2$$

The next two elements, phosphorus and sulphur, are weak reducing agents and weak oxidizing agents. They react slowly with oxygen and chlorine (strong oxidizing agents), moderately with concentrated $HNO_3$ on heating, but they have no reaction with dilute acids (weaker oxidizing agents).

$$S + Cl_2 \longrightarrow SCl_2$$
$$2S + Cl_2 \longrightarrow S_2Cl_2$$
$$S + 2HNO_3 \longrightarrow SO_2 + H_2O + NO + NO_2$$

Phosphorus does not react directly with hydrogen, but molten sulphur will react slowly with hydrogen to form hydrogen sulphide.

$$H_2 + S \longrightarrow H_2S$$

Notice in this case that the sulphur is now acting as an oxidizing agent and the hydrogen as a reducing agent.

Chlorine, at the other extreme to the metals in chemical reactivity, is a strong oxidizing agent. Thus, it has no reaction with $O_2$, dilute acids or concentrated acids which are themselves oxidizing agents, but in sunlight it will react violently with hydrogen (which can act as a reducing agent) to form hydrogen chloride.

$$H_2 + Cl_2 \longrightarrow 2HCl$$

On the extreme right, argon (a noble gas) shows no reactivity with any of these reagents.

Other periods show a similar trend in reactivity to period 3. Excluding the noble gases, elements vary from strong reducing agents on the extreme left to elements which are weak reducing agents and/or, weak oxidizing agents and finally to strong oxidizing agents.

## 3.11 Periodicity in the structure and properties of chlorides

Look closely at table 3.5. It shows the formula, state and boiling point of chlorides of the elements in the first three periods.

**Table 3.5** The formula, state and boiling point of chlorides of the elements in the first three periods

| **Period 1** | | | **H** | | | | | **He** |
|---|---|---|---|---|---|---|---|---|
| Formula of chloride | | | HCl | | | | | No chloride |
| State of chloride (at 20°C) | | | g | | | | | — |
| B.pt. of chloride /°C | | | −85 | | | | | — |
| **Period 2** | **Li** | **Be** | **B** | **C** | **N** | **O** | **F** | **Ne** |
| Formula of chloride | LiCl | $BeCl_2$ | $BCl_3$ | $CCl_4$ | $NCl_3$ | $OCl_2$ ($O_7Cl_2$) | FCl | No chloride |
| State of chloride (at 20°C) | s | s | g | l | l | g (g) | g | — |
| B.pt. of chloride/°C | 1350 | 487 | 12 | 77 | 71 | 2 (decomposes) | −101 | — |
| **Period 3** | **Na** | **Mg** | **Al** | **Si** | **P** | **S** | **$Cl_2$** | **Ar** |
| Formula of chloride | NaCl | $MgCl_2$ | $Al_2Cl_6$ | $SiCl_4$ | $PCl_3$ ($PCl_5$) | $SCl_2$ ($S_2Cl_2$) | $Cl_2$ | No chloride |
| State of chloride (at 20°C) | s | s | s | l | l (s) | l (l) | g | — |
| B.pt. of chloride /°C | 1465 | 1418 | 423 | 57 | 74 (164) | 59 (138) | −35 | — |

- ○ Work out the oxidation numbers of each element in the chlorides shown. How do the oxidation numbers of the elements vary across period 2 and period 3?
- ○ How does the volatility of the chlorides (as indicated by their boiling points) vary across period 2 and period 3?
- ○ The structures of the chlorides change from giant structures on the left of each period to simple molecular structures on the right. Which chlorides in periods 2 and 3 have giant structures and which have simple molecular structures?

We shall look at the trends in properties across a period more fully in chapter 12.

## Summary

1 Most of the credit for arranging the elements in a periodic table is given to Mendeléev. Mendeléev arranged the elements known to him in order of increasing relative atomic mass and showed that elements with similar properties recurred at regular intervals.
2 The vertical columns of similar elements are called groups and the horizontal rows of elements are called periods.
3 In modern periodic tables, all the elements are in strict order of atomic number and the modern periodic law states that the properties of elements are a periodic function of their atomic numbers.
4 Besides a division into vertical groups of similar elements it is also useful to split the periodic table into six blocks of elements with similar properties.

The reactive metals (elements in Groups I and II)
The transition metals
The lanthanides and actinides
The poor metals (metals in Groups III, IV, V and VI)
The non-metals in Groups III, IV, V, VI and VII
The noble gases

5 In spite of its limitations, the classification of elements as metals, metalloids and non-metals is both useful and convenient. The commonest classification is based on electrical conductivity.

Metals are good conductors of electricity.
Metalloids are poor conductors of electricity.
Non-metals are non-conductors.

6 Metals form giant-metallic structures. They have high melting points, high boiling points, high density and high conductivity.
7 Metalloids and diamond form giant molecular structures. They have very high melting points, very high boiling points, high density but poor conductivity.
8 Non-metals (except diamond) form simple molecular structures. They have low melting points, low boiling points, low density and they are non-conductors.
9 From left to right across the periodic table, elements change from being reactive metals, through less reactive metals, metalloids, less reactive non-metals to reactive non-metals. (On the extreme right are the noble gases.)
10 These trends in metal/non-metal character are reflected in the changing redox behaviour across a period. Excluding the noble gases, the elements change from strong reducing agents (e.g. Na, Mg) through elements which are weak reducing agents (e.g. Si), then weak oxidizing agents (e.g. S) to strong oxidizing agents (e.g. $Cl_2$).

## Study questions

1 Consider the elements (Li, Be, B, C, N, O, F, Ne) in the second period.
(a) Which elements
(i) form cations,
(ii) form a chloride of empirical formula, $XCl_3$,
(iii) react together to form a compound of formula, XY,
(iv) exist as diatomic molecules at room temperature?
(b) How do each of the following properties vary for this sequence of elements?
(i) The boiling points of the elements.
(ii) The oxidation numbers of the elements in their chlorides.
(iii) The boiling points of the chlorides of the elements.
(c) Give a brief explanation of the trends in oxidation number in terms of the electronic structures of the atoms of these elements.
(d) How do you account for
(i) the trend in boiling points of the elements,
(ii) the trend in boiling points of the chlorides of the elements?

2 Use a data book to plot (on the same graph) the melting points and boiling points of the elements from H to Ca against atomic number.
(a) Which elements occur at or near the peaks on
(i) the melting point curves,
(ii) the boiling point curves?
(b) Which elements occur in the troughs on
(i) the melting point curves,
(ii) the boiling point curves?
(c) What type of structure is found in those elements which occur
(i) at or near the peaks,
(ii) in the troughs?
(d) Make a note of elements which are liquid over
(i) unusually large temperature ranges,
(ii) unusually small temperature ranges.
(e) What explanation can you give for the fact that some elements have
(i) unusually large liquid ranges,
(ii) unusually small liquid ranges?

3 Table 3.6 shows the specific heat capacity for most of the elements from Li to Ca in the periodic table.

**Table 3.6** The specific heat capacities for elements Li to Ca inclusive (all values are in J $g^{-1}K^{-1}$)

| **Li** | **Be** | **B** | **C** | **N** | **O** | **F** | **Ne** |
|---|---|---|---|---|---|---|---|
| 3.36 | 1.66 | 1.30 | 0.50 | | | | |

| **Na** | **Mg** | **Al** | **Si** | **P** | **S** | **Cl** | **Ar** |
|---|---|---|---|---|---|---|---|
| 1.26 | 1.05 | 0.88 | 0.71 | 0.82 | 0.76 | 0.97 | — |

| **K** | **Ca** |
|---|---|
| 0.76 | 0.71 |

(a) Plot the values of specific heat capacity against atomic number for the elements listed. Does the graph show a periodic variation?
(b) Calculate the specific heat capacities per mole of atoms for those elements for which values are listed in table 3.6. Now plot these values of specific heat capacity per mole against atomic number. Does the graph show a periodic variation?

4 An element A reacts with another element B to form a compound of formula $AB_2$. The element B exists as molecules of formula $B_2$. Some properties of A, $B_2$ and $AB_2$ are tabulated below.

| | A | $B_2$ | $AB_2$ |
|---|---|---|---|
| Melting point | High (in the range 700–1 200°C) | Very low (less than −50°C) | Moderately high (in the range 400–700°C) |
| Electrical conductivity of the solid | High | Very low | Very low |
| Electrical conductivity of molten material | High | Very low | High |
| Electrical conductivity of aqueous solution of the material | | | High |

(a) Which particles will move when a potential difference is applied across a sample of (i) solid A, (ii) molten $AB_2$?

(b) Explain why the electrical conductivity of molten $AB_2$ is high, whereas that of the solid is very low.
(c) Electrolysis of an aqueous solution of $AB_2$ with Pt electrodes gave A at the cathode and $B_2$ at the anode. Suggest possible names for the elements A and B consistent with all the above information.

5 The following table shows the melting point and conductivity of five substances.

| | Melting point /K | Electrical conductivity in solid state | Electrical conductivity in molten state |
|---|---|---|---|
| Magnesium oxide, MgO | 3 173 | poor | good |
| Sodium chloride, NaCl | 1 081 | poor | good |
| Magnesium, Mg | 923 | good | good |
| Carbon dioxide, $CO_2$ | 217 | poor | poor |
| Silicon(IV) oxide, $SiO_2$ | 1 883 | poor | poor |

(a) Why is the electrical conductivity of MgO(l) good, but that of MgO(s) poor?
(b) Why is the melting point of MgO considerably higher than that of NaCl?
(c) Why is the electrical conductivity of magnesium good in *both* solid and liquid states?
(d) Why is the melting point of $SiO_2$ so much higher than that of $CO_2$ in spite of the fact that C and Si are both in Group IV of the periodic table?

6 Look closely at the boiling points of the elements Na to Ar in figure 3.9 or table 3.3.
(a) Why do the boiling points rise from Na $\rightarrow$ Si? Explain the trend in terms of the structures of the elements.
(b) Why are the boiling points of the second four elements (P to Ar) much lower than those of the first four?
(c) Why is there no clear trend in the boiling points of P, S, Cl and Ar? Plot a graph of the boiling point against the relative molecular mass of the simple molecules ($P_4$, $S_8$, $Cl_2$, Ar) of these elements to help you answer this question.

# Atomic Structure 4

## 4.1 Introduction

Only eighty years ago, scientists believed that atoms were solid indestructible particles like minute billiard balls. Since then, they have built up a great deal of evidence concerning the detailed structure of atoms.

Experiments involving electrolysis suggested that certain compounds contained charged particles called ions. The formation of these ions could be explained by the loss or gain of negatively charged particles (electrons) by atoms. This led to the idea that atoms consisted of a small positive nucleus surrounded by negative electrons. You will probably know that the nucleus is composed of two kinds of particle—protons and neutrons. But, what is the evidence for these particles? In this chapter we shall consider those experiments which have been important in shaping our ideas about atomic structure.

## 4.2 Evidence for atomic structure

1897—THOMSON EXPERIMENTS WITH ELECTRONS

In 1874, G. J. Stoney suggested the name 'electron' for the tiny negative particles which made up an electric current. Stoney realized that the experiments involving electrolysis, which Faraday had carried out earlier in the nineteenth century, could be explained in terms of electrons.

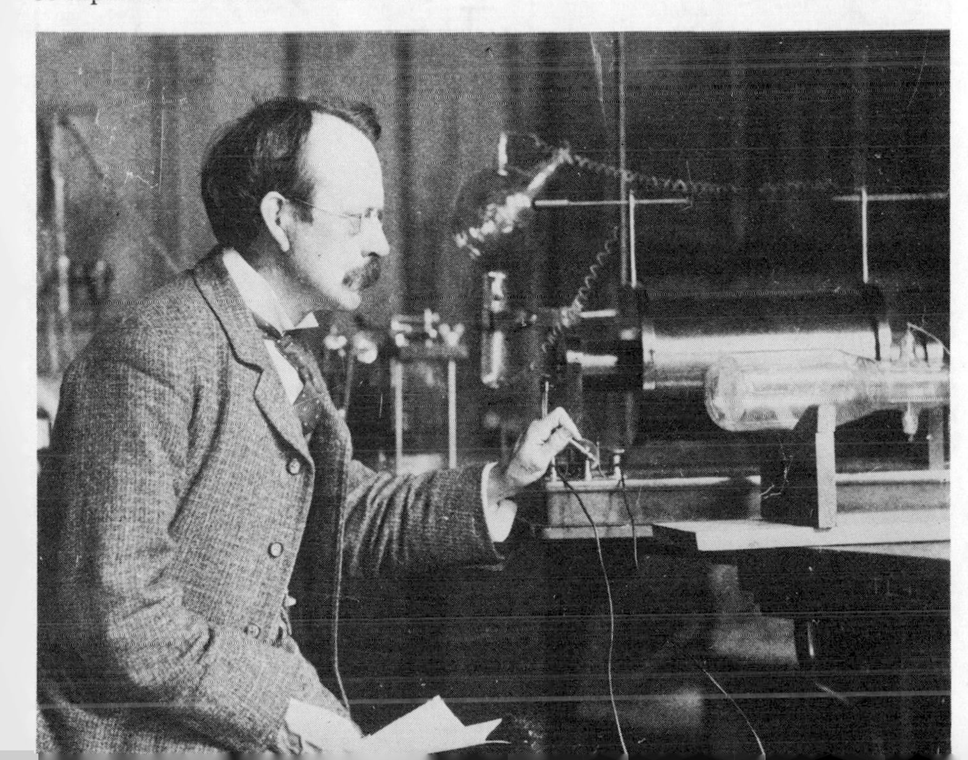

J. J. Thomson (1856–1940) investigating the conductivity of electricity by gases. Using discharge tubes (similar to that on the right of this photograph), Thomson discovered the electron.

*Above* The bright glows produced in discharge tubes are now used for advertising signs and roadside lighting.

However, firm evidence for the existence of electrons was not found until 1897. In that year, J. J. Thomson was investigating the conductivity of electricity by gases at *very low pressure*. At ordinary pressures gases are electrical insulators, but when they are subjected to very high voltages at very low pressures (below 0.01 atm) they 'break down' and conduct electricity. When the gas pressure is lowered to 0.0001 atm and a potential difference of 5 000 volts is applied across the tube containing the gas, the glass container begins to glow. Since then, the bright glows produced have found applications in neon advertising signs and sodium vapour street lamps.

When Thomson applied 15 000 volts across the electrodes of a tube containing a trace of gas a bright green glow appeared on the glass. The luminosity results from the bombardment of the glass by rays travelling in straight lines from the cathode until they strike the anode or the glass walls of the tube. Thomson called these rays **cathode rays** (figure 4.1). Thomson also showed that when cathode rays were deflected onto an electrode of an electrometer, the instrument became negatively charged and the rays could be deflected by a magnetic or by an electric field (figure 4.2).

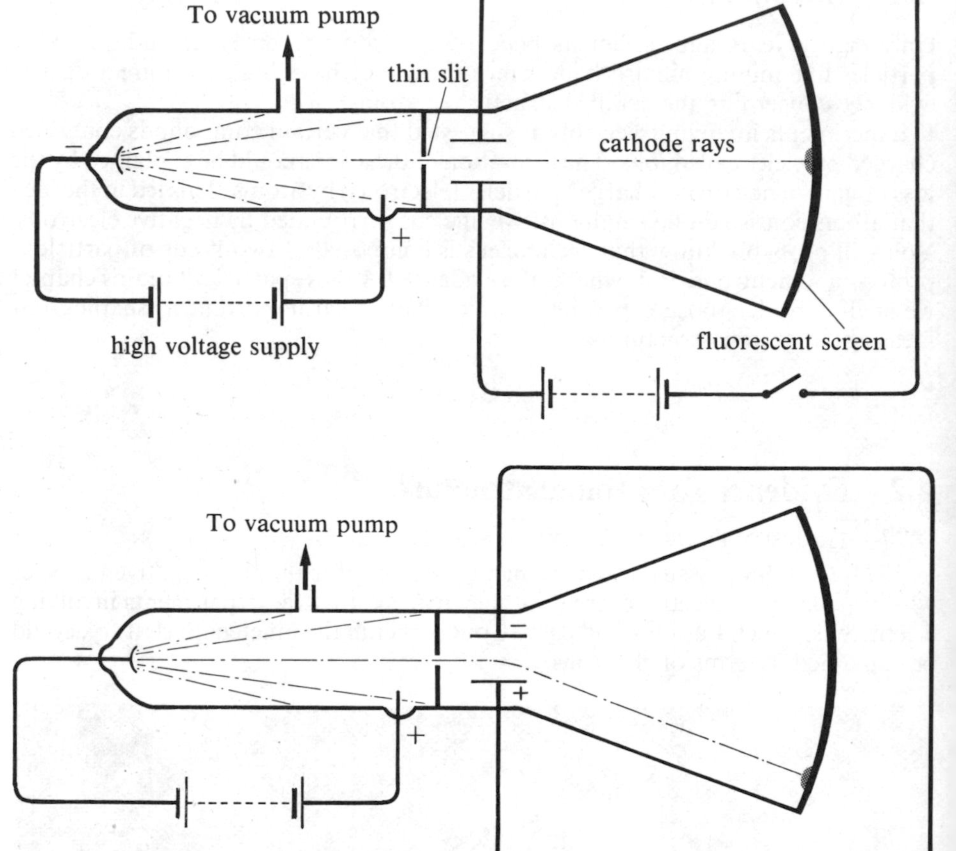


**Figure 4.1** Cathode rays, deflecting plates uncharged.

**Figure 4.2** Cathode rays, deflecting plates charged.

When the rays were deflected by an electric field across a pair of charged plates, the rays moved away from the negative plate towards the positive plate. This suggested that the cathode rays were negatively charged. Thomson studied the bending of a thin beam of cathode rays by magnetic and electric fields and concluded that they consisted of electrons—tiny negatively charged particles. Later, electrons were shown to be 1 840 times lighter than hydrogen atoms.

Now, $6 \times 10^{23}$ H atoms weigh 1g

$$\therefore \text{ 1 H atom weighs } \frac{1}{6 \times 10^{23}}\text{ g}$$

$$\therefore \text{ 1 electron weighs } \frac{1}{6 \times 10^{23}} \times \frac{1}{1\,840}\text{ g} = 9 \times 10^{-28}\text{g}$$

Exactly the same results were obtained with the cathode rays irrespective of the gas present or the materials used to make the tube and the electrodes. This suggested that electrons were present in the atoms of all substances.

- ○ Where do you think the electrons in the cathode rays have come from?
- ○ Why do the cathode rays cause the glass to fluoresce?
- ○ Which piece of Thomson's evidence showed that cathode rays could not be rays of light?
- ○ If the cathode ray beam in figure 4.1 were subjected to a magnetic field at right angles into the page which way would it be deflected?

THOMSON DISCOVERS POSITIVE RAYS

In his experiments with cathode rays, Thomson had noticed a red glow around the cathode but on the opposite side to the anode. To investigate this red glow he designed a discharge tube with a central cathode which had a hole in it (figure 4.3).

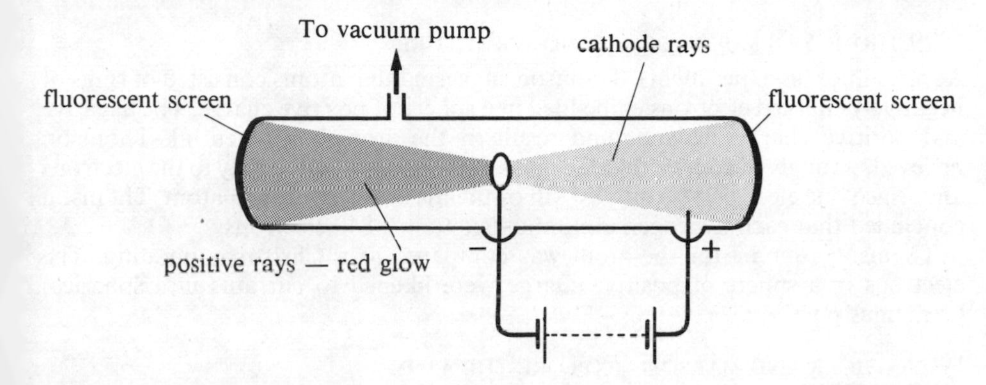


**Figure 4.3** The apparatus used by Thomson to obtain positive rays.

This time he saw a red fluorescence as well as a green fluorescence in the tube. The green fluorescence was caused by cathode rays—electrons. The red glow was caused by rays which were deflected in the opposite direction to electrons by magnetic and electric fields, showing that they contained positively charged particles. Thomson called these rays **positive rays**. They required much larger fields than electrons to cause their deflection and the mass of the particles in them depended on the nature of the gas in the tube.

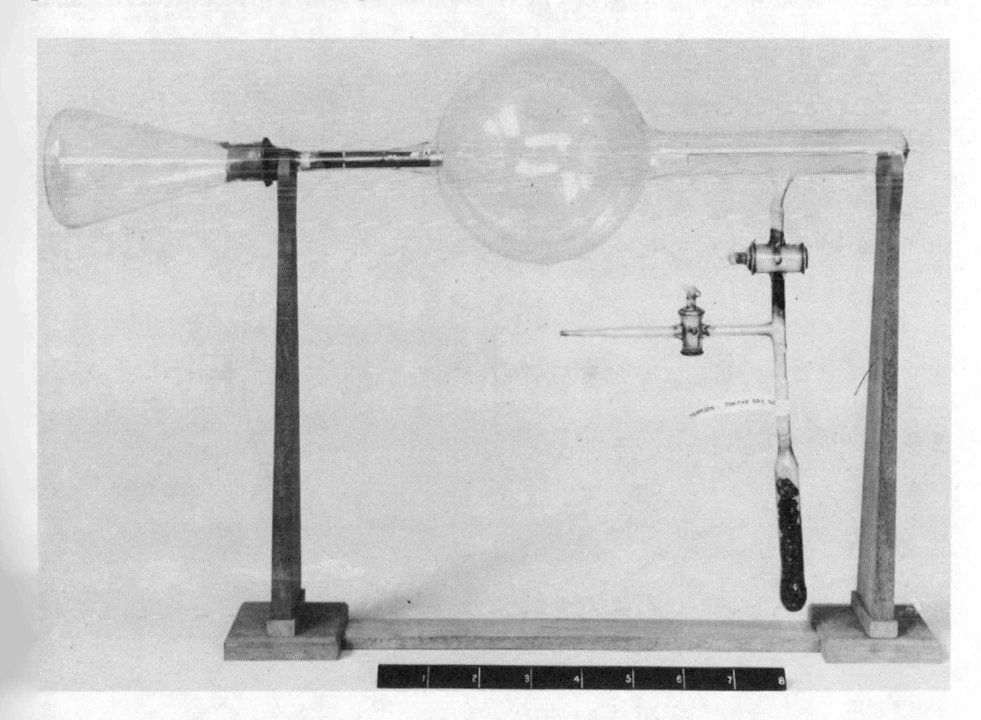


The apparatus with which Thomson investigated positive rays. Compare this with the diagram in figure 4.3.

For example, when the gas in the tube is hydrogen, the mass of the positive particles is almost identical to that of a hydrogen atom. If the gas in the tube is oxygen, the positive particles have a mass almost identical to that of oxygen atoms.

Thomson concluded that the positive rays were produced by collisions between molecules of gas in the tube and electrons in the cathode rays. When a fast-moving electron, streaming away from the cathode, collides with a gas molecule it splits the molecule into atoms and strips off one or more electrons. The atoms are converted into positive ions by the removal of electrons.

$$\underset{\text{fast-moving electron}}{e^-} + \underset{\text{hydrogen molecule}}{H_2} \longrightarrow \underset{\text{retreating electron}}{e^-} + \underbrace{H^+ + H^+}_{\text{hydrogen ions}} + \underbrace{e^- + e^-}_{\text{electrons stripped from H atoms}}$$

The ion is attracted towards the cathode and passes through the hole in it to the space behind.

The lightest positive ions are produced when the gas in the tube is hydrogen. Just as the unit of negative charge is the electron, so the unit of positive charge is that associated with a hydrogen ion, $H^+$. This unit of positive charge is usually called a **proton**.

## 1899 THOMSON'S MODEL OF ATOMIC STRUCTURE

As a result of his experiments, Thomson suggested that atoms consisted of rings of negatively charged electrons embedded in a sphere of positive charge. The negative and positive charges balance and result in the atom being neutral. Thomson believed, wrongly of course, that the mass of the atom was due only to the electrons and, since one electron was only $\frac{1}{1840}$th of the mass of a hydrogen atom, Thomson concluded that each hydrogen atom must contain 1 840 electrons.

Thomson's model for the atom was compared to a Christmas pudding. The electrons in a sphere of positive charge were likened to currants in a spherical Christmas pudding.

## 1909 GEIGER AND MARSDEN EXPLORE THE ATOM

How were the electrons arranged within the atom? Ernest Rutherford, who had been one of Thomson's research students at Cambridge University, had the idea of probing inside the atom using alpha-particles from radioactive substances as 'nuclear bullets'. Alpha-particles are helium ions, $He^{2+}$. They travel a few centimetres in air and produce a tiny pinpoint of light on striking a fluorescent screen.

In order to investigate atomic structure using 'alpha-particle bullets', Geiger and Marsden, working under Rutherford's guidance at Manchester University, built the apparatus shown in figure 4.4. A piece of radium within a protective lead shield

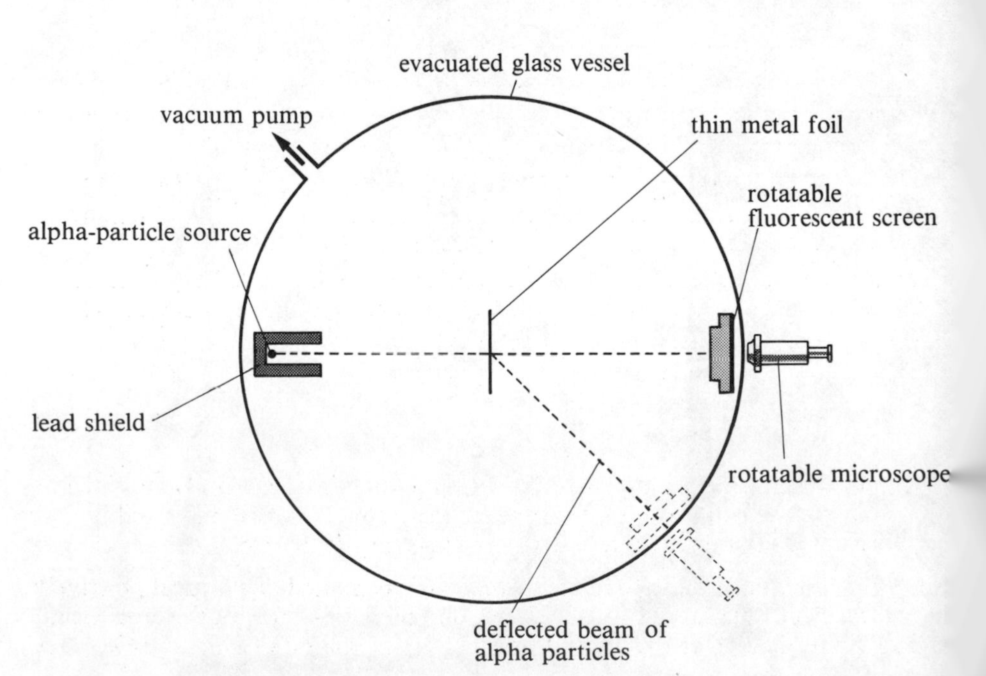


**Figure 4.4** The apparatus used by Geiger and Marsden to investigate the deflection of α-particles by thin metal foil.

was used as the source of alpha-particles. A narrow hole through the lead allowed the alpha-particles to travel in only one direction through the evacuated vessel towards a fluorescent zinc sulphide screen.

When an alpha-particle hits the fluorescent screen a tiny pinpoint of light can be observed through the microscope. Even when a thin metal foil was placed in the path of the alpha-particles, flashes were still seen on the screen.

Using Thomson's model of atomic structure, Rutherford expected that most of the very fast-moving alpha-particles might pass straight through the foil or be deviated only a little. Most of the alpha-particles did pass straight through the foil, but when the detecting screen and microscope were rotated from the straight-on position flashes could still be seen. Clearly, some of the alpha-particles are deflected by the foil, but, to everyone's surprise, one particle in every 10 000 appeared to rebound from the foil.

Here is Rutherford's account of the startling discovery.

> 'I remember Geiger coming to me in great excitement and saying, "We have been able to get some of the alpha-particles coming backwards . . .".
> It was quite the most incredible event that has ever happened to me in my life. It was almost as incredible as if you fired a 15-inch shell at a piece of tissue paper and it came back and hit you. On consideration, I realized that this scattering backwards must be the result of a single collision . . . with a system in which the greater part of the mass of the atom was concentrated in a minute nucleus.'

*Above* Thomson (right) and Rutherford are unquestionably two of the most brilliant scientists of the twentieth century.

## 1911 RUTHERFORD EXPLAINS THE STRUCTURE OF THE ATOM

Since alpha-particles are positively charged, Rutherford suggested that deflections and reflections could only be caused by the particles coming close to a concentrated region of positive charge. Only a very small fraction of the alpha-particles were deflected, so he concluded that the region of positive charge which caused the scattering must occupy a relatively small part of the atom (figure 4.5).

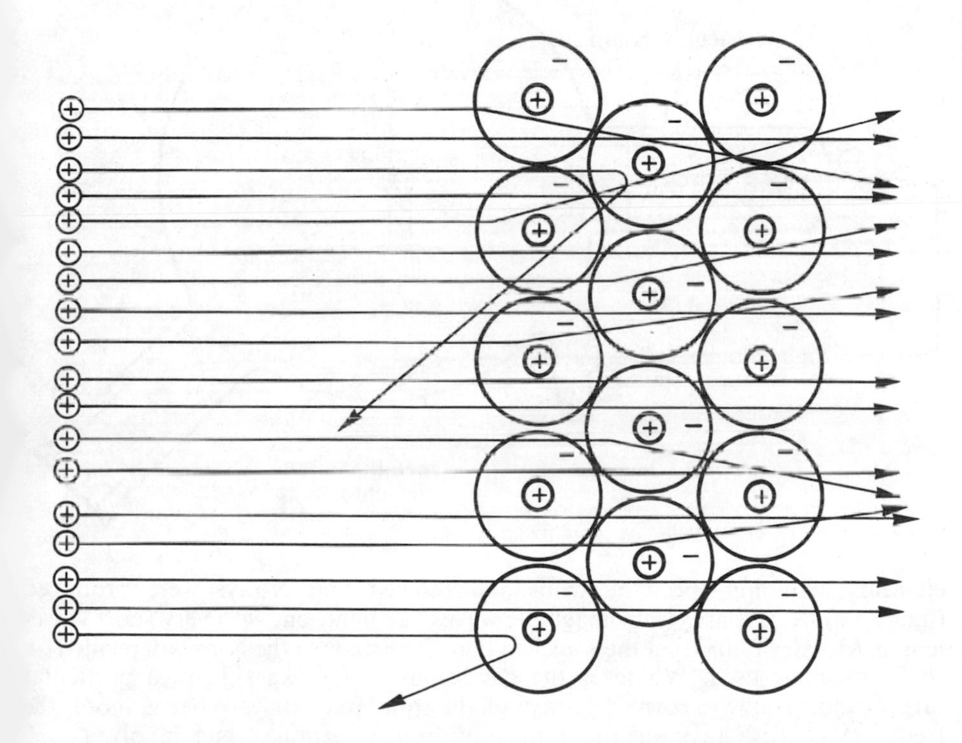

**Figure 4.5** Most alpha-particles pass straight through the foil without coming near a nucleus. Others come close to a nucleus and are deflected. A few alpha-particles approach a nucleus head-on and are repelled back on the same side of the foil as they approached.

- ○ Why do most of the alpha-particles pass straight through the metal foil?
- ○ Why do some alpha-particles appear to rebound from the metal foil?
- ○ What factors will affect the extent of scattering of the alpha-particles?

utherford suggested that atoms in the metal foil consisted of a central positively harged nucleus composed of protons, where the mass of the atom was concen-rated. This nucleus was surrounded by a much larger volume in which the electrons nove.

From the angles through which alpha-particles are deflected Rutherford cal-ulated that the nucleus of an atom would have a radius of about $10^{-14}$m whereas hat of the whole atom is about $10^{-10}$m; the nucleus is about one ten-thousandth f the size of an atom. Thus, if we magnified an atom to the size of a football

If we magnified an atom one million million times to the size of Wembley Stadium, the nucleus would be the size of a pea at the centre of the pitch.

stadium (about 100m across), the nucleus would be represented by a pea placed at the centre of the pitch.

Notice the difference between Rutherford's atomic model and that proposed by Thomson a few years earlier. Rutherford's atomic model has been compared to the solar system; Rutherford pictured each atom as a miniature solar system with electrons orbiting the nucleus, like planets orbiting the sun. He suggested that hydrogen, which had the smallest atoms, would have *one* proton in its nucleus, balanced by *one* orbiting electron. Atoms of helium which were next in size would have *two* protons in their nucleus, balanced by *two* orbiting electrons, etc., etc.

Although Rutherford's idea of planetary electrons has now been discarded, his idea of a small positive nucleus has been supported by many experiments.

## 1913 MOSELEY EXPLORES THE NUCLEUS

Calculations based on the results obtained by Geiger and Marsden showed that the number of unit positive charges (protons) in the nucleus of an atom was *about* half its relative atomic mass. At about the same time, it was noticed that the number of positive charges on the nucleus was equal to the atom's numbered position in the periodic table which had been given the symbol Z. But, was Z merely a number corresponding to the order of the element in the periodic table. Perhaps it could be obtained experimentally from the properties of an element and be shown to be a fundamental characteristic of atomic structure?

In 1913, Moseley, working at Oxford University, bombarded various metallic

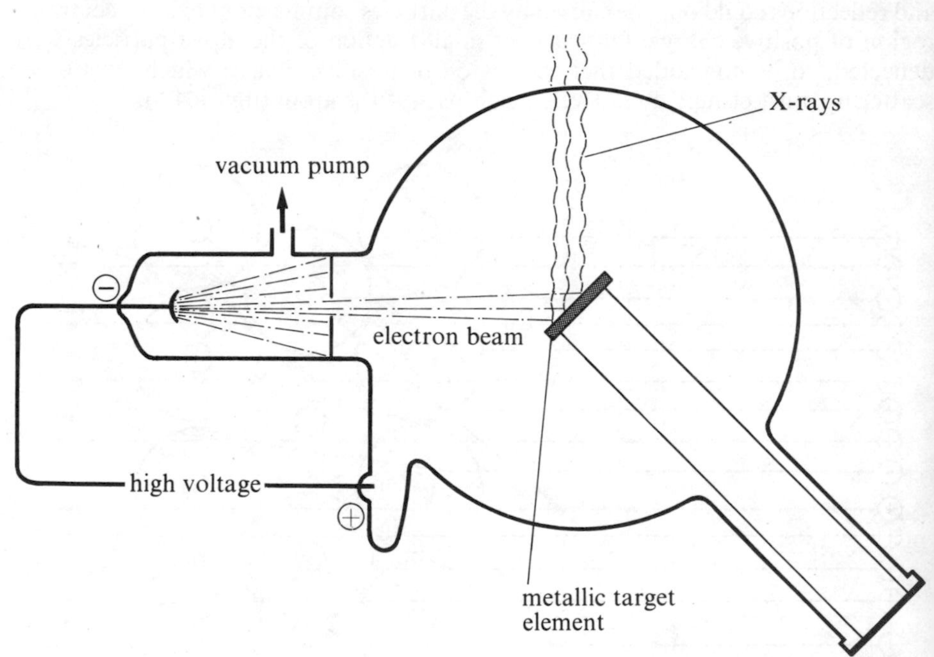


**Figure 4.6** An X-ray tube.

elements with high-speed electrons and showed that X-rays were produced (figure 4.6). X-rays are electromagnetic waves like light, but with very short wavelength. Moseley found that the wavelength or frequency of the X-rays depended on the element he used. Whatever the conditions of his experiment, a particular target element always formed X-rays of the same frequency. What is more, the frequency of the X-rays was found to be given by a formula which involved Z.

$$\sqrt{v} = a(Z - b)$$

($v$ is the frequency of the X-rays, $a$ and $b$ are constants—see figure 4.7).

Thus, Z which was believed to be of no more significance than the atom's order in the periodic table, was shown to be of fundamental significance to the properties of an atom.

Z is known as the **atomic number**; it is the most important feature of an element's individuality. The atomic number represents

(i) the number of protons in the nucleus,
(ii) the number of electrons in the neutral atom,
(iii) the order in which the element appears in the periodic table.

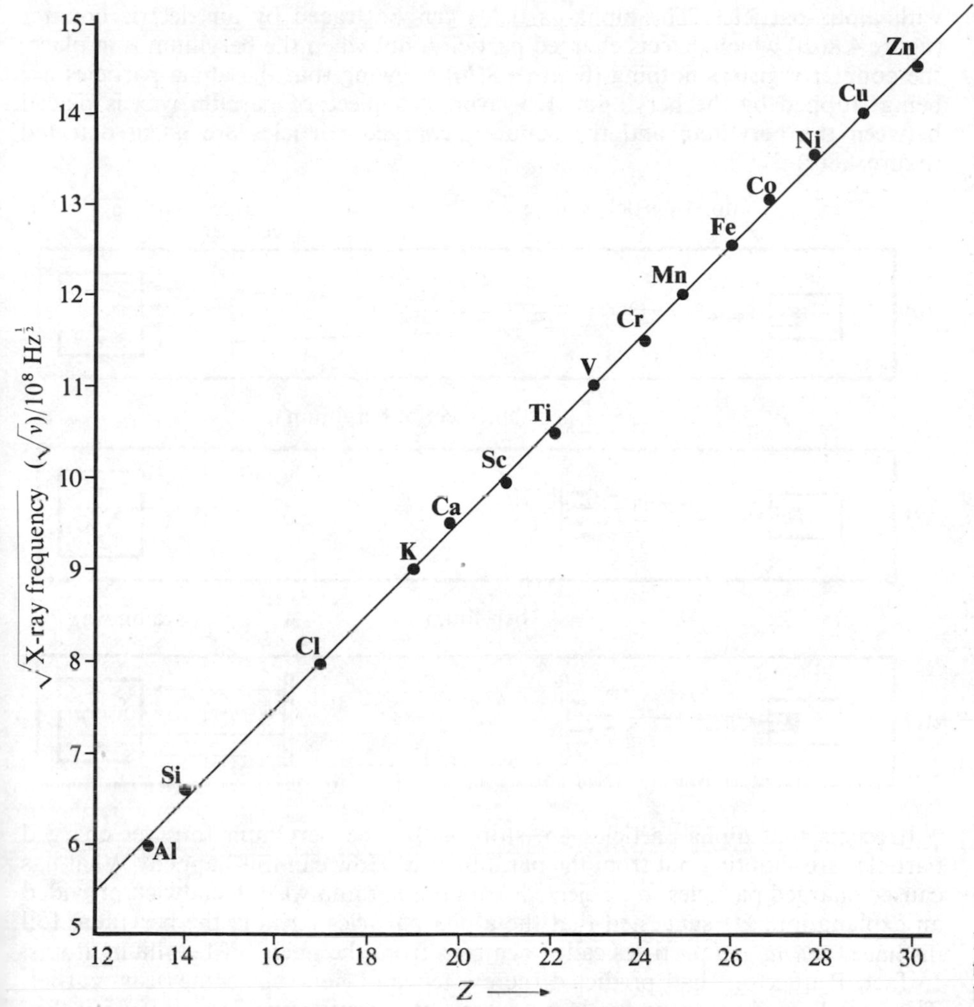


**Figure 4.7** A graph of the square root of the frequency of X-rays against the atomic number of the target element.

The atomic number of an element can therefore be predicted from its order in the periodic table; element number five in the periodic table has an atomic number of five, element 25 has an atomic number of 25 and so on.

Moseley's equation was also important in showing the values of Z for three undiscovered elements. Blank spaces occurred in Moseley's results between molybdenum(Mo) and ruthenium(Ru), between neodymium(Nd) and samarium(Sm) and between tungsten(W) and osmium(Os). Moseley's experiments sparked off a search for these missing elements which were found to occupy the blank spaces.

What are the names of the three missing elements?

Another valuable outcome of Moseley's work was that the characteristic frequency of the X-rays from a particular element provided an easy and certain method of identifying the element.

## 1932 CHADWICK DISCOVERS THE NEUTRON

In spite of the success of Rutherford and Moseley in explaining atomic structure one major problem remained unsolved.

If the hydrogen atom contains one proton and the helium atom contains two protons, then the relative atomic mass of helium should be twice that of hydrogen. Unfortunately, the relative atomic mass of helium is four and not two.

Chadwick, one of Rutherford's collaborators, was able to show where the extra mass in helium atoms came from. Chadwick bombarded a thin sheet of beryllium

Part of the original apparatus with which Chadwick established the existence of neutrons.

with alpha-particles. The alpha-particles can be traced by an electric counter (figure 4.8(*a*)) which detects charged particles, but when the beryllium is in place, the counter registers nothing (figure 4.8(*b*)) showing that the alpha-particles are being stopped by the beryllium. However, if a piece of paraffin wax is placed between the beryllium and the counter, charged particles are again detected (figure 4.8(*c*)).

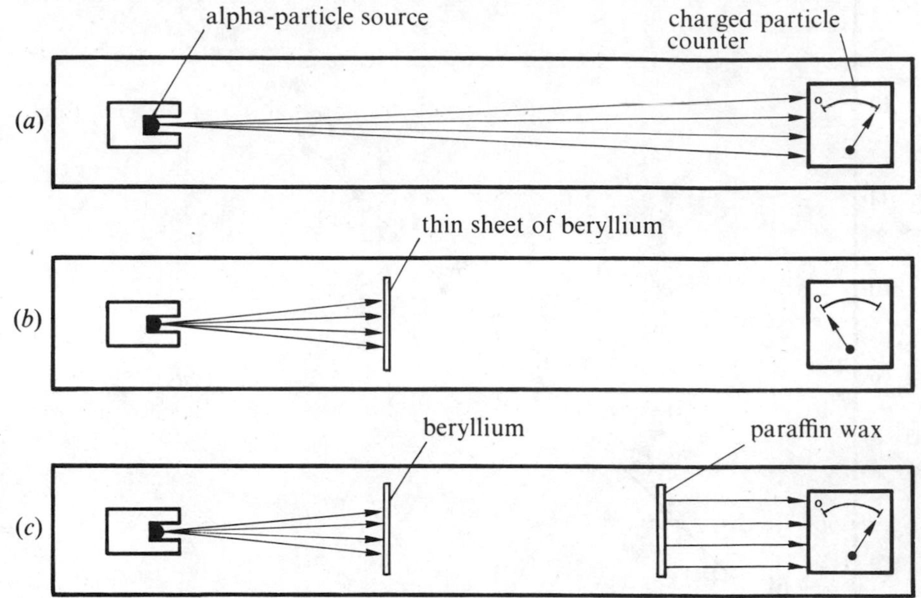


**Figure 4.8** The experiment in which Chadwick detected neutrons.

It seems that alpha-particles are stopped by the beryllium foil, yet charged particles are shooting out from the paraffin wax. How can this happen? What has caused charged particles to be ejected from the paraffin wax? Chadwick provided an explanation. He suggested that the alpha-particles striking the beryllium foil displaced *uncharged* particles called **neutrons** from the nuclei of beryllium atoms. In fact, Rutherford had predicted the existence of neutrons some years earlier. These uncharged neutrons would not affect the charged particle counter, but they could displace positively charged protons from the paraffin wax which would affect the counter. A summary of Chadwick's explanation is shown in figure 4.9. Further experiments showed that neutrons had almost the same mass as protons and Chadwick was able to explain the difficulty concerning the relative atomic masses of hydrogen and helium.

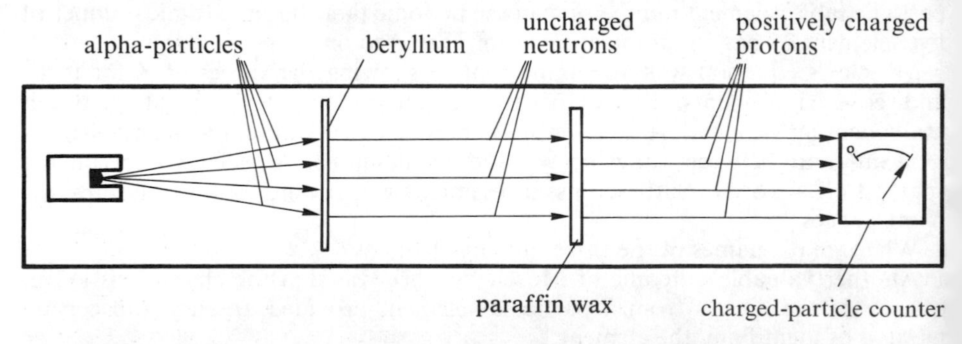


**Figure 4.9** Chadwick's explanation of his experiment to detect neutrons.

Hydrogen atoms have one proton, no neutrons and one electron; helium atoms have two protons, two neutrons and two electrons. Since the mass of the electron is negligible compared to the masses of the proton and neutron, a helium atom is four times as heavy as a hydrogen atom, i.e. the relative atomic mass (atomic weight) of helium = 4.

Lithium is the third element in the periodic table. Its relative atomic mass is very nearly seven.

- ○ What is the atomic number of lithium?
- ○ How many electrons does each lithium atom contain?
- ○ How many neutrons does each lithium atom contain?
- ○ How many protons, neutrons and electrons does one $Li^+$ ion contain?

## 4.3 Fundamental sub-atomic particles

As a result of the experiments described in section 4.2 scientists believed that all atoms were composed of three fundamental particles—protons, neutrons and electrons.

The nucleus of the atom is composed of protons and neutrons each of which has a mass almost equal to that of a hydrogen atom. Virtually all the mass of the atom is concentrated in the nucleus which occupies only a small fraction of the total volume of the atom. The neutron has no charge, whereas the proton carries one positive charge.

The nucleus of a fluorine atom has a diameter of about $10^{-12}$cm and a mass of $3.1 \times 10^{-23}$g

$$\therefore \text{ The density of the fluorine nucleus } = \frac{\text{mass}}{\text{volume}}$$

$$= \frac{3.1 \times 10^{-23}}{\frac{4}{3}\pi(5 \times 10^{-13})^3} \text{ g cm}^{-3}$$

$$= 6 \times 10^{13}\text{g cm}^{-3}.$$

The incredibly high density within the fluorine nucleus suggests that the particles within it are drawn very close together by extremely powerful forces which are capable of overcoming the repulsion of the protons for each other. However, these powerful forces within the nucleus are effective over only a very short range since they do not pull the outer electrons into the nucleus. The electrons have a mass which is only $\frac{1}{1840}$th of that of the proton and they carry one negative charge.

Table 4.1 summarizes the relative masses, the relative charges and the position within the atom of these sub-atomic particles. The atoms of all elements are built up from these three particles, different atoms having different numbers of the three particles. Other particles have been detected during atom-splitting experiments, but all these particles are less stable than protons, neutrons and electrons and only exist under extreme conditions.

**Table 4.1** The relative masses, relative charges and positions within the atom of the three fundamental sub-atomic particles

| **Particle** | **Mass (relative to that of a proton)** | **Charge (relative to that on a proton)** | **Position within the atom** |
|---|---|---|---|
| **Proton** | 1 | +1 | nucleus |
| **Neutron** | 1 | 0 | nucleus |
| **Electron** | $\frac{1}{1840}$ | −1 | shells |

## 4.4 Isotopes

n 1913 Thomson discovered that some elements could have atoms with different nasses. Thomson's equipment was developed by Aston and others into a mass pectrometer which could compare the relative masses of atoms (section 1.2).

Thomson and Aston discovered that some naturally occurring elements conained atoms that were not exactly alike. When atoms of these elements were onized and deflected in a mass spectrometer, the beam of ions separated into two r more paths. Since all the ions in the beam have the same electrical charge, those n the separated paths must have different masses. Consequently, the atoms from hich the ions were formed must also have different masses.

These atoms of the same element with different masses are called **isotopes**. All ne isotopes of one particular element have the *same atomic number* because they ave the same number of protons, but they have *different mass numbers* because ney have different numbers of neutrons.

Figure 4.10 shows a mass spectrometer trace for chlorine. It shows that chlorine onsists of two isotopes; each has an atomic number of 17, but they have different otopic masses, 35 and 37 respectively.

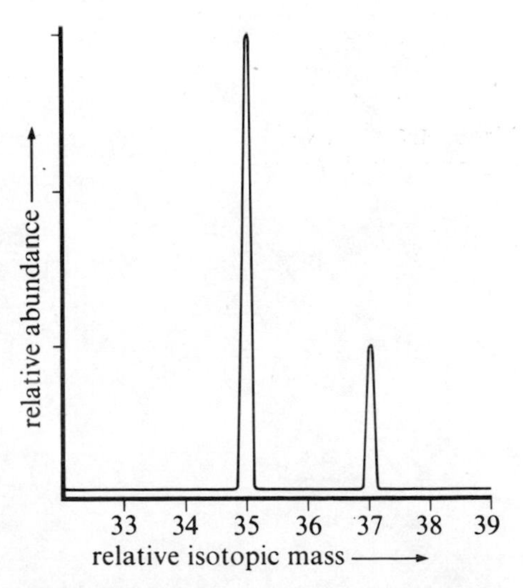


**Figure 4.10** A mass spectrometer trace for chlorine.

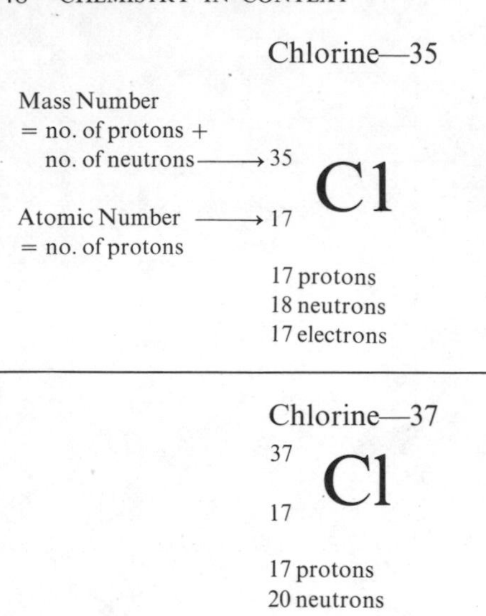


**Figure 4.11** Specifying the mass number and atomic number with the symbol of an isotope.

Atoms of chlorine-35 have 17 protons and 18 neutrons, whereas atoms of chlorine-37 have 17 protons and 20 neutrons. The number of protons + the number of neutrons in an atom is called the **mass number**. Thus, chlorine-35 has a mass number of 35 (17 + 18) and chlorine-37 has a mass number of 37 (17 + 20).

We can write the symbol $^{35}_{17}Cl$ in order to specify the mass number and the atomic number of the chlorine-35 atom. Figure 4.11 shows both the chlorine isotopes represented in this way. The mass number is written at the top and to the left of the symbol; the atomic number is written at the bottom and to the left of the symbol.

Isotopes have the same number of electrons and hence the same chemical properties, since chemical properties depend upon the transfer and redistribution of electrons. However, since isotopes have different numbers of neutrons, they will have different masses and hence different physical properties. For example, pure $^{37}_{17}Cl_2$ will have a higher density, higher melting point and higher boiling point than pure $^{35}_{17}Cl_2$.

## 4.5 Mass numbers, relative isotopic masses and relative atomic masses

The relative mass of a single isotope is called its **relative isotopic mass**. **Relative isotopic masses** relate to the same scale as relative atomic masses on which the isotope $^{12}_{6}C$ is given a relative mass of 12.0000 units. (Relative atomic masses were introduced in section 1.3.) On this scale, the mass of the proton (1.0074 units) is almost the same as that of the neutron (1.0089 units) and the mass of the electron is very small in comparison (0.0005 units). Now since the *relative* masses of the proton and neutron are both very close to one and the electron has a negligible mass, it follows that all relative isotopic masses will be very close to whole numbers. In fact, the relative isotopic mass of an isotope will be very close to its mass number (number of protons + number of neutrons) and the two are assumed to be almost identical in all but the most accurate work.

However, naturally occurring elements often consist of a mixture of isotopes and this results in relative atomic masses which are not close to a whole number, since the relative atomic mass of an element represents the *average* mass of one atom taking into account the different isotopes and their relative proportions. For example, chlorine consists of two isotopes—$^{35}_{17}Cl$ and $^{37}_{17}Cl$ with relative isotopic masses of 35 and 37 respectively. If chlorine consisted of only $^{35}_{17}Cl$, its relative atomic mass would be 35; if it consisted of only $^{37}_{17}Cl$, its relative atomic mass would be 37. A 50:50 mixture of $^{35}_{17}Cl$ and $^{37}_{17}Cl$ would have a relative atomic mass of 36.

Look at figure 4.10. The relative heights of the peaks corresponding to $^{35}_{17}Cl$ and $^{37}_{17}Cl$ indicate that the isotopes occur in the ratio of about 3:1 (i.e. 75% chlorine-35 and 25% chlorine-37). This results in a relative atomic mass for chlorine of 35.5.

**Table 4.2** Relating the relative atomic mass of chlorine to the composition of its isotopic mixture

| $\% ^{35}_{17}Cl$ | 100 | 75 | 50 | 25 | 0 |
|---|---|---|---|---|---|
| $\% ^{37}_{17}Cl$ | 0 | 25 | 50 | 75 | 100 |
| Relative Atomic Mass | 35.0 | 35.5 | 36.0 | 36.5 | 37.0 |

Table 4.2 shows this more clearly. Can you see that *the relative atomic mass is a weighted mean of the relative isotopic masses* of the different isotopes; weighted that is, in the proportions in which they occur?

∴ The relative atomic mass of chlorine $= 35 \times \frac{75}{100} + 37 \times \frac{25}{100}$

(35 = relative isotopic mass of chlorine-35; $\frac{75}{100}$ = proportion of Cl-35; 37 = relative isotopic mass of chlorine-37; $\frac{25}{100}$ = proportion of Cl-37)

$= 26.25 + 9.25$

$= 35.50$

Figure 4.12 shows a mass spectrometer trace for the isotopes of neon. The atomic number of neon is 10.

- ○ What are the relative isotopic masses of the two neon isotopes?
- ○ What are the mass numbers of the two neon isotopes?
- ○ How many protons, neutrons and electrons has each of the neon isotopes?
- ○ What are the percentage abundancies of the two isotopes?
- ○ Calculate the relative atomic mass of neon.

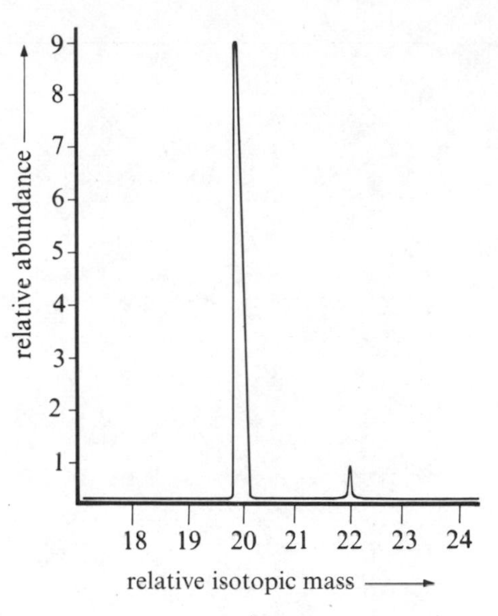


**Figure 4.12** A mass spectrometer trace for neon.

## Summary

1 Cathode rays consist of tiny negatively charged particles called electrons which have a mass of $\frac{1}{1840}$th the mass of a hydrogen atom.
2 Rutherford pictured the atom as a miniature solar system in which electrons orbited the positively charged nucleus like planets orbiting the sun.
3 The nucleus is very small compared with the size of the atom as a whole. The volume of the nucleus is only about one million millionth (i.e. one $10^{-12}$th) of the total volume of the atom.
4 The nucleus is composed of two particles:
protons with a mass of one relative to the hydrogen atom and a relative charge of +1, and neutrons with a mass of one relative to the hydrogen atom but with no charge.
5 Atomic number = number of protons
= order of element in the periodic table.
6 Mass number = number of protons + number of neutrons.
7 Isotopes are atoms with the same atomic number, but different mass numbers. Isotopes have the same number of electrons and therefore the same chemical properties.
Isotopes have different numbers of neutrons (i.e. different masses) and therefore different physical properties.

## Study questions

1 (a) What are the atomic numbers of tellurium (Te) and iodine (I)?
(b) What are the relative atomic masses (atomic weights) of Te and I?
(c) What are the numbers of protons, neutrons and electrons in the commonest isotopes of tellurium (Te-128) and iodine (I-127)?
(d) Why does Te come before I in the periodic table?
(e) Te comes before I in the periodic table, but Te has a larger relative atomic mass than I. Explain.
(f) Look closely at the periodic table and write down the names of two other *pairs* of elements which are placed in the periodic table in the reverse order to their relative atomic masses.

The five main proposals in Dalton's Atomic Theory of matter were:
1 All matter is composed of tiny indestructible particles called atoms.
2 Atoms cannot be created or destroyed.
3 Atoms of the same element are alike in every way.
4 Atoms of different elements are different.
5 Atoms can combine together in small numbers to form molecules.
(a) In the light of modern knowledge about atoms, isotopes, molecules and atomic structure comment on the truth of each of these proposals.
(b) Why is Dalton's Atomic Theory still useful in spite of these limitations?

3 Assume that the fluorine atom ($^{19}_{9}F$) is a sphere of diameter $10^{-10}$m and that its nucleus is a sphere of diameter $10^{-14}$m.
(a) What is (i) the atomic number,
(ii) the mass number of $^{19}_{9}F$?
(b) What is the *actual* mass of the nucleus in *one* $^{19}_{9}F$ atom?
($6 \times 10^{23}$ $^{1}_{1}H$ atoms have a mass of 1g.)
(c) What is the density of the nucleus in a $^{19}_{9}F$ atom?
(d) What does the value in (c) suggest about the forces within the $^{19}_{9}F$ nucleus?
(e) What is the ratio of the volume of the atom to the volume of the nucleus in a $^{19}_{9}F$ atom?

4 Natural silicon in silicon-containing ores contains 92% silicon-28, 5% silicon-29 and 3% silicon-30.
(a) What is the atomic number of silicon?
(b) What are the relative isotopic masses of the 3 silicon isotopes?
(c) What is the relative atomic mass of silicon?
(d) Samples of pure silicon obtained from natural silicon ores, mined in different parts of the world, have slightly different relative atomic masses. Explain.

5 Natural hydrogen contains two isotopes; $^{1}_{1}H$ and $^{2}_{1}H$ (sometimes called deuterium).
(a) Write down the possible formulae of a hydrogen molecule.
(b) What are the possible relative molecular masses for a hydrogen molecule?
(c) 1 $dm^3$ (1 litre) of a deuterium-enriched sample of hydrogen gas was found to weigh 0.10g at s.t.p. What is the isotopic composition of this gas?

6 The accurate relative isotopic masses for five isotopes are shown below.

| $^{1}_{1}H$ | $^{2}_{1}H$ (D) | $^{12}_{6}C$ | $^{14}_{7}N$ | $^{16}_{8}O$ |
|---|---|---|---|---|
| 1.0078 | 2.0141 | 12.0000 | 14.0031 | 15.9949 |

(a) Calculate the accurate relative molecular masses (molecular weights) for
(i) $N_2$ (ii) DCN (iii) CO (iv) $C_2H_4$ (v) $C_2D_2$.
(b) The relative molecular mass of a certain gas in a high resolution mass spectrometer was 28.0171. What gas is probably under observation?

7 A compound containing only carbon, hydrogen and oxygen has peaks in its mass spectrum which correspond to relative masses of 60 (the molecular ion), 43, 31, 29 and 17 relative to the standard $^{12}_{6}C = 12$.
(a) What fragmented particles could be responsible for relative masses of 17, 29 and 31?
(b) Deduce the *two* possible structures for the compound.

8 The mass spectrum of dichloromethane shows peaks corresponding to 84, 86 and 88 atomic mass units (a.m.u.) having relative intensities of 9:6:1
(a) What particles cause the three peaks?
(b) Why are their relative intensities 9:6:1? (H = 1.0; C = 12.0; Cl = 35.5.)

# Electronic Structure 5

## 5.1 Evidence for the electronic structure of atoms

When atoms react, one or more electrons may be transferred from one atom to another or electrons may be shared between the reacting atoms in a different way. Remember that it is *only* electrons which are involved in chemical reactions; protons and neutrons take no part. Furthermore, since chemical reactions involve electrons, the similarities in chemical properties of certain elements (e.g. Na, K and Li; F, Cl and Br) suggest that these elements may have similar electronic structures.

It should be possible to obtain information about the arrangement of electrons in atoms by studying the ease with which atoms lose electrons. It would be useful to know the amount of energy which is needed to remove one electron from an atom. This energy is known as the **ionization energy** and is given the symbol $\Delta H_I$.

*The first ionization energy of an element* is the energy required to remove one electron from each atom in a mole of gaseous atoms producing one mole of gaseous ions with one positive charge. Thus, the first ionization energy of sodium is the energy required for the process

$$Na(g) \longrightarrow Na^+(g) + e^- \qquad \Delta H = \Delta H_{I_1}$$

(The subscript number one in the symbol $\Delta H_{I_1}$ indicates the *first* ionization energy.) Several methods can be used to determine the ionization energy of an element. We shall consider two methods in this chapter. One of these methods involves bombarding the gaseous element with electrons; the other method involves a study of the spectra of the element.

## 5.2 Obtaining ionization energies by electron bombardment

In this experiment the element is used as a gas at very low pressure inside a valve. Certain radio and television valves can be adapted for use in this experiment. Figure 5.1 shows a suitable circuit for measuring the ionization energy of a noble gas.

When the cathode is heated from the 6.3 V a.c. heater, electrons are emitted from its surface. These electrons fall on the grid which is positive and a current flows in the cathode/grid circuit. As electrons stream from the cathode towards the grid they bombard atoms of the noble gas inside the valve.

Using the potentiometer device on the right hand side in figure 5.1 it is possible to alter the voltage between the cathode and the grid. As the potential difference between the cathode and grid increases, the electrons will be accelerated through

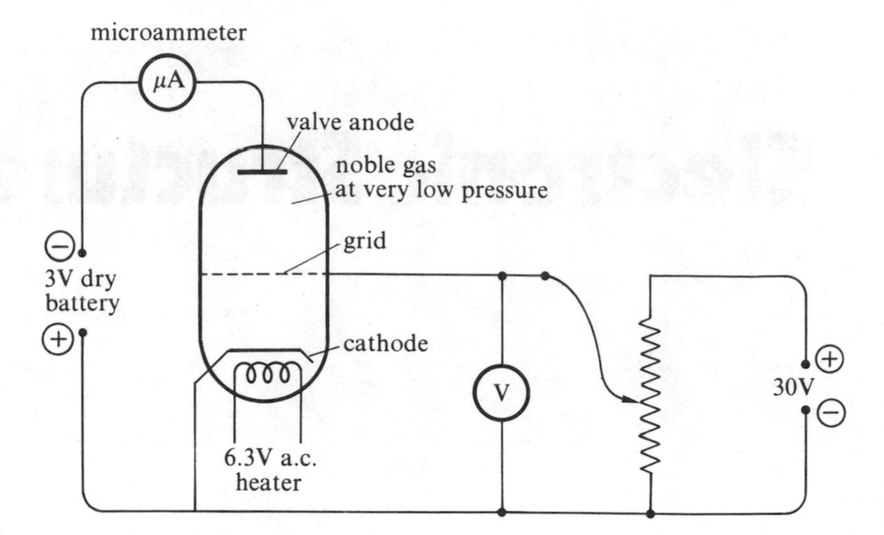


**Figure 5.1** A circuit diagram of the apparatus used to determine the ionization energy of a noble gas.

the valve at greater and greater speeds. (Notice that in this experiment, the *anode* is the most *negative* terminal inside the valve. We have called it the *anode* because it is that part of the valve which is *normally* positive and used as the anode. Although this terminal is negative it cannot give off a stream of electrons because it is not heated.)

If the electrons are travelling slowly between the cathode and the grid, they will collide gently with the noble gas atoms without causing any disruption. However, as the voltage between cathode and grid is gradually increased, the kinetic energy of the bombarding electrons eventually becomes so great that on collision with a gaseous atom, the most loosely bound electron is knocked off the atom forming a positively charged ion.

| $e^-$ | + | $X(g)$ | $\longrightarrow$ | $X^+(g)$ | + | $e^-$ | + | $e^-$ |
|---|---|---|---|---|---|---|---|---|
| fast-moving electron | | gaseous atom | | gaseous ion | | electron knocked off the atom | | retreating electron |

As soon as positive ions ($X^+(g)$) appear they are attracted towards the anode (the most negative terminal inside the valve) and a flow of electrons takes place in the cathode/anode circuit from cathode to anode in order to combine with positive ions appearing on the anode surface. When ionization first occurs, a current suddenly flows in the cathode/anode circuit and a rapid increase appears on the reading of the microammeter. The kinetic energy of the bombarding electrons which is just sufficient to ionize the gaseous atoms inside the valve can be determined from the voltage across the cathode and grid when a sudden increase occurs in the micro-ammeter reading.

Figure 5.2 shows how the current in the microammeter varies with the voltage across the cathode and grid when the experiment is carried out with xenon and then argon.

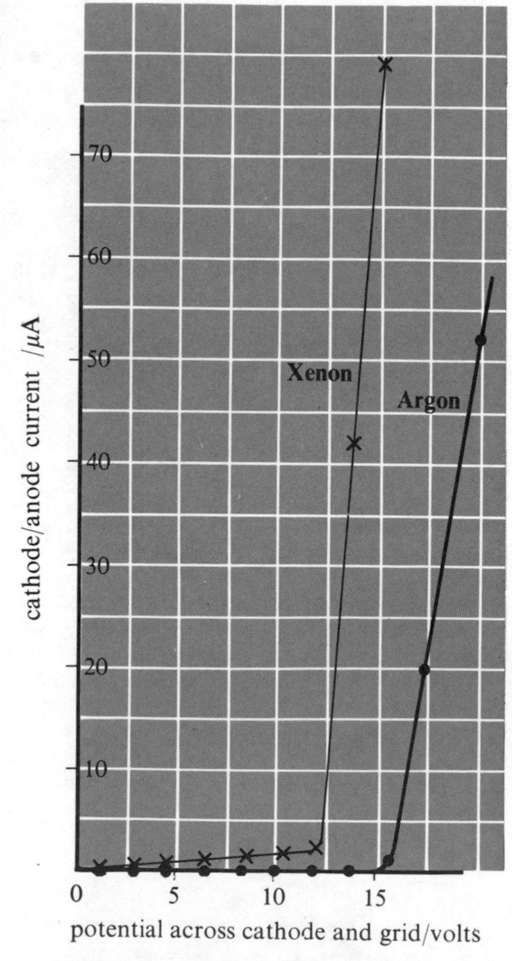


**Figure 5.2** A graph of the current in the anode circuit against the voltage across the cathode and grid in experiments to determine the first ionization energy of two noble gases.

○ At what voltage does the ionization of xenon first occur?
○ At what voltage does the ionization of argon first occur?

A similar experiment was carried out with helium, and the potential difference a which ionization of the gas occurred was 22 volts.

When one electron (with a charge of $1.6 \times 10^{-19}$ C) is accelerated through potential difference of one volt, it acquires $1.6 \times 10^{-19} \times 1$ joules of energy.

When one electron is accelerated through 22 volts, it acquires $1.6 \times 10^{-19} \times 2$ joules of energy. This energy is sufficient to ionize one helium atom.

∴ The first ionization energy of helium

$$= 1.6 \times 10^{-19} \times 22 \text{ J per atom}$$
$$= 1.6 \times 10^{-19} \times 22 \times 6 \times 10^{23} \text{ J mole}^{-1}$$
$$= 211.2 \times 10^{4} \text{ J mole}^{-1}$$
$$= 2112 \text{ kJ mole}^{-1}.$$

## 5.3 Obtaining ionization energies from emission spectra

When sodium chloride is heated strongly in a bunsen, it gives off a brilliant yellow light. Various other substances will also emit light in this way when they are supplied with sufficient energy. No doubt you will have noticed that compounds of one particular element always emit the same colour of light when they are heated strongly in a bunsen flame.
What colour of light is emitted by:

- ○ potassium compounds,
- ○ calcium compounds,
- ○ copper compounds?

*Above* The bright lights! Certain gaseous materials such as neon emit light when they are subjected to high voltages in discharge tubes.

Some gaseous materials emit light when they are subjected to high voltages in electric discharge tubes. Neon advertising signs work in this way and the yellow sodium street lamps are discharge tubes containing sodium vapour.

If the light emitted by these substances is examined using a spectroscope it is found not to consist of a continuous range of colours like the spectrum of white light or the colours in a rainbow. Instead, the light emitted by these substances is composed of separate lines of different colour. This kind of spectrum is called a **line emission spectrum**.

The photographs in figure 5.3 show the line emission spectra of hydrogen, sodium, calcium and barium in the visible region. Notice that each element has its own characteristic set of lines different from those of any other element. Consequently, elements can be identified by a study of their line emission spectra.

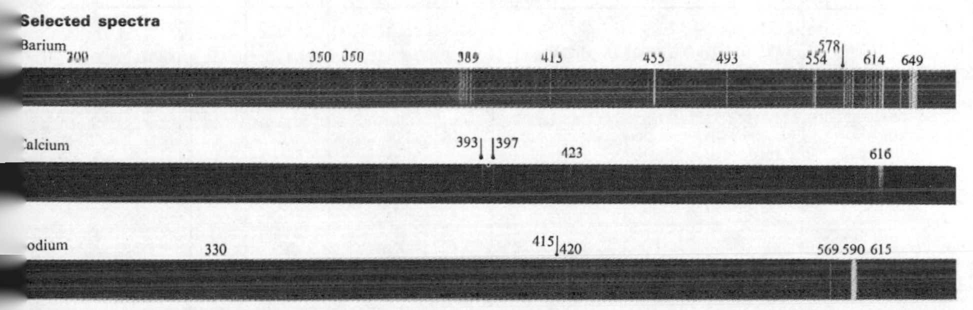


**Figure 5.3** Photographs of the line emission spectra of sodium, calcium and barium in the visible region.

Each line in an emission spectrum corresponds to light of a particular frequency. 'he frequencies of the more prominent lines in the visible region of the spectrum f hydrogen are shown in figure 5.4. This particular series of lines is known as the **almer Series** for hydrogen.

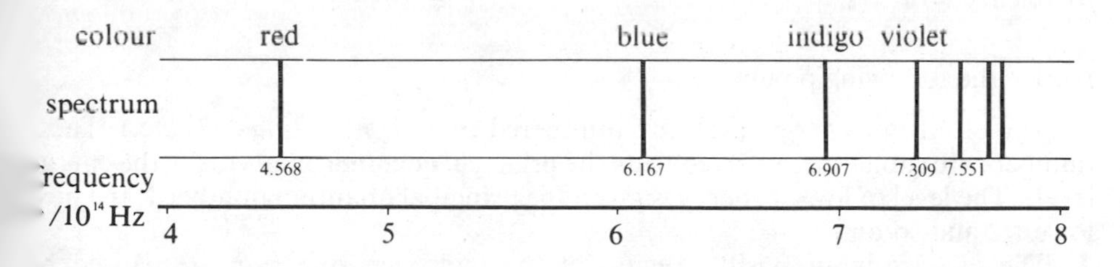


**Figure 5.4** The colours and frequencies of the more prominent lines in the visible region of the atomic hydrogen spectrum.

What do you notice about the spacing of the lines? How do you explain the pacing of the lines? Why does an element only emit light of certain frequencies?

In order to answer these questions it is necessary to assume that the electrons in atom can exist only at certain energy levels. Under normal conditions, the ectrons in an atom or ion will fill up the lowest energy levels first. When sufficient ergy is supplied to an atom, it is sometimes possible to promote (excite) an ectron from a lower energy level to a higher one. This process is called **excitation**. nce the electron is unstable in the higher energy level it will emit the excess ergy as radiation and drop back into the lower energy level. However, the energy fference between the higher and lower energy levels can have only certain fixed lues because the energy levels themselves can only have certain values. Now, since e energy of any radiation is determined by its frequency, it means that the radian emitted when the electron falls from the higher to the lower energy level can ve only certain fixed frequencies (i.e. certain specific colours).

The small amount of radiation emitted by an electron when it falls from a higher to a lower energy level is referred to as a **quantum** of radiation and each electron transition has an associated quantum of radiation.

The relationship between a quantum of energy ($E$) and the frequency ($v$) of its radiation is

$$E = h \times v$$

$h$ is a constant, called Planck's constant. The value of $h$ is $4 \times 10^{-13}$ kJ s mole$^{-1}$ or $4 \times 10^{-13}/6 \times 10^{23} = 6.66 \times 10^{-37}$ kJ s molecule$^{-1}$.

Figure 5.5 shows the energy levels in a hydrogen atom and the electron transitions which correspond to the lines in the visible region of the hydrogen spectrum.

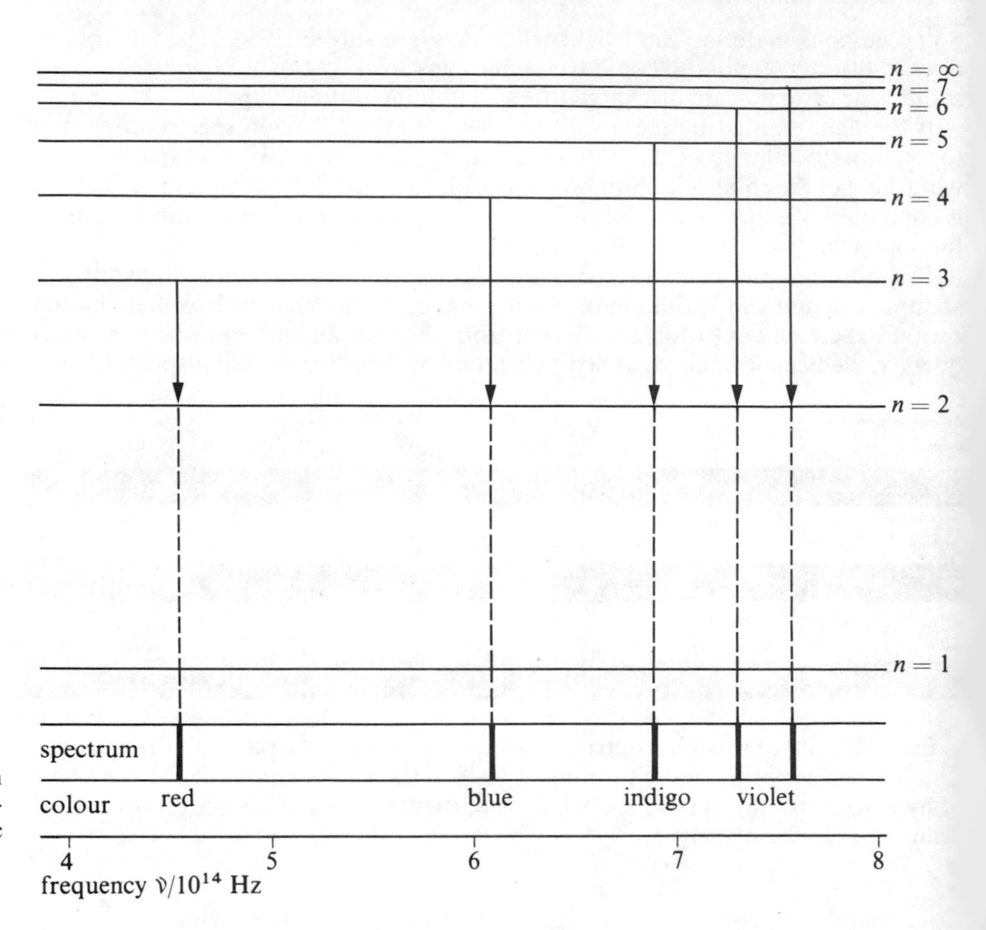


**Figure 5.5** The energy levels in a hydrogen atom and the electron transitions corresponding to lines in the visible region of the atomic hydrogen spectrum.

Notice the following points.

**1** The electronic energy levels are numbered ($n = 1$, $n = 2$, $n = 3$, etc.). Thes numbers are sometimes referred to as the **principal quantum numbers** for the energ levels. The level of lowest energy is given the principal quantum number 1, the nex lowest 2 and so on.

**2** The colours in the visible region of the hydrogen spectrum are caused b electron transitions from higher levels to the level $n = 2$ and not to the lowes energy level ($n = 1$). For example, transitions from $n = 3$ to $n = 2$ result in red line (at frequency $4.568 \times 10^{14}$Hz) in the hydrogen spectrum, while transitior from $n = 4$ to $n = 2$ produce a blue line.

**3** As the energy levels get closer and eventually come together it follows that tl spectral lines will also get closer and eventually come together.

When transitions occur from higher levels to the lowest energy level ($n = 1$) mo energy is released than with transitions to the $n = 2$ level and lines appear at high frequencies (i.e. higher energy range) in the spectrum. In the case of hydroge transitions down to the $n = 1$ level result in lines in the ultraviolet region of tl spectrum.

If sufficient energy is given to an atom to excite an electron just beyond tl highest energy level, then the electron will escape and the atom becomes an io **Ionization** has taken place. Now, in an atom, the highest possible energy level cc responds to the frequency at which the lines in the spectrum come together. Thus, by determining the frequency at which the converging spectral lines come together, we can find the ionization energy of an element. This particular frequency is called the **'convergence limit'**.

The frequencies in table 5.1 relate to electron transitions to the $n = 1$ level in hydrogen. Lines corresponding to these frequencies occur in the ultraviolet region of the hydrogen spectrum. These lines which result from transitions to the lowest energy level in an atom are known as the **Lyman Series** of lines.

- ○ Work out the difference in frequency ($\Delta\nu$) between successive lines in the Lyman Series for hydrogen.
- ○ Plot a graph of $\Delta\nu$ (vertically) against frequency, $\nu$. (Use the value of the lower frequency for plotting $\nu$.)
- ○ Use your graph to estimate the frequency when $\Delta\nu$ becomes 0.
- ○ $\Delta\nu$ becomes 0 when the difference in energy between the electronic energy levels becomes 0. Use the relationship $E = h\nu$ to find the energy which corresponds to the frequency when $\Delta\nu$ becomes 0. This energy is the ionization energy for hydrogen.

An accurate value for the frequency at the convergence limit for hydrogen is $32.7 \times 10^{14}$ Hz. Using $E = h\nu$ the energy of radiation with this frequency

$$= 4 \times 10^{-13} \times 32.7 \times 10^{14} \text{ kJ mole}^{-1}.$$

$$= 1308 \text{ kJ mole}^{-1}$$

$\therefore$ The ionization energy of hydrogen $= 1308$ kJ mole$^{-1}$.

**Table 5.1** Frequencies of the Lyman Series for hydrogen

| Frequency $\nu/10^{14}$Hz | Transition to which frequency corresponds | | |
|---|---|---|---|
| 24.66 | $n = 2$ | to | $n = 1$ |
| 29.23 | $n = 3$ | to | $n = 1$ |
| 30.83 | $n = 4$ | to | $n = 1$ |
| 31.57 | $n = 5$ | to | $n = 1$ |
| 31.97 | $n = 6$ | to | $n = 1$ |
| 32.21 | $n = 7$ | to | $n = 1$ |
| 32.37 | $n = 8$ | to | $n = 1$ |

## 5.4 Using ionization energies to predict electronic structures—evidence for shells

If an atom containing several electrons is provided with sufficient energy it will lose one electron. Additional supplies of energy may result in the removal of a second electron, then a third, then a fourth and so on. A succession of ionizations is possible each of which has its associated ionization energy. For example, the first ionization energy of sodium corresponds to

$$Na(g) \longrightarrow Na^{+}(g) + e^{-} \qquad \Delta H_{I_1} = +494 \text{ kJ mole}^{-1},$$

whereas the second ionization energy of sodium corresponds to

$$Na^{+}(g) \longrightarrow Na^{2+}(g) + e^{-} \qquad \Delta H_{I_2} = +4564 \text{ kJ mole}^{-1}.$$

The successive ionization energies of an element are usually obtained from spectroscopic experiments.

**Table 5.2** The successive ionization energies of beryllium

| Element | Ionization Energy/kJ mole$^{-1}$ | | | |
|---|---|---|---|---|
| | First | Second | Third | Fourth |
| Beryllium | 900 | 1758 | 14905 | 21060 |
| | 2 electrons relatively easy to remove | | 2 electrons very difficult to remove | |

Table 5.2 shows the successive ionization energies of beryllium. Twice as much energy is needed to remove the second electron from beryllium than to remove the first, but eight times as much energy is required to remove the third electron than the second. The values for the different ionization energies in table 5.2 suggest that beryllium has two electrons which are relatively easy to remove and two electrons which are very difficult to remove. Chemists have deduced that the beryllium atom has two electrons in a low energy level ($n = 1$) and therefore very difficult to remove; and two other electrons in a higher energy level ($n = 2$) and therefore easy to remove. This is represented on an energy level diagram in figure 5.6.

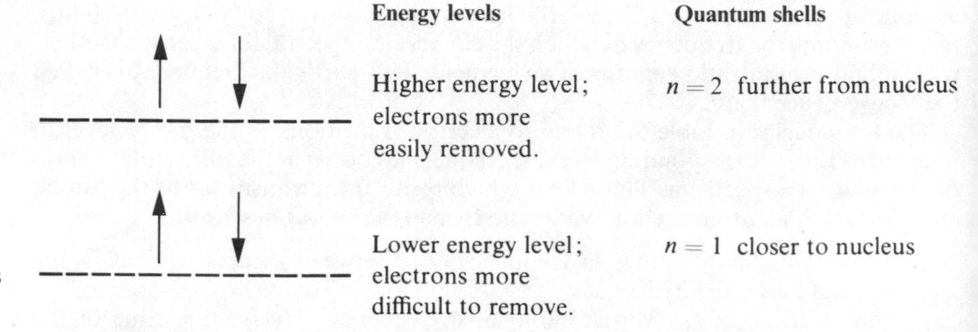


**Figure 5.6** Energy levels and quantum shells for electrons in a beryllium atom.

The electron arrangement in beryllium is written as 2,2.Lithium, with three electrons, has an electron arrangement of 2,1 which shows two electrons in the $n = 1$ energy level and one electron in the $n = 2$ energy level.

In a beryllium atom, the two electrons in the lowest energy level ($n = 1$) spend most of their time closer to the nucleus than do the two electrons in the $n = 2$ level. These two electrons in the $n = 1$ level are said to occupy the first quantum **shell**. The two electrons in the second energy level ($n = 2$) spend most of their time further from the nucleus and are said to occupy the second quantum shell.

Notice that there are two ways of describing the electrons in an atom:

(a) in terms of the **energy levels** which they occupy, $n = 1, 2, 3$, etc.
(b) in terms of their **distance from the nucleus**.

In figure 5.6 the electrons have been represented by arrows. When an energy level is filled the electrons have paired-up with each other and in each of these pairs the electrons are spinning in opposite directions. Evidence that electrons pair up and spin in opposite directions is provided by spectroscopic and magnetic measurements. Chemists believe that paired electrons can only be stable when they spin in opposite directions so that the magnetic attraction which results from their opposite spins can counterbalance the electrical repulsion which results from their identical negative charges.

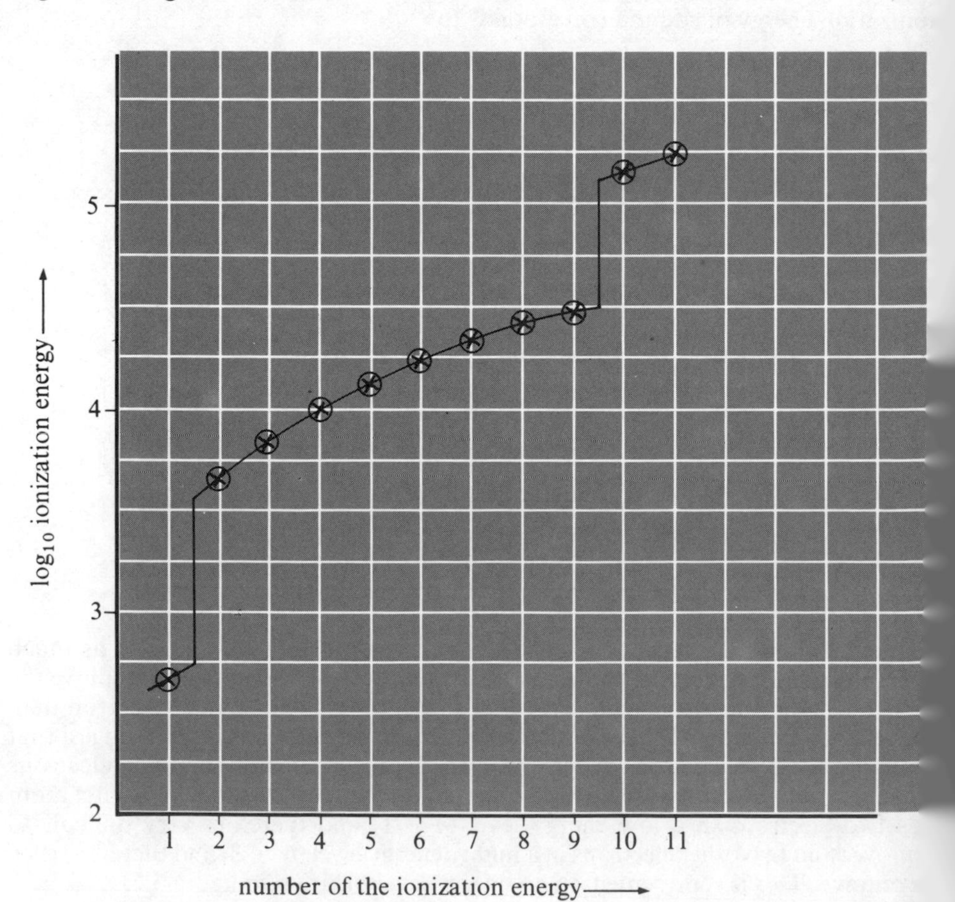


**Figure 5.7** A graph of the logarithm to base ten of the successive ionization energies for sodium.

## 5.5 How are the electrons arranged in larger atoms?

Figure 5.7 shows a graph of the logarithm to base ten ($\log_{10}$) of the successive ionization energies for sodium.

Look closely at figure 5.7.

- ○ How many electrons does sodium have in the first shell close to the nucleus?
- ○ How many electrons does sodium have in the second shell?
- ○ How many electrons does sodium have in the third shell far away from the nucleus?
- ○ Write the electron structure for sodium showing the number of electrons in each shell. (The electron structure for beryllium is 2,2.)
- ○ Draw an energy level diagram similar to figure 5.6 for the electrons in a sodium atom.

Figure 5.8(*a*) shows a sketch graph of the logarithm to base ten of the successive ionization energies for potassium. This graph has been used to construct an energy level diagram (figure 5.8(*b*)) for the electrons in a potassium atom.

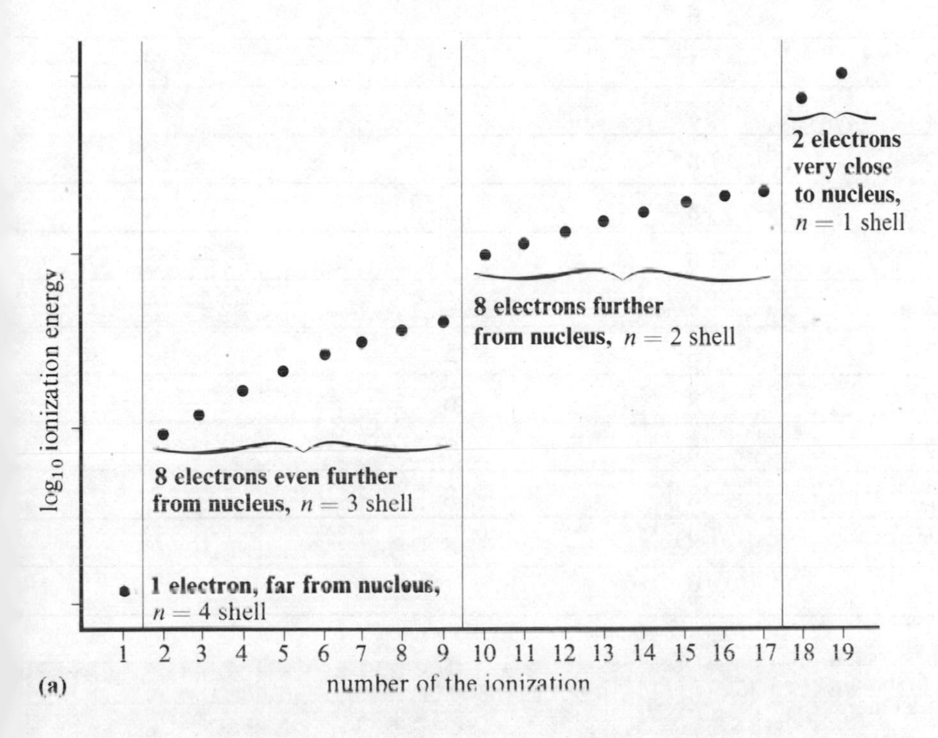


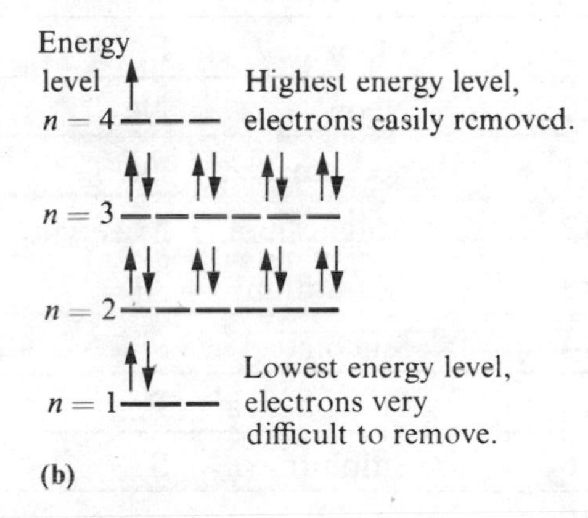


**Figure 5.8** (*a*) *Left* A sketch graph of the logarithm to base ten of the successive ionization energies for potassium.

(*b*) *Above* An energy level diagram for the electrons in a potassium atom.

The electron structure of potassium is written as 2,8,8,1 which shows the number of electrons in each of the quantum shells as we move out from the nucleus. Using ionization energy data, it is possible to predict the electron patterns in other elements. Table 5.3 shows the electron arrangements of the first 20 elements in the periodic table.

Notice that the first shell is filled when it contains two electrons and the second shell is filled when it contains eight electrons. Although the electron structures shown in table 5.3 suggest that the third shell can contain only eight electrons, it is possible for it to hold as many as 18 electrons.

## 5.6 Evidence for sub-shells of electrons

Figure 5.9 shows the first ionization energies of the first 40 elements plotted against atomic number. The graph can be divided into sections ending with a noble (inert) gas whose ionization energy is higher than those of all the elements between it and the previous noble gas. The large ionization energies of noble gases are another indication of their stable electronic structures and unreactivity.

**Table 5.3** Electron arrangements of the first twenty elements in the periodic table

| Atomic Number | Element | Symbol | Number of electrons in the first shell | Number of electrons in the second shell | Number of electrons in the third shell | Number of electrons in the fourth shell |
|---|---|---|---|---|---|---|
| 1 | Hydrogen | H | 1 | — | — | — |
| 2 | Helium | He | 2 | — | — | — |
| 3 | Lithium | Li | 2 | 1 | — | — |
| 4 | Beryllium | Be | 2 | 2 | — | — |
| 5 | Boron | B | 2 | 3 | — | — |
| 6 | Carbon | C | 2 | 4 | — | — |
| 7 | Nitrogen | N | 2 | 5 | — | — |
| 8 | Oxygen | O | 2 | 6 | — | — |
| 9 | Fluorine | F | 2 | 7 | — | — |
| 10 | Neon | Ne | 2 | 8 | — | — |
| 11 | Sodium | Na | 2 | 8 | 1 | — |
| 12 | Magnesium | Mg | 2 | 8 | 2 | — |
| 13 | Aluminium | Al | 2 | 8 | 3 | — |
| 14 | Silicon | Si | 2 | 8 | 4 | — |
| 15 | Phosphorus | P | 2 | 8 | 5 | — |
| 16 | Sulphur | S | 2 | 8 | 6 | — |
| 17 | Chlorine | Cl | 2 | 8 | 7 | — |
| 18 | Argon | Ar | 2 | 8 | 8 | — |
| 19 | Potassium | K | 2 | 8 | 8 | 1 |
| 20 | Calcium | Ca | 2 | 8 | 8 | 2 |

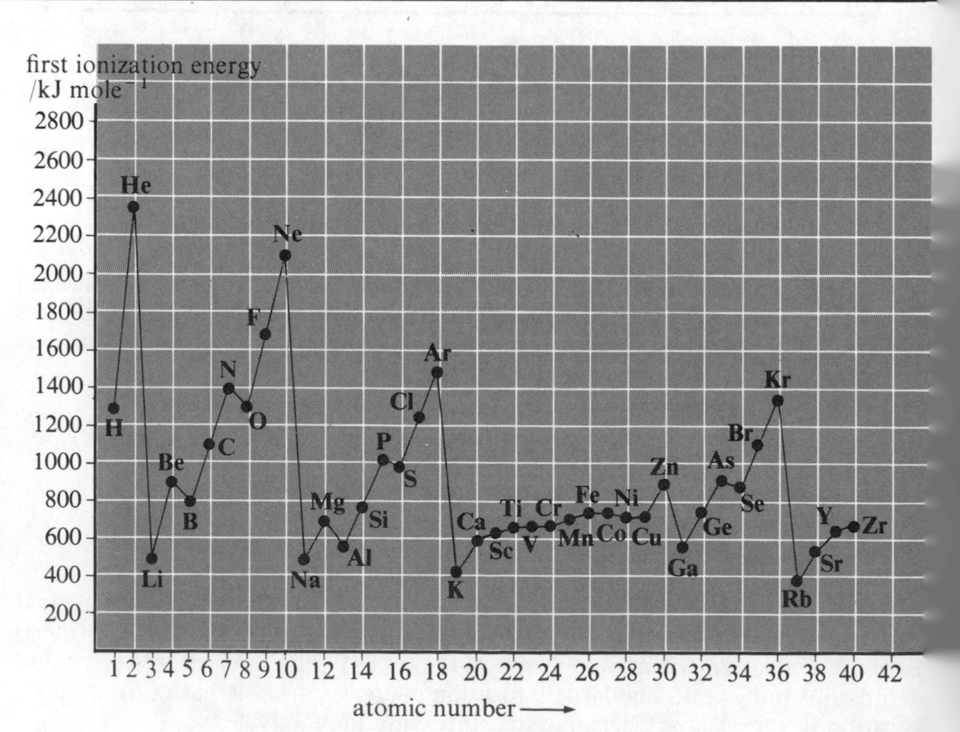


**Figure 5.9** A graph of the first ionization energies of the elements plotted against atomic number.

The graph between one noble gas and the next can also be divided into sub-sections. These sub-sections contain either two, six or ten points on the graph. For example, after both He and Ne there are deep troughs in the graph followed by a small intermediate peak (Li to Be and Na to Mg), i.e. sub-sections of two points. Immediately after Be and Mg, there are groups of six points (B to Ne and Al to Ar). Between Sc and Zn there is a sub-section containing 10 points.

Now, the points on the graph in figure 5.9 between one noble gas and the next correspond to the filling of one shell with electrons. This means that the sub-sections of points correspond to **sub-shells** of electrons. By studying ionization energies and atomic spectra in this way, scientists have concluded that:

the $n = 1$ shell can have 2 electrons in the same sub-level (sub-shell).

the $n = 2$ shell can have 2 electrons in one sub-level and,
6 electrons in a slightly higher sub-level.

the $n = 3$ shell can have 2 electrons in one sub-level,
6 electrons in a slightly higher sub-level and,
10 electrons in a still slightly higher sub-level.

the $n = 4$ shell can have 2 electrons in one sub-level,
6 electrons in a slightly higher sub-level,
10 electrons in a still slightly higher sub-level and,
14 electrons in a still slightly higher sub-level.

The sub-shells (or sub-levels) containing 2 electrons are called *s* sub-shells (*s* electrons).

The sub-shells (or sub-levels) containing 6 electrons are called *p* sub-shells (*p* electrons).

The sub-shells (or sub-levels) containing 10 electrons are called *d* sub-shells (*d* electrons).

The sub-shells (or sub-levels) containing 14 electrons are called *f* sub-shells (*f* electrons).

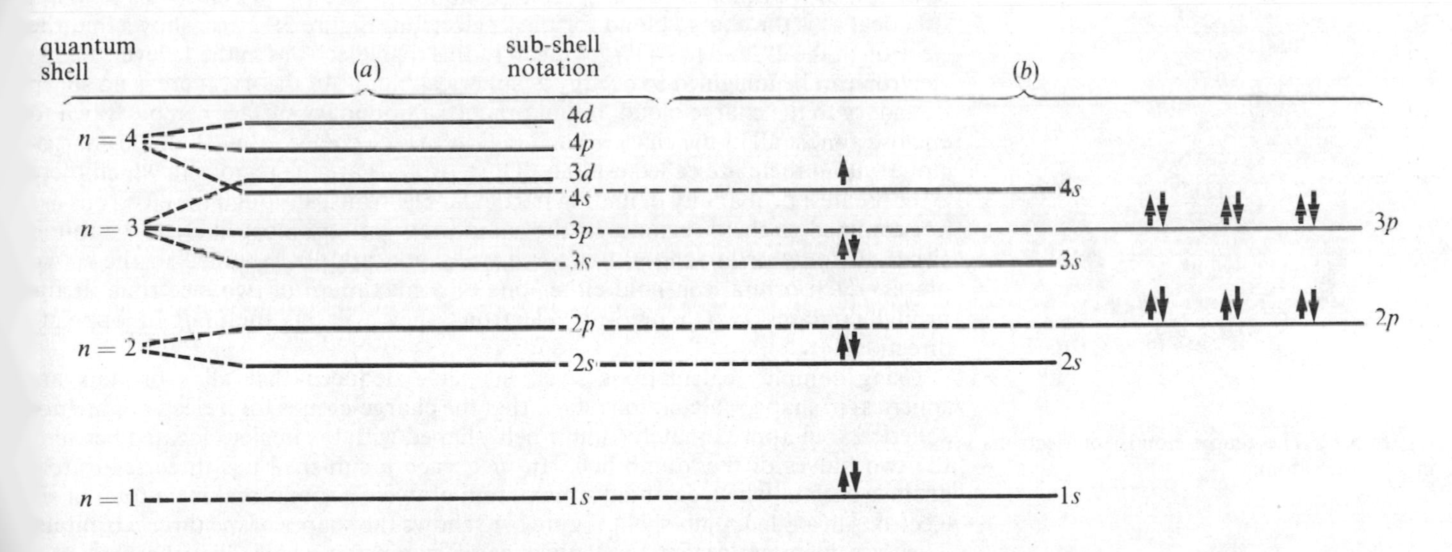


Figure 5.10 (*a*) The arrangement of energy sub-levels in a potassium atom. (*b*) The distribution of electrons amongst the sub-levels in a potassium atom.

he positions of these various sub-shells are shown on the left of figure 5.10 and e distribution of electrons amongst sub-levels in a potassium atom are shown on e right. Compare figure 5.10 with figure 5.8(*b*).

Notice that the *electrons always occupy the lowest available energy sub-levels* and at *electrons 'pair-up' as soon as each sub-level is half-filled.*

hus, potassium has 2 electrons in the 1*s* sub-level (i.e. $1s^2$)
2 electrons in the 2*s* sub-level (i.e. $2s^2$)
6 electrons in the 2*p* sub-level (i.e. $2p^6$)
2 electrons in the 3*s* sub-level (i.e. $3s^2$)
6 electrons in the 3*p* sub-level (i.e. $3p^6$)
and 1 electron in the 4*s* sub-level (i.e. $4s^1$).

This means that the electronic structure for potassium can be written in terms of energy levels as 2, 8, 8, 1; and more precisely in terms of energy sub-levels as $1s^{②}, 2s^{②}\ 2p^{⑥}, 3s^{②}\ 3p^{⑥}, 4s^{①}$.

Figure 5.10 shows that the $3d$ sub-level is just above the $4s$ sub-level, but just below the $4p$ sub-level. This means that once the $4s$ level is filled (at element Ca in the periodic table) further electrons enter the $3d$ level, not the $4p$ level. Thus, scandium (the element after Ca in the periodic table) has the electronic structure $1s^2, 2s^2\ 2p^6, 3s^2\ 3p^6, 4s^2, 3d^1$ and *not* $1s^2, 2s^2\ 2p^6, 3s^2\ 3p^6, 4s^2, 4p^1$. The filling of the $3d$ level is of great significance in the chemistry of the elements from Sc to Zn. These are known as the transition elements and we shall deal with them in some detail in chapters 18 and 19.

○ Which element has the electronic structure $1s^2, 2s^2\ 2p^3$?
○ Write the electronic structure of this element in terms of energy levels.
○ Write the electronic structure of silicon in terms of energy levels.
○ Write the electronic structure of silicon in terms of energy sub-levels.
○ Draw an energy sub-level diagram for the electrons in a silicon atom.

## 5.7 Electrons and orbitals

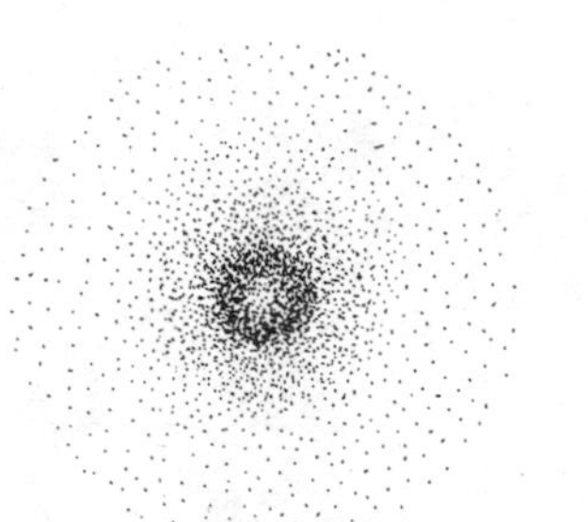

**Figure 5.11** The charge-cloud for the $1s$ electron in a hydrogen atom.

We have described the electrons in atoms as occupying certain quantum shells at increasing distances from the nucleus. But, suppose we could photograph the position of an electron in a hydrogen atom at any given moment. The electron is moving in an unknown path at high speed and if we took a second photograph an instant later the electron would occupy a different position. If we superimposed millions of such photographs they would produce a picture resembling a cloud composed of an enormous number of points, each point representing one position of the electron (figure 5.11). Thus, in the hydrogen atom the electron can be imagined to be a diffused cloud of negative charge. The diffused cloud for the $1s$ electron in hydrogen is spherical in shape, but notice that the density of charge is not uniform throughout the cloud. Figure 5.12 shows the charge-clouds for electrons in a lithium atom. The charge-cloud for the two $1s$ electrons is again spherical as is the charge-cloud for the $2s$ electron. Figure 5.12 also shows that the electron in the $2s$ level has a larger mean radius than electrons in the $1s$ level. The $2s$ electron can be imagined to occupy a 'spherical band'. In theory, there is no sharp boundary to the charge-cloud, but in practice a boundary surface can be drawn to enclose almost all of the charge. Such regions which enclose almost all the charge-cloud within them are called **orbitals**. Thus, orbitals are the regions in which there is the greatest probability of finding particular electrons, although the electrons are not confined to these regions. In the same way, you are most likely to be found either at home or in school, but fortunately you are not confined to these two places. Each orbital can hold either one or a maximum of two electrons. If the orbital contains two 'paired-up' electrons they will be spinning in opposite directions.

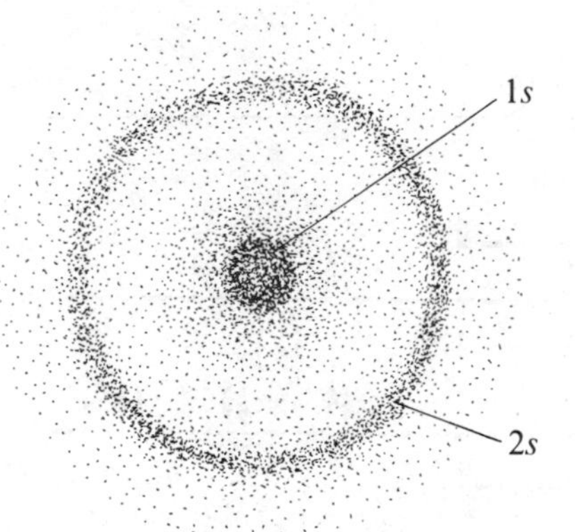


**Figure 5.12** The charge-clouds for electrons in a lithium atom.

Using complex calculations scientists have deduced that all $s$ orbitals are spherical in shape. Calculations show that the charge-clouds for $p$ electrons are not spherical, but approximately 'dumb-bell' shaped with the nucleus located between the two halves of the 'dumb-bell'. In fact, each $p$ sub-shell has three separate orbitals each of which can hold a maximum of two electrons, making a total of six electrons in a filled $p$ sub-shell. Figure 5.13 shows the shapes of the three $p$ orbitals. They are identical except for their axes of symmetry which, like the axes of cartesian co-ordinate system, are mutually at right angles. Thus, it is convenient

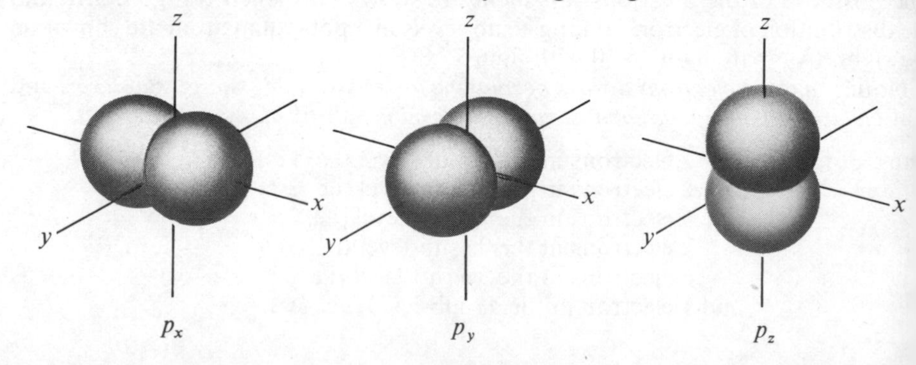


**Figure 5.13** The shapes and relative positions of the three $p$ orbitals in a $p$ sub-shell.

to distinguish between the $p$ orbitals by labelling them $p_x$, $p_y$ and $p_z$. Electrons will, of course, occupy the three $p$ orbitals singly at first because of the natural repulsion that exists between any paired electrons.

The charge-clouds for $d$ electrons are even more complex than those of $p$ electrons and we need not discuss them at this stage.

Figure 5.14 shows how the electronic structures of beryllium, carbon, nitrogen and oxygen can be represented to show the number of electrons in each orbital. Each box represents an orbital.

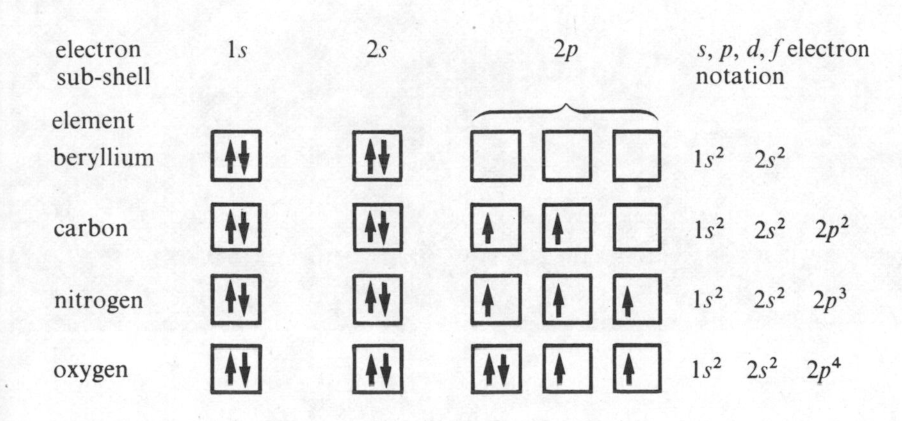


**Figure 5.14** An 'electrons-in-boxes' representation of the electronic structures of beryllium, carbon, nitrogen and oxygen.

## 5.8 Electronic structures and the periodic table

In the periodic table, elements are arranged in groups with similar chemical properties. The chemical similarities amongst elements in the same group arises because of their similar electron configurations. Table 5.4 shows the electron structures of the first 20 elements arranged in periodic table fashion. Note that elements in the same group have similar electron structures.

**Table 5.4** The electron structures of the first twenty elements arranged in periodic table fashion

| | | | | | | | | |
|---|---|---|---|---|---|---|---|---|
| **Period 1** | | | H<br>1 | | | | | He<br>2 |
| **Period 2** | Li<br>2,1 | Be<br>2,2 | B<br>2,3 | C<br>2,4 | N<br>2,5 | O<br>2,6 | F<br>2,7 | Ne<br>2,8 |
| **Period 3** | Na<br>2,8,1 | Mg<br>2,8,2 | Al<br>2,8,3 | Si<br>2,8,4 | P<br>2,8,5 | S<br>2,8,6 | Cl<br>2,8,7 | Ar<br>2,8,8 |
| **Period 4** | K<br>2,8,8,1 | Ca<br>2,8,8,2 | | | | | | |

ROUP O: THE NOBLE (INERT) GASES

ll of the noble gases have completely filled $s$ and $p$ orbitals in their outer shell; elium ($1s^2$), neon ($1s^2$, $2s^2\ 2p^6$), argon ($1s^2$, $2s^2\ 2p^6$, $3s^2\ 3p^6$), etc. Because of their omplete and stable electronic structures noble gases have large first ionization nergies. The atoms of these elements are very stable and exist as monatomic olecules; they do not combine with each other under any conditions to form iatomic molecules nor do they combine readily with the atoms of other elements.

The first noble gas compound was not prepared until 1962 and chemists have ill not succeeded in preparing a single stable compound of helium or neon. The ert character of these elements is attributed to their stable electronic structures ith filled $s$ and $p$ sub-shells.

Crystals of $XeF_2$ on the surface of a glass vessel. The crystals form when Xe and $F_2$ are mixed and exposed to sunlight (*Dr. J. H. Holloway*)

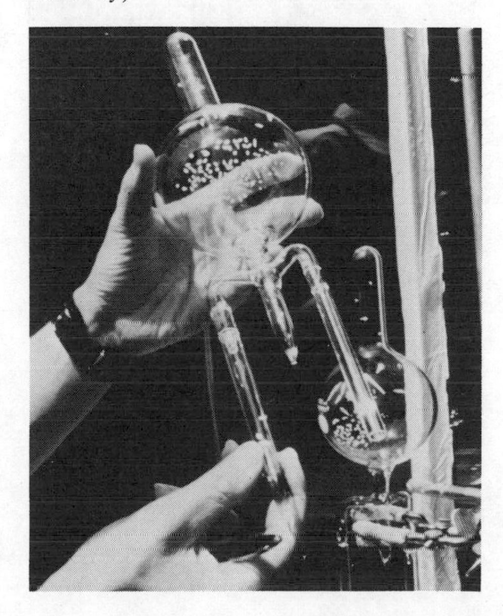

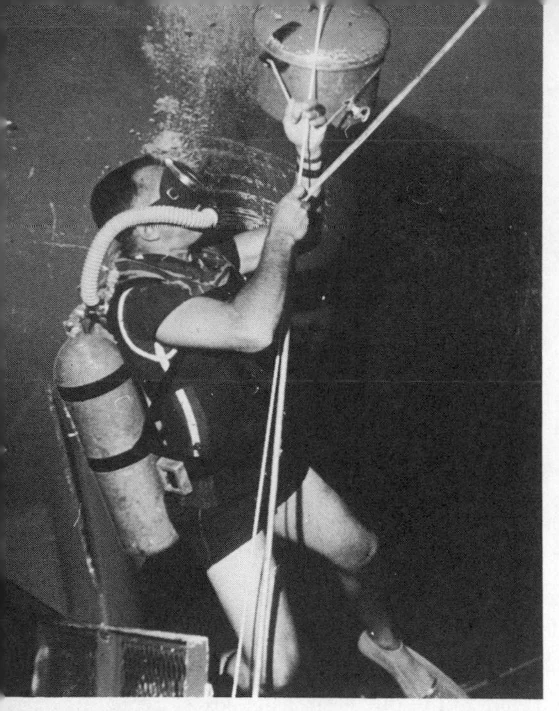

*Above* A diver explores the sea bed. Strapped to his back are cylinders containing a mixture of oxygen/helium rather than oxygen/nitrogen. Helium is much less soluble in the blood than nitrogen and the diver is therefore less likely to suffer from 'bends' (pains caused by gases coming out of the blood) as he rises to the surface.

*Right* Up she goes! An airship filled with helium gently floats upwards, though still held to the ground by ropes.

As a result of such stable electronic structures, the atoms of noble gases have little interaction with each other and the forces of attraction between separate atoms are very small. Thus, noble gases have low melting points and low boiling points. Helium has the lowest melting point and the lowest boiling point of any known substance (3 K and 4 K respectively). Radon, the heaviest inert gas, boils at −62°C (211 K).

GROUP I: THE ALKALI METALS

Each alkali metal follows a noble gas in the periodic table. Thus, alkali metals have one electron in their outermost shell which is fairly easily removed (low first ionization energy). By losing their outermost electron atoms of alkali metals form singly charged positive ions with stable electron structures like the previous noble gas ($Li^+$, $Na^+$, $K^+$, etc.). This means that all the alkali metals have an oxidation number of +1 in their compounds and they are very reactive because of the ease with which they lose the single electron in their outermost shell.

| |
|---|
| Li<br>2,1 |
| Na<br>2,8,1 |
| K<br>2,8,8,1 |
| Rb<br>2,8,18,8,1 |
| Cs |
| Fr |

- ○ How many electrons have the atoms of Group VII elements in the outermost shell?
- ○ Write down the symbols with the correct charge for the ions of three elements in Group VII.
- ○ Why do elements in Group VII form ions with the same charge?
- ○ How many electrons have the atoms of Group II elements in the outermost shell?
- ○ Why do Group II elements have an oxidation number of +2 in all their compounds?

## Summary

1 Each element has a characteristic emission spectrum which can be used to identify it.

2 The electrons in an atom can exist only at certain energy levels. When an atom is provided with sufficient energy, electrons can be promoted to higher energy levels. When such a promoted electron falls from a higher to a lower energy level, a quantum of radiation is emitted and this radiation can be detected as a line in the emission spectrum of the element.

3 The ionization energy of an element can be obtained by electron bombardment in the gas phase or from the convergence limit of the Lyman Series of lines in its emission spectrum.

4 The principal quantum numbers ($n$ = 1, 2, 3, etc.) are used to denote the energy levels in atoms.

5 There are two ways of describing the electrons in an atom:
(a) in terms of the energy levels ($n$ = 1, 2, 3, etc.) which they occupy;
(b) in terms of their distance from the nucleus.
6 In a stable atom the electrons occupy the lowest available energy sub-levels and they begin to 'pair-up' once each sub-level is half-filled.
7 *s* sub-levels can hold a maximum of 2 electrons,
*p* sub-levels can hold a maximum of 6 electrons,
*d* sub-levels can hold a maximum of 10 electrons, and
*f* sub-levels can hold a maximum of 14 electrons.
8 The region in which an electron moves for most of the time is called an orbital. Each orbital can hold a maximum of two electrons. *s* sub-shells contain one orbital; *p* sub-shells contain three orbitals.

## Study questions

1 (a) Draw a sketch graph of the logarithm to base 10 for the successive ionization energies for phosphorus.
(b) What conclusions could you deduce from the graph concerning the electronic structure of phosphorus?
(c) Write the electronic structure of phosphorus using *s*, *p*, *d*, *f* notation.

2 The following table shows the ionization energies (in kJ mole$^{-1}$) of five elements lettered A, B, C, D and E.

| Element | 1st ionization energy | 2nd ionization energy | 3rd ionization energy | 4th ionization energy |
|---|---|---|---|---|
| A | 500 | 4600 | 6900 | 9500 |
| B | 740 | 1500 | 7700 | 10500 |
| C | 630 | 1600 | 3000 | 4800 |
| D | 900 | 1800 | 14800 | 21000 |
| E | 580 | 1800 | 2700 | 11600 |

(a) Which of these elements is most likely to form an ion with a charge of 1+? Give reasons for your answer.
(b) Which two of the elements are in the same group of the periodic table? Which group do they belong to?
(c) In which group of the periodic table is element E likely to occur? Give reasons for your answer.
(d) Which element would require the least energy to convert one mole of gaseous atoms into ions carrying two positive charges?

3 The electron energy levels of a certain element can be represented as $1s^2$, $2s^2$ $2p^6$, $3s^2$ $3p^4$.

(a) Sketch a graph for the first seven ionization energies of the element against the number of the ionization.
(b) What is the atomic number of the element?
(c) What is the name of the element?
(d) Draw an energy level diagram for the electrons in an atom of the element.

4 (a) List three factors which influence the size of the first ionization energy of an element.
(b) The first ionization energy of the elements in Group I are shown below.

| Element | First ionization energy/kJ mole$^{-1}$ |
|---|---|
| Li | 520 |
| Na | 500 |
| K | 420 |
| Rb | 400 |
| Cs | 380 |

Explain the variation in the first ionization energy with atomic number.

(c) Explain, with examples, how ionization energies can provide evidence for the arrangement of electrons in shells.
(d) Explain, with examples, how the electrons in an atom can be described
(i) in terms of their energy levels,
(ii) in terms of their distance from the nucleus.

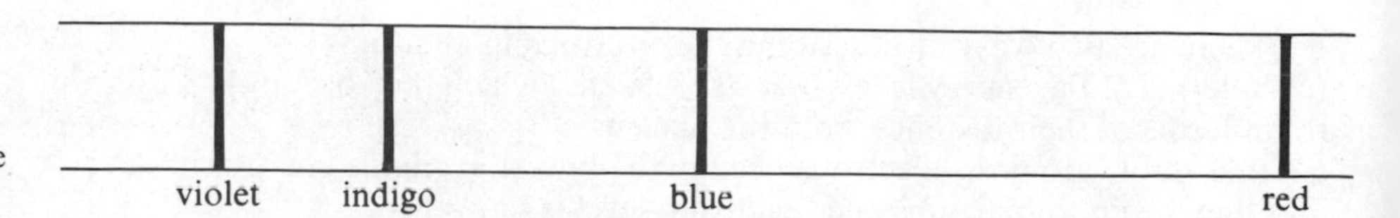


**Figure 5.15** A simplified diagram of the visible line spectrum of the hydrogen atom.

5 Figure 5.15 shows a simplified diagram of the visible line spectrum of the hydrogen atom.
(a) What quantity is represented along the horizontal axis in figure 5.15?
(b) Is this quantity increasing or decreasing from left to right?
(c) Explain why the atomic spectrum consists of a series of lines.
(d) To which energy level do the transitions corresponding to the visible lines in the spectrum of hydrogen relate?
(e) Why is the line spectrum of an element sometimes compared to the 'fingerprint of a criminal'?

6 The graph in figure 5.16 shows the first and second ionization energies of the elements nitrogen to calcium.
(a) Why is there a sudden drop in the ionization energy after neon and after argon?
(b) Why is the first ionization energy of magnesium greater than that for aluminium?
(c) Why is the second ionization energy of each element greater than the corresponding first ionization energy?
(d) Why do the maxima for the two graphs occur at different atomic numbers?

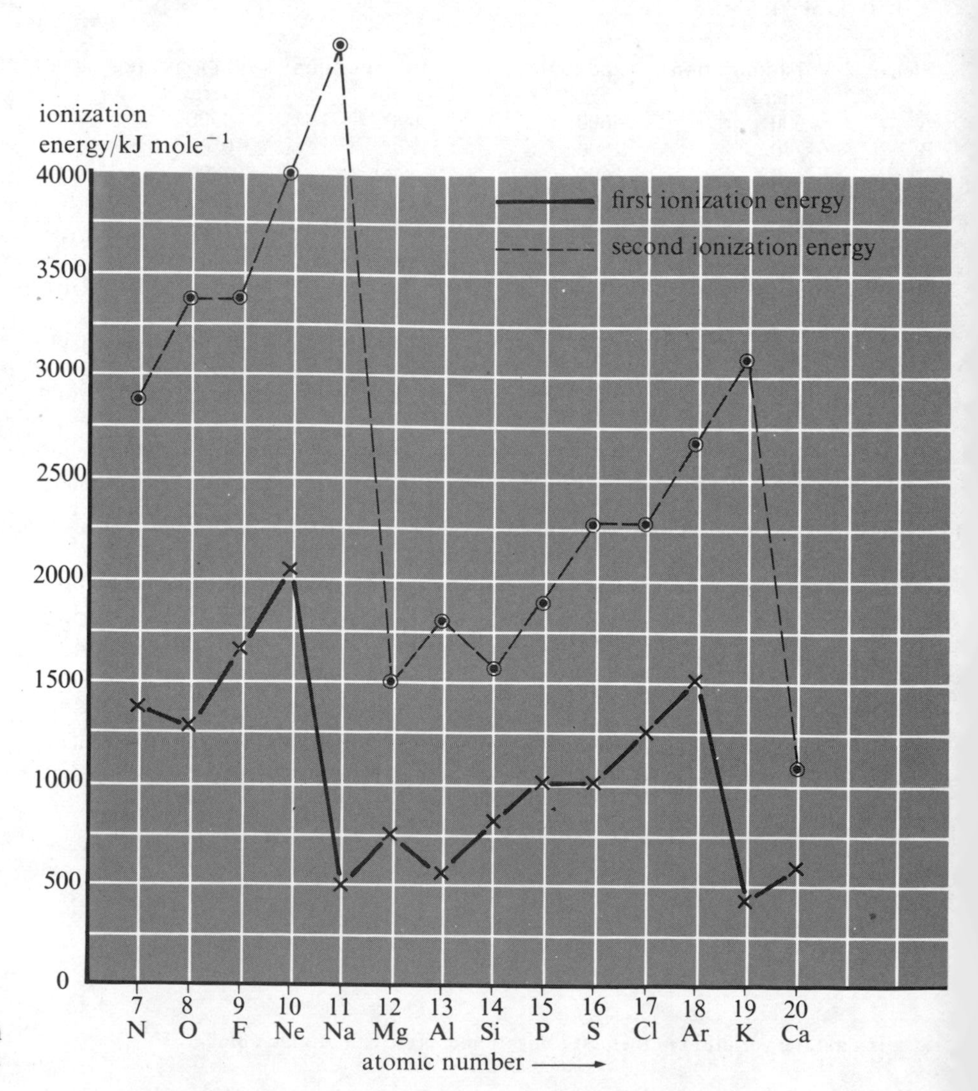


**Figure 5.16** The first and second ionization energies of the elements nitrogen to calcium.

# Nuclear Structure and Radioactivity 6

## 6.1 Introduction

In 1896, the Frenchman Henri Becquerel was investigating the possible production of X-rays from uranium salts exposed to bright sunlight. By chance, he happened to leave some crystals of a uranium salt on a photographic plate wrapped in black paper. Poor sunshine caused him to abandon his intended investigation, but, for interest's sake, he decided to develop the photographic plate. To his surprise, the plate was considerably darkened in the region where the crystals had been. Further experiments showed that the uranium salt was emitting a radiation which had quite different properties to X-rays and Becquerel called this phenomenon **radioactivity**.

Intrigued by Becquerel's discovery, Pierre and Marie Curie examined the radioactivity of uranium salts in some detail. In particular they looked at a uranium ore called pitchblende. They found that pitchblende was much more radioactive than might be expected from its uranium content and after a long and tedious extraction process they isolated two other radioactive elements from the ore, polonium and radium. Radium was found to be about two million times more radioactive than uranium. By 1900, thorium and actinium were also known to be radioactive.

The fogging of a photographic plate by radioactive materials still has practical applications. For example, the metabolism of $CO_2$ by a plant leaf can be monitored by exposing the leaf to an atmosphere of radioactive $^{14}CO_2$ and then holding the leaf against a photographic plate. When the plate is developed the darkest parts indicate the presence of $^{14}C$.

Before very long radioactive substances had been shown to cause another effect which could be used to detect their radiations—scintillation. When a screen coated with zinc sulphide is placed near a radioactive radium salt and examined with a magnifying lens in the dark, tiny flashes of light appear on the surface of the zinc sulphide. Nowadays, crystals of zinc sulphide are combined with a photomultiplier which can detect even the faintest of flashes and measure the radiations falling on the crystal. This arrangement of crystal and photomultiplier, known as a scintillation counter, is used by many research scientists to detect alpha-particles (helium nuclei, $^4_2He^{2+}$).

The flashes of light on the zinc sulphide result from alpha-particles hitting the surface of the crystal and emitting the energy which they possess in the form of light. The alpha-particles are produced during the spontaneous radioactive decay of radium atoms. From time to time, a radium atom 'explodes' and ejects an alpha-particle from its nucleus at great speed.

This spontaneous disintegration of atoms is responsible for radioactivity and those isotopes which decay (break up) in this way are said to be **radioactive**.

*Above* Pierre (1859–1906) and Marie Curie (1867–1934) spent nearly four years isolating the radioactive elements polonium and radium from pitchblende. In 1903, they shared the Nobel Physics Prize with Becquerel and then in 1911, Marie was awarded the Chemistry Prize – the first person ever to win two Nobel Prizes.

## 6.2 Alpha-particles, beta-particles and gamma-rays

When radium-226 decays it emits alpha-particles, but other radioactive substances behave differently when they disintegrate.

For example, potassium-43 emits much lighter beta-particles during decay and other materials radiate gamma-rays during disintegration.

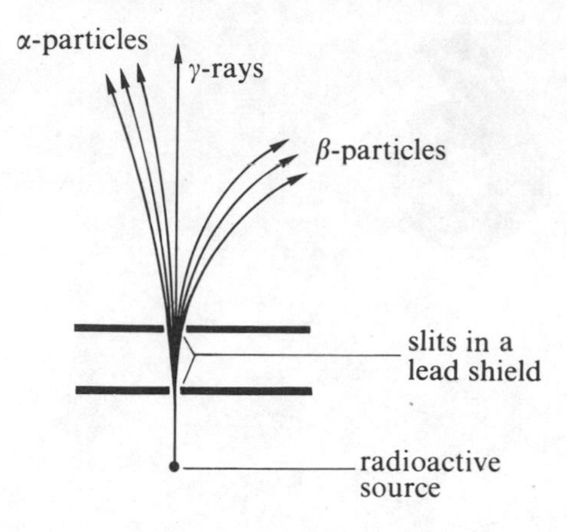


**Figure 6.1** The effect of a magnetic field, perpendicular to the paper, on α-particles, β-particles and γ-rays.

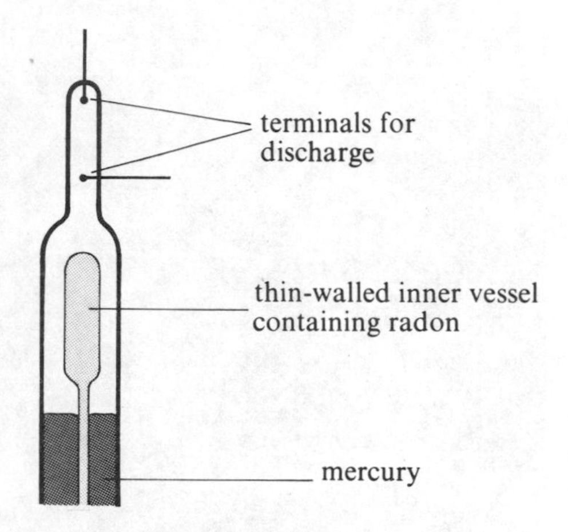


**Figure 6.2** The experiment designed by Rutherford and Royds to investigate the nature of α-particles.

### EVIDENCE FOR THE NATURE OF ALPHA-PARTICLES AND BETA-PARTICLES

Alpha-particles, beta-particles and gamma-rays can be separated easily and conveniently using magnetic or electric fields (figure 6.1).

Alpha-particles are deflected in a direction indicating that they are positively charged; beta-particles are deflected in the opposite direction and must be negatively charged; gamma-rays are unaffected.

Rutherford determined the charge/mass ratio for alpha-particles by measuring their deflection in magnetic and electric fields. The value of charge/mass was half the value for ${}^{1}_{1}H^{+}$ ions (protons) showing that alpha-particles were either singly charged particles of atomic mass two, or doubly charged particles of atomic mass four, or triply charged particles of atomic mass six, or . . . What is correct? The matter was settled by Rutherford and Royds using an ingenious yet remarkably simple experiment (figure 6.2).

Radioactive gaseous radon was allowed to decay for several days in a very thin-walled glass tube within an evacuated thick walled container. Alpha-particles from the decaying radon passed through the thin walls of the inner tube into the evacuated space outside. After a few days, the gas that had accumulated in the outer space was compressed into the top of the apparatus by raising the mercury level and then analysed by passing an electric discharge through it. When the radiation was analysed using a spectrometer it became clear that the gas collecting in the outer space was helium. This confirmed that alpha-particles had a mass of four units and thus a charge of +2. Alpha-particles are therefore doubly charged helium ions, ${}^{4}_{2}He^{2+}$.

Beta-particles were found to have similar properties to cathode rays. Measurement of charge/mass confirmed that beta-particles were electrons. The symbol ${}^{0}_{-1}e$ can be used to represent an electron since its mass number ($= n_p + n_n$) is zero, and it carries a unit charge of $-1$.

### EVIDENCE FOR THE NATURE OF GAMMA-RAYS

Here is a list of the properties of gamma-rays. What conclusion can you make from each piece of evidence?

1. Gamma-rays are unaffected by an electric field.
2. Gamma-rays are unaffected by a magnetic field.
3. Gamma-rays can penetrate several centimetres of lead.
4. Gamma-rays can be diffracted by the lattice of a crystal.
5. When an atom emits gamma-rays there is no change in atomic number or mass.

This and other evidence suggests that gamma-rays are electromagnetic waves similar to light rays and X-rays, but with a wavelength of only $10^{-12}$ metres which is about one-tenth the wavelength of X-rays and about one hundred-thousandth of the wavelength of visible light. The emission of gamma-rays provides a means whereby a nucleus can lose surplus energy. The nature and properties of alpha-particles, beta-particles and gamma-rays are summarized in table 6.1.

**Table 6.1** The nature and properties of alpha-particles, beta-particles and gamma-rays

| Emission | Nature | Relative effect of electric and magnetic fields | Range in air | Relative penetrating power |
|---|---|---|---|---|
| α-particles | helium nuclei (${}^{4}_{2}He^{2+}$) | small deflection | a few cm | 1 |
| β-particles | electrons (${}^{0}_{-1}e$) | large deflection | a few m | 100 |
| γ-rays | electromagnetic waves | no deflection | a few km | 10 000 |

## .3 Nuclear equations

/hen an atom disintegrates by losing an alpha-particle ($^{4}_{2}He^{2+}$), the remaining agment will have a mass number four units less than the original atom and an omic number two units lower. For example, when $^{238}_{92}U$ loses an alpha-particle ontaining two protons and two neutrons), the breakdown product will have a ass number of 234 and an atomic number of 90. All isotopes (section 4.4) with an tomic number of 90 are atoms of thorium, Th. Thus, the products from the nuclear ecay of uranium-238 are $^{234}_{90}Th$ and $^{4}_{2}He$ and the process can be summarized by the ollowing nuclear equation.

$$^{238}_{92}U \longrightarrow {}^{234}_{90}Th + {}^{4}_{2}He$$

*oth atomic mass and atomic number must balance in the nuclear equation.*

lpha-decay is common to most of the radioactive isotopes of elements with tomic number greater than 82 (lead). Isotopes which decay by loss of alpha-articles include radium-226, plutonium-238, polonium-218 and radon-220.

○ Write nuclear equations for the alpha-decay of radium-226 and plutonium-238.

○ Use the periodic table on page 25 to find the symbols of the breakdown products.

lements with an atomic number less than 82 do not usually exhibit alpha-decay, lthough various isotopes of these elements emit beta-particles. For example, arbon-14 ($^{14}_{6}C$) decays by beta-particle emission forming $^{14}_{7}N$. The decay process an be represented by the equation

$$^{14}_{6}C \longrightarrow {}^{14}_{7}N + {}^{0}_{-1}e$$

Jotice that both mass and charge are conserved in beta-decay as in alpha-decay.

○ What happens to the nucleus of the carbon-14 atoms during the beta-emission?

○ Where does the ejected electron come from?

)uring beta-decay, mass number remains constant, so the number of protons + eutrons stays constant. However, the atomic number increases by one which neans that a neutron in the $^{14}_{6}C$ nucleus is converted to a proton + an electron

$$^{1}_{0}n \longrightarrow {}^{1}_{1}p + {}^{0}_{-1}e$$

'he proton remains in the nucleus (now $^{14}_{7}N$), but the electron is ejected as a eta-particle.

Thus, the overall effect of beta-decay causes an increase of one proton in an sotope and a corresponding decrease of one neutron.

○ Write nuclear equations for the beta-decay of $^{90}_{38}Sr$ and $^{131}_{53}I$.

○ How do chemical reactions differ from nuclear reactions?

)uring chemical reactions electrons are redistributed either by transfer from one tom to another or by sharing between atoms. Chemical reactions involve only the uter parts of atoms—the electrons.

During nuclear reactions one element may be converted to another either by adioactive decay or by atomic fission or fusion. Nuclear reactions involve he nucleus—the protons and neutrons. Another important difference between hemical reactions and nuclear reactions is that the energy changes in nuclear eactions are usually much greater than those in chemical reactions.

## 5.4 Stable and unstable isotopes

At the beginning of this century detailed investigations were made of the dis-ntegration of naturally occurring radioactive isotopes. Besides considering the ypes of radioactive change and the products of decay, scientists were also interested n the rate of disintegration and the relative stability of different radioactive sotopes.

Experiments showed that the natural decay of a radioactive isotope could neither be retarded nor accelerated by physical or chemical means. For example, extremes of temperature and pressure had virtually no effect on the rate of decay which was characteristic of the isotope concerned. Scientists found that the most convenient method of expressing the rate of disintegration of a radioactive element was the **half-life period.** The half-life of a radioactive isotope is the time taken for the amount or concentration of the isotope to fall to half of its original value. Since the radioactivity of the isotope will also halve in this time, the half-life can also be defined as the time for the radioactivity of the isotope to be reduced to half its initial value. The half-life remains constant for each radioactive decay process. The half-life for $^{60}_{27}Co$ is 5.2 years. This means that starting with one gram of cobalt-60, only a half gram would remain after 5.2 years. After another 5.2 years, the half gram would have decayed to one-quarter gram and 5.2 years after that (i.e. 15.6 years from the start of the experiment) only one-eighth of a gram would remain. Table 6.2 gives the half-lives of a few radioactive isotopes.

**Table 6.2** The half-lives of some radioactive isotopes

| Radioactive isotope | Half-life |
|---|---|
| Uranium-238, $^{238}_{92}U$ | $4.5 \times 10^9$ years |
| Carbon-14, $^{14}_{6}C$ | $5.7 \times 10^3$ years |
| Radium-226, $^{226}_{88}Ra$ | $1.6 \times 10^3$ years |
| Strontium-90, $^{90}_{38}Sr$ | 28 years |
| Iodine-131, $^{131}_{53}I$ | 8.1 days |
| Bismuth-214, $^{214}_{83}Bi$ | 19.7 minutes |
| Polonium-214, $^{214}_{84}Po$ | $1.5 \times 10^{-4}$ seconds |

The shorter the half-life the faster the isotope decays and the more unstable it is. The longer the half-life the slower the decay process and the more stable the isotope. Thus, the half-life of a radioactive isotope provides a quantitative measure of its relative stability.

Uranium-238 with a half-life of $4.5 \times 10^9$ years might be described as 'almost stable', but polonium-214 is quite the reverse.

Suppose you had one gram of $^{214}_{84}Po$

- ○ How much of it would be left after $1.5 \times 10^{-4}$ seconds?
- ○ How much of it would be left after $3 \times 10^{-4}$ seconds?
- ○ How much of it would be left after $1.5 \times 10^{-3}$ seconds?
- ○ How much of it would be left after 1.5 seconds?

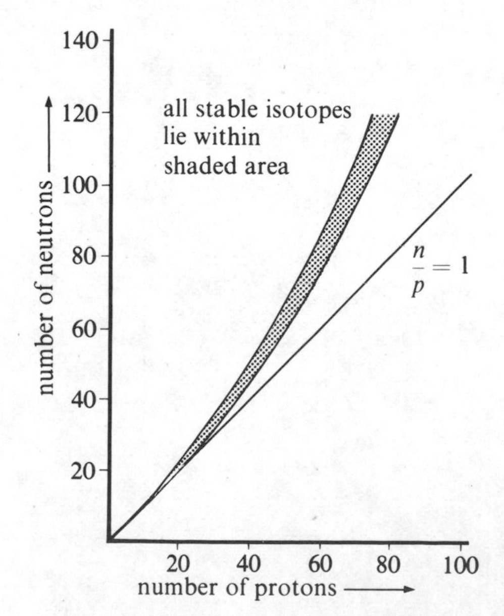


**Figure 6.3** Variation in the number of neutrons with the number of protons for stable, non-radioactive isotopes.

## 6.5 The effect of *n/p* ratio on nuclear stability

For stable isotopes of small mass, the numbers of protons and neutrons are approximately equal, but, as the isotopic mass increases, the number of neutrons becomes increasingly larger than the number of protons.

The shaded area in Figure 6.3 shows the variation in the number of neutrons with the number of protons for stable, non-radioactive isotopes. The straight line corresponds to equal numbers of protons and neutrons ($n/p = 1$). Figure 6.3 shows that if a nucleus is to be stable its $n/p$ ratio must lie within a restricted range for any given atomic number. How does the $n/p$ ratio for stable isotopes vary as atomic number increases?

### ALPHA-DECAY FOR ISOTOPES WITH N/P RATIO BELOW THE STABLE STATE

As a result of alpha-particle emission, the ratio of neutrons to protons will usually increase. For example, $^{226}_{88}\mathrm{Ra}$ has a $n/p$ ratio of $\frac{138}{88}$ (= 1.568), whereas its alpha-decay product, $^{222}_{86}\mathrm{Rn}$, has a $n/p$ ratio of $\frac{136}{86}$ (= 1.581). Thus, isotopes which are unstable due to an excess of protons over neutrons ($n/p <$ stable state) can increase their stability by alpha-particle emission. In many cases, these isotopes are unstable because their atomic mass is just too large. They become stable simply by losing protons and neutrons which make up the ejected alpha-particle.

One other method, besides alpha-decay, by which nuclei with a low $n/p$ ratio may stabilize themselves is **electron capture**. In this case, the unstable nucleus captures an electron from its innermost shell which combines with a proton to form a neutron.

$$^{0}_{-1}\mathrm{e} + ^{1}_{1}p \longrightarrow ^{1}_{0}n$$

In this way, the number of neutrons rises by one and the number of protons falls by one, thus the $n/p$ ratio increases. At the same time, energy is emitted as gamma-radiation.

### BETA-DECAY FOR ISOTOPES WITH N/P RATIO ABOVE THE STABLE STATE

During beta-decay, a neutron splits up forming a proton which remains in the nucleus and an electron which is ejected.

$$^{1}_{0}n \longrightarrow ^{1}_{1}p + ^{0}_{-1}\mathrm{e}$$

As a result of this, the number of neutrons in an isotope decreases by one and the number of protons increases by one. Thus isotopes which are unstable due to an excess of neutrons over protons can try to stabilize themselves by beta-particle emission.

Isotopes which show beta-particle loss have more neutrons than a stable isotope of the same element. Thus, the isotopes of an element with the highest mass number are the ones most likely to emit beta-particles in order to increase their stability. Consequently, carbon-14 decays by beta-emission, but carbon-12 is stable. Hydrogen-3 is beta-active, but hydrogen-1 is stable.

Many of the nuclei synthesized artificially by neutron bombardment have a $n/p$ ratio above the stable value and consequently these isotopes emit beta-particles. For example, radioactive $^{32}_{15}\mathrm{P}$, obtained by neutron bombardment of $^{31}_{15}\mathrm{P}$, is a beta-emitter,

$$^{1}_{0}n + ^{31}_{15}\mathrm{P} \longrightarrow ^{32}_{15}\mathrm{P} \longrightarrow ^{32}_{16}\mathrm{S} + ^{0}_{-1}\mathrm{e}$$

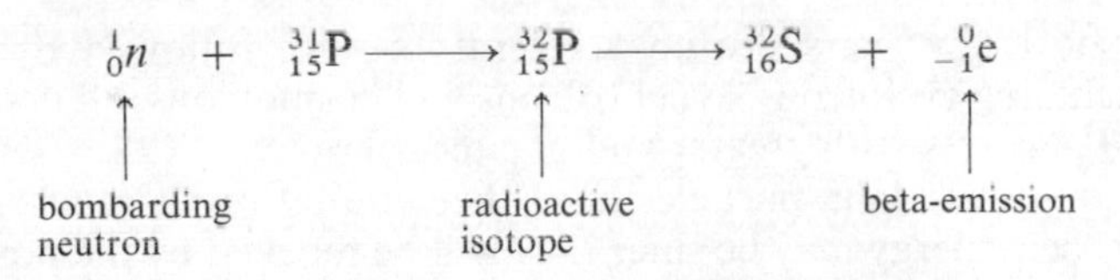


## 6.6 An alchemist's dream—producing radioactive isotopes

Many of the alchemists who practised during the sixteenth and seventeenth centuries were little better than charlatans or magicians. Preoccupied, as they were, in searching for a universal solvent which would dissolve all matter or a means of turning lead into gold, they seemed doomed to failure. Their attempts at transmuting one element into another led to nothing more than gloom and despair, yet curiously enough natural radioactive transmutations were happening all the time in nature.

The earliest evidence for the disintegration of an atomic nucleus was obtained in 1919 when Rutherford bombarded nitrogen gas with alpha-particles and discovered that protons were ejected. Later Blackett showed that the ejection of a proton was accompanied by the formation of an oxygen-17 isotope.

$$^{14}_{7}\mathrm{N} + ^{4}_{2}\mathrm{He} \longrightarrow ^{1}_{1}p + ^{17}_{8}\mathrm{O}$$

*Right* The alchemists of the sixteenth and seventeenth centuries were the first scientists to attempt the transmutation of elements.

Within a few years, Rutherford, Chadwick and their collaborators had demonstrated several other transmutations including F → Ne, Na → Mg and Al → Si, using alpha-particles as projectiles.

A few years later, Cockcroft and Walton working under Rutherford at Cambridge, developed equipment for bombarding atoms with protons rather than alpha-particles. Hydrogen gas was ionized to form electrons and protons. The protons were then accelerated to immense speeds by attracting them through a potential of more than 300 000 volts. In 1932, Cockcroft and Walton used their accelerator to bombard lithium atoms with protons and produce helium.

$$^{1}_{1}p + ^{7}_{3}\mathrm{Li} \longrightarrow ^{4}_{2}\mathrm{He} + ^{4}_{2}\mathrm{He}$$

During the next few years proton accelerators were followed by a variety of machines including cyclotrons, synchrotrons and cosmotrons, all designed to provide charged particles with greater and greater energy.

Both protons and alpha-particles have their limitation as 'nuclear bullets' no matter what their energy may be since they will be repelled by the positive nucleus of any target atom they approach. Fortunately, the discovery of the neutron opened up new possibilities for nuclear transmutations.

One of the first cyclotrons designed by Lawrence. Positively charged particles enter the cyclotron through tube X and are then accelerated around the machine in an outward-spiralling path by a series of voltage kicks. Eventually, the particles emerge from the tube at the top as high-energy projectiles.

- ○ What disadvantages do protons and alpha-particles have as nuclear projectiles?
- ○ What advantages do protons have over alpha-particles as nuclear projectiles?
- ○ Why are electrons ineffective as nuclear projectiles?
- ○ Why are neutrons superior to protons, alpha-particles and electrons as nuclear projectiles?

Radioactive $^{32}_{15}\mathrm{P}$ is obtained by neutron bombardment of $^{31}_{15}\mathrm{P}$.

$$^{1}_{0}n + ^{31}_{15}\mathrm{P} \longrightarrow ^{32}_{15}\mathrm{P}$$

The $^{32}_{15}\mathrm{P}$, being heavier than the stable isotope, undergoes beta-decay to $^{32}_{16}\mathrm{S}$.

$$^{32}_{15}\mathrm{P} \longrightarrow ^{32}_{16}\mathrm{S} + ^{0}_{-1}\mathrm{e}$$

Radioactive $^{32}_{15}\mathrm{P}$ has been used to study the uptake and metabolism of phosphoru by plants from phosphate fertilizers.

Another radioactive isotope obtained by neutron bombardment is $^{14}_{6}\mathrm{C}$.

$$^{1}_{0}n + ^{14}_{7}\mathrm{N} \longrightarrow ^{14}_{6}\mathrm{C} + ^{1}_{1}\mathrm{H}$$

$^{14}_{6}\mathrm{C}$ has been extensively used in the elucidation of carbon pathways includin photosynthesis and protein synthesis in plants and animals.

The reconstructed 600 MeV synchro-cyclotron of the European Organization for Nuclear Research at Geneva. To the left is the rotary capacitor (No. 2) which controls the frequency of the accelerating electric field. Behind is the 2 500-tonne electromagnet between the poles of which protons circulate *in vacuo* and behind the technicians one of the two vacuum pumps.

At first artificial transmutations involved elements with low relative atomic masses, but in 1940 McMillan and Abelson succeeded in synthesizing heavier elements from uranium, the heaviest naturally occurring element. Uranium-238 was bombarded with neutrons producing uranium-239

$$^{1}_{0}n + ^{238}_{92}\mathrm{U} \longrightarrow ^{239}_{92}\mathrm{U}$$

However, $^{239}_{92}\mathrm{U}$ is radioactive, losing beta-particles in two stages to form, first neptunium, and then plutonium.

$$^{239}_{92}\mathrm{U} \longrightarrow ^{239}_{93}\mathrm{Np} + ^{0}_{-1}\mathrm{e}$$

$$^{239}_{93}\mathrm{Np} \longrightarrow ^{239}_{94}\mathrm{Pu} + ^{0}_{-1}\mathrm{e}$$

By 1955, another seven elements with atomic numbers from 95–101 had been synthesized by Seaborg and his co-workers, and in the following 20 years evidence was obtained for the next four elements with atomic numbers 102–105.

All these relatively new elements beyond uranium in the periodic table are radioactive and most of them disintegrate rapidly into smaller atoms.

## 6.7 Atomic fission—nuclear (atomic) energy

Prior to 1938, scientists believed that nuclear reactions consisted solely in removing a relatively small particle such as an alpha-particle, an electron or a proton from a nucleus. In 1938, however, two German scientists, Hahn and Strassman, realized that isotopes of intermediate mass were formed when uranium was bombarded with neutrons. In this case uranium nuclei had split into two similar halves.

This type of splitting during radioactive decay is very different from the previous examples we have studied in which one large fragment, and one very small fragment (either $^{4}_{2}\mathrm{He}$, $^{0}_{-1}\mathrm{e}$ or $^{1}_{1}p$) were produced. Thus, the term **atomic fission** or **nuclear fission** was used to describe the disintegration of an atom into two large fragments.

Hahn and Strassman noticed one other important feature of atomic fission—it could produce vast quantities of energy. Calculations suggested that one gram of natural uranium could provide as much energy as 10 000 grams of coal. Could the energy released during uranium fission be harnessed for the service of humanity? The answer, of course, was yes, but not without despair and destruction. In 1942, the first atomic reactor was built in America by the Italian scientist Enrico Fermi.

Enrico Fermi (1901–1954) who built the first nuclear reactor in a Chicago squash court in 1942.

The possibility of nuclear warfare became obvious and the first 'atomic bombs' were dropped on Hiroshima and Nagasaki in Japan in 1945. The effect was shattering—almost incredible. Vast areas of land were destroyed and thousands of inhabitants were killed, many of them not by the explosion itself but by the nuclear radiation emitted. Such nuclear warfare created worldwide unrest and indignation and, during the 1950s and 1960s, large numbers of British scientists became supporters of a Nuclear Disarmament Campaign which opposed the increased testing and development of nuclear weapons by Britain, America and Russia. Fortunately, these countries had begun to exploit the peaceful uses of nuclear energy. In 1954, the first atomic power station supplying electrical energy was opened in Russia and in 1956, Calder Hall in Cumberland, the world's first industrial-scale nuclear power station came into operation.

Hiroshima after the bomb.

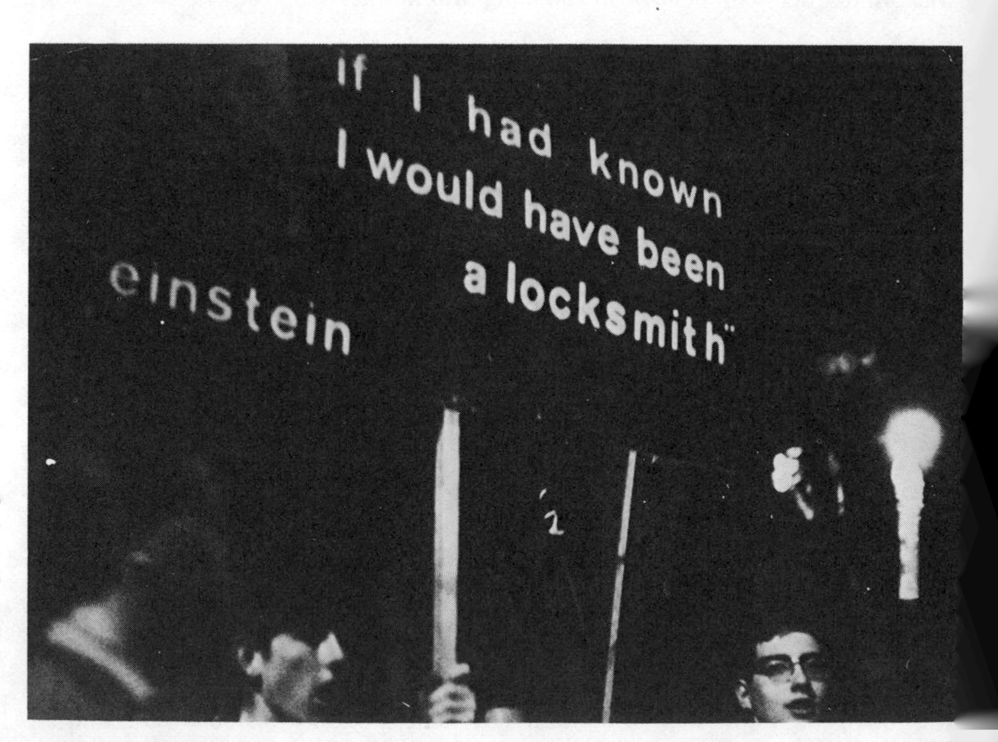


During the 1950's and 1960's, large numbers of British scientists and students became supporters of a Nuclear Disarmament Campaign. In the photograph *right*, students parade Einstein's thoughts on what he would have done had he known what nuclear science would do for mankind. The students were parading through the West End of London during a candlelight procession.

URANIUM FISSION

Natural uranium contains two isotopes, $^{238}_{92}U$ and the much rarer $^{235}_{92}U$. Experiments showed that the rarer isotope of mass number 235 is mainly responsible for the vast release of energy during fission. When natural uranium is bombarded by neutrons the $^{235}_{92}U$ isotopes absorb neutrons and then split up violently releasing two smaller nuclei, two or three neutrons and a large quantity of energy.

For example:

$$^{1}_{0}n + {}^{235}_{92}U \longrightarrow [{}^{236}_{92}U] \longrightarrow {}^{90}_{38}Sr + {}^{143}_{54}Xe + 3^{1}_{0}n$$

Although U-238 does not undergo fission during neutron bombardment, it absorbs neutrons and disintegrates to neptunium and plutonium.

$$^{1}_{0}n + ^{238}_{92}U \longrightarrow ^{239}_{92}U + \text{gamma-rays}$$

$$^{239}_{92}U \longrightarrow ^{239}_{93}Np + ^{0}_{-1}e$$

$$^{239}_{93}Np \longrightarrow ^{239}_{94}Pu + ^{0}_{-1}e$$

The plutonium-239 which is formed undergoes atomic fission similar to U-235 and the newer atomic power stations, such as the one at Dounreay, use Pu-239 as fuel.

## 6.8 Chain reactions, atomic bombs and nuclear reactors

All of the neutrons released during the fission of U-235 are capable of splitting another uranium nucleus. Figure 6.4 shows how an exploding chain reaction can occur during uranium fission.

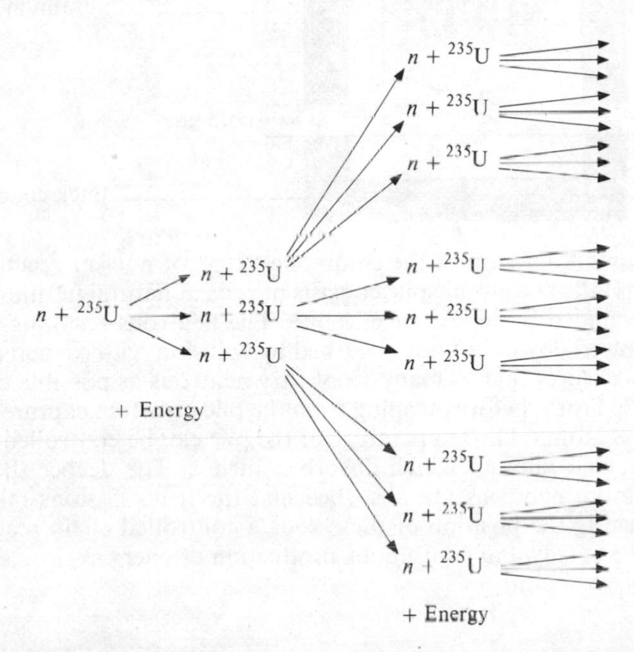


**Figure 6.4** A simplified diagram to illustrate an exploding chain reaction during uranium fission.

One reaction releases 3 neutrons, 3 neutrons release 9, then 27, 81, 243 etc. Provided the chain can be built up rapidly and completely there is an enormous release of energy—an atomic bomb.

This will only occur if the sample of U-235 is large enough. If it is too small the secondary neutrons will escape from the sample before they have caused further fissions and the chain reaction will be broken (figure 6.5(*a*)). Thus, a small piece of $^{235}U$ is quite safe, but a larger piece, within which the secondary neutrons cause further fissions before escaping, may explode (figure 6.5(*b*)). There is, therefore, a **critical size** below which $^{235}U$ is safe, but above which it may explode.

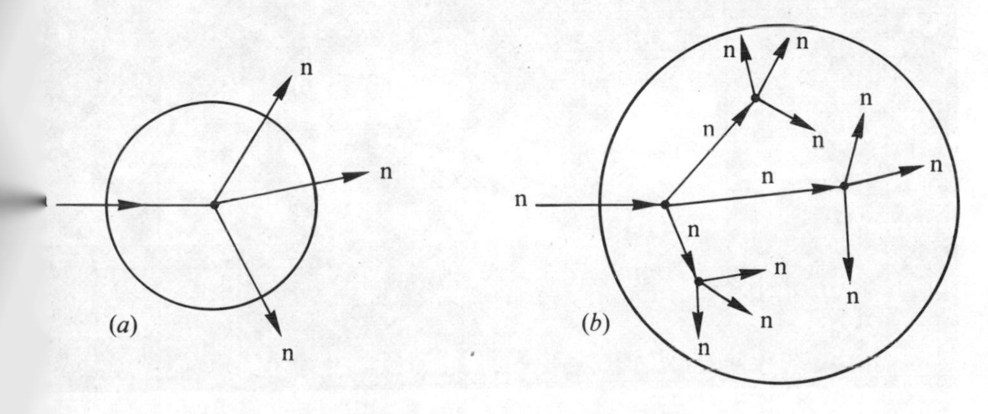


**Figure 6.5** A small piece of $^{235}U$ (as in (*a*)) is quite safe, but a piece larger than the critical size (as in (*b*)) will explode.

In an atomic bomb two pieces of pure U-235 or Pu-239 each below the critical size are brought together to form one piece larger than the critical size. Fission of U-235 or Pu-239 then proceeds in an uncontrolled manner as an exploding chain reaction.

- ○ Why must the U-235 used in an atomic bomb be pure?
- ○ How can the fission of U-235 be controlled in an atomic reactor so that the energy is converted into electricity or heat?

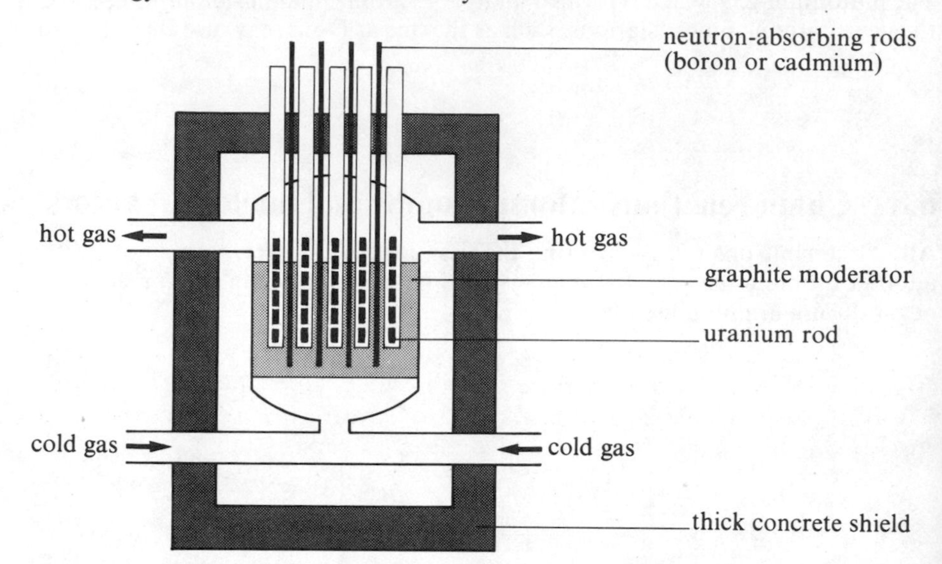


**Figure 6.6** A simplified diagram of a nuclear reactor.

Figure 6.6 shows a diagram of the commonest type of nuclear reactor similar to that at Calder Hall. The atomic pile consists of rods of natural uranium inserted in channels surrounded by blocks of graphite. The neutrons resulting from fission of $^{235}U$ are slowed down, but not absorbed by collision with carbon atoms in the graphite. This ensures that as many secondary neutrons as possible cause further fission of U-235 atoms before escaping from the pile or before capture by the more plentiful U-238 atoms. The temperature of the pile can be controlled by movable rods of boron or cadmium which absorb neutrons. The deeper these rods are inserted, the more neutrons are absorbed and the fewer fissions take place. By carefully adjusting the position of these rods a controlled chain reaction can be obtained with a steady and continuous production of energy.

Atomic energy: controlled or uncontrolled! Britain's first nuclear explosion superimposed above the Dounreay Prototype Fast Reactor and the Dounreay Fast Reactor on the right of the picture.

The heat produced during the fission process is removed by a stream of circulating carbon dioxide. The hot gas can be used to make steam and then drive a turbine.

○ Why must the uranium rods be stored well apart before use in the reactor?
○ Why do fewer fissions occur when the rods are removed from the pile even though they are the same distance apart?
○ Why is it necessary to surround the pile with a thick concrete shield?

Besides liberating energy, the atomic reactor also produces plutonium as a result of the neutron bombardment of U-238.

$$^{1}_{0}n + ^{238}_{92}\text{U} \longrightarrow ^{239}_{92}\text{U}$$
$$^{239}_{92}\text{U} \longrightarrow ^{239}_{93}\text{Np} + ^{0}_{-1}\text{e}$$
$$^{239}_{93}\text{Np} \longrightarrow ^{239}_{94}\text{Pu} + ^{0}_{-1}\text{e}$$

The plutonium can be separated from the uranium and used as the atomic fuel in nuclear reactors such as the one at Dounreay.

Since plutonium requires faster neutrons for fission than U-235, no graphite moderator is necessary. The central core of plutonium is surrounded by natural uranium which captures any neutrons escaping from the central core and becomes converted to plutonium. By careful design and control, it is possible to produce more plutonium in the surrounding uranium than is used up in the central core. Consequently, this type of reactor is called a **fast-fission breeder reactor** since it uses fast neutrons and 'breeds' plutonium from uranium.

Although such reactors produce more plutonium than they use, we are not getting 'something for nothing'. There is no contradiction of the laws of Conservation of Matter and Energy since uranium and plutonium are being converted into more plutonium. Check this yourself using the last set of equations.

The energy from disintegrating plutonium that destroyed Nagasaki is being tamed. For more than 20 years nuclear scientists have been trying to harness plutonium as the fuel for the future in 'fast-breeder' reactors. Many people, however, consider the dangers in using plutonium are too great. Atom bombs in the hands of terrorists or unscrupulous governments might prove catastrophic. On the other hand, the benefits of plutonium seem irresistible—cheap, abundant electricity for hundreds of years. Will our descendants curse us or bless us for the legacy of plutonium?

Technicians loading cadmium absorbers into a small-scale test reactor at Harwell. The top shield plate of the reactor has been removed and is in the background.

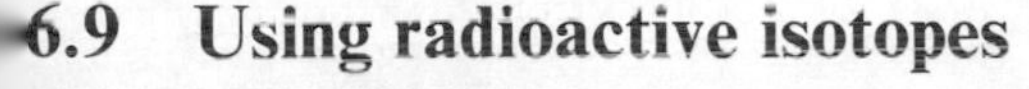

## 6.9 Using radioactive isotopes

TREATMENT OF CANCER

The structure of DNA within the genes of animals and plants can be changed by radiation. For reasons which are not yet clear, cancer cells are more susceptible than normal cells to destruction by radiation. Because of this, gamma-rays can be used in the treatment of cancer. Penetrating gamma-radiation from cobalt-60 ($^{60}_{27}$Co) is used in treating inaccessible growths, whereas superficial cancers, such as skin cancer, can be treated by less penetrating radiation from $^{32}_{15}$P or $^{90}_{38}$Sr in plastic sheets strapped on the affected area.

STUDYING METABOLIC PATHWAYS

Radioactive isotopes can be used to trace the uptake and metabolism of various elements by animals and plants. For example, the uptake of phosphate and the metabolism of phosphorus by plants can be studied using a fertilizer containing $^{32}_{15}$P. Radioactive tracer studies using $^{14}_{6}$C have helped in the elucidation of photosynthesis and protein synthesis. $^{131}_{54}$I has been used in the diagnosis and treatment of thyroid diseases and in research into thyroid gland functioning.

THICKNESS GAUGES AND EMPTY PACKET DETECTORS

The radiation passing through a material decreases as the material gets thicker. Thus, the amount of penetrating beta- or gamma-radiation can be used to estimate the thickness of various materials such as paper, metal or plastic. The advantage of using radiation for measuring the thickness of materials is that there need be no

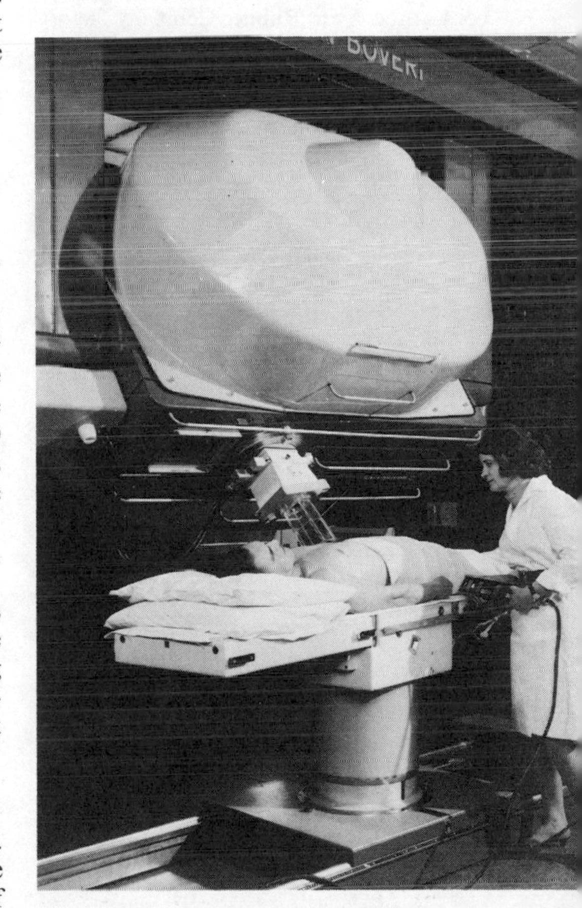

A patient being treated by radiotherapy.

A radiation thickness gauge using strontium-90 being used to control the thickness of tyre cord at the Avon Rubber Company factory at Melksham.

touching, marking or tearing of the article concerned. Thus, radiation thickness gauges can be used to control the thickness of sheet steel emerging from a high-speed rolling mill. Beta-rays can be used for material thicknesses up to about 0.2cm of steel whilst gamma-rays can be used with steel up to 10cm thick.

A similar use of radioactive isotopes is made in level gauges and empty-packet detectors. The level of liquid in a closed vessel can be found by placing a source on one side of the container and a detector on the other. A sudden decrease in the detected radiation, when source and detector are moved down the vessel simultaneously, shows the level of liquid within the vessel. Level gauges of this type are used to measure the amount of liquid in fire extinguishers and gas cylinders. In a similar fashion, empty-packet detectors can be set to reject empty or insufficiently filled packets of biscuits or cigarettes.

## Summary

1 Radioactivity results from the spontaneous disintegration of nuclei.

2 Naturally occurring elements which undergo radioactive decay emit alpha-particles ($^{4}_{2}He^{2+}$ ions), beta-particles (electrons) and gamma-rays (electromagnetic radiation).

3 Gamma-radiation is far more penetrating than alpha- and beta-radiation. It can penetrate thick metal sheets and is responsible for most of the damage caused by the disintegration of radioactive substances.

4 The half-life period of a radioactive isotope is the time taken for the mass or concentration of the isotope (or for its rate of decay) to fall to half its initial value. The half-life is virtually unaffected by external conditions and can be used as a measure of its relative stability.

5 The $n/p$ ratio for stable isotopes lies within a very restricted range for any given atomic number.
Isotopes which are unstable due to an excess of protons over neutrons ($n/p <$ stable value) can increase their stability by alpha-particle emission.
Isotopes which are unstable due to an excess of neutrons over protons ($n/p >$ stable value) can increase their stability by beta-particle emission.

6 Atomic (nuclear) fission is the disintegration of an isotope into two large fragments.

7 Chemical reactions involve electrons on the outer parts of atoms, whereas nuclear reactions involve protons and neutrons in the nucleus.

## Study questions

1 (a) What are the relative masses and charges of alpha- and beta-particles?
(b) What is the nature of gamma-radiation?
(c) What evidence can you provide to support your statements concerning the nature of alpha-, beta- and gamma-radiation?

2 A radioactive source with a very long half-life emits alpha-particles of energy $5 \times 10^{-1}$ J at the rate of $4.2 \times 10^{6}$ per second. The source is embedded in material sufficient to absorb the alpha-particles. The material loses heat at the rate of 8.4J per minute for every 10°C its temperature is above that of its surroundings and eventually a steady temperature is attained.
(a) What conditions apply when the steady temperature is attained?
(b) How much has the temperature of the material risen above its surroundings when the steady state is achieved?
(c) Mention two consequences or practical applications of this radioactive heating effect

3 (a) How do the atomic number and the mass number of an isotope change when its nucleus loses
(i) an alpha-particle, (ii) a beta-particle,
(iii) gamma-rays, (iv) a neutron?
(b) $^{232}_{90}Th$ emits a total of six alpha-particles and four beta-particles in its natural decay sequence. What is the atomic number, mass number and symbol of the final product
(c) Write nuclear equations to summarize the following changes:
(i) when a $^{7}_{3}Li$ nucleus absorbs a colliding proton the product disintegrates into two exactly similar fragments.
(ii) $^{232}_{90}Th$ atoms undergo alpha-decay to form radium atoms.

(d) Scintillation counting shows that 1mg of polonium emits $3 \times 10^{18}$ alpha-particles in the course of complete decay.
What is the relative atomic mass of polonium? (State the assumptions in your calculation.)

4 $^{214}_{83}Bi$ has a half-life of 20 minutes.
(a) Plot a graph of the percentage of the $^{214}_{83}Bi$ remaining (vertical) against time.
What percentage of the $^{214}_{83}Bi$ remains after 70 minutes?
(b) Rewrite the following nuclear equations substituting symbols and numbers for the question marks.
(i) $^{14}_{?}? + ^{?}_{?}He \rightarrow ^{?}_{?}O + ^{1}_{1}?$
(ii) $^{?}_{?}H + ^{56}_{?}Fe \rightarrow ^{57}_{?}Co + ^{1}_{0}?$

5 Outline the importance or uses of two of the following isotopes.
$^{60}_{27}Co$, $^{32}_{15}P$, $^{131}_{54}I$, $^{239}_{94}Pu$

6 (a) How is carbon-14 produced in nature?
(b) Write an equation for the beta-decay of $^{14}_{6}C$.
(c) How is $^{14}_{6}C$ used in 'dating' archaeological remains?
(d) How was radioactive 'dating' used to confirm the validity of the Dead Sea Scrolls and the forgery of Piltdown Man?

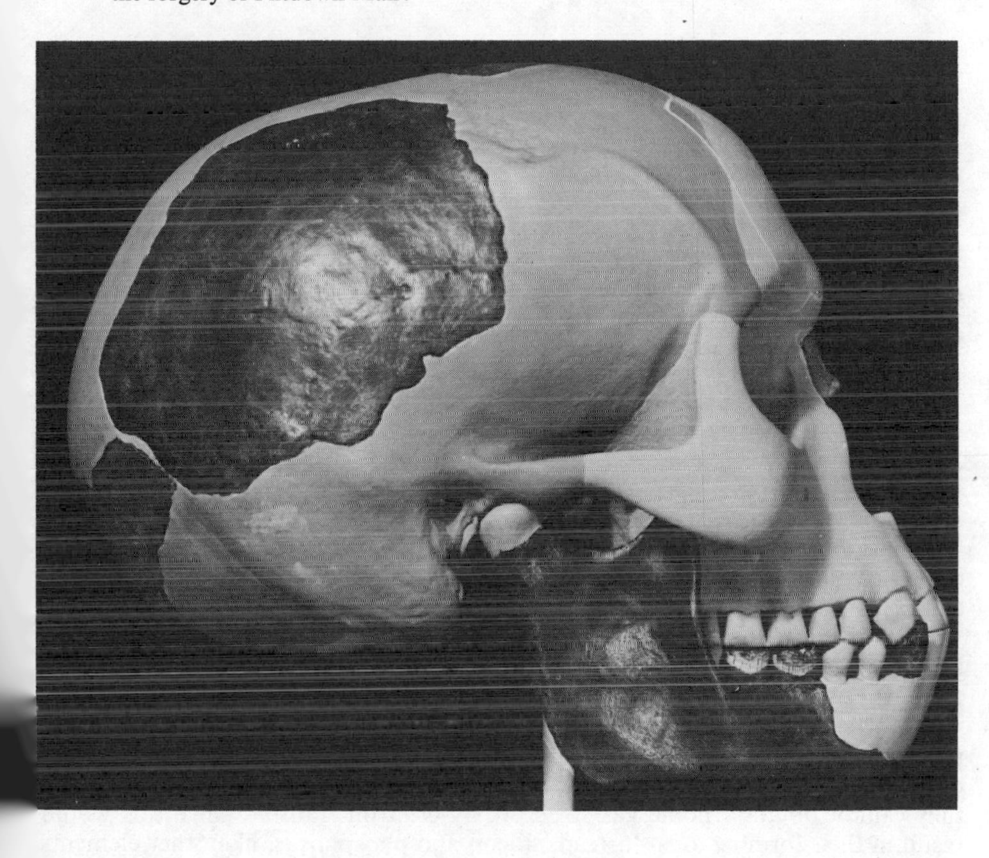

The reconstructed skull of Piltdown Man, a forgery uncovered by radioactive dating.

# 7 The Electronic Theory and Chemical Bonding

## 7.1 Introduction

Look at the information in table 7.1 relating to the electronic structures of the atoms and ions of the elements in period 3.

**Table 7.1** Electronic structures of the atoms and ions of the elements in period 3

| **Element** | Na | Mg | Al | Si | P | S | Cl | Ar |
|---|---|---|---|---|---|---|---|---|
| **Electronic structure of the atom** | 2,8,1 | 2,8,2 | 2,8,3 | 2,8,4 | 2,8,5 | 2,8,6 | 2,8,7 | 2,8,8 |
| **No. of electrons in outermost shell of atom** | 1 | 2 | 3 | 4 | 5 | 6 | 7 | 8 |
| **Ion formed** | $Na^+$ | $Mg^{2+}$ | $Al^{3+}$ | — | — | $S^{2-}$ | $Cl^-$ | — |
| **Electronic structure of the ion** | 2,8 | 2,8 | 2,8 | — | — | 2,8,8 | 2,8,8 | — |

The first three elements in the period (Na, Mg and Al) *lose* electrons from their outermost shell to form positively charged ions ($Na^+$, $Mg^{2+}$ and $Al^{3+}$) with an electron structure like neon, the previous noble gas.

Elements at the end of the period in Groups VI and VII (S and Cl) *gain* electrons to form negatively charged ions ($S^{2-}$ and $Cl^-$) with the same electron structure as the next noble gas, argon.

But, what about elements in the middle of the period such as silicon and phosphorus? These elements do not usually form ions in their compounds, but do they obtain an electronic structure similar to that of a noble gas? The answer is 'yes', but they cannot obtain a noble gas structure by electron transfer, since this would result in their forming ions. Instead, silicon and phosphorus, like other elements in Groups IV and V, achieve electron structures like noble gases by sharing electrons with other atoms in their compounds.

In 1916, Kossel and Lewis realized that all of the noble gases, with the exception of helium, had an outer shell containing eight electrons. They suggested that this arrangement was responsible for the stability and inertness of the noble gases. Thus, *when elements form compounds they either lose or gain or share electrons so as to achieve stable (low energy) electron configurations similar to the next higher or lower noble gas in the periodic table.* This simple idea forms the basis of the electronic theory of bonding.

Since Kossel and Lewis put forward their ideas in 1916, the noble gases have been shown to be more reactive than expected and we now know of many compounds in which the elements do not have a noble gas structure. For example, most of the ions of transition metals (e.g. $Fe^{2+}$, $Fe^{3+}$, $Cu^{2+}$) do not have an electron structure like a noble gas nor does the sulphur atom in $SO_2$ or $SO_3$. Nevertheless, Kossel and Lewis's ideas still form the basis of modern theories of bonding which explain the formulae and structure of most compounds and the forces holding particles together.

## 7.2 Transfer of electrons—electrovalent (ionic) bonding

The most typical ionic compounds are formed when a metal element from Group I or Group II reacts with a non-metal from Group VI or Group VII. When the reaction occurs electrons are transferred from the metal to the non-metal until the outer electron shells of the resulting ions are identical to those of the nearest noble gases.

Figure 7.1 shows how the transfer of electrons from lithium to oxygen forms ions in lithium oxide. In figure 7.1, the nucleus of each atom is represented by its symbol and the electrons in each shell are represented by groups of dots or crosses around the symbol. Ions are shown in square brackets with the charge at the top right hand corner.

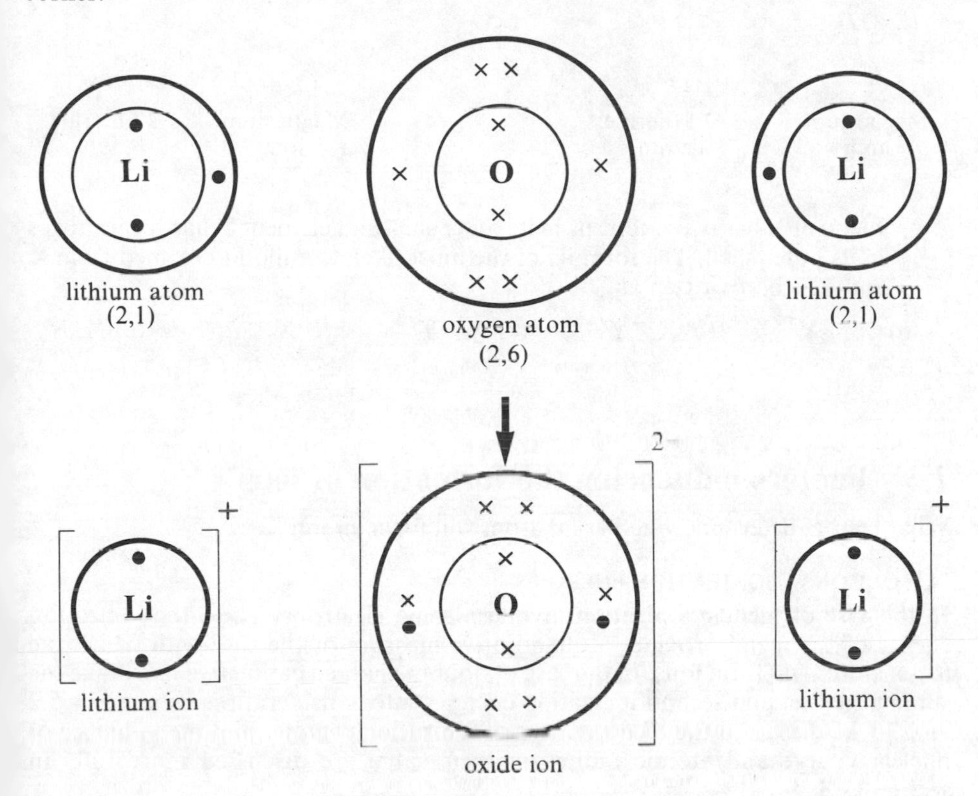


**Figure 7.1** Transfer of electrons from lithium to oxygen in the formation of lithium oxide

- ○ What is the electron structure of: (i) the lithium ion, (ii) the oxide ion?
- ○ Which noble gas has an electron structure like $Li^+$?
- ○ Which noble gas has an electron structure like $O^{2-}$?
- ○ Why is it that two lithium atoms react with only one oxygen atom?

Although the electrons of the different atoms in figure 7.1 are shown by dots and crosses, you must not think that electrons of lithium are any different from those of oxygen. All electrons are identical; they are shown differently in the diagram so that we can follow their transfer more easily.

Figure 7.2 shows the electron transfers which take place during the formation of sodium chloride and magnesium fluoride. Only those electrons in the outer shell of each atom are represented by dots or crosses round their symbol, although the full electron structures are shown in brackets below the symbols.

Draw similar 'dot/cross' diagrams to represent the electron transfers which take place in the formation of magnesium sulphide and potassium oxide.

Remember when writing 'dot/cross' diagrams that the dots and crosses are simply a means of counting electrons; they cannot show the location of electrons within the atom since electrons are distributed in space as diffuse negative charge-clouds.

The formation of ions in such compounds as lithium oxide, sodium chloride and magnesium fluoride involves the *complete transfer* of electrons. In the crystals of these substances ions are held together by **ionic (electrovalent) bonds**—an electrostatic attraction between oppositely charged ions. The structure, properties and bonding in ionic compounds are discussed in sections 9.7 and 9.8.

$$\text{Na}\bullet \quad + \quad {}^{\times}_{\times}\overset{\times\times}{\underset{\times\times}{\text{Cl}}}{}^{\times}_{\times} \longrightarrow [\text{Na}]^{+} \quad [{}^{\times}_{\bullet}\overset{\times\times}{\underset{\times\times}{\text{Cl}}}{}^{\times}_{\times}]^{-}$$

| (2, 8, 1) | (2, 8, 7) | | (2, 8) | (2, 8, 8) |
|---|---|---|---|---|
| Sodium atom | Chlorine atom | | Sodium ion | Chloride ion |

$$\text{Mg}\mathbf{:} \quad + \quad 2\ {}^{\times}_{\times}\overset{\times\times}{\underset{\times}{\text{F}}}{}^{\times}_{\times} \longrightarrow [\text{Mg}]^{2+} \quad 2\,[{}^{\times}_{\times}\overset{\times\times}{\underset{\times\bullet}{\text{F}}}{}^{\times}_{\times}]^{-}$$

| (2, 8, 2) | (2, 7) (2, 7) | | (2, 8) | (2, 8) (2, 8) |
|---|---|---|---|---|
| Magnesium atom | 2 Fluorine atoms | | Magnesium ion | 2 Fluoride ions |

**Figure 7.2** Electron transfers during the formation of sodium chloride and magnesium fluoride.

○ Element X has two electrons in its outer shell and element Y has six electrons in its outer shell. The formula of the most likely compound formed from X and Y is best written as

(*a*) $(X^+)_3Y^{3-}$ (*b*) $X^{2+}Y^{2-}$ (*c*) $X^{2-}Y^{2+}$ (*d*) $X_2{}^+Y_2{}^-$ (*e*) $X^+Y^-$

## 7.3 Factors influencing the formation of ions

What factors determine whether an atom will form an ion?

### (A) CATIONS—IONIZATION ENERGIES

In the case of metals, ionization involves losing electrons. Thus, the ionization energy of the metal provides a quantitative measure of the ease with which the metal atoms will form ions. In the case of doubly-charged cations we must take the sum of the first and second ionization energies into consideration. In sections 5.2 and 5.3 we discussed the measurement of ionization energies and the influence of nuclear charge and atomic radius on their value are discussed more fully in section 12.3.

The lower the ionization energy of an atom or ion the more easily will it lose an electron. The further an electron is from the nucleus, the less firmly it is held and the easier it can be lost. Thus, in Group I, caesium which has the largest atoms, forms ions most easily in spite of having the largest positive nuclear charge. In fact, caesium loses electrons so easily that it is used in photoelectric cells (section 18.6(*b*)). In general, it becomes easier to form positive ions on passing down a group from atoms of smaller to those of larger relative atomic mass.

Once an electron has been lost from an atom, the overall positive charge can hold the remaining electrons more firmly. Thus, although many metals form doubly-charged cations, those with three units of charge are less common and ions with four positive charges are very rare. In fact, it requires four times as much energy to form $Mg^{2+}(g)$ from Mg(g) than to form $Na^+(g)$ from Na(g) and the process $Al(g) \rightarrow Al^{3+}(g)$ requires ten times as much energy as $Na(g) \rightarrow Na^+(g)$.

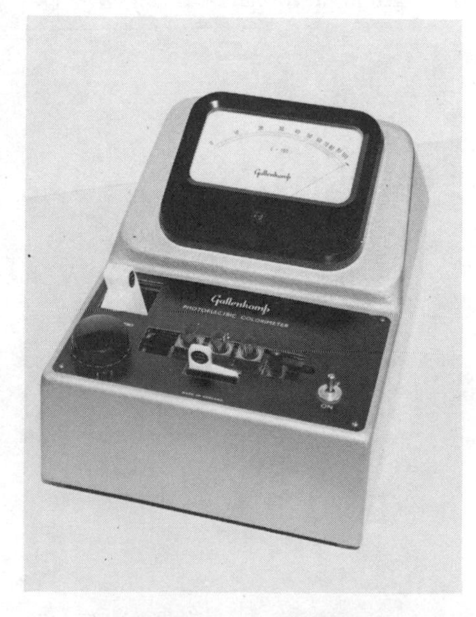


A photoelectric colorimeter which can be used to determine the concentration of a coloured solution. There is an explanation of the working of the instrument in section 18.6(b)(i). This often involves the ionization of caesium atoms.

### (B) ANIONS—ELECTRON AFFINITY

Ionization energies provide an indication of the ease with which atoms or ions can lose electrons. They are concerned with the formation of cations and are of no use in considering the formation of anions which involves the gain of an electron or electrons. A more appropriate measurement in the case of anions is the electron affinity, which gives the energy change for the process:

$$X(g) + e^- \longrightarrow X^-(g)$$

Notice that the electron affinity for X(g) is simply the reverse of the first ionization energy for $X^-(g)$. Electron affinities can be measured by valve measurements or spectroscopic methods similar to those used in the determination of ionization energies (p. 51). Table 7.2 shows some values of electron affinities. The more negative (i.e. more exothermic) the electron affinity, the more stable is the anion formed. Thus, $Cl^-$ is more stable than $Br^-$ with respect to their corresponding atoms and $Br^-$ is more stable than $I^-$.

**Table 7.2** Electron affinities for some atoms and ions (Values are in kJ mole$^{-1}$)

| | | | |
|---|---|---|---|
| $H \longrightarrow H^-$ <br> −72 | | | |
| $F \longrightarrow F^-$ <br> −333 | $Cl \longrightarrow Cl^-$ <br> −364 | $Br \longrightarrow Br^-$ <br> −342 | $I \longrightarrow I^-$ <br> −295 |
| $O \longrightarrow O^-$ <br> −141 | $O^- \longrightarrow O^{2-}$ <br> +791 | $S \longrightarrow S^-$ <br> −200 | $S^- \longrightarrow S^{2-}$ <br> +649 |

- ○ Why do you think the electron affinities of halogen atoms are more exothermic than those of the oxygen atom or the sulphur atom?
- ○ Why do you think the electron affinities of $O^-$ and $S^-$ are endothermic?

In general, electron affinities become more exothermic as a period is crossed from left to right because the incoming electron is attracted more strongly by the progressively smaller atoms with an increasing positive charge in their nucleus.

### (C) LATTICE ENERGIES

Although ionization energies and electron affinities provide information about the ease with which atoms form ions, these processes are often endothermic. For example, the process of converting solid sodium to gaseous sodium ions ($Na(s) \rightarrow Na^+(g) + e^-$) is endothermic. This suggests that $Na^+(g)$ is less stable than Na(s). However, the gaseous sodium ion ($Na^+$) is stabilized when it comes into contact with $Cl^-$ ions to form the solid crystal, because the process:

$$Na^+(g) + Cl^-(g) \longrightarrow Na^+Cl^-(s)$$

is very exothermic ($\Delta H = -781$ kJ mole$^{-1}$). The heat change for this reaction is known as the **Lattice Energy** for sodium chloride.

The more exothermic the lattice energy the more stable the ionic compound formed. Lattice energies are discussed in more detail in section 11.12.

## 7.4 Sharing electrons—covalent bonding

Look closely at figure 7.3 which shows an electron density map for the hydrogen molecule. (This can be compared with the electron density map for sodium chloride, a typical ionic compound in figure 9.20.) Lines on the map join points of

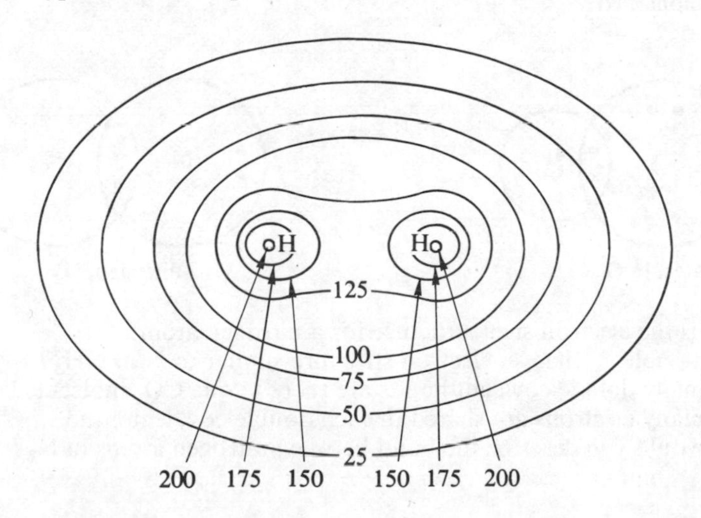


**Figure 7.3** Electron density map for hydrogen (units for the contours are electrons per nm$^3$) After C. A. Coulson, *Proc. Cam. Phil. Soc.*, **34**, 210.

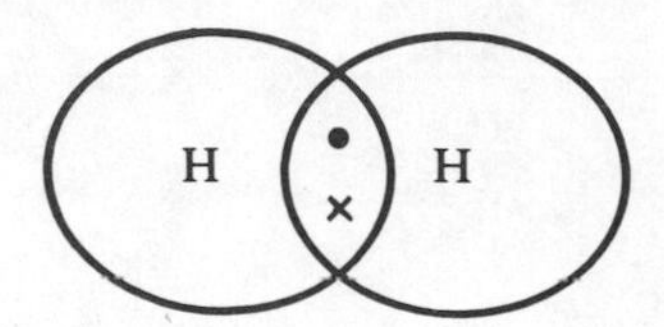


**Figure 7.4** A 'dot/cross' representation of the electron structure of a hydrogen molecule.

equal electron density in the same way that contours on a map join points at the same height above sea-level. Notice that, although the highest concentration of electrons is near each nucleus, there is also a high concentration of electrons between the two nuclei. This suggests that in molecules such as $H_2$, electrons are shared by the two hydrogen atoms. The two atoms are held together by the strong attractions of their nuclei for the electrons in between. Each hydrogen atom has only one electron and is very reactive. If, however, two hydrogen atoms come sufficiently close together their electron orbitals can overlap so that the pair of electrons is attracted to each nucleus and shared by each atom. A schematic 'dot/cross' representation of this is shown in figure 7.4. Each hydrogen atom now has two electrons which is the same electron structure as helium. Thus, the molecule, $H_2$ is much more stable than the H atom.

*The sharing of a pair of electrons between two atoms constitutes a covalent bond.*

In a normal covalent bond, each atom contributes one electron to the shared pair which then go towards filling the outermost shell of both atoms.

Covalent bonding can be used to explain the structure and formulae of molecules of non-metals (e.g. $Cl_2$, $P_4$, $S_8$) and also of non-metal/non-metal compounds (e.g. HCl, $CH_4$, $CO_2$). In these substances, *each atom usually gains a noble-gas electron structure* as a result of electron sharing. Figure 7.5 shows how this happens

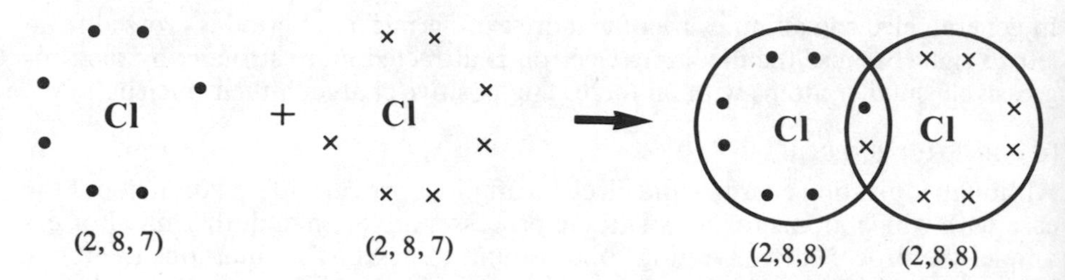


**Figure 7.5** The formation of a covalent bond between two chlorine atoms.

in the case of chlorine, $Cl_2$. Each chlorine atom has the electronic structure 2, 8, 7 with seven electrons in its outermost shell. By sharing one pair of electrons they acquire an electron structure similar to argon, (2,8,8). Figure 7.6 shows the electron 'dot/cross' diagrams for ammonia, $NH_3$; water, $H_2O$; carbon dioxide, $CO_2$ and nitrogen, $N_2$.

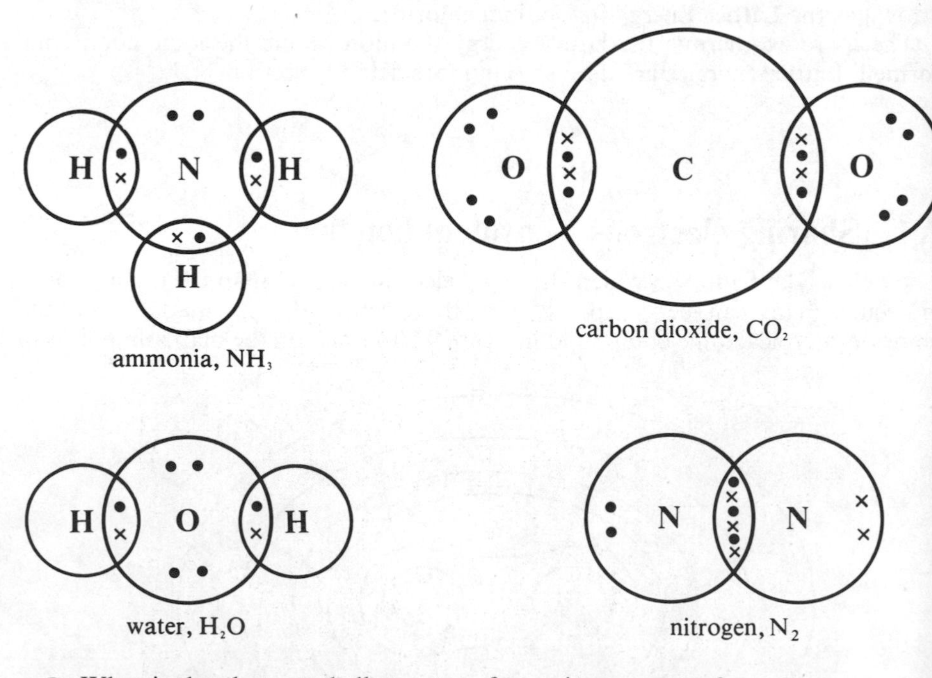


**Figure 7.6** Electron 'dot/cross' diagrams for ammonia, water, carbon dioxide and nitrogen.

- ○ What is the electron shell structure for a nitrogen atom?
- ○ Which noble gas has an electron structure similar to N in $NH_3$?
- ○ How many double covalent bonds are there in one $CO_2$ molecule?
- ○ How many electrons are shared in each double covalent bond?
- ○ How would you describe the bond between nitrogen atoms in $N_2$?

In carbon dioxide, double bonds are formed by the sharing of four electrons, two contributed by each atom. Similarly, triple bonds can be formed by the sharing of six electrons, three contributed by each atom (see nitrogen in figure 7.6). Remember that 'dot/cross' diagrams are merely a convenient method for counting electrons. They do not represent the positions of electrons, nor do they give any indication of the shape of the molecules.

## 7.5 Determining the structures of simple molecules

When a beam of electrons is passed through a gas or vapour at low pressure, the nuclei of the atoms in the gaseous molecules scatter (diffract) the electrons. From the pattern of the diffracted electrons, it is possible to determine the structure of the gas causing the diffraction. Figure 7.7 shows the experimental arrangement for electron diffraction studies. Electrons ejected by a heated filament are accelerated through about 50 000 volts and collimated into a thin beam. The gas under investigation is admitted through a jet towards a cold surface which will condense it. As the electron beam passes through the gas, diffraction takes place and the scattered electrons strike a photographic plate.

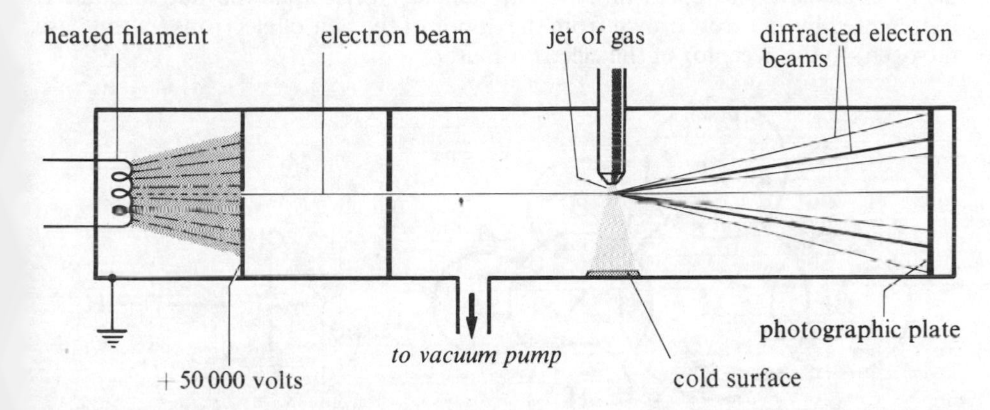


**Figure 7.7** The experimental arrangement for electron diffraction studies.

The diffraction photograph shows a series of concentric circles. Each circle corresponds to the distance between atoms within one molecule or between atoms in different molecules. By comparing the observed diffraction pattern of the gas with diffraction patterns calculated for various assumed structures it is possible to deduce the correct structure of particles in the gas. Furthermore, the diffraction pattern gives information from which bond lengths and bond angles can be determined.

Electron diffraction may be used with simple molecules in the gaseous or vapour state. It gives bond lengths to an accuracy of $\pm 0.001$ nm or better. (The structure of materials in the crystalline state can be investigated by X-ray diffraction (section 9.2).)

*Left* An electron diffraction pattern produced by electrons passing through a very thin film of silver.

*Above* An electron diffraction pattern produced by electrons passing through a very thin film of gold.

## 7.6 Co-ordinate (dative covalent) bonding

In a normal covalent link, each atom provides one electron for the bond. However, in a few compounds a bond is formed by the sharing of a pair of electrons both of which originate from one atom. Since both electrons in the bond are donated by one atom, the bonding is known as **dative covalent** or **co-ordinate**. Dative covalent bonding plays an important part in the formation of the addition compound, $NH_3.AlCl_3$.

In the vapour phase at high temperatures, aluminium chloride consists of simple molecules of $AlCl_3$ with covalent bonds between the aluminium and chlorine atoms (figure 7.8). Notice that aluminium has only six electrons in its outer shell—two short of the noble gas structure for neon. Now look at the 'dot/cross' structure of ammonia in figure 7.6. The nitrogen atom has eight electrons in its outer shell, but two of these electrons are not shared with any other atom. When ammonia gas and aluminium chloride vapour are mixed they react rapidly to form the solid $NH_3.AlCl_3$. In this compound, the $NH_3$ and $AlCl_3$ have formed a bond because the N atom in $NH_3$ can donate its unshared pair of electrons towards the Al atom in $AlCl_3$, enabling the Al atom to achieve a noble gas structure. Figure 7.9 shows the electron 'dot/cross' structure for the compound and also the simpler notation using single lines to represent covalent bonds. Notice that the dative bond is represented by an arrow drawn from the donor of the pair of electrons (in this case nitrogen) to the acceptor of the electron-pair.

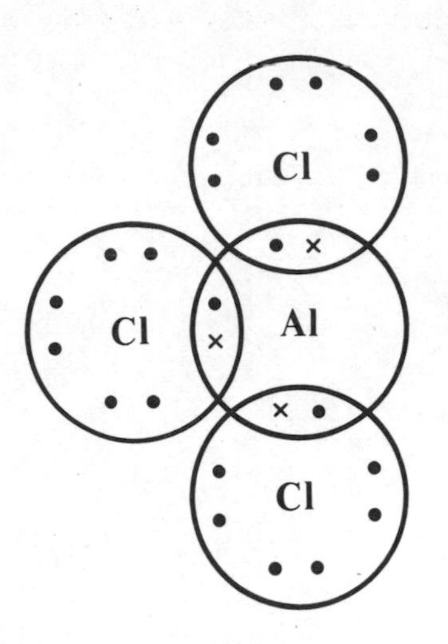


**Figure 7.8** An electron 'dot/cross' diagram for $AlCl_3$.

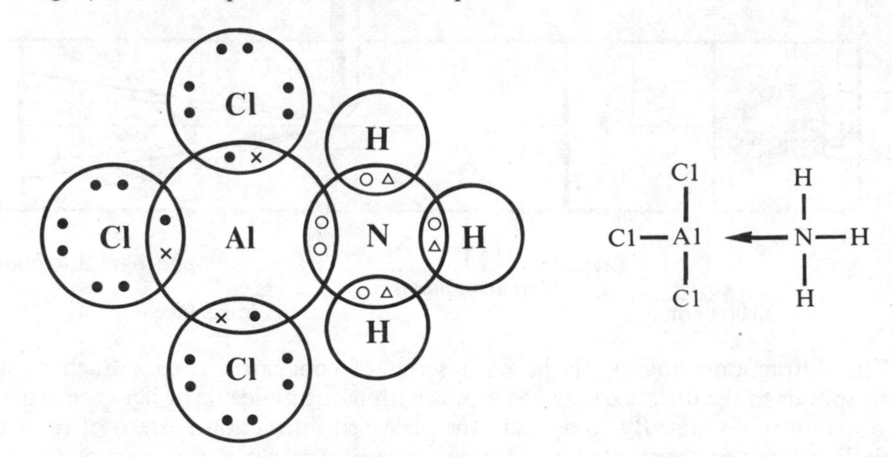


**Figure 7.9** The electronic structure and bonding in $AlCl_3.NH_3$.

When gaseous aluminium chloride is cooled, $AlCl_3$ molecules dimerize to form molecules of $Al_2Cl_6$. Monomers of $AlCl_3$ are held together in the $Al_2Cl_6$ dimer by dative covalent bonding (figure 7.10).

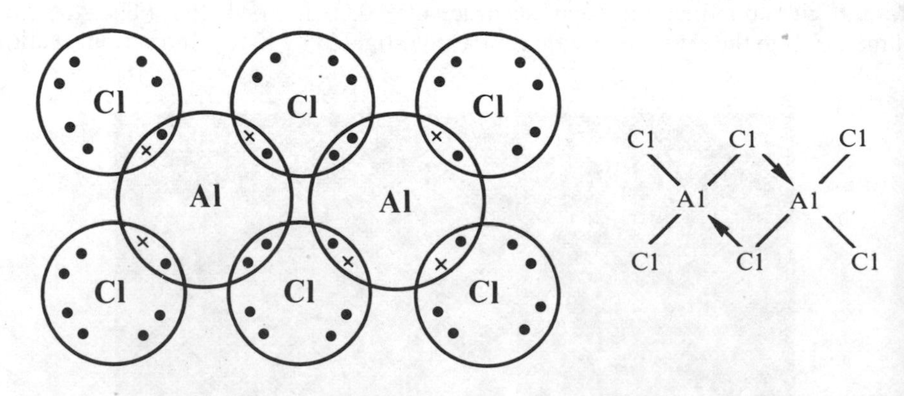


**Figure 7.10** The electronic structure and bonding in $Al_2Cl_6$.

Compounds containing unshared pairs of electrons readily form dative covalen bonds. For example, when ammonia is mixed with gaseous hydrogen chloride white clouds of ammonium chloride form. The reaction can be thought to resul from the formation of a dative covalent bond between the unshared pair of electron on the N atom in $NH_3$ and an $H^+$ ion from the HCl (figure 7.11). A similar reactio to this occurs when hydrogen chloride dissolves in water (figure 7.12).

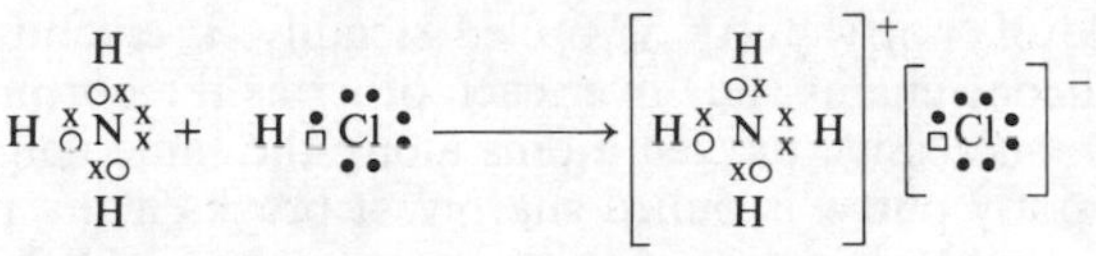

$$H_3N + HCl \longrightarrow [NH_4]^+ [Cl]^-$$

Figure 7.11 An electron 'dot/cross' representation for the formation of $NH_4Cl$ from $NH_3$ and HCl.

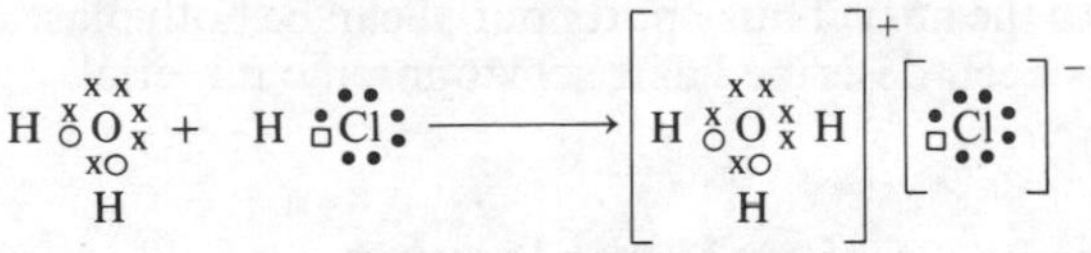

$$H_2O + HCl \longrightarrow [H_3O]^+ [Cl]^-$$

Figure 7.12 An electron 'dot/cross' representation for the reaction between hydrogen chloride and water.

Notice that both the ammonium ion and the oxonium ion ($H_3O^+$) have an overall charge of +1 which originates from the $H^+$ ions. In these ions this charge is distributed over all the structure; it is not localized on any one atom.

Once a dative bond has formed, it is indistinguishable from a covalent bond—the only difference between the two lies in the way we pictured the bond forming. Thus, although the $NH_4^+$ and $H_3O^+$ ions are sometimes represented as,

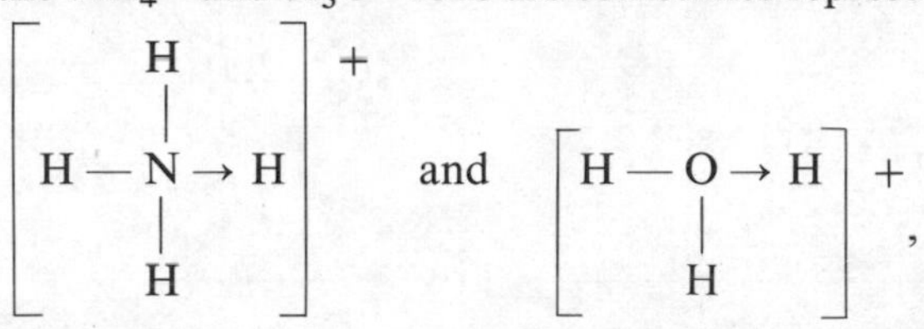

$$\left[ H-N(H)(H) \rightarrow H \right]^+ \quad \text{and} \quad \left[ H-O(H) \rightarrow H \right]^+,$$

showing the co-ordinate bond as an arrow, it is important to remember that all four N–H bonds in $NH_4^+$ are identical as are the three O H bonds in $H_3O^+$. There is no difference between any of the four hydrogen atoms in $NH_4^+$ and all four N–H bonds have the same length and the same strength.

The idea of dative covalent bonding has also been used to explain the existence of more than one oxide of the same non-metal. The structure of carbon dioxide can be explained purely in terms of covalent bonds, but that of carbon monoxide must involve a dative covalent bond if both atoms are to attain a 'noble-gas' structure.

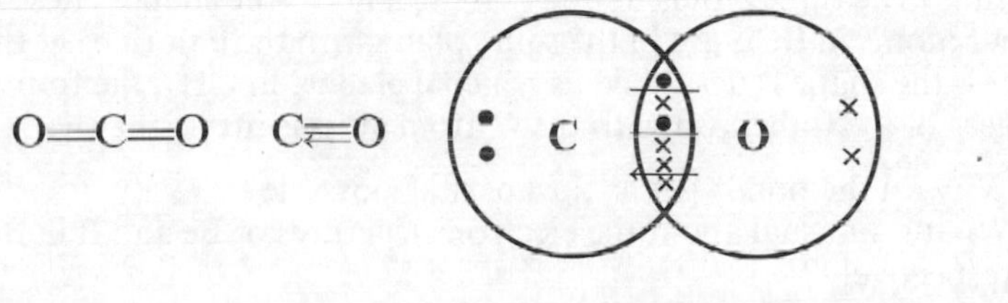

$$O{=}C{=}O \qquad C \leftleftarrows O$$

The strange properties of silicone bouncing putty ('potty putty') can be explained in terms of co-ordinate bonds. The structure of 'potty putty' resembles a silicone in which some silicon atoms have been replaced by boron (figure 7.13). These boron atoms have only six electrons in their outermost shell and readily form dative covalent bonds with the oxygen atoms in neighbouring silicone chains.

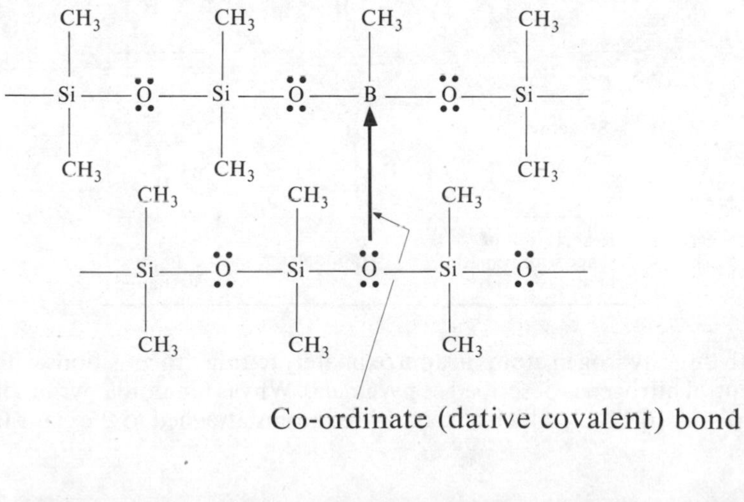

$$—Si(CH_3)_2—O—Si(CH_3)_2—O—B(CH_3)—O—Si(CH_3)_2—$$
$$—Si(CH_3)_2—O—Si(CH_3)_2—O—Si(CH_3)_2—O—$$

Co-ordinate (dative covalent) bond

Figure 7.13 Co-ordinate bonding in 'potty-putty'.

When a sample of 'potty putty' is pulled steadily, it extends like a piece of plasticine. The silicone chains slide over each other as the boron atoms form co-ordinate bonds to successive oxygen atoms along the same neighbouring chain. However, if the 'potty putty' is pulled sharply, it breaks like a piece of crumbly cheese because the boron atoms are unable to progress smoothly by dative-bonding from one oxygen to the next. Thus, 'potty putty' can be both plastic and brittle—no wonder that it has been described as a schizophrenic material.

## 7.7 The shapes of simple molecules

Look closely at the structures and 'dot/cross' diagrams of the simple molecules in figure 7.14. These structures have been determined by electron diffraction studies.

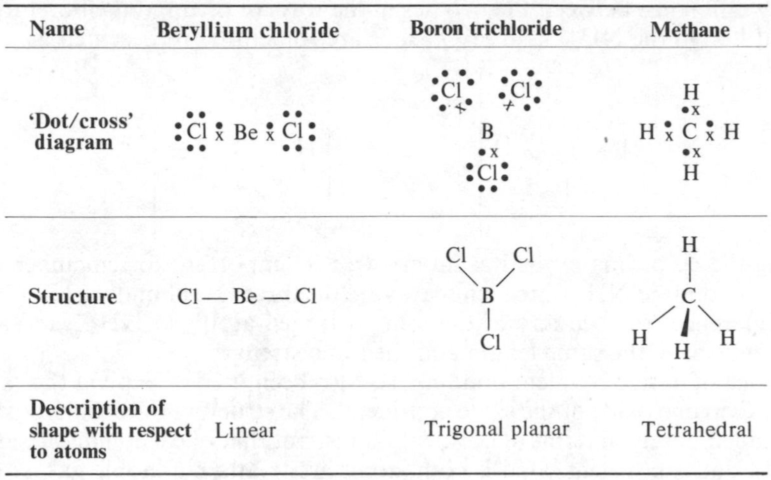

| Name | Beryllium chloride | Boron trichloride | Methane |
|---|---|---|---|
| 'Dot/cross' diagram | :Cl ⁕ Be ⁕ Cl: | (dot/cross diagram) | H x C x H |
| Structure | Cl — Be — Cl | (structure) | (structure) |
| Description of shape with respect to atoms | Linear | Trigonal planar | Tetrahedral |

Figure 7.14 Electron 'dot/cross' diagrams and structures for $BeCl_2$, $BCl_3$ and $CH_4$.

Notice that the bonds in $BeCl_2$, $BCl_3$ and $CH_4$ spread out so as to be as far apart as possible. The three atoms in $BeCl_2$ are in a line—the shape is described as **linear**. The four atoms in $BCl_3$ are in the same plane with the chlorines at the corners of a triangle—the shape is described as **trigonal planar**. In $CH_4$, the four H atoms lie at the apices of a tetrahedron with the C atom at its centre—the shape is **tetrahedral**.

- ○ Why do the bonds get as far apart as possible?
- ○ What is unusual about the electron structures of Be and B in $BeCl_2$ and $BCl_3$ respectively?

Now look at the 'dot/cross' diagrams and structures of $NH_3$ and $H_2O$ in figure 7.15.

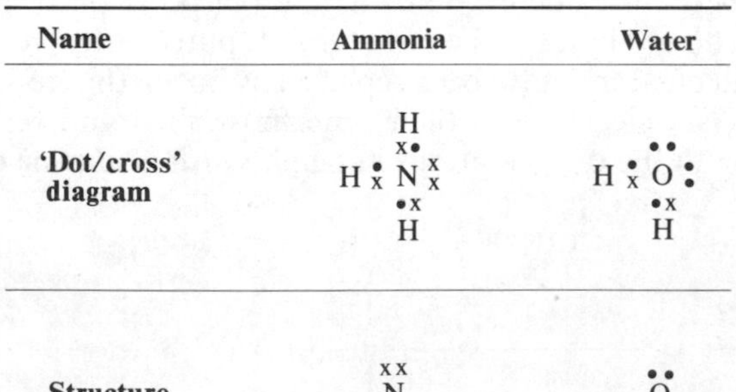

| Name | Ammonia | Water |
|---|---|---|
| 'Dot/cross' diagram | (dot/cross diagram) | (dot/cross diagram) |
| Structure | (structure) | (structure) |
| Description of shape with respect to atoms | Pyramidal | Bent (V-shaped) |

Figure 7.15 Electron 'dot/cross' diagrams and structures for ammonia and water.

$NH_3$ with three hydrogen atoms in approximately tetrahedral positions with respect to the central nitrogen is described as **pyramidal**. Why is ammonia pyramidal? Why is it not planar like $BCl_3$ which also has three atoms attached to the central boron

The answer lies in the non-bonded **lone pair of electrons** on the nitrogen atom. This lone pair on the nitrogen occupies the fourth tetrahedral position around the nitrogen atom in the $NH_3$ molecule. Each of the N–H bonds in ammonia is composed of a region of negative charge similar to the lone pair. The nitrogen atom is therefore surrounded by four regions of negative charge. These four negative-charge clouds repel each other as far apart as possible and consequently the shape of ammonia, though pyramidal with respect to atoms, can be described as tetrahedral with respect to negative centres around the central nitrogen atom.

What about the water molecule? Can its **bent** or **V-shaped** structure be explained in a similar manner?

- ○ How many lone pairs does the O atom in water possess?
- ○ How many centres of negative charge are there around the O atom in water?
- ○ What is the shape of water with respect to negative centres around the central O atom?

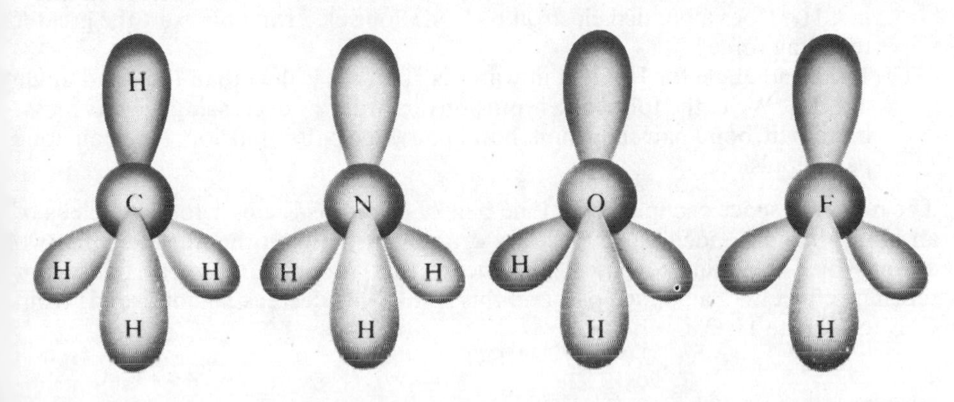


**Figure 7.16** Electron charge-cloud models of $CH_4$, $NH_3$, $H_2O$ and HF.

Figure 7.16 shows the electron charge-cloud models of $CH_4$, $NH_3$ and $H_2O$. You will realize from this discussion that it is the number of regions of negative charge (not the number of bonds) around the central atom(s) which dictates the shape of a molecule. Thus, $CO_2$ (figure 7.6) and HCN (figure 7.17) with two negative centres around their central C atoms are both linear with respect to negative centres and linear with respect to atoms. In $CO_2$, each double covalent bond counts as a single negative centre as does the triple covalent bond in HCN. In contrast to $CO_2$, $SO_2$ has three centres of negative charge around the S atom, and is V-shaped with respect to atoms (figure 7.17). Our discussions concerning the shapes of molecules are summarized in table 7.3.

H—C≡N

O=S=O

**Figure 7.17** The electronic structure and bonding in HCN and $SO_2$.

**Table 7.3** The number of negative centres, number of lone pairs and shapes of some simple molecules

| No. of negative centres around central atom | Angular separation of negative centres | Shape w.r.t. negative centres | Example | Shape w.r.t. atoms | No. of lone pairs |
|---|---|---|---|---|---|
| 2 | 180° | linear | $BeCl_2$, $CO_2$ | linear | 0 |
| 3 | 120° | trigonal planar | $BCl_3$, $BF_3$ | trigonal planar | 0 |
| 3 | 120° | trigonal planar | $SO_2$ | bent or v-shaped | 1 |
|  | 109° | tetrahedral | $CH_4$, $NH_4^+$, $BH_4^-$, $CCl_4$ | tetrahedral | 0 |
|  | 109° | tetrahedral | $NH_3$, $H_3O^+$ | pyramidal | 1 |
|  | 109° | tetrahedral | $H_2O$, $F_2O$ | bent or V-shaped | 2 |
|  | 90° | octahedral | $SF_6$ | octahedral | 0 |

THE FINE STRUCTURE OF METHANE, AMMONIA AND WATER

Look closely at table 7.4. Molecules of $CH_4$ are perfectly symmetrical. Consequently, the bond angle for H–C–H is that of a regular tetrahedron, 109°.

**Table 7.4** Bond angles in methane, ammonia and water.

| | Methane $CH_4$ | Ammonia $NH_3$ | Water $H_2O$ |
|---|---|---|---|
| HXH bond angle (X = C, N or O) | 109° | 107° | 105° |

- ○ The bond angle for H–N–H in $NH_3$ is 107°, i.e. 2° less than the bond angle in $CH_4$. Does a bonded electron pair or a lone electron pair exert the greater repelling force?
- ○ The bond angle for H–O–H in water is 105°, i.e. 2° less than the bond angle in $NH_3$. Write the following repulsions in order of decreasing effectiveness: bond pair/bond pair repulsion, bond pair/lone pair repulsion, lone pair/lone pair repulsion.

The region in space occupied by a lone pair of electrons is closer to the nucleus of an atom than a bonded pair which are drawn out between the nuclei of the two atoms which they bind together. This means that a lone pair can exert a greater repelling effect than a bonded pair and this results in a decreasing bond angle from $CH_4$ to $NH_3$ to $H_2O$.

## Summary

1 In many cases, when elements form compounds they either lose, gain or share electrons so as to achieve stable electron configurations similar to the next highe or lower noble gas in the periodic table.

2 An electrovalent bond involves the complete transfer of electrons from one atom to another forming oppositely charged ions. The ions are held together in the solid crystal by the electrostatic attraction between their opposite charges.

3 In general, cations form easiest when:
(a) the resulting charge is small
(b) the radius of the atom is large.

4 In general, anions form easiest when:
(a) the resulting charge is small
(b) the radius of the atom is small.

5 A covalent bond involves the sharing of a pair of electrons between two atoms each atom contributing one electron to the shared pair. A double covalent bon involves the sharing of four electrons, two contributed by each atom.

6 A co-ordinate (dative covalent) bond involves the sharing of a pair o electrons between two atoms, both electrons in the bond being donated by on atom.

7 The shape of a simple molecule is dictated by the number of regions c negative charge around the central atom. The regions of negative charge coul be a non-bonded (lone) pair of electrons, a shared pair of electrons, four share electrons in a double covalent bond or even six electrons in a triple bond.

8 A lone pair of electrons exerts a greater repelling effect than a bonded pai

## Study questions

1 Draw 'dot/cross' electron structures showing electrons in the outermost shell of each ato in the following compounds. Show the overall charge on each ion in the ionic compound

(a) $Cl_2$, $H_2O_2$, $O_2$, $C_2H_4$, $C_2H_2$, HCN
(b) LiF, $CaCl_2$, $Na_3P$, $Al_2S_3$
(c) $PCl_3$, $Cl_2O$, $HNO_2$, $H_2CO_3$, $C_2H_5OH$
(d) $SO_2$, undissociated $HNO_3$, undissociated $H_2SO_4$.

2 (a) How is the type of bonding in the chlorides of the elements Na, Mg, Al, Si, P and S related to
(i) their position in the periodic table,
(ii) the number of electrons in the outermost shells of these elements?
(b) Explain the terms ionic, covalent and co-ordinate as applied to bonds in compounds. Illustrate your answer by clear 'dot/cross' diagrams to show the positions of all electrons in the outer shells of each atom or ion in four of the following compounds: hydrogen chloride, ammonia, ammonium chloride, aluminium chloride, sodium hydride.

3 Draw 'dot/cross' diagrams and predict the shapes with respect to atoms for molecules of the following compounds:

$$SF_6,\ POCl_3,\ SOCl_2,\ H_3O^+,\ BF_3,\ C_2H_2.$$

4 (a) State the electronic configurations of the following atoms, (e.g. Be would be 2,2);

C, N, O, F.

(b) Draw a series of 'dot/cross' diagrams to show the structures of the simplest hydrides formed by carbon, nitrogen, oxygen and fluorine.
(c) Sketch and describe the shapes of the molecules depicted and discuss the influence of any lone pairs of electrons present on the shapes adopted.
(d) What shape would you predict for the following:

$$NH_4^+,\ NH_3,\ NH_2^-?$$

5 X, Y and Z represent elements of atomic numbers 9, 19 and 34.
(a) Write the electronic structures for X, Y and Z, (e.g. Be would be 2, 2).
(b) Predict the type of bonding which you would expect to occur between
(i) X and Y, (ii) X and Z, (iii) Y and Z.
(c) Draw 'dot/cross' diagrams for the compounds formed, showing only the electrons in the outermost shell for each atom.
(d) Predict, giving reasons, the relative
(i) volatility,
(ii) electrical conductance,
(iii) solubility in water
you would expect for the compound formed between X and Y compared to that formed between X and Z.

6 Explain the following:
(a) Tin and lead form the ions $Sn^{4+}$ and $Pb^{4+}$ respectively, but carbon and silicon, elements in the same group do not form $C^{4+}$ and $Si^{4+}$ ions.
(b) The carbon-oxygen bond length in methoxymethane (dimethyl ether, $CH_3.O.CH_3$) is 0.14nm, while that in carbon dioxide is 0.12nm and that in carbon monoxide is only 0.11nm.
(c) The C O bond lengths in $CO_3^{2-}$ are all identical.
(d) Aluminium fluoride has a much higher melting point than aluminium chloride.

The first electron affinities for the elements in period 3 are given below.

| Element | Na | Mg | Al | Si | P | S | Cl | Ar |
|---|---|---|---|---|---|---|---|---|
| First electron affinity/kJ mole$^{-1}$ | −20 | +67 | −30 | −135 | −60 | −200 | −364 | — |

(a) What is the general trend in the value of the electron affinity from Na to Cl?
(b) Explain the general trend noticed in part (a).
(c) Why is the first electron affinity of Mg more positive than one might expect from the general trend in the values above?
(d) Why is the first electron affinity of silicon more exothermic than that of phosphorus?

# 8 Intermolecular Forces

## 8.1 Polar and non-polar molecules

Fill a burette with water. Open the tap and bring a charged ebonite rod close to the stream of water issuing from the jet of the burette. The water is deflected from its vertical path towards the charged rod (figure 8.1). Why is this?

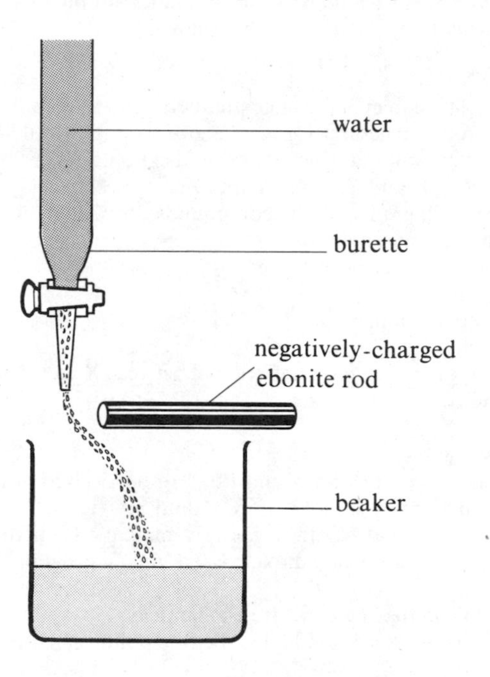


**Figure 8.1** The effect of a charged rod on a thin stream of water.

The ebonite rubbed with fur has a negative charge. When the ebonite rod is replaced by a positively-charged rod, the water is again deflected towards the rod. Why is this?

The results in table 8.1 show what happens when water is replaced by other liquids. Those liquids which are affected are always deflected *towards* the charged rod.

**Table 8.1** Testing the deflection of a jet of liquid using a charged rod

| Liquids showing a marked deflection | Liquids showing no deflection |
|---|---|
| Water | |
| Trichloromethane | Tetrachloromethane |
| Propanone (acetone) | |
| Ethoxyethane (ether) | |
| Nitrobenzene | Benzene |
| Cyclohexene | Cyclohexane |
| Ethanol | |

- Why is trichloromethane deflected, but not tetrachloromethane?
- Why is nitrobenzene deflected, but not benzene?
- Why is cyclohexene deflected, but not cyclohexane?

Look at the structures of these six molecules in figure 8.2. Notice that those molecules which are unaffected by a charged rod are more symmetrical than those which are deflected. This lack of symmetry, in $CHCl_3$ for example, means that the centre of positive charge does not coincide exactly with the centre of negative charge and the molecule is affected by an electrostatic field from the charged rod.

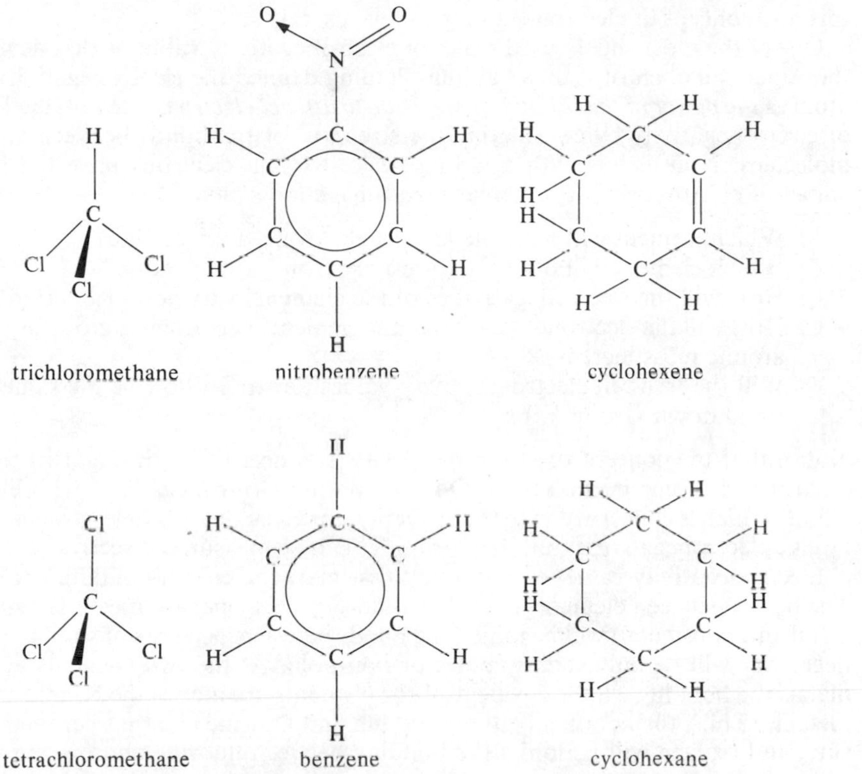

trichloromethane nitrobenzene cyclohexene

tetrachloromethane benzene cyclohexane

**Figure 8.2** The structures of some simple molecules.

When two different atoms are joined by a covalent bond, their attractions for the onding electrons will not be the same. For example, in a molecule of HCl the onding electrons will not be shared equally by the hydrogen and chlorine atoms. ı fact, the chlorine atom has a greater attraction for the electrons in the covalent ond. In chemical language, we say that chlorine is more **electronegative** than ydrogen. Consequently, the centre of negative charge in the HCl molecule is rawn towards the chlorine atom and it is closer than the centre of positive charge the nucleus of the chlorine atom (figure 8.3).

The overall distortion of charge in molecules such as HCl, which results from ıequal sharing of electrons, is known as **polarization** and the molecules are said to **polar**. The separation of charge in the molecule is referred to as a **dipole**.

Molecules, such as tetrachloromethane, benzene and cyclohexane, containing a mmetrical distribution of similar atoms and in which equal dipoles cancel each her exactly are **non-polar**.

Molecules, such as water, trichloromethane, nitrobenzene and cyclohexene, in ıich dipoles do not cancel each other are polar. When these liquids stream from a ırette past a charged rod, molecules are attracted towards the charged rod and the is deflected. When a positive rod is used it is the negative ends of dipoles in the lar molecules that are attracted towards the rod. With a negative rod, positive ds of the dipoles are attracted.

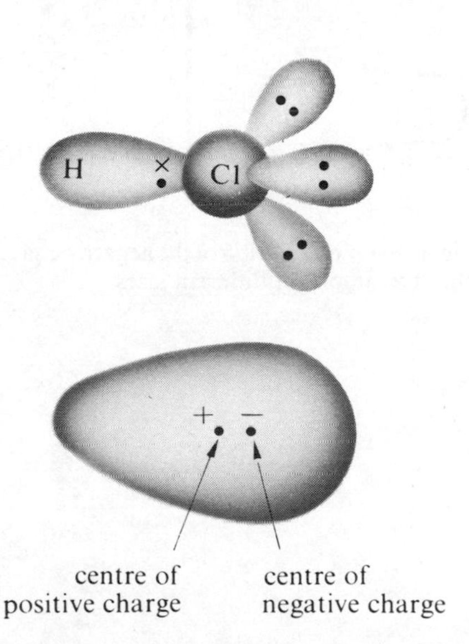


**Figure 8.3** The centres of positive and negative charge in a molecule of HCl.

## 8.2 Electronegativity, polarity and bonding

In non-symmetrical molecules such as HCl and $H_2O$, the centre of positive charge and the centre of negative charge do not coincide. The polarity in these molecules results from the presence of highly electronegative elements such as Cl and O.

Various attempts have been made to quantify the electronegativity or electron-attracting power of an atom within a molecule. Obviously, the electron-attracting power of an atom will be different in different compounds since it will depend on the other atoms to which the atom under consideration is attached. For this reason, it is difficult to estimate values of electronegativity with real accuracy. In spite of this, the concept of electronegativity is very useful.

One of the most widely used scales of electronegativity values is that devised by the American chemist, Linus Pauling. Pauling defined the electronegativity of an atom as *the power of that atom in a molecule to attract electrons*. He obtained values of electronegativity by considering the strengths of the bonds between atoms in molecules. Non-metals with a strong desire to gain electrons have the highest values of electronegativity whereas metals have low values.

- ○ Which element will have the highest electronegativity value?
- ○ Which element will have the lowest electronegativity value?
- ○ How will the electronegativities of the elements vary across a period?
- ○ How will the electronegativities of the elements vary down a group as relative atomic mass increases?
- ○ Will the trend in electronegativity values down Group I be the same as the trend down Group VII?

Notice that the concept of electronegativity is concerned with the attraction for electrons of atoms in molecules. Do not confuse electronegativity with electron affinity which is concerned with the attraction for electrons of single gaseous atoms. Unlike electronegativities, electron affinities can be measured directly.

Electronegativity values can be used to estimate the polarity of different bonds The bonds between elements of widely differing electronegativities (i.e. between a metal and non-metal) will be ionic. The bonds between elements of similar electro-negativity will be only slightly polar or non-polar. If the two elements are non-metals the bonding will be covalent; if the elements are metals the bonding will be metallic. Thus, the bonding between sodium and fluorine like that between potas-sium and oxygen will be ionic; the bonding between fluorine and oxygen will be covalent and that between potassium and sodium will be metallic. Of course, two metals would not react with each other, but two liquid metals could mix intimatel with each other and there would be metallic bonding between different meta atoms in the mixture.

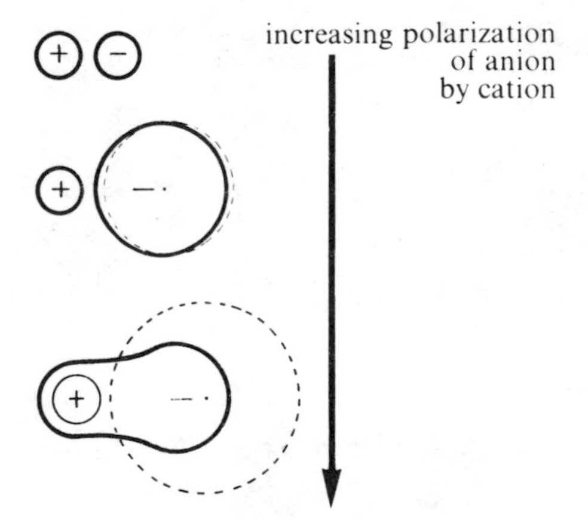


**Figure 8.4** Polarization of the negative charge-cloud in anions of different sizes.

The existence of a dipole confers partial ionic character on a polar molecule. A the polarity of the molecule increases so does the extent of its ionic character.

As a result of differing electronegativities there is a distortion of the equal sharin which one would expect in a pure covalent bond. Similarly, chemists believe tha there is also distortion of the charge in ionic substances since cations will attrac the negative charge-cloud of the anions with which they are associated (figure 8.4 This polarization of the ions confers partial covalent character on the ionic bondin Although it is convenient to regard bonds as either ionic or covalent, it is as well t remember that the wholly ionic and wholly covalent bonds described in chapter are extreme types. All ionic bonds have some covalent character and most covale bonds have some ionic character.

## 8.3 Intermolecular forces

In the last chapter, we considered the bonding between ions and that betwe atoms. We must now consider the forces between molecules. We know, for exampl that within a molecule of $CHCl_3$ the three Cl atoms and the H atom are joined the central carbon atom by strong covalent bonds, but what kind of forces hold t various $CHCl_3$ molecules together? What are the forces like between one CHC molecule and its neighbours?

(A) DIPOLE-DIPOLE ATTRACTIONS

In section 8.1 we investigated the deflection of a jet of $CHCl_3$ by a charged ebonite rod. Our observations led us to conclude that $CHCl_3$ molecules were polar owing to the non-symmetrical distribution of charge within each molecule. The interactions between permanent dipoles explain the attraction between neighbouring $CHCl_3$ molecules. These attractions are called **permanent dipole-permanent dipole attractions** (figure 8.5). The existence of dipole-dipole attractions will explain the

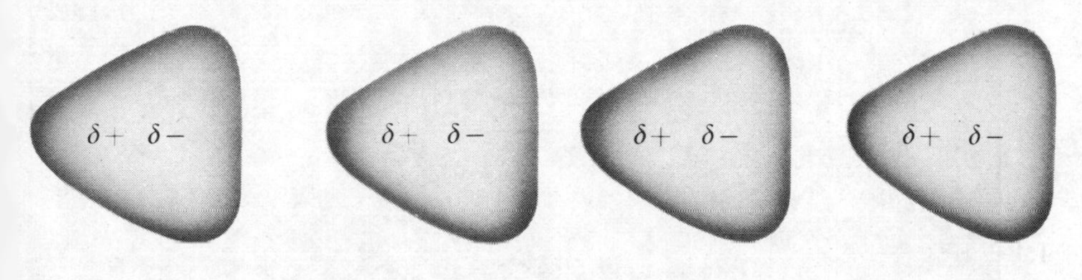


**Figure 8.5** Permanent dipole–permanent dipole attractions in polar molecules.

forces holding together *polar* molecules in liquids such as trichloromethane ($CHCl_3$), propanone (acetone, $CH_3COCH_3$) and nitrobenzene ($C_6H_5NO_2$). But, what about *non-polar* molecules in liquids such as tetrachloromethane ($CCl_4$) and benzene ($C_6H_6$)? How are we to account for the forces between molecules in these substances which have no permanent dipole? What holds $CCl_4$ molecules together in liquid $CCl_4$?

(B) EVIDENCE FOR INTERMOLECULAR FORCES BETWEEN NON-POLAR MOLECULES—VAN DER WAALS' FORCES

*(i) The properties of noble gases*

The noble gases are monatomic, existing as single atoms in the gas phase at room temperature. These symmetrical, non-polar atoms have no permanent dipole and do not form any normal bonds, yet all the noble gases will condense to liquids and ultimately form solids provided the temperature is reduced sufficiently. The possible liquefaction and solidification of noble gases would suggest the existence of intermolecular forces in these non-polar substances which can hold the molecules together in the solid and liquid state.

Furthermore, energy is required both to melt the solid and to boil the liquid noble gases showing that cohesive forces are operating between molecules. For example, the energy of sublimation for solid xenon is 14.9 kJ $mole^{-1}$.

*(ii) The non-ideal behaviour of gases*

According to the kinetic theory of gases, the molecules of an ideal gas

occupy negligible volume, and

exert no forces on one another.

Using this model, the properties of ideal gases can be summarized in **the ideal gas equation**,

$$pV = nRT \text{ (see section 10.7)}$$

For one mole of gas $pV = RT$ and in this case $pV/RT = 1$.

Although this equation represents the behaviour of gases accurately at low pressures (i.e. $\leqslant$ atmospheric pressure) and high temperatures (i.e. well above their condensation point when the gas is least like a liquid), substantial and increasing deviations occur at low temperatures and high pressures.

- ○ What is the value of $pV/RT$ for one mole of an ideal gas?
- ○ Would $pV/RT$ increase, remain constant or decrease as $p$ increased for an ideal gas?
  Look closely at figure 8.6.

○ How does $pV/RT$ vary as $p$ increases for $N_2$ (a real gas) at 673 K?
○ How does $pV/RT$ deviate from ideal behaviour as temperature is reduced?

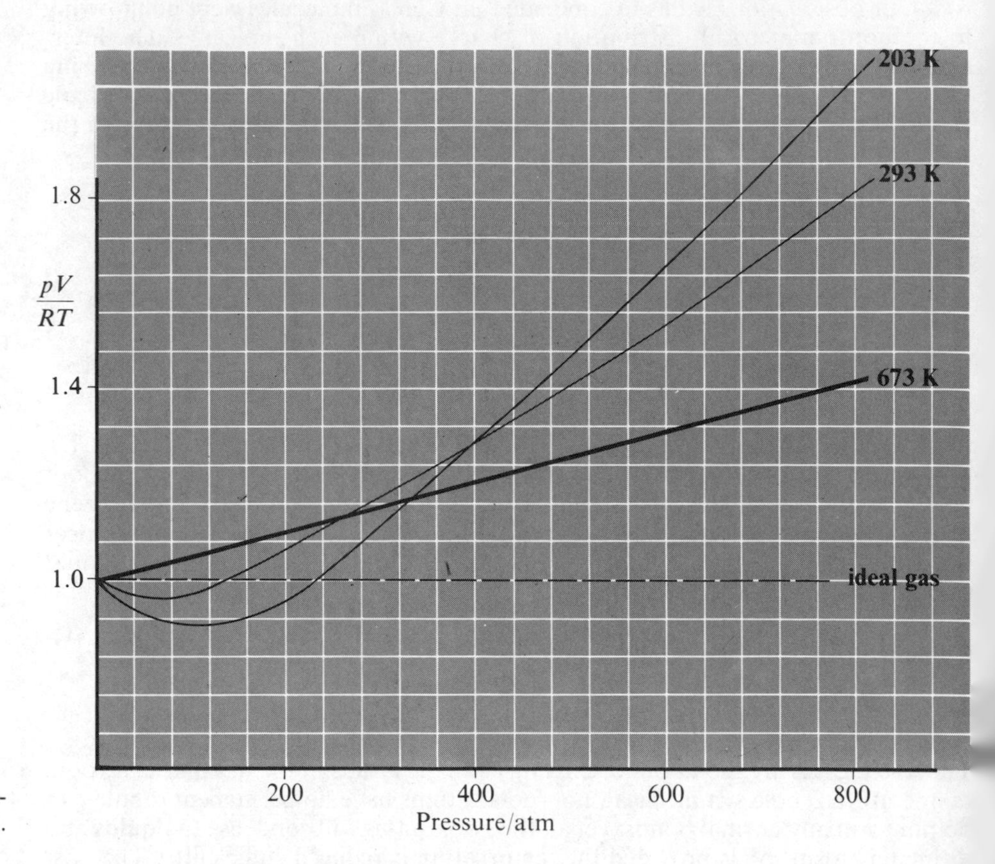


**Figure 8.6** Variation of $pV/RT$ with increasing pressure for $N_2$ at 673 K, 293 K and 203 K.

Figure 8.6 shows how the properties of nitrogen deviate from ideal behaviour at low temperature and high pressure. These deviations can be understood by reconsidering the two fundamental assumptions of the kinetic theory of gases stated at the beginning of this sub-section.

1 *In deriving the ideal gas equation, it is assumed that the molecules are point masses moving around in the whole volume of the container*. If the molecules are not points, but are of finite size, each must exclude a certain volume of the container from all the others. If we call this 'excluded' volume $b$, then the 'true' volume available for molecular motion is $(V - b)$.

Thus, the equation $pV = nRT$ should be written as $P(V - b) = nRT$ which means that the effect of finite molecular size causes the observed pressure for a particular volume to be greater than that predicted by the ideal gas equation.

The constant $b$ has a value of about 30cm$^3$ mole$^{-1}$ for many gases. Since one mole of many gases occupies approximately 30cm$^3$ on liquefaction, this would suggest that $b$ is roughly the same as the volume of the molecules.

2 *In deriving the ideal gas equation, it is assumed that the molecules exert no force on each other*. Unfortunately intermolecular forces cannot be neglected; this is particularly so at high pressure. The pressure of a gas results from particles of the gas bombarding the walls of their container. Within the bulk of the gas, intermolecular forces cancel each other out, but those molecules near the walls experience an overall net force tending to pull them back into the bulk. In effect, attractive forces cause molecules near the walls to transfer some of their momentum to other gas molecules rather than to the walls and this results in a measured pressure lower than the 'true' pressure. The magnitude of the 'pressure reduction' will be proportional to both the concentration of molecules near the wall ($\propto \frac{1}{V}$) and the concentration of molecules within the bulk (also $\propto \frac{1}{V}$). Thus, for one mole of gas the 'pressure reduction' can be written as $a/V^2$ where $a$ is a constant. By adding the term $a/V^2$ to the actual pressure, we obtain the corrected pressure term $[p + (a/V^2)]$.

If we replace $p$ and $V$ in the ideal gas equation by their corrected expressions we obtain

$$[p + (a/V^2)].(V - b) = nRT$$

which is known as the **Van der Waals' Real Gas Equation**. This equation was first used by the Dutch physicist Van der Waals in 1873.

Why does this Real Gas Equation reduce to $pV = nRT$ at low pressure and high temperature?

The properties of noble gases and the non-ideal behaviour of real gases provide evidence for the existence of cohesive forces between non-polar molecules. These weak, short-range forces of attraction, independent of normal bonding forces between molecules are known as **Van der Waals' forces**. Van der Waals' bonds are, of course, much weaker than covalent and ionic bonds. For example, the energy of sublimation for solid chlorine (i.e. the energy required to overcome the Van der Waals' forces between one mole of $Cl_2$ molecules) is only 25 kJ mole$^{-1}$, whereas the bond energy of chlorine (i.e. the energy required to break one mole of Cl—Cl covalent bonds) is 244 kJ mole$^{-1}$. Roughly speaking, Van der Waals' bonds are between one-tenth and one-hundredth the strength of covalent bonds.

## 8.4 How do Van der Waals' forces arise?

The electrons in a molecule are in continual motion. At any particular moment, the electron charge cloud around the molecule will not be perfectly symmetrical. There is more negative charge on one side of the molecule than on the other and it possesses an instantaneous electric dipole. This dipole will induce dipoles in neighbouring molecules and if the positive end of the original dipole is pointing towards a neighbouring molecule, then the induced dipole will have its negative end pointing towards the positive of the original dipole. In this way, weak **induced dipole-induced dipole attractions** exist between molecules.

Obviously, these induced dipoles will act first one way, then another way and continually arise and disappear as a result of electron movement, but notice that the force between the original dipole and the induced dipole will always be an attraction. Consequently, even though the average dipole on each molecule over a period of time is zero, the resultant forces between molecules at any instant are not zero.

As the size of a molecule increases, the number of constituent electrons increases too, and the induced dipole-induced dipole attractions become stronger.

The increase in boiling point for the elements in Group VII ($F_2$, $Cl_2$, $Br_2$, and $I_2$—figure 8.7) and the increase in boiling point for the homologous series of alkanes ($CH_4$, $C_2H_6$, $C_3H_8$, etc. in figure 25.3) result from stronger Van der Waals' attractions with increasing relative molecular mass.

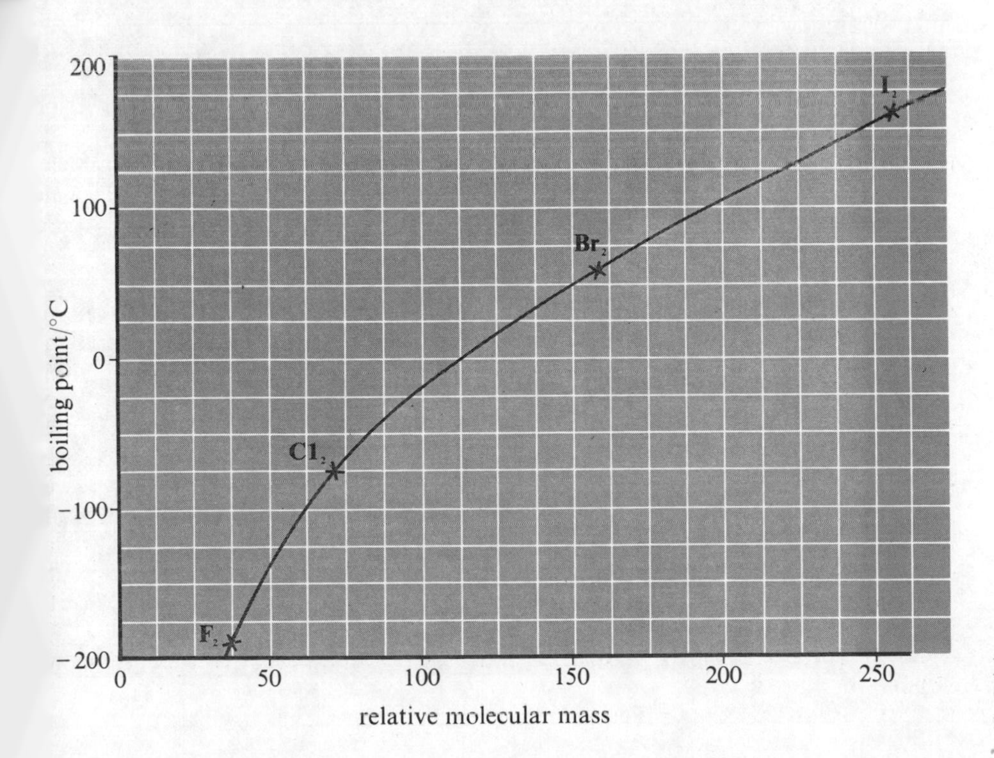


**Figure 8.7** Boiling points for the elements in Group VII plotted against relative molecular mass.

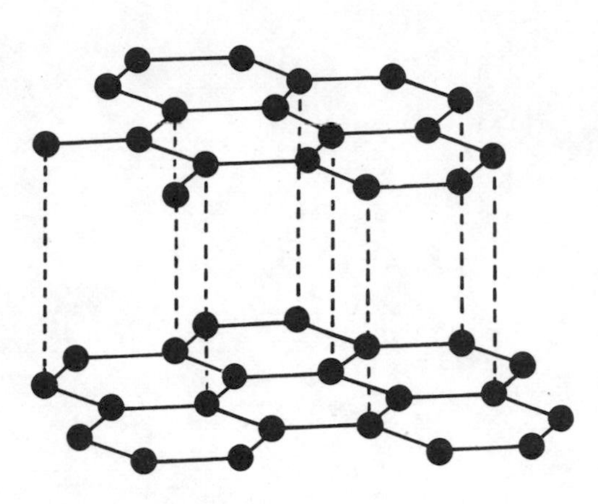

Figure 8.8 The structure of graphite.

Although the Van der Waals' forces between simple molecules such as $CCl_4$, $Cl_2$ and $CH_4$ are very small, the total Van der Waals' forces between the molecules of a large polymer with many contacts can be very significant indeed. The strength of non-polar polymers such as poly(ethene) and poly(propene) is dependent on significant Van der Waals' bonding between parallel molecules. Furthermore, experiments have shown that the tensile strength of high density poly(ethene) which has tightly packed parallel molecules is three times as large as that of low density poly(ethene) which is packed less tightly and therefore has weaker Van der Waals' attractions.

Van der Waals' forces also account for the properties of graphite. Crystals of graphite are composed of parallel layers of hexagonally arranged carbon atoms (figure 8.8). Within each layer, carbon atoms are linked by strong covalent bonds, whereas the parallel layers are held together by Van der Waals' forces. The C—C distance within each layer is 0.14nm but the distance between adjacent layers is 0.34nm. The Van der Waals' bonding between the layers is strong enough to hold the layers together, but weak enough to allow them to slide over each other. Because of this, graphite is soft and acts as a solid lubricant.

## 8.5 Hydrogen bonding

Look closely at the graphs in figure 8.9 showing the boiling points of hydrides in Group IV, Group V, Group VI and Group VII. Notice that the boiling points of the Group IV hydrides decrease with decreasing relative molecular mass from $SnH_4$ to $CH_4$. Is there a similar decrease for the hydrides of Groups V, VI and VII?

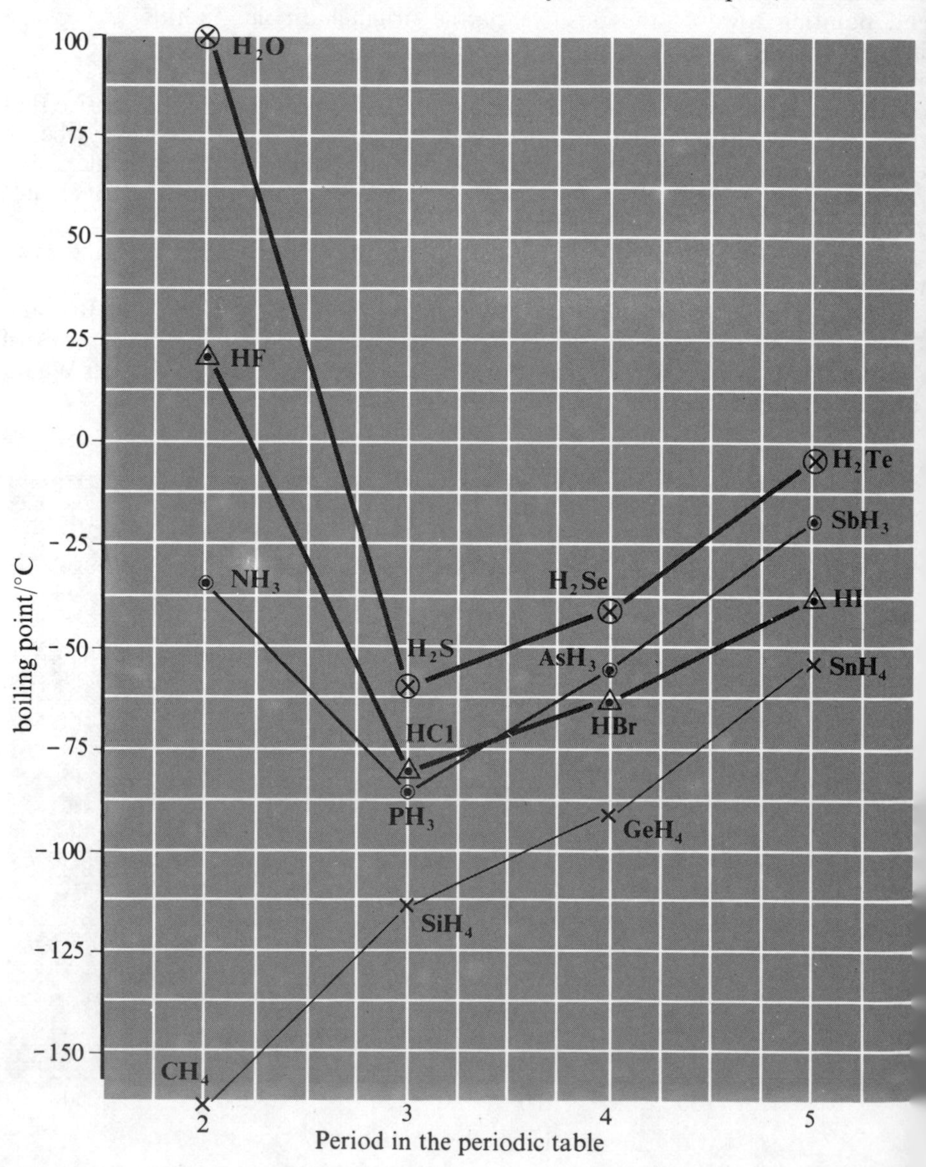


Figure 8.9 Variation of the boiling points of the hydrides in Groups IV, V, VI and VII.

Figure 8.9 shows the expected decrease in boiling point from $H_2Te$ through $H_2Se$ to $H_2S$, but $H_2O$ has a much higher boiling point than one would expect. A similar pattern appears with the hydrides in Group VI and those in Group VII where we find a much higher boiling point for $NH_3$ and HF than extrapolation of the graphs would suggest.

How can we account for these unusually high boiling points for $H_2O$, $NH_3$ and HF?

In water, liquid ammonia and liquid hydrogen fluoride there must be unusually strong intermolecular forces. Why is this?

$H_2O$, $NH_3$ and HF are all very polar since they contain the three most electronegative elements, oxygen, nitrogen and fluorine linked directly to hydrogen which is weakly electronegative. This results in exceptionally polar molecules with stronger intermolecular forces than usual. These particularly strong intermolecular forces are known as **hydrogen bonds.**

**Table 8.2** Data for the hydrides of elements in Group VI

| **Compound** | **Melting point /K** | **Molar heat of fusion, $\Delta H^{\ominus}_{fus}$ /kJ mole$^{-1}$** | **Molar heat of vaporization, $\Delta H^{\ominus}_{vap}$ /kJ mole$^{-1}$** |
|---|---|---|---|
| $H_2O$ | 273 | 6.02 | 40.7 |
| $H_2S$ | 188 | 2.39 | 18.7 |
| $H_2Se$ | 207 | 2.51 | 19.3 |
| $H_2Te$ | 225 | 4.18 | 23.2 |

○ Look closely at the data in table 8.2. Do you think the melting point, the molar heat of fusion and the molar heat of vaporization of water are influenced by hydrogen bonding?

○ Which other physical properties of a substance, besides those already considered, will be influenced by the existence of hydrogen bonds? Why can H atoms attached to N, O and F atoms form such strong intermolecular forces?

## 8.6 What is a hydrogen-bond?

N, O and F are the three most electronegative elements. When they are bonded to an H atom, the shared pair of electrons in the covalent bond is drawn towards the electronegative atom. Now the H atom has no electrons other than its share of those in this covalent bond which are being pulled away from it by the more electronegative N, O or F.

Since the H atom has no inner shell of electrons, the single proton in its nucleus is unusually 'bare' and readily available for any form of dipole-dipole attraction. Thus, H atoms attached to N, O or F are able to interpose themselves between two of these atoms exerting an attractive force on them and bonding them together (figure 8.10). The two larger atoms are drawn closer with an H atom

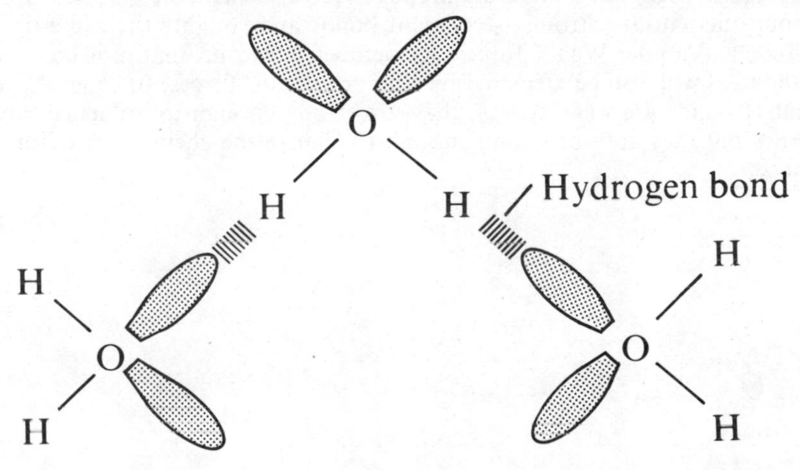


**Figure 8.10** Hydrogen bonding in water.

effectively buried in their electron clouds. *H-bonds are therefore extra strong intermolecular, permanent dipole-permanent dipole attractions.*

The essential requirements for an H-bond are:

1 a hydrogen atom attached to a highly electronegative atom;
2 an unshared pair of electrons on the electronegative atom.

In the $NH_3$ molecule, there are three N—H bonds and one non-bonded electron pair which means that there can be an average of one H-bond per molecule.

In water, however, there are two O—H bonds and two unshared electron pairs per molecule. This means that each $H_2O$ molecule can form two hydrogen bonds and this helps to explain the three dimensional lattice structure in ice (section 8.8(a)).

○ How many unshared electron pairs are there in each HF molecule?
○ What is the maximum possible number of H-bonds per HF molecule?

## 8.7 Estimating the strength of hydrogen bonds in water

Which one of the following values could be used to estimate the strength of an H-bond in water?

○ The strength of an O—H bond in $H_2O$.
○ The heat evolved when one mole of $H_2O$ forms from its elements.
○ The heat of vaporization of water.
○ The melting point of ice.

H-bonds in water are intermolecular forces: they are the forces which hold the molecules together. When water is vaporized the $H_2O$ molecules, which are relatively close to each other in the liquid, are pulled much further apart in forming the vapour. The energy provided to vaporize the water is needed to overcome the forces holding the water molecules together in the liquid state. Thus, the heat of vaporization of water will give a rough idea of the strength of H-bonds between $H_2O$ molecules. But, the heat of vaporization overcomes Van der Waals' bonds between $H_2O$ molecules as well as the H-bonds. How can we estimate the strength of H-bonding alone?

Figure 8.11 shows a graph of the molar heats of vaporization for the hydrides of elements in Group VI plotted against relative molecular mass (molecular weight). Now, if we assume that $H_2S$, $H_2Se$ and $H_2Te$ have intermolecular forces due *only* to Van der Waals' bonds (negligible H-bonding) we can find a value for the strength of Van der Waals' forces in water by extrapolating the curve through the values of $\Delta H_{vap}$ for $H_2S$, $H_2Se$ and $H_2Te$. This is indicated by the thick circled cross in figure 8.11; it gives a predicted heat of vaporization for water of 18.5 kJ mole$^{-1}$.

Total strength of intermolecular forces (Van der Waals' bonds + H-bonds) in water = 40.7 kJ mole$^{-1}$.

Estimated strength of Van der Waals' forces in water = 18.5 kJ mole$^{-1}$
∴ Approximate strength of H-bonds in water = 40.7 − 18.5 = 22.2 kJ mole$^{-1}$

Usually, the strength of H-bonds are in the range 5 to 40 kJ mole$^{-1}$. Thus H-bonds are about one-tenth as strong as covalent bonds and roughly the same order of magnitude as Van der Waals' forces. Remember, however, that molecules which are H-bonded will also be attracted by Van der Waals' forces. In general, we can say that H-bonds are weak forces; they are strong enough to influence physical properties but they are not strong enough to change the chemical reactions of a substance.

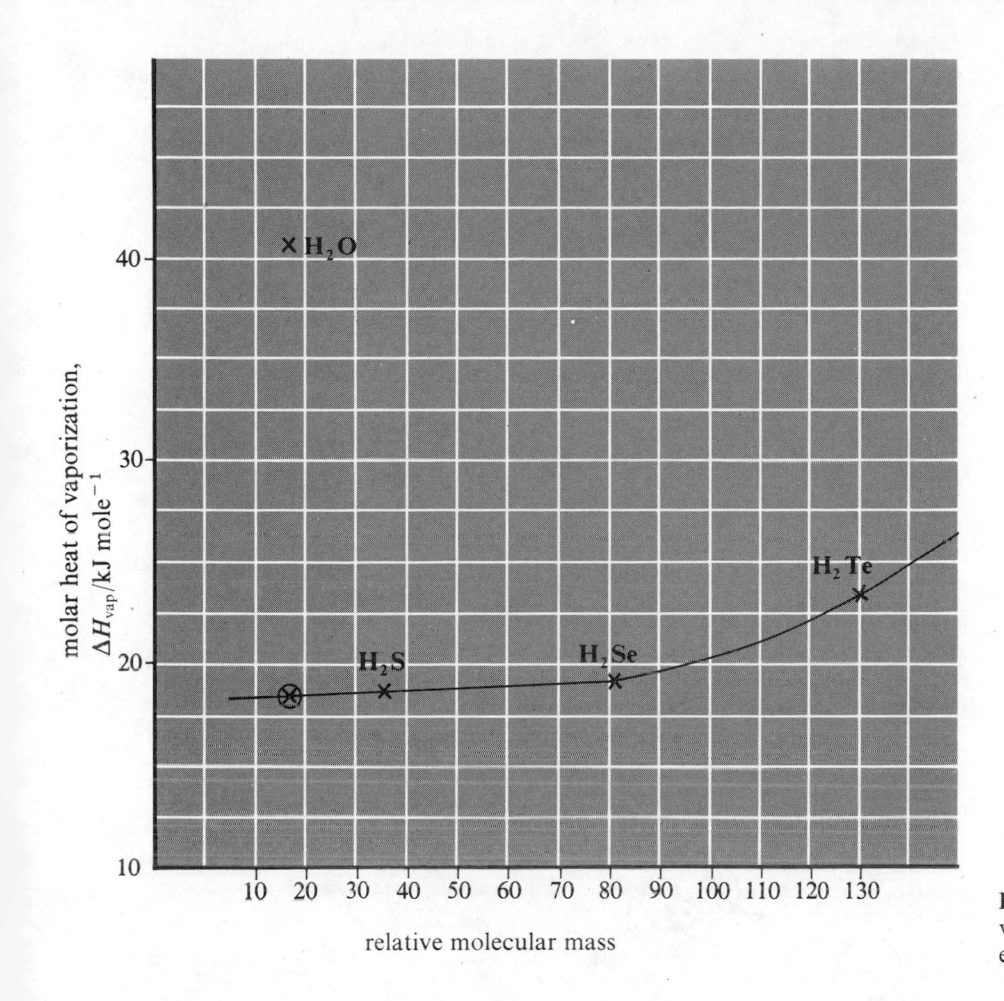


**Figure 8.11** Variation of the molar heats of vaporization for the hydrides of Group VI elements.

## 8.8 The influence and importance of hydrogen-bonding

(A) THE STRUCTURE AND PROPERTIES OF WATER AND ICE

The melting point, boiling point and molar heats of vaporization and fusion of water have already been mentioned. The unexpectedly high values result from the association of water molecules due to the formation of H-bonds. This extra bonding between $H_2O$ molecules also causes high surface tension and high viscosity. Because of the high surface tension of water, which provides a sort of 'skin effect', it is possible for small but relatively dense articles such as razor blades and beetles to float on an undisturbed water surface.

The presence of *two* hydrogen atoms and *two* lone electron pairs in each water molecule results in a three dimensional tetrahedral structure in ice. Each oxygen atom in ice is surrounded tetrahedrally by four others with hydrogen bonds linking each pair of oxygen atoms (figure 8.12).

The distance between adjacent oxygen atoms is 0.276nm, more than twice the O—H distance of 0.096nm in gaseous water molecules. This suggests that the hydrogen atom linking the two oxygen atoms in ice is probably not midway between them.

The arrangement of water molecules in ice creates a very open structure which accounts for the fact that ice is less dense than water at 0°C. When ice melts, the regular lattice breaks up and the water molecules can pack more closely and so the liquid has a higher density. In liquid water, the strong hydrogen bonding results in some ordered packing of water molecules over a short range. However, there is no long-range order as the H-bonds are continually being broken and formed and the regions of short-range order are continually changing.

The anomalous physical properties of ice and water which result from H-bonding have a staggering influence biologically and environmentally.

*Above* Walking on water! It's all done by hydrogen-bonding and surface tension.

**Figure 8.12** The structure of ice.

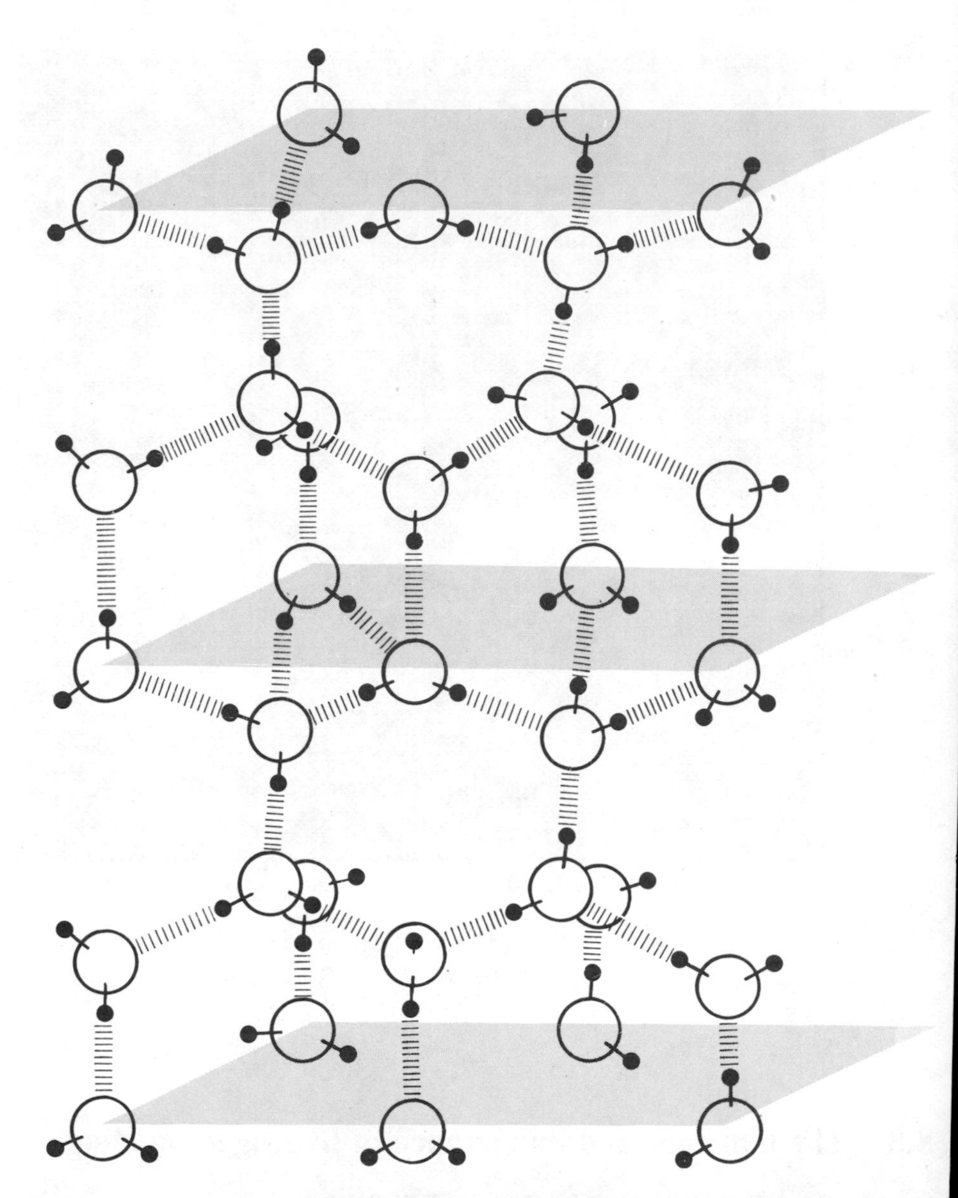

*Below* Ice floats on water. The arrangement of water molecules in ice creates a very open structure which is, in fact, less dense than water.

The fact that ice is less dense than water at 0°C means that ponds and lakes freeze from the surface downwards and the layer of ice insulates the water below, preventing complete solidification. If ice were more dense than water, water would freeze from the bottom upwards and ponds would freeze completely killing fish, aquatic plants and other water-living creatures.

Water has an unusually high boiling point owing to hydrogen bonds. Without the operation of these H-bonds, water would probably be a gas under normal atmospheric conditions; oceans, lakes and rivers would never exist and it would never rain!

Earlier in this section, we mentioned the exceptionally high surface tension of water. Were it not for this, water would never rise through the capillary tubes in the roots and stems of plants.

The high polarity of water means that polar and ionic substances are usually soluble in water. Plants can obtain the salts which they require for growth by absorption of these materials into their bloodstream from the aqueous solution in roots.

Animals obtain the essential ionic and polar substances which they need by absorption of these materials into their bloodstream from the aqueous solution in their stomachs and intestines.

### (B) THE DIMERIZATION OF CARBOXYLIC ACIDS

When carboxylic acids such as ethanoic acid (acetic acid) are dissolved in benzene, the solute particles appear to have a relative molecular mass twice as large as expected (table 8.3).

**Table 8.3** The relative molecular masses of ethanoic acid and benzoic acid

| **Carboxylic acid** | **Formula** | **Relative molecular mass, $M_r$** | **$M_r$ measured by depression of F.Pt. or elevation of B.Pt. in benzene** |
|---|---|---|---|
| Ethanoic acid (acetic acid) | $CH_3COOH$ | 60 | 120 |
| Benzoic acid | $C_6H_5COOH$ | 122 | 244 |

In order to explain these results, it is believed that the carboxylic acid forms hydrogen-bonded dimers in the benzene. The dimers are particularly stable since each pair of acid molecules is linked by two H-bonds, not just one (figure 8.13).

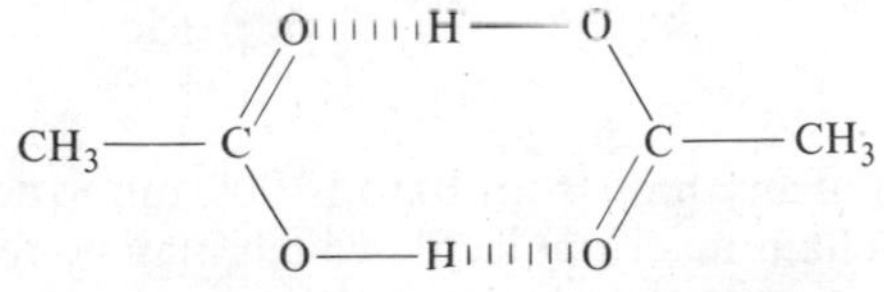

**Figure 8.13** Hydrogen bonding in carboxylic acid dimers.

The presence of dimerized pairs of carboxylic acid molecules in benzene has been confirmed by electron diffraction measurements. Furthermore, measurements of the relative molecular mass of ethanoic acid in the vapour state also suggests the presence of dimers.

- ○ One measurement of the relative molecular mass of ethanoic acid in the vapour state gave a value of 90. To what extent is the acid dimerized?
- ○ Ethanoic acid could, in theory, form trimers, tetramers, etc. in benzene. Why are dimers formed in preference?
- ○ When ethanoic acid is dissolved in water dimers do not form. Why is this?
- ○ Name one other solvent, besides benzene, in which ethanoic acid will form H-bonded dimers.

The effect of dimerization on the boiling points of carboxylic acids is discussed in section 31.1.

### (C) THE HARDNESS OF IONIC CRYSTALS

Crystals of anhydrous calcium sulphate(VI) (anhydrite, $CaSO_4$) are very hard and very difficult to cleave. On the other hand, crystals of hydrated calcium sulphate(VI) (gypsum, $CaSO_4 . 2H_2O$) are soft and easily cleaved. Why is there such a difference in hardness between apparently similar materials?

The structure of gypsum consists of layers containing both $Ca^{2+}$ and $SO_4^{2-}$ ions separated by layers of water molecules. Within the $Ca^{2+}/SO_4^{2-}$ layers the ions are held together by strong electrovalent bonds, but these separated $Ca^{2+}/SO_4^{2-}$ layers are linked by relatively weaker H-bonds from $SO_4^{2-}$ ions in alternate layers to water molecules in the intermediate region. Consequently, the gypsum can be readily cleaved and scratched along the layer of water molecules.

In contrast, anhydrite has a completely ionic structure involving only $Ca^{2+}$ and $SO_4^{2-}$ ions: it is therefore, much harder than gypsum and cannot be easily cleaved.

### (D) THE STRUCTURES AND PROPERTIES OF BIOLOGICAL MOLECULES

H-bonds are present in the structures of proteins, carbohydrates and nucleic acids and the structure, properties and functions of these biological molecules are dependent on their H-bonding.

#### *Proteins*

Proteins are composed of a long sequence of amino acids joined by the **peptide bond**. The general formula of an amino acid is:

$$\mathrm{H_2N{-}C(H)(R){-}C({=}O){-}O{-}H}$$

The $-\mathrm{C}({=}\mathrm{O})\mathrm{OH}$ group of one amino acid molecule can react with the $-\mathrm{N}\langle^{\mathrm{H}}_{\mathrm{H}}$ group of a second amino acid to eliminate water and form a peptide bond.

$$\mathrm{H_2N{-}C(H)(R){-}C({=}O){-}OH} + \mathrm{H_2N{-}C(H)(R){-}C({=}O){-}OH} \longrightarrow \mathrm{H_2N{-}C(H)(R){-}\underbrace{C({=}O){-}N(H)}_{\text{peptide link}}{-}C(H)(R){-}C({=}O){-}OH} + H_2O$$

Proteins are composed of anything from 10 to 10 000 amino acids linked by peptide bonds to form a long-chain macromolecule which may be represented as:

$$\mathrm{H_2N{-}C(H)(R){-}C({=}O){-}N(H){-}C(H)(R){-}C({=}O){-}N(H){-}C(H)(R){-}C({=}O)}\ldots\ldots\mathrm{N(H){-}C(H)(R){-}C({=}O){-}OH}$$

The sequence of amino acids in the protein chain is usually known as **the primary structure** of the protein whilst the detailed configuration of the chain is called **the secondary structure**. One of the commonest secondary structures in proteins is a spiral arrangement known as the α-helix form.

The amino acid units are arranged in a coiled helix which is stabilized by H-bonds between the $\rangle$N—H group of one amino acid unit and the $\rangle$C=O group of the fourth unit further along the chain (figure 8.14).

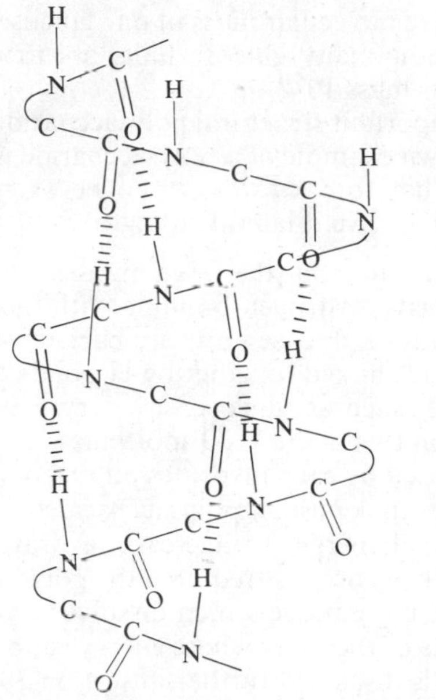

**Figure 8.14** Hydrogen bonding in the coiled helical structure of proteins.

Consequently, H-bonding plays an important part in maintaining the structure of proteins. The precise position and sequence of amino acids in the protein structure means that proteins can carry coded information and control growth or metabolism in plants and animals. In fact, most biological catalysts (which are usually known as **enzymes**) are proteins, so that life itself is very dependent on H-bonding.

The presence of polar $>C=O$ and $>N-H$ groups in proteins which can H-bond with water means that proteins are water soluble. They are, therefore, readily absorbed into the blood and into cellular fluids for easy metabolism by plants and animals.

The properties of amino acids and the peptide link are discussed further in section 32.8.

*Carbohydrates (polysaccharides)*

Most common polysaccharides (e.g. starch, cellulose, glycogen) are built up from the single monosaccharide, glucose.

The structure of glucose is:

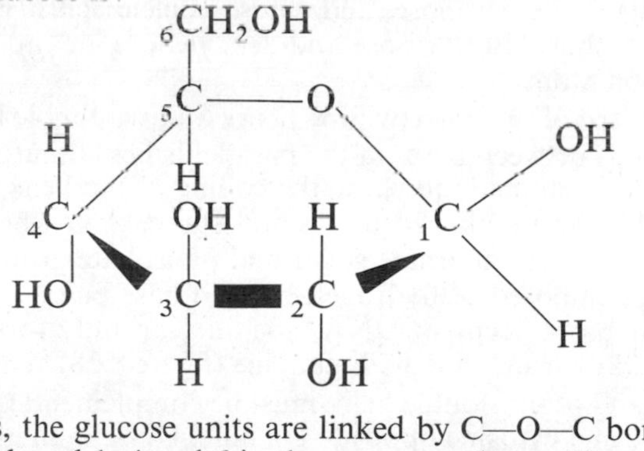

In polysaccharides, the glucose units are linked by C—O—C bonds between the carbon atoms numbered 1, 4 and 6 in the structure above. Figure 8.15 shows the structure of cellulose in which the carbon atom labelled $_1C$ in one glucose unit is linked to $_4C$ in the next. The relative molecular masses of natural polysaccharides range from $10^5$ to $10^7$.

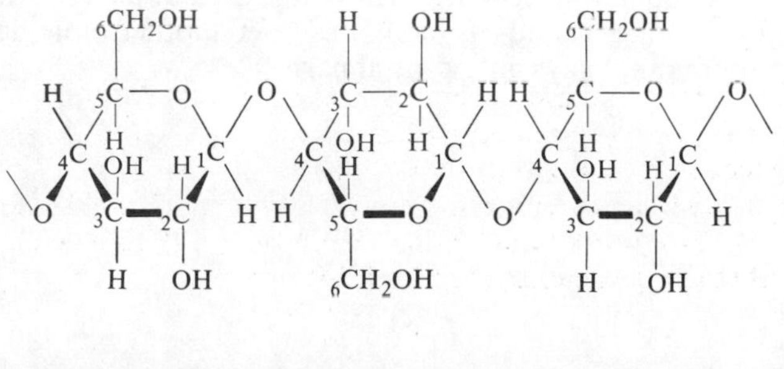

**Figure 8.15** The structure of cellulose.

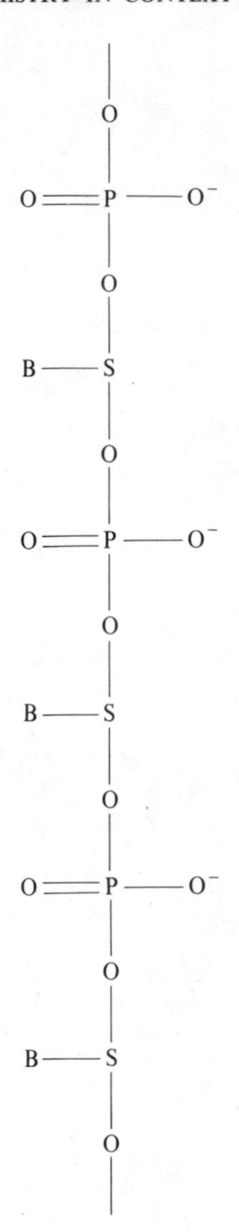


**Figure 8.16** The simplified structure of a single-stranded nucleic acid molecule.

*Above* An X-ray diffraction photograph of DNA. The arrangement of dots in the form of a cross is typical of a double helix structure.

- ○ What is the relative molecular mass of one glucose unit in a polysaccharide?
- ○ Approximately how many glucose units are there in a polysaccharide of relative molecular mass $10^6$?
- ○ Cellulose is an important structural polysaccharide in the cell walls of plants. It is the best known example of a polysaccharide with considerable mechanical strength. What forces can operate between neighbouring chains of cellulose in order to give it lateral strength?

Carbohydrates are non-ionic, but the large number of —OH groups give rise to the possibility of extensive hydrogen-bonding with themselves or with water. In cellulose, the long chains of glucose units are packed very close forming a strong bundle of parallel strands linked together by H-bonds.

Small carbohydrates, such as glucose, are very water soluble owing to their ability to hydrogen-bond with the $H_2O$ molecules. This, coupled with the large amount of energy released by carbohydrates on reaction with oxygen, means that both animals and plants make use of polysaccharides as energy storage materials.

In plants, the principal energy storage carbohydrate is starch whereas that in animals is glycogen. When energy is required the polysaccharides (starch or glycogen) break down releasing glucose which dissolves easily in the blood for rapid transport to those parts of the body where energy supplies are needed. The chemistry of carbohydrates is discussed further in section 30.7.

### *Nucleic acids*

The monomer units in nucleic acids are nucleotides which consist of a complex organic base, B, a sugar, S, and a phosphate group as shown below:

$$\mathrm{B{-}S{-}O{-}\overset{\overset{\displaystyle O}{\|}}{P}{-}O^-} \quad (\text{with } \mathrm{P{-}O{-}} \text{ below})$$

In the nucleic acid molecule a chain of nucleotides are linked together through their sugar/phosphate groups (figure 8.16).

Essentially, there are two types of nucleic acids, deoxyribose nucleic acids (DNA) in which the sugar is deoxyribose, and ribose nucleic acids (RNA) in which the sugar molecule is ribose. Both ribose and deoxyribose are monosaccharides containing five carbon atoms.

DNA is composed of two intertwining helices: it is a double helix, held together by hydrogen-bonds between bases in the parallel helices (figure 8.17).

No attempt has been made to show the coiling of the helix in figure 8.17. The structure of DNA can be likened to a spiral staircase in which the sides of the staircase are composed of alternate sugar and phosphate groups and the steps of the staircase are composed of hydrogen-bonded base pairs.

Although four bases occur in DNA, adenine can only form hydrogen-bonds with thymine, and cytosine only with guanine (figure 8.18). Thus, the sequence of bases in each strand of the double helix must be complementary; adenine must be opposite thymine and cytosine opposite guanine. DNA is an essential constituent of the nuclei of cells; it makes up the genes which are responsible for the transmission of characteristics from one generation to the next. In a healthy plant or animal, cells are continually dividing and replacing those which die away. During cell division, the relatively weak hydrogen bonds between the two strands in the double helix break and new helices are then synthesized in the two 'daughter' cell using the separated single strands as templates—yet another biological process in which hydrogen-bonding plays an essential part.

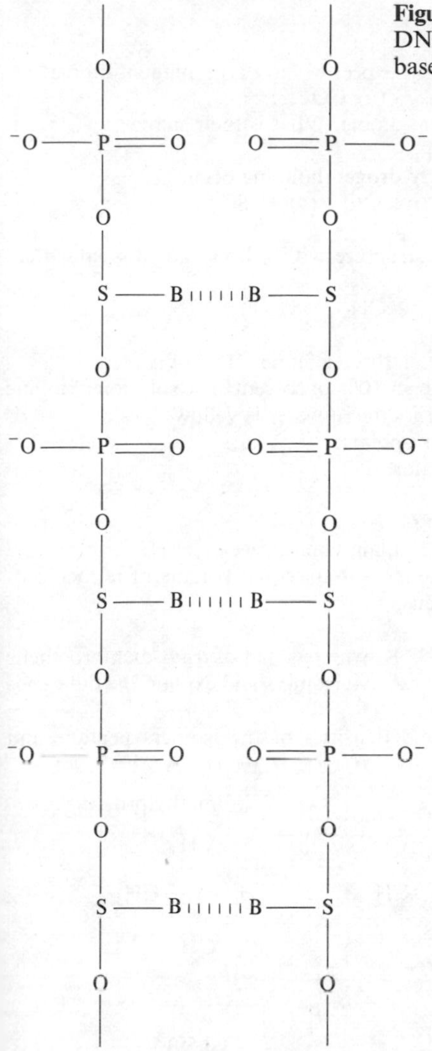

**Figure 8.17** *Left* The simplified structure of DNA showing the hydrogen bonds between bases in the parallel chains.

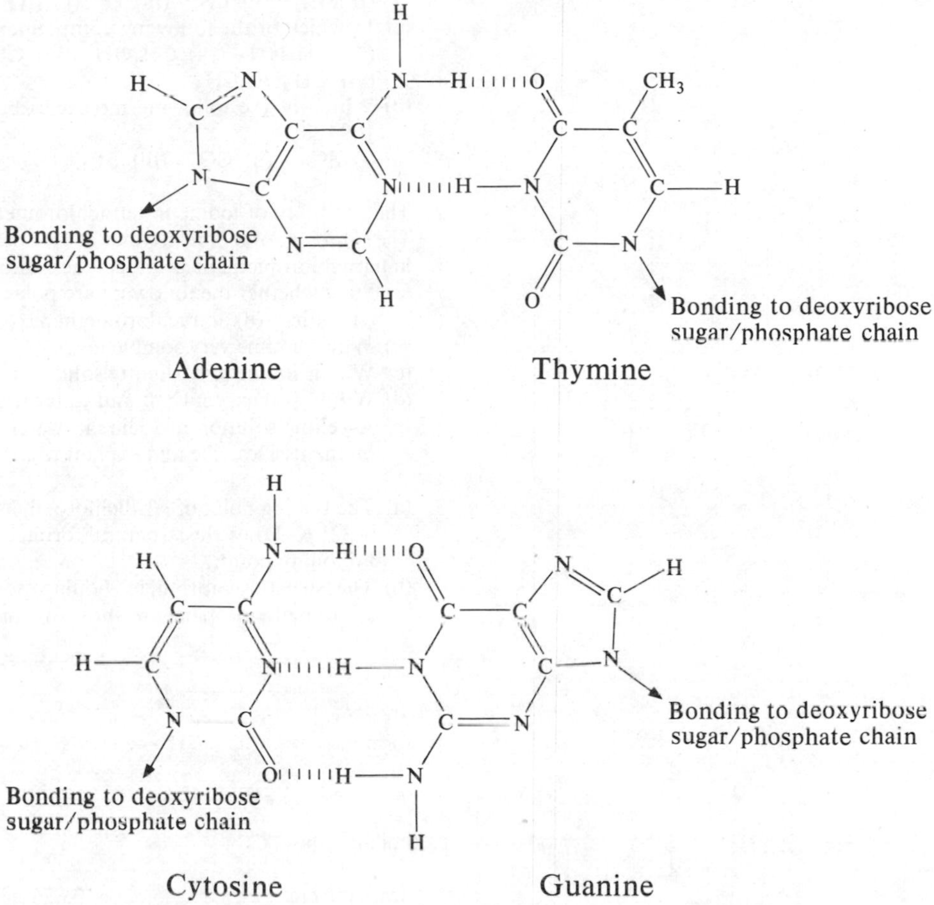

Adenine

Thymine

Cytosine

Guanine

Bonding to deoxyribose sugar/phosphate chain

**Figure 8.18** Hydrogen-bonded base pairs in DNA. (→ = bonds to deoxyribose in the sugar phosphate chain.)

# Summary

1 When two different atoms are joined by a covalent bond, their attractions for the bonding electrons will not be the same. The distortion of charge which results from the unequal sharing of electrons is known as polarization and molecules with an overall distortion of charge are said to be polar.
2 The electronegativity of an atom has been defined as the power of that atom in a molecule to attract electrons.
3 Do not confuse electronegativity with electron affinity. Electronegativity is concerned with the attraction for electrons of atoms in molecules; electron affinity is concerned with the attraction for electrons of single gaseous atoms.
4 Van der Waals' forces are weak, short range forces of attraction between molecules. They are independent of normal bonding forces. Van der Waals' forces are essentially induced dipole-induced dipole attractions.
5 Hydrogen bonds are extra strong intermolecular, permanent dipole-permanent dipole attractions.
6 Hydrogen-bonding can occur when
(i) a hydrogen atom is attached to a highly electronegative atom (usually N, O or F) and
(ii) this highly electronegative atom has an unshared pair of electrons.
7 Hydrogen-bonding in water is responsible for the unexpectedly high values of its melting point, boiling point, latent heats, surface tension and viscosity.
8 Hydrogen-bonds are present in the structures of proteins, carbohydrates and nucleic acids and the biological properties and functions of these molecules in living things are very dependent on their H-bonding.

# Study questions

1 (a) Which of the following molecules would you expect to have a permanent dipole?
(i) $GeH_4$ (ii) ICl (iii) $SiF_4$ (iv) $CH_2Cl_2$ (v) $CO_2$
(b) The following molecules have no permanent dipole. What is their shape?
(i) $BCl_3$ (ii) $CS_2$ (iii) $C_2Cl_2$ (iv) $CBr_4$
(c) In which of the following compounds will hydrogen-bonding occur?
(i) $C_2H_5NH_2$ (ii) $CH_3OH$ (iii) $CH_3I$ (iv) $CF_4$ (v) $H_2SO_3$
(vi) $CH_3.O.CH_3$
(d) Which of the following molecules have a structure with a bond angle greater than 109° 28′?
(i) $SCl_2$ (ii) $CO_2$ (iii) $BF_3$ (iv) $NF_3$ (v) $CH_4$

2 The solubility of iodine in tetrachloromethane at 10°C is 3g per 100g of solvent. The solubility of iodine in water at 10°C is 0.02g per 100g of solvent. The solution of iodine in tetrachloromethane is violet; the solution of iodine in water is yellow.
(a) State whether the following are polar or non-polar:
(i) iodine, (ii) tetrachloromethane, (iii) water.
(b) Why is iodine very soluble in $CCl_4$?
(c) Why is iodine only slightly soluble in water?
(d) Would you expect $CCl_4$ and water to mix? Explain your answer.
(e) A yellow solution of iodine in water is shaken with an equal volume of tetrachloromethane. Describe and explain what happens.

3 (a) The boiling point of *cis*-dichloroethene is 333 K, whereas that of *trans*-dichloroethene is 321 K. Draw the structural formulae of these two isomers and explain the difference in boiling point.
(b) The structural formulae, boiling points and densities of the isomers pentane and 2,2-dimethylpropane are shown below.

| | pentane | 2,2-dimethylpropane |
|---|---|---|
| Structural formula | $CH_3—CH_2—CH_2—CH_2—CH_3$ | $CH_3—C(CH_3)_2—CH_3$ |
| boiling point/°C | 36 | 9 |
| density/g cm$^{-3}$ | 0.626 | 0.591 |

(i) Why does pentane have a higher boiling point than 2,2-dimethylpropane?
(ii) Why does pentane have a higher density than 2,2-dimethylpropane?
(iii) 2-methylbutane is an isomer of pentane and 2,2-dimethylpropane. How do you think its boiling point and density will compare with these two substances? Explain your answer.

4 The relative molecular masses and molar heats of vaporization for three of the hydrides of elements in Group V are given below.

| Compound | Relative molecular mass | $\Delta H^{\ominus}_{vap}$/kJ mole$^{-1}$ |
|---|---|---|
| $NH_3$ | 17 | 23.4 |
| $PH_3$ | 34 | 14.6 |
| $AsH_3$ | 78 | 17.5 |

(a) Plot a graph of $\Delta H_{vap}$ against relative molecular mass for the three hydrides.
(b) Why is the value of $\Delta H_{vap}$ for $NH_3$ unexpectedly high?
(c) Use your graph to estimate a value for the $\Delta H_{vap}$ of $NH_3$ assuming that it has only Van der Waals' bonds.
(d) Predict a value for the strength of H-bonds in $NH_3$.
(e) The strength of H-bonding in water is approximately 22 kJ mole$^{-1}$. How do you explain the fact that H-bonding in $NH_3$ is only about half the strength of that in $H_2O$?

5 (a) Water, in its state of lowest potential energy, exists as a regular, crystalline lattice, but it usually exists as a liquid at room temperature and pressure. Reconcile these facts with the statement that 'all systems tend to a state of minimum potential energy'.
(b) The strength of the H-bond in ice is 22 kJ mole$^{-1}$. How much energy is required to break one mole of hydrogen-bonds in ice?
(c) What percentage of the hydrogen-bonds are broken when ice melts, assuming that all the energy involved in the heat of fusion is used to break hydrogen-bonds? (The heat of fusion of ice is 6 kJ mole$^{-1}$.)

(d) What is the shape of the $H_2O$ molecule? Explain why the $H_2O$ molecule is polar.
(e) What shape would you predict for the hypothetical molecule $H_2X$ if it were non-polar? Explain your prediction.

6 Experiments were carried out with the three liquids, A, B and C, (each with the empirical formula CHF) in order to find their polarity and relative molecular mass. The results of these experiments are given in the table below.

| Compound | Empirical formula | Effect of a charged rod on a thin stream of the liquid issuing from a burette | Approx. mass of 1 $dm^3$ of gas at s.t.p./g |
|---|---|---|---|
| A | CHF | nil | 3 |
| B | CHF | liquid is attracted to charged rod | 3 |
| C | CHF | nil | 6 |

(a) Which of the three liquids is (are) polar?
(b) What is the relative molecular mass of
(i) A, (ii) B, (iii) C?
(c) Draw *one* possible structural formula for A. (H = 1, C = 12, F = 19)
(d) Draw *two* possible structural formulae for B.
(e) Draw *three* possible structural formulae for C.

7 Proteins are polypeptides with relative molecular masses in the range $10^3$ to $10^6$. They are composed of a sequence of amino acids joined by the peptide link.
(a) Write a general formula for the monomer amino acid units which make up proteins.
(b) What is the approximate relative molecular mass for *one* amino acid unit *in* the protein?
(c) About how many amino acid units will there be in a protein of relative molecular mass $10^5$?
(d) Draw a section of a protein structure.
(e) How does hydrogen-bonding explain:
(i) the water solubility of proteins,
(ii) the precise configurations of protein enzymes,
(iii) the elasticity of natural protein fibres such as hair, wool and silk?

8 Suggest reasons for the following:
(a) The boiling points of water, ethanol and ethoxyethane (diethyl ether) are in the reverse order of their relative molecular masses unlike those of their analogous sulphur compounds $H_2S$, $C_2H_5SH$ and $C_2H_5SC_2H_5$.
(b) $BF_3$ is non-polar, but $NF_3$ is polar.

# 9 Structure, Bonding and Properties – The Solid State

## 9.1 Introduction

One of the major achievements of chemistry is the way in which new materials, such as plastics and alloys with predetermined and desirable properties, have been synthesized by chemists. In order to design these new substances, it is necessary to know how the structure of materials can affect their properties. Thus, it is not surprising that one of the most important aspects of chemistry is the investigation of the structure of materials and the bonds holding the particles together.

In this chapter, we shall begin by looking at the methods used to investigate the structures of solid materials and then turn to a consideration of the properties of these materials and the way in which properties are dictated by structure and bonding.

## 9.2 Evidence for the structure of materials

Many of the physical properties of a particular material can provide evidence of some sort for the structure of that material. For example, the diffraction of X-rays by a solid provides evidence for the arrangement of particles within it; the amount of heat required to melt the solid yields information concerning the forces of attraction between these particles, whilst the effect of an electric current on the molten material can tell us something about the nature of the constituent particles.

Each of these phenomena gives different information about the material, but together they provide a detailed picture of its structure and bonding.

### EVIDENCE FROM X-RAY DIFFRACTION

Look through a piece of thin stretched cloth (possibly your handkerchief) at a bright light. The pattern you see is caused by the deflection of light when it passes through the regularly spaced threads of the fabric. This deflection of light is called **diffraction** and the patterns produced are **diffraction patterns**.

When the cloth is rotated in front of the light, the diffraction pattern rotates. If the cloth is stretched so that the strands of the fabric get closer, then the pattern spreads out further. If the strands are arranged differently the pattern changes. From the diffraction pattern which we observe, it is possible to deduce the pattern of the strands in the fabric. The same idea was used to determine the pattern in which particles pack in a crystal.

In 1912, the German scientist, von Laue, obtained a photograph of the diffraction pattern produced by passing X-rays (which have a much smaller wavelength than ordinary light rays) through a crystal. Using a similar technique W. L. Bragg (now Sir Lawrence Bragg) was able to determine the simple cubic structure of sodium chloride. Nowadays, the structure of extremely complex substances such as proteins and nucleic acids can be investigated using X-ray diffraction studies.

*Left* An early version of the X-ray spectrometer designed by W. L. Bragg. X-rays generated in the tube A pass through the slit at B on to the inclined face of the crystal at C. The reflection of the X-rays from the crystal face is measured in the ionization chamber D.

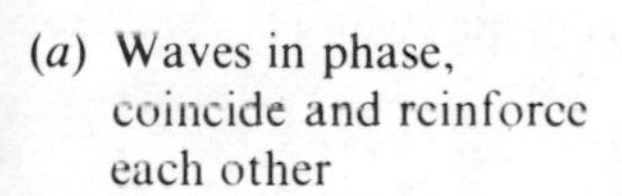


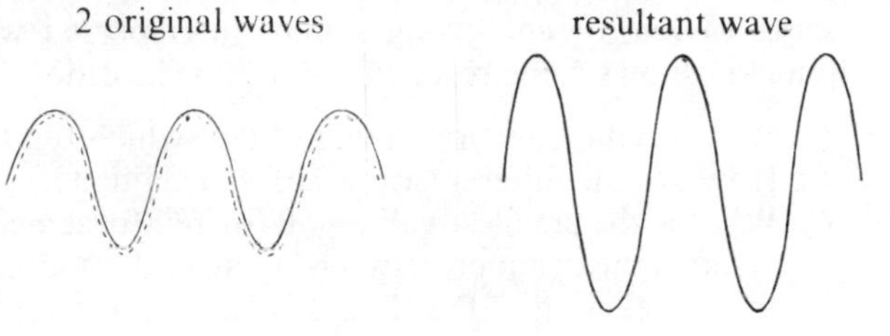


(*b*) Waves out of phase, interfere and destroy each other

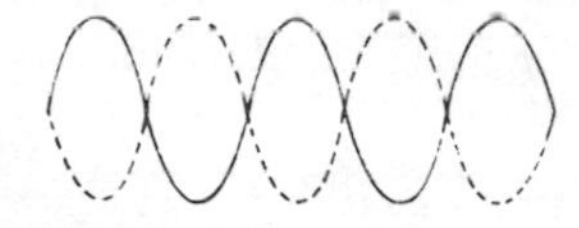

no signal

**Figure 9.1** (*a*) Reinforcement of waves in phase. (*b*) Interference of waves out of phase.

When a beam of X-rays falls on a crystal composed of regularly spaced atoms or ions the X-rays will be reflected. In most instances, waves from the reflected X-rays will interfere with and destroy each other. However, it is possible for the X-rays to be reflected by the particles so that their waves coincide and reinforce each other (figure 9.1). Under what conditions will reinforcement occur? Figure 9.2 shows a beam of X-rays being directed on to a crystal such as sodium chloride. Only three layers of the crystal are represented.

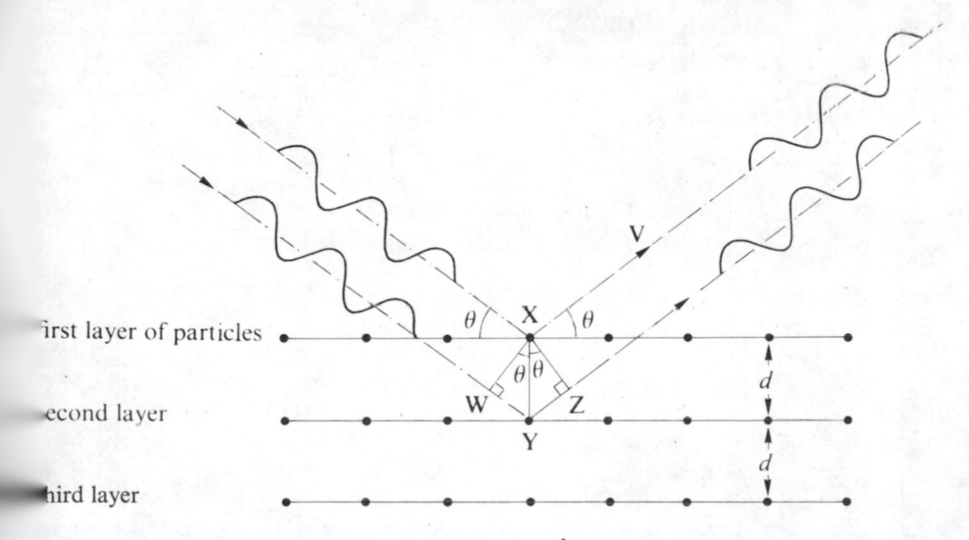


**Figure 9.2** The diffraction of two waves of X-rays by the regularly spaced particles in a crystal.

○ Figure 9.2 shows two waves in phase approaching the crystal. After reflection, the ray emerging along XV is ahead of the ray emerging along YZ. Why is this?

○ How much further does the ray reflected at the second layer travel?

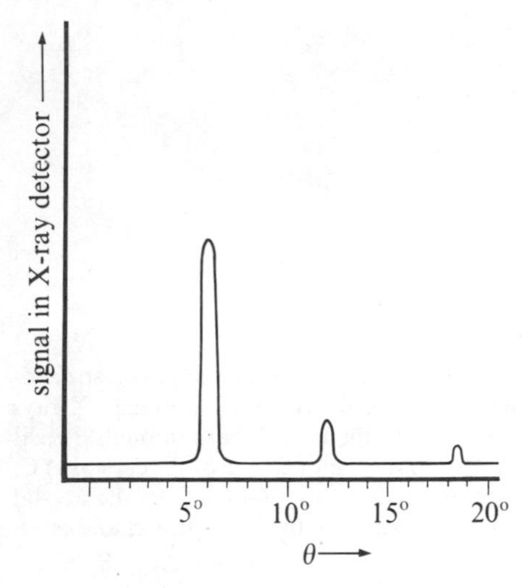

**Figure 9.3** Signals from a crystal of sodium chloride as the angle ($\theta$) between the direction of the X-ray beam and the crystal face increases during diffraction studies.

If the waves are to be in phase again after reflection, the difference in distance travelled by the two rays (i.e. the **path difference**) must equal a whole number of wavelengths, $n\lambda$.

$$\text{Path difference} = WY + YZ$$
$$= XY \sin\theta + XY \sin\theta$$
$$= 2XY \sin\theta = 2d \sin\theta$$

where $d$ = distance between layers.

Thus, for reinforcement of the reflected X-rays

$$2d \sin\theta = n\lambda.$$

which is known as the **Bragg equation**. For X-rays of a particular wavelength, $\lambda$, and for a given value of $d$, maximum reinforcement will occur only at certain values of $\theta$. If $\theta$ is measured and $\lambda$ is known, then the distance, $d$, between the layers in the crystal can be obtained.

By slowly changing the value of $\theta$ from 0° to 90°, an experimenter would obtain a series of reinforced, strong signals interspersed with regions of cancellation. Figure 9.3 shows some results for sodium chloride.

- ○ Why are there several values of $\theta$ at which reinforcement occurs?
- ○ How are the different values of $\theta$ explained?
- ○ What is the smallest value of $\theta$ for reinforcement?
- ○ What is the distance between layers in the sodium chloride crystal? (Assume the wavelength of the X-rays is $5.8 \times 10^{-2}$nm.)

**Figure 9.4** *Below right* An X-ray goniometer.

**Figure 9.5** *Below* The goniometer head.

### THE X-RAY GONIOMETER

Crystal structures are determined by the method described above using an X-ray goniometer (figures 9.4 and 9.5).

X-rays are directed through a narrow collimating channel (at the top left-hand corner of figure 9.4) towards the flat face of a single crystal mounted on the projection from the goniometer head at the centre of the instrument.

The goniometer head (figure 9.5) can be rotated in two arcs at right angles so that the crystal faces can be set in the correct orientation relative to the direction of the X-ray beam. The microscope on the right-hand side of the instrument is used for observing the crystal whilst setting its orientation on the goniometer head. A cylindrical drum is placed over the central well of the instrument and supports the X-ray film around the crystal.

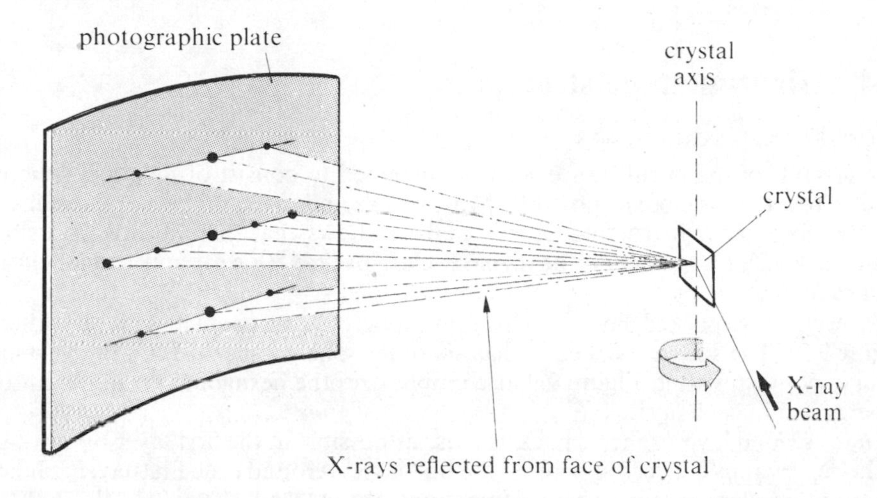


**Figure 9.6** X-rays reflected from the face of a crystal produce a pattern of dots on the photographic plate.

Figure 9.6 shows how a pattern of dots is produced during X-ray diffraction. As the crystal is rotated through 360° on the goniometer head, different planes in the crystal come to the correct angle for reinforcement and an X-ray beam is diffracted towards the photographic plate where a spot appears. If a flat film were used, the spots would occur along curves. Usually the film is bent into a cylinder with the crystal at its centre, so that the spots appear in straight lines. Figure 9.7 shows an X-ray photograph of raw cotton fibres. From the distances between spots, the relative brightness of the spots and, of course, the angles of diffraction, it is possible to deduce a great deal of information about the structure of a crystal.

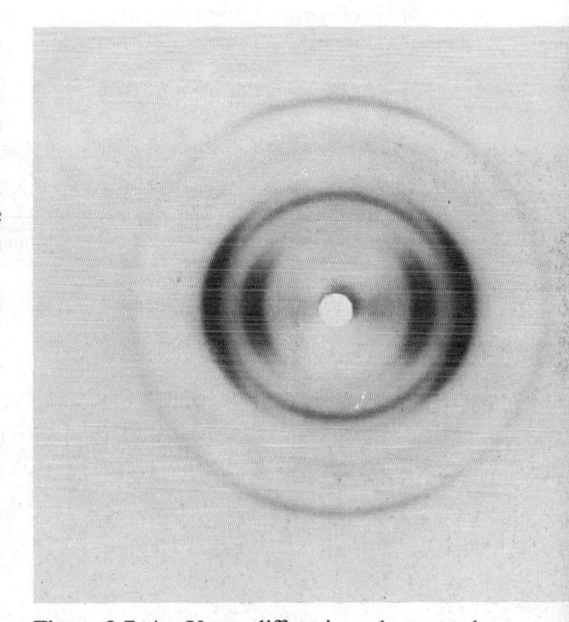

**Figure 9.7** An X-ray diffraction photograph of raw cotton fibres.

## 9.3 Electron density maps

When X-rays strike a crystal, they are diffracted by the electrons in the atoms or ions. Consequently, the larger the atom, the more electrons it possesses and the brighter the spot will be on the diffraction pattern. By analysing both the positions and the intensities of the spots on a diffraction pattern using a computer, it is possible to determine the charge density of electrons within the crystal. The charge density is measured in terms of electrons per cubic nanometer (or electrons per cubic Ångström). Points of equal density in the crystal are joined by contours giving an electron density 'map' similar to the one in figure 9.8. By mounting such

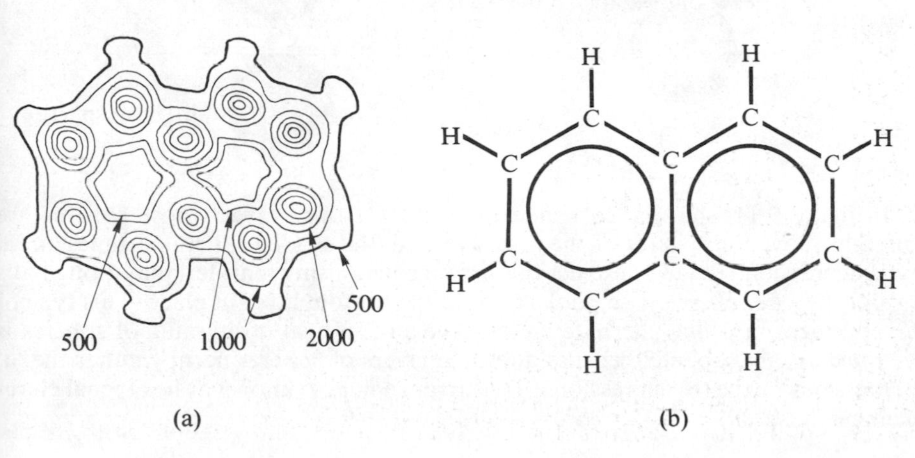


**Figure 9.8** (*a*) Electron density map of naphthalene (contours in electrons per $nm^3$). (*b*) Structural formula of naphthalene. How many atoms can be located from the electron density map? Which atoms are they? Why are the other atoms not evident from the electron density map?

maps on clear plastic or perspex it has been possible to assemble accurate three-dimensional structures of complex substances such as proteins and nucleic acids and then to see the likely positions of different atoms within them. There are many different arrangements in which atoms can be packed in repeating units to form crystal lattices. A few of these arrangements which occur quite frequently in natural crystals are considered in the following sections.

## 9.4 Giant metallic structures

CLOSE-PACKED STRUCTURES

The crystals of many substances can be imagined to consist of identical spheres packed together as close as possible. Most metals and many molecular substances have a close-packed structure. X-ray analysis shows that there are two possible close-packed arrangements: **hexagonal close-packed (h.c.p)** and **cubic close-packed (c.c.p.)**.

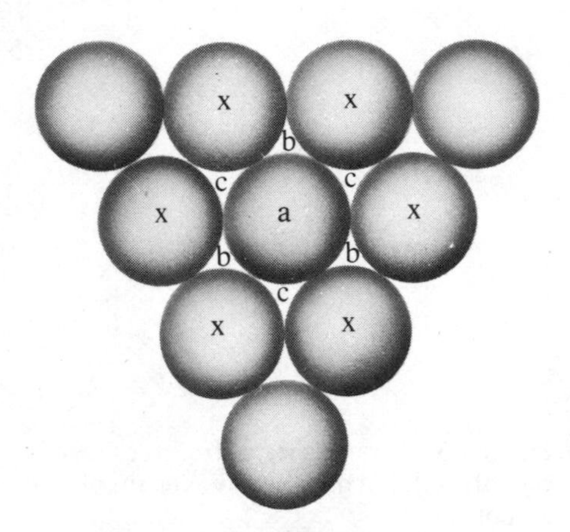


Figure 9.9 A layer of close-packed atoms.

In each close-packed layer, each atom (sphere) is in contact with six others (figure 9.9). The sphere marked 'a' has six other spheres (marked 'x') in the same layer in contact with it. Figure 9.9 also emphasizes the hexagonal arrangement of spheres within each layer.

In the second layer, spheres pack as close as possible to the first layer by 'sitting' in the depressions between spheres in the first layer. Around each first-layer sphere, there are six depressions. These depressions are marked alternately 'b' and 'c' around the sphere 'a'. If the depressions marked 'b' are used those marked 'c' cannot be. (Figure 9.10 shows the second layer of spheres (dashed) in place. Have the 'b' or 'c' depressions been used?) Notice that three spheres in the second layer touch each sphere in the first layer.

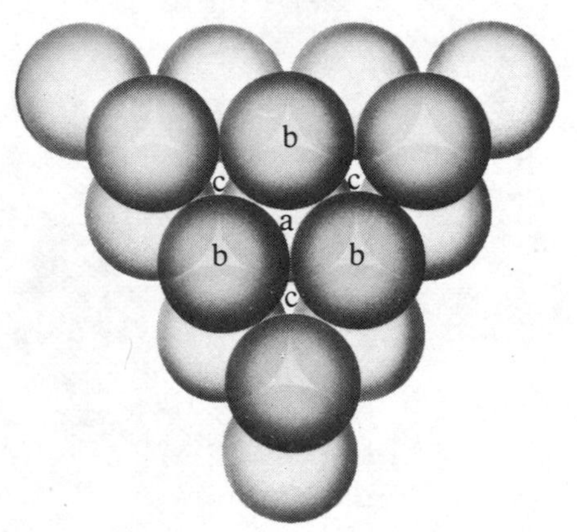


Figure 9.10 First and second layers of close-packed spheres (atoms).

This means that any one sphere (atom) touches 12 others in these close-packed arrangements—six in its own layer, three in the layer above and three in the layer below. This is summarized by saying that its **co-ordination number** is 12.

The third layer of spheres, however, may be added in two quite distinct ways. There are two types of depression available for spheres in the third layer, but the two types of depression are not exactly the same. One type of depression, labelled 'a' in figure 9.10, lies directly above the centre of sphere 'a' in the first layer. The other type of depression, denoted by 'c', lies directly over a hole in the first layer.

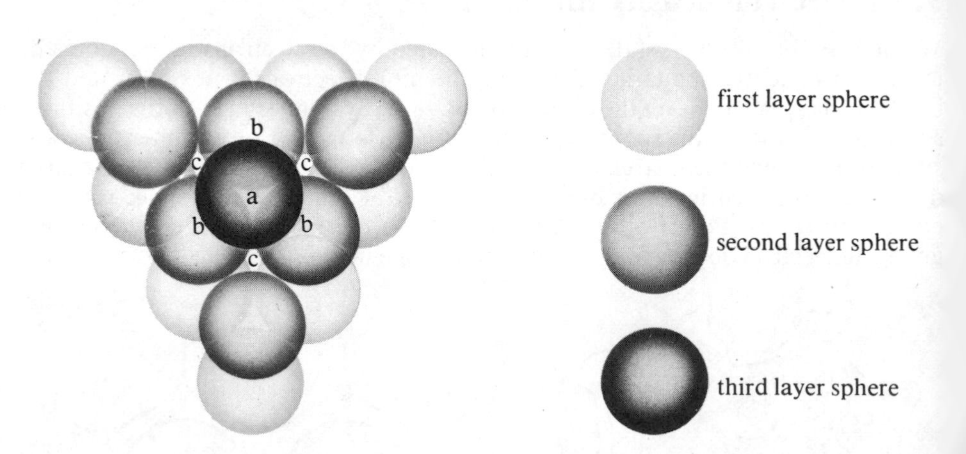


Figure 9.11 First, second and third layers of close-packed spheres (ababab arrangement).

In figure 9.11, the shaded sphere is part of a third layer whose spheres are directly above the spheres of the first layer, i.e. the shaded sphere occupies an 'a' type depression. Try to construct this arrangement using marbles, ping-pong balls or polystyrene spheres. Use books to hold the bottom layer in place. This type of arrangement in which alternate layers have an identical positioning of spheres is denoted as 'ababab' etc. because alternate layers of spheres occupy either the 'a' depressions or the 'b' depressions. This arrangement is known as **hexagonal close-packing (h.c.p.)**.

The diagrams in figure 9.12 show the arrangement of spheres (atoms) in the h.c.p. structure. (The spheres are shaded differently to show the packing of layers with respect to each other more clearly. There is, of course, no difference in the atoms in the different layers of a metal.) Zinc and magnesium are examples of metals with hexagonal close-packed structures.

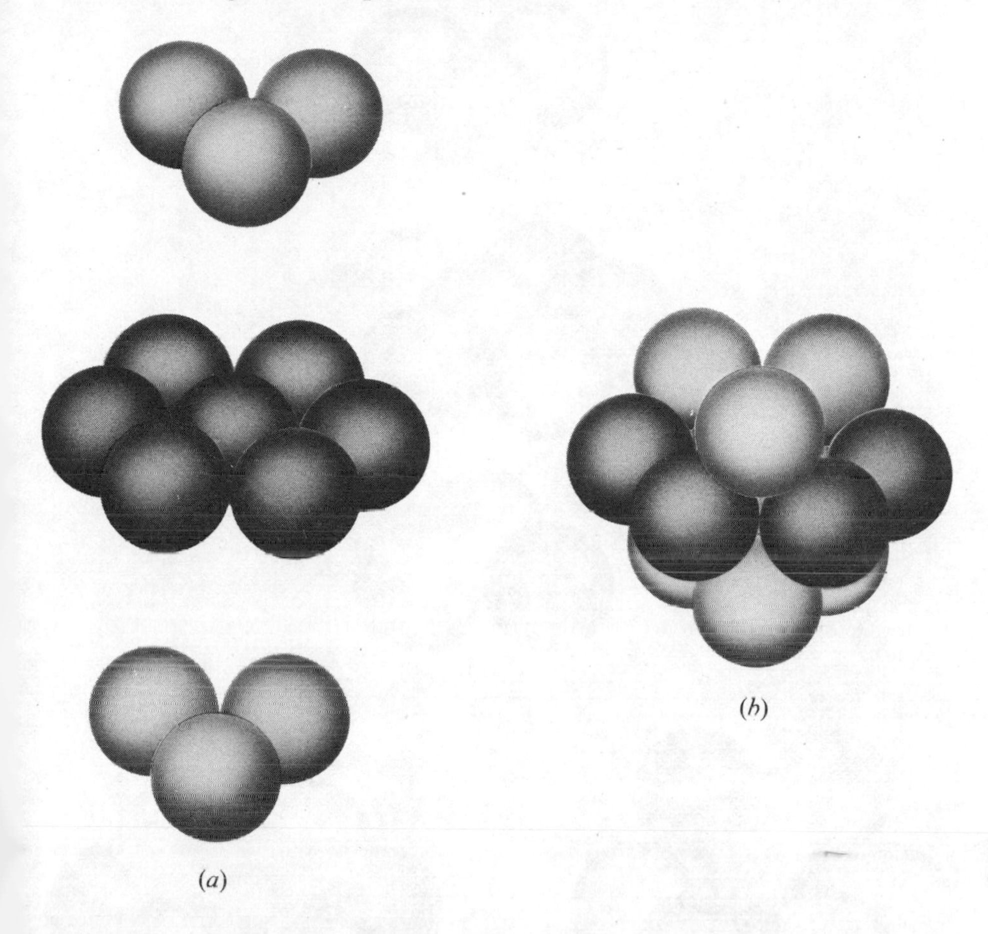


**Figure 9.12** Hexagonal close-packed spheres (*a*) exploded view, (*b*) normal view.

An alternative close-packed arrangement occurs when spheres in the third layer occupy 'c' depressions, above neither the first layer nor the second layer spheres. This arrangement is shown in figure 9.13.

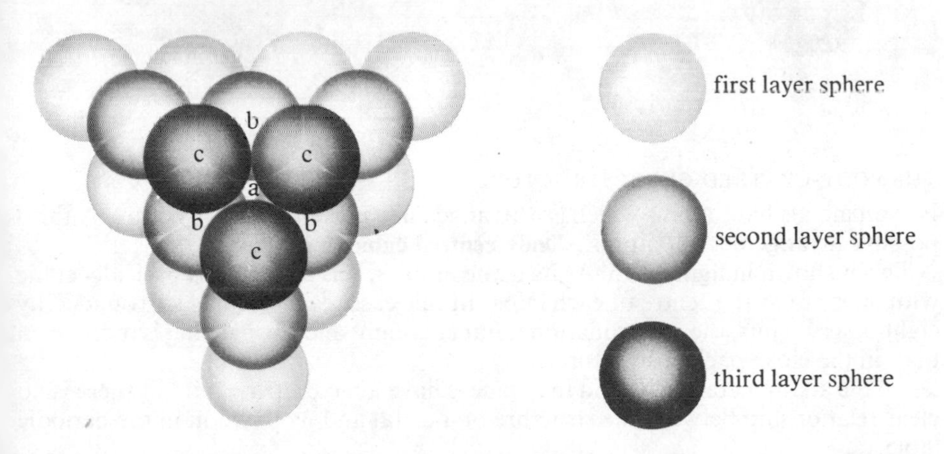


**Figure 9.13** First, second and third layers of close-packed spheres (abcabc arrangement).

This type of arrangement in which the first layer of spheres occupy 'a' depressions, the second layer of spheres occupy 'b' depressions and the third layer of spheres occupy 'c' depressions can be repeated to give a sequence of layers denoted as 'abcabc' etc. This arrangement is known as **cubic close-packing (c.c.p.)**.

The spheres are arranged hexagonally within each layer, but this sequence of layers also has cubic symmetry. Hence, it is described as a *cubic* close-packed structure. The diagrams in figure 9.14 show the arrangement of three layers with respect to each other. Some of the spheres have been shaded so that the cubic symmetry is more obvious. Figure 9.14(*c*) which shows the cubic symmetry most clearly has been obtained by rotation of the second diagram.

Aluminium, copper, lead, silver, gold and platinum are examples of metals with a cubic close-packed structure.

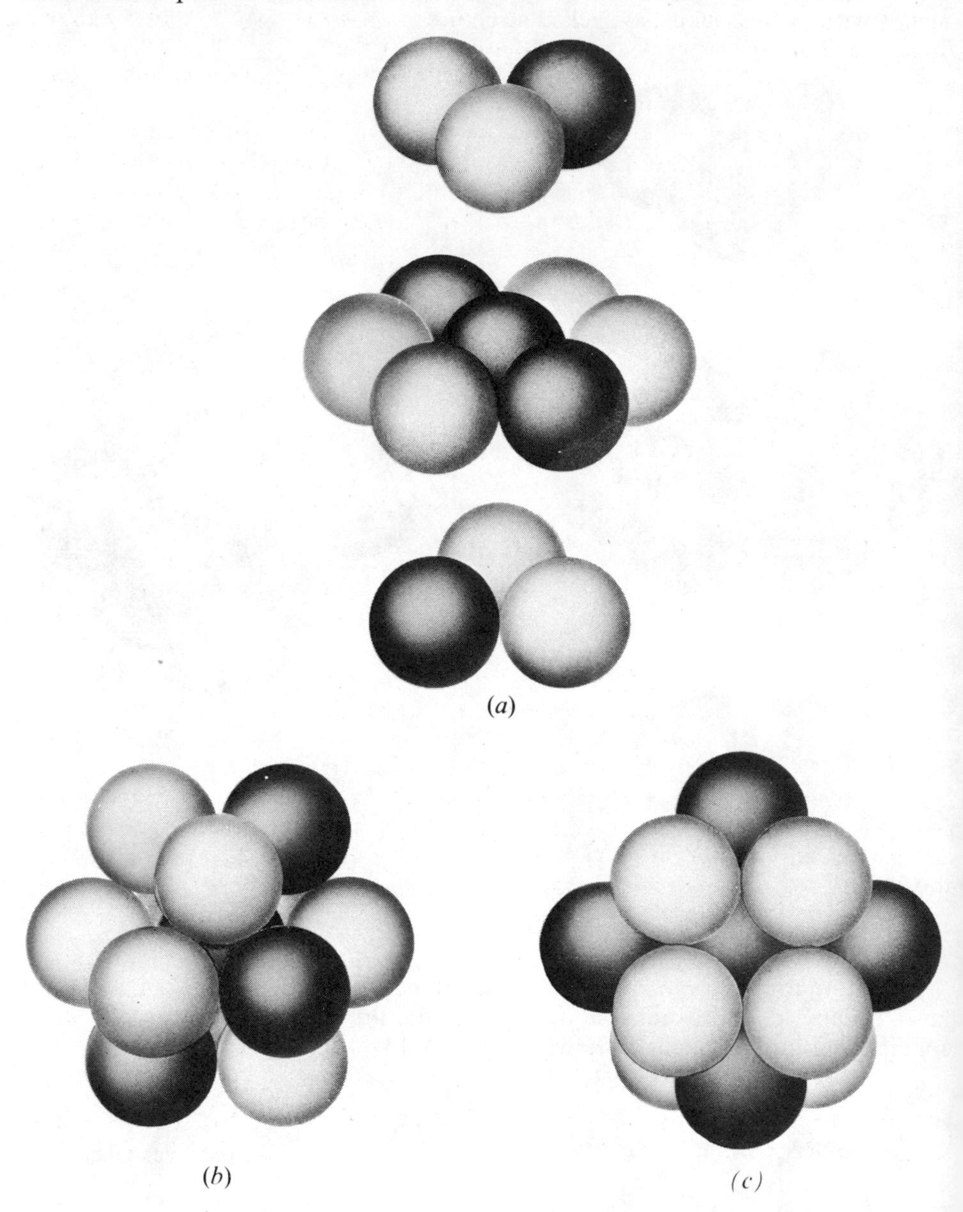


**Figure 9.14** Cubic close-packed spheres. (*a*) exploded view, (*b*) normal view, (*c*) rotation of the normal view to show cubic symmetry more clearly.

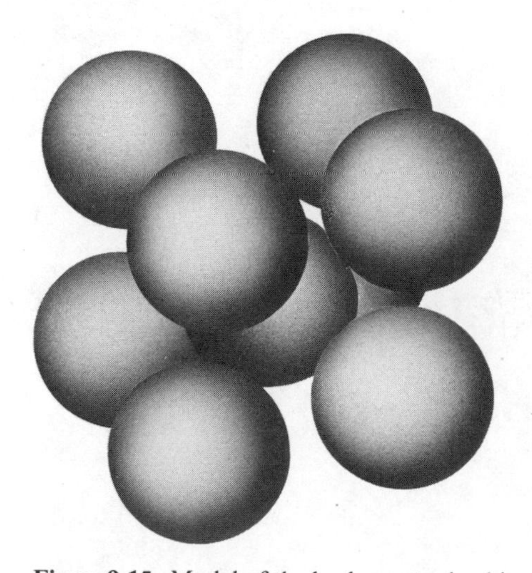

**Figure 9.15** Model of the body-centred cubic structure of metal atoms.

### THE BODY-CENTRED CUBIC STRUCTURE

Not all metals have atoms which are arranged in a close-packed structure. A third possible arrangement of atoms is **body-centred cubic (b.c.c.)**.

This is shown in figure 9.15. As its name implies, this structure is basically cubic with an atom at the centre of each cube. In this case, each atom is surrounded by eight others. Thus, the co-ordination number is eight and the packing is more open than in the close-packed structures.

All the alkali metals, iron and manganese have a b.c.c. structure, but there is no clear relationship between the structure of a metal and its position in the periodic table.

## 9.5 The properties of metals

If the surface of unpolished tinplate is examined closely, it is possible to see small irregularly shaped areas clearly separated from each other by distinct boundaries. These irregularly shaped boundaries are known as **grains**. X-ray analysis shows that the metal atoms are packed regularly within the grains. Thus, these grains are small irregularly shaped crystals of the metal pushed tightly together and containing millions and millions of atoms in a giant structure.

The properties of a metal depend on its crystal structure (i.e. close-packed or body-centred cubic) and also on the size of crystals (grains). Metals usually have high densities, high melting points, high boiling points and high molar heats of fusion and vaporization. In addition, they are good conductors of heat and electricity, they are malleable and ductile.

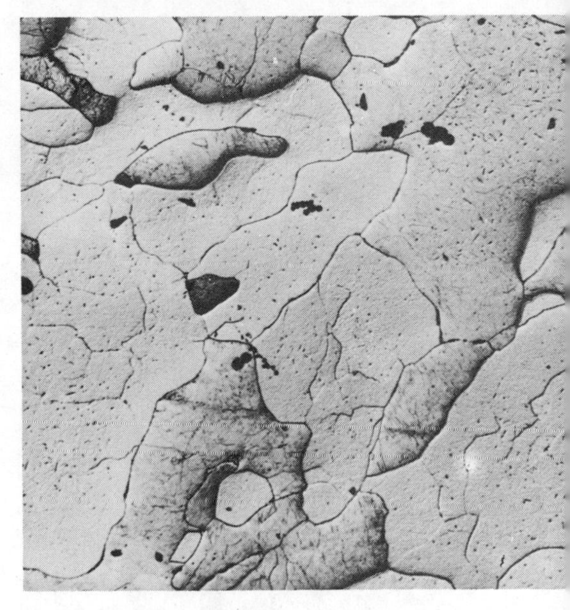

*Above* Crystal grains on the surface of cast iron.

Look back at section 3.9 (p. 32).

- ○ How do chemists visualize the strong forces between metal atoms?
- ○ Why do most metals have a high density?
- ○ Why do most metals have high melting points, high boiling points and high latent heats?
- ○ Why are metals good conductors of electricity?

Chemists believe that the outermost electrons in metal atoms move about freely within the lattice. Thus, the metal consists of positive ions surrounded by a 'sea' of moving electrons. The outer electrons are said to be **delocalized** as they move from one place to another in the crystal, rather than staying in one locality. The negatively charged electrons attract *all* the positively charged ions and bind the nuclei together.

The relatively low affinity for electrons results in two other properties common to metals—low electronegativity and low ionization energy.

There are no rigid, directed bonds in the metal, so that layers of atoms can slide over each other when a force is applied to the metal. This relative movement of layers in the metal lattice is called '**slip**'. After 'slipping' the atoms settle into new positions and the crystal structure is restored. Thus, a metal can be hammered into different shapes or drawn out into a wire; i.e. it is malleable and ductile. Figure 9.16 shows what happens when 'slip' occurs.

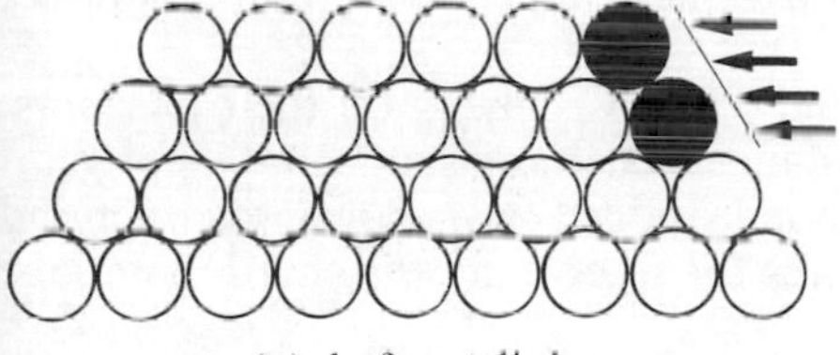

(*a*) before 'slip'

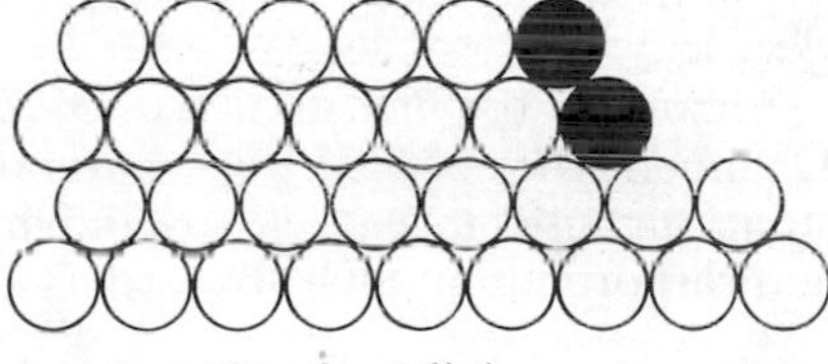

(*b*) after 'slip'

**Figure 9.16** The arrangement of metal atoms (*a*) before and, (*b*) after 'slip'.

When a metal is placed under stress, it will behave elastically provided the stress is not too great. This is the way an engineer would like a metal to behave in his machines and structures. However, a point is reached when the stress is so great that the metal behaves plastically and the changes are permanent and irreversible—this is 'slip'.

Can 'slip' be prevented so as to improve the elastic behaviour of the metal? Metallurgists can increase the strength of metals by inserting barriers in the metal lattice which prevent 'slip' occurring.

Two important methods of strengthening metals are:
(a) reducing the size of the crystal grains,
(b) alloying.

Slip does not readily take place across grain boundaries. Thus, metals with small grain size are harder to deform, less malleable and less ductile than metals with larger grains. By careful choice of metallurgical techniques and casting processes the grain size can be reduced.

Metals will readily form alloys since the metallic bond is non-specific. The presence of small quantities of a second element in the metal frequently increases its strength. For example, brass—an alloy of zinc and copper—is much stronger than either pure copper or pure zinc. Atoms of the second metal are different in size to those of the original metal. These differently-sized atoms interrupt the orderly arrangement of atoms in the lattice and prevent them sliding over each other.

## 9.6 Giant molecular structures

Elements, such as carbon and silicon, which have medium electronegativity values and the ability to form four covalent bonds, can form giant molecular structures of covalently bonded atoms.

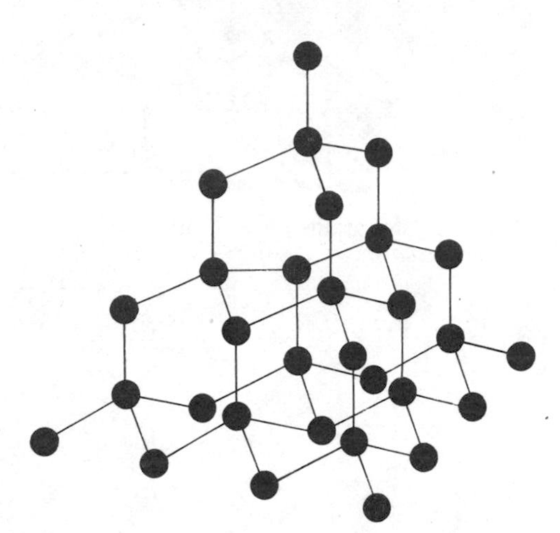

**Figure 9.17** The structure of diamond.

### THE STRUCTURE OF DIAMOND

In diamond, each carbon atom can be imagined to be at the centre of a regular tetrahedron surrounded by four other carbon atoms whose centres are at the corners of the tetrahedron (figure 9.17). Each carbon atom forms four covalent bonds by sharing electrons with each of its four nearest neighbours.

- ○ What is the co-ordination number of carbon atoms in diamond?
- ○ Will the carbon atoms on the outside of the diamond form four covalent bonds?

Silicon (section 3.9, p. 33) and silicon carbide (SiC, used in the abrasive, 'carborundum') exist in a similar crystal structure to diamond. Silicon(IV) oxide ($SiO_2$) is a further example of a compound with a giant molecular structure. The simplest way to appreciate the structure of $SiO_2$ is to picture the structure of pure silicon and imagine that each Si—Si bond is bridged by an oxygen atom. In each of these structures, atoms are linked by localized electrons in strong covalent bonds throughout the whole three-dimensional arrangement.

It is therefore very difficult to distort a covalently-bonded crystal since this would involve breaking covalent bonds. Consequently, diamond, SiC and $SiO_2$ are hard and brittle with very high melting points and very high boiling points. Furthermore, the localization of the electrons within covalent bonds prevents their moving freely in an applied electric field and thus these materials do not conduct electricity.

Almost all the industrial uses of diamond and SiC depend on their hardness. Diamond is one of the hardest natural substances. Diamonds, unsuitable for gemstones, are used in glass cutters and in diamond studded saws. Powdered diamond and carborundum (SiC) are used as abrasives for smoothing very hard materials.

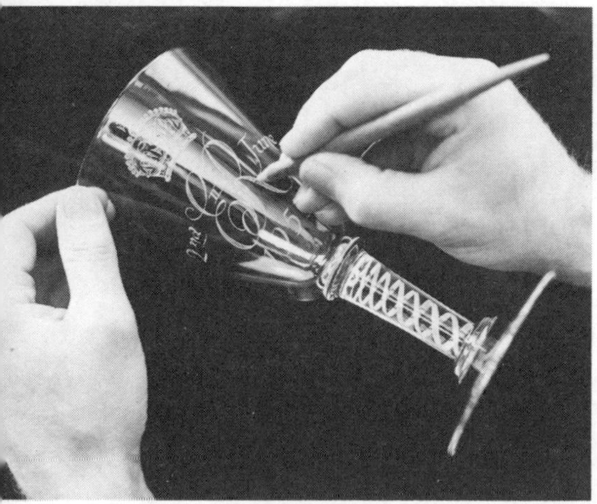

A glass engraver using a diamond-tipped 'pencil' to produce a design on a glass goblet.

A workman uses a diamond cutter on a large piece of plate glass.

A workman uses a drill tipped with diamonds to extract granite samples from the Eleanor Cross monument at Charing Cross, London.

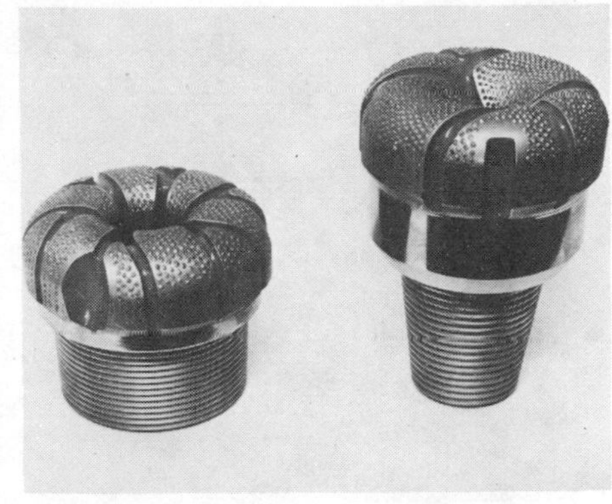

Natural diamonds set in the surface of a drill bit.

THE STRUCTURE OF GRAPHITE

In graphite, the carbon atoms are arranged in flat, parallel layers. Each layer contains millions of hexagonally-arranged carbon atoms (figure 9.18). Each carbon atom is covalently bonded to three other atoms in its layer, so that each layer can be viewed as a two-dimensional sheet polymer or layer lattice. The carbon-carbon bond length within the layer is 0.142nm, suggesting some multiple bond character, since this is intermediate in length between a single and a double carbon-carbon bond. The strong covalent bonds within the layers account for the very high melting point (3 730°C) of graphite. Owing to such a high melting point graphite is used to make crucibles for molten metals and a special form of heat-resistant graphite, pyrographite, is used for the exhaust cones of rockets.

The distance between the layers in graphite is 0.335nm which is much longer than a single carbon-carbon bond. Thus, the bonding between the layers is restricted to relatively weak forces and the layers can slide over each other easily.

This accounts for the softness of graphite and its use as a lubricant.

- Why is it better to lubricate a car lock or a zip-fastener with a pencil than with oil?
- Why is graphite often more suitable than oil for lubricating the moving parts of machinery?
- How do you think the proportion of graphite to clay affects the hardness of a pencil?

**Figure 9.18** The structure of graphite.

The bonding in graphite can be pictured as three trigonally-arranged covalent bonds formed by three of the four valence electrons of carbon whilst the fourth electron is delocalized over the whole layer. This delocalization of electrons similar to that in metals results in graphite conducting electricity and in its shiny appearance. The electrical conductivity of graphite enables it to be used as electrodes in electric furnaces and during electrolyses.

## 9.7 Giant ionic structures

onic structures are formed when atoms with large differences in electronegativity orm compounds. Electrons are transferred from atoms of low electronegativity o those of high electronegativity and the oppositely-charged ions which result are eld together by strong electrostatic forces of attraction. The electrical force binding the ions together is described as an ionic or electrovalent bond.

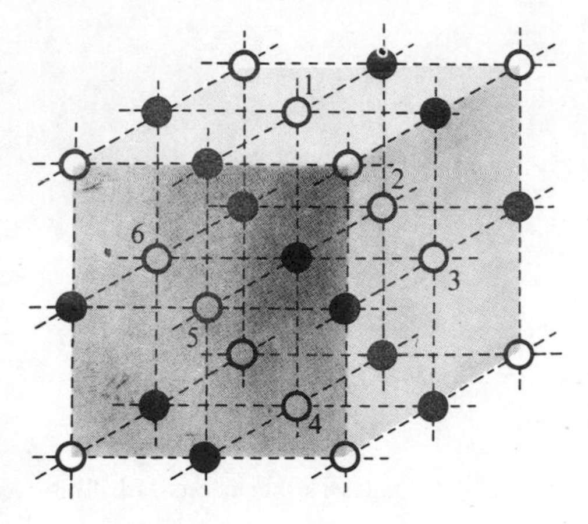


○ indicates the centre of a sodium ion, $Na^+$

● indicates the centre of a chloride ion, $Cl^-$

**Figure 9.19** The structure of sodium chloride.

X-ray analysis shows that the particles in different ionic structures can be arranged in different patterns. The simplest structure amongst ionic compounds is the **simple cubic structure**. Sodium chloride, potassium bromide and calcium oxide all have a simple cubic structure. Figure 9.19 shows a diagram of the structure of sodium chloride. Notice how the ions are arranged in a cubic pattern. Figure 9.19 shows only a few $Cl^-$ and $Na^+$ ions, but remember that there will be millions and millions of ions in even the smallest visible crystal of sodium chloride.

Each positive sodium ion is surrounded by six $Cl^-$ ions and each negative chloride ion is surrounded by six $Na^+$ ions. The six $Na^+$ ions round the central $Cl^-$ ion in figure 9.19 have been numbered. Four of the $Na^+$ ions (numbered 2,3,5 and 6) are in the same horizontal layer of the crystal as the central $Cl^-$ ion, one $Na^+$ ion (number 1) is in the layer above and another $Na^+$ ion (number 4) is in the layer below.

○ What is the co-ordination number of $Cl^-$ ions in NaCl?
○ What is the co-ordination number of $Na^+$ ions in NaCl?
○ What properties are usually associated with giant ionic crystals?

MEASURING THE SIZE OF IONS

X-ray measurements on ionic solids can be presented in the form of electron density maps in order to determine the size of different ions. Figure 9.20 shows an electron density map for sodium chloride. The circular contours suggest that the electron distribution in these ions is spherical and the spacing of the contour lines enables us to distinguish particles with different numbers of electrons.

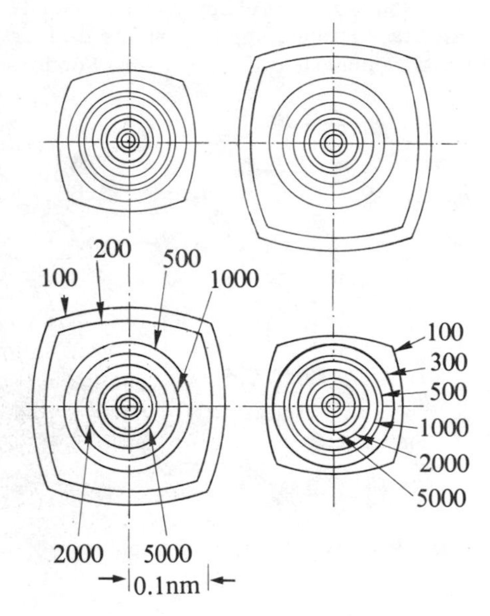


**Figure 9.20** Electron density map for sodium chloride. (Electron densities are expressed as electrons per cubic nanometre.)

Look closely at figure 9.20.

○ Which are the sodium ions?
○ What is the interionic distance between neighbouring $Na^+$ and $Cl^-$ ions?
○ What is the ionic radius of:

(i) $Na^+$, (ii) $Cl^-$?

Using electron density maps of this kind, it is possible to compile tables of ionic radii. But remember, the size of a particular ion can vary slightly depending on the size and charge of other ions in the crystal and so values of ionic radii for the same ion do not always agree. The trends in the size of ionic radii in relation to the periodic table are discussed in section 12.3.

## 9.8 The properties of ionic crystals

Crystals of ionic solids are:
1 Hard and brittle,
2 Involatile with high melting points and high boiling points,
3 Good conductors of electricity when molten, but non-conductors when solid,
4 Soluble in polar solvents such as water, but insoluble in non-polar solvents such as tetrachloromethane.

In an ionic solid each ion is held in the crystal by strong attractions from the oppositely charged ions around it. Consequently, ionic solids like sodium chloride are hard and difficult to cut. However, they are also very brittle and may be split cleanly (**cleaved**) using a sharp edged razor. When the crystal is tapped sharply along a particular plane it is possible to displace one layer of ions relative to the next. As a result of this displacement, ions of similar charge come together and repel each other forcing apart the two portions of the crystal (figure 9.21).

Why is it easier to cleave ionic solids in certain directions rather than others? The forces of attraction between oppositely-charged ions in an ionic lattice are so strong that large quantities of heat energy must be supplied to the crystal before the ions vibrate vigorously enough to overcome the forces of attraction between one another and move away from their relative positions. Thus, ionic solids have high melting points and high molar heats of fusion. Even in the molten ionic liquid there will still be strong forces between the oppositely-charged ions. Consequently, ionic solids have high boiling points and high molar heats of vaporization.

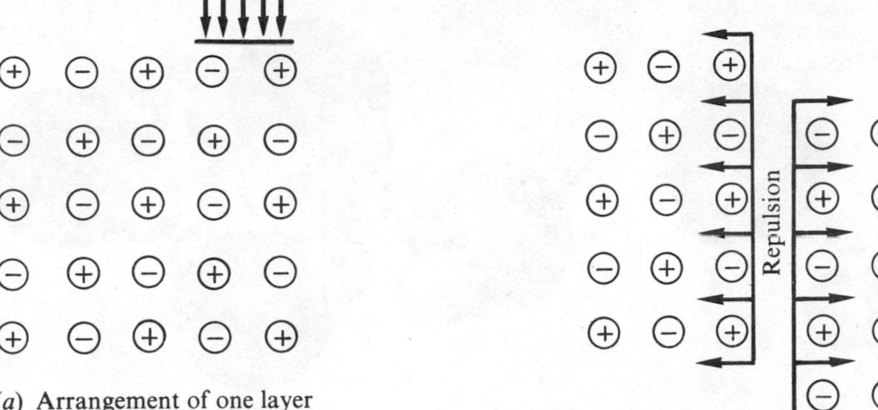


(*a*) Arrangement of one layer of ions before displacement

(*b*) Arrangement of ions after displacement

**Figure 9.21** Crystal cleavage in ionic compounds.

- ○ Why do molten ionic compounds conduct electricity?
- ○ Why are solid ionic compounds non-conductors?
- ○ Why are ionic compounds such as sodium chloride soluble in water, but insoluble in tetrachloromethane?

### WHAT HAPPENS WHEN AN IONIC SOLID DISSOLVES IN WATER?

When sodium chloride dissolves in water, the crystal lattice is broken up forming separated $Na^+$ and $Cl^-$ ions in aqueous solution. Where does the energy required to separate the oppositely-charged ions come from? We have seen already that water contains highly polar molecules. The positive ends of polar water molecules are attracted to negative ions in the crystal and negative ends of the water molecules are attracted to positive ions. The formation of ion-solvent bonds results in a release of energy sufficient to cause the detachment of ions from the lattice (figure 9.22). Thus, ionic crystals will often dissolve in polar solvents such as water, ethanol and propanone (acetone).

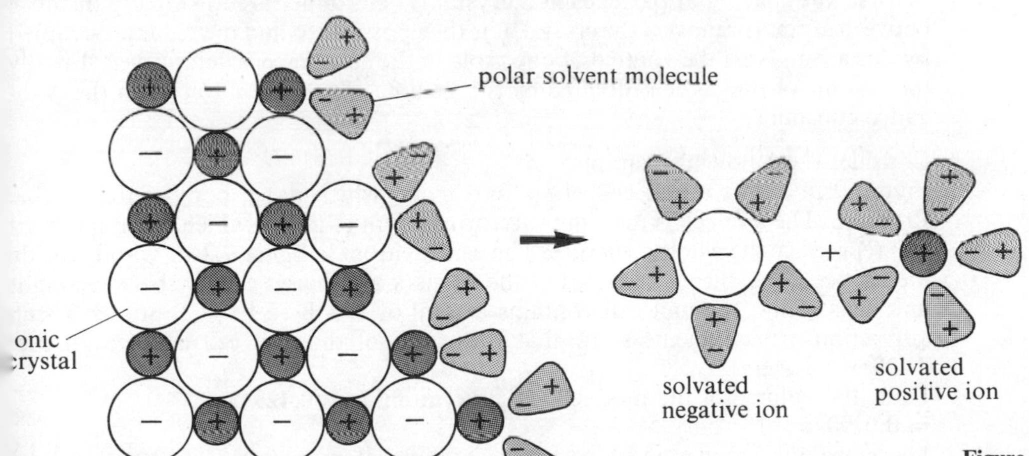


**Figure 9.22** Polar solvent molecules dissolving an ionic solid.

The attachment of polar solvent molecules to ions is known generally as **solvation**. Chemists say that the ions have been **solvated**. Very often the solvent is water and in this specific case the process is called **hydration** and the ions are said to be **hydrated**. Figure 9.23 shows the structures of hydrated positive and negative ions.

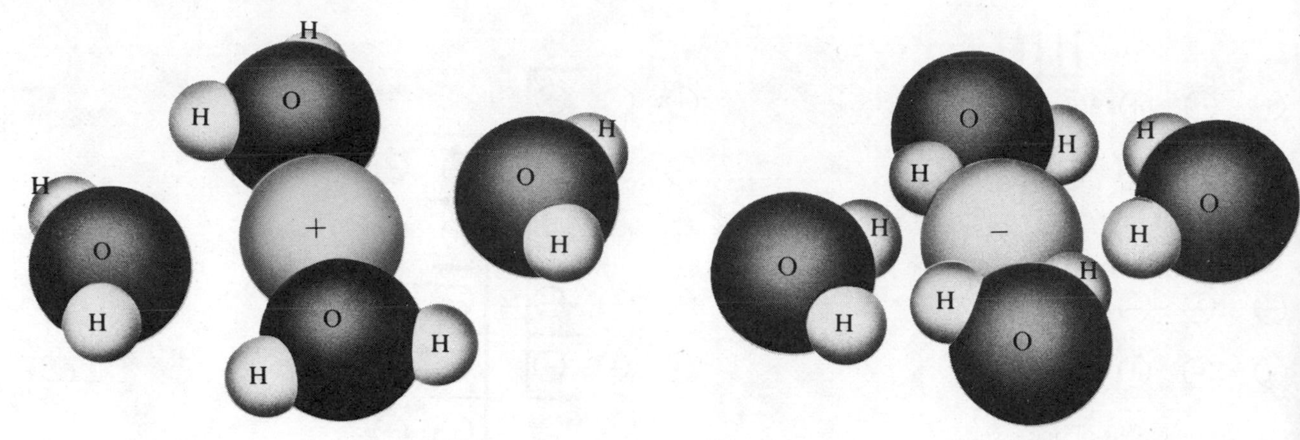


**Figure 9.23** The structures of hydrated positive and negative ions. Only those water molecules in the horizontal plane are shown.

Frequently the attraction of the central ion extends beyond the first layer of polar solvent molecules to a whole sheath of solvent molecules which may be several layers thick. This envelope of attracted solvent molecules reduces the movement of solvated ions which therefore diffuse much slower than expected and move less easily towards electrodes during electrolysis.

Non-polar liquids such as tetrachloromethane, benzene and hexane will not solvate ionic solids. Why is this? Non-polar molecules are held together by weak intermolecular (Van der Waals') forces much smaller in magnitude than the forces between ions in an ionic crystal. The ion-ion attractions are much stronger than either the solvent-solvent interactions or the ion-solvent interactions in this case and so the non-polar solvent molecules cannot penetrate the ionic lattice. Thus, sodium chloride is virtually insoluble in tetrachloromethane and benzene.

## 9.9 Determining the Avogadro constant from X-ray studies

One of the most valuable results of X-ray diffraction studies has been the accurate determination of the Avogadro constant, $L$. By this means, $L$ can be found to an accuracy of one in ten thousand.

First, the spacing of particles in a crystal is determined. Knowing the distance between atoms (or ions) in the crystal, it is then possible to find the volume occupied by one atom. Next the volume of one mole of the substance is determined. Finally the volume of one mole is divided by the volume of one atom to obtain the Avogadro constant.

Consider the following example:

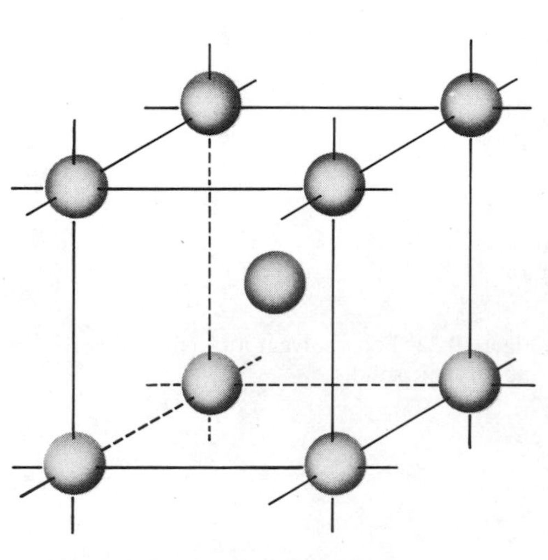

**Figure 9.24** A unit cell of sodium.

Figure 9.24 shows a **unit cell** of sodium metal which has a body-centred cubic structure. The unit cell is the simplest arrangement of atoms which when repeated will reproduce the whole structure. The central atom in figure 9.24 is wholly inside the unit cell, but the eight atoms at the corners are shared equally between eight unit cells. Thus, the unit cell contains a total of $1 + 8 \times \frac{1}{8}$ (= 2) atoms. X-ray diffraction measurements show that the width of the unit cell is 0.429nm (i.e. $0.429 \times 10^{-7}$cm).

Thus, the volume of the unit cell (i.e. two atoms) = $(0.429 \times 10^{-7})^3\text{cm}^3$
$= 0.0790 \times 10^{-21}\text{cm}^3$

Therefore, the volume occupied by one sodium atom = $0.0395 \times 10^{-21}\text{cm}^3$

The relative atomic mass of sodium = 22.99

The density of sodium = $0.97\text{g cm}^{-3}$

Using the equation, volume $= \frac{\text{mass}}{\text{density}}$,

the volume of one mole ($L$ atoms) of sodium $= \frac{22.99}{0.97}\text{cm}^3$

$= 23.70\text{cm}^3$

$$\therefore \text{ The Avogadro constant, } L = \frac{23.70}{0.0395 \times 10^{-21}} = 6.0 \times 10^{23}\text{ mole}^{-1}.$$

## 9.10 Simple molecular structures

Non-metal elements and compounds of non-metals containing only elements with a high electronegativity such as Cl, O and N usually exist as simple molecules. Substances like iodine, carbon dioxide, water and naphthalene ($C_{10}H_8$) are composed of molecules. In these molecular substances, the atoms are joined together *within* the molecule by strong covalent bonds, but the separate molecules are attracted to each other by much weaker Van der Waals' bonds.

X-ray diffraction measurements on crystals of molecular substances have been used to determine both the distance between atoms within each molecule and the distance between separate molecules. Figure 9.25 shows the arrangement of $I_2$ molecules in solid iodine. The arrangement of molecules in the crystal lattice is described as **face-centred cubic**—the molecules are arranged in a *cube* with a molecule at each corner and a molecule at the *centre* of each *face*. Notice that the relative positions of molecules with 12 nearest neighbours in the face-centred cubic structure is the same as the relative positions of metal atoms in the cubic close-packed structure.

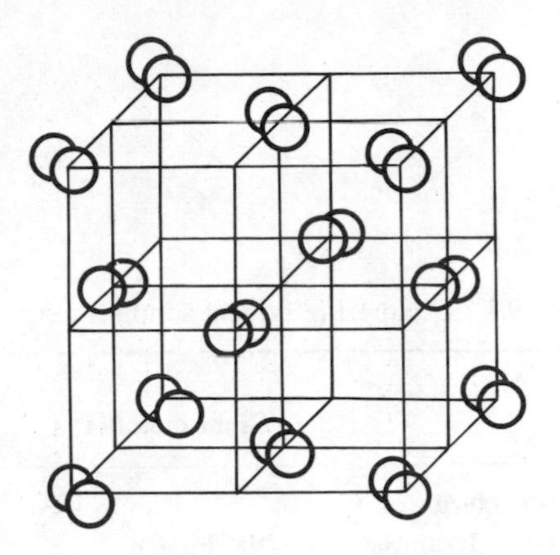


**Figure 9.25** The crystal structure of iodine.

Analysis of the X-ray diffraction pattern for solid iodine suggests that the distance between two iodine atoms in a molecule is 0.166nm. Thus, the I—I covalent bond length is 0.166nm. The distance between the centres of iodine atoms in adjacent molecules is 0.43nm which gives a value of 0.215nm for the Van der Waals' radius for iodine.

The packing of molecules in crystal structures is often more complicated than the packing of simple atoms or ions, but each substance is found to have a characteristic and uniform lattice arrangement.

### THE PROPERTIES OF SIMPLE MOLECULAR SUBSTANCES

Crystals of molecular substances such as iodine, naphthalene ($C_{10}H_8$), sucrose (sugar, $C_{12}H_{22}O_{11}$) and ice are usually soft with low melting points and low boiling points. Molecular compounds do not conduct electricity when molten or dissolved in water. In fact, non-polar molecular compounds such as iodine and naphthalene are almost insoluble in water, though they are usually very soluble in non-polar solvents such as benzene and tetrachloromethane.

- ○ How does the bonding in ice and sugar differ from that in non-polar simple molecules like naphthalene and iodine?
- ○ Why are ice and sugar harder than iodine and naphthalene?
- ○ Why is sugar soluble in water unlike iodine and naphthalene?

The properties of molecular compounds can be explained in terms of their structure which consists of simple molecules. Although the atoms within these molecules are joined by strong covalent forces, the separate molecules are held together by very weak Van der Waals' forces. These weak inter-molecular bonds permit the molecules to be separated easily. Hence the crystals of simple molecular substances are usually soft with low melting points and low boiling points.

Molecular compounds contain neither delocalized electrons (as do metals) nor ions (as do ionic compounds). Thus, they cannot conduct electricity when molten or when dissolved in water.

Why are non-polar substances such as iodine insoluble in water, but soluble in benzene?

In liquids of high polarity such as water, there are strong water-water attractions which are considerably stronger than either iodine-iodine attractions or iodine-water attractions. Consequently, iodine molecules cannot penetrate the water structure and there is little tendency for water molecules to solvate uncharged iodine molecules. Iodine is therefore almost insoluble in water.

However, in non-polar liquids, such as benzene and tetrachloromethane, there are weak intermolecular forces. Thus benzene-benzene attractions are similar in strength to iodine-benzene and iodine-iodine attractions. Thus, it is easy for benzene molecules to penetrate into the iodine crystal and solvate the iodine molecules. Thus, iodine dissolves in benzene.

## 9.11 Comparing typical solid structures

The structure, bonding and properties of the four common solid structures (giant metallic, giant molecular, giant ionic and simple molecular) are summarized and compared in table 9.1. Use the information in table 9.1 to answer the following question.

**Table 9.1** Comparing typical solid structures

| | Giant metallic | Giant molecular (Giant covalent) | Giant ionic | Simple molecular |
|---|---|---|---|---|
| 1. **Structure** | | | | |
| (i) Examples | Na, Fe, Cu | Diamond, SiC, $SiO_2$ | $Na^+Cl^-$, $Ca^{2+}O^{2-}$, $(K^+)_2SO_4^{2-}$ | $I_2$, $S_8$, $C_{10}H_8$, HCl, $CH_4$ |
| (ii) Constituent particles | Atoms | Atoms | Ions | Molecules |
| (iii) Type of substance | Metal element with low electro-negativity | Non-metal element in Group IV or its compound | Metal/non-metal compound (A compound of elements with a large difference in electro-negativity) | Non-metal element or non-metal/non-metal compound (Elements with high electro-negativity) |
| 2. **Bonding** in the solid | Attraction of outer mobile electrons for positive nuclei binds atoms together by *strong* metallic bonds | Atoms are linked through the whole structure by *very strong* covalent bonds from one atom to the next | Attraction of positive ions for negative ions—*strong* ionic bonds | Strong covalent bonds hold atoms together within the separate molecules; separate molecules are held together by *weak* Van der Waals' bonds |
| 3. **Properties** | | | | |
| (i) Volatility | Non volatile<br>High m.pt., high b.pt., high latent heats | Non volatile<br>Very high m.pt., very high b.pt., very high latent heats | Non volatile<br>High m.pt., high b.pt., high latent heats | Volatile<br>Low m.pt., low b.pt., low latent heats |
| State at room temp. | Usually solid | Solid | Solid | Usually gases or volatile liquids |
| (ii) Hardness/malleability | Hard, yet malleable | Very hard and brittle | Hard and brittle | Soft |
| (iii) Conductivity | Good conductors when solid or liquid | Non-conductors (Graphite is an exception) | Non-conductors when solid. Good conductors when molten or in aqueous solution —*electrolytes* | Non-conductors when solid, liquid and in aqueous solution. (A few (e.g. HCl) react with water to form electrolytes.) |
| (iv) Solubility | Insoluble in polar and non-polar solvents, but soluble in liquid metals | Insoluble in all solvents | Soluble in polar solvents (e.g. $H_2O$), insoluble in non-polar solvents (e.g. $CCl_4$) | Insoluble in polar solvents (e.g. $H_2O$), soluble in non-polar solvents (e.g. $CCl_4$) |

Consider each of the following in the solid state: Cu, Si, $NH_3$, NaI, Xe.

Which solid would

○ be a good electrical conductor?
○ be a poor electrical conductor, but conduct on melting?
○ be hard and brittle and insoluble in water?
○ be easy to cleave?
○ have strong hydrogen bonds between molecules?
○ have the lowest melting point?

## Summary

1 When X-rays are directed at a crystalline solid they are diffracted by particles in the lattice. The diffracted X-rays can be detected using an X-ray film and the diffraction pattern which is obtained can be used to determine the arrangement of particles in the crystal.
2 For reinforcement of reflected X-rays,

$$2d \sin \theta = n\lambda,$$

which is known as the Bragg equation.
($d$ = distance between layers in the crystal, $\theta$ = angle between direction of X-rays and the crystal face, $n$ = 1, 2, 3, etc., $\lambda$ = wavelength of X-rays).
3 The structure and bonding of a substance dictate its properties. Table 9.1 summarizes the structure, bonding and properties of the four typical solid structures (giant metallic, giant molecular, giant ionic and simple molecular).
4 The co-ordination number of an atom or ion is the number of its nearest neighbours.
5 Metals form either close-packed structures (co-ordination number = 12) or body-centred structures (co-ordination number = 8).
6 Electrons which are not held tightly between atoms in directed bonds, but can move freely from one atom to another are said to be delocalized.
7 In diamond, each carbon atom is joined by four tetrahedrally-spaced covalent bonds to four other atoms (co-ordination number = 4).
8 In graphite, carbon atoms are arranged hexagonally in flat, parallel layers. It is described as a two-dimensional sheet polymer or layer lattice.
9 Sodium chloride forms a simple cubic structure in which the co-ordination numbers of both $Na^+$ and $Cl^-$ ions are six.
10 The attachment of solvent molecules to the particles of a solute during the dissolving process is known as solvation. Very often the solvent is water and in this particular case the process is called hydration.

## Study questions

(a) Draw a three-dimensional model of the sodium chloride crystal.
(b) Name the constituent particles and explain their positions relative to each other.
(c) What holds the crystal together?
(d) Why does sodium chloride have a high melting point and a high boiling point?
(e) Why is sodium chloride soluble in water?
(f) Explain why sodium chloride will not conduct in the solid state, but is a good conductor when molten.

(a) What do you understand by the terms:
(i) atom, (ii) molecule, (iii) ion?
(b) Use one or more of these terms to describe the following structures.
(i) copper, (ii) solid carbon dioxide, (iii) graphite.
(c) Explain how the properties of copper and graphite are related to their structure and bonding.

Explain in terms of the structure and bonding in diamond and graphite why:
(a) diamond is more dense than graphite,
(b) diamond is an insulator, whereas graphite is a conductor,
(c) diamond has a larger heat of combustion than graphite, and
(d) diamond is used as an abrasive, whereas graphite is used as a lubricant.

4 Consider the following five types of crystalline solids:
A Metallic,
B Ionic,
C Giant molecular (macromolecular),
D Composed of monatomic molecules, and
E Composed of molecules containing a small number of atoms.
Select the letter (A–E) corresponding to the structure most likely to show the following properties.
(a) An element which conducts electricity and boils at 1 600°C to form gaseous monatomic atoms.
(b) A solid which melts at −250°C.
(c) A solid with a very high molar heat of vaporization which does not conduct when liquid.
(d) A hard, brittle solid which easily cleaves.
(e) A substance which boils at −50°C and decomposes at high temperatures.

5 Consider the following five substances in the solid state:
A sodium
B silicon
C tetrachloromethane
D argon
E potassium bromide
Select the letter (A–E) corresponding to the substance most likely to show the structure or property described in each of the following questions.
(a) A monatomic substance held together by Van der Waals' forces.
(b) A solid of low melting point composed of polyatomic molecules.
(c) A network solid of covalently bonded atoms.
(d) A non-conducting solid which melts to form a liquid which conducts electricity.
(e) A substance which exists as a liquid over a temperature range of only 3 K.
(f) A substance which is decomposed by an electric current in the liquid state.

6 Diamond is one of the hardest natural substances. Before the production of carborundum, powdered diamond was the most widely used abrasive. Carborundum (silicon carbide) is made by heating coke with sand (silicon(IV) oxide) at 2 500°C.
(a) What is an abrasive?
(b) What are the main uses for abrasives?
(c) Write an equation for the manufacture of carborundum from coke and sand.
(d) Why do you think carborundum has superseded diamond as the most widely used abrasive?
(e) 'Diamonds are a girl's best friend'! Why? Carborundum has not superseded diamonds in this context. Why not?
(f) It has been said that the discovery of carborundum enabled the industrial revolution to occur swiftly during the late nineteenth and early twentieth centuries. Why was this?

7 Yellow phosphorus contains simple molecules of formula $P_4$ whereas red phosphorus forms a giant molecular layer lattice.
(a) Draw diagrams to illustrate the structures of yellow and red phosphorus.
(b) Explain the bonding in these structures by drawing simple electronic (dot/cross) diagrams.
(c) State two differences in properties between these allotropes.
(d) How are these differences in properties between yellow and red phosphorus explained by their different structures?

8 Solid iodine has a dark purple lustrous appearance. In addition, solid iodine displays a small electrical conductivity. When liquid iodine chloride (ICl) is electrolysed iodine is liberated at the cathode.
Discuss and explain each of these observations, explaining whether they are consistent with the expected structure of the substances involved and the position of iodine in the periodic table.

9 Discuss and explain the following:
(a) The ionic nature of $MgCl_2$ is greater than that of $AlCl_3$, which in turn is greater than that of $SiCl_4$.
(b) Silicon(IV) oxide is a solid at room temperature which does not melt until 1 973 K whereas carbon dioxide (m.pt. 217 K) is a gas at room temperature.
(c) Both calcium oxide and sodium chloride have a simple cubic structure and similar interionic distances yet calcium oxide melts at 2 973 K whereas sodium chloride melts at 1 074 K.
(d) Glucose ($C_6H_{12}O_6$) is much more soluble in water than in benzene, but cyclohexane ($C_6H_{12}$) is much more soluble in benzene than in water.

10 When X-rays of wavelength 0.1537nm are directed towards a crystal of KCl, reinforcement of the X-rays first occurs when the angle between the direction of the rays and the cube face is 14°.
(a) Using the Bragg equation, calculate the distance between layers of ions in the crystal.
(b) At what angle would the second reinforcement of X-rays occur with the crystal in the same orientation?

(c) Figure 9.26 shows a unit cube for potassium chloride.
  (i) Use your result in part (a) to calculate the volume of the unit cube.
  (ii) How many $K^+$ $Cl^-$ 'ion pairs' does the unit cube contain? (Remember that ions at the corners of a cube are shared equally amongst eight cubes.)
  (iii) What is the volume occupied by one $K^+$ $Cl^-$ 'ion pair'?
(d) The density of KCl is 2.0g $cm^{-3}$ and its relative formula mass (formula weight) is 74.6. Using these values, calculate the total volume of one mole of $K^+Cl^-$.
(e) Using the volume of one mole of $K^+Cl^-$ 'ion pairs' (obtained in part (d)) and the volume of one $K^+Cl^-$ 'ion pair' (obtained in part (c) (iii)), calculate a value for the Avogadro constant, $L$.

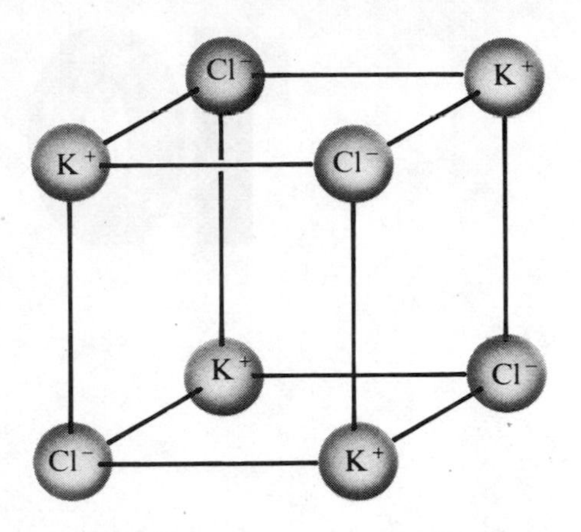


**Figure 9.26** A unit cube of potassium chloride.

11 (a) What is meant by the term co-ordination number?
(b) Name two metals with a close-packed structure and two metals with a body-centred cubic structure.
(c) What is the co-ordination number of atoms
  (i) in a close-packed structure,
  (ii) in a body-centred cubic structure?
(d) Suppose one particular metal can have either a close-packed or a body-centred cubic structure. In which of these two forms would you expect it to have:
  (i) the highest density,
  (ii) the highest melting point,
  (iii) the highest electrical conductivity,
  (iv) the greatest malleability?
  Give reasons for your choice in each case.

12 The following data apply to the compounds $XCl_x$ and $YCl_y$.

| | **Melting point** /°C | **Boiling point** /°C | **Solubility in water** /g per 100g | **Solubility in benzene** /g per 100g |
|---|---|---|---|---|
| $XCl_x$ | 801 | 1 443 | 37 | 0.063 |
| $YCl_y$ | −22.6 | 76.8 | 0.08 | miscible with benzene in all proportions. |

(a) What types of bond(s) are present in these two chlorides?
(b) Explain clearly how the bonding in each chloride leads to the great differences in volatility and solubility shown in the data above.

# 10 The Gaseous State

## 10.1 Evidence for moving particles

The Roman poet Lucretius (*c.* 98–55 B.C.)

Even the Greeks believed in particles. As early as 60 B.C., Lucretius, in his poem, *The Nature of the Universe* suggested that matter existed in the form of invisible particles.

Unfortunately, the Greeks rarely performed experiments so their theories remained nothing more than good ideas. They had, however, a vast range of everyday experience which supported their beliefs that matter was particulate. They knew, for instance, that a small amount of flavouring such as pepper or ginger could give a whole dish a strong, distinctive taste. This suggested that tiny particles in the pepper could spread throughout the stew to give it a particular flavour.

When dyes such as Tyrian purple were dissolved in water, a tiny amount of the dye could colour an enormous volume of water, suggesting that there must be many particles of the purple pigment in only a little solid and that the particles must be very small.

Use the idea of particles to explain the following:

- ○ It is possible to smell the perfume a girl is wearing from some distance.
- ○ Solid blocks of air freshener provide a pleasant smell in bathrooms and disappear after a time without leaving any liquid.

Particles of vapour from the scent and the air freshener which have a distinctive smell have *mixed* with air particles and *moved* to other parts of the room. This mixing and movement of matter which results from the kinetic energy of moving particles is called **diffusion**.

The idea that particles in gases and liquids are moving has been confirmed by many other experiments involving diffusion.

One other phenomenon which provides strong evidence for moving particles is **Brownian Motion**.

During the 1820's, Robert Brown, a botanist, was carrying out a study of pollen grains. At first, Brown believed that he could observe the pollen grains more effectively through his microscope if they were suspended in water, but to his annoyance he found that the pollen continually jittered around in the water in a random manner. As a result of Brown's observation, the random movement of solid particles suspended in a liquid or in a gas is called Brownian Motion.

- ○ What particles are present in water? How are these particles moving?
- ○ What caused the pollen grains to jitter about in Brown's experiment? Why do they move so randomly and haphazardly?

## 10.2 Examining the Brownian Motion of smoke particles in air

Although Brown's observation of pollen grains was the first recorded example of this phenomenon, it is possible to watch Brownian Motion more conveniently using smoke particles in air. Figure 10.1 shows a suitable arrangement which you could use for this experiment.

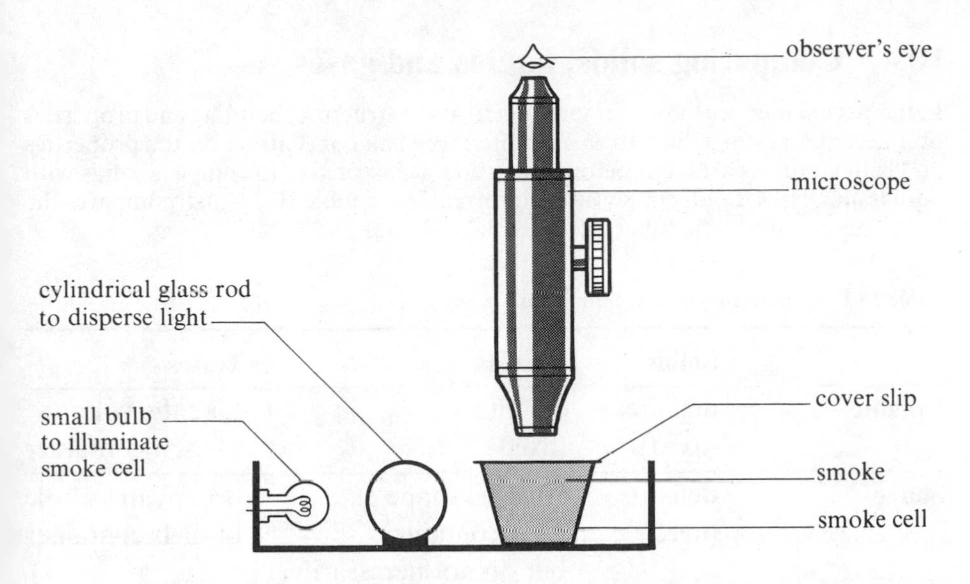


**Figure 10.1** An arrangement for examining the Brownian motion of smoke particles in air.

Use a teat pipette to inject some smoke from a smouldering piece of string into the small glass smoke cell and close the cell with a cover slip. When the illuminated cell is viewed through the microscope the smoke particles look like tiny jittering pin points of light. Why is this?

The smoke particles which you see through the microscope are under constant bombardment from even tinier and invisible molecules in the air. Each individual smoke particle is bombarded first on one side and then on another side and so it moves first this way and then that way in a random jittery motion. The movement of the smoke particles which you can see provides strong evidence for the movement of air molecules which, of course, you cannot see. What do you think will happen if the heat from the small illuminating lamp causes the temperature of the air and smoke in the cell to rise significantly?

## 10.3 The Kinetic Theory of Matter

Our ideas about the movement of particles in solids, liquids and gases are summarized in the **Kinetic Theory of Matter**. The word kinetic is derived from the Greek word *kineo* which means 'I move'. The main points in the Kinetic Theory can be summarized as follows:

1 All matter is composed of tiny, invisible particles.

2 Particles of different substances are different in size.

3 In solids, the particles are relatively close together and have smaller amounts of energy than the same particles in the liquid and gaseous states at higher temperatures. Consequently, solid particles cannot overcome the strong forces of attraction holding them together and they can only vibrate about fixed positions in the solid crystal. They have vibrational and rotational motion, but no translational motion.

4 In liquids, the particles are slightly further apart than in solids and possess larger amounts of energy. Thus, they are able to overcome the forces between each other to some extent and can move freely around each other whilst in close proximity. The liquid particles have vibrational, rotational and translational motion.

5 In gases, the particles are much more widely separated than those in solids and liquids and have much larger amounts of energy. The gas particles have sufficient energy to overcome the forces of attraction between each other almost completely and move rapidly, randomly and haphazardly into any space available. According to the Kinetic Theory, an ideal gas can be imagined to consist of a collection of point mass particles in random motion.

An increase in temperature causes an increase in the average kinetic energy of particles. The kinetic energy is manifested as vibrational, rotational and translational energy in gases and liquids and as vibrational and rotational energy in solids.

## 10.4 Comparing solids, liquids and gases

In the last chapter, we looked in some detail at the structure, bonding and properties of different types of solids. In this chapter, we shall concentrate on the properties and behaviour of gases, but before doing so it is instructive to compare solids with liquids and gases. Look closely at the information in table 10.1 which compares the bulk (macroscopic) characteristics of solids, liquids and gases.

**Table 10.1** Comparing solids, liquids and gases

| | **Solids** | **Liquids** | **Gases** |
|---|---|---|---|
| **Volume** | definite (fixed) | definite (fixed) | take the volume of their container |
| **Shape** | definite (fixed) | take the shape of their container, but do not necessarily occupy all of it | occupy the whole of their container |
| **Relative compressibility** | nil | almost nil | great |
| **Relative density** | great | great | small |

Notice the similarities between solids and liquids, but the considerable differences between solids and gases. Solids and liquids have fixed volumes, they have similar densities and are almost non-compressible.

The particles in solids and liquids are very close, but in gases they are more widely spaced with much empty space in between. The closeness of the particles in solids and liquids means that these materials have relatively large densities and negligible compressibilities; there are strong forces of attraction between the particles and so the material is held together as a fixed volume.

In contrast, the widely-spaced particles in a gas result in much lower densities and much larger compressibilities compared to solids and liquids. The forces between widely-spaced gas molecules are very weak so that the particles readily move away from each other and occupy the whole volume of their container.

In terms of their bulk characteristics, solids and liquids have certain features in common yet they are quite distinct. Liquids unlike solids are fluid, a property which they share with gases. However, the strong forces of attraction between the liquid particles holds them tightly together in a fixed volume even though they have sufficient energy to move freely around each other. It is this freedom of movement of the particles which enables a liquid to flow freely and smoothly on pouring and to take the shape of its container.

In a solid, particles are confined to fixed places in the lattice and it is possible to predict the positions of particles in the structure with some accuracy—the solid is said to have **long-range order**. In a liquid, particles are not confined to fixed positions; they can move freely around each other and it is difficult to predict the positions of particles around an initial reference molecule with any accuracy—the liquid is said to have **short-range order**.

The particles in a gas are much more widely spaced than those in a liquid or solid. The forces between such widely-spaced particles are so weak that the gas particles readily move away from each other and occupy the whole volume of their container.

## 10.5 Gay-Lussac's law and Avogadro's theory

During the later part of the eighteenth century, chemists were particularly interested in the reactions between different gases. In 1781, Henry Cavendish found that whenever he exploded hydrogen with oxygen, a given volume of hydrogen would always react with half of its own volume of oxygen. In 1805, Joseph Louis Gay-Lussac repeated this experiment and confirmed that two volumes of hydrogen always reacted with one volume of oxygen provided that the volumes of hydrogen and oxygen were measured under the same conditions of temperature and pressur

Struck by the simplicity of the 2:1 ratio, Gay-Lussac examined the volumes in which other gases reacted together. Provided that all the volumes in each experiment were measured at the same temperature and pressure, he found that:

1 volume of hydrogen + 1 volume of chlorine → 2 volumes of hydrogen chloride;

1 volume of nitrogen + 2 volumes of oxygen → 2 volumes of nitrogen dioxide;

1 volume of hydrogen chloride
+ 1 volume of ammonia → solid ammonium chloride.

In 1808, Gay-Lussac published his **law of combining volumes**, now known as **Gay-Lussac's law**: *when gases react, the volumes of the reacting gases and the volumes of any gaseous products are in the ratio of small whole numbers provided the volumes are measured at the same temperature and pressure.*

As soon as Gay-Lussac had announced his law of combining volumes, chemists began to look for an interpretation of it in terms of Dalton's Atomic Theory of Matter published in 1803.

Chemists realized that since $100cm^3$ of hydrogen just reacted with $100cm^3$ of chlorine, the number of hydrogen and chlorine molecules in these volumes must be simply related. Otherwise some molecules of hydrogen or chlorine would be left over after the reaction.

In 1811, the Italian scientist Amedeo Avogadro suggested an important hypothesis which later became known as **Avogadro's theory**: *equal volumes of all gases at the same temperature and pressure contain equal numbers of molecules.*

This suggestion by Avogadro explained beautifully the results leading to Gay-Lussac's Law.

For example, experiments show that:

1 volume of hydrogen + 1 volume of chlorine → 2 volumes of hydrogen chloride.

If we assume that one volume of hydrogen contains $n$ molecules, then by Avogadro's theory, one volume of chlorine contains $n$ molecules and two volumes of hydrogen chloride contain $2n$ molecules.

$\Rightarrow$ $n$ molecules of hydrogen
+ $n$ molecules of chlorine → $2n$ molecules of hydrogen chloride

$\Rightarrow$ $\frac{1}{2}$ molecule of hydrogen
+ $\frac{1}{2}$ molecule of chlorine → 1 molecule of hydrogen chloride

Notice that half of a molecule of hydrogen goes to form one molecule of hydrogen chloride. Now, since one molecule of hydrogen chloride must contain at least one hydrogen atom, half a molecule of hydrogen is at least one atom, so the simplest formula for one hydrogen molecule is $H_2$.

By the same argument, the simplest formula for the chlorine molecule is $Cl_2$.

This means that one atom of hydrogen and one atom of chlorine form one molecule of hydrogen chloride.

Therefore, the formula of hydrogen chloride is HCl and the last equation can be rewritten using formulae as:

$\frac{1}{2}$ molecule of $H_2$ + $\frac{1}{2}$ molecule of $Cl_2$ → 1 molecule of HCl.

Notice that by applying Avogadro's theory to the volumes of reacting gases it is possible to deduce the formulae of simple gases such as $H_2$, $Cl_2$ and HCl.

○ What empirical formula would have been deduced for hydrogen chloride if we had assumed that the formula of the hydrogen molecule was $H_4$?
○ Why can the formula of the hydrogen molecule *not* be $H_3$?
○ When hydrogen chloride reacts with metals only one series of salts can be formed and none of these salts contain hydrogen, i.e. there are no salts corresponding to $NaHSO_4$ from $H_2SO_4$. What does this suggest about the number of hydrogen atoms in one molecule of hydrogen chloride?

## 10.6 Obtaining relative molecular masses (molecular weights) using Avogadro's theory

A further consequence of Avogadro's theory was that it introduced a means of obtaining the relative masses of molecules in gases and vapours.

If, for example, $1 dm^3$ of a certain gas, X, is 20 times as heavy as $1 dm^3$ of hydrogen at the same temperature and pressure, it follows from Avogadro's theory that molecules of X must be 20 times as heavy as hydrogen molecules and therefore 40 times as heavy as hydrogen atoms.

We can summarize this reasoning in the following way:

By Avogadro's theory,

$$\frac{\text{mass of } x \text{ dm}^3 \text{ of X}}{\text{mass of } x \text{ dm}^3 \text{ of hydrogen}} = \frac{\text{mass of } n \text{ molecules of X}}{\text{mass of } n \text{ molecules of hydrogen}}$$

$$\therefore \frac{\text{mass of } x \text{ dm}^3 \text{ of X}}{\text{mass of } x \text{ dm}^3 \text{ of hydrogen}} = \frac{\text{mass of 1 molecule of X}}{\text{mass of 1 molecule of hydrogen}} \quad \text{(dividing by } n\text{)}$$

Now, since one hydrogen molecule contains two atoms, we can write:

$$\Rightarrow \frac{\text{mass of } x \text{ dm}^3 \text{ of X}}{\text{mass of } x \text{ dm}^3 \text{ of hydrogen}} = \frac{\text{mass of 1 molecule of X}}{\text{mass of 2 atoms of hydrogen}}$$

$$= \tfrac{1}{2} \times \frac{\text{mass of 1 molecule of X}}{\text{mass of 1 atom of hydrogen}}$$

$$\Rightarrow \frac{\text{mass of } x \text{ dm}^3 \text{ of X}}{\text{mass of } x \text{ dm}^3 \text{ of hydrogen}} = \tfrac{1}{2} \times \text{relative molecular mass of X}$$

if we assume that the relative atomic mass of H = 1.

Thus, by assigning a relative mass of one unit to the hydrogen atom, it became possible for chemists to determine relative molecular masses by simply comparing the mass of a given volume of a particular gas with the mass of the same volume of hydrogen. This profound, yet fundamentally simple realization led to increased understanding and experimentation in chemistry.

Having obtained relative molecular masses (molecular weights), chemists were able to find the volumes occupied by one mole of gas (table 10.2). The results of a large number of experiments show that one mole of all gases at 25°C and one atmosphere pressure occupies $24.4 \pm 0.1 dm^3$ (or $22.4 \pm 0.1 dm^3$ at s.t.p.).

**Table 10.2** Volumes occupied by 1 mole of various gases at 25°C and 1 atmosphere pressure.

| Gas | Mass of empty ($1 dm^3$) flask /g | Mass of flask + gas /g | Mass of $1 dm^3$ of gas /g | Relative molecular mass of gas | Volume of 1 mole of gas/$dm^3$ |
|---|---|---|---|---|---|
| $O_2$ | 161.45 | 162.76 | 1.31 | 32 | 24.5 |
| $N_2$ | 161.45 | 162.60 | 1.15 | 28 | 24.3 |
| CO | 161.45 | 162.60 | 1.15 | 28 | 24.3 |
| $CO_2$ | 161.45 | 163.26 | 1.81 | 44 | 24.3 |

This volume is known as the **Molar Volume** (Gram Molecular Volume). It is not surprising that one mole of all gases occupies the same volume at the same temperature and pressure, since they contain the same number of molecules.

Gay-Lussac's Law and Avogadro's Theory also enabled chemists to obtain the formulae of more complex gases than say $H_2$, $Cl_2$ and HCl.

The following example illustrates this.

In order to find the formula of a hydrocarbon, $10 cm^3$ of the gaseous hydrocarbon were mixed with $33 cm^3$ of oxygen which was in excess. The mixture was exploded and after cooling to room temperature, the residual volume of gas occupied $28cm^3$. On adding concentrated potassium hydroxide the volume decreased to $8cm^3$. This remaining gas is excess oxygen.

Volume of hydrocarbon reacting = $10cm^3$
Volume of oxygen reacting = $33 - 8 = 25cm^3$
Volume of carbon dioxide (absorbed by potassium hydroxide) = $28 - 8 = 20cm^3$

If we give the hydrocarbon a hypothetical formula of $C_xH_y$, we can write a general equation for combustion as:

$$C_xH_y + \left(x + \frac{y}{4}\right)O_2 \longrightarrow xCO_2 + \frac{y}{2}H_2O$$

$$1 \text{ mole} \quad \left(x + \frac{y}{4}\right) \text{moles} \quad x \text{ moles} \quad \frac{y}{2} \text{moles}$$

Applying Avogadro's theory:

$$\Rightarrow 1 \text{ volume } C_xH_y + \left(x + \frac{y}{4}\right) \text{volumes } O_2 \longrightarrow x \text{ volumes } CO_2 + \text{negligible volume } H_2O \ldots (1)$$

(liquid at room temperature)

$10cm^3 C_xH_y$ $\quad 25cm^3 O_2$ $\quad 20cm^3 CO_2$

$\Rightarrow 1cm^3 C_xH_y$ $\quad 2.5cm^3 O_2$ $\quad 2cm^3 CO_2$

$\Rightarrow 1 \text{ volume } C_xH_y + 2.5 \text{ volumes } O_2 \longrightarrow 2 \text{ volumes } CO_2 \ldots (2)$

By comparing equations (1) and (2)

$$x = 2 \text{ and } x + \frac{y}{4} = 2.5; \quad \therefore y = 2$$

∴ The molecular formula of the hydrocarbon is $C_2H_2$

By applying Gay-Lussac's Law and Avogadro's Theory, it also became possible to analyse gas mixtures.

A sample of air was analysed by explosion with excess hydrogen in order to determine its percentage by volume of oxygen. $25cm^3$ of air was mixed with $100cm^3$ of hydrogen and exploded. After cooling to room temperature, the final volume was $110cm^3$.

- ○ Write an equation to represent the reaction of hydrogen with oxygen. Put in state symbols and remember that the measurements are all made at room temperature.
- ○ What are the relative volumes of hydrogen and oxygen which react?
- ○ What is the decrease in volume on reaction?
- ○ What proportion of the decrease in volume is due to oxygen?
- ○ How many $cm^3$ of oxygen are present in the original air sample?
- ○ What is the percentage by volume of oxygen in the air sample?

Robert Boyle (1627-1691), one of the foremost scientists of the seventeenth century. Boyle carried out investigations into many areas of science including combustion, besides his research into the volume and pressure of gases.

## 10.7 Important Gas Laws and the Ideal Gas Equation

During the seventeenth, eighteenth and nineteenth centuries, a great deal of scientific research involved the physical and chemical properties of gases. Careful investigations were carried out in order to discover how the volume of a gas varied with changes in temperature and pressure.

As early as 1662, Robert Boyle had discovered that *the volume of a fixed mass of gas is inversely proportional to its pressure, provided the temperature remains constant.*

This is known as **Boyle's Law**, which can be expressed mathematically as

$$V \propto \frac{1}{p}$$

About a century later, in 1787, Charles showed that *the volume of a fixed mass of gas is directly proportional to its absolute temperature, provided the pressure remains constant*.

This is known as **Charles's law** which could be expressed mathematically as

$$V \propto T$$

By combining Boyle's Law and Charles's Law we get

$$V \propto \frac{T}{p}$$

$$\therefore pV \propto T$$

$$\text{and } \frac{pV}{T} = \text{constant}$$

$$\text{so } \frac{p_1V_1}{T_1} = \frac{p_2V_2}{T_2}$$

This last equation enables the volume of a gas to be obtained under any conditions of temperature and pressure (say $T_1$ and $p_1$), provided its volume ($V_2$) is known under some other conditions of temperature ($T_2$) and pressure ($p_2$).

What is the value of the constant in the expression $pV/T$ = constant? The value will, of course, depend on the amount of gas taken. Let us calculate the constant for one mole of gas. It is given the symbol $R$ and called **the gas constant.**

We know that one mole of gas occupies a volume of 22.4dm$^3$ at one atmosphere pressure and 273 K.

$$\therefore \text{ using } R = \frac{pV}{T} = \frac{1 \times 22.4}{273}$$

$$\Rightarrow \mathbf{R = 0.082 \text{ atm dm}^3 \text{ K}^{-1} \text{ mole}^{-1}}$$

Operating in S.I. units, the volume of one mole of gas is 0.0224m$^3$ at a pressure of 101 325 N m$^{-2}$ (i.e. 101 325 Pa*) and 273 K.

$$\therefore \text{ In S.I. units, } R = \frac{pV}{T} = \frac{101\,325 \times 0.0224}{273}$$

$$= 8.31 \text{ N m K}^{-1} \text{ mole}^{-1}$$

$$\Rightarrow \mathbf{R = 8.31 \text{ J K}^{-1} \text{ mole}^{-1}}$$

So for one mole of gas, $pV = RT$
and for $n$ moles of gas, $pV = nRT$ and this is known as the **Ideal Gas Equation**.

Consequently, a gas which obeys this equation exactly is called a 'Perfect' or an 'Ideal' gas. In practice, real gases obey the equation very closely at low pressure and high temperature (i.e. under conditions which make a gas most like a gas and least like a liquid, section 8.3(b)(ii).)

Suppose the air inside a room is at one atmosphere pressure and just suppose the sample of air is exactly one-fifth by volume of oxygen and four-fifths by volume of nitrogen.

- ○ What pressure is exerted by the oxygen?
- ○ What pressure is exerted by the nitrogen?

Now suppose you have two identical cylinders, joined by a tap, one containing pure oxygen at one atmosphere pressure, the other containing pure nitrogen at one atmosphere pressure. If all the oxygen in the first cylinder is forced through the tap into the second cylinder:

- ○ What is the total pressure in the second cylinder?
- ○ What is the pressure of nitrogen in the second cylinder?
- ○ What is the pressure of oxygen in the second cylinder?

In order to answer the last five questions you will have assumed **Dalton's law of partial pressures** which says that *the total pressure of a mixture of gases is equal to the sum of the pressures that each gas would exert if it occupied the space alone.*

* The internationally accepted unit of pressure is the pascal, Pa. One pascal is defined as a pressure of one newton per square metre, N m$^{-2}$

This can be expressed mathematically as

$$p_{\text{total}} = p_1 + p_2 + p_3 \ldots p_n$$

$p_{\text{total}}$ is the total pressure of the mixture, whilst $p_1$ is the partial pressure of gas 1, $p_2$ is the partial pressure of gas 2, etc.

## 10.8 Rationalizing the gas laws with the kinetic theory of gases

Using the simple model for gases as randomly moving small particles, it is possible to predict the laws which we have discussed in the last three sections. The kinetic theory of gases makes three fundamental assumptions:

1 The gas molecules have negligible volume.
2 The gas molecules exert no forces upon each other between collisions.
3 All collisions are perfectly elastic so there is no loss of energy during collisions.

Consider a cube of side, $l$ containing $n$ molecules each of mass, $m$. Suppose the velocity of a typical molecule is $c$ with component velocities of $u$, $v$ and $w$ along the co-ordinate axes (figure 10.2). (i.e. the component velocity of the molecule in the horizontal plane towards face A of the cube is $u$).

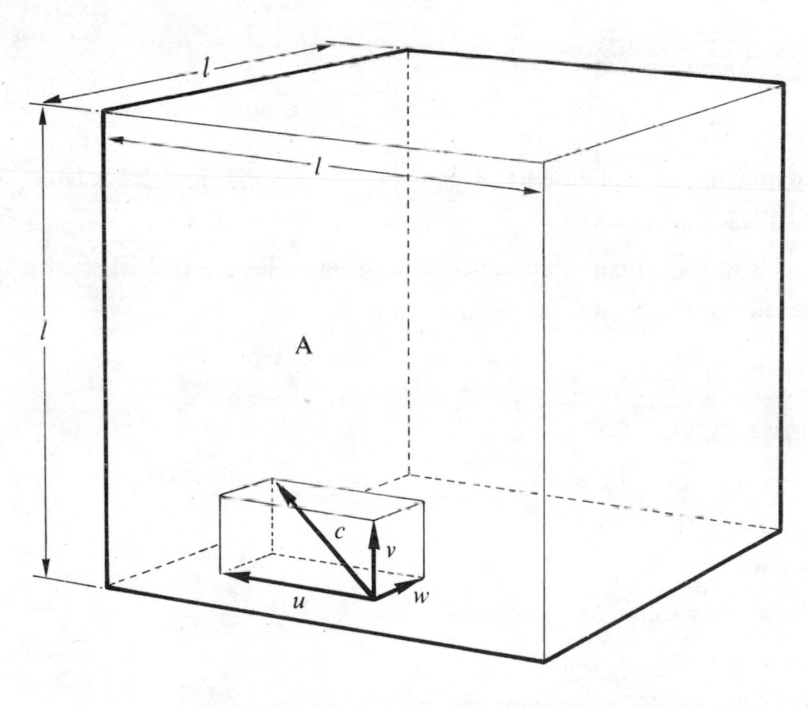


**Figure 10.2** Examining the velocity of a typical molecule in order to develop a simple model for the kinetic theory of gases.

$$\therefore \; c^2 = u^2 + v^2 + w^2$$

Now, momentum change of one molecule on rebounding from face A $= mu - (-mu) = 2mu$
and number of collisions of one molecule with A per second $= u/2l$
$\therefore$ change of momentum per second for this molecule at face A $= 2mu \,.\, u/2l = mu^2/l$

If we assume that the average value of $u^2$ for all $n$ molecules is $\overline{u^2}$ then,

$$\overline{u^2} = \frac{u_1^2 + u_2^2 + u_3^2 \ldots u_n^2}{n}$$

$\therefore$ Average change of momentum per second on face A for $n$ molecules $= n \,.\, \dfrac{m\overline{u^2}}{l}$

$\Rightarrow$ Force on A = change of momentum per second

$$= \frac{nm\overline{u^2}}{l}$$

$$\Rightarrow \text{pressure on face A} = \frac{\text{force}}{\text{area}} = \frac{nm\overline{u^2}}{l} \cdot \frac{1}{l^2} = \frac{nm\overline{u^2}}{l^3}$$

$$\therefore p = \frac{nm\overline{u^2}}{V} \text{ (since } l^3 = \text{vol, } V\text{)}$$

But, the pressure on each wall of the container is the same

$$\therefore \frac{nm\overline{u^2}}{V} = \frac{nm\overline{v^2}}{V} = \frac{nm\overline{w^2}}{V}$$

i.e. $\overline{u^2} = \overline{v^2} = \overline{w^2}$

But $\overline{c^2} = \overline{u^2} + \overline{v^2} + \overline{w^2} = 3\overline{u^2}$

Where $\overline{c^2}$ is the mean square velocity.

Now, since $\overline{u^2} = \dfrac{\overline{c^2}}{3}$

$$\Rightarrow p = \frac{nm}{V} \cdot \frac{\overline{c^2}}{3}$$

$$\therefore pV = \tfrac{1}{3} nm\overline{c^2}$$

We can now use the equation $pV = \frac{1}{3}nm\overline{c^2}$ to explain the various gas laws.

**The Ideal Gas Equation**

The kinetic energy of $n$ gas molecules of mass, $m = \frac{1}{2}m\overline{c^2}.n = \frac{1}{2}mn\overline{c^2}$

Now kinetic energy $\propto$ absolute temperature, $T$

i.e. $mn\overline{c^2} \propto T$

Since $pV = \frac{1}{3}mn\overline{c^2}$

$\Rightarrow pV = \text{constant} \times T$

and for 1 mole $pV = RT$

**Boyle's law**

We have already seen that:

$pV = \text{constant} \times T$

$\therefore pV = \text{constant}$ if $T$ is constant and mass is constant.

**Charles's law**

$pV = \text{constant} \times T$

$\therefore V \propto T$ if $p$ is constant and mass is constant.

**Avogadro's theory**

Consider gas A and gas B, then

$$p_A V_A = \tfrac{1}{3}m_A n_A \overline{c^2}_A \text{ and } p_B V_B = \tfrac{1}{3}m_B n_B \overline{c^2}_B.$$

For equal volumes of A and B at the same pressure,

$p_A = p_B$ and $V_A = V_B$

$\therefore p_A V_A = p_B V_B$

$\Rightarrow m_A n_A \overline{c^2}_A = m_B n_B \overline{c^2}_B$

If the gases are at the same temperature, then the kinetic energies of their particles are equal:

i.e. $\frac{1}{2} m_A \overline{c^2}_A = \frac{1}{2} m_B \overline{c^2}_B$

$\Rightarrow n_A = n_B$

i.e., equal volumes of gases at the same temperature and pressure contain equal numbers of molecules (Avogadro's theory).

## 10.9 Investigating the distribution of molecular speeds in gases—the Zartmann experiment

Clearly the molecules in a sample of gas will not all have the same velocity. At any particular moment, some will be almost stationary whilst others will be moving very rapidly and they will be moving in very different directions as a result of their haphazard motion. During the early 1930's, Zartmann and Ko developed an ingenious technique for determining the distribution of molecular speeds. Their apparatus is shown in figure 10.3.

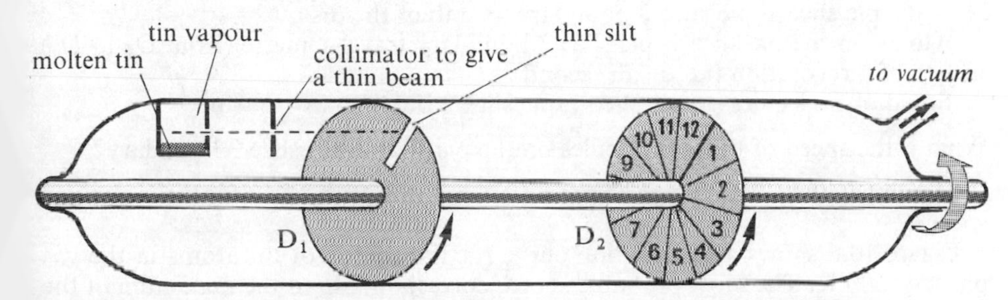


Figure 10.3 The Zartmann experiment.

$D_1$ and $D_2$ are discs which rotate rapidly and at the same speed on a common axle, within an evacuated container. In front of $D_1$ is an oven containing molten tin at a controlled temperature.

Tin vapour streams from a small hole in the oven and strikes $D_1$. Tin atoms pass through $D_1$ in small bursts each time the slit in $D_1$ is in line with the opening in the oven and the collimator. Effectively, $D_1$ acts as a 'starting gate' for repeated 'molecular races' between $D_1$ and $D_2$.

No more molecules can get past $D_1$ until the disc has completed a full rotation and the slit is opposite the opening in the oven again. Those molecules passing $D_1$ travel towards the second rotating disc, $D_2$ and spread throughout its pie slices.

$D_1$ lets through burst after burst of tin atoms and a layer of tin builds up on $D_2$. The faster molecules will hit the plate before the slower ones, since they all started from the slit simultaneously. Figure 10.4 shows how the deposited tin spreads over the pie slices. The distribution of the tin deposited on $D_2$ depends on the distribution of the speeds of the tin atoms. In order to find the distribution of tin on $D_2$, it is cut up into pie slices which are weighed to determine the amount of solid deposited on each slice.

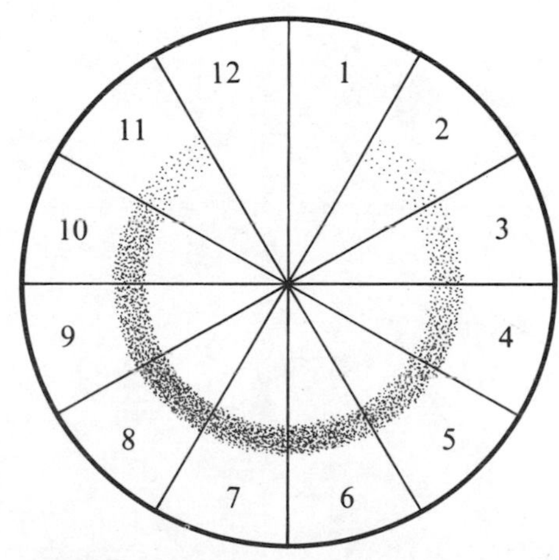


Figure 10.4 The distribution of tin over the pie slices of the target disc in the Zartmann experiment.

- ○ Why is no tin deposited on pie slice 1?
- ○ Why is no tin deposited on pie slice 12?
- ○ What happens to the distribution of tin amongst the pie slices if the speed of rotation of $D_1$ and $D_2$ is increased?
- ○ Why must the space between $D_1$ and $D_2$ be continuously evacuated?

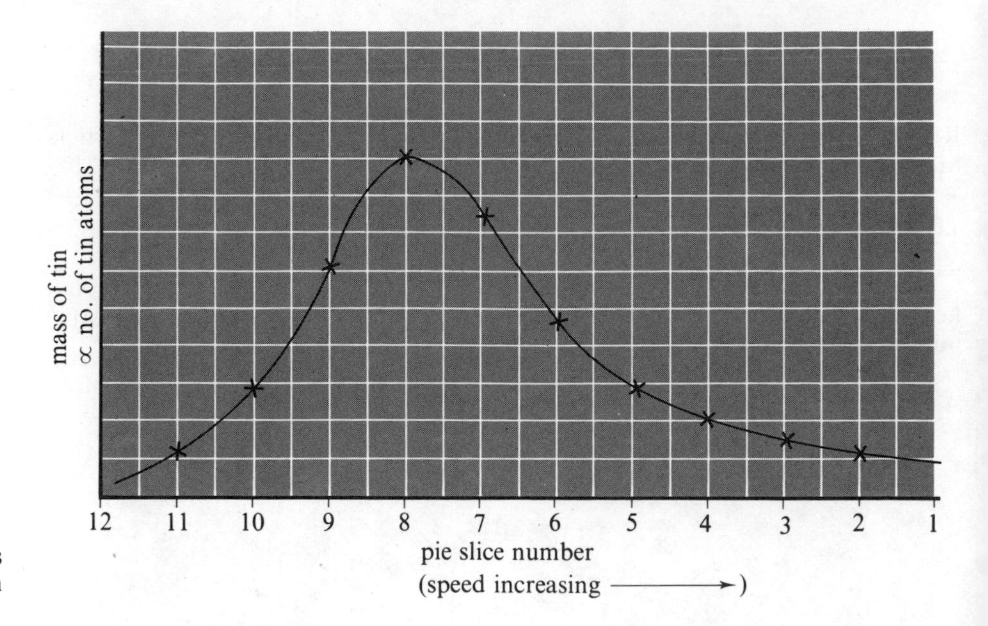


**Figure 10.5** A distribution curve for the mass of tin deposited on the different pie slices in the Zartmann experiment.

Figure 10.5 shows a distribution curve for the mass of tin deposited on each pie slice.

If we know the distance between $D_1$ and $D_2$ and the rate of rotation of the discs we can calculate the velocity of those molecules hitting a particular point on $D_2$.

For example:
suppose the distance from $D_1$ to $D_2$ is one metre and the discs rotate at 50 revolutions per second.
1 revolution of the discs takes $\frac{1}{50}$th second.
Now the pie slice 1–pie slice 2 boundary $= \frac{1}{12}$th of the disc.
$\therefore$ Atoms on the pie slice 1–pie slice 2 boundary travel 1 metre (from $D_1$ to $D_2$) in $\frac{1}{12}$th of a revolution (i.e. $\frac{1}{600}$th second).
$\therefore$ Speed of molecules on pie slice 1–pie slice 2 boundary $= 600 \mathrm{m\,s^{-1}}$.

What is the speed of those molecules on the pie slice 2–pie slice 3 boundary?

Proceeding in this way, it is possible to obtain a distribution curve for the molecular speeds.

Figure 10.6 shows a distribution curve for the speeds of tin atoms in the gas phase at 500 K. The most probable speed, corresponding to the maximum of the curve, (i.e. more molecules possess this speed than any other) is 250 metres per second. The root mean square speed at this temperature is about 325 metres per second.

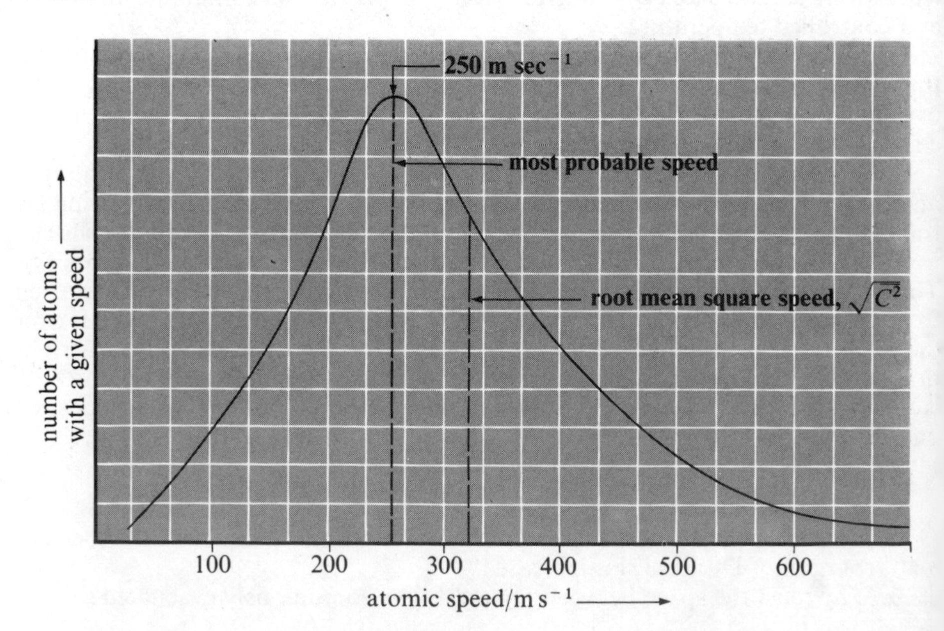


**Figure 10.6** A distribution curve for the speeds of tin atoms in the gas phase at 500 K.

What is the effect of increased temperature on the distribution of molecular and atomic speeds?

Figure 10.7 shows the distribution curves for the speeds of tin atoms at 500 K and 1 000 K. There are some important points to notice about the effect that increased temperature has had on the distribution and sizes of the atomic speeds.

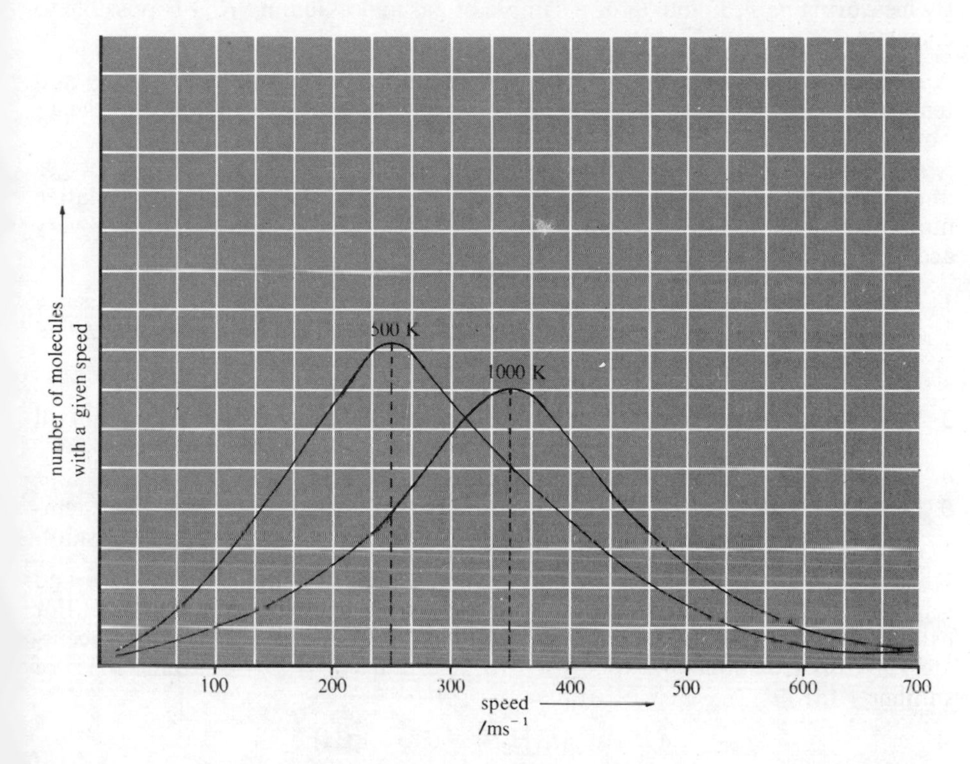


**Figure 10.7** Distribution curves for the speeds of tin atoms at 500 K and 1 000 K.

**1** The most probable speed increases as the temperature increases. At 500 K, the most probable speed for tin atoms is 250m s$^{-1}$, but at 1 000 K the most probable speed has increased to 360m s$^{-1}$.

**2** The distribution curve becomes flatter as the temperature rises. This means that there are fewer molecules with the most probable speed, but there is a greater proportion of high speed molecules.

It is important to remember that the graph in figure 10.7 is really a histogram. Thus, the area beneath the curve is proportional to the total number of molecules involved and since we are dealing with the *same total* number of molecules at 500 K and 1 000 K, the areas beneath the two curves must be identical.

## 10.10 Determining the relative molecular masses of gases and volatile liquids

Nowadays, the most accurate and convenient method of measuring the relative molecular mass of a substance is by mass spectrometry. Using this method, relative molecular masses can be obtained in a few minutes to an accuracy of one part in a million.

Originally, the relative molecular masses of gases and volatile substances were obtained by methods dependent on the gas laws discussed earlier in this chapter. Two of the more common methods employed were those designed by Regnault and Victor Meyer.

FINDING THE RELATIVE MOLECULAR MASS OF A GAS BY DIRECT WEIGHING—REGNAULT'S METHOD.

Using the Ideal Gas Equation $pV = nRT$, and $n = \dfrac{m}{M}$

where $m$ is the mass of gas whose relative molar mass is $M$,

We can write $pV = \frac{m}{M}.RT$

$\Rightarrow M = \frac{m}{pV}.RT$

By measuring $m$, $p$, $V$ and $T$ for a sample of gas and assuming $R$, it is possible to calculate $M$.

A container of known volume ($V$) is weighed full of the gas at pressure, $p$ and temperature, $T$. The same container is then weighed after evacuation in order to obtain the mass of gas inside ($m$).

Alternatively, Regnault's method can be done by comparing the mass of the gas under investigation with the mass of the same volume of a gas of known relative molecular mass. Although the method is limited to gaseous substances, it is very accurate provided the following precautions are taken.

1 The container must be large, so as to give an appreciable volume of gas, since gases are so light that any slight error in weighing would be greatly magnified if the mass of the substance weighed is very small.
2 The gas used must be perfectly pure; in particular perfectly dry.
3 The container should be filled and emptied several times to make sure that all the air or the previous gas has been driven out.
4 The container must be evacuated as completely as possible.
5 The container must be weighed, both filled and evacuated, at the same temperature and pressure so that the upthrust on the container remains constant.

Regnault's method has been used to determine relative molecular masses and relative atomic masses to an accuracy of more than one part in a thousand. For example, the relative molecular mass of $H_2S$ ($M_r(H_2S)$) could be obtained accurately by this technique and then used to determine the relative atomic mass of sulphur ($A_r(S)$).

$$A_r(S) = M_r(H_2S) - 2 \times A_r(H)$$

FINDING THE RELATIVE MOLECULAR MASS OF A VOLATILE LIQUID

This method, for use with volatile liquids, is a development of an earlier technique used by Victor Meyer. Although it gives only an approximate value for the relative molecular mass, it enables an accurate value to be obtained if the empirical formula of the substance can be determined.

Essentially, the principle of the experiment is the same as that described for gases earlier in this section.

$$pV = nRT$$

$$\text{and since } n = \frac{m}{M}$$

$$\therefore pV = \frac{m}{M}.RT$$

$$\Rightarrow M = \frac{m}{pV}.RT$$

Thus, the relative molecular mass ($M$) of the liquid can be calculated by finding the volume of vapour ($V$) formed from a known mass of liquid ($m$) at a measured temperature ($T$) and pressure ($p$) provided we assume the value of the gas constant ($R$).

Figure 10.8 shows a suitable apparatus to use. Draw a few $cm^3$ of air into the graduated syringe and fit the self-sealing rubber cap over the nozzle. Now pass steam through the outer jacket until the thermometer reading and the volume of air in the syringe become steady. Continue to pass steam through the jacket and record the temperature ($T$) and the volume of air in the syringe. Now fill the hypodermic syringe with about $1 cm^3$ of the liquid under investigation, ensuring that all air is expelled from the needle and the latter is completely filled with liquid. Weigh the hypodermic syringe and its contents and then put the needle through the self-sealing cap of the graduated syringe so that it is well clear of the nozzle. Inject about $0.2 cm^3$ of liquid into the larger syringe, withdraw the hypodermic syringe and immediately reweigh the latter and its contents.

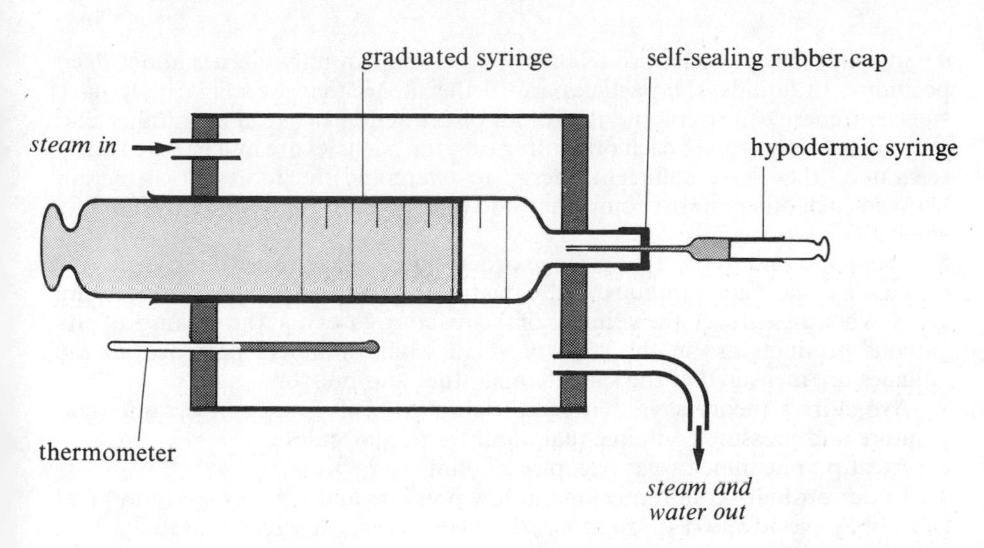


**Figure 10.8** Determination of the relative molecular mass of a volatile liquid.

The liquid in the graduated syringe will evaporate and the final volume of air plus vapour in the graduated syringe should be recorded when the volume becomes steady. Finally record the atmospheric pressure.

Here are some typical results.

Mass of liquid, X, vaporized, $m = 0.16\text{g}$

Initial volume of air in graduated syringe $= 10\text{cm}^3$

Final volume of air + vapour in syringe $= 56\text{cm}^3$

$\therefore$ Volume of X vaporized $= 46\text{cm}^3$

$= 46 \times 10^{-6}\text{m}^3$

Temperature of X $= 100°\text{C} = 373\ \text{K}$

Pressure of X $= 1\text{atm} = 101\,325\ \text{N m}^{-2}$

Gas constant, $R = 8.31\ \text{J K}^{-1}\ \text{mole}^{-1}$

Using $M = \dfrac{m}{pV}.RT$

$$M = \frac{0.16}{101\,325 \times 46 \times 10^{-6}} \times 8.31 \times 373 = 106.4\ \text{g mole}^{-1}$$

i.e. Relative molecular mass of X $\approx 106$

- ○ Why is this method unsuitable for liquids which boil above 80°C?
- ○ Why should the hypodermic syringe be handled as little as possible between weighings?
- ○ What are the main sources of error in this experiment?
- ○ The substance X, for which the typical results were given earlier, has the percentage composition by mass of 22.0% C, 4.6% H and 73.4% Br.
  (i) Calculate the empirical formula of X.
  (ii) What is the molecular formula of X?
  (iii) What is the accurate relative molecular mass of X?

## Summary

1 Diffusion is the mixing and movement of matter which results from the kinetic energy of moving particles.
2 Brownian motion is the random movement of particles of solid suspended in a liquid or in a gas.
3 All matter is composed of tiny, invisible particles.

4 In solids, the particles are relatively close and can only vibrate about fixed positions. In liquids, the particles are further apart than in solids; they have sufficient energy to overcome the forces of attraction between each other and they move freely around each other. In gases, the particles are much more widely separated; they have sufficient energy to overcome the forces of attraction between each other almost completely and move rapidly and randomly into any space available.
5 Solids are said to have long-range order, liquids have short-range order.
6 Gay-Lussac's law summarizes the relationship between volumes of reacting gases: when gases react, the volumes of the reacting gases and the volumes of any gaseous products are in the ratio of small whole numbers provided all the volumes are measured at the same temperature and pressure.
7 Avogadro's theory says that equal volumes of all gases at the same temperature and pressure contain equal numbers of molecules.
8 At s.t.p., one mole of gas occupies 22.4dm³.
9 Under ordinary conditions (and at low pressure and high temperature) real gases obey the Ideal Gas Equation, $pV = nRT$, very closely.

## Study questions

1 (a) Draw sketch graphs of (i) $p$ against $V$, (ii) $p$ against $1/V$, (iii) $pV$ against $p$, (iv) $pV$ against $V$, for a constant number of moles of a perfect gas at constant temperature.
(b) Plot on separate sketch graphs, (i) $V$ against $T$ (°C), (ii) $V$ against $T$ (K), for a constant number of moles of a perfect gas at constant pressure.
(c) Plot $pV$ against $T$ and $pV/T$ against $p$ for a constant number of moles of ideal gas.
(d) Draw a sketch graph to show how the pressure of a constant mass of ideal gas will vary as the temperature rises from absolute zero in a container of constant volume. How will the graph change if the gas tends to dissociate ($X_2 \rightarrow 2X$) as temperature increases? (For each graph in sections (a), (b), (c) and (d), label the axes and show the zero for each axis.)

2 A balloon can hold 1 000cm³ of air before bursting. The balloon contains 975cm³ of air at 5°C. Will it burst when it is taken into a house at 25°C? Assume that the pressure of the gas in the balloon remains constant.

3 What would you expect to be the effect of a change in
(i) pressure, (ii) volume, (iii) temperature,
on the rate of diffusion of a particular gas? Explain your answer in terms of the kinetic theory.

4 (a) A mixture of two gases in a container exerts a pressure of 800mm Hg and occupies a volume of 400cm³. If one of these gases (A) occupies a volume of 300cm³ under the same conditions of temperature and pressure, what pressure does the other gas (B) exert in the mixture?
(b) When gaseous argon is allowed to diffuse, it separates into a lighter and heavier fraction. What does this suggest about the nature of argon?

5 (a) Describe, and explain in terms of molecular theory, what happens when
(i) a small quantity of a volatile liquid is introduced into the space above mercury in a barometer.
(ii) a solute in a solvent is shaken with a second solvent, which is immiscible with the first solvent.
(b) 0.50g of a volatile liquid was introduced into a globe of 1 000cm³ capacity. The globe was heated to 91°C so that all the liquid vaporized. Under these conditions the vapour exerted a pressure of 190mm mercury. What is the relative molecular mass of the liquid?

6 An industrial chemist who works for a firm which manufactures ethyne (acetylene) for use in oxyacetylene welding discovers a very cheap way of producing ethyne. Unfortunately his ethyne is contaminated with ethene (ethylene). Unless the mixture contains at least 50% by volume of ethyne, it is useless. In order to determine the relative proportions of the two in the mixture, he exploded 10cm³ of the mixture with 30cm³ of oxygen. After absorbing the residual $CO_2$ in KOH, the uncombined $O_2$ occupied 2cm³. What was the composition of the mixture by volume?
(*Hint*: let vol. of one gas be $x$ cm³. Write an equation for the combustion of that gas and hence work out in terms of $x$ how many cm³ of $O_2$ it reacts with. Repeat this with the other gas. Find total volume $O_2$ used in terms of $x$. Find actual vol. of $O_2$ used from data, hence find $x$.)

7 The extent of combustion of petrol vapour in an internal combustion engine can be shown in terms of the percentage of $CO_2$ in the exhaust gases. $60cm^3$ of exhaust gas (assumed to contain only CO, $CO_2$ and $N_2$) were mixed with $20cm^3$ of $O_2$ (excess) and exploded. On cooling, there had been a decrease to $70cm^3$. On adding KOH, the volume decreased to $35cm^3$. What is the composition by volume of the mixture and hence what is the percentage by volume of $CO_2$ in the exhaust fumes?

8 On decomposition with electric sparks, two volumes of a gaseous compound containing only phosphorus and hydrogen gave a deposit of phosphorus (solid) and three volumes of hydrogen at the same temperature and pressure.
(a) Apply Avogadro's theory and deduce the number of atoms of hydrogen in one molecule of the phosphorus hydride. (Assume hydrogen is diatomic.)
(b) Write an empirical formula for the phosphorus hydride in as much detail as you are able.
(c) What do you think its molecular formula is?
(d) What measurement would you make to try to confirm this?
(e) Predict two physical and two chemical properties of this compound via the periodic table.

9 It is required to find the composition by volume of a sample of Town gas which contains only hydrogen, carbon monoxide and nitrogen. $40cm^3$ of the Town gas were carefully exploded with $40cm^3$ of oxygen (known to be in excess), so that only hydrogen and carbon monoxide would react with oxygen. On cooling to room temperature, the volume was $51cm^3$ and on adding concentrated KOH, the volume decreased further to $41cm^3$.
(a) Write equations (including state symbols) for the reactions which occur on explosion. (Remember that water will be liquid, as the measurements are made at room temperature.)
(b) What are the relative volumes of gaseous reactants and products in these reactions?
(c) What volume of $CO_2$ is produced?
(d) What is the volume of CO in the original $40cm^3$ of Town gas?
(e) What is the total decrease in volume as a result of the explosion?
(f) What decrease in volume is caused by
(i) CO (ii) $H_2$ on explosion?
(g) What are the volumes of
(i) $H_2$ (ii) $N_2$ in the original $40cm^3$ of Town gas?

10 $5.2cm^3$ of a gaseous hydrocarbon were exploded with an excess of oxygen. After cooling to the original room temperature and pressure, a contraction in volume of $7.8cm^3$ was observed. A further contraction of $10.4cm^3$ was noted after treatment with concentrated KOH.
What is the molecular formula of the hydrocarbon?

11 When $20cm^3$ of a gaseous hydrocarbon, A, are exploded with $150cm^3$ of oxygen, the residual gases occupied $110cm^3$. After shaking the residual gases with aqueous sodium hydroxide, the final volume was $30cm^3$. (All volumes were measured at the same temperature and pressure.)
(a) What is the molecular formula of A?
(b) Write six possible structural formulae for A and name each of these structures.

# 11 Energy Changes and Bonding

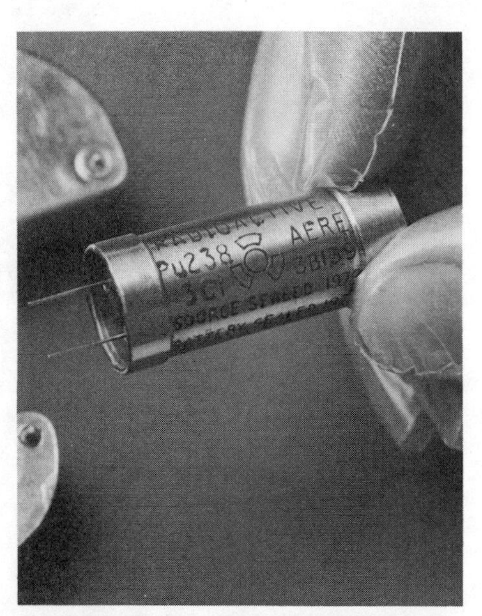


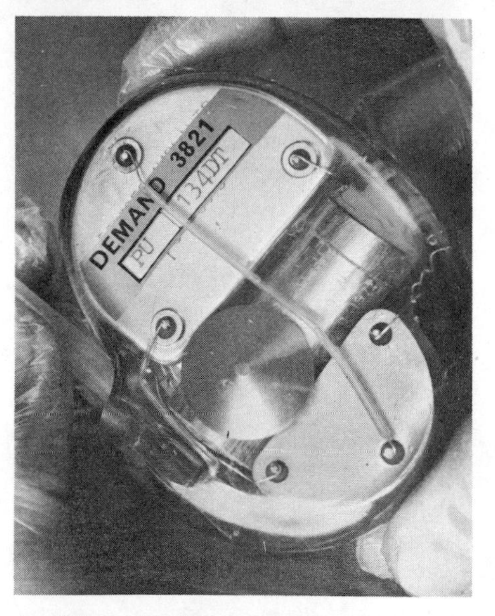


A miniature nuclear battery for a heart pacemaker, (*a*) before construction, (*b*) after construction. The nuclear battery (35mm long × 15mm across) uses heat from the radioactive decay of plutonium-238 to generate electricity in a miniature thermopile. The encapsulation of the plutonium ensures that there is no hazard from radiation. The batteries are designed to last 20 years.

## 11.1 Energy and energy changes

Energy is the most precious commodity we have. Without it there could be no life, no warmth, no movement. Energy gives us the power to do work, and in every country people's living standards are closely related to the availability of energy.

From the earliest times, man worshipped the sun. This is not surprising, for through the process of photosynthesis, the sun provides us with most of our food and, over millions of years, it has created our supplies of the fossil fuels (coal, oil and natural gas). More than ninety per cent of our present energy needs are supplied by these fossil fuels—valuable reserves of energy that cannot last forever.

For several years, Britain and the rest of Western Europe has been dependent on the Middle East for its increasing demands for oil. After the outbreak of the Arab/Israeli war in 1973, the price of oil rose alarmingly and created serious inflation in the oil-importing countries such as Britain. Recently, however, the prospects for Europe have brightened with the discovery of first natural gas, and then vast oilfields, below the North Sea. During the 1980's, Britain hopes, not only to supply its own needs for oil and gas, but to have a surplus available for export.

The transfer of energy to or from chemicals plays a crucial part in chemical processes in industry and in living things. Consequently, the study of these energy changes are as important to us as the study of the changes in the materials themselves.

Our present-day living conditions rely heavily on the ready availability of energy in its various forms. Chemical energy is converted to heat energy when fuels such as gas, oil and coal are burnt in our homes and in industry. Just think of the vast quantities of chemical energy converted to mechanical energy each day from the petrol burnt in our vehicles!

Within our own bodies, energy changes are vital. Foods such as fats and carbohydrates are important biological fuels. During metabolism, the chemical energy in these foods can be converted to heat energy to keep us warm, to mechanical energy in our muscles and to electrical energy in the signals within our nerve fibres. Chemical energy is also converted to electrical energy when the materials in cells and batteries are used to generate electricity.

All these important processes involve energy changes. This in itself would be a sufficient answer to the question 'Why study energy changes?', but quite apart from their importance to industry and society the energy changes in chemical reactions, as we shall see, can lead to a better understanding of the fundamental chemistry involved.

## 11.2 The ideas and language of thermochemistry

Very often chemical changes are accompanied by changes in the **heat conten** **(enthalpy, *H*)** of the materials which are reacting and the change in the heat conten is shown by a change in temperature. Indeed, the change in temperature whe substances react often provides evidence that a chemical change has taken place.

When an exothermic reaction occurs, heat is given out and the temperature o the products rises above room temperature. Eventually, the temperature of th products falls to room temperature again as the heat produced is lost to the sur roundings (figure 11.1(*a*)). Thus, the heat content (enthalpy) of the products ($H_2$) i

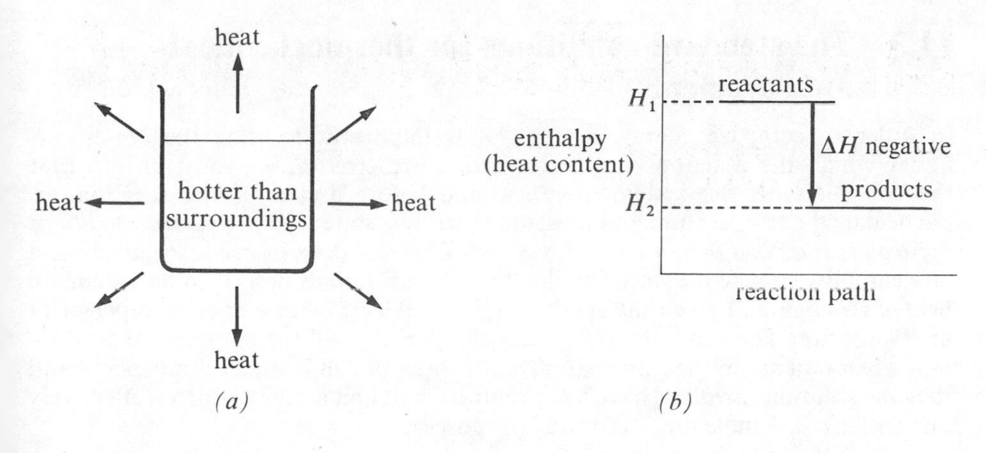


**Figure 11.1** In an exothermic reaction, heat is lost to the surroundings and $\Delta H$ is negative.

less than that of the reactants ($H_1$). Since the materials have *lost* enthalpy, we can see that the enthalpy change for the reaction, $\Delta H$, (sometimes called the **heat of reaction**) is negative (figure 11.1 (*b*)).

For example, when magnesium reacts with oxygen, heat is evolved.

$$2Mg(s) + O_2(g) \longrightarrow 2MgO(s) \qquad \Delta H = -1204 \text{ kJ}$$

Chemical energy in the magnesium and in the oxygen is partly transferred to chemical energy in the magnesium oxide and partly evolved as heat. Thus, the magnesium oxide has less energy than the starting materials, magnesium and oxygen. The value of $\Delta H$, the heat of reaction, relates to the amounts shown in the equation—2 moles of Mg atoms, 1 mole of $O_2$ molecules and 2 moles of MgO (figure 11.2).

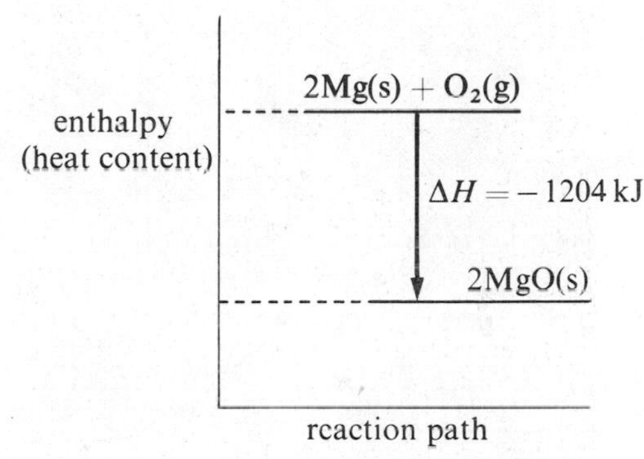


**Figure 11.2** An energy level diagram for the exothermic reaction of magnesium with oxygen.

When an endothermic reaction occurs, the heat required for the reaction is taken from the reacting materials and the temperature of the products falls below the initial temperature. Eventually, the temperature of the products rises to room temperature again as heat is *absorbed* from the surroundings. In this case, the heat content of the products is greater than that of the reactants and the enthalpy change (heat of reaction), $\Delta H$, is positive (figure 11.3).

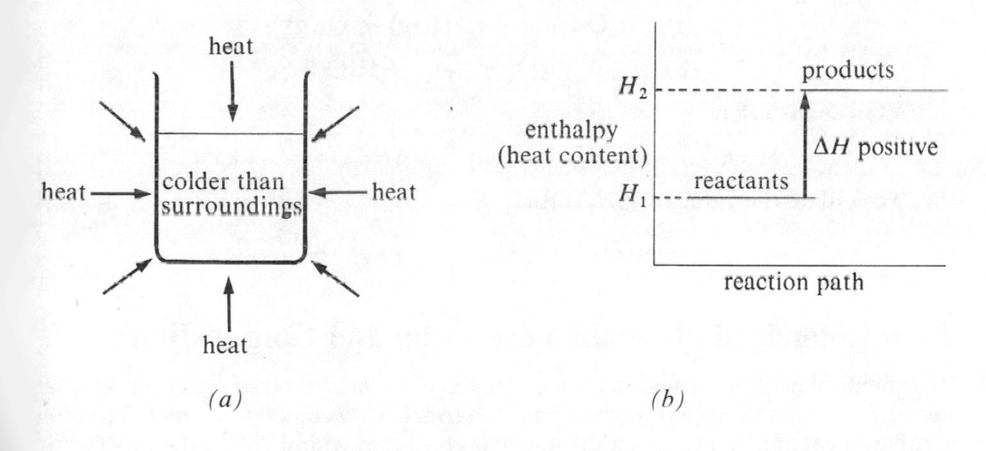


**Figure 11.3** In an endothermic reaction, heat is gained from the surroundings and $\Delta H$ is positive.

We can summarize these ideas as:

$$\text{Enthalpy change} = \text{heat of reaction}$$

$$\Delta H = H_2 - H_1$$

Since the enthalpy change manifests itself as heat, the term 'heat of reaction' is often used in place of 'enthalpy change of reaction'.

Remember that $\Delta H$ refers only to the energy change *for the reacting materials.* The surroundings will necessarily gain whatever heat the reacting materials lose and vice versa. Thus, the *total amount* of energy is unchanged during a chemical reaction. *Energy may be exchanged between the materials and the surroundings but the total amount of energy of the materials and the surroundings remains constant.* This important concept is known as the **Conservation of Energy**.

## 11.3 The standard conditions for thermochemical measurements

In order to compare energy changes, it is important to state the conditions under which the reaction was performed. Furthermore, we must ensure that the conditions of the system are the same before and after the reaction. In particular, the temperature and pressure should be stated. *The standard conditions for temperature and pressure are 298 K (25°C) and 1 atmosphere* respectively and any enthalpy change measured under these conditions is described as a **standard heat of reaction** and given the symbol $\Delta H^{\ominus}$ or $\Delta H^{\ominus}_{298}$ with a special superscript and subscript. The symbol, $\Delta H^{\ominus}_{298}$, also implies that all the substances involved in the reaction are in their normal physical states at 298 K and 1 atmosphere and that any solutions involved have a concentration of unit activity which is effectively one molar (i.e. 1 mole $dm^{-3}$) for our purposes.

Thus, $\Delta H^{\ominus}_{298}$ for the reaction

$$2H_2(g) + O_2(g) \longrightarrow 2H_2O(1)$$

must relate to gaseous hydrogen, gaseous oxygen and liquid water (not steam).

In the case of elements which exist as different allotropes and compounds which exist as different polymorphs, the most stable form at 298 K and 1 atm, is chosen as the standard. Consequently, $\Delta H^{\ominus}_{298}$ values for reactions involving carbon should relate to the allotrope graphite rather than diamond.

We can therefore define the standard heat of a reaction as: *the amount of heat absorbed or evolved when the molar quantities of reactants as stated in the equation react together under standard conditions.*

i.e., at a pressure of 1 atmosphere,
at a temperature of 298 K,
with substances in their physical states normal under these conditions
and solutions having unit activity (effectively 1 M).

Consider the reaction

$$2H_2(g) + O_2(g) \longrightarrow 2H_2O(l) \qquad \Delta H^{\ominus}_{298} = -575 \text{ kJ}$$

- ○ Is the reaction endothermic or exothermic?
- ○ Which has the greater enthalpy, the products or the reactants?
- ○ What is the value of $\Delta H^{\ominus}_{298}$ for

$$\text{(i) } 2H_2O(l) \longrightarrow 2H_2(g) + O_2(g)$$

$$\text{(ii) } H_2(g) + \tfrac{1}{2}O_2(g) \longrightarrow H_2O(l)$$

- ○ Given also that

$$H_2O(l) \longrightarrow H_2O(g) \qquad \Delta H^{\ominus}_{298} = +44 \text{ kJ},$$

calculate the value of $\Delta H_{298}$ for

$$2H_2(g) + O_2(g) \longrightarrow 2H_2O(g)$$

## 11.4 Standard Heats of Formation and Combustion

*The* **standard heat of formation** *of a substance is the heat evolved or absorbed when one mole of the substance is formed from its elements under standard conditions.* The standard heat of formation of a substance is given the symbol $\Delta H_f^{\ominus}$; the superscript $^{\ominus}$ indicates standard conditions and the subscript $_f$ refers to the formation reaction.

Thus, the statement $\Delta H_f^{\ominus}(MgO(s)) = -602 \text{ kJ mole}^{-1}$ relates to the formation of 1 mole of magnesium oxide from 1 mole of Mg atoms and $\frac{1}{2}$ mole of $O_2$ molecules, i.e.

$$Mg(s) + \tfrac{1}{2}O_2(g) \longrightarrow MgO(s)$$

One important consequence of the definition of standard heat of formation is that the heat of formation of an element in its physical state normal under standard conditions is zero since no heat change is involved when an element is formed from itself.

$$Cu(s) \longrightarrow Cu(s) \qquad \Delta H_f^{\ominus}(Cu(s)) = 0$$

$$H_2(g) \longrightarrow H_2(g) \qquad \Delta H_f^{\ominus}(H_2(g)) = 0$$

Obviously, $\Delta H^{\ominus}_{298}$ for the process $H_2(g) \rightarrow H_2(g)$ is zero, but this is not so for the process $H_2(g) \rightarrow 2H(g)$ which involves atomization, i.e. the conversion of $H_2$ molecules to single H atoms. In fact, the **standard heat of atomization** of an element is defined as *the enthalpy change when one mole of gaseous atoms is formed from the element under standard conditions.* Thus, the standard heat of atomization of hydrogen ($\Delta H^{\ominus}_{at}$ ($H_2(g)$)) refers to the process:

$$\tfrac{1}{2}H_2(g) \longrightarrow H(g) \qquad \Delta H^{\ominus}_{at}(H_2(g)) = +218 \text{ kJ mole}^{-1}$$

A wide range of experimental techniques is available for determining the atomization energies of elements. One technique widely used with gaseous diatomic elements is to find the minimum frequency of radiation ($\nu$) required to dissociate the gaseous molecules into atoms.
For example:

$$I_2(g) \xrightarrow{E = h\nu} I(g) + I(g) \qquad \Delta H = h\nu$$

Since the energy ($E$) of the radiation is given by $h\nu$, where $h$ is Planck's Constant:

$$\Rightarrow \Delta H^{\ominus}_{at}(I_2(g)) = \frac{\Delta H}{2} = \frac{h\nu}{2}$$

The heats of atomization of solid and liquid elements are more difficult to measure. For many elements, they can be obtained using specific heat capacities and molar heats of fusion and vaporization.

In discussing the heat changes in chemical reactions, it is useful to visualize each compound as having a definite heat content or enthalpy. For convenience, all elements are assigned a heat content of zero under standard conditions and the standard heat of formation of a compound then provides a measure of the heat content of the compound relative to its constituent elements. Remember, however, that these enthalpy values are *only* relative values. They give no information about the absolute energy content of a substance which will depend on the potential energy inherent in the electrical and nuclear interactions of the constituent particles plus the kinetic energy possessed by the atoms and sub-atomic particles.

The standard heat of formation of water ($\Delta H^{\ominus}_f$ ($H_2O(l)$)) = $-286$ kJ mole$^{-1}$) tells us that, under standard conditions, water has a lower energy content than the hydrogen and oxygen from which it is formed (figure 11.4). On an atomic scale, we can imagine that some of the potential energy in the molecules of $H_2$ and $O_2$ has been converted to kinetic energy in the molecules of $H_2O$ and this has resulted in a temperature rise. We can compare this with the conversion of potential energy to kinetic energy and heat when a stone rolls down a hill.

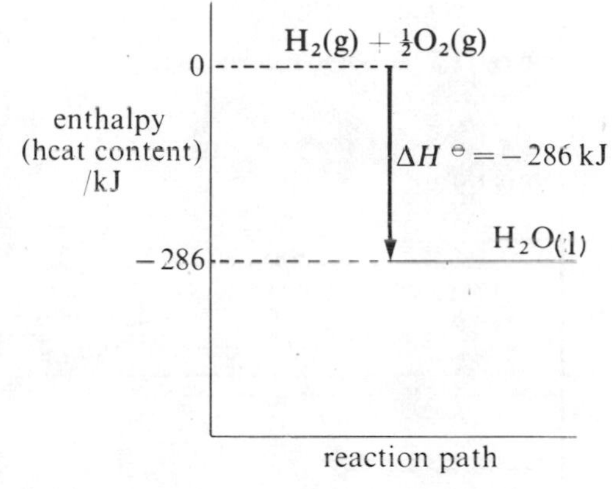


**Figure 11.4** An energy level diagram for the standard heat of formation of water.

One other enthalpy change of considerable importance is the standard heat of combustion which is particularly relevant to the study of fuels and foods.

The **standard heat of combustion** of a substance, $\Delta H^{\ominus}_c$, is defined as *the enthalpy change when one mole of the substance is completely burnt in oxygen under standard conditions.* For example, the standard heat of combustion of ethanol ($\Delta H^{\ominus}_c$ ($C_2H_5OH(l)$)) is $-1\,368$ kJ mole$^{-1}$.

i.e. $C_2H_5OH(l) + 3\tfrac{1}{2}O_2(g) \longrightarrow 2CO_2(g) + 3H_2O(l) \qquad \Delta H^{\ominus} = -1\,368$ kJ

- ○ Why is it necessary to say 'completely burnt in oxygen' in the definition of $\Delta H^{\ominus}_c$?
- ○ Draw the heats of combustion of both graphite and diamond on the same energy level diagram.

$$\Delta H^{\ominus}_c \text{ (C (graphite))} = -393 \text{ kJ mole}^{-1}$$
$$\Delta H^{\ominus}_c \text{ (C (diamond))} = -395 \text{ kJ mole}^{-1}$$

- ○ Which allotrope has the larger enthalpy (energy content)?
- ○ Which allotrope is the more stable?
- ○ What is the enthalpy change for the process

$$\text{C (graphite)} \longrightarrow \text{C (diamond)?}$$

Study question 1 (page 163) shows a very simple (one might say crude) apparatus in which the heat of combustion of a liquid fuel could be determined. The apparatus in figure 11.5 shows an instrument, known as a bomb calorimeter, used for accurate thermochemical determinations. Scientists use equipment such as this to determine the energy values of foods and fuels. The apparatus is specially designed to avoid

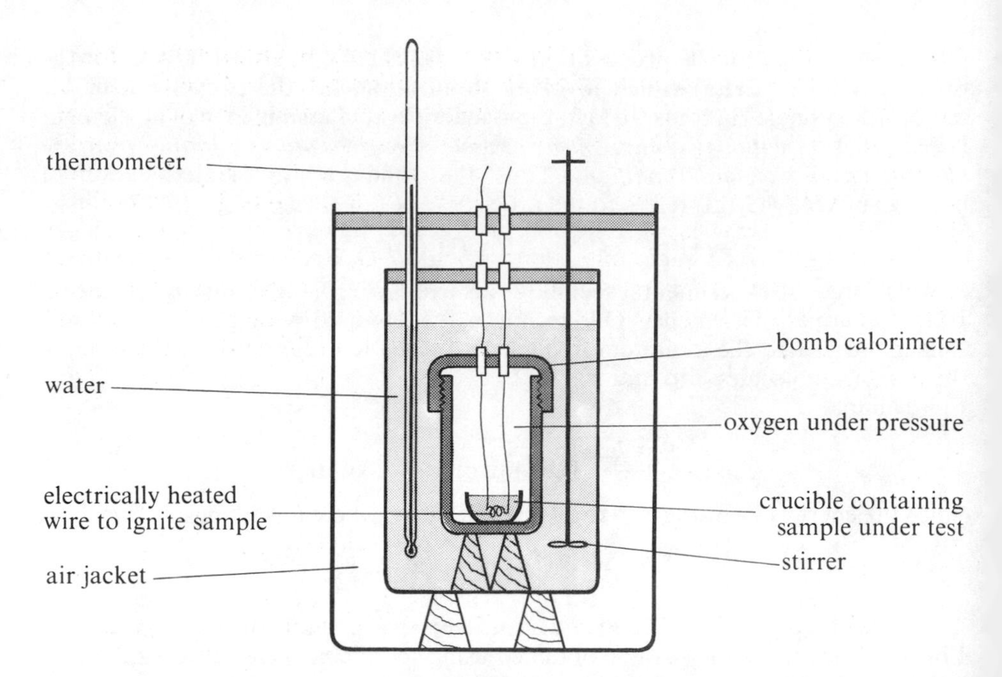


**Figure 11.5** A bomb calorimeter.

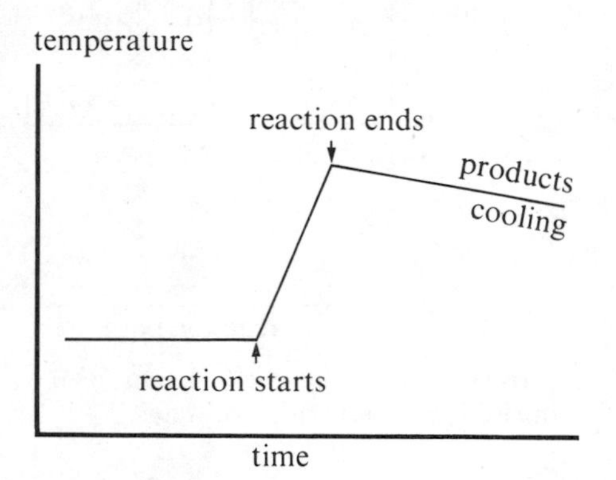


**Figure 11.6** A graph of temperature against time for a reaction in a bomb calorimeter.

heat losses by completely surrounding the 'bomb' with water. Furthermore, heat losses can be eliminated altogether if the thermochemical investigation is coupled with an electrical calibration. First of all, the chemical reaction is carried out in the calorimeter and the temperature is plotted against time before, during and after the reaction (figure 11.6). The experiment is now repeated, but this time an electrical heating coil replaces the reactants. The current in the coil is carefully adjusted so as to give a temperature/time curve identical to that obtained in the chemical reaction. By recording the current during the time of this electrical calibration, it is possible to calculate the electrical energy supplied with great accuracy. This electrical energy is exactly the same as the energy change in the chemical reaction and since it includes both the heat absorbed by the system and the heat lost from the system, it eliminates the need for a heat loss correction.

The practical significance of heats of combustion is clear enough. Fuel engineers and dieticians call them energy values. The prices of fuels are closely related to their energy values and energy-providing foods, such as sugar, are classified by the amount of heat they liberate when they are metabolized in the body.

## 11.5 Measuring standard heats of formation

The standard heats of formation of carbon dioxide, magnesium oxide and many other oxides can be measured directly by using a bomb calorimeter similar to that discussed in the last section.

However, there are many compounds for which $\Delta H_f^\ominus$ cannot be measured directly. Consider, for example, tetrachloromethane ($CCl_4$); graphite and chlorine do not combine readily, nor does $CCl_4$ decompose easily into its constituent elements. In other words, neither the reaction

$$C\text{ (graphite)} + 2Cl_2(g) \longrightarrow CCl_4(l),$$

nor the reaction

$$CCl_4(l) \longrightarrow C\text{ (graphite)} + 2Cl_2(g)$$

can be carried out in a calorimeter.

Boron(III) oxide ($B_2O_3$) and aluminium oxide ($Al_2O_3$) provide a different prob lem in attempting to measure their standard heats of formation. In their case, it i difficult to burn the elements, boron and aluminium, completely in oxygen becaus a protective layer of oxide coats the unreacted element.

- ○ Write an equation relating to the standard heat of formation of carbo monoxide.
- ○ Why is it impossible to obtain $\Delta H_f^\ominus$ (CO(g)) directly?

As a result of the problems involved in these measurements, chemists have had to look for methods of obtaining standard heats of formation indirectly. One indirect method which has been widely used involves measuring the heat of combustion of the compound and the heats of combustion of its constituent elements. These values can be linked in an energy cycle or an energy level diagram with the heat of formation of the compound. The energy cycle in figure 11.7 shows how this can be done for carbon monoxide. Notice that figure 11.7 effectively shows two routes for converting graphite and oxygen to $CO_2$. One of these is the direct route straight from graphite and oxygen to $CO_2$, whilst the alternative route proceeds via CO. It would seem reasonable that the overall enthalpy change for the conversion of graphite to carbon dioxide is independent of the route taken, so we can write,

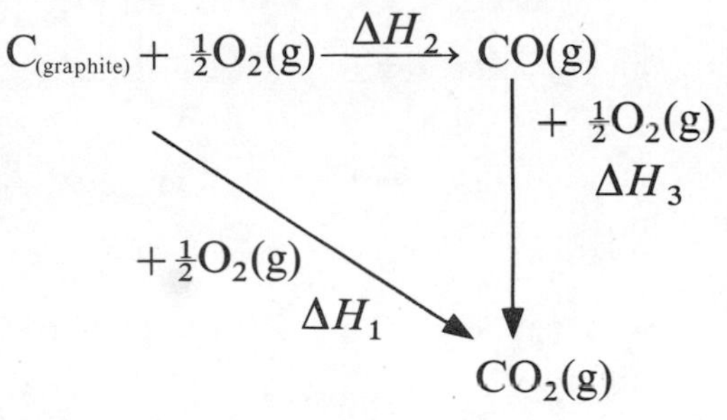

$$C_{(graphite)} + \tfrac{1}{2}O_2(g) \xrightarrow{\Delta H_2} CO(g)$$

$+ \frac{1}{2}O_2(g)$ $\Delta H_3$

$+ \frac{1}{2}O_2(g)$ $\Delta H_1$

$CO_2(g)$

**Figure 11.7** An energy cycle incorporating the heat of formation of carbon monoxide.

$$\Delta H_1 = \Delta H_2 + \Delta H_3 \qquad \text{Equation 1}$$

$$\text{now } \Delta H_1 = \Delta H_c^\ominus \text{ (C (graphite))} = -393 \text{ kJ mole}^{-1},$$

$$\text{and } \Delta H_3 = \Delta H_c^\ominus \text{ (CO(g))} = -283 \text{ kJ mole}^{-1},$$

$$\text{so } \Delta H_2 = \Delta H_f^\ominus \text{ (CO(g))} = \Delta H_1 - \Delta H_3$$

$$= -393 - (-283)$$

$$= -110 \text{ kJ mole}^{-1}$$

The chemical processes and the different routes involved are represented in an energy level diagram in figure 11.8.

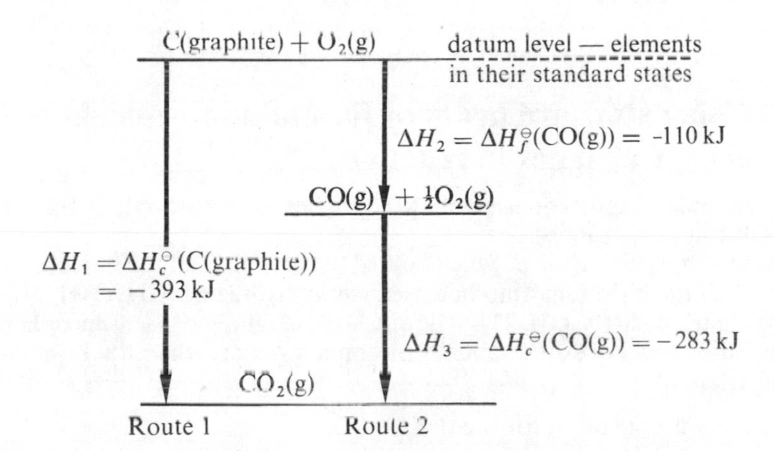


**Figure 11.8** An energy level diagram incorporating the heat of formation of carbon monoxide.

The argument we have just used in obtaining equation 1 is a specific example of **Hess's Law of Constant Heat Summation** which says that *the energy change in converting reactants, A + B, to products, X + Y, is the same, regardless of the route by which the chemical change occurs provided the initial and final conditions are the same.*

Of course, Hess's Law is simply an application of the more fundamental law of conservation of energy.

- ○ Why must Hess's Law hold if the law of conservation of energy is to be obeyed?
- ○ Suggest experiments which would enable you to verify Hess's Law.

We have used Hess's Law to find the standard heat of a reaction which cannot be measured directly. Before moving on, let us look at a more complex example involving butane ($C_4H_{10}$). Carbon and hydrogen will not react directly to produce butane. Therefore, the formation of butane, represented by the equation

$$4C \text{ (graphite)} + 5H_2(g) \longrightarrow C_4H_{10}(g)$$

cannot be carried out directly in a calorimeter. However, the standard heat of formation of butane can be obtained by measuring the standard heats of combustion of butane and its constituent elements, carbon and hydrogen.

An energy level diagram for the reactions concerned is shown in figure 11.9.

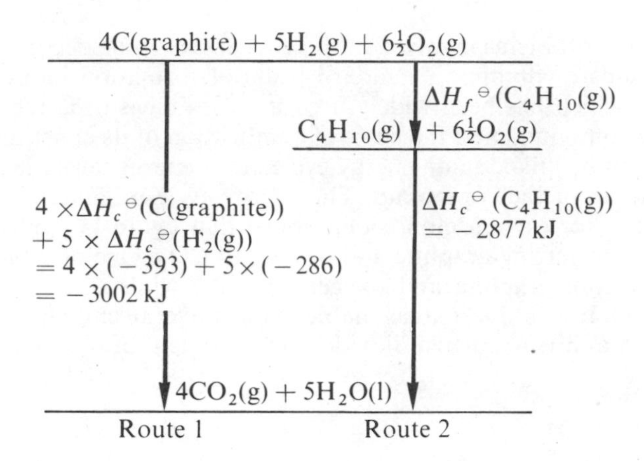

Figure 11.9

**Figure 11.9** An energy level diagram used to obtain the heat of formation of butane indirectly.

Notice that route 1 involves four times the standard heat of combustion of graphite and five times the standard heat of combustion of hydrogen.

By Hess's Law,
Enthalpy change for route 1 = Enthalpy change for route 2

$\therefore\ -3002 = \Delta H_f^{\ominus}\,(C_4H_{10}(g)) - 2877$

$\Rightarrow$ Standard heat of formation of butane, $\Delta H_f^{\ominus}\,(C_4H_{10}(g)) = -125$ kJ mole$^{-1}$

*Above* The booster stage for the Saturn V rocket, used to lift the American astronauts into orbit during the Apollo Project, being hoisted into position at the Kennedy Space Centre, Florida. During its initial stage, Saturn V burns 15 tonnes of kerosene/oxygen mixture every second for the first 2½ minutes.

## 11.6 Using standard heats of formation to calculate the energy changes in reactions

One of the most important uses of $\Delta H_f^{\ominus}$ values is in calculating the enthalpy changes in chemical reactions.

During the *Apollo 11* project which landed the first man on the moon on 21 July 1969, the engines of the lunar module used methylhydrazine ($CH_3.NH.NH_2$) and dinitrogen tetraoxide ($N_2O_4$). These liquids were carefully chosen since they ignite spontaneously and very exothermically on contact. What is the enthalpy change for their reaction?

First we write the equation for the reaction.

$$4CH_3.NH.NH_2(l) + 5N_2O_4(l) \longrightarrow 4CO_2(g) + 12H_2O(l) + 9N_2(g)$$

Now draw an energy cycle by adding the formation equations from the same elements to both sides of the equation under consideration (figure 11.10).

By Hess's Law, the total enthalpy change for the formation of carbon dioxide, water, and nitrogen will be the same whether they are formed directly from their elements or via the intermediates, $CH_3.NH.NH_2$ and $N_2O_4$.

$$\therefore\ 4\,\Delta H_f^{\ominus}(CH_3.NH.NH_2(l)) + 5\,\Delta H_f^{\ominus}(N_2O_4(l)) + \Delta H^{\ominus}_{reaction}$$
$$= 4\,\Delta H_f^{\ominus}(CO_2(g)) + 12\,\Delta H_f^{\ominus}(H_2O(l))$$

The standard heats of formation are:

$$\Delta H_f^{\ominus}(CH_3.NH.NH_2(l)) = +53 \text{ kJ mole}^{-1}$$
$$\Delta H_f^{\ominus}(N_2O_4(l)) = -20 \text{ kJ mole}^{-1}$$
$$\Delta H_f^{\ominus}(CO_2(g)) = -393 \text{ kJ mole}^{-1}$$
$$\Delta H_f^{\ominus}(H_2O(l)) = -286 \text{ kJ mole}^{-1}$$

$$\Rightarrow 4(+53) + 5(-20) + \Delta H^{\ominus}_{reaction} = 4(-393) + 12(-286)$$
$$\text{so, } \Delta H^{\ominus}_{reaction} = -5116 \text{ kJ}$$

Thus, $4CH_3.NH.NH_2(l) + 5N_2O_4(l) \longrightarrow 4CO_2(g) + 12H_2O(l) + 9N_2(g)$
$\Delta H^{\ominus} = -5116$ k

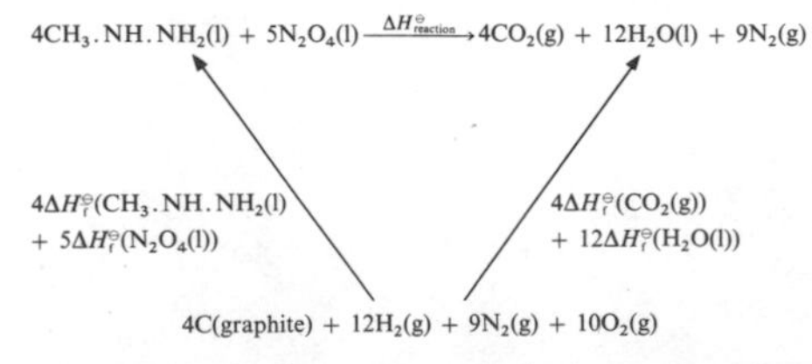


**Figure 11.10** An energy cycle incorporating the reaction of methylhydrazine with dinitrogen tetraoxide.

## 11.7 Using standard heats of formation to predict the relative stabilities of compounds

Most compounds are formed exothermically from their elements. Thus, the standard heats of formation of water, carbon dioxide, aluminium oxide and many other compounds are negative. These compounds are therefore at a lower energy level than their constituent elements, i.e. the compounds are energetically more stable than the elements from which they are formed. But just consider the following problem.

The standard heat of formation of hydrogen peroxide is $-188$ kJ mole$^{-1}$. From this, we would expect $H_2O_2$ to be stable, yet it decomposes fairly readily into water and oxygen. How do you explain this?

The answer, and the important point, lies in the fact that $\Delta H_f^{\ominus}$ ($H_2O_2(l)$) only describes the stability of hydrogen peroxide *relative to its elements.*

$$H_2(g) + O_2(g) \longrightarrow H_2O_2(l) \qquad \Delta H = -188 \text{ kJ}$$

$H_2O_2$ is obviously more stable than its elements, but on decomposition it produces, not $H_2(g) + O_2(g)$ for which $\Delta H^{\ominus}$ is $+188$ kJ, but $H_2O(l) + \frac{1}{2}O_2(g)$ for which $\Delta H^{\ominus}$ is $-98$ kJ, i.e.

$$H_2O_2(l) \longrightarrow H_2O(l) + \tfrac{1}{2}O_2(g) \qquad \Delta H = -98 \text{ kJ}$$

Thus, hydrogen peroxide is energetically stable with respect to its elements, but unstable with respect to water and oxygen. *This example shows just how important it is to specify with respect to what substances a compound is stable or unstable.*

A few compounds, such as ethyne ($C_2H_2$), trioxygen (ozone, $O_3$), carbon disulphide ($CS_2$), and nitrogen oxide (NO) are formed endothermically from their elements.

$$2C(\text{graphite}) + H_2(g) \longrightarrow C_2H_2(g) \qquad \Delta H^{\ominus} = +227 \text{ kJ}$$

$$\tfrac{1}{2}N_2(g) + \tfrac{1}{2}O_2(g) \longrightarrow NO(g) \qquad \Delta H^{\ominus} = +90 \text{ kJ}$$

These compounds have positive standard heats of formation and they are therefore energetically unstable with respect to their elements. Why, then, don't these compounds decompose instantaneously into their constituent elements?

Ethyne(acetylene), trioxygen(ozone), carbon disulphide and nitrogen oxide can all be stored for long periods at room temperature and pressure in the absence of a catalyst. These compounds do, however, begin to decompose at high temperatures or in the presence of a catalyst. In order to explain the unexpected stability of these compounds, we must distinguish between *energetic* stability and *kinetic* stability.

Thus, ethyne and nitrogen oxide are energetically unstable with respect to their elements, but at low temperatures and pressures the decomposition reactions are so slow that both ethyne and nitrogen oxide are kinetically stable. The kinetic stability of these energetically unstable compounds can be compared to the situation of a stone resting on a hillside. The stone is energetically unstable. Given the opportunity, it would roll to the bottom of the hill where it would come to rest in a position of lower energy. Resting on the hillside, stuck behind a tuft of grass, the stone's movement has been prevented—it is kinetically stable in spite of its energetic instability.

Diamond provides another example of an energetically unstable, yet kinetically stable, substance. At normal temperatures and pressures, diamond is unstable with respect to its allotrope, graphite.

$$C\ (\text{diamond}) \longrightarrow C\ (\text{graphite}) \qquad \Delta H^{\ominus} = -2 \text{ kJ}$$

Fortunately, the rate of transformation of diamond to graphite is immeasurably slow at room temperature and so the diamond is kinetically stable. 'Diamonds are for ever', or are they?

The kinetic stability of nitrogen oxide with respect to its elements shows, yet again, how important it is to state clearly with respect to what a substance is stable. Obviously, nitrogen oxide is kinetically stable with respect to nitrogen and oxygen, but in the presence of air or oxygen it is energetically and kinetically unstable with respect to nitrogen dioxide. Hence, nitrogen oxide reacts rapidly with oxygen to form brown fumes of nitrogen dioxide.

$$2NO(g) + O_2(g) \longrightarrow 2NO_2(g)$$

## 11.8 Predicting whether reactions will occur

The enthalpy change of a reaction is sometimes used as a rough guide to the likelihood that the reaction will occur. If $\Delta H^{\ominus}$ for a reaction is negative, energy is lost when the reaction occurs and the products are more stable than the reactants.

Thus, exothermic reactions are more likely to occur than endothermic reactions and those reactions which occur spontaneously are often very exothermic.

*Although the value of $\Delta H^{\ominus}$ can provide some indication of the likelihood of a reaction, there are limitations on its use and it is important to bear these in mind.*

**1** $\Delta H^{\ominus}$ shows the relative energetic stabilities of the reactants and products for a reaction, but it says nothing about the *kinetic* stability of the products relative to the reactants. Thus, $\Delta H^{\ominus}$ *is no guide to the rate of a reaction*, it cannot tell us whether the reaction is fast or slow. A reaction may be enormously exothermic, yet nothing happens because the reaction rate is immeasurably slow and the reactants are kinetically stable with respect to the products (e.g. a mixture of hydrogen and oxygen at room temperature). We shall be studying reaction rates in more detail in chapter 23.

**2** In order to make accurate predictions concerning the relative energy levels of reactants and products, it is necessary to consider not only the energy lost or gained by the reacting system, but also any energy changes inside that system. For example, when a gas is produced in a reaction or a solid dissolves in a liquid, there is a marked increase in the disorder of the system itself and an increase in the number of ways in which the energy is distributed in the system. Consequently, in predicting the likelihood of any reaction we should take into account the change in order (or disorder) introduced into the system. At normal temperatures, this additional disorder is usually unimportant, but it explains why certain endothermic reactions happen so spontaneously. For example, many solids dissolve in water readily and easily in spite of the fact that the process of dissolving is endothermic. The reason for this is that, although the enthalpy change is positive, there is an enormous increase in disorder (or entropy) as the solid dissolves, causing the reaction to occur spontaneously.*

**3** Predictions from $\Delta H_f^{\ominus}$ values relate only to standard conditions, i.e. 298 K and atmospheric pressure. The situation may be very different under different conditions or in the presence of a catalyst.

- ○ The heat of combustion of octane (in petrol) is $-5513$ kJ mole$^{-1}$. Would you expect petrol to react with oxygen?
- ○ Why are petrol/oxygen mixtures stable at room temperature before a spark is applied?
- ○ How does the spark initiate a reaction between petrol and oxygen?

## 11.9 Heats of combustion and molecular structure

Look closely at the information in table 11.1. Notice the similarity in the heats of combustion for butane and methylpropane. Perhaps the similarity is not surprising after all they contain the same atoms and the same bonds. This might, however suggest that each kind of bond makes a fixed contribution to the total enthalpy change. This idea is confirmed when we inspect the heats of combustion of the simpler alkanes (table 11.2). Two important points emerge from the data in table 11.2.

First, there is a regular difference in structure of a —$CH_2$— group from one alkane to the next. Secondly, there is a similar difference of about 650 kJ in the value of $\Delta H_c^{\ominus}$ from one alkane to the next. This would suggest that each additional —$CH_2$— group is responsible for a fixed increment in the heat of combustion and that each bond makes a characteristic contribution to the overall energy content of a substance.

What happens to the different bonds when an alkane burns? During combustion the bonds between the atoms in the alkane molecule are broken and new bonds form between carbon and oxygen in $CO_2$ and between hydrogen and oxygen in $H_2O$. Notice that each alkane has one C—C bond and two C—H bonds more than the previous alkane. Check this for yourself.

* For a fuller discussion of the change in order (or disorder) in chemical systems consult a more advanced text under the topic 'entropy'.

**Table 11.1** The formulae, bonds and heats of combustion of butane and methylpropane.

| | **Butane** | **Methylpropane** |
|---|---|---|
| Molecular formula | $C_4H_{10}$ | $C_4H_{10}$ |
| Structural formula | H H H H<br>\| \| \| \|<br>H—C—C—C—C—H<br>\| \| \| \|<br>H H H H | H H H<br>\| \| \|<br>H—C—C—C—H<br>\| \| \|<br>H \| H<br>H—C—H<br>\|<br>H |
| Bonds | 3 C—C<br>10 C—H | 3 C—C<br>10 C—H |
| $\Delta H_c^\ominus$/kJ mole$^{-1}$ | −2877 | −2869 |

**Table 11.2** The formulae and heats of combustion of the simpler alkanes

| **Alkane** | **Molecular formula** | **Structural formula** | $\Delta H_c^\ominus$ /kJ mole$^{-1}$ | **Difference in $\Delta H_c^\ominus$** /kJ |
|---|---|---|---|---|
| Methane | $CH_4$ | H<br>\|<br>H—C—H<br>\|<br>H | −890 | |
| | | | | 670 |
| Ethane | $C_2H_6$ | H H<br>\| \|<br>H—C—C—H<br>\| \|<br>H H | −1560 | |
| | | | | 660 |
| Propane | $C_3H_8$ | H H H<br>\| \| \|<br>H—C—C—C II<br>\| \| \|<br>II II H | −2220 | |
| | | | | 657 |
| Butane | $C_4H_{10}$ | H H H H<br>\| \| \| \|<br>H—C—C—C—C—H<br>\| \| \| \|<br>H H H H | −2877 | |
| | | | | 643 |
| Pentane | $C_5H_{12}$ | H H H H H<br>\| \| \| \| \|<br>H—C—C—C—C—C—H<br>\| \| \| \| \|<br>H H H H H | −3520 | |

Obviously, energy must be supplied in order to break a bond between two atoms, whilst energy is released when a bond forms. Thus, *bond-breaking is an endothermic process whereas bond-making is exothermic*. Now, since chemical reactions involve bond-breaking followed by bond-making, it means that the heat of a reaction is the energy difference between bond-breaking and bond-making processes (figure 11.11). In section 11.11, we shall see how information concerning the strengths of bonds can be used to calculate the heats of reactions.

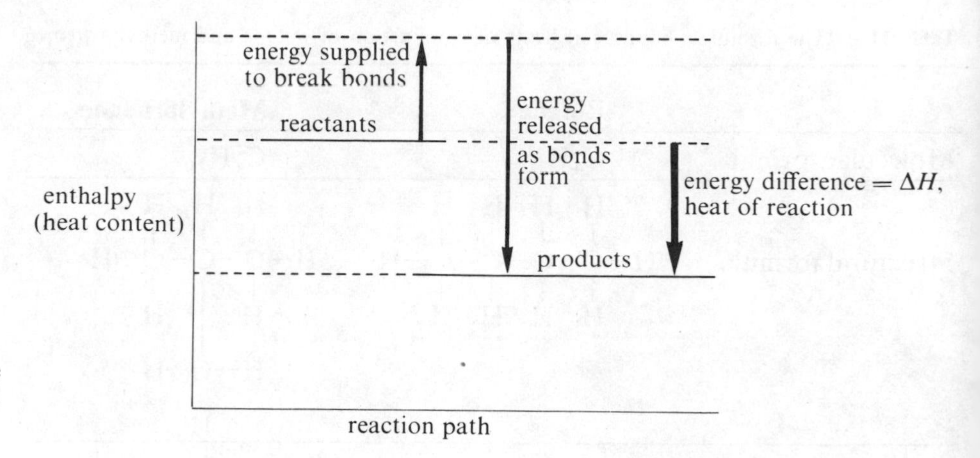


**Figure 11.11** The heat of a reaction is the energy difference between bond-breaking and bond-making processes.

As the alkane molecules get larger, more energy is required to break the bonds between their carbon and hydrogen atoms, but even larger amounts of energy are released as these atoms form carbon dioxide and water. On burning, each alkane molecule forms one $CO_2$ molecule and one $H_2O$ molecule more than the previous alkane.

$$CH_4 + 2O_2 \longrightarrow CO_2 + 2H_2O$$
$$C_2H_6 + 3\tfrac{1}{2}O_2 \longrightarrow 2CO_2 + 3H_2O$$
$$C_3H_8 + 5O_2 \longrightarrow 3CO_2 + 4H_2O$$

Thus, the heat of combustion gets larger by a constant amount as we progress along the homologous series of alkanes. The exothermic nature of the reactions indicates that the energy released on making the bonds in $CO_2$ and $H_2O$ is greater than the energy required to break the bonds in the alkanes and $O_2$. An exothermic change is evidence for the formation of stronger bonds, whereas an endothermic change is evidence for the formation of weaker bonds. Indeed, the energy we get from burning such fuels as coal, oil and natural gas results from the formation of strong C=O and H—O bonds in $CO_2$ and $H_2O$ respectively.

## 11.10 Finding the strength of the C—H bond and the C—C bond

From our discussions in section 11.9, it would seem that a definite quantity of energy (known as the **bond energy**) may be associated with each type of bond. This energy is absorbed when the bond is broken and evolved when the bond is formed. In order to find the strength of the C—H bond, it would seem appropriate to study methane which contains C—H bonds only. Consider the following equation which relates to the heat of atomization of methane

$$CH_4(g) \longrightarrow C(g) + 4H(g)$$

- ○ How many C—H bonds are broken during the atomization of 1 mole of methane?
- ○ Suppose $\Delta H$ for this process is $x$ kJ mole$^{-1}$. What is the bond energy of one C—H bond in kJ mole$^{-1}$?
- ○ Why is equation 2 not simply the reverse of the heat of formation of methane?

Although the standard heat of formation of methane does not relate directly to bond energies, we can relate the C—H bond energy to the heat of formation of methane from its monatomic gaseous elements since this is merely the reverse of equation 2.

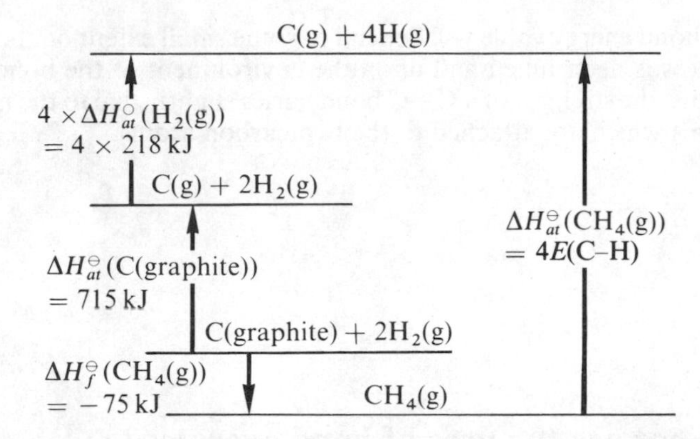


**Figure 11.12** An energy level diagram which can be used in determining the C—H bond energy.

Thus, in order to obtain the C—H bond energy, usually written $\bar{E}$ (C—H), we must first find the standard heats of atomization for the elements present in methane as well as the standard heat of formation of methane. An energy diagram connecting the relevant quantities for methane is shown in figure 11.12. From this diagram we can write:

$$4E\,(\text{C—H}) = 4 \times 218 + 715 + 75\ \text{kJ}$$
$$\Rightarrow 4E\,(\text{C—H}) = 1\,662\ \text{kJ}$$

∴ **The average C—H bond energy in methane, $E$(C—H) = 415.5 kJ mole$^{-1}$**

*Although average bond energies are usually referred to simply as bond energies, they are sometimes called* **bond energy terms** *and denoted by the symbol* $\bar{E}$.

If we assume that the C—H bonds in ethane have the same strength as those in methane, we can now calculate the carbon–carbon bond energy, $E$ (C—C). Figure 11.13 shows the appropriate enthalpy changes in an energy diagram.

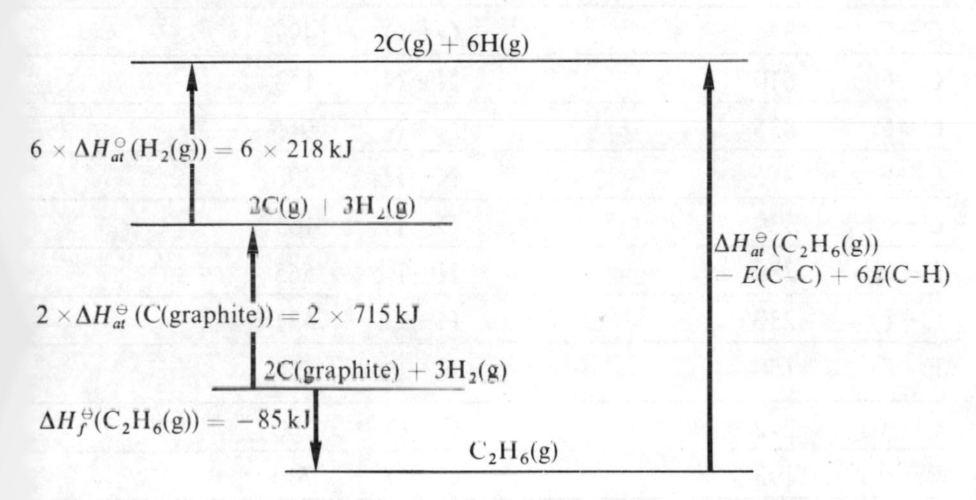


**Figure 11.13** An energy level diagram used in determining the C—C bond energy.

Equating the two routes from $C_2H_6$(g) to 2C(g) + 6H(g) we can write:

$$E\,(\text{C—C}) + 6E\,(\text{C—H}) = -(-85) + 2 \times 715 + 6 \times 218\ \text{kJ}$$
$$\Rightarrow E\,(\text{C—C}) + 6E\,(\text{C—H}) = 2\,823\ \text{kJ}$$

Assuming that $E$ (C—H) = 415.5 kJ;

$$E\,(\text{C—C}) + 6 \times 415.5 = 2823$$
$$\Rightarrow E\,(\text{C—C}) = 2\,823 - 2\,493 = 330\ \text{kJ}$$

∴ **The C—C bond energy in ethane, $E$(C—C) = 330 kJ mole$^{-1}$**

Proceeding in this way we could calculate the strength of other bonds. The idea that bond energies are additive and transferable from one molecule to another holds very well in most cases, but results show small deviations in the strength of one particular type of bond in different molecules.

For example, the bond energy for C—C has been determined using several different compounds giving values which range from 330 to 346 kJ mole$^{-1}$. This shows that a bond energy value will depend to some small extent on the compound from which it was determined and upon the environment of the bond within the molecule. Thus, the strength of a C—C bond varies slightly due to the nature of the different atoms which are attached to the two carbon atoms.

## 11.11 What are the uses of bond energies?

Although the concept of bond energies is rather artificial, corresponding to no physically realizable process, it is nevertheless very useful. Four of its important uses are:

(a) comparing the strengths of bonds,
(b) understanding structure and bonding,
(c) estimating the enthalpy changes in reactions, and
(d) understanding the mechanisms of chemical reactions.

Table 11.3 shows a list of some bond energies.

**Table 11.3** Average bond energies (bond energy terms)

| Bond | Bond energy, *E*(X—Y) /kJ per mole of bonds | Bond | Bond energy, *E*(X—Y) /kJ per mole of bonds |
|---|---|---|---|
| C—H | 413 | O—O | 146 |
| C—C | 346 | O=O | 497 |
| C=C | 610 | N—N | 163 |
| C≡C | 835 | N≡N | 945 |
| C—F | 495 | N—H | 390 |
| C—Cl | 339 | O—H | 463 |
| C—Br | 280 | H—F | 565 |
| C—I | 230 | H—Cl | 431 |
| F—F | 158 | H—Br | 365 |
| Cl—Cl | 242 | C—O | 360 |
| Br—Br | 193 | C=O | 740 |
| I—I | 151 | Si—O | 464 |
| Si—Si | 226 | | |

Probably the most important application of bond energies is their use in estimating the enthalpy changes in chemical reactions. These estimations are particularly useful when calorimetric or other experimental measurements cannot be made. Consider the following example:

Hydrazine is often used as a rocket fuel as it can be stored conveniently as a liquid and it reacts very exothermically with oxygen forming purely gaseous products.

$$N_2H_4(g) + O_2(g) \longrightarrow N_2(g) + 2H_2O(g); \qquad \Delta H^\ominus = -622 \text{ kJ}$$

It has been suggested that hydrazine/fluorine mixtures might be more exothermic than hydrogen/oxygen mixtures. Using bond energies from table 11.3, we can find $\Delta H$ for the reaction of hydrazine with fluorine.

$$N_2H_4(g) + 2F_2(g) \longrightarrow N_2(g) + 4HF(g)$$

The following equation shows clearly those bonds which are broken and those which form.

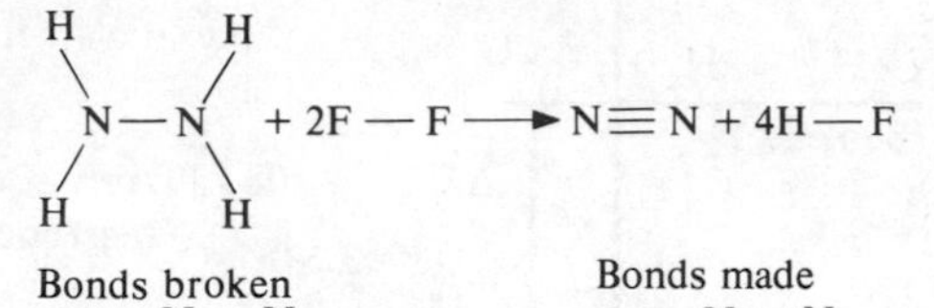

$$\mathrm{H_2N{-}NH_2} + 2\mathrm{F{-}F} \longrightarrow \mathrm{N{\equiv}N} + 4\mathrm{H{-}F}$$

| Bonds broken | | Bonds made | |
|---|---|---|---|
| one | N—N | one | N≡N |
| four | N—H | four | H—F |
| two | F—F | | |

- ○ Calculate the energy required to break the bonds in this reaction.
- ○ Calculate the energy evolved when the bonds form in the products.
- ○ What is the value of $\Delta H^{\ominus}$ for this reaction per mole of hydrazine?
- ○ Is the reaction with fluorine more or less exothermic than the reaction with oxygen?
- ○ If you had to decide whether to use oxygen or fluorine for the rocket flight, what other factors besides the exothermicity of the reaction would influence your final decision?

For most reactions, the values of $\Delta H$ estimated via bond energy terms agree closely with experimentally determined values and this has further established the usefulness of bond energies. Occasionally, the estimated values for $\Delta H$ are in strong disagreement with the experimental results and this has resulted in a re-examination of the structure of compounds and an improvement in our understanding of bonding. A striking example of this is provided by benzene and other aromatic compounds. At one time, molecules of benzene were thought to contain hexagons of carbon atoms linked by alternate double and single bonds (figure 11.14).

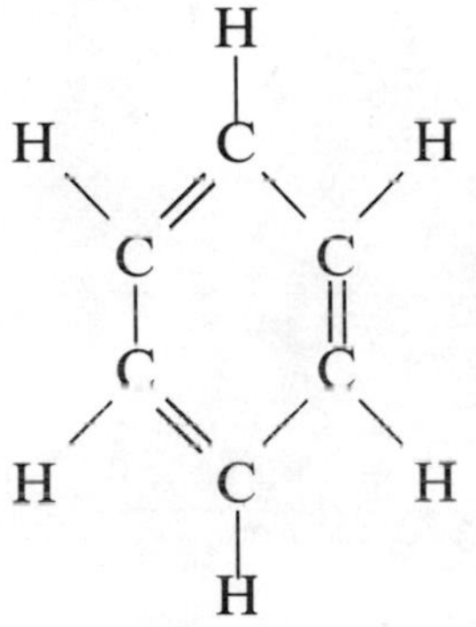

Figure 11.14 The Kekulé structure for benzene.

Now, when cyclohexene, containing one (C=C) bond, undergoes an addition eaction with hydrogen, the enthalpy change is $-120$ kJ mole$^{-1}$.

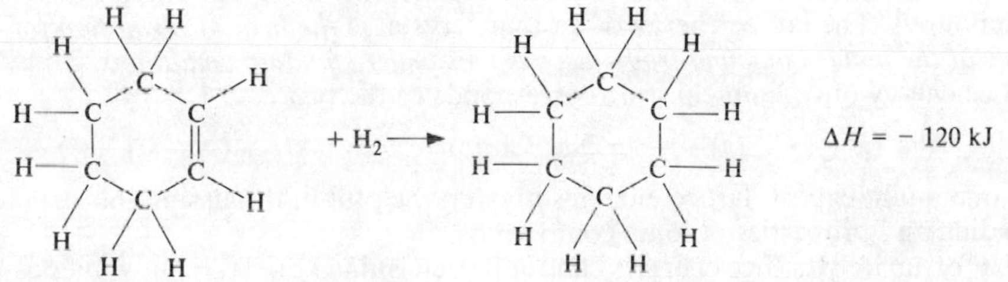

$$\text{cyclohexene} + \mathrm{H_2} \longrightarrow \text{cyclohexane} \qquad \Delta H = -120 \text{ kJ}$$

his would suggest that the hydrogenation of 'Kekulé's' benzene, containing hree (C=C) bonds should have an enthalpy change of $-360$ kJ. However, when xperiments are performed, the heat of hydrogenation of benzene is found to be ·208 kJ mole$^{-1}$ and not $-360$ kJ mole$^{-1}$.

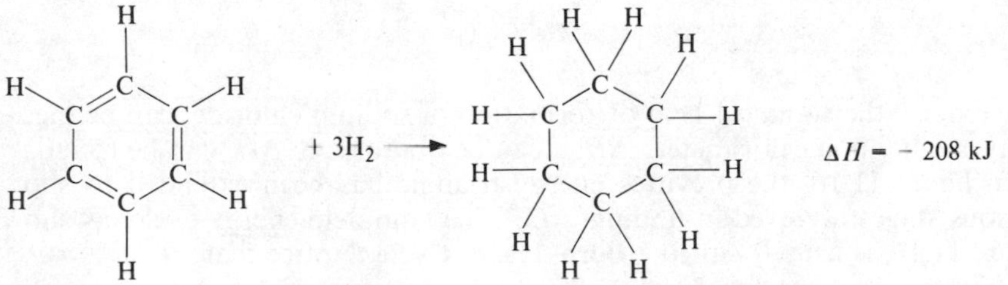

$$\text{benzene} + 3\mathrm{H_2} \longrightarrow \text{cyclohexane} \qquad \Delta H = -208 \text{ kJ}$$

would seem that one mole of benzene is 152 kJ more stable than the Kekulé ructure would suggest and this increased stability is emphasized in figure 11.15.

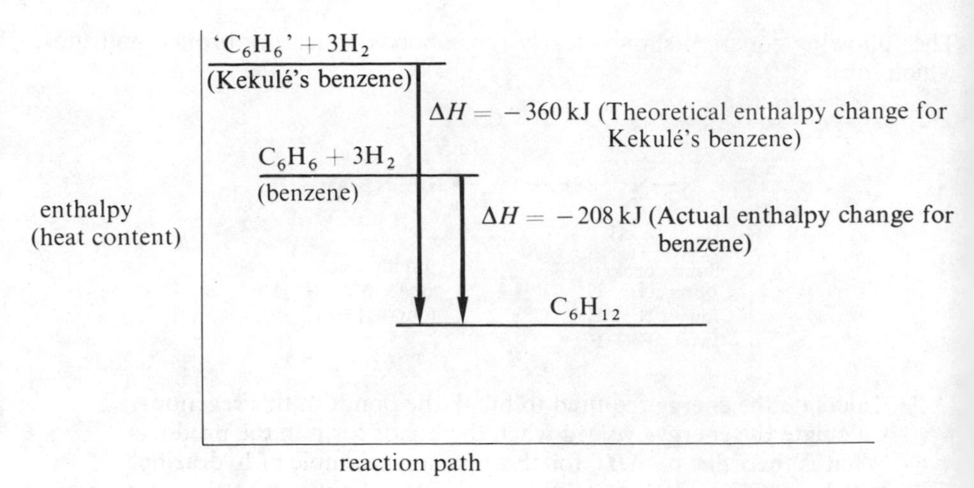


**Figure 11.15** An energy level diagram to emphasize the unexpected stability of benzene.

As a result of this discrepancy, chemists were forced to re-examine the structure of benzene. It is now known that all six C—C bonds in its molecule are equal in length and in strength. The molecule does not contain alternate double and single bonds as Kekulé had proposed. Furthermore, this new structure for benzene has greatly improved our understanding of its properties and reactions and those of other aromatic compounds. The structure of benzene is discussed further in section 27.2.

## 11.12 Energy changes in forming ionic substances

In the last section, we found that bond energies were extremely useful in helping us to understand the properties and reactions of molecular substances. Bond energies for molecular substances can be compared with **lattice energies** for ionic compounds. The lattice energy of an ionic crystal is *the heat of formation for one mole of the ionic compound from gaseous ions under standard conditions*. Thus, the lattice energy of sodium chloride corresponds to the process:

$$Na^+(g) + Cl^-(g) \longrightarrow Na^+Cl^-(s) \qquad \Delta H_{latt}\,(Na^+Cl^-(s))$$

As one might expect, lattice energies are very helpful in discussing the structure bonding and properties of ionic compounds.

Unfortunately, lattice energies cannot be determined directly, but values can be obtained indirectly by means of an energy cycle connecting the ionic solid, the gaseous ions and the elements in their standard states.

$$Na(s) + \tfrac{1}{2}Cl_2(g) \xrightarrow{\Delta H_x} Na^+(g) + Cl^-(g)$$

$\Delta H_f^{\ominus}\,(Na^+Cl^-(s))$ ↘ $Na^+Cl^-(s)$ ↙ $\Delta H_{latt}(Na^+Cl^-(s))$

Since $\Delta H_f^{\ominus}$, the standard heat of formation of sodium chloride, can be measure conveniently in a calorimeter, $\Delta H_{latt}$ can be obtained if $\Delta H_x$ can be found.

In figure 11.16, the previous energy triangle has been extended to show th various stages involved in finding $\Delta H_x$. The complete energy cycle, as shown figure 11.16, is usually called a **Born–Haber Cycle**. Notice that:

$$\Delta H_x = \Delta H_{at}(Na(s)) + \Delta H_i(Na(g)) + \Delta H_{at}(Cl_2(g)) + \Delta H_e(Cl)$$

The first two stages in this process involve atomizing and then ionizing sodiur The heat of atomization of sodium

$$Na(s) \xrightarrow{\Delta H_{at}(Na(s))} Na(g)$$

can be obtained from values of its heat of fusion, heat of vaporization and speci heat capacity.

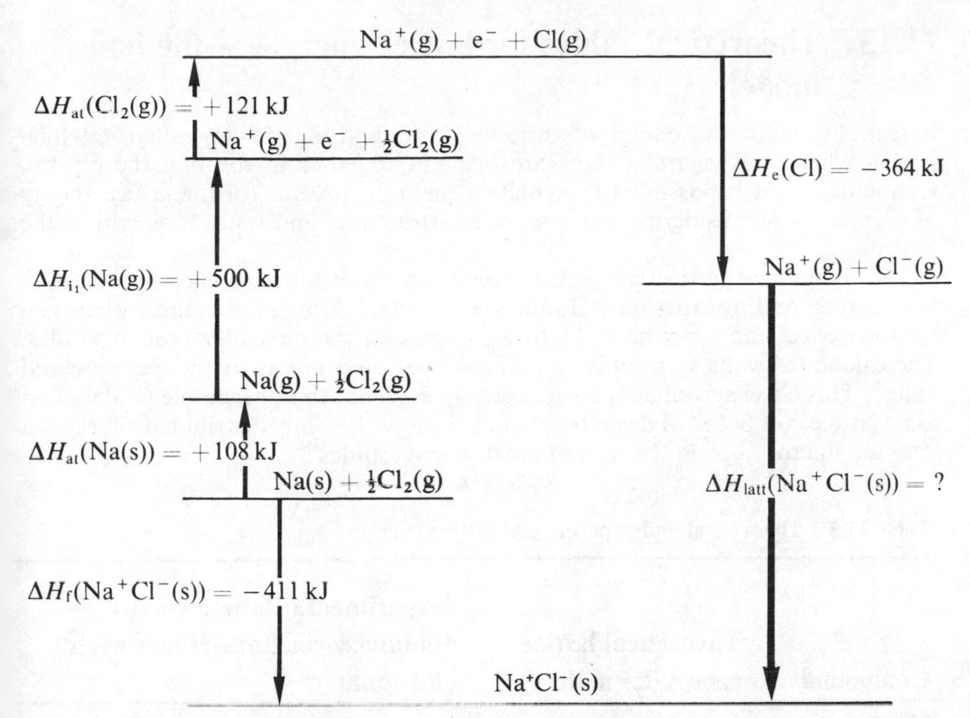


**Figure 11.16** The Born–Haber cycle for sodium chloride.

The first ionization energy of sodium,

$$Na(g) \xrightarrow{\Delta H_{i_1}(Na(g))} Na^+(g) + e^-,$$

can be determined spectroscopically or by valve measurements (sections 5.2, 5.3). The third and fourth stages, in the expression for $\Delta H_x$ above, involve the atomization of chlorine and the conversion of chlorine atoms to chloride ions respectively. The latter process is, of course, called the electron affinity for chlorine (section 7.3, page 80).

The heat of atomization of chlorine,

$$\tfrac{1}{2}Cl_2(g) \xrightarrow{\Delta H_{at}(Cl_2(g))} Cl(g)$$

can be obtained from spectroscopic studies (section 11.4), whilst the electron affinity for chlorine,

$$Cl(g) + e^- \xrightarrow{\Delta H_e(Cl)} Cl^-(g)$$

can be found by methods similar to the valve measurements discussed on page 52.

The lattice energy for sodium chloride can now be obtained since all other values in the Born–Haber cycle can be determined experimentally.

Using the values in figure 11.16,

$$\Delta H_{latt}(Na^+Cl^-(s)) = (364 - 121 - 500 - 108 - 411)\ \text{kJ mole}^{-1}$$
$$= -776\ \text{kJ mole}^{-1}.$$

This lattice energy gives us some idea of the force of attraction between $Na^+$ and $Cl^-$ ions in crystalline sodium chloride. The lattice energies of various other ionic crystals are given in table 11.4.

- ○ Why does the lattice energy become progressively more exothermic along the series NaI, NaBr, NaCl, NaF?
- ○ The interionic distance between $Na^+$ and $F^-$ ions in NaF is very similar to that between $Mg^{2+}$ and $O^{2-}$ ions in MgO. Why, then, is the lattice energy of MgO about four times more exothermic than that of NaF?

Notice the similar order of magnitude between the lattice energies quoted in table 11.4 and the bond energies of covalent bonds in table 11.3. This would suggest that ionic bonds and covalent bonds are roughly similar in strength. Although discussion is often restricted to the two extreme types of bond, it is important to realize their underlying similarity; they both, of course, result from electrostatic forces, but involve different distributions of electrons.

**Table 11.4** Lattice energies of some ionic solids

| Compound | Lattice energy /kJ mole$^{-1}$ |
|---|---|
| NaF | −915 |
| NaCl | −776 |
| NaBr | −742 |
| NaI | −699 |
| $MgCl_2$ | −2489 |
| MgO | −3933 |

## 11.13 Theoretical values for lattice energies—the ionic model

Essentially, the lattice energy of an ionic compound is the energy change which occurs when well-separated ions are brought together in forming the crystal. Consequently, it is possible to calculate a theoretical value for the lattice energy of a crystal by considering the interionic attractions and repulsions within the lattice.

The theoretical lattice energies of some ionic substances are compared with their corresponding experimental values in table 11.5. Notice the similarity between the theoretical and experimental lattice energies for the three alkali metal halides. The calculated values are only one or two per cent less than the experimental values. This close agreement provides strong evidence that the simple model of an ionic lattice composed of discrete spherical ions with evenly distributed charge is a very satisfactory one in the case of alkali metal halides.

**Table 11.5** Theoretical and experimental lattice energies

| Compound | Theoretical lattice energy/kJ mole$^{-1}$ | Experimental lattice energy (obtained via Born-Haber cycle) /kJ mole$^{-1}$ |
|---|---|---|
| NaCl | −766 | −776 |
| NaBr | −731 | −742 |
| NaI | −686 | −699 |
| AgCl | −768 | −890 |
| AgBr | −759 | −877 |
| AgI | −736 | −867 |

Now, look at the theoretical and experimental lattice energies for the silver halides in table 11.5. For these compounds the theoretical values are approximately 130 kJ mole$^{-1}$ (i.e. about 15 %) less than the experimental values. In this case, the simple ionic model is not completely satisfactory and requires some modification.

When the difference in electronegativity between the ions in the crystal is large as in the case of alkali metal halides, the ionic model is satisfactory. However, when the difference in electronegativity gets smaller, as in the case of silver halides, there is a significant disagreement between experimental results and those calculated in terms of a simple ionic model. Experiments show that the bonding between ions is stronger than the ionic model predicts. The explanation of this anomaly is that the bonding in this case is not purely ionic but is intermediate in character between ionic and covalent. The partly covalent nature of the bonds can be interpreted by saying that the ionic bonds have been polarized (section 8.2) or by suggesting that the electrons are incompletely transferred in forming the ions. Once again, this emphasizes the fact that ionic and covalent bonds are simply extreme types and that the bonding in most substances is intermediate in character between purely ionic and purely covalent.

## 11.14 Solution, hydration and lattice energy

When ionic solids dissolve in water, heat is usually evolved or absorbed. Why is this? Can we explain the enthalpy changes in terms of the processes taking place at a molecular and ionic level?

When sodium chloride is dissolved in water, the overall change can be represented as:

$$NaCl(s) + (aq) \longrightarrow Na^+(aq) + Cl^-(aq)$$

For one mole of the solute and the formation of an infinitely dilute solution, this process is described as the **heat of solution** of sodium chloride, $\Delta H_{soln}$ (NaCl(s)) $= +5$ kJ

In order to understand this enthalpy change, the overall process of solution can be divided into two distinct stages which are shown in figure 11.17. First, the separation of the solid ionic crystal into monatomic gaseous ions, i.e. the reverse of the lattice energy process,

$$NaCl(s) \longrightarrow Na^+(g) + Cl^-(g) \qquad -\Delta H_{latt} = +776 \text{ kJ}.$$

Second, the solvation (hydration) of these gaseous ions by water molecules which is known as the **hydration energy** $\Delta H_{hyd}$ (section 9.8).

$$Na^+(g) + Cl^-(g) + (aq) \longrightarrow Na^+(aq) + Cl^-(aq)$$

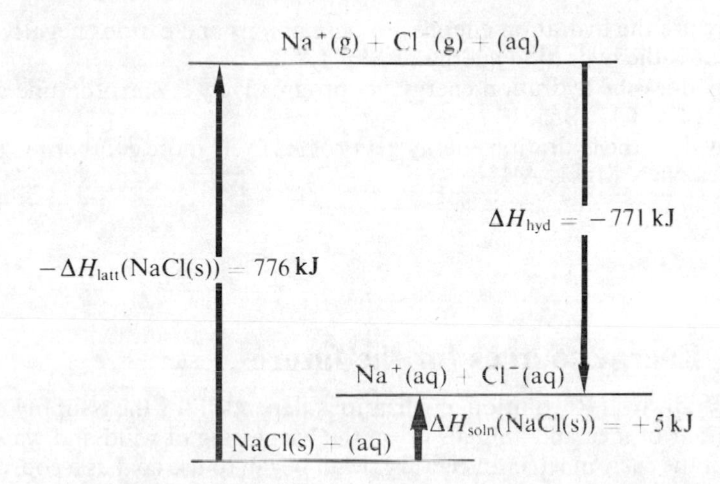


**Figure 11.17** The relationship between lattice energy, hydration energy and heat of solution.

The first of these stages is, of course, always endothermic since it involves separating the ions of the solute. The second stage, on the other hand, is always exothermic since it involves the attraction of ions in the solute for solvent molecules. The overall enthalpy change on solution will depend on whether the endothermic or the exothermic stage is the larger.

In the case of sodium chloride, the endothermic process is marginally greater than the exothermic process so the heat of solution has a small positive value. The lattice energy and the hydration energy for ionic substances are nearly always large values and the heat of solution, which is the difference between these two values, is positive in some cases and negative in others. The relationship between these three quantities is:

$$\Delta H_{soln} = -\Delta H_{latt} + \Delta H_{hyd}$$

Check this for yourself in figure 11.17.

Notice that the hydration energy used in figure 11.17 is really the sum of the separate hydration energies of $Na^+$ and $Cl^-$. Clearly, the individual solvation energy for $Na^+$ cannot be measured directly since sodium ions always exist in combination with some anion or other. Nevertheless, it is often very useful to know the individual solvation energies for particular ions and various attempts have been made to estimate these values from the overall hydration energy of the ionic compound. The convention we shall follow in this book is to accept $-1\,075$ kJ as the hydration energy for $H^+$. All other enthalpies of hydration can then be obtained from the overall hydration energies of the compounds.

A few individual hydration energies are given in table 11.6. It does, of course, follow from what we have just said that the hydration energies of compounds can be obtained by addition of the hydration energies of their constituent ions.

**Table 11.6** Enthalpies of hydration of some ions

| **Ion** | $\Delta H_{hyd}$ **/kJ mole$^{-1}$** | **Ion** | $\Delta H_{hyd}$ **/kJ mole$^{-1}$** |
|---|---|---|---|
| $H^+$ | −1075 | $F^-$ | −457 |
| $Li^+$ | −499 | $Cl^-$ | −381 |
| $Na^+$ | −390 | $Br^-$ | −351 |
| $K^+$ | −305 | $I^-$ | −307 |
| $Mg^{2+}$ | −1891 | | |
| $Ca^{2+}$ | −1562 | | |
| $Al^{3+}$ | −4613 | | |

Look closely at table 11.6.

- ○ Why are the hydration energies of *both* anions and cations negative?
- ○ What is the hydration energy of $MgCl_2$?
- ○ Why does the hydration energy get progressively less exothermic along the series $F^-$, $Cl^-$, $Br^-$, $I^-$?
- ○ Why does the hydration energy get progressively more exothermic along the series $Na^+$, $Mg^{2+}$, $Al^{3+}$?

## 11.15 Energy sources for the future

Until the Industrial Revolution, civilizations depended for their supply of energy on the labour of men and animals or on the harnessing of wind and water.

Then, in the early nineteenth century, Man began to use coal as a source of heat and power in his homes and industries as he progressed beyond the level of a simple agrarian society. At first, the use of coal was very small, but as the Industrial Revolution gained momentum towards the middle of the century, the consumption of coal began to double every fifteen years.

Early in the present century, oil began to make an impact. Discoveries multiplied fast and oil became plentiful, cheap and a source of many other products. Without oil, the internal combustion engine would have been impossible and the revolution in land, sea and air transport could never have taken place. Oil became the most convenient fuel for many industrial and domestic purposes and the basic raw material for the organic chemicals industry. Its consumption rose by leaps and bounds. By 1960, oil had outstripped coal as the major source of energy in most industrialized countries.

A decade later, the first danger signals appeared as the rate of oil consumption increased faster than the discovery of new reserves.

Unfortunately, there is only a finite amount of oil on the Earth. From 1870 to 1970 we consumed only 15%, but in a single decade of the seventies we shall consume another 15%. If we continue at this rate, we shall exhaust our supplies of oil well before the year 2050.

The situation with coal is much less alarming since its rate of consumption is lower and reserves are possibly twenty times greater than those of oil. Nevertheless, it has been estimated that 80% of our coal reserves will have disappeared by the year 2500.

It is vital that we start to conserve our resources now, turning our thoughtless over-consumption to a more intelligent use of fuel and avoiding such frivolous wastes of energy as electrically operated tooth brushes and excessive heating an

*Above* Cutting peat, another form of fossil fuel, for domestic heating. Peat is still used as a source of fuel in some areas. It consists of plant fibres and mud and is probably the first stage in the formation of coal.

lighting in homes, offices, schools and factories. For example, it has been shown that more than 20 % of the energy used in heating buildings could be saved by better insulation of roofs and walls, double glazing of windows and the acceptance of slightly lower, but still comfortable, temperatures.

One of the most wasteful users of fuel is the private car in which only 10–20 % of the energy in the fuel ultimately goes towards moving the car. Hopefully, batteries will one day supersede the inefficient internal combustion engine. Vast quantities of precious oil are also being squandered on industrial and domestic heating and on the generation of electric power. Oil is much too valuable for this purpose. It should be conserved as feedstock for essential chemicals such as plastics, paints, pesticides and pharmaceuticals. A more suitable fuel for power generation is coal and either coal or natural gas can be used for heating. The increasing demands for electricity, however, will call for an increase in nuclear power generation. Furthermore, the electricity from nuclear power stations is likely to become cheaper than that from oil-powered stations over the next few years. By the year 2000, it is expected that the percentage of electricity generated from nuclear energy will have risen to about 70 % from its present level of 12 %.

*Left* The huge turbine in the nuclear power station at Dounreay. Heat produced from nuclear reactions generates steam to drive the turbine.

Eventually, it is hoped to develop the so called 'fast-breeder' reactor which can transform the energy in uranium 150 times more efficiently than present-day reactors. Unfortunately, the high capital cost of these new reactors and the problems of handling and storing large quantities of highly toxic waste with complete safety are preventing their construction and development.

Faced with these problems of dwindling energy resources and limited capital resources, it is essential to examine all possibilities for future energy provision. At the present time, hydro-electric schemes supply less than 1 % of the world's electricity, yet there are possibilities for expansion in many areas. The production of hydro-electricity is essentially limited in its location and has only been developed so far in those areas which are both mountainous and industrially advanced. It has been estimated that the full development of all suitable areas could increase the production of hydro-electricity by twenty times.

A less obvious source of water power lies in the rise and fall of the tides. The enormous movement of water between high and low tides around the world results from the moon's gravitational pull on the seas. By damming tidal basins, it has been possible to generate electricity using low speed turbines. One such installation is across the estuary of the Rance in Northern Brittany. There are many other places in the world, including the Bristol Channel, with potential for such development.

By far the biggest source of our energy income is the radiation received directly from the sun. Scientists have calculated that one tenth of the solar energy falling on the Arizona Desert could provide enough energy for the whole of the USA, yet in practice very little direct use of solar energy is made. This is because technology has not yet provided an efficient and economic method of transforming solar radiation into useable energy. The most promising possibility for solar energy is

*Above* Power from the wind – the Admiral's Cup inshore race at Cowes.

the heating of buildings. Although sunlight is intermittent and unreliable in many areas, developments in the storage of solar energy may help to solve this problem. Alternatively, solar energy could be used in conjunction with conventional heating systems and this would still save considerable quantities of our fossil fuels.

The sun's inexhaustible source of energy comes from nuclear fusion. The origin of this energy is exactly the same as that of the hydrogen bomb. Matter, below the sun's surface, is crushed together under a pressure millions upon millions of times greater than atmospheric pressure. Under this immense pressure atoms collide and undergo nuclear fusion. Since only the smallest atoms can undergo fusion, the largest proportion of the sun's energy results from the fusion of heavy-hydrogen (deuterium) atoms forming helium.

$$^{2}_{1}H + ^{2}_{1}H \longrightarrow ^{3}_{2}He + ^{1}_{0}n + \text{Energy}$$

As a result of this fusion process, the energy radiated from the sun's surface in one year is equivalent to 250 thousand million million million tonnes of oil. If only man could harness the energy from such a fusion process on earth, he would have solved his energy problems for ever. Sea water, for example, contains 34 grams of deuterium per tonne, energy equivalent to 200 tonnes of oil. Another advantage of nuclear fusion is that the products are stable non-radioactive elements, unlike the products of nuclear fission breeder reactors which pose considerable disposal problems and safety hazards. It is impossible to predict when, or indeed if ever, controlled fusion will be achieved. The major problem is that it requires temperatures of about 100 000 000°C before atomic collisions are vigorous enough to cause fusion. Although such elevated temperatures have not yet been reached, experiments have achieved temperatures close to those required to sustain the fusion reaction and it is conceivable that full reaction conditions could be achieved within the next twenty years.

In the meantime, however, a world of limitless energy is still only a dream and we must learn to use and conserve our finite resources more wisely and more efficiently.

## Summary

1 The transfer of energy to or from chemicals plays a crucial part in chemical processes in industry and in living things.

2 In an exothermic reaction, heat is lost from the reacting materials; in an endothermic reaction, heat is gained by the reacting materials.

3 The standard heat of a reaction is the amount of heat absorbed or evolved when the molar quantities of reactants as stated in the equation react together under standard conditions.

4 Standard conditions for thermochemical measurements are:
a pressure of one atmosphere,
a temperature of 298 K,
substances in their normal physical states for these conditions,
concentrations of unit activity.

5 The standard heat of formation of a substance, $\Delta H_f^\ominus$, is the heat evolved or absorbed when one mole of the substance is formed from its elements under standard conditions.

6 The heat of atomization of an element, $\Delta H_{at}^\ominus$, is the enthalpy change when one mole of gaseous atoms is formed from the element under standard conditions.

7 The standard heat of combustion of a substance, $\Delta H_c^\ominus$, is the enthalpy change when one mole of the substance is completely burnt in oxygen under standard conditions.

8 Hess's Law of Constant Heat Summation says that the energy change in converting reactants, A + B, to products, X + Y, is the same regardless of the route by which the chemical change occurs.

9 A substance is said to be stable if it tends neither to decompose nor to react spontaneously.

10 The heat of formation of a compound provides a measure of the energetic stability of that compound relative to its elements. If $\Delta H_f^\ominus$ is negative, the compound is more stable than its elements.

11 It is important to distinguish between energetic stability and kinetic stability. Compounds such as ethyne ($C_2H_2$) are energetically unstable with respect to their elements, yet they do not change at ordinary temperatures and pressures because their reaction rate is immeasurably slow—they are kinetically stable.
12 Bond-breaking is an endothermic process whereas bond-making is exothermic.
13 A definite quantity of energy can be associated with each type of covalent bond. This energy is absorbed when the bond is broken and evolved when the bond is formed. Thus, the bond energy for the C—Cl bond, $E$(C—Cl), is defined as the energy required to break one mole of C—Cl bonds forming uncharged products under standard conditions.
14 Bond energies for molecular substances can be compared with lattice energies for ionic compounds which give the heat of formation of one mole of the ionic crystal from gaseous ions under standard conditions.
15 The close agreement between calculated and experimental lattice energies for many ionic compounds provides strong evidence that the simple model of an ionic lattice composed of discrete spherical ions with evenly distributed charge is generally very satisfactory.

## Study questions

1 Figure 11.18 shows a diagram of the apparatus used by a student to determine the heat of combustion of ethanol. The heat produced by the burning fuel warms a known mass of water. By measuring the mass of fuel burnt and the temperature rise of the water, it is possible to obtain an approximate value for the heat of combustion of the fuel.

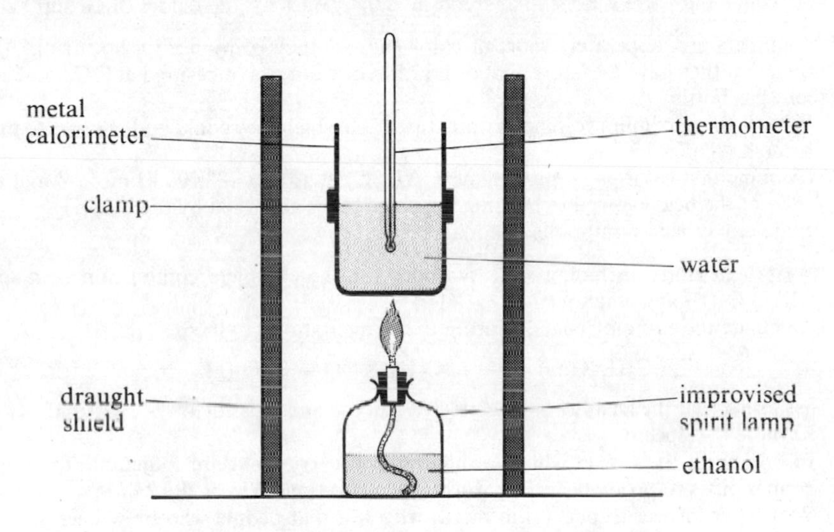


**Figure 11.18** A simple apparatus to determine the heat of combustion of a liquid fuel.

Volume of water in calorimeter = 400cm$^3$
Initial temperature of water = 12°C
Final temperature of water = 22°C
Mass of ethanol burnt = 0.92g
(Specific heat capacity of water = 4.2 J g$^{-1}$ K$^{-1}$)

(a) How much heat is required to raise the temperature of the water from 12°C to 22°C? (This is the amount of heat produced when 0.92g of ethanol burn.)
(b) How much heat would be produced when 1 mole of ethanol burns? (Call this the heat of combustion of ethanol.)
(c) Why is the answer to (b) not described as the *standard* heat of combustion of ethanol?
(d) An accurate value for $\Delta H_c^\ominus(C_2H_5OH(l))$ is $-1368$ kJ mole$^{-1}$. Mention four serious errors in the simple experiment which could be responsible for the poor result.

2 Ethanol ($C_2H_5OH$) cannot be prepared directly from its elements so the standard heat of formation of ethanol must be obtained indirectly.
(a) Define the terms: $\Delta H_f^\ominus$ and $\Delta H_c^\ominus$.
(b) Write an equation for the formation of ethanol from its elements in their standard states.
(c) Draw an energy cycle linking the heat of formation of ethanol with its heat of combustion and the heat of combustion of its constituent elements.

(d) Calculate $\Delta H_f^\ominus(C_2H_5OH(l))$.

$$\Delta H_c^\ominus(C(graphite)) = -393 \text{ kJ mole}^{-1}$$
$$\Delta H_c^\ominus(H_2(g)) = -286 \text{ kJ mole}^{-1}$$
$$\Delta H_c^\ominus(C_2H_5OH(l)) = -1\,368 \text{ kJ mole}^{-1}$$

3 Table 11.7 shows the standard heats of formation of the oxides of elements in Group I and Group IV.

**Table 11.7** Standard heats of formation of the oxides of elements in Group I and Group IV

| Group I | $\Delta H_f^\ominus$ /kJ mole$^{-1}$ | Group IV | $\Delta H_f^\ominus$ /kJ mole$^{-1}$ |
|---|---|---|---|
| $Li_2O$ | −596 | $CO_2$ | −393 |
| $Na_2O$ | −416 | $SiO_2$ | −909 |
| $K_2O$ | −361 | $GeO_2$ | −551 |
| $Rb_2O$ | −330 | $SnO_2$ | −581 |
| $Cs_2O$ | −318 | $PbO_2$ | −277 |

(a) Why does $\Delta H_f^\ominus$ for the oxides of Group I elements get less exothermic as the atomic numbers of the elements increase?
(b) Why is the $\Delta H_f^\ominus$ of $SiO_2$ much more exothermic than that for $CO_2$?
(c) Suggest reasons why there is a steady trend in $\Delta H_f^\ominus$ values for the oxides of Group I elements, but there is no similar trend in $\Delta H_f^\ominus$ values for the oxides of Group IV.

4 Two campers are desperately short of calor gas, yet they badly need a hot drink. They estimate that they have the equivalent of 1.12dm$^3$ of calor gas (measured at 0°C and 1 atm) in their 'gas bottle'.
(a) What is the maximum volume of water (at 20°C) which they could boil in order to make some hot coffee?
(Assume that calor gas is pure butane, $\Delta H_c(C_4H_{10}(g)) = -3\,000$ kJ mole$^{-1}$ and that 75% of the heat evolved in burning the calor gas is absorbed by the water.)
(b) State any other assumptions you make.

5 (a) Write equations, including state symbols, for the complete combustion of glucose ($C_6H_{12}O_6(s)$) and ethanol ($C_2H_5OH(l)$).
(b) Calculate the enthalpy change during the fermentation of glucose,

$$C_6H_{12}O_6(s) \longrightarrow 2C_2H_5OH(l) + 2CO_2(g)$$

assuming that the heats of combustion of glucose and ethanol are −2 820 and −1 368 kJ mole$^{-1}$ respectively.
(c) In breweries, the vats in which fermentation is carried out are sometimes fitted with copper pipes to promote cooling during fermentation. Why is this?
(d) With your answer to part (c) in mind, why is it that people who brew beer at home usually put their fermentation vessel in a warm place (such as an airing cupboard) during fermentation?

6 A possible mechanism for the reaction of fluorine with methane is

$$CH_4 + F_2 \xrightarrow{\text{slow}} CH_3\cdot + HF + F\cdot$$

$$CH_3\cdot + F\cdot \xrightarrow{\text{fast}} CH_3F$$

(a) Use the bond energies in table 11.3 to calculate the enthalpy change in each step of the reaction mechanism.
(b) Is the suggested mechanism viable? Explain your answer.
(c) Write the equations for a reaction between $CH_4$ and $Cl_2$ assuming a similar mechanism to that for $CH_4$ and $F_2$ and calculate the enthalpy change in each step.
(d) Is such a mechanism viable for the $CH_4/Cl_2$ reaction? Explain your answer.
(e) Why is it that fluorine will react with methane in the dark, whereas chlorine only reacts appreciably with methane in sunlight?

7 (a) A simple rule regarding the solubility of solutes in solvents is 'Like dissolves like'. In general, non-polar solutes dissolve readily in non-polar solvents but not in polar ones. On the other hand, polar solutes dissolve readily in polar solvents but are insoluble in non-polar solvents. Discuss and explain this pattern in solubility, mentioning specific examples and referring to solute-solute, solvent-solvent and solute-solvent attractions.
(b) Explain the terms lattice energy, hydration energy and heat of solution with reference to the hypothetical substance, $X^+Y^-$.
(c) Draw an energy diagram relating the three terms in part (b).
(d) Calculate the hydration energy of potassium iodide, assuming that its heat of solution is $+21$ kJ mole$^{-1}$ and its lattice energy is $-642$ kJ mole$^{-1}$.

8 (a) Explain what is meant by the terms
(i) ionization energy,
(ii) atomization energy,
(iii) lattice energy.
(b) Draw a complete, fully-labelled Born-Haber Cycle for the formation of potassium bromide.
(c) Using the information in the table below, calculate the lattice energy of potassium bromide.

| Reaction | $\Delta H$/kJ mole$^{-1}$ |
|---|---|
| $K(s) + \frac{1}{2}Br_2(l) \longrightarrow K^+Br^-(s)$ | $-392$ |
| $K(s) \longrightarrow K(g)$ | $+90$ |
| $K(g) \longrightarrow K^+(g) + e^-$ | $+420$ |
| $\frac{1}{2}Br_2(l) \longrightarrow Br(g)$ | $+112$ |
| $Br(g) + e^- \longrightarrow Br^-(g)$ | $-342$ |

(d) The values of the lattice energies of the other potassium halides are:

| Compound | KF | KCl | KI |
|---|---|---|---|
| Lattice energy/kJ mole$^{-1}$ | $-813$ | $-710$ | $-643$ |

What explanation can you give for the trend in these values?

9 The enthalpy changes involved in the synthesis of calcium oxide are represented in a Born-Haber cycle below. (The numerical values printed beside the cycle are in kJ mole$^{-1}$.)

$Ca^{2+}(g) + O^{2-}(g)$

$+790 = \Delta H_e(O^-)$

$Ca^{2+}(g) + e^- + O^-(g)$

$-141 = \Delta H_e(O)$

$Ca^{2+}(g) + 2e^- + O(g)$

$+249 = \Delta H_{at}(O)$

$Ca^{2+}(g) + 2e^- + \frac{1}{2}O_2(g)$

$+1\,100 = \Delta H_{i_2}(Ca)$

$Ca^+(g) + e^- + \frac{1}{2}O_2(g)$

$+590 = \Delta H_{i_1}(Ca)$

$Ca(g) + \frac{1}{2}O_2(g)$

$+177 = \Delta H_{at}(Ca)$

$Ca(s) + \frac{1}{2}O_2(g)$

$-636 = \Delta H_f(CaO(s))$

$Ca^{2+}O^{2-}(s)$

$\Delta H_{latt}(CaO(s))$

(a) Calculate the lattice energy for calcium oxide, $\Delta H_{latt}(CaO(s))$.
(b) Why is $\Delta H_{i_2}(Ca)$ greater than $\Delta H_{i_1}(Ca)$?
(c) Why is $\Delta H_e(O)$ negative whereas $\Delta H_e(O^-)$ is positive?
(d) State and explain how the value of
(i) the first ionization energy and
(ii) the heat of atomization
of magnesium would compare with the corresponding values for calcium.

# 12 Patterns Across the Periodic Table

## 12.1 Periodic properties

In chapter 3 we noticed that a very large number of chemical and physical properties of the elements vary periodically with atomic number. *The idea of periodicity embodied in the periodic table is one of the major unifying themes in chemistry.* It provides an organized structure to our knowledge and understanding of inorganic chemistry and it reduces the need for memorizing isolated pieces of factual information. In this chapter, we shall look more closely at the trends and gradations in properties of the elements in periods 2 and 3 in the periodic table and their more important compounds. Several of the bulk physical properties of the elements, such as melting point and density, depend upon their structure and bonding and we discussed these relationships in chapter 3. However, the structure and bonding of the elements are in turn related to atomic properties such as electron structures, ionization energies and atomic radii and it is therefore profitable to discuss the trends in these atomic properties and then relate them to physical and chemical properties. Table 12.2 provides a summary of the structure, bonding and properties of the elements in periods 2 and 3.

## 12.2 Atomic structure and the periodic table

Chemists now realize that the atomic number of an element dictates the number and arrangement of its electrons.

Table 12.1 shows the electronic (shell and sub-shell) structures for the elements in periods 2 and 3.

**Table 12.1** The electronic structures of the elements in periods 2 and 3

| **Period 2** | Li | Be | B | C | N | O | F | Ne |
|---|---|---|---|---|---|---|---|---|
| **Electronic shell structure** | 2,1 | 2,2 | 2,3 | 2,4 | 2,5 | 2,6 | 2,7 | 2,8 |
| **Electronic sub-shell structure** | $1s^22s^1$ | $1s^22s^2$ | $1s^22s^22p^1$ | $1s^22s^22p^2$ | $1s^22s^22p^3$ | $1s^22s^22p^4$ | $1s^22s^22p^5$ | $1s^22s^22p^6$ |
| **Period 3** | Na | Mg | Al | Si | P | S | Cl | Ar |
| **Electronic shell structure** | 2,8,1 | 2,8,2 | 2,8,3 | 2,8,4 | 2,8,5 | 2,8,6 | 2,8,7 | 2,8,8 |
| **Electronic sub-shell structure** | $1s^22s^22p^6$ $3s^1$ | $1s^22s^22p^6$ $3s^2$ | $1s^22s^22p^6$ $3s^23p^1$ | $1s^22s^22p^6$ $3s^23p^2$ | $1s^22s^22p^6$ $3s^23p^3$ | $1s^22s^22p^6$ $3s^23p^4$ | $1s^22s^22p^6$ $3s^23p^5$ | $1s^22s^22p^6$ $3s^23p^6$ |

**Table 12.2** A summary of the structure, bonding and properties of the elements in periods 2 and 3

| | | | |
|---|---|---|---|
| **Period 2** | Li Be | B C | N O F Ne |
| **Period 3** | Na Mg Al | Si | P S Cl Ar |
| **Type of element** | metal | metalloid | non-metal |
| **Structure** | Metallic<br>Close-packed or body-centre cubic arrangement of atoms | Giant-molecular<br>Infinite lattice structure | Simple-molecular<br>Discrete small molecules (e.g. $P_4$, $Cl_2$, $O_2$, Ne) |
| **Bonding** | Metallic: strong forces of attraction of positive ions for mobile outer electrons. | Covalent: very strong forces of attraction between atoms due to the attraction of nuclei for shared electrons. | Molecular (Van der Waals'): weak forces between molecules (strong covalent forces hold atoms together within the molecule). |
| **Co-ordination number** | 8 or 12 | 4 or less | 1 or 2 usually |
| **Properties** | | | |
| m.pt./b.pt. | High | Very high | Low |
| $\Delta H_{fus}/\Delta H_{vap}$ | High | Very high | Low |
| Molar volume | High/Moderate | Low | High |
| Conductivity | Good | Poor | Nil |

*Elements in the same group of the periodic table have similar electron configurations.* For example, Group I elements (Li to Fr) have one electron in their outer shell ($ns^1$). Group II elements (Be to Ra) have two electrons in their outer shell ($ns^2$) and Group VII elements (F to At) have seven outer shell electrons ($ns^2np^5$).

Since electron configurations recur in a periodic manner, it is not surprising that properties show a similar periodic recurrence. We shall investigate the similarities in the properties of elements within a *group* more thoroughly in chapters 14, 16, and 17. Meanwhile, we can investigate the variation of atomic properties across a period.

## 12.3 Atomic properties and the periodic table

### ATOMIC AND IONIC SIZE

Since the electron cloud of an atom has no definite limit, the size of an atom cannot be defined simply and easily. However, *the radius of an atom is often defined as the distance of closest approach to another identical atom.* This means that the size of an atom is determined by the effective volume of the outer electrons. But, when we say the atomic radius of iodine in the solid is 0.133nm, do we mean half the distance between the two iodine atoms in an $I_2$ molecule or half the distance between two iodine atoms in adjacent molecules which are not chemically bonded (figure 12.1)? In order to clarify this situation, *half the distance between two covalently bonded iodine atoms is defined as the covalent radius for iodine and half*

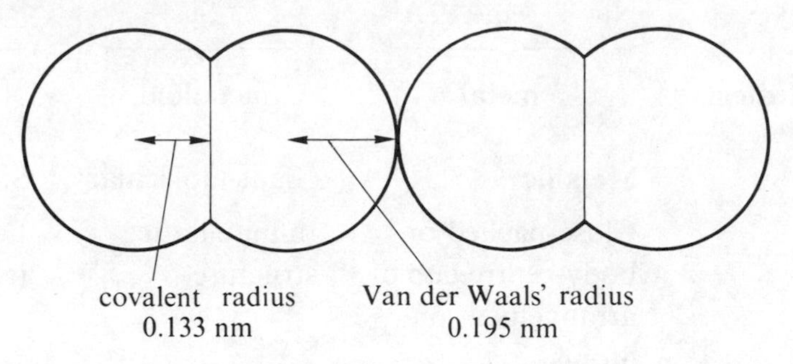


**Figure 12.1** Covalent (atomic) and Van der Waals' radii for iodine.

*the distance between two iodine atoms which are not chemically bonded is defined as the Van der Waals' radius.* The former gives a measure of the length of a covalent bond and the latter gives a measure of the length of a Van der Waals' bond. The covalent radius for iodine is 0.133nm and the Van der Waals' radius is 0.195nm. Clearly, the type of bonding greatly influences the size of an atom. When chemists refer to *atomic radii* they are usually thinking of *covalent radii.*

The size of atoms in metallic elements can be obtained by determining the internuclear distance in the metallic crystal using X-ray diffraction and then dividing by two to give an atomic radius.

Figure 12.2 shows the atomic radii (covalent radii) of the elements Li–F and Na–Cl in the periodic table. Notice that along each period there is a gradual decrease in atomic size as the outer electron shell is being filled.

| | Li | Be | B | C | N | O | F |
|---|---|---|---|---|---|---|---|
| covalent radius /nm | 0.123 | 0.089 | 0.082 | 0.077 | 0.070 | 0.066 | 0.064 |

| | Na | Mg | Al | Si | P | S | Cl |
|---|---|---|---|---|---|---|---|
| covalent radius /nm | 0.156 | 0.136 | 0.125 | 0.117 | 0.110 | 0.104 | 0.099 |

**Figure 12.2** Covalent (atomic) radii of the elements Li to F and Na to Cl.

- ○ What causes the general decrease in size across a period?
- ○ How is the attraction of the positive nucleus for the outer electrons changing across the period?

Moving from one element to the next across a period, electrons are being added to the same shell at about the same distance from the nucleus and protons are being added to the nucleus. Therefore, the electrons are attracted and pulled towards the nucleus with an increasing positive charge and so the radius of the atom decreases. However, the rate of decrease in the radius becomes smaller as the atoms get heavier since the addition of one more proton to the 11 already present in sodium causes a greater proportional change in nuclear attractive power than the addition of one more proton to the 16 already present in sulphur. Thus, the atomic radius falls by 0.020nm from Na to Mg, but by only 0.005nm from S to Cl.

So far we have considered only atomic sizes. Ionic sizes are also useful in explaining and understanding chemical properties since they provide a measure of the space occupied by an ion in the crystal lattice. Figure 12.3 shows the radii of the most stable ions for some of the first 20 elements in the periodic table. Look for the following general patterns in figure 12.3.

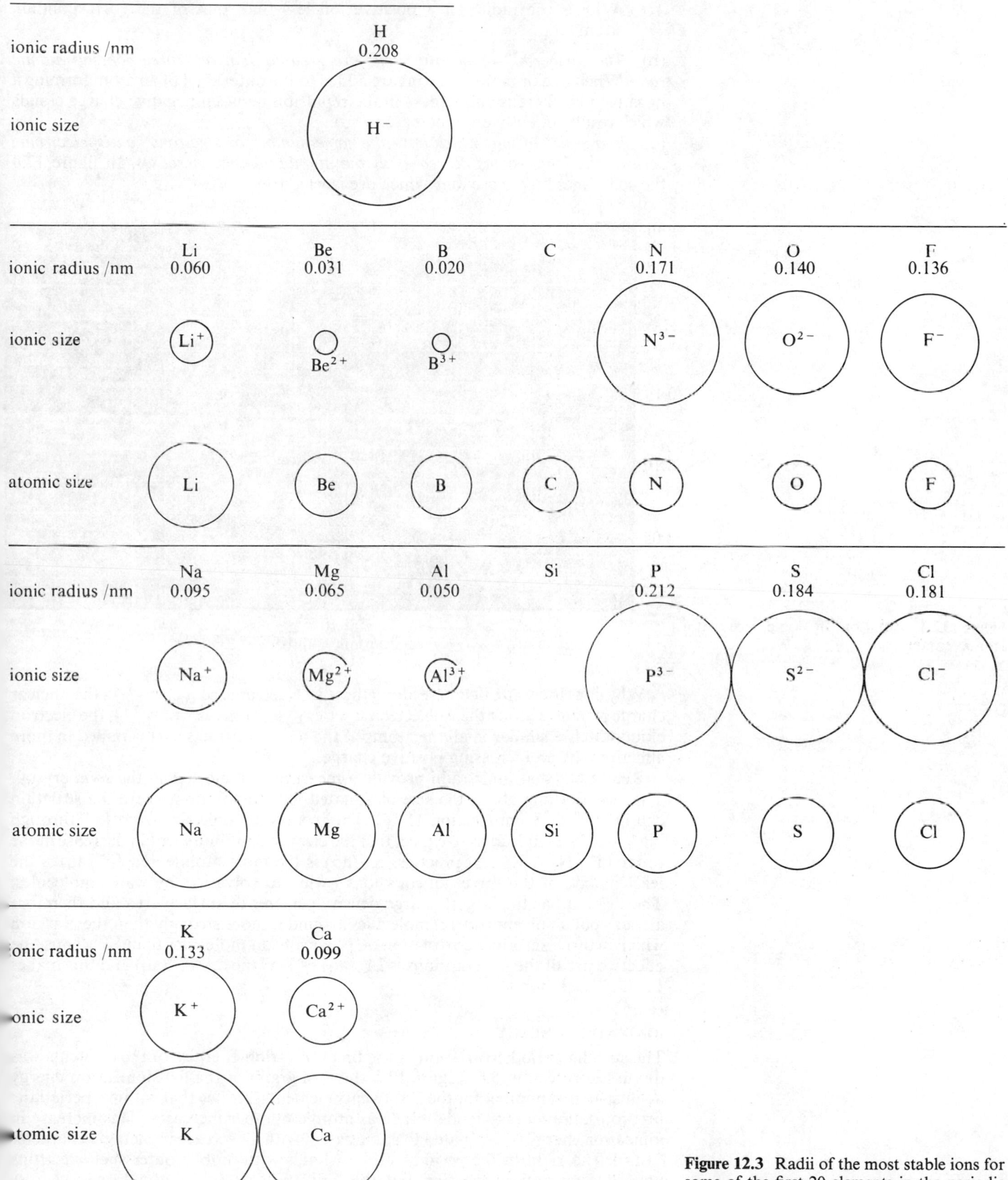


**Figure 12.3** Radii of the most stable ions for some of the first 20 elements in the periodic table.

(a) *The ionic radii of positive ions are smaller than the corresponding atomic radii.*

- ○ Which electrons are usually removed from metals in forming their positive ions?
- ○ How does the electronic structure of a positive ion (e.g. $Na^+$ or $Mg^{2+}$) compare with that of the corresponding metal atom?
- ○ Why is the radius of a positive ion less than that of the corresponding atom?

(b) *The ionic radii of negative ions are greater than the corresponding atomic radii.* When one or more electrons are added to the outer shell of an atom forming a negative ion, there is an increase in the repulsion between negative charge clouds which results in an overall increase in size.

(c) *In a series of ions which have the same number of electrons (an isoelectronic series), the ionic radius decreases as the atomic number increases.* In figure 12.4 the solid lines link those ions which are isoelectronic.

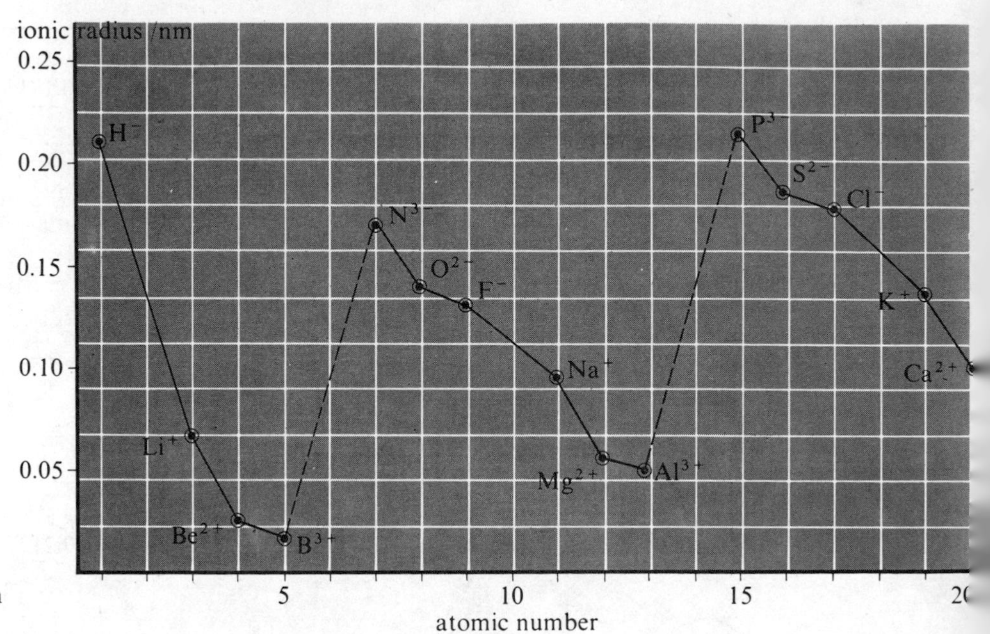


**Figure 12.4** Variation in ionic size with atomic number.

Why does ionic size decrease along these isoelectronic sequences? As the nuclear charge increases along the isoelectronic series (e.g. from $N^{3-}$ to $Al^{3+}$), the electron cloud, which is similar in all the atoms of the series, contracts as it is pulled in more effectively by an increasing positive charge.

Remember that ionic radii provide a measure of ionic size in the *solid* crystal. They say nothing about the size of aquated ions in solution where the situation can be very different (section 11.14). The increasing ionic size from $Li^+$ through $Na^+$ to $K^+$ might lead us to expect that the electrical mobility would decrease in the order $Li^+$, $Na^+$, $K^+$. In practice, $K^+$(aq) is the most mobile and $Li^+$(aq) is the least mobile of the three aqueous ions owing to solvation by water molecules. The $Li^+$ ion has the largest charge density per unit of surface area and therefore attracts polar solvent (water) molecules around it more strongly than the $Na^+$ ion which in turn exerts more attraction on polar solvent molecules than $K^+$. Thus, the effective size of the aqueous ions is $Li^+(aq) > Na^+(aq) > K^+(aq)$ and this makes $Li^+$(aq) least mobile.

## IONIZATION ENERGY

The striking periodic variation in the first ionization energies of the elements was discussed in section 5.6. Figure 12.5 shows a graph of the first ionization energy against atomic number for the first twenty elements. Notice that within a period the first ionization energy tends to rise as atomic number increases. This increase in ionization energy is associated to some extent with a decrease in metallic character from left to right in the periodic table. Metals with mobile outer-shell electrons have low ionization energies, but the ionization energy gradually rises across the period as metallic character disappears and electrons are more tightly held

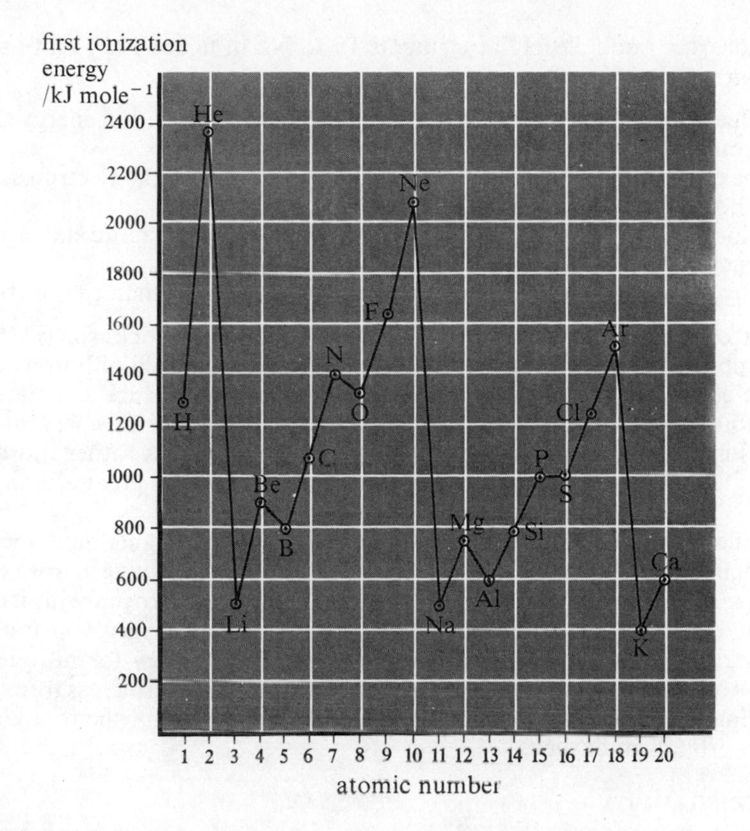


**Figure 12.5** The first ionization energies plotted against atomic number for the elements H to Ca

in covalent bonds. But, how are ionization energies related to atomic structure? What factors do you think will affect its size?

The ionization energy of an atom is influenced mainly by three atomic parameters.

1 *The distance of the outermost electron from the nucleus*
As the distance increases, the attraction of the positive nucleus for the negative electron will decrease and consequently the ionization energy will decrease.

2 *The size of the positive nuclear charge*
As the nuclear charge gets more and more positive, its attraction for the outermost electron increases and consequently the ionization energy increases.

3 *The screening (shielding) effect of inner electrons*
The outermost electrons are repelled by all the other electrons in the atom besides being attracted by the positive nucleus. Chemists say that the outermost electron is **screened** (shielded) from the attraction of the nucleus by the repelling effect of inner electrons. For example, the outermost electron in an atom with an atomic number of $Z$ (i.e. $Z$ protons in the nucleus) experiences an attraction by the nucleus of charge $Z$ and a repulsion from $(Z - 1)$ electrons. Thus, it might appear that the $(Z - 1)$ electrons could cancel all but one charge on the nucleus. In practice, the outermost electron occupies an orbital which is not completely outside the orbitals of other inner electrons and so the screening (shielding) effect is much less than perfect. This means that the electron experiences an overall effective nuclear attraction which is much greater than that from one proton. In general, however, *the screening effect by inner electrons is more effective the closer these inner electrons are to the nucleus*. Thus,

(a) Electrons in shells of lower principal quantum number are more effective shields than electrons in shells of higher quantum number.

(b) Electrons in the same shell exert a negligible shielding effect on each other. This means that we need only consider inner shells of electrons when we discuss the screening effect on an outermost electron.

*Moving from left to right across any period, there is a general increase in the first ionization energy*. This is due to the fact that the nuclear charge is increasing across the period from one element to the next and the increasing nuclear charge causes a decrease in the atomic radius (see page 167) and thus a decrease in the distance of the outermost electron from the nucleus. The screening effect remains almost the same from one element to the next across a period since electrons are added successively to the same shell and such electrons screen each other very little from the increasing nuclear charge. Look closely at elements Li to Ne in figure 12.5. Note that the ionization energy of beryllium is higher than that of boron.

- ○ Which other element in period 2 has a higher first ionization energy than the element immediately after it?
- ○ The electron structure of beryllium is $1s^2 2s^2$. Write out the electron structure of boron in the same notation.
- ○ Which of the two elements, beryllium or boron, has the more stable electron structure?
- ○ Why is the first ionization energy of beryllium higher than that of boron?

The electron configuration of beryllium is $1s^2 2s^2$, whereas that of boron is $1s^2 2s^2 2p^1$. All the sub-shells in beryllium are filled, but the outer sub-shell of boron contains only one electron. In the same way that filled electron shells are associated with extra stability, *there is also some extra energetic stability associated with filled sub-shells*. This means that the electron structure in beryllium is rather more stable than we might have expected and its first ionization energy is greater than that of boron.

A similar situation arises with nitrogen which has a higher first ionization energy than oxygen. The electron structures of nitrogen and oxygen are $1s^2 2s^2 2p^3$ and $1s^2 2s^2 2p^4$ respectively. The half filled $2p$ sub-shell in nitrogen with its evenly distributed charge is more stable than the $2p$ sub-shell in oxygen which contains four electrons. This results in a higher first ionization energy for nitrogen than oxygen. Alternatively, we can visualize the greater ease of electron loss from oxygen as resulting from the extra repulsion involved in *pairing* two of the four electrons in the $2p$ orbitals of oxygen.

## ELECTRONEGATIVITY

*Elements that tend to acquire electrons in their chemical interactions are said to be* **electronegative** and the electronegativity of an atom provides a numerical measure of the power of that atom in a molecule to attract electrons (section 8.2). The electronegativity values for main group elements (based on Pauling's scale) are given in table 12.3 and the values for the first 20 elements in the periodic table are plotted graphically against atomic number in figure 12.6.

**Table 12.3** Electronegativities of the main-group elements.

| | | | | | | | | |
|---|---|---|---|---|---|---|---|---|
| | | H 2.1 | | | | | | He — |
| Li 1.0 | Be 1.5 | | B 2.0 | C 2.5 | N 3.0 | O 3.5 | F 4.0 | Ne — |
| Na 0.9 | Mg 1.2 | | Al 1.5 | Si 1.8 | P 2.1 | S 2.5 | Cl 3.0 | Ar — |
| K 0.8 | Ca 1.0 | | Ga 1.6 | Ge 1.8 | As 2.0 | Se 2.4 | Br 2.8 | Kr — |
| Rb 0.8 | Sr 1.0 | | In 1.7 | Sn 1.8 | Sb 1.9 | Te 2.1 | I 2.5 | Xe — |
| Cs 0.7 | Ba 0.9 | | Tl 1.8 | Pb 1.8 | Bi 1.9 | Po 2.0 | At 2.2 | Rn — |
| Fr 0.7 | Ra 0.9 | | | | | | | |

*Notice that electronegativities decrease down a group, but increase across a perioc* As expected, the most electronegative elements are the reactive non-metals in th top right hand corner of the periodic table, whereas the least electronegativ elements are the reactive metals in the bottom left hand corner. Why does electrc negativity increase from left to right across the periodic table? Moving from on element to the next across a period, the nuclear charge increases by one unit an one electron is added to the outer shell. As the positive charge on the nucleus rise the atom has an increasing electron-attracting power and therefore an increasin electronegativity.

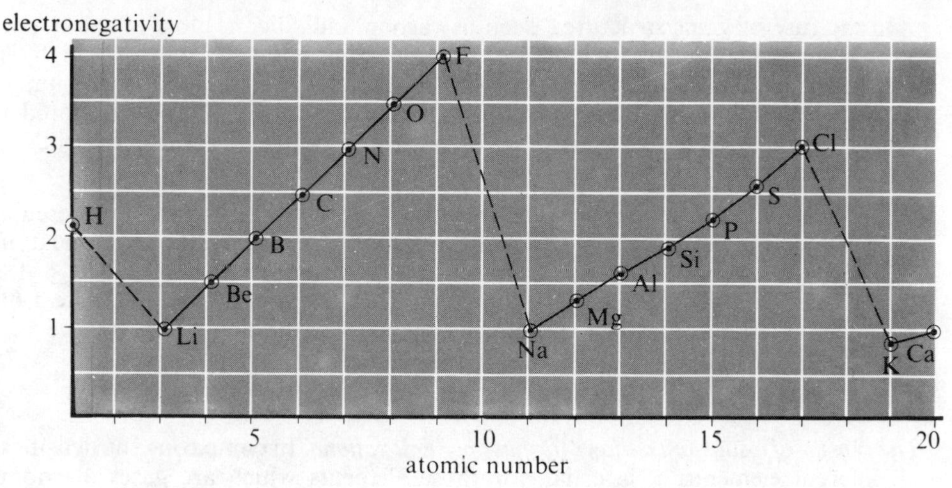


**Figure 12.6** Variation in electronegativity with atomic number.

## 12.4 Physical properties and the periodic table

### MELTING POINT

*The melting point represents the temperature at which a pure solid is in equilibrium with its pure liquid at atmospheric pressure.* Moving across a period from left to right, the melting point rises through the metals and metalloids and then drops abruptly to low values for the non-metals (see figure 3.8). Although there is some evidence of periodicity in the melting points of elements, it is not so obvious as the periodicity of other properties. This is partly because there is more than one factor involved in the determination of a melting point. Melting points depend on both the structure (packing) and on the bonding in a substance. Since there are abrupt changes in bonding and structure across a period there are also abrupt changes in the melting point. For example with the period Na → Ar, the melting point rises sharply from the last metallic element aluminium to the giant molecule silicon and then falls even more abruptly from silicon to the simple molecule phosphorus. In the period Li → Ne, the melting point rises steeply from the metallic element beryllium to the giant molecule boron and then falls sharply from the giant molecule carbon to the simple molecule nitrogen.

### MOLAR HEAT OF FUSION

*The molar heat of fusion represents the amount of energy required to melt one mole of the solid at its melting point.* Clearly, there will be a close similarity in the periodicity of the heats of fusion and the periodicity of the melting points (figure 3.8). When a marked change of structure and bonding occurs from one element to the next across a period there is a sharp change in the $\Delta H_{fus}$ in the same way that we noticed sharp changes in the melting point in the last sub-section.

Consider the changes in $\Delta H_{fus}$ across the third period from Na to Ar. In metals, (Na, Mg, Al) a good deal of the bonding remains when the metal liquefies and since only restricted bond breaking has occurred, the heat of fusion is only moderate. In giant covalent molecules such as silicon, nearly all the bonds must be broken before the solid melts and therefore the heat of fusion is very large. In simple molecules ($P_4$, $S_8$, $Cl_2$ and Ar), the heats of fusion are very small because melting in these elements involves only the breaking of molecular (Van der Waals') bonds.

### BOILING POINT AND MOLAR HEAT OF VAPORIZATION

The periodic trends in this case can be related to those in the last two sub-sections. However, since structure (packing) has been disrupted in forming the liquid, the size of the boiling point and the $\Delta H_{vap}$ depends very much on the strength of the bonds which are broken in converting the liquid to a vapour. In the case of metals, most of the metallic bonding still exists in the liquid state, but the atoms are separated to a considerable distance in forming the vapour and this involves complete breakage of the metal bonds. Thus, the boiling points of metals are much higher than their melting points.

In the case of giant structures, such as carbon and silicon, most of the bonds have been broken during melting and so the boiling point, though very high, is not much higher than the melting point. The boiling of elements with simple molecular structures, such as $P_4$, and $S_8$ and $Cl_2$, involves breaking weak Van der Waals' bonds. This is relatively easy for most of these elements and the boiling points occur at low temperatures, not much above the melting point.

- ○ What simple kinetic process does the molar heat of vaporization represent?
- ○ Why is the molar heat of vaporization usually larger than the molar heat of fusion?
- ○ Use the values in tables 3.2 and 3.3 to plot a graph of the molar heat of vaporization for at least 10 elements against their boiling point. Is there any relationship between these two quantities?

DENSITY

*The density of a substance shows its mass per unit volume*. In comparing the densities of different elements, it is usual, for those elements which are gases at room temperature, to quote the density of the element as a liquid at its boiling point. What factors will influence the density of an element—atomic mass, atomic radius (which determines the volume of each atom) and crystal structure (which determines how close to each other the atoms are packed)? Although each of these factors varies from one element to the next across a period there is a periodic trend in density (see figure 3.10).

Across a period there is a general increase in density until a maximum is reached at Group IV for periods 2 and 3 or at the end of the transition metals for the later periods. This increase in density can be related to the increasing atomic mass and decreasing atomic radius. A maximum is reached when the atomic radius reaches a low value and the strength of the metallic or giant-covalent bonding is at a maximum. In spite of having higher atomic masses and lower atomic radii, elements at the right hand side of each period have relatively low densities because they form simple molecular structures in which molecules are only weakly bonded by Van der Waals' forces.

MOLAR VOLUME

*The molar volume of an element is the volume occupied by* 1 *mole* ($6 \times 10^{23}$ *atoms*) *of the element*. It is very dependent on atomic radius and structure (packing). Suppose the relative atomic mass of the element is $A_r$ and its density is $\rho$ g cm$^{-3}$.

$$6 \times 10^{23} \text{ atoms of the element have a mass of } A_r \text{ g}$$

$$\therefore 6 \times 10^{23} \text{ atoms of the element have a volume of } \frac{A_r}{\rho} \text{ cm}^3$$

$$\therefore \text{Molar volume} = \frac{A_r}{\rho} = \frac{\text{Relative atomic mass in grams}}{\text{density}}$$

Since molar volume is *inversely related to density*, it is not surprising that trends in molar volume are the *reverse* of trends in density. Thus molar volumes gradually fall across a period to a minimum value and then rise again (figure 3.10). However, the major factor affecting volume is structure rather than atomic radius and there tend to be abrupt changes in molar volume across a period where there are obvious changes in structure. For example, the atomic radius of phosphorus is less than that of silicon as expected, but phosphorus has the larger molar volume because it contains weakly bonded $P_4$ molecules whereas silicon is a strongly bonded giant molecule.

## 12.5 Patterns in the formulae of compounds

Table 12.4 shows the formulae of the oxides, chlorides, and hydrides of the elements in periods 2 and 3.

Look at the formulae of the chlorides.

- ○ What is the oxidation number of each element in its chloride?
- ○ How do these oxidation numbers vary across periods 2 and 3?
- ○ Why do the oxidation numbers vary in this way?

**Table 12.4** Formulae of the chlorides, oxides and hydrides of the elements in periods 2 and 3.

| **Period 2** | Li | Be | B | C | N | O | F | Ne |
|---|---|---|---|---|---|---|---|---|
| **Empirical formula of chloride** | LiCl | $BeCl_2$ | $BCl_3$ | $CCl_4$ | $NCl_3$ | $Cl_2O$ | ClF | — |
| **Formula of oxide** | $Li_2O$ | BeO | $B_2O_3$ | CO<br>$CO_2$ | $N_2O$<br>NO<br>$NO_2$<br>$N_2O_3$<br>$N_2O_4$<br>$N_2O_5$ | $O_2$ | $OF_2$ | — |
| **Empirical formula of simplest hydride** | LiH | $BeH_2$ | $BH_3$ | $CH_4$ | $NH_3$ | $H_2O$ | HF | — |
| **Period 3** | Na | Mg | Al | Si | P | S | Cl | Ar |
| **Empirical formula of chloride** | NaCl | $MgCl_2$ | $AlCl_3$ | $SiCl_4$ | $PCl_3$<br>$PCl_5$ | $SCl_2$<br>SCl | $Cl_2$ | — |
| **Formula of oxide** | $Na_2O$<br>$Na_2O_2$ | MgO | $Al_2O_3$ | $SiO_2$ | $P_2O_3$<br>$P_2O_5$ | $SO_2$<br>$SO_3$ | $Cl_2O$<br>$Cl_2O_7$ | — |
| **Empirical formula of hydride** | NaH | $MgH_2$ | $AlH_3$ | $SiH_4$ | $PH_3$ | $H_2S$ | HCl | — |

The oxidation numbers of the elements in their oxides are always positive (except for fluorine) since oxygen is the most electronegative element apart from fluorine. The oxidation numbers of the elements from Li to Ar are represented graphically in figure 12.7. Oxidation numbers of the elements in compounds with oxygen are indicated by an 'x'. The pattern in these oxidation numbers provides further evidence of periodic properties. The maximum oxidation number of each element (apart from fluorine) is the same as its group number. For example, lithium in Group I has oxidation number +1, boron in Group III has oxidation number +3 and nitrogen in Group V has a maximum oxidation number of +5. Thus, the maximum oxidation number corresponds to the number of electrons in the outermost shell of the atoms of each element. The oxidation numbers of the elements in their hydrides (indicated by an '**o**') are also shown in figure 12.7. These have been obtained by assuming that the oxidation number of hydrogen in non-metal compounds is +1, whilst its oxidation number in metal hydrides is −1. Another interesting periodic pattern emerges this time. The oxidation numbers of the elements with hydrogen rise from +1 to +3 for the metals in Groups I, II, and III, plunge to −4 for the elements in Group IV and then rise through −3, −2, and −1 for the elements in Groups V, VI and VII.

Many of the oxidation numbers correspond to the atom's losing or gaining enough electrons to obtain a completely-filled outer shell of electrons of the type $ns^2np^6$. This tendency is certainly the case for elements in Groups I, II and III which contain 1, 2 and 3 electrons respectively in their outer shell. In forming compounds, the elements in these three groups lose 1, 2 or 3 electrons respectively forming ions such as $Na^+$, $Mg^{2+}$ and $Al^{3+}$ with oxidation numbers of +1, +2 and +3. In contrast, the elements in Groups V, VI and VII gain 3, 2 and 1 electrons respectively to achieve stable electron structures in ions such as $P^{3-}$, $S^{2-}$ and $Cl^-$. Thus, the elements in these groups have oxidation numbers of −3, −2 and −1 respectively.

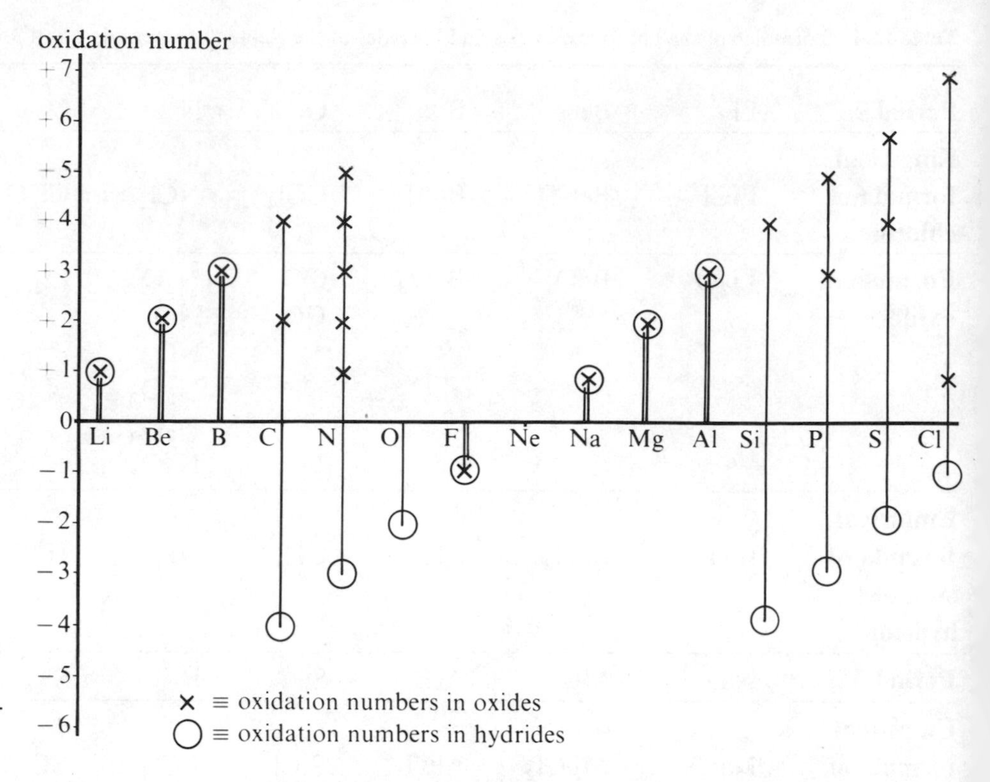


**Figure 12.7** Oxidation numbers of the elements from Li to Ar.

## 12.6 Patterns in the properties of chlorides

Table 12.5 shows various properties of the chlorides of the elements in the third period.

- ○ How do the states of the chlorides at 20°C vary across the third period?
- ○ How do the boiling points of the chlorides vary across the period?
- ○ How do the conductivities of the molten chlorides vary across the period?

### STRUCTURE AND BONDING

Each of the variations to which the above three questions point can be explained in terms of the structure and bonding in the chlorides concerned.

NaCl and $MgCl_2$ are giant structures composed of oppositely charged ions attracted to each other by strong electrostatic forces in ionic bonds. This means that the melting points and boiling points of these compounds are high, but the molten substances will conduct electricity because the ions which they contain can move towards the electrode of opposite charge.

All the other chlorides are simple molecular structures composed of small discrete molecules attracted to each other by relatively weak Van der Waals' forces. This means that the melting points and boiling points of these compounds are low and as liquids they will not conduct electricity.

### HEATS OF FORMATION

Elements on the left of the periodic table are highly electropositive metals keen to give up electrons, whereas those on the right are highly electronegative non-metals (excluding the noble gases). Between these extremes, the electronegativity slowly increases from left to right. Consequently, there is a gradual decrease in the heat evolved when the elements react with 1 mole of chlorine (Cl). Sodium and chlorine react violently liberating 411 kJ per mole of Cl.

$$2Na(s) + Cl_2(g) \longrightarrow 2(Na^+Cl^-(s)) \qquad \Delta H = -822\ \text{kJ}$$

The strong ionic bonding in $Na^+Cl^-$ results in a highly stable product and a reaction which is very exothermic. On the other hand, the reaction between sulphur and chlorine (two electronegative elements) is relatively feeble and much less exothermic.

$$2S(s) + Cl_2(g) \longrightarrow S_2Cl_2 \qquad \Delta H = -60\ \text{kJ}$$

**Table 12.5** Various properties of the chlorides of the elements in the third period

| **Formula of chloride** | $NaCl$ | $MgCl_2$ | $Al_2Cl_6$ | $SiCl_4$ | $PCl_3$ ($PCl_5$) | $S_2Cl_2$ | $Cl_2$ |
|---|---|---|---|---|---|---|---|
| **State of chloride (at 20°C)** | s | s | s | l | l (s) | l | g |
| **b.pt. of chloride/°C** | 1465 | 1418 | 423 | 57 | 74 (164) | 136 | −35 |
| **Conduction of electricity by molten or liquid chloride** | good | good | v. poor | nil | nil | nil | nil |
| **Structure of chloride** | Giant structures | | Simple molecular structures | | | | |
| **Heat of formation of chloride at 298 K/kJ mole$^{-1}$** | −411 | −642 | −1408 | −640 | −320 | −60 | 0 |
| **Heat of formation of chloride at 298 K per mole of Cl atoms/kJ** | −411 | −321 | −235 | −160 | −107 | −30 | 0 |
| **Effect of adding chloride to water** | solid dissolves readily | | chloride reacts with water, HCl fumes are produced. | | | | some $Cl_2$ reacts with water |
| **pH of aqueous solution of chloride** | 7 | 6.5 | 3 | 2 | 2 | 2 | 2 |

REACTION WITH WATER

When the ionic chlorides are added to water, there is an immediate attraction of polar water molecules for ions in the chloride. The solid dissolves forming single aquated ions such as $Na^+$(aq) and $Cl^-$(aq) (i.e. separate metal and non-metal ions surrounded by polar water molecules).

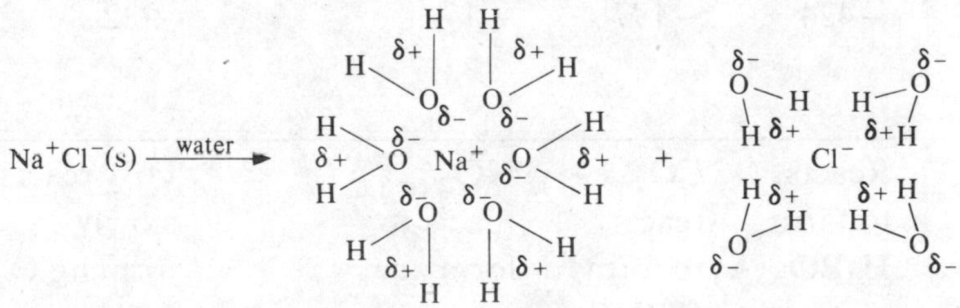

$Na^+Cl^-(s) \xrightarrow{\text{water}}$ $Na^+$ (surrounded by $\overset{\delta-}{O}$ of six $\overset{\delta+}{H}$–O–$\overset{\delta+}{H}$ molecules) + $Cl^-$ (surrounded by $\overset{\delta+}{H}$ of four water molecules)

The solution of sodium chloride is neutral (pH = 7).

In contrast to this, the solution obtained when aluminium chloride is added to water is very acidic (pH = 3). The acidic nature of the solution can be explained in terms of the relatively small but highly charged $Al^{3+}$ ion which draws electrons away from its surrounding water molecules and causes them to give up $H^+$ ions (see also section 15.5).

$$[Al(H_2O)_6]^{3+}(aq) \longrightarrow [Al(H_2O)_5(OH)]^{2+}(aq) + H^+(aq)$$

The non-metal chlorides in period 3 all react with water forming acidic solutions.

$$SiCl_4(l) + 2H_2O(l) \longrightarrow SiO_2(s) + \underbrace{4H^+(aq) + 4Cl^-(aq)}_{\text{hydrochloric acid}}$$

$$PCl_3(l) + 3H_2O(l) \longrightarrow \underbrace{H_3PO_3(aq)}_{\text{phosphonic (phosphorous) acid}} + \underbrace{3H^+(aq) + 3Cl^-(aq)}_{\text{hydrochloric acid}}$$

$$2S_2Cl_2(l) + 2H_2O(l) \longrightarrow 3S(s) + SO_2\,(aq) + \underbrace{4H^+\,(aq) + 4Cl^-\,(aq)}_{\text{hydrochloric acid}}$$

A similar pattern is found in the chlorides of period 2 (study question 2).

## 12.7 Patterns in the properties of oxides

Tables 12.6 and 12.7 show various properties of the oxides of the elements in period 2 and period 3 respectively. There are marked periodic patterns in the structure, bonding and properties of these oxides.

**Table 12.6** Properties of the oxides of elements in period 2.

| **Formula of oxide** | $Li_2O$ | $BeO$ | $B_2O_3$ | $CO_2$ ($CO$) | $N_2O$ $NO$ $NO_2$ $N_2O_4$ $N_2O_5$ | $O_2$ | $OF_2$ |
|---|---|---|---|---|---|---|---|
| **State of oxide (at 20°C)** | solid | solid | solid | gases | gases (except $N_2O_5$, a solid) | gas | gas |
| **Conduction of electricity by molten or liquid oxide** | good | moderate | v. poor | nil | nil | nil | nil |
| **Structure of oxide** | **Giant structures** | | | **Simple molecular structures** | | | |
| **Heat of formation of oxide at 298 K/kJ mole$^{-1}$** | −596 | −611 | −1273 | −394 ($CO_2$) | +33 ($NO_2$) | | +22 |
| **Heat of formation of oxide at 298 K per mole of O/kJ** | −596 | −611 | −424 | −197 | +17 | | +22 |
| **Effect of adding oxide to water** | Reacts to form LiOH(aq) alkaline solution | BeO does not react with water but it is amphoteric | Reacts to form $H_3BO_3$, a very weak acid | $CO_2$ reacts to form $H_2CO_3$, a weak acid | $NO_2$ reacts to form an acid solution of $HNO_3$ and $HNO_2$ | | $OF_2$ reacts slowly forming $O_2$ and an acidic solution of HF |
| **Nature of oxide** | **Basic (alkaline)** | **Amphoteric** | **Acidic** | **Acidic** | **Acidic** | | **Acidic** |

**Table 12.7** Properties of the oxides of elements in period 3

| **Formula of oxide** | $Na_2O$ | $MgO$ | $Al_2O_3$ | $SiO_2$ | $P_4O_{10}$ ($P_4O_6$) | $SO_3$ ($SO_2$) | $Cl_2O_7$ ($Cl_2O$) |
|---|---|---|---|---|---|---|---|
| **State of oxide (at 20°C)** | solid | solid | solid | solid | solid (solid) | liquid (gas) | liquid (gas) |
| **Conduction of electricity by molten or liquid oxide** | good | good | good | v. poor | nil | nil | nil |
| **Structure of oxide** | **Giant structure** | | | | **Simple molecular structure** | | |
| **Heat of formation of oxide at 298 K/kJ mole$^{-1}$** | −416 | −602 | −1676 | −910 | −2984 ($P_4O_{10}$) | −395 ($SO_3$) | +80 ($Cl_2O$) |
| **Heat of formation of oxide at 298 K per mole of O/kJ** | −416 | −602 | −559 | −455 | −298 | −132 | +80 |
| **Effect of adding oxide to water** | Reacts to form NaOH(aq) alkaline solution | Reacts to form $Mg(OH)_2$ weakly alkaline solution | Does not react with water but it is amphoteric | Does not react with water, but it is acidic | $P_4O_{10}$ reacts to form $H_3PO_4$- acid solution | $SO_3$ reacts to form $H_2SO_4$- acid solution | $Cl_2O_7$ reacts to form $HClO_4$- acid solution |
| **Nature of oxide** | **Basic (alkaline)** | **Basic (weakly alkaline)** | **Amphoteric** | **Acidic** | **Acidic** | **Acidic** | **Acidic** |

STRUCTURE AND BONDING

*In each period, the oxides of metals and metalloids have giant structures, whereas the oxides of non-metals are composed of simple molecules.* Thus, lithium oxide, $Li_2O$; beryllium oxide, $BeO$; sodium oxide, $Na_2O$; magnesium oxide, $MgO$; and aluminium oxide, $Al_2O_3$, may be regarded as ionic structures. Consequently, they are solids at room temperature, with high melting points and boiling points. These ionic oxides will conduct electricity in the molten state.

The metalloids, boron and silicon, form oxides with giant molecular structures. Thus, in boron oxide, $B_2O_3$, boron and oxygen atoms are arranged in layers with covalent bonds linking one atom to the next in a giant sheet structure. The giant structures, held together by strong covalent bonds, result in solids with high melting points and boiling points. The melting point of $B_2O_3$ is 577°C and that of $SiO_2$ is 1700°C. Unlike ionic solids, giant covalent solids do not conduct electricity in the molten state.

Carbon dioxide ($CO_2$), nitrogen dioxide ($NO_2$), oxygen difluoride ($OF_2$), phosphorus(V) oxide ($P_4O_{10}$), sulphur(VI) oxide($SO_3$) and dichlorine heptoxide ($Cl_2O_7$) consist of discrete small molecules. These simple molecular oxides are much more volatile than the ionic metal oxides and the giant covalent metalloid oxides. They have low melting points and low boiling points and they do not conduct electricity in the liquid state.

Notice the gradations in structure and bond type across each period from ionic oxides and chlorides to simple molecular oxides and chlorides. These gradations can be correlated with changes in electronegativity across the period from low values on the left to high values on the right. Thus, atoms of low electronegativity, such as Na, Mg and Li, form compounds in which they have given up electrons either to very electronegative chlorine atoms in chlorides or to very electronegative oxygen atoms in oxides. In contrast to this, compounds formed between the more electronegative atoms (such as Si, P, S and F) and either chlorine or oxygen exist as discrete molecules (e.g. $SiCl_4$, ClF) or as giant covalent structures (e.g. $SiO_2$).

Thus, the bonding in the oxides and chlorides becomes less ionic and more covalent as we move across a period from atoms of low to atoms of high electronegativity.

### HEATS OF FORMATION

Tables 12.6 and 12.7 show the standard heats of formation of some of the oxides in periods 2 and 3. Notice that most of the oxides have a negative heat of formation. This means that oxides as a set are very stable compounds. Indeed, only fluorides are generally more stable than oxides.

The standard heats of formation of the oxides vary from very large negative values to relatively small positive values. At first sight, there appears to be no distinct pattern in the standard heats of formation of the oxides. However, when we compare the heats of formation *per mole of oxygen atoms* (i.e. a value which represents the stability of bonds in the oxide to one mole of oxygen atoms) an obvious pattern appears (tables 12.6 and 12.7 and figure 12.8). The heat of for-

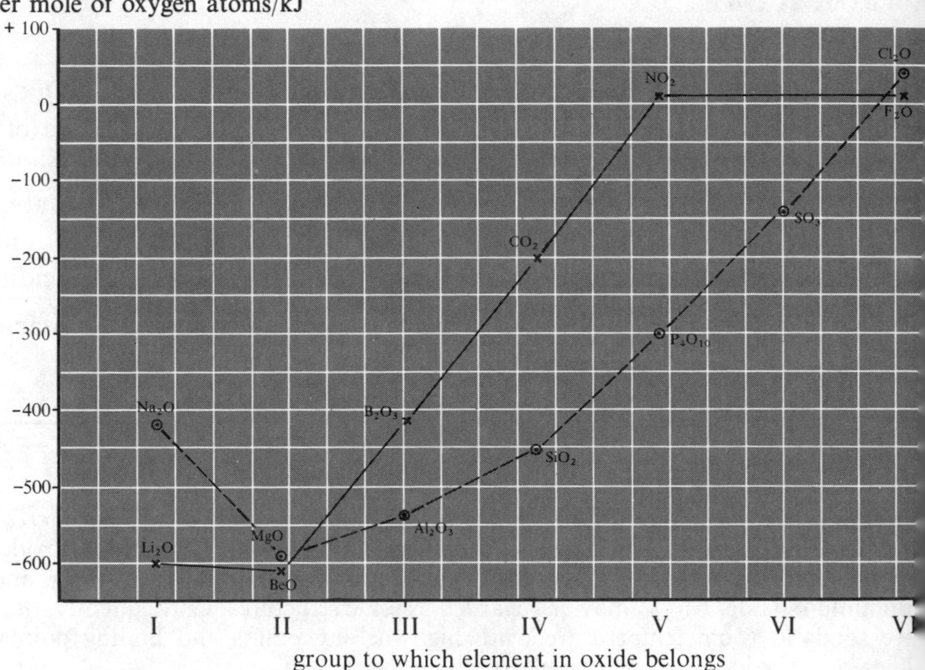


**Figure 12.8** Variation in the heats of formation of oxides in periods 2 and 3 per mole of oxygen atoms/kJ per mole of O atoms.

mation per mole of oxygen atoms tends to decrease (i.e. become less negative) as the atomic number increases across a period. This means that, in general, oxygen forms its most stable compounds with elements such as Li, Na, Mg and Al which are furthest removed from it in the periodic table. This is generally the case with other pairs of elements also.

### REACTION WITH WATER—ACID-BASE CHARACTER OF OXIDES

- ○ Write an equation for the reaction of the following oxides with water: $Na_2O$, MgO, $P_4O_{10}$, $SO_3$.
- ○ How does the acid-base character of the oxides vary across periods 2 and 3?

As we pass across a period from left to right there is a steady change in the structure of the oxide from ionic, through giant covalent to simple molecular. This change in structure leads to a profound difference in the way in which the oxides react with water, acids and alkalis.

The ionic oxides contain $O^{2-}$ ions in the crystal lattice. Thus, $O^{2-}$ ions in $Li_2O$ and $Na_2O$, the oxides of Group I metals, react vigorously with water to form alkaline solutions.

$$O^{2-} + H_2O \longrightarrow 2OH^-$$
$$Li_2O(s) + H_2O(l) \longrightarrow 2Li^+(aq) + 2OH^-(aq)$$
$$Na_2O(s) + H_2O(l) \longrightarrow 2Na^+(aq) + 2OH^-(aq)$$

These oxides would react even more vigorously with acids forming a solution containing cations of the metal.

$$Na_2O(s) + 2H^+(aq) \longrightarrow 2Na^+(aq) + H_2O(l)$$

The oxides of Group II metals do not react so readily with water or acids since the large charge density on the Group II cation holds the $O^{2-}$ ions more firmly.

Thus, MgO is only slightly soluble in water, although it reacts readily with acids to form a solution of magnesium ions.

$$MgO(s) + H_2O(l) \longrightarrow Mg^{2+}(aq) + 2OH^-(aq)$$
$$MgO(s) + 2H^+(aq) \longrightarrow Mg^{2+}(aq) + H_2O(l)$$

BeO is insoluble in water, but it shows basic properties by dissolving in acids to form $Be^{2+}$ salts.

$$BeO(s) + 2H^+(aq) \longrightarrow Be^{2+}(aq) + H_2O(l)$$

However, BeO also resembles acidic oxides by reacting with alkalis to form beryllates.

$$BeO(s) + 2OH^-(aq) + H_2O(l) \longrightarrow \underset{\text{beryllate}}{Be(OH)_4{}^{2-}(aq)}$$

Oxides such as BeO, which show *both basic and acidic properties* are called **amphoteric oxides**.

Aluminium oxide is also amphoteric. It does not react with water, but it will react with both $H^+$ and $OH^-$ ions.

$$Al_2O_3(s) + 6H^+(aq) \longrightarrow 2Al^{3+}(aq) + 3H_2O(l)$$
$$Al_2O_3(s) + 2OH^-(aq) + 3H_2O(l) \longrightarrow \underset{\text{aluminate}}{2Al(OH)_4{}^-(aq)}$$

The remaining oxides of the elements in periods 2 and 3 are all acidic except CO, $OF_2$, $N_2O$, NO, and $ClO_2$ which are described as **neutral** since they show *neither acidic nor basic character*.

Boron(III) oxide, $B_2O_3$, reacts with water to form boric(III) acid, $H_3BO_3$, a very weak acid.

$$B_2O_3(s) + 3H_2O(l) \longrightarrow 2H_3BO_3(aq)$$

$CO_2$ dissolves in water and reacts slightly to form weak carbonic acid, $H_2CO_3$.

$$CO_2(g) + H_2O(l) \rightleftharpoons H_2CO_3(aq)$$

$CO_2$ will react with the $OH^-$ ions in alkalis to form first hydrogencarbonate, $HCO_3{}^-$ and then carbonate, $CO_3{}^{2-}$.

$$CO_2(g) + OH^-(aq) \longrightarrow HCO_3{}^-(aq)$$
$$CO_2(g) + 2OH^-(aq) \longrightarrow CO_3{}^{2-}(aq) + H_2O(l)$$

Silicon(IV) oxide, $SiO_2$, does not react with water, but it reacts with concentrated alkalis forming silicate(IV) ions, $SiO_3{}^{2-}$.

$$SiO_2(s) + 2OH^-(aq) \longrightarrow SiO_3{}^{2-}(aq) + H_2O(l)$$

$NO_2$ reacts with water to form a mixture of two acids, $HNO_2$ and $HNO_3$.

$$2NO_2(g) + H_2O(l) \longrightarrow HNO_2(aq) + HNO_3(aq)$$

The oxides of P, S and Cl (except $ClO_2$) react readily with water to form strong acids.

$$\underset{\text{phosphorus(V) oxide (phosphorus pentoxide)}}{P_4O_{10}(s)} + 6H_2O(l) \longrightarrow \underset{\text{phosphoric(V) acid}}{4H_3PO_4(aq)}$$

$$\underset{\text{sulphur(VI) oxide (sulphur trioxide)}}{SO_3(g)} + H_2O(l) \longrightarrow \underset{\text{sulphuric(VI) acid}}{H_2SO_4(aq)}$$

$$Cl_2O_7(l) + H_2O(l) \longrightarrow 2HClO_4(aq)$$

dichlorine heptoxide — chloric(VII) acid (perchloric acid)

$$Cl_2O(g) + H_2O(l) \longrightarrow 2HClO(aq)$$

dichlorine oxide — chloric(I) acid (hypochlorous acid)

Notice that *as we cross the periodic table, we move from the ionic oxides of metals, which are basic, to the oxides of metalloids with giant covalent structure, which are weakly basic, weakly acidic or amphoteric, and finally to the simple molecular oxides of non-metals, which are acidic.*

## Summary

1 Periodicity as embodied in the periodic table is one of the major unifying themes in chemistry.
2 Elements in the same group of the periodic table have similar electron configurations and this results in similar chemical properties.
3 Moving from left to right across a period there is a decrease in atomic radius since the increasing positive charge in the nucleus pulls the electrons closer.
4 Moving from left to right across a period, there is a general increase in the first ionization energy. This results from an increasing nuclear charge and a decreasing atomic radius across the period causing the outermost electron to be held more tightly.
5 The electronegativity of an atom provides a numerical measure of the power of that atom in a molecule to attract electrons.
6 Moving from left to right across a period there is an increase in the electronegativities of elements.
7 In forming compounds, the elements in Groups I, II, and III form positive ions with charges of $+1$, $+2$, and $+3$ respectively. In contrast, the elements in Groups V, VI and VII gain electrons forming negative ions with charges of $-3$, $-2$ and $-1$ respectively.
8 Moving from left to right across a period,
(i) the chlorides of the elements change from ionic, involatile metal chlorides to simple molecular, volatile non-metal chlorides.
(ii) the hydrides of the elements change from ionic, involatile metal hydrid to simple molecular, non-metal hydrides. The hydrides also become more acid across the period.
(iii) the oxides of the elements change from being ionic, involatile, and basic in metal oxides, through giant-molecular, involatile, amphoteric, metalloid oxides to simple molecular, volatile, and acidic non-metal oxides.

## Study questions

1 Use the information in the following table to explain the statements below.

| | Na | Mg | Al | Si | P | S | Cl |
|---|---|---|---|---|---|---|---|
| Atomic radius /nm | 0.156 | 0.136 | 0.125 | 0.117 | 0.110 | 0.104 | 0.099 |
| Ionic radius /nm | 0.095 | 0.065 | 0.050 | | | 0.184 | 0.181 |
| First ionization energy/kJ mole$^{-1}$ | 492 | 743 | 579 | 791 | 1060 | 1003 | 1254 |

(a) The atomic radius decreases across the period from Na to Cl.
(b) The ionic radii of $Na^+$, $Mg^{2+}$ and $Al^{3+}$ are less than their respective atomic radi whereas the ionic radii of $Cl^-$ and $S^{2-}$ are greater than their respective atomic radii
(c) The first ionization energies show a general increase from Na to Cl.
(d) The first ionization energy of Al is less than that for Mg.
(e) The first ionization energy of S is less than that for P.

**Table 12.8** Various properties of the chlorides of the elements in period 2

| **Formula of chloride** | LiCl | $BeCl_2$ | $BCl_3$ | $CCl_4$ | $NCl_3$ | $Cl_2O$ | ClF |
|---|---|---|---|---|---|---|---|
| **State at 20°C** | s | s | g | l | l | g | g |
| **B.pt. of chloride/°C** | 1350 | 487 | 12 | 77 | 71 | 2 | −101 |
| **Conduction of electricity by liquid chloride** | good | v. poor | nil | nil | nil | nil | nil |
| **Structure** | giant structures | | simple molecular structures | | | | |

2 Table 12.8 shows various properties of the chlorides of the elements in period 2.
(a) Explain the variation in state and boiling point of the chlorides across period 2.
(b) Both LiCl and $BeCl_2$ are giant structures; LiCl(l) is a good conductor of electricity, but $BeCl_2$(l) is only a poor conductor of electricity.
What type of structure does LiCl have?
Why does LiCl conduct electricity?
Suggest a possible structure for $BeCl_2$(s).
(c) How would you expect LiCl and $BCl_3$ to behave when added to water?
(d) How is the type of bonding in the chlorides of the elements in period 2 related to their electronegativity?

3 The behaviour of the hydrides of the elements Na–Ar with water is summarized below.

| NaH | $MgH_2$ | $AlH_3$ | $SiH_4$ | $PH_3$ | $H_2S$ | HCl |
|---|---|---|---|---|---|---|
| react forming $H_2$(g) and an alkaline soln | | | no reaction | reacts to form a slightly alkaline solution | reacts to form a slightly acidic solution | reacts to form an acidic solution |

(a) Write equations to summarize the reactions of NaH and $MgH_2$ with water.
(b) Suggest a reason why $SiH_4$ has no reaction with water.
(c) Write an equation to account for the formation of a slightly alkaline solution when $PH_3$ reacts with water.
(d) Write an equation to account for the formation of an acidic solution when HCl reacts with water.
(e) Explain the trends and differences in the reactions of these hydrides with water in terms of fundamental atomic properties.

4 Consider the elements Li, Be, B, N, F, Ne.
(a) Which exist as diatomic molecules in the gaseous state at room temperature?
(b) Which has the highest boiling point?
(c) Which form a chloride of formula $XCl_3$?
(d) Which has the largest first ionization energy?
(e) Which has the smallest second ionization energy?
(f) Which form hydrides which dissolve in water to give an alkaline solution?

5 (a) Draw dot/cross diagrams to show the electronic structures of
(i) LiF, (ii) $CF_4$, (iii) $NF_3$.
(b) Predict the shapes with respect to atoms of molecules of
(i) $BF_3$, (ii) $CF_4$, (iii) $NF_3$, (iv) $OF_2$.
(c) Why is it not meaningful to talk about the shape of LiF?

6 (a) What is the nature of the bonds in the oxides formed when Na, Mg, Al and S react with excess oxygen?
(b) How do these oxides react with
(i) water, (ii) dilute acids, (iii) alkali?
(c) Magnesium chloride is a high melting point solid, aluminium chloride is a solid which sublimes readily at about 180°C and silicon tetrachloride is a volatile liquid. Explain the nature of the chemical bonding in these chlorides and show how this accounts for the above differences in volatility.

7 (a) Write down the symbols of the elements in the third period of the periodic table (ending with the noble gas, argon) in order of increasing atomic number.
(b) Which of these elements are
(i) '*s*-block' elements,
(ii) '*d*-block' elements?

(c) (i) Write the empirical formula of the chloride formed by the element of atomic number 13.
(ii) Describe briefly how you could prepare a sample of this chloride.

(d) The chloride of element X is represented as $XCl_n$. Suppose you are asked to find the value of *n*.
(i) Describe the experiments you would carry out.
(ii) List the measurements you would take.
(iii) Explain how you would calculate your result.

8 Element Y has an atomic number of 31. Use your knowledge of chemical periodicity to answer the following questions about Y.
(a) Write the electronic (*s*, *p*, *d*, *f*) configuration for Y.
(b) Is element Y in the '*s*-block', '*p*-block' or '*d*-block'?
(c) What is the principal oxidation number of Y?
(d) What is the probable formula of the oxide of Y?
(e) Is the oxide of Y likely to be acidic, amphoteric or basic?
(f) Write equations for any reactions which the oxide of Y would undergo with
(i) water,
(ii) dilute nitric(V) acid,
(iii) dilute sodium hydroxide.

# Competition Processes 13

## 13.1 Fundamental reactions in inorganic chemistry

The reactions of inorganic substances can be divided into four major classes, all of which involve competition in one way or another.

1. Redox
2. Acid-base
3. Precipitation
4. Complexing

Other processes such as decomposition and synthesis usually fit into one of these four categories.

REDOX REACTIONS

Redox processes have already been discussed in Chapter 2. It is worth recalling the following important types of redox reaction.

(a) *The reaction of metals with non-metals*

e.g. $2Na(s) + Cl_2(g) \longrightarrow 2Na^+Cl^-(s)$

$2Mg(s) + O_2(g) \longrightarrow 2Mg^{2+}O^{2-}(s)$

(b) *The reaction of metals with water and steam*

e.g. $Ca(s) + 2H_2O(l) \longrightarrow Ca^{2+}(OH^-)_2(s) + H_2(g)$

(c) *The reaction of metals with acids*

e.g. $Mg(s) + 2H^+(aq) \longrightarrow Mg^{2+}(aq) + H_2(g)$

(d) *Reactions at the electrodes during electrolysis*

e.g. during the electrolysis of molten sodium chloride,

Anode (+) $2Cl^-(l) \longrightarrow Cl_2(g) + 2e^-$

Cathode (−) $2Na^+(l) + 2e^- \longrightarrow 2Na(s)$

Rows of mercury cells used in the electrolytic manufacture of chlorine and sodium hydroxide from brine.

(e) *Disproportionation reactions* (see section 16.10)

Which one of the following processes involves redox? (Remember that the best check for redox is to consider the oxidation numbers of the elements involved in the reaction.)

- ○ $Ag^+ + 2NH_3 \longrightarrow [Ag(NH_3)_2]^+$
- ○ $H^+ + NH_3 \longrightarrow NH_4^+$
- ○ $Ba^{2+} + CrO_4^{2-} \longrightarrow BaCrO_4$
- ○ $2Al + 3Cl_2 \longrightarrow Al_2Cl_6$
- ○ $CuCO_3 \longrightarrow CuO + CO_2$

ACID-BASE REACTIONS

You will have met many examples of acid–base reactions already. These will have included reactions of acids with insoluble bases (metal oxides and hydroxides)

$$CuO(s) + 2H^+(aq) \longrightarrow Cu^{2+}(aq) + H_2O(l)$$

$$Cu(OH)_2(s) + 2H^+(aq) \longrightarrow Cu^{2+}(aq) + 2H_2O(l)$$

and reactions of acids with alkalis

$$Na^+(aq) + OH^-(aq) + H^+(aq) \longrightarrow Na^+(aq) + H_2O(l)$$

as well as less obvious acid–base reactions such as that between aqueous ammonia and dilute acid

$$NH_3(aq) + H^+(aq) \longrightarrow NH_4^+(aq).$$

In the course of this chapter we shall extend our ideas of acid–base processes to include such reactions as acid + carbonate.

PRECIPITATION REACTIONS

*Precipitation occurs when the mixing of aqueous solutions leads to the formation of an insoluble substance.* For example, a white precipitate of silver chloride forms when aqueous solutions of silver nitrate(V) and sodium chloride are mixed.

$$Ag^+(aq) + NO_3^-(aq) + Na^+(aq) + Cl^-(aq) \longrightarrow Ag^+Cl^-(s) + Na^+(aq) + NO_3^-(aq)$$

REACTIONS INVOLVING COMPLEXES

The formation of complexes is of great importance in the chemistry of transition metals (section 18.6, page 267) and indeed in the chemistry of most aqueous metal ions. In aqueous solution, most cations ($M^{2+}$) are surrounded by water molecules to form a hydrated ion of the form $[M(H_2O)_n]^{2+}$. This aquated ion can be regarded as a complex ion. Other important complexes form when cations are surrounded by anions or molecules other than water, e.g. when aluminium hydroxide dissolves in excess NaOH(aq):

$$Al(OH)_3(s) + OH^-(aq) \longrightarrow [Al(OH)_4]^-(aq)$$

tetrahydroxoaluminate(III)
(aluminate)

The $[Al(OH)_4]^-$ ion consists of an $Al^{3+}$ ion surrounded by four $OH^-$ ions.

## 13.2 Redox—competition for electrons

Most redox reactions involve a transfer of electrons from one reactant to another. When metals react with non-metals, electrons are transferred from the metal to the non-metal with the subsequent formation of ions.

$$2Na \longrightarrow 2Na^+ + 2e^-$$

$$Cl_2 + 2e^- \longrightarrow 2Cl^-$$

The metal, which loses electrons, is oxidized whilst the non-metal, which gains electrons, is reduced. The oxidizing agent (oxidant) is the electron acceptor and the reducing agent (reductant) is the electron donor. In this case, the reactions occur because most non-metals are eager to gain electrons whilst metals will part with their electrons quite readily. In the competition for electrons, the non-metal wins 'hands down'. However, the competition for electrons in redox reactions is no always so clear cut. Consequently, chemists have developed the idea of electrode potentials to check whether (and in which direction) a redox reaction is occurring between two systems.

In chapter 2, we discovered that electron transfer occurred when a $Zn(s)/Zn^{2+}(aq)$ half-cell was connected by a salt bridge to a $Cu(s)/Cu^{2+}(aq)$ half-cell provided the two metals are joined through a pea bulb. The bulb lights showing that electrons are flowing through the wire (figure 13.1).

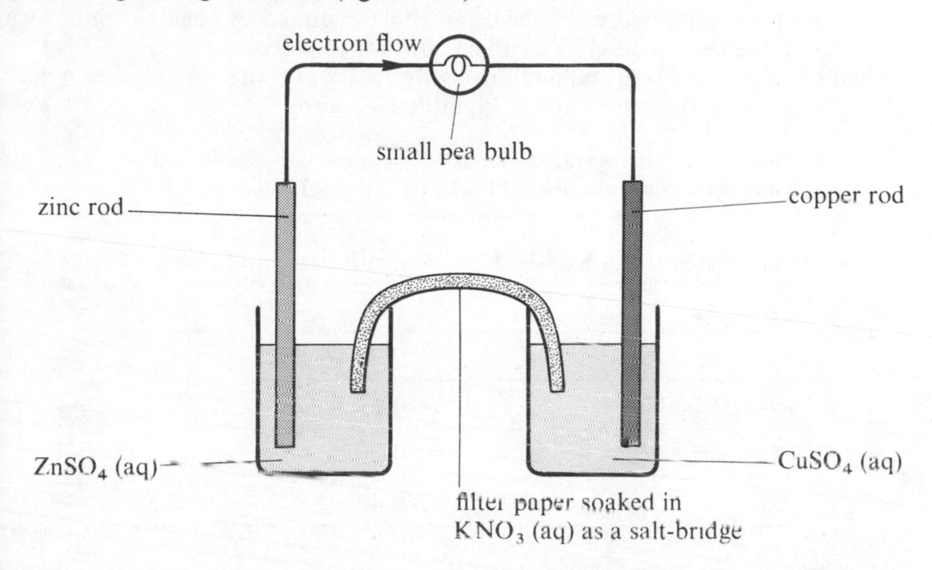


**Figure 13.1** Electron transfer between two half-cells connected by a salt-bridge.

Zinc dissolves from the zinc rod which loses weight, whilst copper is deposited on the copper rod. The reactions at the terminals can be represented by the following half equations.

$$Zn(s) \longrightarrow Zn^{2+}(aq) + 2e^-$$

$$Cu^{2+}(aq) + 2e^- \longrightarrow Cu(s)$$

Electrons flow from the negative zinc terminal through the external circuit to the positive copper terminal.

If the pea bulb is replaced by a high resistance voltmeter (e.g. a valve voltmeter) it is possible to measure the potential difference set up between the two half cells.

By using a high resistance voltmeter, the current flowing in the external circuit is virtually zero and the cell registers its maximum potential difference usually called the **electromotive force (e.m.f.)**. This e.m.f. gives a quantitative measure of the likelihood of the redox reaction taking place in the cell.

When the pea bulb in figure 13.1 is replaced by a valve voltmeter, a voltage of 1.10 volts is recorded.

This combination of half-cells and the resulting cell e.m.f. can be summarized in a diagram as

| $Zn(s)$ | \| | $Zn^{2+}(aq)$ | ⁝ | $Cu^{2+}(aq)$ | \| | $Cu(s)$ | $E = 1.10V$ |
|---|---|---|---|---|---|---|---|
| ↑ | | ↑ | | ↑ | | ↑ | ↑ |
| metal conducting terminal on the left | | materials in contact with metal terminal on left | ↑ Salt bridge or porous partition | materials in contact with metal terminal on right | | metal conducting terminal on the right | Cell e.m.f., i.e. potential of metal terminal on right relative to that on the left (for zero current). |

The voltage shown in the summary above relates to changes indicated by the order of materials in the diagram, i.e.

$$Zn(s) \longrightarrow Zn^{2+}(aq) + 2e^-$$

$$\text{and } Cu^{2+}(aq) + 2e^- \longrightarrow Cu(s)$$

So, for the change,

$$Zn(s) + Cu^{2+}(aq) \longrightarrow Zn^{2+}(aq) + Cu(s) \qquad E = 1.10V$$

and for the reverse process,

$$Zn^{2+}(aq) + Cu(s) \longrightarrow Zn(s) + Cu^{2+}(aq) \qquad E = -1.10V$$

Notice that this convention leads to a positive value of $E$ if the reaction actually takes place in that direction when the cell is short-circuited. A negative value of $E$ indicates that the cell reaction as written cannot take place.

When the $Zn(s)/Zn^{2+}(aq)$ half-cell in figure 13.1 was replaced by other metal/metal ion half-cells, the results shown in table 13.1 were obtained.

**Table 13.1** The overall cell e.m.f. developed between various metal/metal ion half-cells and the $Cu(s)/Cu^{2+}(aq)$ half-cell

| Metal/metal ion half-cell | Overall cell e.m.f./Volts |
|---|---|
| $Zn(s)/Zn^{2+}(aq)$ | +1.10 |
| $Fe(s)/Fe^{2+}(aq)$ | +0.78 |
| $Pb(s)/Pb^{2+}(aq)$ | +0.47 |
| $Cu(s)/Cu^{2+}(aq)$ | 0.00 |
| $Ag(s)/Ag^{+}(aq)$ | −0.46 |

The first three metals (zinc, iron, and lead) are all negative with respect to copper, i.e. these metals are stronger reducing agents than copper and go into solution as their ions when the cells are short-circuited. On the other hand, the $Ag(s)/Ag^{+}(aq)$ half-cell produces a negative e.m.f. relative to the other three half-cells. This means that the silver electrode is positive with respect to copper and in this case copper is acting as the reducing agent. The cell reaction can be summarized as:

$$Ag(s)|Ag^{+}(aq) \,\vdots\, Cu^{2+}(aq)|Cu(s) \qquad E = -0.46V$$

As we might expect, the reaction,

$$2Ag(s) + Cu^{2+}(aq) \longrightarrow 2Ag^{+}(aq) + Cu(s),$$

has a negative e.m.f. of −0.46 volts since it is the reverse reaction which actually occurs when the cell is short-circuited. Figure 13.2 shows diagrammatically the relative potentials for the copper, zinc and silver half-cells.

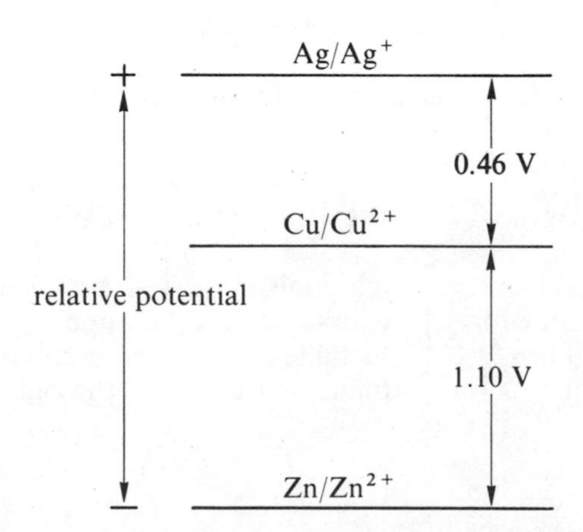


**Figure 13.2** Relative cell potentials. What would be the e.m.f. of a $Zn(s)|Zn^{2+}(aq) \vdots Ag^{+}(aq)|Ag(s)$ cell?

- ○ Why is it impossible to measure the electrode potential for a single half-cell
- ○ What other factors besides the particular metals and ions used will affect th size of an electrode potential?
- ○ Look at figure 13.2. What would be the e.m.f. of a $Zn(s)/Zn^{2+}(aq) \vdots Ag^{+}(aq$ $Ag(s)$ cell? Write half equations for the reactions which take place in the tw half-cells of this combination when it is short-circuited.

Notice that the e.m.f. values in table 13.1 can be used to compare the relative tendencies of the metals in metal/metal ion systems to release electrons and form ions. Thus zinc, at the head of the table is a strong reducing agent, releasing electrons readily, whilst silver at the bottom of the table is a very poor reducing agent. The e.m.f. values give a quantitative record of the position of the metal in the electrochemical series (E.C.S.).

## 13.3 Standard electrode potentials

So far, we have seen that the magnitude of cell e.m.f.'s enable metals to be placed in order of their relative ability as reducing agents. Conversely, we can also draw up a list of metal ions showing their ability as oxidizing agents. Thus, $Ag^+$(aq) is a stronger oxidizing agent than $Cu^{2+}$(aq), i.e. it is a better competitor for electrons.

You will have realized already that it is impossible to obtain the electrode potential for a single half-cell because e.m.f.'s can only be measured for a completed circuit with two electrodes. In other words, *only differences in potential are measurable*. Nevertheless, it would be extremely useful if we could summarize all e.m.f. data by giving each half-cell a characteristic value. This can be done by arbitrarily assigning an electrode potential of zero to one particular system and then comparing all other systems with this standard.

*The standard chosen for electrode potentials is not a metal but hydrogen.* The so called **standard hydrogen half-cell** (sometimes loosely referred to as the standard hydrogen electrode) is shown in figure 13.3.

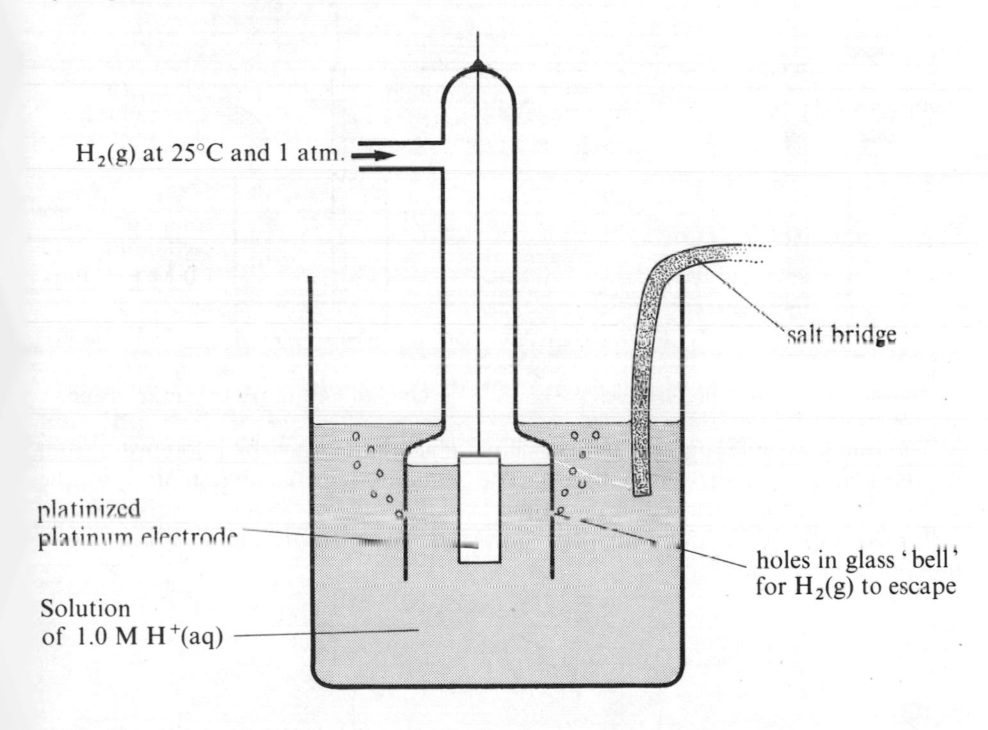


**Figure 13.3** The standard hydrogen half-cell.

This consists of $H_2$ gas at one atmosphere pressure and 25°C bubbling around a platinized platinum electrode which is immersed in a 1.0M solution of $H^+$ ions. In this way, the platinum electrode is alternately bathed in first $H^+$(aq) and then $H_2$(g). Hydrogen is adsorbed on the platinum and an equilibrium is established between the adsorbed layer of $H_2$ and $H^+$ ions in the solution.

$$\tfrac{1}{2}H_2(g) \rightleftharpoons H^+(aq) + e^-$$

The function of the platinized platinum electrode is threefold.

1 It acts as an inert metal connection to the $H_2/H^+$ system. (There is no tendency for Pt to form ions itself.)

2 It allows $H_2$ gas to be adsorbed on to its surface.

3 It is covered by a loosely deposited layer of finely divided platinum (i.e. it is

*platinized*) to increase its surface area so that it can establish an equilibrium between $H_2(g)$ and $H^+(aq)$ as rapidly as possible.

The standard electrode potential ($E^\ominus$) for this reference half-cell is taken as zero, i.e.

$$\tfrac{1}{2}H_2(g) \longrightarrow H^+(aq) + e^- \qquad E^\ominus = 0.00V$$

## 13.4 Measuring standard electrode potentials

The potential of the platinum electrode in a standard hydrogen half-cell will depend on the temperature, the concentration of $H^+$ ions and the pressure of the $H_2(g)$.

Consequently, *in measuring and comparing electrode potentials we must choose the same standard conditions for all measurements.* The standard conditions chosen are similar to those for thermochemical measurements.

1 All ionic species at a concentration of 1.0 M
2 Any gases involved at a pressure of 1 atmosphere
3 A temperature of 25°C (298 K)
4 Platinum used as the electrode when the half-cell system does not include a metal

Figure 13.4 shows a standard $Cu^{2+}(aq)/Cu(s)$ half-cell and a standard $Fe^{3+}(aq)/Fe^{2+}(aq)$ half-cell. The latter system does not involve a metal so a platinum electrode must be used as the electrical connection. Notice also that the solution is 1.0 M with respect to both $Fe^{3+}(aq)$ and $Fe^{2+}(aq)$.

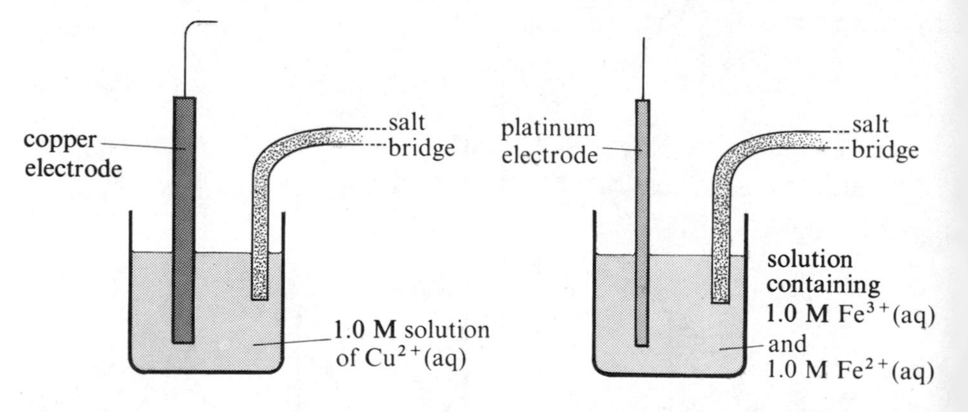


**Figure 13.4** Two standard half-cells.

The standard potentials of all other half-cells are obtained by reference to that of the standard hydrogen half-cell. The standard electrode potential of the $Cu^{2+}(aq)/Cu(s)$ half-cell is thus the potential difference between the electrodes of a cell consisting of the standard hydrogen half-cell and the standard $Cu^{2+}(aq)/Cu(s)$ half-cell (figure 13.5).

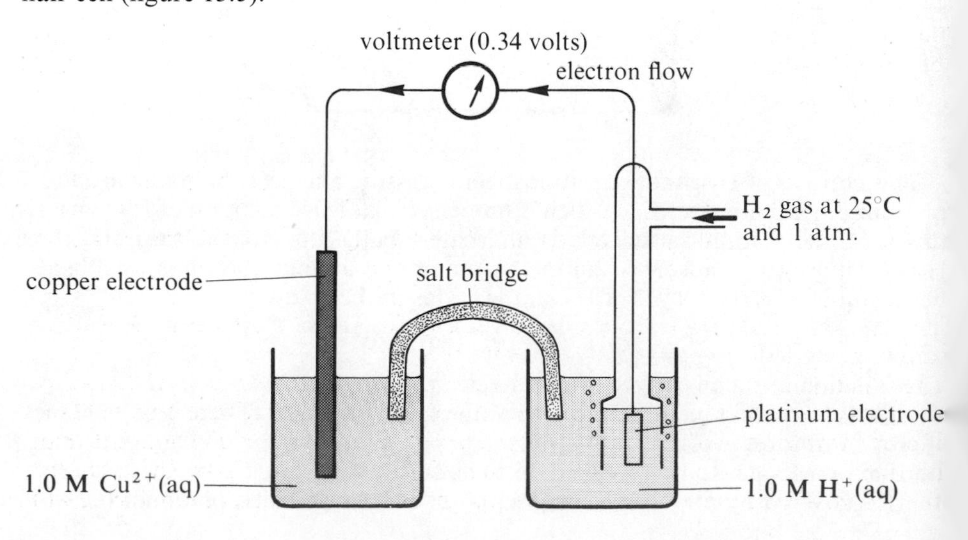


**Figure 13.5** Measuring the standard electrode potential of the $Cu^{2+}(aq)/Cu(s)$ half-cell.

When this cell is set up the potential is 0.34 volts and copper is the positive terminal. The reactions occurring at the electrodes are

$$H_2 \longrightarrow 2H^+ + 2e^- \text{ and}$$
$$Cu^{2+} + 2e^- \longrightarrow Cu.$$

Now, since the standard electrode potential for the system $H_2 \rightarrow 2H^+ + 2e^-$ is arbitrarily taken as zero, the standard electrode potential for the system $Cu^{2+} + 2e^- \rightarrow Cu$ is +0.34 volts, the plus sign being used because the copper electrode is positively charged with respect to the standard hydrogen electrode, i.e.

$$Cu^{2+} + 2e^- \longrightarrow Cu \qquad E^\ominus = +0.34\text{V}$$
$$Pt(s)|H_2(g), H^+(aq) \,\vdots\, Cu^{2+}(aq)|Cu(s) \qquad E^\ominus = +0.34\text{V}$$

*By convention, the oxidized form is written first when a particular redox half-equation and its standard electrode potential are being referred to.* Thus, $Cu^{2+}(aq)/Cu(s)$, $E^\ominus = +0.34$ volts means that the half-cell reaction

$$Cu^{2+}(aq) + 2e^- \longrightarrow Cu(s)$$

has a standard electrode potential of +0.34 volts. Because of this, standard electrode potentials are sometimes called **standard reduction potentials** since they relate to the *reduction of the more oxidized species.*

When inert electrodes are present (as in the hydrogen half-cell), the least oxidized form of the components of the system should be written next to the electrode. Hence, the standard hydrogen electrode is written as $H^+(aq)$, $H_2(g)/Pt(s)$ or as $Pt(s)/H_2(g)$, $H^+(aq)$, but not as $Pt(s)/H^+(aq)$, $H_2(g)$.

We are now in a position to define the *standard electrode (reduction) potential, $E^\ominus$, of a standard half-cell as the potential of that half-cell relative to a standard hydrogen half-cell under standard conditions.*

When a standard zinc half-cell is connected to a standard hydrogen half-cell, the e.m.f. produced is 0.76 volts, but in this case the zinc electrode is negative. Thus, the standard electrode potential for the $Zn^{2+}(aq)/Zn(s)$ half-cell is −0.76 volts, i.e.

$$Zn^{2+}(aq) + 2e^- \longrightarrow Zn(s) \qquad E^\ominus = -0.76\text{V}$$

In this case, the reactions occurring at the electrodes are

$$Zn(s) \longrightarrow Zn^{2+}(aq) + 2e^- \text{ and}$$
$$2H^+(aq) + 2e^- \longrightarrow H_2(g)$$

Notice that the reaction at the hydrogen electrode is now in the opposite direction to that which occurred when a $Cu^{2+}(aq)/Cu(s)$ half-cell was connected to a standard hydrogen half-cell.

True or false: The statement $Cu^{2+}(aq)/Cu(s)$, $E^\ominus = +0.34$ volts means that:

1 copper is more positive than $Cu^{2+}$ ions.
2 a solution of $Cu^{2+}(aq)$ is 0.34 volts more positive than a Cu electrode immersed in it.
3 the reaction $Cu^{2+}(aq) + 2e^- \rightarrow Cu(s)$ has a standard electrode potential of +0.34 volts.
4 the copper electrode of a standard $Cu^{2+}(aq)/Cu(s)$ half-cell is 0.34 volts more positive than the platinum electrode in a standard hydrogen half-cell to which it is connected.

## 13.5 Relative strengths of oxidizing agents and reducing agents

The values of standard electrode potentials provide a direct measure of the relative oxidizing and reducing power of different species.

The standard electrode potentials of the zinc, hydrogen and copper half-cells have been tabulated in table 13.2 in order to emphasize the relative strengths of the different oxidizing and reducing agents. $Cu^{2+}$ is a stronger oxidizing agent than either $H^+$ or $Zn^{2+}$ since $Cu^{2+}$ can oxidize both $H_2$ to $H^+$ and Zn to $Zn^{2+}$. On the other hand, Zn is a more powerful reducing agent than either $H_2$ or Cu since it will

**Table 13.2** Standard electrode potentials of the zinc, hydrogen and copper half-cells

| | Oxidizing agent | | Reducing agent | | $E^{\ominus}$ /volts |
|---|---|---|---|---|---|
| increasing strength of oxidizing agent ↑ | $Cu^{2+}(aq) + 2e^-$ | ⟶ | $Cu(s)$ | increasing strength of reducing agent ↓ | +0.34 |
| | $H^+(aq) + e^-$ | ⟶ | $\frac{1}{2}H_2(g)$ | | 0.00 |
| | $Zn^{2+}(aq) + 2e^-$ | ⟶ | $Zn(s)$ | | −0.76 |

reduce $H^+$ to $H_2$ and $Cu^{2+}$ to Cu. Notice that *the relative size of $E^{\ominus}$ gives a measure of the strengths of both oxidants and reductants.* $Cu^{2+}$, the most powerful oxidizing agent in this table, has the largest (most positive) value for $E^{\ominus}$ whereas $Zn^{2+}$, the least powerful oxidizing agent, has the smallest (most negative) value for $E^{\ominus}$.

On the other hand, the value of $E^{\ominus}$ for the reaction

$$Zn(s) \longrightarrow Zn^{2+}(aq) + 2e^-$$

is +0.76 volts. So, Zn, the most powerful reducing agent, has the most positive value for $E^{\ominus}$. It is worth remembering that *the more positive the value of $E^{\ominus}$, the more energetically favourable is the reaction.*

The standard electrode potentials for a large number of redox half-reactions are shown in table 13.3. The table is arranged so that the strongest oxidizing agent, $F_2$, which has the most positive value for $E^{\ominus}$, is at the top of the list and the weakest oxidizing agent, $K^+$, with the most negative value for $E^{\ominus}$, is at the bottom. Thus, fluorine is a better competitor for electrons than any other oxidizing agent in the list, whilst potassium ions have the weakest tendency to accept electrons. Conversely, $F^-$ is the weakest reducing agent with least tendency to donate electrons, whilst K is the most powerful reducing agent in the list. This illustrates the recipro-

**Table 13.3** Standard electrode potentials

| | Oxidizing agent | | Reducing agent | | $E^{\ominus}$/volts |
|---|---|---|---|---|---|
| Strongest oxidizing agent | $F_2(g) + 2e^-$ | ⟶ | $2F^-(aq)$ | Weakest reducing agent | +2.87 |
| increasing strength of oxidizing agent ↑ | $H_2O_2(aq) + 2H^+(aq) + 2e^-$ | ⟶ | $2H_2O(l)$ | increasing strength of reducing agent ↓ | +1.77 |
| | $MnO_4^-(aq) + 4H^+(aq) + 3e^-$ | ⟶ | $MnO_2(s) + 2H_2O(l)$ | | +1.70 |
| | $2HClO(aq) + 2H^+(aq) + 2e^-$ | ⟶ | $Cl_2(aq) + 2H_2O(l)$ | | +1.59 |
| | $MnO_4^-(aq) + 8H^+(aq) + 5e^-$ | ⟶ | $Mn^{2+}(aq) + 4H_2O(l)$ | | +1.51 |
| | $Cl_2(aq) + 2e^-$ | ⟶ | $2Cl^-(aq)$ | | +1.36 |
| | $MnO_2(s) + 4H^+(aq) + 2e^-$ | ⟶ | $Mn^{2+}(aq) + 2H_2O(l)$ | | +1.23 |
| | $Br_2(aq) + 2e^-$ | ⟶ | $2Br^-(aq)$ | | +1.09 |
| | $NO_3^-(aq) + 2H^+(aq) + e^-$ | ⟶ | $NO_2(g) + H_2O(l)$ | | +0.80 |
| | $Ag^+(aq) + e^-$ | ⟶ | $Ag(s)$ | | +0.80 |
| | $Fe^{3+}(aq) + e^-$ | ⟶ | $Fe^{2+}(aq)$ | | +0.77 |
| | $2H^+(aq) + O_2(g)$ | ⟶ | $H_2O_2(aq)$ | | +0.68 |
| | $I_2(aq) + 2e^-$ | ⟶ | $2I^-(aq)$ | | +0.54 |
| | $Cu^{2+}(aq) + 2e^-$ | ⟶ | $Cu(s)$ | | +0.34 |
| | $2H^+(aq) + 2e^-$ | ⟶ | $H_2(g)$ | | 0.00 |
| | $Pb^{2+}(aq) + 2e^-$ | ⟶ | $Pb(s)$ | | −0.13 |
| | $Fe^{2+}(aq) + 2e^-$ | ⟶ | $Fe(s)$ | | −0.44 |
| | $Zn^{2+}(aq) + 2e^-$ | ⟶ | $Zn(s)$ | | −0.76 |
| | $Al^{3+}(aq) + 3e^-$ | ⟶ | $Al(s)$ | | −1.66 |
| | $Mg^{2+}(aq) + 2e^-$ | ⟶ | $Mg(s)$ | | −2.37 |
| | $Na^+(aq) + e^-$ | ⟶ | $Na(s)$ | | −2.71 |
| Weakest oxidizing agent | $K^+(aq) + e^-$ | ⟶ | $K(s)$ | Strongest reducing agent | −2.92 |

cal or conjugate character of an oxidizing agent and its corresponding reducing agent in a redox half-equation. *The stronger the oxidizing agent, the weaker is its corresponding or conjugate reducing agent.*

## 13.6 Using standard electrode potentials

TO PREDICT THE POSSIBILITY OF REACTIONS

Why is it that $Cu^{2+}(aq)$ can oxidize Zn(s), but $Zn^{2+}(aq)$ cannot oxidize Cu(s)?

As we have seen already, $E^{\ominus}$ values provide an indication of the relative strengths of oxidizing agents and reducing agents.

The value of $E^{\ominus}$ for the reaction

$$Cu^{2+}(aq) + 2e^- \longrightarrow Cu(s) \text{ is } +0.34 \text{ volts,}$$

and that for the reaction

$$Zn(s) \longrightarrow Zn^{2+}(aq) + 2e^- \text{ is } +0.76 \text{ volts.}$$

Consequently, the overall potential for the reaction

$$Cu^{2+}(aq) + Zn(s) \longrightarrow Cu(s) + Zn^{2+}(aq) \text{ is } 1.10 \text{ volts}$$

i.e.

$$Cu^{2+}(aq) + 2e^- \longrightarrow Cu(s) \qquad + 0.34 \text{ volts}$$
$$Zn(s) \longrightarrow Zn^{2+}(aq) + 2e^- \qquad + 0.76 \text{ volts}$$

Add: $$Cu^{2+}(aq) + Zn(s) \longrightarrow Cu(s) + Zn^{2+}(aq) \qquad + 1.10 \text{ volts}$$

The overall positive value for the reaction potential suggests that the process is energetically feasible.

Conversely, the overall potential for the reaction

$$Cu(s) + Zn^{2+}(aq) \longrightarrow Cu^{2+}(aq) + Zn(s) \text{ is } -1.10 \text{ volts}$$

and the negative value suggests that the reaction is unlikely to occur. In general, *reactions with an overall positive potential are energetically feasible whereas those with an overall negative value are not so.* Relating this to table 13.3, it means that any oxidizing agent on the left will oxidize any reducing agent below it on the right. In other words,

stronger oxidizing agent + stronger reducing agent → weaker reducing agent + weaker oxidizing agent

Use table 13.3 to predict whether the following substances will react with one another.

- ○ $F_2$ and Na
- ○ $F^-$ and $Na^+$
- ○ $F_2$ and $Na^+$
- ○ $Cl_2$ and $Fe^{2+}$
- ○ $I_2$ and $Br^-$

In predicting whether a reaction will occur, it is important to remember that $E^{\ominus}$ *relates to the probability of the reaction, not to the quantity of materials reacting.* The reaction is just as likely (or unlikely) to occur whether we have a milligram or a kilogram, a thimbleful or a bucketful, one mole or two moles and the electrode e.m.f. is independent of the number of electrons being transferred.

Thus, if $E^{\ominus}$ for the reaction $Fe^{3+}(aq) + e^- \rightarrow Fe^{2+}(aq)$ is +0.77 volts, $E^{\ominus}$ for the reaction $2Fe^{3+}(aq) + 2e^- \rightarrow 2Fe^{2+}(aq)$ is also +0.77 volts, and *not* +1.54 volts.

Although an overall positive value of $E^{\ominus}$ for a redox pair suggests that the reaction should take place, in practice the reaction may be too slow to occur. The important point is that $E^{\ominus}$ relates to the relative stabilities of reactants and products and, therefore, it indicates the feasibility of the reaction from an energetic standpoint. However, $E^{\ominus}$ gives no information whatsoever about *the rate of a reaction or its kinetic feasibility*.

Thus, $E^\ominus$ values predict that $Cu^{2+}(aq)$ should oxidize $H_2(g)$ to $H^+(aq)$, i.e.

$$Cu^{2+}(aq) + 2e^- \longrightarrow Cu(s) \qquad +0.34$$

$$H_2(g) \longrightarrow 2H^+(aq) + 2e^- \qquad 0.00$$

Add: $$Cu^{2+}(aq) + H_2(g) \longrightarrow Cu(s) + 2H^+(aq) \qquad +0.34$$

But, nothing happens when hydrogen is bubbled into copper(II) sulphate(VI) solution because the reaction rate is effectively zero.

Another important point to remember in using $E^\ominus$ values is that they *relate only to standard conditions*. Changes in temperature, pressure, and concentration will affect the values of electrode potentials. In particular, all electrode potentials become more positive when the concentration of reactant ions is increased, since the reduction of these ions is then more likely to occur, and less positive when the concentration of reactant ions is reduced.

For example, the standard electrode potential of copper is $+0.34$ volts, but the electrode potential of copper in contact with 0.1M $Cu^{2+}(aq)$ is only $+0.31$ volts. Similarly, the standard electrode potential of zinc is $-0.76$ volts, but the electrode potential of zinc in contact with 0.1M $Zn^{2+}(aq)$ is reduced to $-0.79$ volts.

The standard electrode potential for the half-reaction

$$Fe^{3+}(aq) + e^- \longrightarrow Fe^{2+}(aq) \text{ is } +0.77 \text{ volts.}$$

What happens to the electrode potential when the concentration of

- ○ $Fe^{3+}(aq)$ rises,
- ○ $Fe^{3+}(aq)$ falls,
- ○ $Fe^{2+}(aq)$ rises,
- ○ $Fe^{2+}(aq)$ falls,
- ○ both $Fe^{3+}(aq)$ and $Fe^{2+}(aq)$ are doubled?

The effect of conditions on electrode potentials is nicely illustrated by the reaction between $MnO_2(s)$ and HCl(aq). Under standard conditions, $MnO_2$ will not oxidize 1.0M HCl(aq) to $Cl_2$, i.e.

$$MnO_2(s) + 4H^+(aq) + 2e^- \rightarrow Mn^{2+}(aq) + 2H_2O(l) \qquad +1.23 \text{ volts}$$

$$2Cl^-(aq) \rightarrow Cl_2(g) + 2e^- \qquad -1.36 \text{ volts}$$

$$MnO_2(s) + 4H^+(aq) + 2Cl^-(aq) \rightarrow Mn^{2+}(aq) + 2H_2O(l) + Cl_2(g) \qquad -0.13 \text{ volts}$$

However, when manganese(IV) oxide is warmed with *concentrated* hydrochloric acid the electrode potentials for each half-equation become more positive, and the overall electrode potential becomes positive. Hence, chlorine is produced.

### TO CALCULATE THE E.M.F.'S OF CELLS

This can be done in a similar manner to the prediction of likely reactions by adding together the electrode potentials of the two half-cells. One of the first practical cells to be used was **the Daniell Cell**, invented in 1836 (figure 13.6). Notice how closely

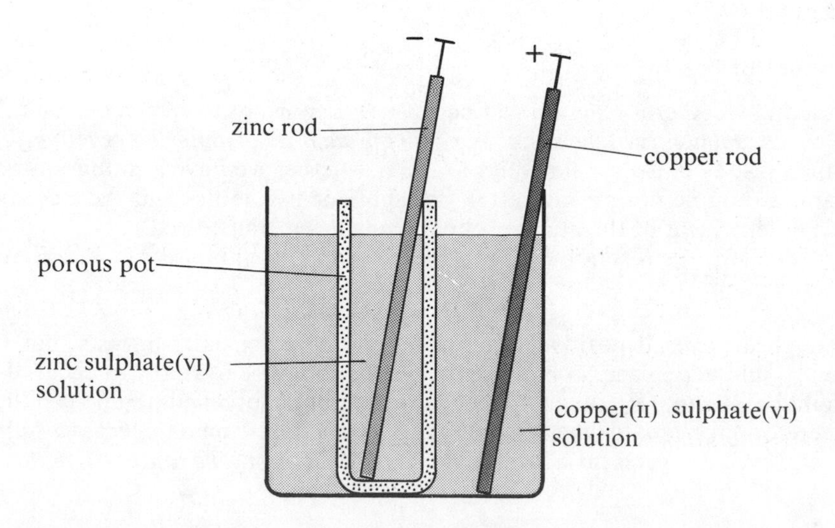


**Figure 13.6** The Daniell Cell.

the Daniell cell resembles the arrangement in figure 13.1 by using $Zn^{2+}(aq)/Zn(s)$ and $Cu^{2+}(aq)/Cu(s)$ half-cells. The only difference is that the Daniell cell has a rigid porous pot which allows movement of ions between the two solutions in place of a filter paper soaked in potassium nitrate(v) solution.

When the Daniell cell is operating, zinc goes into solution as zinc ions and copper deposits on the copper electrode. Under standard conditions, the e.m.f. of the cell is 1.10 volts, i.e.

$$Zn(s) \longrightarrow Zn^{2+}(aq) + 2e^- \qquad +0.76 \text{ volts}$$

$$Cu^{2+}(aq) + 2e^- \longrightarrow Cu(s) \qquad +0.34 \text{ volts}$$

$$\text{Add:} \quad Cu^{2+}(aq) + Zn(s) \longrightarrow Zn^{2+}(aq) + Cu(s) \qquad +1.10 \text{ volts}$$

The arrangement in the Daniell cell can be summarized as:

$$Zn(s)|Zn^{2+}(aq) \;\vdots\; Cu^{2+}(aq)|Cu(s) \qquad E = 1.10 \text{ volts}$$

TO SHOW THE RELATIVE OXIDIZING AND REDUCING POWER OF DIFFERENT SPECIES and to show the relative reactivity of metals and non-metals (see sections 13.2 and 13.5).

## 13.7 Commercial cells

Cells and batteries provide a useful and economic way of obtaining energy from chemical reactions. The Daniell cell, discussed in the last section, has now been replaced by cheaper and more convenient cells.

THE LECLANCHÉ DRY CELL

Probably the commonest, cheapest, and most convenient cell in use at the present time is the Leclanché dry cell (figure 13.7). It is used in a wide range of small electrical appliances such as torches, bicycle lamps, radios and electric bells.

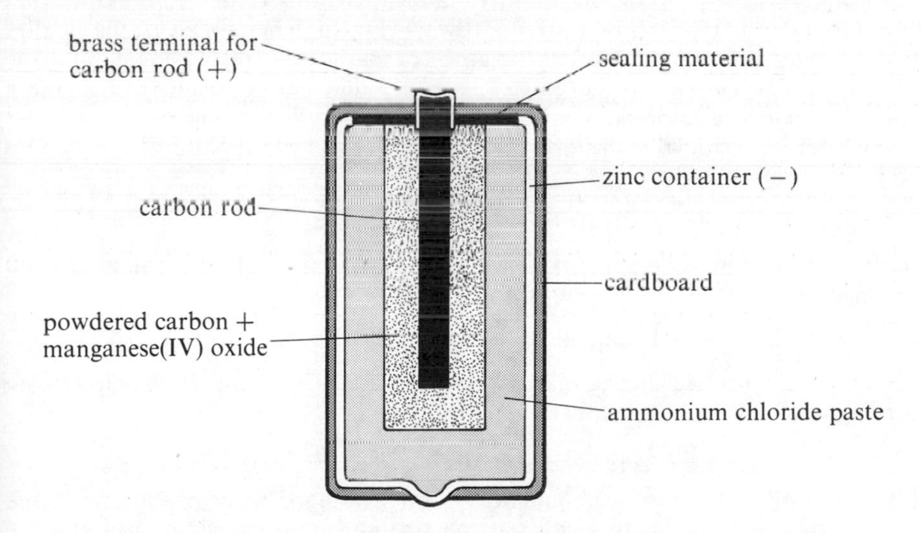


**Figure 13.7** The Leclanché Dry Cell.

- ○ Why is the Leclanché dry cell cheaper than the Daniell cell?
- ○ Why is the Leclanché dry cell more convenient and more portable than the Daniell cell?
- ○ Why is the ammonium chloride used as a paste rather than as a dry solid?

The Leclanché cell can be summarized as:

$$Zn(s)|Zn^{2+}(aq) \;\vdots\; NH_4^+(aq), [2NH_3(g) + H_2(g)]|C \text{ (graphite)} \qquad E = 1.5 \text{ volts}$$

The negative terminal in the cell is zinc, as in the Daniell cell.

$$Zn(s) \longrightarrow Zn^{2+}(aq) + 2e^-$$

The carbon rod is the positive terminal at which ammonium ions are converted to ammonia and hydrogen.

$$2NH_4^+(aq) + 2e^- \longrightarrow 2NH_3(g) + H_2(g)$$

The surface area of the positive terminal is increased by surrounding the carbon rod with a mixture of powdered charcoal and manganese(IV) oxide. The purpose of the manganese(IV) oxide is to oxidize hydrogen produced at the electrode to water thus preventing any bubbles of the gas from coating the carbon terminal which would reduce its efficiency. The ammonia gas causes no such problems because it dissolves rapidly in the water of the paste. It is necessary to use a paste since dry solid ammonium chloride would not conduct electricity.

A single dry cell can produce a potential difference of 1.5 volts, but batteries of these cells giving 100 volts and more are frequently used.

When the cell is in use, the casing is eaten away as the zinc is converted to $Zn^{2+}$ ions. In time, the zinc case disintegrates and the paste oozes through the outer cardboard.

ACCUMULATORS (SECONDARY CELLS)

Obviously, the dry cell cannot be used to provide continuous supplies of electrical energy indefinitely, nor is there any means of restoring or *recharging* the cell so that it can be used again. Cells such as this which can be used once only and cannot be recharged are called **primary cells** in contrast to **secondary cells** or **accumulators** which can be recharged and used again.

When an accumulator is recharged, an electric current is passed through it in the opposite direction to that current which the cell generates itself and chemical reactions occur at the terminals as the original substances reform. Thus, recharging is an example of electrolysis and is the reverse of discharge.

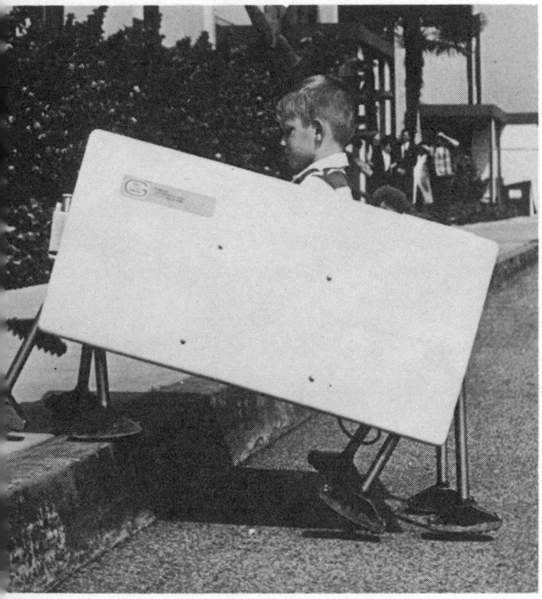

This chin-controlled robot has been designed to aid severely-handicapped children. The robot is powered by four motor-cycle batteries and can work for twelve hours without being recharged.

THE LEAD-ACID ACCUMULATOR

This is certainly one of the most commonly used types of accumulator. Virtually all car batteries are composed of either three or six lead-acid cells in series giving a total potential difference of approximately 6 or 12 volts respectively.

Most electric milk floats run on lead-acid accumulators since they have only a limited mileage and can be conveniently recharged for each day's work. In recent years, the escalating cost of petrol has made the electrically powered motor car an increasingly viable proposition. Unfortunately, attempts to develop and market an economic vehicle of this type are still thwarted by high battery costs, the smaller power/weight ratio of the 'battery' engine compared to a petrol engine and the limited mileage between recharging. It is likely that battery cars will become a reality when a cheap, low weight, rechargeable battery is developed.

The negative terminal in the lead-acid accumulator is a lead plate which gives up electrons forming lead(II) ions during discharge.

$$Pb(s) \longrightarrow Pb^{2+}(aq) + 2e^-$$

The positive terminal is lead(IV) oxide which reacts with $H^+$ ions in the sulphuric(VI) acid electrolyte also forming lead(II) ions.

$$PbO_2(s) + 4H^+(aq) + 2e^- \longrightarrow Pb^{2+}(aq) + 2H_2O(l)$$

The $Pb^{2+}$ ions formed during discharge react with $SO_4^{2-}$ ions in the electrolyte forming insoluble lead(II) sulphate(VI).

$$Pb^{2+}(aq) + SO_4^{2-}(aq) \longrightarrow PbSO_4(s)$$

It is important that discharge should not be too prolonged or the precipitate of fine lead sulphate will change to a coarser, inactive and non-reversible form and the accumulator will become much less efficient.

## 13.8 Acids and bases

During the nineteenth century, the Swedish chemist Arrhenius put forward the idea that *acids were substances which dissociated in water to produce hydrogen ions, $H^+$*.

$$HCl \longrightarrow H^+ + Cl^-$$

Strong acids were thought to be completely dissociated in aqueous solution whereas weak acids were only partially dissociated with a high proportion of the acid remaining in the undissociated form. Thus, a solution of ethanoic (acetic) acid could be represented by the following equilibrium.

$$CH_3COOH \rightleftharpoons CH_3COO^- + H^+$$

With the development of knowledge concerning atomic structure, chemists realized that it was highly improbable that any ion so small as $H^+$, which is simply a proton, could exist independently in aqueous solution. Indeed, the $H^+$ ion has a radius 50 000 times less than that of a lithium ion which itself is regarded as a very small ion.

Thus, $H^+$ ions with their high charge density were believed to associate with polar water molecules in aqueous solution as $H_3O^+$ ions, known as oxonium (hydronium) ions. Experimental evidence for the hydration of $H^+$ ions in aqueous solution was obtained by observing the movement of $H^+$ ions during electrolysis.

Consequently, the dissociation of acids in water can be represented more accurately as

$$HCl + H_2O \longrightarrow H_3O^+ + Cl^-$$

$$CH_3COOH + H_2O \rightleftharpoons H_3O^+ + CH_3COO^-$$

Two important advantages of this refinement to Arrhenius's theory were immediately apparent.

1 It recognized the essential role of the 'solvent' water molecules in the dissociation of acids.

2 It explains why substances such as hydrogen chloride and ethanoic acid only show their usual acidic properties in the presence of water.

Pure dry ethanoic acid and solutions of hydrogen chloride or ethanoic acid in organic solvents such as toluene contain no $H^+$ or $H_3O^+$ ions. Thus, they are non-electrolytes, they do not affect dry litmus paper and they do not react with carbonates to produce $CO_2$.

Although $H^+$ ions exist in aqueous solution as $H_3O^+$, for the sake of simplicity it is conventional to write $H^+$ rather than $H_3O^+$ in equations.

According to the Arrhenius theory, *bases were regarded as substances which reacted with $H^+$ ions to form water.* For example,

$$CuO + 2H^+ \longrightarrow Cu^{2+} + H_2O$$

$$Cu(OH)_2 + 2H^+ \longrightarrow Cu^{2+} + 2H_2O$$

Alkalis, soluble bases, were regarded as substances which dissociated in water producing $OH^-$ ions.

$$NaOH(s) \xrightarrow{\text{water}} Na^+(aq) + OH^-(aq)$$

- ○ Write an equation for the reaction of sodium oxide with dilute hydrochloric acid.
- ○ Write an equation for the reaction of sodium sulphide with dilute hydrochloric acid.
- ○ Compare the two equations you have just written. Do you regard the second equation as an acid-base reaction? Explain your answer.

A life-saving acid/base reaction. During the Apollo 13 space project, the astronauts discovered that carbon dioxide was building up in their crippled spacecraft as they travelled towards the Earth. By an ingenious use of lithium hydroxide, they were able to repair the air purification equipment in their spacecraft. The photograph shows test pilot Scott MacLeod holding one of the lithium hydroxide containers.

## 13.9 The Brønsted–Lowry theory of acids and bases

The definitions of acids and bases in the last section are somewhat unsatisfactory since they can only be applied to aqueous solutions.

Furthermore, the Arrhenius definition of a base is restricted to those substances which react with $H^+$ ions *to form water*.

In order to broaden the scope of acid-base reactions to include non-aqueous systems and to include a wide range of bases, Brønsted and Lowry independently suggested the following definitions in 1923.

*An acid is a proton donor.*
*A base is a proton acceptor.*

Thus, the relationship between an acid and its corresponding base is

$$\underset{\substack{\text{acid}\\\text{(proton donor)}}}{HB} \rightleftharpoons \underset{\text{proton}}{H^+} + \underset{\substack{\text{base}\\\text{(proton acceptor)}}}{B^-}$$

HB and $B^-$ are said to be *conjugate* and to form a **conjugate acid-base pair**. HB is the conjugate acid of $B^-$ and $B^-$ is the conjugate base of HB.

○ What is the conjugate acid of
(i) $CH_3COO^-$, (ii) $HSO_4^-$, (iii) $NH_3$, (iv) $OH^-$?

○ What is the conjugate base of
(i) HCl, (ii) $H_3O^+$, (iii) $HSO_4^-$, (iv) $NH_3$?

According to the Brønsted-Lowry theory, acid-salts (such as $NaHSO_4$) and ammonium ions are recognized as acids along with hydrochloric acid and ethanoic acid. What is more, the definition of a base now includes all anions, water and ammonia as well as oxide and hydroxide ions.

| Acid | | Base | | Base | | Acid |
|---|---|---|---|---|---|---|
| $HSO_4^-$ | + | $OH^-$ | $\longrightarrow$ | $SO_4^{2-}$ | + | $H_2O$ |
| $NH_4^+$ | + | $OH^-$ | $\longrightarrow$ | $NH_3$ | + | $H_2O$ |
| $2H_3O^+$ | + | $S^{2-}$ | $\longrightarrow$ | $2H_2O$ | + | $H_2S$ |
| $H_3O^+$ | + | $NH_3$ | $\longrightarrow$ | $H_2O$ | + | $NH_4^+$ |

Notice in the equations above that *water can act both as a base and as an acid* and does so simultaneously during its dissociation.

$$\underset{\text{acid}}{H_2O} + \underset{\text{base}}{H_2O} \rightleftharpoons \underset{\text{base}}{OH^-} + \underset{\text{acid}}{H_3O^+}$$

## 13.10 Relative strengths of acids and bases

When the acid HB dissociates in water the following equilibrium is established (sections 22.5 and 22.9).

$$HB + H_2O \rightleftharpoons B^- + H_3O^+$$

*Acids differ in the extent to which they donate protons to water in aqueous solution.* Those which donate all their protons to water molecules are known as strong acids and the equilibrium in the equation above is well over to the right. Thus, the conjugate base of a strong acid is weak and vice versa.

If HB is a weak acid, relatively few molecules will donate $H^+$ ions to water and the equilibrium will be well over to the left.

Using the idea of an acid as a proton donor and a base as a proton acceptor, we can arrange all acids and bases in a 'league table' showing their order of relative strengths (table 13.4).

If an acid is weak (i.e. has little tendency to donate protons) it follows that its conjugate base is strong since it must have a strong affinity for protons. For example, hydrogen sulphide is a weak acid, but the sulphide ion is a strong base. Thus, in table 13.4 the acids increase in strength down the page, but their conjugate bases gradually become weaker. The relative strengths of acids are discussed further in sections 22.5 and 22.9.

## 13.11 Acid-base reactions—competition for protons

When dilute hydrochloric acid is added to a solution of sodium sulphide, hydrogen sulphide is produced.

$$S^{2-} + \underbrace{2H^+ + 2Cl^-}_{\text{hydrochloric acid}} \longrightarrow H_2S + 2Cl^-$$

**Table 13.4** Relative strengths of acids and bases

| Name of acid | Acid | | $H^+$ + Base | Name of base |
|---|---|---|---|---|
| ethanol | $C_2H_5OH$ | ⇌ | $H^+ + C_2H_5O^-$ | ethoxide |
| water | $H_2O$ | ⇌ | $H^+ + OH^-$ | hydroxide |
| ammonium | $NH_4^+$ | ⇌ | $H^+ + NH_3$ | ammonia |
| hydrogen sulphide | $H_2S$ | ⇌ | $2H^+ + S^{2-}$ | sulphide |
| ethanoic acid | $CH_3COOH$ | ⇌ | $H^+ + CH_3COO^-$ | ethanoate |
| sulphuric(IV) acid | $H_2SO_3$ | ⇌ | $2H^+ + SO_3^{2-}$ | sulphate(IV) |
| oxonium | $H_3O^+$ | ⇌ | $H^+ + H_2O$ | water |
| sulphuric(VI) acid | $H_2SO_4$ | ⇌ | $2H^+ + SO_4^{2-}$ | sulphate(VI) |
| hydrochloric acid | HCl | ⇌ | $H^+ + Cl^-$ | chloride |
| chloric(VII) acid | $HClO_4$ | ⇌ | $H^+ + ClO_4^-$ | chlorate(VII) |

(acid strength increases ↓; base strength decreases ↓)

This reaction emphasizes that hydrochloric acid is a stronger acid than hydrogen sulphide since it donates protons to sulphide ions in forming $H_2S$. Alternatively, we might say that sulphide is a stronger base than chloride and in competition for protons the sulphide wins convincingly.

Thus, *acid-base reactions involve competitions between bases for protons and, in this respect, they can be compared to redox reactions which involve competitions between oxidizing agents for electrons.*

When dilute hydrochloric acid or sulphuric(IV) (sulphurous) acid is added to a solution of sodium benzoate a white precipitate of benzoic acid appears.

$$\underset{\text{benzoate ion}}{C_6H_5COO^-(aq)} + \underbrace{H^+(aq) + Cl^-(aq)}_{\text{hydrochloric acid}} \longrightarrow \underset{\text{benzoic acid}}{C_6H_5COOH(s)} + Cl^-(aq)$$

$$\underset{\text{benzoate ion}}{C_6H_5COO^-(aq)} + \underset{\text{sulphuric(IV) acid}}{H_2SO_3(aq)} \longrightarrow \underset{\text{benzoic acid}}{C_6H_5COOH(s)} + HSO_3^-(aq)$$

Hence, both hydrochloric acid and sulphuric(IV) acid are stronger than benzoic acid.

However, when ethanoic acid is added to a solution of sodium benzoate no apparent reaction occurs. Ethanoic acid does not protonate benzoate ions, because ethanoic acid is a weaker acid than benzoic acid. Thus, we could place benzoic acid between ethanoic acid and sulphuric(IV) acid in the 'league' showing relative strengths of acids in table 13.4.

○ When three drops of Universal Indicator solution were added to separate portions of
(i) ammonium chloride solution,
(ii) a dilute solution of phenol ($C_6H_5OH$) in water and
(iii) water,
solutions (i) and (ii) produced an orange colour, whereas (iii) was yellow-green. What can you conclude from this about the relative acidities of the three solutions?

○ When a few magnesium turnings were added to the three solutions mentioned in the last paragraph, solution (i) evolved hydrogen most vigorously and (iii) least vigorously. What can you conclude from this about the relative acidities of (i), (ii) and (iii)?

○ Where would you place phenol ($C_6H_5OH$) in the 'league' of relative acidity in table 13.4?

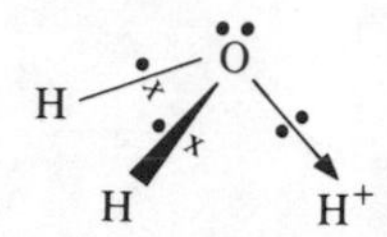


**Figure 13.8** Co-ordinate bonding in an aqueous $H^+$ ion.

## 13.12 Complex ions

In the last few sections, we have been discussing the reactions of acids with bases. Acids donate protons ($H^+$ ions) and in aqueous solution these $H^+$ ions are strongly attracted to polar water molecules by co-ordinate bonds forming $H_3O^+$ ions (figure 13.8).

In the same way, other cations besides $H^+$ (e.g. $M^{2+}$) will also exist in aqueous solution as hydrated ions (e.g. $[M(H_2O)_n]^{2+}$). Since the $H^+$ ion (a bare proton) is more than 50 000 times smaller than any other cation, the bonding between $H_2O$ and $H^+$ is much stronger than that between $H_2O$ molecules and other cations. However, the larger size of other cations enables them to associate with two, four or even six water molecules, whereas $H^+$ associates with only one.

The number of water molecules associated with a given cation has been firmly established in some cases, and in other cases it has been established more tentatively through indirect evidence. Thus, the ion $Mg^{2+}$(aq) is represented as $[Mg(H_2O)_6]^{2+}$, $Cu^{2+}$(aq) is represented as $[Cu(H_2O)_4]^{2+}$ whilst $Ag^+$(aq) can be represented as $[Ag(H_2O)_2]^+$.

In these ions, the water molecules can be considered to be bound to the central cation by co-ordinate bonds (figure 13.9).

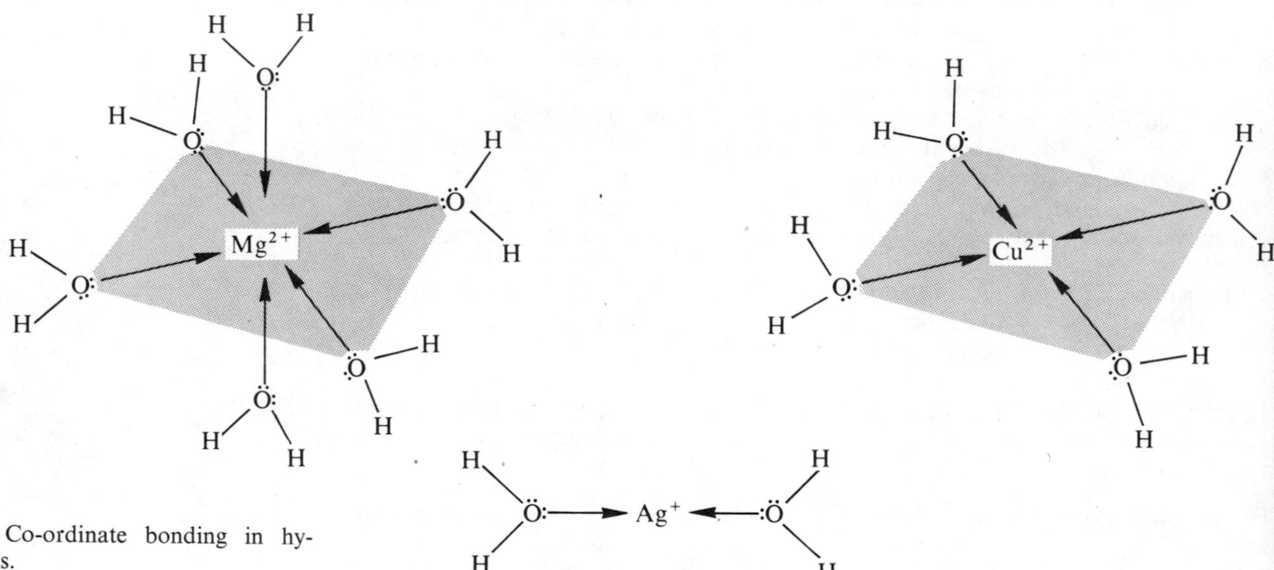


**Figure 13.9** Co-ordinate bonding in hydrated cations.

Other polar molecules, besides water, can also co-ordinate to metal cations. Thus, in ammonia solution $Cu^{2+}$ exists principally as $[Cu(NH_3)_4]^{2+}$ whilst $Ag^+$ exists as $[Ag(NH_3)_2]^+$.

It is not only polar molecules but also anions that can associate with cations in this way. Thus, when anhydrous copper(II) sulphate(VI) is added to concentrated hydrochloric acid, the solution becomes yellow-green due to the formation of $[CuCl_4]^{2-}$ ions.

Ions such as $[Cu(H_2O)_4]^{2+}$, $[Cu(NH_3)_4]^{2+}$ and $[CuCl_4]^{2-}$, in which a metal ion is associated with a number of anions or neutral molecules are known as **complex ions**. The anions and molecules firmly bonded to the central cation are called **ligands**. Each ligand contains at least one atom bearing a lone pair of electrons, which can be donated to the central cation forming a co-ordinate (dative) bond. The ligand is said to be co-ordinated to the central ion.

Water is by far the most common ligand. The strength of the co-ordinate bonds between different cations and water molecules can be compared by determining their **hydration energies** (sections 9.8 and 11.14), which relate to the process

$$M^{x+}(g) + nH_2O(l) \longrightarrow [M(H_2O)_n]^{x+}(aq).$$

Look closely at table 13.5.

- ○ How does the hydration energy change as ionic radius increases?
- ○ How does the hydration energy change as the charge on an ion increases?
- ○ In general, how will the strength of co-ordinate bonds from ligands to a central cation vary with the size of the cation and with the charge on the cation?

**Table 13.5** Hydration energies of some cations

| Ion | Ionic radius /nm | Hydration energy /kJ mole$^{-1}$ |
|---|---|---|
| $Li^+$ | 0.068 | −499 |
| $Na^+$ | 0.098 | −390 |
| $K^+$ | 0.133 | −305 |
| $Rb^+$ | 0.148 | −281 |
| $Cs^+$ | 0.167 | −248 |
| $Mg^{2+}$ | 0.065 | −1891 |
| $Ca^{2+}$ | 0.094 | −1562 |
| $Sr^{2+}$ | 0.110 | −1413 |
| $Ba^{2+}$ | 0.134 | −1273 |
| $Al^{3+}$ | 0.045 | −4613 |
| $Ga^{3+}$ | 0.062 | −4650 |

## 13.13 Naming complex ions

The systematic naming of complex ions is based on four simple rules.

1 *Number the ligands* using Greek prefixes; mono-, di-, tri-, etc.

2 *Identify the ligands* using names ending in -o, e.g. $NH_3$ -ammino, $CN^-$ -cyano, $Cl^-$ -chloro, $OH^-$ -hydroxo, $H_2O$ -aquo.

3 *Name the cation* using the English name for the cation in a positively-charged complex, but the Latinized name ending with the suffix *-ate* for the central metal atom in a negatively-charged complex, i.e. aluminate for aluminium, cuprate for copper, ferrate for iron, plumbate for lead, zincate for zinc, etc.

4 *Indicate the oxidation number of the central cation* using Roman numerals (I, II, III, etc.)

The following examples show you how to apply these rules.

| Formula of complex ion | 1 Number the ligands | 2 Name the ligand | 3 Name the central cation (suffix-*ate* for anions) | 4 Indicate oxidation number of central cation |
|---|---|---|---|---|
| $[Cu(NH_3)_4]^{2+}$ | tetra...... | .ammino..... | copper ....... | (II) |
| $[CuCl_4]^{2-}$ | tetra...... | .chloro ..... | cuprate....... | (II) |
| $[Fe(CN)_6]^{3-}$ | hexa....... | cyano ..... | ferrate ....... | (III) |
| $[Cu(NH_3)_2]^+$ | di ....... | ammino..... | copper ....... | (I) |

Now try to name the following ions:

$[Al(OH)_4]^-$, $[Zn(NH_3)_4]^{2+}$, $[Fe(CN)_6]^{4-}$, $[CrCl_2(H_2O)_4]^+$.

## 13.14 Polydentate ligands and chelation

Most ligands can form only one co-ordinate bond with a cation. They are said to be **unidentate** since they have only 'one tooth' with which to attach themselves to the central cation in a complex (the word *dens* in Latin means tooth). In some cases, each ligand molecule or anion can form more than one link with the metal ion and these are said to be **polydentate** ('many teeth'). Examples of bidentate ligands are:

ethanedioate (oxalate), 1,2-diaminoethane, and 2-hydroxybenzoate (salicylate)

$$^{-}O(O{=})C{-}C({=}O)O^{-} \qquad H_2\ddot{N}CH_2.CH_2\ddot{N}H_2 \qquad C_6H_4(COO^{-})(\ddot{O}H)$$

Some ligands such as edta (ethylenediaminetetraacetate) can form as many as six dative bonds with the central ion and are known as hexadentate ligands.

The complex ions which form between polydentate ligands and cations are known as **chelates** or **chelated complexes** from a Greek word *chelos* meaning 'a crab's claw', since the ligand forms a clawing pincer-like grip on the metal ion. Figure 13.10 shows the chelate formed between iron(III) and ethanedioate (oxalate) ions.

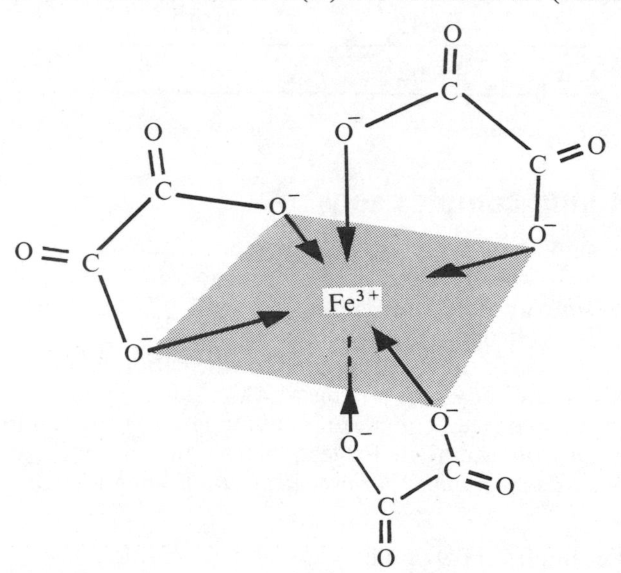


**Figure 13.10** The structure of the iron(III)/ethanedioate chelate, $[Fe(C_2O_4)_3]^{3-}$

In general, polydentate ligands are more powerful and more versatile than simple unidentate ligands and the stability of a complex is much enhanced by chelation as the pincer-like grip of the polydentate ligand can hold the central cation more securely.

## 13.15 Complexing—competition for cations

In section 13.11 we noticed that *acid-base reactions involve competitions between different bases for $H^+$ ions*. In a similar fashion, *complexing reactions usually involve competitions between different ligands for metal cations*. Thus, when excess ammonia solution is added to aqueous copper(II) sulphate(VI), ammonia molecules displace water molecules from hydrated copper(II) ions forming $[Cu(NH_3)_4]^{2+}$ and the colour changes from pale blue to a much deeper blue.

$$[Cu(H_2O)_4]^{2+}(aq) + 4NH_3(aq) \longrightarrow [Cu(NH_3)_4]^{2+}(aq) + 4H_2O(l)$$

In this case, ammonia acts as a stronger ligand than water.

In general, it can be shown that a more powerful ligand will displace a less powerful ligand from a cation complex. Thus, when 2-hydroxybenzoate ions (salicylate $HOC_6H_4COO^-$) are added to aqueous iron(III) chloride the colour changes from yellow to a deep purple as 2-hydroxybenzoate ions displace water molecules from hydrated iron(III) ions forming purple $[Fe(HOC_6H_4COO)_3]$.

$$[Fe(H_2O)_6]^{3+}(aq) + 3HOC_6H_4COO^-(aq) \rightarrow [Fe(HOC_6H_4COO)_3](s) + 6H_2O(l$$

When edta (an even stronger ligand than 2-hydroxybenzoate) is added to the purple suspension, the edta displaces 2-hydroxybenzoate ions and the colour changes from purple to pale yellow which is the colour of the iron(III)/edta complex.

One of the most fascinating and intriguing demonstrations of the relative strengths of different ligands for a particular cation is outlined in the sequence of reactions shown in table 13.6. Going down the table, the ligands are increasing in strength and becoming stronger competitors for $Ag^+$ ions, i.e.

$$edta^{2-} > S^{2-} > CN^- > I^- > S_2O_3^{2-} > Br^- > NH_3 > Cl^- > H_2O.$$

**Table 13.6** The relative strengths of different ligands for $Ag^+$

| | Ligand | Colour and state of complex |
|---|---|---|
| $[Ag(H_2O)_2]^+$ | $H_2O$ | clear solution |
| add ↓ $NaCl(aq)$ | | |
| $AgCl(s)$ | $Cl^-$ | white precipitate |
| add ↓ $NH_3(aq)$ | | |
| $[Ag(NH_3)_2]^+$ | $NH_3$ | clear solution |
| add ↓ $KBr(aq)$ | | |
| $AgBr(s)$ | $Br^-$ | cream precipitate |
| add ↓ $Na_2S_2O_3(aq)$ | | |
| $[Ag(S_2O_3)_2]^{3-}$ | $S_2O_3^{2-}$ | clear solution |
| add ↓ $KI(aq)$ | | |
| $AgI(s)$ | $I^-$ | yellow precipitate |
| add ↓ $KCN(aq)$ | | |
| $[Ag(CN)_2]^-$ | $CN^-$ | clear solution |
| add ↓ $Na_2S(aq)$ | | |
| $Ag_2S(s)$ | $S^{2-}$ | black precipitate |
| add ↓ sodium edta(aq) | | |
| $[Ag(edta)]^-$ | $edta^{2-}$ | clear solution |

Notice that some reactions in the sequence result in the precipitation of insoluble solids. These precipitates can be regarded in some respects as neutral complexes. Being uncharged, they are less readily hydrated by polar water molecules than the charged complexes and so they are less likely to dissolve in water.

## 3.16 Investigating the stoichiometry of complex ions

Essentially, a determination of the stoichiometry of a complex ion involves a measurement of the number of ligands complexing with one metal cation. This can be investigated by various methods, the most important of which are:

titration methods involving competitive complexing (described below), and

colorimetric methods requiring measurement of the colour intensity of the mixture as the proportion of metal cation to ligand is varied (see section 18.6, page 267).

### DETERMINATION OF THE STOICHIOMETRY OF THE IRON(III)/EDTA COMPLEX BY COMPETITIVE COMPLEXING

When sodium 2-hydroxybenzoate (sodium salicylate) is added to a solution of $Fe^{3+}$ ions a deep purple complex ion is formed. As edta is added to this purple solution, the colour slowly fades being replaced by a pale yellow iron(III)/edta complex. Of course, the purple colour remains as long as there are more than enough $Fe^{3+}$ ions for the edta, since these excess $Fe^{3+}$ ions will form a purple complex with the 2-hydroxybenzoate. When the purple colour just disappears, the quantities of $Fe^{3+}$ and edta just balance and neither is in excess. Thus, 2-hydroxy-benzoate ions can act as indicators in deciding when just sufficient edta has been added to react with all the $Fe^{3+}$.

In a particular experiment, 20cm$^3$ of 1.0M iron(III) chloride was pipetted into a flask and a few drops of sodium 2-hydroxybenzoate were added. 0.1M edta was then added from a burette and the solution became a clear yellow colour after the addition of 20.1cm$^3$.

- ○ How many moles of $Fe^{3+}$ were taken?
- ○ How many moles of edta reacted with the $Fe^{3+}$ taken?
- ○ How many moles of edta react with one mole of $Fe^{3+}$?
- ○ What is the formula of the $Fe^{3+}$/edta complex?

## 13.17 The use and importance of complex ions

Complex ions are important in industry, in the laboratory and in biology.

### COMPLEX IONS OF BIOLOGICAL IMPORTANCE

Two essential biological macromolecules composed of complex ions are **chlorophyll** and **haemoglobin.** Chlorophyll is the green pigment in plant cells which is responsible for absorbing the radiant energy of sunlight and converting it into chemical energy in the bonds of carbohydrate molecules synthesized by the plant.

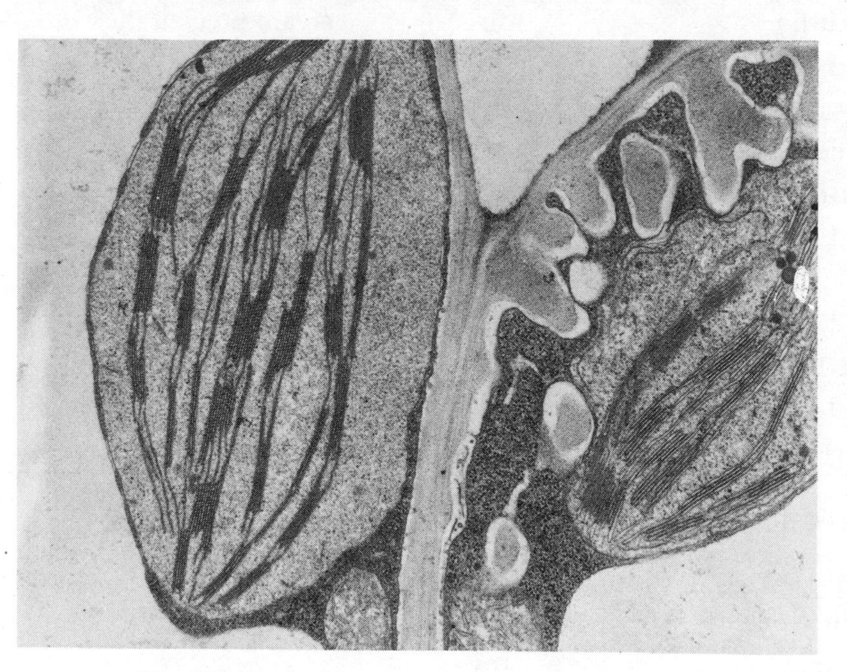

Dark parallel lines of chlorophyll in two chloroplasts from a leaf of flax. The magnification is 46 000 times.

The chlorophyll molecule is composed of a complicated cyclic tetradentat ligand containing four nitrogen atoms surrounding magnesium in a square plana arrangement (figure 13.11).

Sunlight reaching the earth's surface has a maximum intensity in the blue-gree region of the spectrum in the wavelength range 450 to 550nm. Curiously, th chlorophyll molecule has its weakest absorption in this portion of the visibl spectrum. The maximum absorption peaks for chlorophyll are, in fact, at 680nn in the red region of the spectrum and at 440nm in the violet (figure 13.12). In spit of this inefficiency in absorbing the radiation of greatest intensity, chlorophyll ha other properties that make it particularly suitable as a photosynthetic pigment. I can receive energy both directly from light and indirectly from other pigments i plants such as carotenoids; it can store radiant energy and transfer it to othe molecules, and its ring structure of alternating double and single bonds enables it t function in the transference of hydrogen atoms from one molecule to another.

Figure 13.11 also shows the structure of haemoglobin, the red pigment preser in red blood cells. Haemoglobin acts as the transporter of oxygen in the blood. is composed of the complex cyclic tetradentate haem ligand attached to the protei globin. Notice the striking similarity of the haem portion of the complex to tl

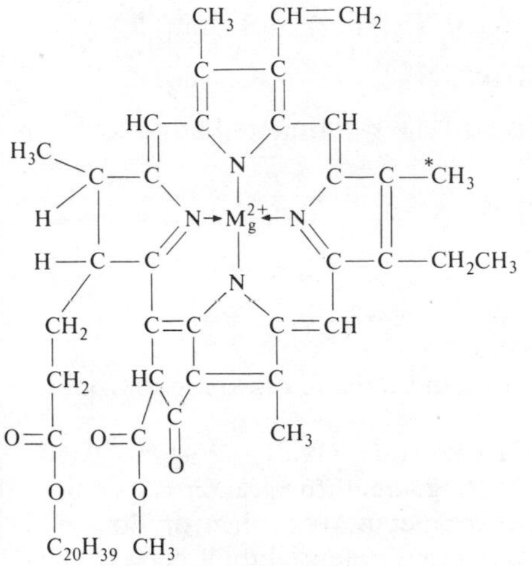

Chlorophyll a
(In chlorophyll b, the methyl group marked by an asterisk is replaced by a —CHO group)

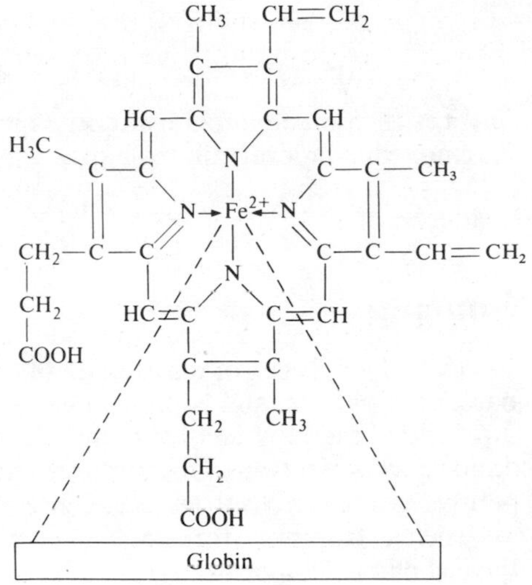

Haemoglobin

**Figure 13.11** Complex ions of biological importance

chlorophyll molecule, suggesting that the two may have been adapted from the same original substance during the course of evolution. In haemoglobin, the central metal ion is iron(II), not magnesium. The active part of the haemoglobin complex is the $Fe^{2+}$ ion which is co-ordinated to four nitrogen atoms in the haem ligand and also to two nitrogen atoms in the globin. One of the latter two positions can be occupied weakly and reversibly by oxygen and it is this property which enables haemoglobin to transport oxygen in the bloodstream from the lungs to other parts of the body. Unfortunately, much stronger ligands than oxygen, such as cyanide and carbon monoxide, can also occupy this position. Unlike oxygen, they attach themselves to the haem group irreversibly so they act as acute poisons by eliminating the oxygen-transporting ability of haemoglobin.

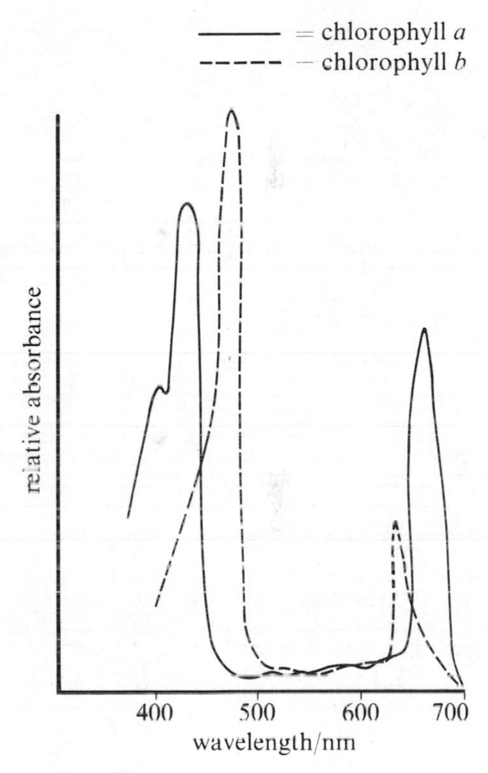


**Figure 13.12** The absorption spectrum of chlorophyll.

## COMPLEX IONS OF INDUSTRIAL IMPORTANCE

Complex ions play an important part in the methods used to soften hard water and in the extraction of metals from their ores.

Soap contains the sodium salts of long chain carboxylic acids such as sodium hexadecanoate (sodium palmitate) and sodium octadecanoate (sodium stearate) (section 31.3). When hard water, containing calcium and magnesium ions, is mixed with soap, it reacts with the anions in these salts forming insoluble compounds which we see as 'scum' on the surface of the water.

$$Ca^{2+}(aq) + 2CH_3(CH_2)_{16}COO^- \longrightarrow Ca(CH_3(CH_2)_{16}COO)_2(s)$$

| $Ca^{2+}(aq)$ | + | $2CH_3(CH_2)_{16}COO^-$ | ⟶ | $Ca(CH_3(CH_2)_{16}COO)_2(s)$ |
|---|---|---|---|---|
| ions in hard water | | octadecanoate (stearate) ions in soap | | insoluble ppte. of calcium octadecanoate (stearate) in 'scum' |

Various ligands will react with the calcium and magnesium ions to form soluble and stable complex ions. As a result of this, the calcium and magnesium ions are unable to react with the anions in soap and the water is softened. Thus, sodium polyphosphate ($Na_6P_6O_{18}$), which contains the powerful polyphosphate ligand ($P_6O_{18}{}^{6-}$) is sold for domestic and industrial water softening under the trade name Calgon, a term derived from the expression 'calcium gone'. edta and its sodium salt, known under the trade names Sequestrol and Versene, are also used in water softening. However, these two substances are poisonous and cannot be used in softening drinking water. Consequently, their use is restricted to the softening of water required for industrial processes such as the dyeing of textiles.

Complex formation frequently plays an essential part in the purification of metal ores and the subsequent extraction of the metal. An excellent example of this is provided by the purification of aluminium oxide from bauxite (section 15.6) by conversion to the complex ion, $[Al(OH)_4]^-$. The extraction of gold and silver involve the formation of complex cyanides.

Impure silver ores such as argentite (containing $Ag_2S$) and horn silver (containing AgCl) are first mixed with a solution of sodium cyanide to form the soluble complex cyanide.

$$Ag_2S(s) + 4CN^-(aq) \longrightarrow 2[Ag(CN)_2]^-(aq) + S^{2-}(aq)$$
$$AgCl(s) + 2CN^-(aq) \longrightarrow [Ag(CN)_2]^-(aq) + Cl^-(aq)$$

Silver is then precipitated by adding zinc dust to the solution and any excess zinc is removed by adding dilute acid.

$$2[Ag(CN)_2]^-(aq) + Zn(s) \longrightarrow [Zn(CN)_4]^{2-}(aq) + 2Ag(s)$$

## Summary

1 The major classes of reaction involving inorganic substances are redox, acid-base, precipitation and complexing.
2 Redox reactions involve competition for electrons. Oxidizing agents (oxidants) accept electrons. The stronger the oxidizing agent, the greater is its competitive desire for electrons. Reducing agents (reductants) are electron donors.
3 The electromotive force of a cell is the maximum potential difference which the cell can generate.
4 The standard conditions for electrochemical ($E^\ominus$) measurements are similar to those for thermochemical measurements:- 25°C, gases at 1atm. pressure, aqueous solutions 1.0M, platinum as the electrode when the half-cell system does not include a metal.
5 The standard electrode (reduction) potential, $E^\ominus$, of a standard half-cell is the potential of that half-cell relative to a standard hydrogen half-cell under standard conditions. $E^\ominus$ which relates to the reduction half-equation has the same sign as the particular half-cell relative to the standard hydrogen half-cell.
6 The size and sign of $E^\ominus$ values provide a measure of the relative strength of oxidizing agents and their conjugate reducing agents.
7 An overall positive $E^\ominus$ for a redox pair suggests that a reaction should take place. In practice:
(i) likely reactions may be too slow to occur,
(ii) reactions which are unlikely under standard conditions (to which $E^\ominus$ values refer) may take place readily under different conditions of temperature, pressure and concentration.
8 Arrangements which generate electric currents from chemical reactions are called cells. A battery is composed of several cells.
9 Cells which can be used only once and which cannot be recharged are called primary cells. Cells which can be recharged and used again and again are called secondary cells or accumulators.
10 According to the Brønsted-Lowry theory of acids and bases, an acid is a proton donor, whereas a base is a proton acceptor.
11 Acid-base reactions involve competitions between bases for protons ($H^+$ ions). The stronger the base the greater is its competitive desire for protons.
12 Water can act both as a base and as an acid and does so simultaneously during its dissociation.

$$\underset{\text{acid}}{H_2O} + \underset{\text{base}}{H_2O} \rightleftharpoons \underset{\text{base}}{OH^-} + \underset{\text{acid}}{H_3O^+}$$

13 A complex ion is an ion in which a central metal cation is associated with a number of anions or neutral molecules.
14 The anions and molecules bound to the central cation in a complex ion are called ligands.
15 Chelates or chelated complexes are complex ions in which each ligand forms more than one link with the central cation.
16 Complexing reactions involve competitions between ligands for metal cations. The stronger the ligand the greater is its competitive desire for cations.

## Study questions

1 (a) How did Arrhenius define (i) an acid (ii) a base?
(b) How did Brønsted and Lowry define (i) an acid (ii) a base?
(c) Show, by writing appropriate equations how:
(i) the $HSO_4^-$ ion can act as an Arrhenius acid,
(ii) the $HSO_4^-$ ion can act as a Brønsted-Lowry base.

(d) Explain, using examples, how the Brønsted-Lowry definition allows a much wider range of substances to be classified as bases.

2 2.5g of a hydrate of sodium carbonate, $Na_2CO_3.xH_2O$, were dissolved in 250cm$^3$ of water. 25cm$^3$ of the resulting solution required 17.5cm$^3$ of 0.1M hydrochloric acid to neutralize it completely.
(a) How many moles of sodium carbonate react with one mole of hydrochloric acid?
(b) How many moles of hydrochloric acid are there in 17.5cm$^3$ of a 0.1M solution?
(c) How many moles of sodium carbonate react with 17.5cm$^3$ of 0.1M hydrochloric acid?
(d) How many moles of sodium carbonate were there in the original 2.5g?
(e) How many moles of water of crystallization must there have been in the original 2.5g?
(f) What is the value of $x$? [Na = 23, C = 12, O = 16, H = 1]

3 (a) Explain what you understand by the terms oxidation and reduction.
(b) Illustrate your answer to part (a) by writing separate half equations involving electrons for the oxidation and reduction processes when the following substances react:
(i) Mg with $Cl_2$,
(ii) $I_2$ with $S_2O_3^{2-}$,
(iii) $MnO_2$ with conc.HCl,
(iv) acidified $KMnO_4$ with $SO_2$,
(v) $H_2O_2$ with $I^-$,
(vi) $Fe^{3+}$ with $I^-$,
(vii) acidified $BrO_3^-$ with $Br^-$.
(c) Suppose you were required to determine the standard electrode potential of a $Zn^{2+}$(aq)/Zn(s) half-cell.
(i) List all the essential materials and equipment required for this determination.
(ii) Draw a diagram showing how the apparatus would be assembled for the determination.

4 Use oxidation numbers to decide which elements are oxidized or reduced in the following conversions.
(a) $Cr^{3+} \longrightarrow Cr_2O_7^{2-}$
(b) $CrO_4^{2-} \longrightarrow Cr_2O_7^{2-}$
(c) $S_2O_3^{2-} \longrightarrow S_4O_6^{2-}$
(d) $C_2O_4^{2-} \longrightarrow CO_2$
(e) $IO_3^- \longrightarrow I_2$

5 (a) Draw a fully labelled diagram of a Daniell cell and show the direction in which electrons flow in the external circuit when it is used to generate electricity.
(b) Write equations for the reactions at each electrode when the cell produces an electric current.
(c) Deduce a value for the e.m.f. of the cell operating under standard conditions. ($Cu^{2+}$(aq)/Cu(s) = +0.34V, $Zn^{2+}$(aq)/Zn(s) = −0.76V).
(d) How would the e.m.f. of the cell be affected if:
(i) the concentration of $Cu^{2+}$ ions was increased,
(ii) the concentration of $Zn^{2+}$ ions was increased?
Give an explanation of your reasoning in each case.

6 The rusting of iron is an oxidation process.
(a) Explain, with appropriate equations, what happens when iron rusts.
(b) Iron may be protected from rusting by coating with zinc or tin. By reference to the following data, explain why zinc protects iron more effectively than tin, once the protective metal coating has been scratched to expose the iron below.

| | $E^\ominus$/volts |
|---|---|
| $Zn \longrightarrow Zn^{2+} + 2e^-$ | +0.76 |
| $Fe \longrightarrow Fe^{2+} + 2e^-$ | +0.44 |
| $Sn \longrightarrow Sn^{2+} + 2e^-$ | +0.14 |

(c) Why do you think tin is used rather than zinc to coat and protect the inside of 'tin' cans, whereas zinc is used rather than tin to galvanize and protect buckets?

(a) The element nitrogen can form compounds in various oxidation states from −3 to +5.
(i) Construct an oxidation number chart for nitrogen in the following common compounds:

$NH_3$, $NH_4Cl$, $HNO_3$, $HNO_2$, $NO_2$.

(ii) What are the principal oxidation states of nitrogen?
(iii) What are the oxidation numbers of nitrogen in the following compounds?
hydrazine, $N_2H_4$;
dinitrogen oxide, $N_2O$;
nitrogen oxide, NO;
dinitrogen tetraoxide, $N_2O_4$;
ammonium nitrate(v), $NH_4NO_3$.

(b) Nitrogen forms a variety of bonds in different compounds—ionic, covalent, co-ordinate and hydrogen bonds.
Draw dot/cross diagrams to show the electronic structures of atoms and/or ions in the following compounds and indicate the type of bonds involved:
(i) magnesium nitride, $Mg_3N_2$;
(ii) ammonia, $NH_3$;
(iii) nitrogen trichloride, $NCl_3$;
(iv) ammonium chloride, $NH_4Cl$;
(v) sodium nitrate(v), $NaNO_3$.

8 The following table gives the standard electrode potentials for a number of half-reactions.

| | $E^\ominus$/volts |
|---|---|
| $Zn^{2+} + 2e^- \longrightarrow Zn$ | $-0.76$ |
| $Fe^{2+} + 2e^- \longrightarrow Fe$ | $-0.44$ |
| $I_2 + 2e^- \longrightarrow 2I^-$ | $+0.54$ |
| $Fe^{3+} + e^- \longrightarrow Fe^{2+}$ | $+0.77$ |
| $Ce^{4+} + e^- \longrightarrow Ce^{3+}$ | $+1.61$ |

(a) Relative to which half-equation are these electrode potentials expressed?
(b) Which of the substances listed in the above table is:
(i) the strongest oxidizing agent,
(ii) the strongest reducing agent?
(c) Which substance(s) in the table could be used to convert iodide ions to iodine? Write balanced equations for any possible conversions.
(d) A half-cell is constructed by putting a platinum electrode in a solution which is 1.0M with respect to both $Fe^{2+}$ and $Fe^{3+}$ ions. This half-cell is then connected by means of a 'salt bridge' to another half-cell containing an iron electrode in a 1.0M solution of $Fe^{2+}$ ions.
(i) What is the e.m.f. of this cell?
(ii) If the two electrodes are connected externally, what reactions take place in each half-cell?
(iii) In which direction do electrons flow in the external circuit?
(e) Write an equation for the reaction you would expect to occur when an iron nail is placed in a solution of iron(III) sulphate(VI).

# Groups I and II–The Alkali Metals and the Alkaline-Earth Metals

# 14

## 14.1 Introduction

The alkali metals and the alkaline-earth metals are members of Groups I and II respectively in the periodic table (table 14.1). Whilst you are studying these elements, look for the following important features:

1. The similarities between the elements in Group I.
2. The similarities between the elements in Group II.
3. The differences between the elements in Group I and those in Group II.
4. The gradual trends in properties of the elements and their compounds in Group I and the similar trends in Group II.

**Table 14.1** The members of Group I and Group II in the periodic table.

| **Group I** | **Group II** |
|---|---|
| Li | Be |
| Na | Mg |
| K | Ca |
| Rb | Sr |
| Cs | Ba |
| **The alkali metals** | **The alkaline-earth metals** |

In all these metals, the only electrons in their outermost shell occupy an *s*-orbital—hence these elements are sometimes called the **s-block elements**.

- ○ Write the electronic structures of Na and K in terms of *s, p, d, f* notation. (The electronic structure of Li is $1s^2\ 2s^1$).
- ○ Write the electronic structures of Be, Mg and Ca in terms of *s, p, d, f* notation.

You will now realize that *all Group I metals have one outermost s-electron, whereas Group II metals have two outermost s-electrons*. These *s*-electrons, situated much further from the nucleus than all other electrons in a metal, are only weakly held by the positive nucleus. Consequently, the atoms of Group I metals readily lose their single outermost *s*-electron to form stable ions, with a noble gas electron structure, carrying one positive charge, e.g.

$$Na \longrightarrow Na^+ + e^-$$

Similarly, the atoms of Group II metals readily lose two electrons to form stable ions, with a noble gas structure, carrying two positive charges, e.g.

$$Ca \longrightarrow Ca^{2+} + 2e^-$$

## 14.2 Oxidation numbers and *s*-block elements

Tables 14.2 and 14.3 summarize various physical properties of the elements in Groups I and II respectively.

**Table 14.2** Physical properties of the elements in Group I.

| **Element** | Li | Na | K | Rb | Cs |
|---|---|---|---|---|---|
| **Electron structure** | (He)$2s^1$ | (Ne)$3s^1$ | (Ar)$4s^1$ | (Kr)$5s^1$ | (Xe)$6s^1$ |
| **First ionization energy/kJ mole$^{-1}$** | 520 | 500 | 420 | 400 | 380 |
| **Second ionization energy/kJ mole$^{-1}$** | 7300 | 4600 | 3100 | 2700 | 2400 |
| **Atomic radius/nm (metallic radius)** | 0.15 | 0.19 | 0.23 | 0.25 | 0.26 |
| **Melting point/°C** | 180 | 98 | 64 | 39 | 29 |
| **Boiling point/°C** | 1330 | 892 | 760 | 688 | 690 |
| **Density/g cm$^{-3}$** | 0.53 | 0.97 | 0.86 | 1.53 | 1.90 |
| **Standard electrode potential, $M^+(aq)$ /M(s)/Volts** | −3.05 | −2.71 | −2.93 | −2.92 | −2.92 |

Look closely at the first and second ionization energies for Group I elements. The first ionization energy is much lower than the second. In the case of Na, it is nine times easier to remove the first electron than the second. The first electron, in an *s*-orbital, can be removed from the element easily, but the second electron must be removed from a noble gas core which is closer to the nucleus, and therefore requires much more energy. Thus, sodium readily forms $Na^+$ ions, but never forms $Na^{2+}$ ions. This means that sodium and the other elements in Group I have only the one oxidation state of +1 in their compounds.

Look closely at the first, second and third ionization energies of Group II elements in table 14.3.

**Table 14.3** Physical properties of the elements in Group II.

| **Element** | Be | Mg | Ca | Sr | Ba |
|---|---|---|---|---|---|
| **Electron structure** | (He)$2s^2$ | (Ne)$3s^2$ | (Ar)$4s^2$ | (Kr)$5s^2$ | (Xe)$6s^2$ |
| **First ionization energy/kJ mole$^{-1}$** | 900 | 740 | 590 | 550 | 500 |
| **Second ionization energy/kJ mole$^{-1}$** | 1800 | 1450 | 1150 | 1060 | 970 |
| **Third ionization energy/kJ mole$^{-1}$** | 14 800 | 7700 | 4900 | 4200 | — |
| **Atomic radius/nm (metallic radius)** | 0.11 | 0.16 | 0.20 | 0.21 | 0.22 |
| **Melting point/°C** | 1280 | 650 | 838 | 768 | 714 |
| **Boiling point/°C** | 2770 | 1110 | 1440 | 1380 | 1640 |
| **Density/g cm$^{-3}$** | 1.85 | 1.74 | 1.55 | 2.6 | 3.5 |
| **Standard electrode potential, $M^{2+}(aq)$ /M(s)/Volts** | −1.85 | −2.37 | −2.87 | −2.89 | −2.91 |

- ○ What is the ratio of the first ionization energy: second ionization energy: third ionization energy for Mg?
- ○ Why is there a greater increase from the second to the third ionization energy than from the first to the second?
- ○ What is the most stable oxidation state for Group II elements?

## 14.3 Physical properties

The atomic radii of alkali metals are included in table 14.2 and those of the alkaline-earth metals are given in table 14.3. As one might expect, the atomic radii increase with atomic number down the groups since each succeeding element has electrons in one more shell than the previous element.

The outermost *s*-electrons in these metals are held very weakly by the nucleus. Thus, the outer electrons can drift further from the nucleus than in most other atoms, and the elements in Groups I and II have larger atomic radii than those elements which follow them in their respective periods. The large atomic size results in weaker forces between neighbouring atoms since there is a reduced attraction of the nuclear charge for the shared mobile outer electrons as these electrons get further away.

Consequently, the metals in Groups I and II have lower melting points and lower boiling points than we would normally associate with metals. Apart from beryllium, all their melting points are less than 840°C and all the Group I metals melt below 200°C. In contrast, most of the transition metals melt at temperatures above 1 000°C.

The reduced interatomic forces in these metals makes them relatively soft. All the Group I metals can be cut with a pen-knife and, although the Group II metals are harder than those in Group I, they are generally softer than transition metals.

As we have seen, *s*-block elements have larger atomic radii than transition metals of approximately the same relative atomic mass. Thus, the *s*-block elements will be less dense and have larger molar volumes. Of the ten elements in tables 14.2 and 14.3, only barium is more dense than aluminium, a metal normally associated with low density. Their densities vary from approximately 1g $cm^{-3}$ to 3.5g $cm^{-3}$ for barium. Most transition metals have a density greater than 7g $cm^{-3}$.

## 14.4 Chemical properties

*All the metals in Groups I and II are high in the activity (electrochemical) series.* The standard electrode potential (tables 14.2 and 14.3) for the conversion of each metal *to its ions* is positive and greater than 2.0 volts for all metals except beryllium.

Hence, *these metals are very good reducing agents.* They all react vigorously with water, reducing it to hydrogen.

$$M(s) \longrightarrow M^+(aq) + e^-$$

$$H_2O(l) + e^- \longrightarrow \tfrac{1}{2}H_2(g) + OH^-(aq)$$

Excluding lithium, which reacts slower than all the other elements of Group I, the reactivity of the elements with water closely follows the value of $E^\ominus$.

In Group I, for example, sodium reacts vigorously, fizzing and skating about on the water surface. Potassium reacts even more vigorously giving small cracks and pops as the hydrogen explodes and producing a lilac flame as the hydrogen burns. Rubidium and caesium explode violently in contact with water.

- ○ Do $E^\ominus$ values relate to reaction rates or to equilibria? Do they indicate how fast a reaction will go, or simply the direction in which it should go?
- ○ Why do you think lithium reacts more slowly with water than other Group I metals, even though it has the largest $E^\ominus$?

Although electrode potentials relate only to aqueous solutions, they also provide a guide to the general reactivity of a substance. All the *s*-block elements are such good reducing agents that they can react with chlorine, bromine, sulphur, hydrogen and oxygen on heating. They all, of course, tarnish rapidly in air, forming a layer of oxide, and the reactivity of Group I metals is such that lithium, sodium and potassium are usually stored under oil.

## 14.5 Thermal stability of salts

The stability of the salt $M^+X^-(s)$ is dependent on both the size of its ions and the charge on these ions. The greater the charge, the stronger will be the attraction between the ions and the more stable will be $M^+X^-(s)$. The smaller the ions become, the closer they can approach each other in the solid crystal and the more stable is $M^+ X^-(s)$.

Thus, we might expect $M^+ X^-$ to become more stable as ionic charge increases or as ionic radius decreases.

There is however, one other very important factor to bear in mind during any discussion of thermal stability. When large anions such as $CO_3^{2-}$ decompose on heating to form smaller anions such as $O^{2-}$, the crystal containing the latter will generally be more stable than the carbonate, since the charge density on a small $O^{2-}$ ion will be greater than that on a larger $CO_3^{2-}$ ion, and the former will be more strongly attracted to the cations in the crystal. Furthermore, the smaller $O^{2-}$ ions can get closer to the cations than the larger $CO_3^{2-}$ ions.

This point has considerable importance in considering the thermal stability of nitrates(V), carbonates and hydroxides of the *s*-block elements.

### NITRATES(V)

Group I nitrates(V) (except $LiNO_3$) decompose on heating in the bunsen flame to form their corresponding nitrates(III) (nitrites) which are stable.

$$NaNO_3(s) \longrightarrow NaNO_2(s) + \tfrac{1}{2}O_2(g)$$

However, Group II nitrates(V) and $LiNO_3$ decompose on heating to form their corresponding oxide.

$$Mg(NO_3)_2(s) \longrightarrow MgO(s) + 2NO_2(g) + \tfrac{1}{2}O_2(g)$$

$$2LiNO_3(s) \longrightarrow Li_2O(s) + 2NO_2(g) + \tfrac{1}{2}O_2(g)$$

This suggests that the decrease in size from $NO_3^-$ to $NO_2^-$ stabilizes the Group I nitrates(V) to heat, but that the Group II nitrates(V) can only achieve thermal stability by decomposing from their nitrate to form the much smaller oxide ion.

### CARBONATES

The pattern of thermal stability amongst the carbonates is very similar to that of the nitrates. The carbonates of Group I (except $Li_2CO_3$) are stable at the temperature of the bunsen flame, but those of Group II decompose at this temperature to form the corresponding oxide.

$$MgCO_3(s) \longrightarrow MgO(s) + CO_2(g)$$

Here again the Group II compounds are achieving thermal stability by forming their oxide.

How do you think the thermal stability of $MgCO_3$ will compare with that of $BaCO_3$? Explain your answer.

### HYDROXIDES

Yet again, the pattern is repeated. The hydroxides of Group I (except LiOH*) are stable, but those of Group II decompose on heating.

$$Mg(OH)_2(s) \longrightarrow MgO(s) + H_2O(g)$$

## 14.6 The solubility of salts

Table 14.4 shows the results obtained when 0.1 M solutions of $Mg^{2+}$, $Ca^{2+}$ $Sr^{2+}$ and $Ba^{2+}$ are treated with various molar solutions.

Notice that the sulphates(VI), carbonates, ethanedioates(oxalates) and chromates(VI) of the Group II metals become more insoluble as atomic number increases, whereas their hydroxides become more soluble. These results are confirmed by the values of solubilities quoted for the sulphates(VI), carbonates chromates(VI), nitrates(V), and hydroxides in table 14.5.

* LiOH starts to decompose at about 650°C, but NaOH does not decompose until temperatures well above this and above the highest temperature obtainable with a bunsen.

**Table 14.4** Results obtained when 0.1 M solutions of $Mg^{2+}$, $Ca^{2+}$, $Sr^{2+}$ and $Ba^{2+}$ are treated with various molar solutions.

| **0.1 M solution of Group II cation** | Reaction with 1 M $Na_2SO_4$ | Reaction with sat'd. $CaSO_4$ | Reaction with 1 M $Na_2CO_3$ | Reaction with 1 M $Na_2C_2O_4$ | Reaction with 1 M $K_2CrO_4$ | Reaction with 1 M NaOH |
|---|---|---|---|---|---|---|
| $Mg^{2+}$ | no ppte | no ppte | white ppte | no ppte | no ppte | thick white ppte |
| $Ca^{2+}$ | thin white ppte | no ppte | white ppte | thin white ppte | no ppte | white ppte |
| $Sr^{2+}$ | white ppte | white ppte | thick white ppte | white ppte | pale yellow ppte | white ppte |
| $Ba^{2+}$ | thick white ppte | white ppte | very thick white ppte | thick white ppte | thick pale yellow ppte | thin white ppte |
| | Group II sulphates(VI) become more insoluble ↓ | | Group II carbonates become more insoluble ↓ | Group II ethanedioates (oxalates) become more insoluble ↓ | Group II chromates(VI) become more insoluble ↓ | Group II hydroxides become more soluble ↓ |
| **General equations for precipitations** | $M^{2+} + SO_4^{2-} \rightarrow MSO_4(s)$ | $M^{2+} + SO_4^{2-} \rightarrow MSO_4(s)$ | $M^{2+} + CO_3^{2-} \rightarrow MCO_3(s)$ | $M^{2+} + C_2O_4^{2-} \rightarrow MC_2O_4(s)$ | $M^{2+} + CrO_4^{2-} \rightarrow MCrO_4(s)$ | $M^{2+} + 2OH^- \rightarrow M(OH)_2(s)$ |

When an ionic compound dissolves in water, the following process occurs:

$$M^+X^-(s) + (aq) \longrightarrow \overset{+}{M}(aq) + X^-(aq)$$

We can picture this process as the result of two stages. First, the separation of ions in the solid, i.e.

$$M^+X^-(s) \longrightarrow M^+(g) + X^-(g)$$

which is the reverse of the lattice formation process and requires the *input* of the lattice energy; followed by the hydration of these separated ions by water, i.e.

$$M^+(g) + X^-(g) + (aq) \longrightarrow \overset{+}{M}(aq) + X^-(aq)$$

As the ionic radii of $M^+$ and $X^-$ increase, the heat change in both of these processes decreases. The reverse lattice energy process becomes less endothermic and the hydration process becomes less exothermic. Consequently, a decrease in the endothermic process may be cancelled by a decrease in the exothermic process and it is, therefore, difficult to make predictions about changes in solubility from changes in ionic size (section 11.14).

There are, however, one or two general patterns and trends in the solubility of Group I and Group II salts.

*All common Group I salts are soluble.*
*Salts of Group II cations containing anions with a charge of −1 (e.g. chlorides, nitrates*(V)*) are generally soluble, except for the hydroxides.*

Look at the solubilities of the various magnesium salts (all of which contain an anion with charge −1) in table 14.6.

- ○ Approximately, how much more soluble are the other compounds in table 14.6 compared to $Mg(OH)_2$?
- ○ Why are the solubilities given in moles per 100g of water rather than g/100g of water?

*Salts of Group II cations containing anions with a charge of −2 (e.g. sulphates*(VI)*, carbonates) are generally insoluble* (except for some magnesium and a few calcium salts).

Furthermore, there is a distinct decrease in solubility as the atomic number of the metal rises. Thus, beryllium and magnesium sulphates are very soluble in water, calcium sulphate is sparingly soluble, but strontium and barium sulphates are virtually insoluble. The solubilities of some compounds of the Group II metals are shown in table 14.5.

Why do the sulphates(VI), carbonates, ethanedioates(oxalates), and chromates(VI) of the alkaline-earth metals show similar trends in solubility? The solubility of all these salts decrease as the atomic number of the metal ion increases and this pattern is opposite to that observed for the hydroxides and fluorides.

**Table 14.5** The solubilities (at 25°C) of some compounds of Group II metals

| **Compound** | **Solubility/moles per 100g water** |
|---|---|
| $MgSO_4$ | $3600 \times 10^{-4}$ |
| $CaSO_4$ | $11 \times 10^{-4}$ |
| $SrSO_4$ | $0.62 \times 10^{-4}$ |
| $BaSO_4$ | $0.009 \times 10^{-4}$ |
| $MgCO_3$ | $1.3 \times 10^{-4}$ |
| $CaCO_3$ | $0.13 \times 10^{-4}$ |
| $SrCO_3$ | $0.07 \times 10^{-4}$ |
| $BaCO_3$ | $0.09 \times 10^{-4}$ |
| $MgCrO_4$ | $8500 \times 10^{-4}$ |
| $CaCrO_4$ | $870 \times 10^{-4}$ |
| $SrCrO_4$ | $5.9 \times 10^{-4}$ |
| $BaCrO_4$ | $0.011 \times 10^{-4}$ |
| $Mg(OH)_2$ | $0.2 \times 10^{-4}$ |
| $Ca(OH)_2$ | $16 \times 10^{-4}$ |
| $Sr(OH)_2$ | $330 \times 10^{-4}$ |
| $Ba(OH)_2$ | $240 \times 10^{-4}$ |
| $Mg(NO_3)_2$ | $4.9 \times 10^{-1}$ |
| $Ca(NO_3)_2$ | $7.7 \times 10^{-1}$ |
| $Sr(NO_3)_2$ | $3.4 \times 10^{-1}$ |
| $Ba(NO_3)_2$ | $0.35 \times 10^{-1}$ |

**Table 14.6** The solubilities (at 25°C) of various magnesium salts containing singly-charged anions.

| **Compound** | **Solubility/moles per 100g water** |
|---|---|
| $MgCl_2$ | $5.6 \times 10^{-1}$ |
| $MgBr_2$ | $5.5 \times 10^{-1}$ |
| $Mg(OH)_2$ | $2.0 \times 10^{-5}$ |
| $Mg(CH_3COO)_2$ | $5.6 \times 10^{-1}$ |
| $Mg(NO_3)_2$ | $4.9 \times 10^{-1}$ |

Although it is difficult to relate solubilities accurately to such fundamental properties as lattice energy and hydration energy, it is possible to make a semi-quantitative analysis in these terms.

The lattice energies of the sulphates(VI) of Group II metals change relatively little from $MgSO_4$ to $BaSO_4$ since their values are determined largely by the reciprocal of the sum of the ionic radii, $[1/(r_+ + r_-)]$. The ionic radius of the sulphate(VI) ion, $r_-$, is so large that the changing size of the much smaller cation makes very little difference to $1/(r_+ + r_-)$.

However, the changing size of the cation does cause considerable differences in hydration energy. From $Mg^{2+}$ to $Ba^{2+}$, the increasing size of the cation results in smaller hydration energies and the decreasing exothermicity of the hydration process causes a decreasing solubility of their compounds.

The solubilities of the carbonates, ethanedioates (oxalates), chromates(VI) and nitrates(V) of Group II metals vary in the same way as those of their sulphates because the anions in all these compounds are relatively large compared to the constituent cations.

However, this trend of decreasing solubility with atomic number for the alkaline-earth compounds is reversed in the case of their hydroxides and fluorides. In these compounds, the anions ($OH^-$ and $F^-$) are much smaller and lattice energies are no longer approximately constant, but decrease with increasing cation size.

Furthermore, the change in lattice energies is greater than the change in hydration energies, and consequently hydroxides and fluorides become more soluble as the group is descended.

## 14.7 Occurrence of the *s*-block elements

The alkali metals occur in nature only as +1 ions. Sodium and potassium are, o course, the commonest Group I metal ions, being sixth and seventh most abundan of all the elements in the earth's crust (figure 14.1). Lithium is found in trace amount

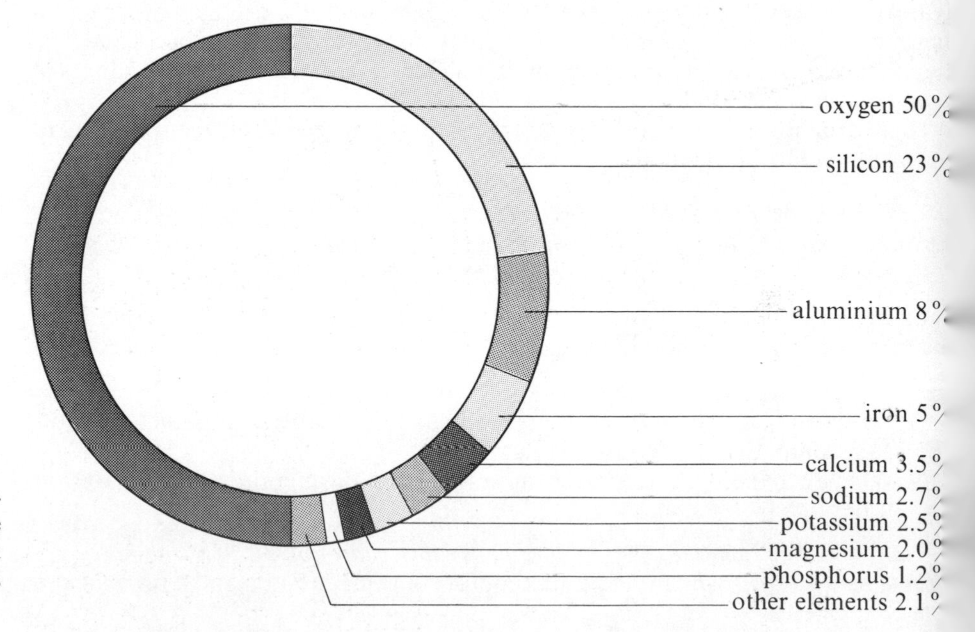


**Figure 14.1** The relative abundance of elements on the earth. (The figures given indicate the % by mass of the particular element in the earth's crust to a depth of 40 kilometres, including the atmosphere.)

in virtually all rocks and clays, but rubidium and caesium are very rare. Franciur the last element in Group I, does not exist naturally—it has only been obtaine in tiny amounts during nuclear reactions, and even these amounts disappe rapidly as the radioactive francium disintegrates and forms more stable elemen

Since many of the compounds of alkali metals are soluble in water, it is n surprising that they are found in relatively large quantities in sea water and as s deposits which have formed by the evaporation of brine. However, many insolub clays also contain alkali metal ions combined as complex silicates and aluminat

The alkaline-earth metals occur in nature only as +2 ions. As these +2 ions react with −2 ions to form insoluble compounds, it is not surprising that Group II elements exist in nature principally as their carbonates, silicates and sulphates(VI).

The only beryllium mineral found in any quantity is the silicate, beryl ($3BeSiO_3 . Al_2(SiO_3)_3$). Emeralds are, in fact, crystals of beryl, coloured green by traces of chromium.

Magnesium is the eighth most abundant element in the earth's crust. It is found extensively as the carbonate in magnesite ($MgCO_3$) and dolomite ($MgCO_3 . CaCO_3$), as the silicate in minerals such as asbestos ($3MgSiO_3 . CaSiO_3$) and as $Mg^{2+}$ ions in sea water.

Calcium is the fifth most abundant element in the earth's crust and the most abundant *s*-block element. Vast quantities of calcium occur as the carbonate in chalk, limestone and marble. Each of these minerals is believed to have been formed by the effect of high temperature and high pressure on the accumulated skeletal remains of marine animals.

Smaller quantities of calcium occur as the sulphate(VI) in anhydrite ($CaSO_4$) and gypsum ($CaSO_4 . 2H_2O$). These sulphate ores are thought to have resulted from the action of sulphuric(VI) acid (produced from the oxidation of sulphide minerals) on limestone.

Strontium and barium are much rarer elements. The commonest ore of strontium is strontianite ($SrCO_3$) and that of barium is barite ($BaSO_4$).

A large crystal of rock salt.

*Above left* The white horse carved in the chalk downland at Uffington, Berkshire.

*Left* An Ethiopian peasant carries harvested salt to the shore.

*Far left* An Ethiopian workman at the Lake Assale salt works shapes slabs of salt into conveniently-sized bars.

The average abundance of radium (the last element in Group II) in the earth's crust has been estimated at less than 1 part per million million. Radium exists naturally as a radioactive element formed during the nuclear disintegration of heavier elements such as uranium.

## 14.8 Manufacture of the *s*-block elements

In order to extract the *s*-block elements from their naturally-occurring compounds, it is necessary to reduce their positive ions, i.e.

$$M^+(aq) + e^- \longrightarrow M(s) \text{ or}$$
$$M^{2+}(aq) + 2e^- \longrightarrow M(s).$$

This reduction can be done either chemically or electrolytically, but the large negative electrode potentials of the *s*-block elements mean that chemical methods are virtually impossible since they would require a reducing agent stronger than the alkali or alkaline-earth metals themselves.

Consequently, these elements are usually extracted by electrolysis. The electrolytes used are always fused compounds rather than aqueous solutions, since water in the solution would be discharged at the cathode in preference to the *s*-block ion. For example,

$$Na^+ + e^- \longrightarrow Na \qquad E^\ominus = -2.71 \text{ volts}$$
$$H_2O + e^- \longrightarrow \tfrac{1}{2}H_2 + OH^- \qquad E^\ominus = -0.42 \text{ volts}$$

The more positive (less negative) electrode potential for the lower reaction means that it will occur in preference to the one above.

*Of the metals in Groups I and II, sodium and magnesium are in much higher demand than the others.* Each of these metals is obtained by electrolysis of its molten chloride.

Look closely at figure 14.2 which shows a simplified diagram of the Downs cell in which sodium is manufactured.

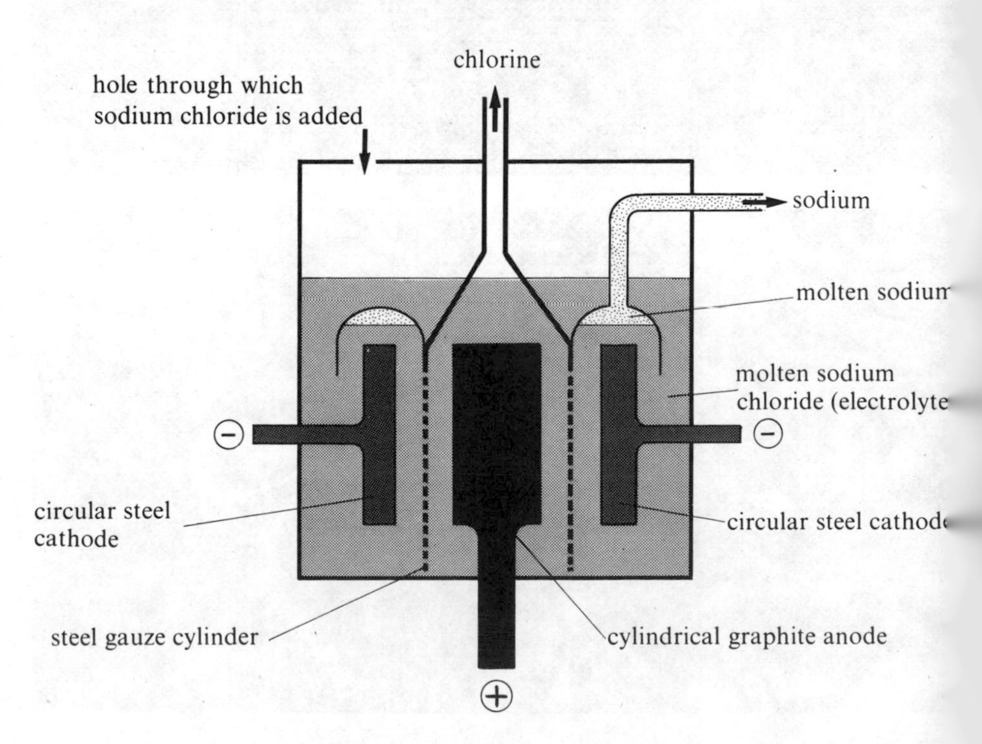


**Figure 14.2** The extraction of sodium in the Downs Cell.

- The electrolyte contains $Na^+(l)$ and $Cl^-(l)$. Write equations for the process at each electrode during the manufacture of sodium.
- In the Downs Cell, the cathode is made of steel, but the anode is made of graphite. Why is the anode made of graphite in spite of the fact that steel would be a better conductor?
- Why is it necessary to incorporate hoods and a steel gauze cylinder in order to prevent the product at the anode from mixing with the product at the cathode?

Since the extraction of these *s*-block elements must involve fused compounds rather than aqueous solutions, large quantities of energy are required not only for the electrolytic process itself, but also to melt the electrolyte.

- Why do you think that so many electrolytic manufacturing plants are located near supplies of hydro-electricity?
- In these electrolyses, the naturally-occurring raw material is usually mixed with some other compound. For example, in the extraction of sodium, the electrolyte of NaCl(l) contains some calcium chloride. How does the addition of calcium chloride reduce the energy expenditure?
- In view of the calcium chloride added to the electrolyte, what substance will contaminate the sodium produced at the cathode?

## 14.9 Uses of the *s*-block elements and their compounds

In spite of being so reactive, sodium, magnesium and the other *s*-block elements have many important uses. Probably the most important use of sodium is as a liquid coolant in nuclear reactors.

Fast nuclear reactors, such as that at Dounreay, operate at a temperature of about 600°C. Water, which boils at 100°C is therefore unsuitable as a cooling liquid in this case. But sodium, which melts at 98°C and boils at 892°C, is ideal for this purpose. Furthermore, sodium is a much better thermal conductor than water and almost as mobile a fluid. It can therefore be circulated easily around the hot reactor core and will transfer the heat away even more effectively than water.

Assuming that the sodium can be prevented from coming into contact with air or moisture, the only major problem in its use as a liquid coolant is that it can act as a solvent for various metals which might otherwise be used for the reactor core and the cooling system. Fortunately, iron and other transition metals are virtually insoluble in the liquid and so they can be used in the construction of the reactor core and the cooling pipes.

Sodium's high electrical conductivity and low density have led to extensive testing to see whether sodium could be used in electricity cables. Of course, sodium wire could never be used because of its reaction with air and water and its poor mechanical strength. However, a method has been devised to overcome these problems by extruding a tube of poly(ethene) (polythene) and simultaneously filling it with sodium.

Another important use of sodium is as an alloy with lead in the manufacture of tetraethyllead(IV) from chloroethane.

$$\underbrace{Pb + 4Na}_{\text{lead-sodium alloy}} + 4C_2H_5Cl \longrightarrow \underbrace{(C_2H_5)_4Pb}_{\text{tetraethyllead(IV)}} + 4NaCl$$

Liquid sodium is used as coolant in the Prototype Fast Reactor at Dounreay. The photograph shows the bottom end of one of the sodium pumps with the central inlet pipe and two outlet pipes.

Tetraethyllead(IV) (T.E.L.) is used as an essential 'antiknock' additive for petrol. Every gallon of petrol contains about 2cm³ of T.E.L. (section 25.6).

Sodium is also used in the reduction of titanium(IV) chloride to titanium by the Goldschmidt process and in the familiar yellow sodium vapour street lamps. In the metals industry, sodium cyanide is used in the extraction of gold and in the hardening of steel.

Sodium hydroxide is the cheapest industrial alkali. Thousands of tonnes are used annually in the manufacture of rayon (cellulose acetate), paper, and soap. Sodium polyphosphate ($Na_6P_6O_{18}$) is used as a water softener and sodium nitrate(III), $NaNO_2$, is used as a preservative in food.

The other metals in Group I are used much less extensively than sodium and its compounds. Potassium is used to make potassium superoxide, $KO_2$, which reacts with water to give oxygen.

$$4KO_2 + 2H_2O \longrightarrow 4KOH + 3O_2$$

$KO_2$ can therefore be used as an emergency source of oxygen in mines and in submarines. Potassium chloride, KCl, is used as a fertilizer and potassium nitrate is an essential constituent of gunpowder.

Owing to its very low first ionization energy, caesium is used as the light-sensitive surface in photocells—devices used to convert light signals to electrical signals. Figure 14.3 shows a simplified diagram of a photocell circuit. The evacuated tube

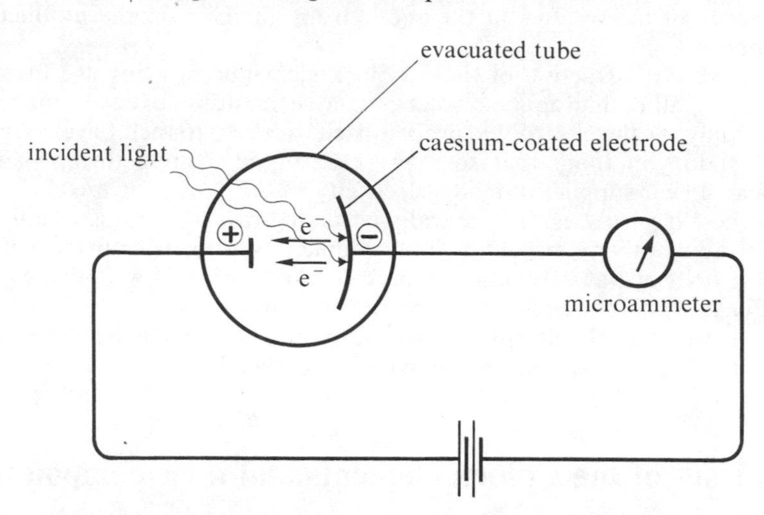


**Figure 14.3** A simplified photocell circuit.

contains two electrodes, one of which is coated with caesium or an alloy of caesium, antimony and silver. In the absence of light, the photocell does not conduct electricity. However, when the caesium coated electrode is struck by light, it emits electrons which are attracted to the positive electrode and an electric current flows in the circuit. Television pickup devices and photographers' lightmeters incorporate the use of photocells.

Magnesium is by far the most commonly used metal in Group II. It is used extensively in the preparation of lightweight alloys with a high tensile strength and particularly in the construction of aircraft parts and household goods. Although pure magnesium has poor structural strength, this can be increased by alloying it with aluminium, zinc and manganese. The aluminium increases the tensile strength of the alloy, the zinc improves its machining properties and the manganese reduces corrosion.

When magnesium burns in air it emits a bright, intense light. Because of this, magnesium powders and ribbons are used as light sources in photography. Flash bulbs contain magnesium (or aluminium) foil in an atmosphere of oxygen. When the bulb is fired, an electric current passes through the metal which is heated up and then suddenly flares into reaction with oxygen.

The use of magnesium flares in photography being demonstrated at an early meeting of the British Association in Birmingham.

The most commonly used compounds of magnesium are magnesium oxide (MgO), magnesium hydroxide ($Mg(OH)_2$), and hydrated magnesium sulphate(VI) ($MgSO_4 . 7H_2O$). Magnesium oxide is used as a lining for high temperature furnaces because of its exceptionally high melting point and low reactivity. Magnesium hydroxide is a very weak alkali. Because of this, it is an important constituent of toothpastes (it neutralizes acids that would otherwise cause decay) and a suspension of $Mg(OH)_2$ in water (commonly called milk of magnesia) is used to treat acid indigestion.

Beryllium is too rare and costly for any large scale uses, but it is used for hardening alloys and other metals such as copper and nickel. Beryllium-copper alloys are particularly resistant to sea-water corrosion and are therefore used for small marine parts.

Calcium, strontium and barium have very few uses owing to their high reactivity. The affinity of these elements for oxygen results in their use as deoxidizers in steel production and as 'getters' (i.e. removers of traces of oxygen) inside electron tubes and radio valves. By far the most commercially important compounds of calcium are lime (CaO) and limestone ($CaCO_3$). Limestone, of course, is the source of lime. Many thousands of tonnes of lime are used annually in counteracting soil acidity and in the production of calcium hydroxide ($Ca(OH)_2$), the cheapest alkali. Enormous quantities of lime are also used in the manufacture of plaster, mortar and cement.

*Left* Quarrying limestone.

The salts of calcium, strontium, and barium produce beautiful flame colours /hen heated to high temperatures in a bunsen flame; calcium salts—red, strontium alts—crimson, and barium salts—pale green. As a result of this, volatile compounds f these elements are used in the production of coloured military signals and in reworks.

*Above* Volatile compounds of calcium, strontium and barium are often used in fireworks.

## ummary

1 In the alkali and alkaline-earth metals, the only electrons in the outermost shell occupy an *s*-orbital.

2 The atoms of these *s*-block metals readily lose their outermost *s*-electrons to form stable ions; $M^+$ if the elements are in Group I, $M^{2+}$ if the elements are in Group II.

3 The metals in Groups I and II show only one oxidation number in their compounds; +1 for Group I elements, +2 for Group II elements.

4 The *s*-block metals have lower melting points, lower boiling points, lower densities and they are softer than transition metals.

5 The *s*-block metals are high in the activity series. They will all reduce water to hydrogen and they all tarnish rapidly in air forming a layer of oxide.

6 The compounds of Group I metals are thermally more stable than those of Group II metals. The carbonates and hydroxides of Group I metals (except $Li_2CO_3$ and LiOH) do not decompose at 1000°C but those of Group II metals decompose to their oxides on heating.
7 All common Group I salts are soluble.
8 Salts of Group II cations containing anions with a charge of $-1$ are generally soluble except for the hydroxides. Salts of Group II cations containing anions with a charge of $-2$ are generally insoluble.
9 The *s*-block elements are manufactured by electrolysis of their molten compounds.
10 Sodium and magnesium are the most widely used *s*-block elements. Sodium is used as a liquid coolant in nuclear reactors and as an alloy with lead in the manufacture of tetraethyllead(IV) ('antiknock' in petrol). Magnesium is used extensively in the manufacture of low density, high strength alloys.

# Study questions

1 (a) State how each of the following properties of the elements in Group II will change with increasing atomic number.
(i) Atomic radius
(ii) Ionization energy
(iii) Strength as reducing agents
(iv) Vigour of reaction with chlorine
(v) Electropositivity
(b) In each case, explain why the five properties change in the way you have suggested.

2 Group II consists of the elements Be, Mg, Ca, Sr, Ba.
(a) Are these elements metals or non-metals? Give one reason for your answer.
(b) What ion or ions are formed by calcium?
(c) Describe briefly five properties that show a general gradation as the group is descended from Be to Ba.
(d) Briefly explain how ionization energies and redox potentials are related to the reactivity of these elements.
(e) Explain why the ion $Mg^+$ is not found in compounds even though less energy is required to remove one electron from Mg and form this ion than is required to remove two electrons to form $Mg^{2+}$.
(f) Beryllium shows certain properties that are not typical of the rest of the group. Mention two of these properties and suggest reasons why the difference should occur.
(g) Group II elements frequently form hydrated salts, while the corresponding compounds of Group I elements are anhydrous. Suggest reasons for this difference.

3 The elements in Group II of the periodic table are barium (Ba), beryllium (Be), calcium (Ca), magnesium (Mg), radium (Ra) and strontium (Sr).
(a) Arrange these elements in order of increasing relative atomic mass.
(b) Write the electronic structures of any two of these elements except beryllium (e.g. Be would be $1s^2\ 2s^2$.)
(c) Draw a sketch graph of ionization energy against number of electrons removed for the successive ionization energies of magnesium.
(d) Why do all the metals in Group II have an oxidation number of $+2$?
(e) Why do the elements in Group II not have an oxidation number of $+1$ or $+3$?
(f) How and why does the first ionization energy of the metals vary as the group is descended?
(g) When solutions of strontium nitrate(V) and potassium chromate(VI) are mixed, a yellow precipitate is formed which dissolves in dilute hydrochloric acid. Explain these observations.

4 Group I consists of the elements Li, Na, K, Rb, Cs.
(a) The first member of a group often shows anomalous properties. Give two respects in which the behaviour of Li is anomalous.
(b) Describe briefly five properties that show a general gradation as the group is descended from Li to Cs.
(c) Explain the meaning of the term 'ionization energy'.
(d) How will successive ionization energies of Na vary?
(e) Why is the ion $Na^+$ formed in normal chemical reactions rather than the ion $Na^{2+}$
(f) How are ionization energies related to the reactivity of these elements?

5 The isotope, $^{42}_{19}K$ decays by beta-particle emission to a stable nuclide. The rate of emission of $\beta$ particles from $^{42}_{19}K$ was followed by a suitable method.
(a) Draw a sketch graph showing the rate of emission of $\beta$ particles (vertically) against time.
(b) 0.02g of $^{42}_{19}K$ were allowed to decay.
(i) Write a nuclear equation (balanced for mass number and atomic number) for the decay process.
(ii) What is the stable nuclide produced in the decay process?
(iii) If the half-life of the potassium isotope is 12.5h, what mass of the stable nuclide will have been formed after 25h?

6 Draw up a table to compare the first three elements in Group II of the periodic table; beryllium, magnesium and calcium. Your table should include:
(a) electronic structures,
(b) relative values of ionization energies,
(c) methods of extraction of elements,
(d) action of water, acids and alkalis on the elements,
(e) the basic strength of oxides and hydroxides,
(f) the volatility and hydrolysis of chlorides.
(If necessary, you should use your knowledge of the periodic table to predict properties for beryllium and its compounds.)

7 Magnesium and calcium occur naturally in the mineral dolomite, $MgCO_3.CaCO_3$, a mixture of insoluble magnesium and calcium carbonates which can be used to produce calcium sulphate(VI) and magnesium sulphate(VI). Calcium sulphate(VI) is used in the manufacture of building materials such as plaster-board. Magnesium sulphate(VI) is used in fireproofing fabrics and as a purgative (Epsom salts).
(a) Describe carefully how you would prepare pure samples of $MgSO_4.7H_2O$ and $CaSO_4$ from dolomite. You may find the following information useful.

| **Compound** | **Solubility/g per 100g water at 20°C** |
|---|---|
| $MgCO_3$ | 0.01 |
| $CaCO_3$ | 0.0014 |
| $MgSO_4$ | 33.0 |
| $CaSO_4$ | 0.21 |

(b) How would you obtain pure $MgSO_4$ from crystals of $MgSO_4.7H_2O$?

8 Sodium sulphate(IV) (sodium sulphite, $Na_2SO_3$) is sometimes added to sausage meat to act as a preservative. The amount of $Na_2SO_3$ present can be determined by boiling a sample of the meat with acid and then determining the quantity of sulphur dioxide produced by titration against iodine.
100g of sausage meat was boiled with 500cm³ of M HCl. The sulphur dioxide evolved was dissolved in water and found to require 12.00cm³ of 0.025M $I_2$ solution in order to oxidize the $SO_2$ as in the equation below.

$$SO_2 + 2H_2O + I_2 \longrightarrow 4H^+ + SO_4^{2-} + 2I^-$$

In order to check the results of the titration, excess barium chloride is added to the final solution after the titration and the resulting precipitate is collected and weighed.

(Na = 23, S = 32, O = 16)

(a) How many moles of $SO_2$ are evolved from 100g of the sausage meat?
(b) How many grams of $Na_2SO_3$ are present in 100g of the sausage meat?
(c) Government scientists often express the amount of $Na_2SO_3$ in meat as parts per million (p.p.m.).

(1p.p.m. = 1g of $Na_2SO_3$ in $10^6$g of meat)

Express the amount of $Na_2SO_3$ in the sausage meat in p.p.m.
(d) Write an equation for the reaction which occurs when $BaCl_2$(aq) is added to the solution at the end of the titration.
(e) Calculate the mass of precipitate formed when excess $BaCl_2$(aq) is added to the solution at the end of the titration.

Calcium sulphate(VI) is found naturally in two forms: anhydrous calcium sulphate(VI), ($CaSO_4$— anhydrite) and hydrated calcium sulphate(VI), ($CaSO_4.2H_2O$—gypsum). When anhydrite is heated with coke, sulphur dioxide is obtained which can then be used to manufacture sulphuric(VI) acid.
If gypsum is heated at about 125°C it dehydrates partially forming plaster of Paris, $(CaSO_4)_2.H_2O$. When this is mixed with water it changes back to $CaSO_4.2H_2O$. The paste expands slightly as it hardens and sets quickly to a firm solid.
When gypsum is heated to 200°C, it loses all its water of crystallization and the anhydrous salt which forms sets only very slowly when mixed with water.

Intersecting tunnels in an anhydrite mine.

(a) Explain what is meant by the terms:
  (i) anhydrous, (ii) hydrated, (iii) water of crystallization.
(b) Write equations for the reactions which occur when:
  (i) anhydrite is heated with coke,
  (ii) plaster of Paris is mixed with water,
  (iii) gypsum is heated at 200°C.
(c) Draw a diagram of the apparatus *you* would use to make plaster of Paris from gypsum.
(d) Explain why plaster of Paris is so suitable for:
  (i) immobilizing broken limbs,
  (ii) making models from moulds.
(e) Why is anhydrite not a suitable alternative for the uses mentioned in (d)?

# Aluminium

# 15

## 15.1 Introduction

Aluminium is by far the most important element in Group III (table 15.1). Although the metal itself is not found free in nature, its compounds are so widespread that it is the third most abundant element (after oxygen and silicon) in the earth's crust. Economically, the most important ore is bauxite (hydrated aluminium oxide, $Al_2O_3 . 2H_2O$) from which the metal is obtained, but there are also considerable and widespread deposits of aluminosilicates in clays, mica and other minerals and extensive deposits of cryolite (sodium hexafluoroaluminate(III), $Na_3AlF_6$) are found in Greenland.

**Table 15.1** The symbols, atomic numbers and electron structures of the elements in Group III.

| Element | Symbol | Atomic number | Electron structure |
|---|---|---|---|
| Boron | B | 5 | $(He)2s^2\ 2p^1$ |
| Aluminium | Al | 13 | $(Ne)\ 3s^2\ 3p^1$ |
| Gallium | Ga | 31 | $(Ar)3d^{10}\ 4s^2\ 4p^1$ |
| Indium | In | 49 | $(Kr)4d^{10}\ 5s^2\ 5p^1$ |
| Thallium | Tl | 81 | $(Xe)4f^{14}\ 5d^{10}\ 6s^2\ 6p^1$ |

Mining bauxite.

**Table 15.2** Values of charge, ionic radius and charge/radius ratio for some common cations.

| Cation | Unit charge | Ionic radius/nm | Charge/radius ratio/unit charge $nm^{-1}$ |
|---|---|---|---|
| $Na^+$ | +1 | 0.098 | 10 |
| $Mg^{2+}$ | +2 | 0.065 | 31 |
| $Al^{3+}$ | +3 | 0.048 | 63 |
| $Zn^{2+}$ | +2 | 0.074 | 27 |
| $Cu^{2+}$ | +2 | 0.069 | 29 |

Much of the chemistry of aluminium compounds is dictated by *the high charge and small radius of the $Al^{3+}$ ion* which results in an unusually large charge density. The high charge density of $Al^{3+}$ is reflected in its large charge/radius ratio which is compared with the values for some other common cations in table 15.2.

## 15.2 Bonding in aluminium compounds

Aluminium and the other elements in Group III are, of course, predominantly trivalent showing an oxidation state of +3. They form ions with a charge of 3+ by losing the three electrons from their outermost shell.

$$Al \longrightarrow Al^{3+} + 3e^-$$

However, the high charge density associated with the $Al^{3+}$ ion will cause some distortion (polarization) of the electron cloud around any ion in contact with it. If the polarization is sufficiently large, there is an effective electron density in the region between the aluminium ion and its neighbouring anion which comprises partial covalent bonding. With large anions such as $Br^-$ and $I^-$, in which polarization occurs more easily, the distortion can be so great that the bonding is best described as covalent.

- ○ Why are large anions polarized more easily than smaller anions?
- ○ Which anion do you think will be least polarized by cations?
- ○ Will $Ga^{3+}$ have a greater or smaller polarizing effect than $Al^{3+}$ on anions? Explain.

The influence of polarization on bonding, structure and properties is beautifully illustrated by the aluminium halides (figure 15.1).

| | Aluminium fluoride | Aluminium chloride | Aluminium bromide |
|---|---|---|---|
| **Bonding:** | Ionic | Intermediate between ionic and covalent | Covalent |
| | $F^-$ ion is too small to be polarized. | $Cl^-$ ion is intermediate in size and partly polarized. | $Br^-$ ion is easily polarized forming a covalent bond to Al. |
| **Structure:** | 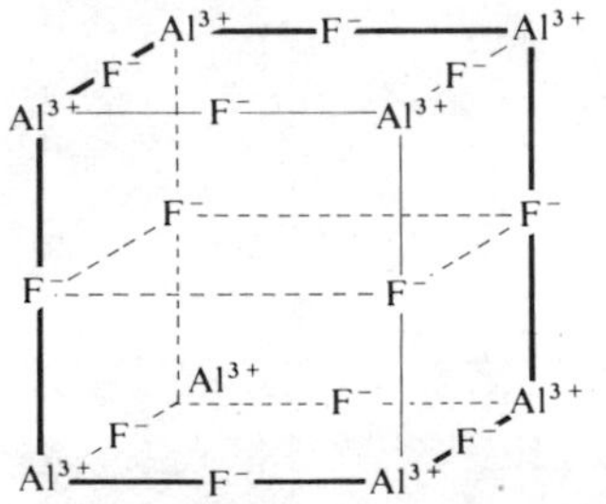 [cubic lattice diagram of $Al^{3+}$ and $F^-$ ions] | Solid shows intermediate bonding, but this sublimes to yield $Al_2Cl_6$ dimers with the same structure as $Al_2Br_6$. At high temps. $AlCl_3$ mols. form. | 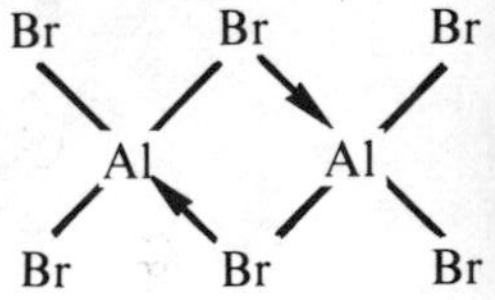 [$Al_2Br_6$ dimer diagram] |
| | Symmetrical ionic crystal. | | Simple molecular (dimer) structure. (The outer shell of each Al atom is completed by a co-ordinate bond from a Br atom in the associated $AlBr_3$ molecule—hence the dimer.) |
| **Melting point:** | 1265°C. | Sublimes at 180°C. | 97°C. |



**Figure 15.1** The influence of polarization on the bonding, structure and properties of aluminium halides.

## 15.3 The bonding and properties of aluminium oxide

*Aluminium oxide is normally regarded as ionic containing $Al^{3+}$ and $O^{2-}$ ions. However, the oxide ions are polarized by the $Al^{3+}$ ions and the bonds have a marked degree of covalent character*. This factor and the small size of the $Al^{3+}$ and $O^{2-}$ ions combine to make the bonding in aluminium oxide very strong indeed.

Aluminium oxide may be prepared by heating powdered aluminium in oxygen or by heating aluminium hydroxide.

$$2Al + \tfrac{3}{2}O_2 \longrightarrow Al_2O_3$$

$$2Al(OH)_3 \longrightarrow Al_2O_3 + 3H_2O$$

The strongly bound structure of $Al_2O_3$ results in its insolubility in water and very high melting point (2 050°C), whilst its intermediate (ionic/covalent) bonding is reflected in its amphoteric character—it dissolves slowly in both dilute acids and dilute alkalis.

$$\underset{\text{acid}}{Al_2O_3(s) + 6H^+(aq)} \longrightarrow 2Al^{3+}(aq) + 3H_2O$$

$$\underset{\text{alkali}}{Al_2O_3(s) + 2OH^-(aq)} + 3H_2O(l) \longrightarrow \underset{\substack{\text{tetrahydroxoaluminate(III)}\\ \text{(aluminate)}}}{2Al(OH)_4^-(aq)}$$

Aluminium oxide occurs naturally in both hydrated and anhydrous forms. The hydrated form (bauxite) has already been mentioned, but the anhydrous forms, emery and corundum, are less well known. Corundum is a hard, crystalline form which may be coloured by traces of red impurity to form *ruby* or by traces of blue impurity to form *sapphire*. The extremely hard, stable structure of $Al_2O_3$ explains why it is used in cement and in refractory-furnace bricks and as an inert surface catalyst in cracking petroleum fractions.

## 15.4 Properties and uses of aluminium

The position of aluminium in the electrochemical series and its electrode potential suggest that it should react readily with oxygen and dissolve rapidly in dilute acids, liberating hydrogen.

$$Al(s) \longrightarrow Al^{3+}(aq) + 3e^- \qquad E^{\ominus} = +1.66V$$

$$3H^+(aq) + 3e^- \longrightarrow \tfrac{3}{2}H_2(g) \qquad E^{\ominus} = 0.00V$$

However, the rapid formation of a thin layer of oxide, no more than $10^{-8}$m thick, prevents further attack by oxygen and retards reaction of the aluminium with dilute acids. Furthermore, this extremely thin oxide layer is virtually non-porous to water, so that the aluminium is completely protected from further oxidation. This is very different from iron, which forms a porous oxide layer (rust) which is readily penetrated by water, allowing the process of corrosion to continue beneath the superficial layer of rust.

Most of the uses of aluminium are, of course, only possible because of the protective oxide coating—saucepans, aircraft and vehicle bodywork, window frames, etc.

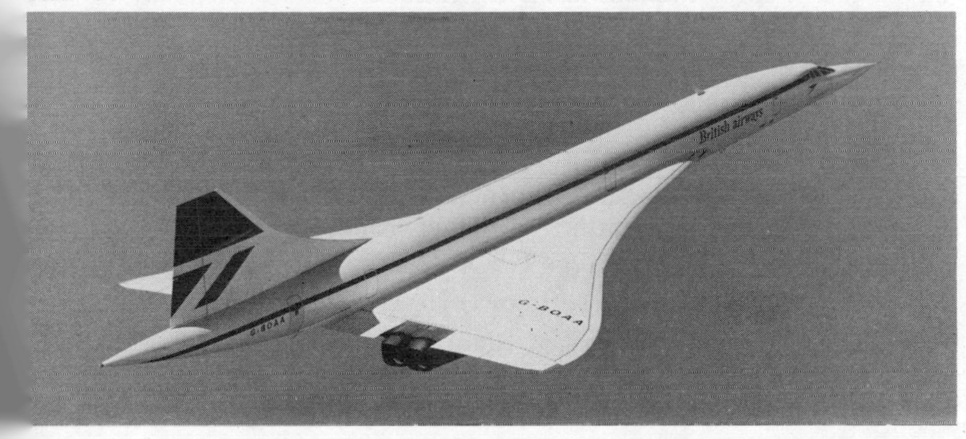


Concorde (206) in flight, much of the aircraft structure is aluminium.

The bodywork of a car emerges from its electrolytic bath after completion of the anodizing process.

In order to protect the aluminium even more, it is possible to increase the thickness of the natural oxide layer to about $10^{-5}$m by a process known as **anodizing**. The aluminium is anodized by making it the anode during the electrolysis of sulphuric(VI) acid. Oxygen, released at the anode, combines with the aluminium thereby thickening the oxide layer. By carrying out the electrolytic anodizing process in the presence of dyes which are adsorbed by the oxide layer, the anodized material can be coloured attractively for use in domestic and personal articles.

- What properties of aluminium (besides its resistance to corrosion) make it suitable for use as:
  (i) milk bottle tops, (ii) baking foil,
  (iii) aircraft bodywork, (iv) electricity cables,
  (v) kettles and pans?
- Why does aluminium not corrode away like iron?
- Aluminium and iron are the two most widely used metals. What advantages are there in using iron rather than aluminium?

Once the oxide layer is removed or penetrated, aluminium reacts readily with many reagents. When strongly heated in air, it will burn to form the oxide and a little nitride.

$$2Al + \tfrac{3}{2}O_2 \longrightarrow Al_2O_3$$

$$2Al + N_2 \longrightarrow 2AlN$$

Aluminium dissolves in hot dilute hydrochloric and sulphuric(VI) acids, evolving hydrogen.

$$Al(s) + 3H^+(aq) \longrightarrow Al^{3+}(aq) + \tfrac{3}{2}H_2(g)$$

With concentrated sulphuric(VI) acid and both concentrated and dilute nitric(V) acid, however, the metal remains unreactive. This so-called 'passivity' of the aluminium is caused by the oxidizing agents reacting with the metal to form an inert, impervious oxide layer.

Aluminium dissolves in hot dilute sodium hydroxide, forming hydrogen and sodium tetrahydroxoaluminate(III).

$$Al(s) + OH^-(aq) + 3H_2O(l) \longrightarrow Al(OH)_4{}^-(aq) + \tfrac{3}{2}H_2(g)$$

The reactions with dilute acids and alkalis are usually very slow at first, but become increasingly vigorous as the oxide layer is removed and the solutions become hotter. Owing to the reaction of aluminium with alkali, aluminium pans should never be cleaned with washing soda which forms an alkaline solution.

## 15.5 Reactions of aqueous $Al^{3+}$ ions

When aluminium salts are added to water, $Al^{3+}$ ions are immediately attracted to the negative end of polar water molecules forming the hexaaquaaluminium(III) ion, $[Al(H_2O)_6]^{3+}$. This is often written simply as $Al^{3+}(aq)$.

However, the electric field associated with a small, highly-charged $Al^{3+}$ ion is so intense that it draws electrons in the O—H bonds of water molecules towards itself, enabling the water molecules to become proton donors. In aqueous solution, free water molecules act as bases and the following equilibrium is established.

$$[Al(H_2O)_6]^{3+} + H_2O \rightleftharpoons [Al(H_2O)_5(OH)]^{2+} + H_3O^+$$

Thus, solutions of $Al^{3+}$ salts are acidic—in fact, as acidic as vinegar. The salts of other divalent and trivalent ions (e.g. $Mg^{2+}$, $Cu^{2+}$, $Fe^{3+}$) behave in a similar fashion to $Al^{3+}$. When bases stronger than water, such as $S^{2-}$ or $CO_3{}^{2-}$, are added to aqueous aluminium salts further $H^+$ ions are removed from $[Al(H_2O)_6]^{3+}$ and insoluble aluminium hydroxide precipitates.

$$2[Al(H_2O)_6]^{3+}(aq) + 3S^{2-}(aq) \longrightarrow 2[Al(H_2O)_3(OH)_3](s) + 3H_2S(g)$$

A similar reaction occurs when NaOH(aq), an even stronger base, is added to aqueous aluminium salts.

$$[Al(H_2O)_6]^{3+}(aq) + 3OH^-(aq) \longrightarrow [Al(H_2O)_3(OH)_3](s) + 3H_2O(l)$$

With excess NaOH(aq), further protons are removed from the insoluble precipitate and an anion forms which is soluble in water.

$$\underset{\text{white ppte}}{[Al(H_2O)_3(OH)_3](s)} + OH^-(aq) \longrightarrow \underset{\text{colourless soln.}}{[Al(H_2O)_2(OH)_4]^-(aq)} + H_2O(l)$$

These reactions of the hydrated $Al^{3+}$ ion are best regarded as the progressive removal of $H^+$ ions from $[Al(H_2O)_6]^{3+}$ by bases of gradually increasing strength. The effects of different bases are summarized in the following table.

$$[Al(H_2O)_6]^{3+}(aq) \rightleftharpoons [Al(H_2O)_5(OH)]^{2+}(aq) + H^+(aq) \rightleftharpoons [Al(H_2O)_3(OH)_3](s) + 3H^+(aq) \rightleftharpoons [Al(H_2O)_2(OH)_4]^-(aq) + 4H^+(aq)$$

| $[Al(H_2O)_5(OH)]^{2+}(aq) + H^+(aq)$ | $[Al(H_2O)_3(OH)_3](s) + 3H^+(aq)$ | $[Al(H_2O)_2(OH)_4]^-(aq) + 4H^+(aq)$ |
|---|---|---|
| Even $H_2O(\rightarrow H_3O^+)$ is a strong enough base to remove one $H^+$. Thus, aqueous $Al^{3+}$ salts are acidic | $NH_3(\rightarrow NH_4^+)$, $S^{2-}$ $(\rightarrow H_2S)$ and $CO_3^{2-}$ $(\rightarrow H_2CO_3)$ are strong enough bases to remove $3H^+$ ions, thus precipitating $[Al(H_2O)_3(OH)_3](s)$ | Only very strong bases such as $OH^-(\rightarrow H_2O)$ can remove $4H^+$ ions |

The reactions of aqueous aluminium ions with NaOH(aq) are frequently represented more simply but less correctly as:

$$Al^{3+}(aq) + 3OH^-(aq) \longrightarrow \underset{\text{white ppte}}{Al(OH)_3(s)}$$

$$Al(OH)_3(s) + OH^-(aq) \longrightarrow \underset{\text{clear soln.}}{Al(OH)_4^-(aq)}$$

Aluminium hydroxide, like aluminium oxide, is amphoteric dissolving in alkalis, as just described, to form aluminates and in acids to give aluminium salts.

When aluminium hydroxide is freshly prepared, the occluded water molecules make the precipitate very gelatinous. Because of this, aluminium hydroxide is used as a *mordant* in dyeing (Latin *mordere*—to bite). A fine precipitate of aluminium hydroxide is deposited in the fibres of the cloth. The gelatinous nature of the precipitate enables the dye to be easily absorbed and held by the $Al^{3+}$ ions. In effect, the aluminium hydroxide acts as a *mordant* by 'biting' both the dye and the cloth.

- ○ Aluminium hydroxide is precipitated in the fibres of the cloth by soaking the material in aluminium sulphate(VI) and then adding alkali. Why is this preferable to adding freshly prepared aluminium hydroxide?
- ○ Aluminium sulphate(VI) can be used for coagulating organic material in water and for coagulating blood. How do $Al^{3+}$ ions in this substance act as coagulating agents?

## 15.6 The extraction of aluminium

Aluminium is obtained industrially by the electrolysis of molten aluminium oxide. This is obtained from bauxite—hydrated aluminium oxide containing impurities such as iron(III) oxide and silicon(IV) oxide. Clearly, the first stage in the production of aluminium must be to obtain pure aluminium oxide from the bauxite (figure 15.2).

**Figure 15.2** The production of pure aluminium oxide from bauxite.

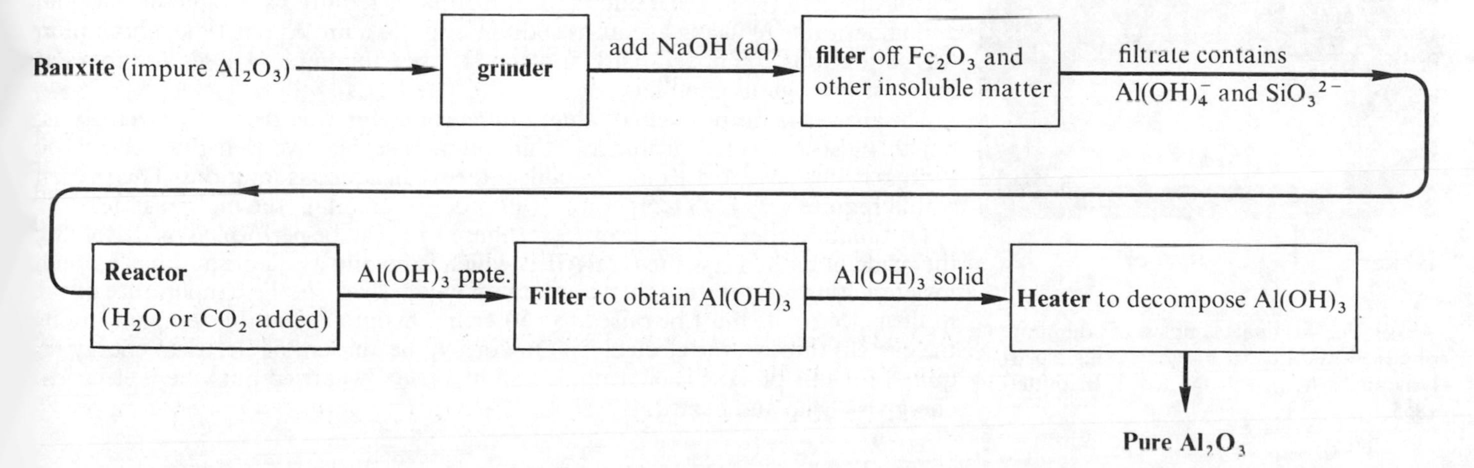

*Right* Rows of electrolytic cells at Alcan's aluminium smelter at Kitimat, British Columbia. What is the white powder on the floor? What are the different workmen doing?

The amphoteric nature of $Al_2O_3$ is an essential feature on which the purification is based.

When the impure bauxite is treated with concentrated sodium hydroxide, $Al_2O_3$ and $SiO_2$ dissolve, but $Fe_2O_3$ and other basic materials remain insoluble.

$$Al_2O_3(s) + 2OH^-(aq) + 3H_2O(l) \longrightarrow 2[Al(OH)_4]^-(aq)$$

$$SiO_2(s) + 2OH^-(aq) \longrightarrow SiO_3^{2-}(aq) + H_2O(l)$$

After filtering, the solution is diluted with water or treated with carbon dioxide to precipitate aluminium hydroxide:

$$2[Al(OH)_4]^- + CO_2 \longrightarrow 2Al(OH)_3(s) + CO_3^{2-}(aq) + H_2O(l)$$

which is then heated to obtain pure $Al_2O_3$.

$$2Al(OH)_3 \longrightarrow Al_2O_3 + 3H_2O$$

Owing to the vast, ubiquitous deposits of clay throughout the world, many attempts have been made to obtain aluminium oxide from clays. However, its separation and purification from large quantities of impurity in the clay is very difficult and, at present, uneconomical.

During the nineteenth century, attempts were made to reduce aluminium oxide chemically. However, the oxide is so stable that it required such unconventional and undesirable reducing agents as sodium or potassium. At one time, aluminium was so precious and novel that Napoleon III used aluminium knives and forks at his most prestigious banquets.

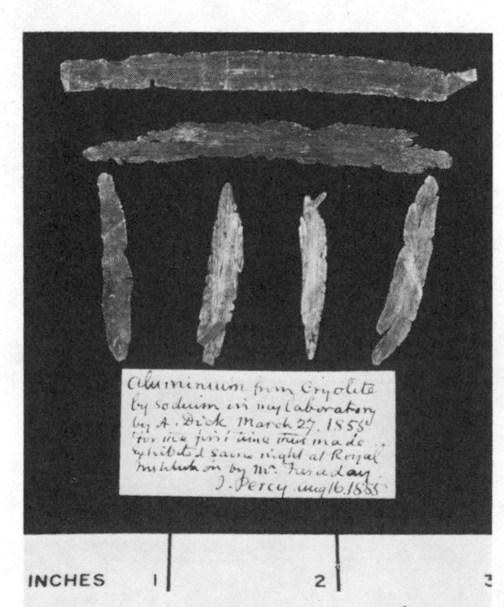


*Above* The original samples of aluminium obtained by Michael Faraday using electrolysis and exhibited at the Royal Institute in 1855.

The answer to the problem of aluminium production was obviously electrolysis, but this also presented problems. Aluminium is so reactive that the electrolytic process would have to use molten solids rather than aqueous solutions. This in turn would require very high temperatures in order to maintain the molten state.

Fortunately, the electrolytic process (figure 15.3) can be performed by dissolving the oxide in molten cryolite ($Na_3AlF_6$) which is readily available and has a much lower melting temperature than aluminium oxide. Even so, the temperature of the molten electrolyte must be raised to 850°C and maintained at this temperature by the current through the electrolyte. Obviously, the amount of electrical energy required for this process is enormous, and it is usually carried out where electrical energy is cheap and plentiful.

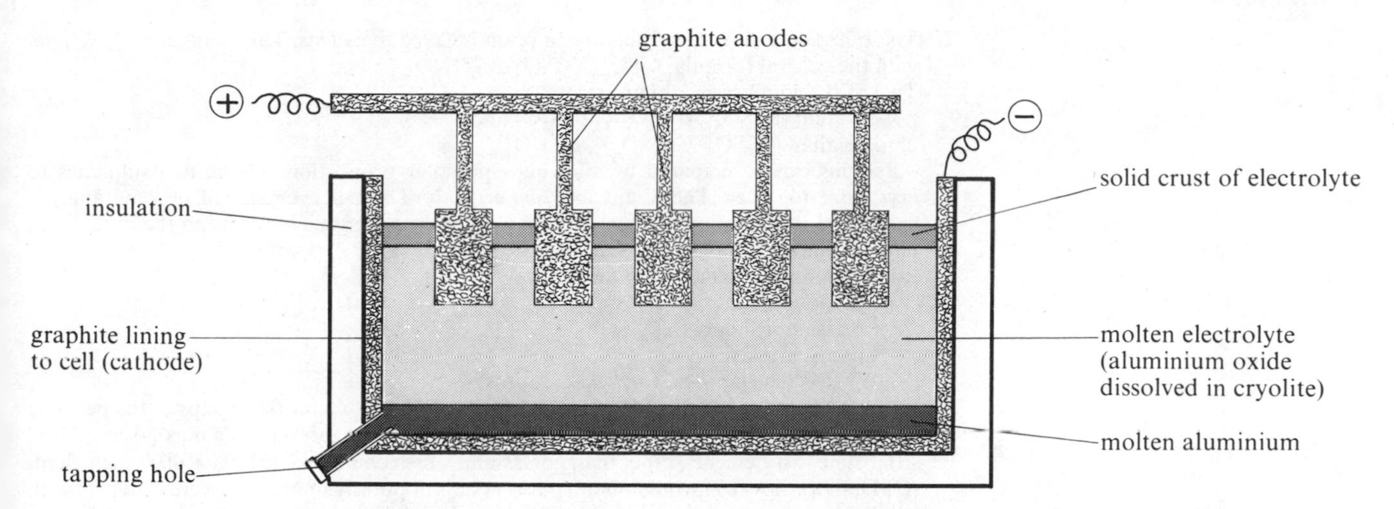


**Figure 15.3** Electrolytic manufacture of aluminium.

Aluminium is discharged at the graphite lining of the cell which acts as the cathode.

$$Al^{3+} + 3e^{-} \longrightarrow Al$$

The molten aluminium collects at the bottom of the cell and is tapped off periodically. Oxygen is evolved at the anodes.

$$2O^{2-} \longrightarrow O_2 + 4e^{-}$$

The oxygen reacts with the carbon of the anode to form oxides of carbon at the high temperatures involved. As a result, the anodes are gradually burnt away and must be replaced from time to time.

*Above* Unloading carbon anode blocks for fitting to their overhead hangers in one of the four 500-metre-long cell rooms at British Aluminium's smelter at Invergordon, Ross and Cromarty. Each anode weighs about 650 kilograms.

## Summary

1 Aluminium is obtained industrially by the electrolysis of aluminium oxide in molten cryolite. The pure oxide is obtained from bauxite.

2 Aluminium is a reactive metal but its reactivity is suppressed by a thin coating of inert oxide.

3 The high charge density of $Al^{3+}$ is responsible for:

(i) the intermediate ionic/covalent bonding of aluminium oxide and its amphoteric character;

(ii) the polarization of anions with which it is associated resulting in the partially covalent nature of aluminium chloride and other aluminium salts;

(iii) the hydration of $Al^{3+}$ ions in aqueous aluminium salts.

4 Aluminium oxide and aluminium hydroxide are amphoteric, dissolving in acids to form aluminium salts and in alkalis to form aluminates.

5 Although pure aluminium is a reactive metal, its compounds have little resemblance to the corresponding compounds of the reactive metals in Groups I and II. The compounds of aluminium are more like those of such 'poor' metals as tin and lead.

## Study questions

The superficial oxide coating on aluminium is effectively removed by washing the metal in mercury(II) chloride solution which is slightly acidic. The solution reacts with the oxide layer and then with the aluminium below. Mercury is displaced by aluminium and then forms an amalgam (alloy) with more aluminium. Oxide does not adhere to this amalgam and if the metal is left in moist air 'feather-like' growths appear on the aluminium as it reacts with oxygen and water vapour in the atmosphere.
Write equations for the reaction of:
(a) $H^+$ ions in the acidic solution with the aluminium oxide layer,
(b) the $HgCl_2(aq)$ with aluminium,
(c) aluminium with oxygen and with water vapour in producing 'feather-like' growths.

2 This question concerns a group of compounds called the *alums*. The alums are *double salts* with the general formula $X_2SO_4.Y_2(SO_4)_3.24H_2O$.
Two of the commonest alums are:
potash alum ($K_2SO_4.Al_2(SO_4)_3.24H_2O$) and
chrome alum ($K_2SO_4.Cr_2(SO_4)_3.24H_2O$).
All alums can be prepared by allowing equimolar proportions of the two sulphates to crystallize together. The alums are *isomorphous*. Thus, if a crystal of chrome alum is suspended in a solution of potash alum, a colourless *overgrowth* forms on the dark violet chrome alum crystal and the similarity in structure is very apparent.
(a) Explain what is meant by the terms:
(i) double salt,
(ii) isomorphous,
(iii) overgrowth.
(b) What is the empirical formula of alums?
(c) What can you conclude about the charges on X and Y and the groups in the periodic table to which X and Y may belong from the general formula of alums given above?
(d) Name two elements, other than potassium, which could take the place of X in an alum.
(e) Describe how you would obtain pure crystals of potash alum if you were provided with $KOH(aq)$, $Al_2O_3(s)$ and $H_2SO_4(aq)$.

3 The density of anhydrous aluminium chloride vapour was measured at 200, 600, and 800°C at atmospheric pressure and the results are given below.

| Temp/°C | 200 | 600 | 800 |
|---|---|---|---|
| density/g $dm^{-3}$ | 6.9 | 2.7 | 1.5 |

(a) Calculate the relative molecular mass of anhydrous aluminium chloride vapour at each temperature. (The gas constant, R = 8.31 J $K^{-1}$ $mole^{-1}$ (0.082 litre atm $K^{-1}$ $mole^{-1}$).
(b) What are the probable molecular formulae of aluminium chloride vapour at 200° and 800°C? (Al = 27.0, Cl = 35.5).
(c) Comment on the state of aluminium chloride vapour at 600°C.
(d) Outline the laboratory preparation of solid anhydrous aluminium chloride from aluminium.

4 Describe and explain with equations what happens in each of the following.
(a) Anhydrous aluminium chloride is added to water and aqueous sodium hydroxide is then added slowly until it is present in excess.
(b) Lithium is heated in dry hydrogen and the excess of the product is added to a suspension of aluminium chloride in ethoxyethane (diethyl ether).
(c) A mixture of aluminium fluoride, ammonium fluoride and sodium nitrate(v) is heated.

5 When 10g of a compound, A, were analysed they contained 5.0g of carbon, 1.25g of hydrogen and 3.75g of aluminium.
(i) 0.120g of A reacts with excess water at s.t.p. to evolve 0.112 $dm^3$ of a gas, B and form a white precipitate, C.
(ii) C dissolves in dilute sodium hydroxide solution and in dilute hydrochloric acid.
(iii) 10 $cm^3$ of B require 20 $cm^3$ of oxygen for complete combustion, producing only carbon dioxide and water.
(a) Identify A, B and C. (H = 1.0; C = 12.0; O = 16.0; Al = 27.0.)
(b) Write equations for the reactions in (i), (ii) and (iii) above.
(c) Write a structural formula for A.

# Group VII – The Halogens 16

## 16.1 Introduction

The halogens are the elements in Group VII of the periodic table—fluorine (F), chlorine (Cl), bromine (Br), iodine (I) and astatine (At) (table 16.1). They are known as the *halogens*—a name derived from Greek meaning 'salt formers'—because they combine readily with metals to form salts.

Generally speaking, the halogens comprise the most reactive group of non-metals and in this chapter we shall contrast their properties with those of the reactive metals in Groups I and II of the periodic table. We shall notice that *the halogens are strong oxidizing agents* whilst the alkali and alkaline-earth metals in Groups I and II are strong reducing agents. Furthermore, we shall see that the halogens exhibit various oxidation numbers in their compounds unlike the *s*-block elements which have only one oxidation number in their compounds.

Although the halogens show distinct trends in behaviour down their group, they emphasize, like the alkali and alkaline-earth metals, the remarkable similarities which exist amongst the elements within one particular group in the periodic table. These similarities in their properties and reactions result very largely from their similar electron structures. Each of the halogens has an outer shell containing seven electrons (i.e. $s^2p^5$).

- ○ Write the electron shell structure for fluorine, chlorine and bromine. (For example; that for beryllium would be $1s^2$, $2s^2$.)
- ○ Why are these elements strong oxidizing agents?
- ○ What is the most likely oxidation number of these elements in their compounds?

**Table 16.1** Group VII (The Halogens).

| |
|---|
| F |
| Cl |
| Br |
| I |
| At |

## 16.2 Sources of the halogens

*The halogens are so reactive that they cannot exist free in nature.* Indeed, fluorine is reactive enough to combine directly with almost all the known elements including some of the noble gases, whilst chlorine reacts directly with all elements except carbon, nitrogen, oxygen and the noble gases.

Consequently, the halogens always occur naturally as compounds with metals in which they are present as negative ions; fluoride ($F^-$), chloride ($Cl^-$), bromide ($Br^-$) and iodide ($I^-$). The last element in the Group, astatine (At), does not occur naturally at all. It is a very unstable, radioactive element which was first synthesized by chemists in the U.S.A. in 1940. It has never been obtained in anything other than minute amounts and the most stable isotope, $^{210}_{85}At$, has a half-life of only 8.3 hours. The name astatine is derived from the Greek word *astatos* meaning 'unstable'.

Fluorine and chlorine are by far the most abundant halogens. The most widespread compounds of fluorine are fluorspar (fluorite), $CaF_2$, such as Derbyshire 'Blue John' and cryolite, $Na_3AlF_6$. Unfortunately these extensive deposits of fluoride are dispersed very thinly over the earth's surface and only a few sources can be worked economically.

The commonest chlorine-compound is, of course, sodium chloride (NaCl) which occurs in sea water and in rock salt. Each kilogram of sea water contains about 30g of sodium chloride.

*Above* Obsolete machines in a Blue John mine in Derbyshire. Why do you think the calcium fluoride is blue?

Bromides and iodides occur in much smaller amounts than either fluorides or chlorides. Sea water contains only small concentrations of bromide, about 70 parts per million by weight. But the extraction of bromine from brine is still economically feasible.

Iodides are even scarcer than bromides. Sea water contains traces of iodide (0.05 parts per million by weight), but the laminarian seaweeds concentrate iodide from the sea to such an extent that fresh wet weed can contain up to 800 parts per million of iodine. Biologists are still perplexed by this capacity of laminaria to hoard such large concentrations of iodide. Curiously, this is not the only problem posed by naturally occurring iodine compounds. The main source of iodine is sodium iodate(v) ($NaIO_3$) which is found only in Chile, mixed with larger proportions of sodium nitrate(v). Why does iodine occur as iodate(v) in such high concentration in only one part of the world, and why does the iodine exist in such a high oxidation state? These two questions still remain unanswered.

Chilean workmen drilling holes in the layers of caliche prior to blasting.

## 16.3 Obtaining the halogens

*The halogens are usually obtained by oxidation of halide ions.*

$$2Hal^- \longrightarrow Hal_2 + 2e^-$$

Fluorine is such a powerful oxidizing agent that fluoride ions cannot be oxidized by any of the common chemical oxidizing agents such as conc.$H_2SO_4$, $MnO_2$ and $KMnO_4$. Consequently, fluorine is obtained commercially by electrolysis. Furthermore, electrolysis cannot be carried out in aqueous solution, since water would be discharged in preference to $F^-$ and even if fluorine were produced it would react with water. The electrolyte is usually potassium fluoride dissolved in liquid anhydrous hydrogen fluoride and the electrodes consist of a graphite anode and a steel cathode.

- ○ Write the symbols, with charges, of all ions present in the electrolyte.
- ○ Write an equation to summarize the formation of fluorine at the anode.
- ○ Write an equation to summarize the process at the cathode.

The other halogens are much less reactive than fluorine and they can be obtained in the laboratory by oxidizing the appropriate halide ions using $MnO_2$ or $KMnO_4$.

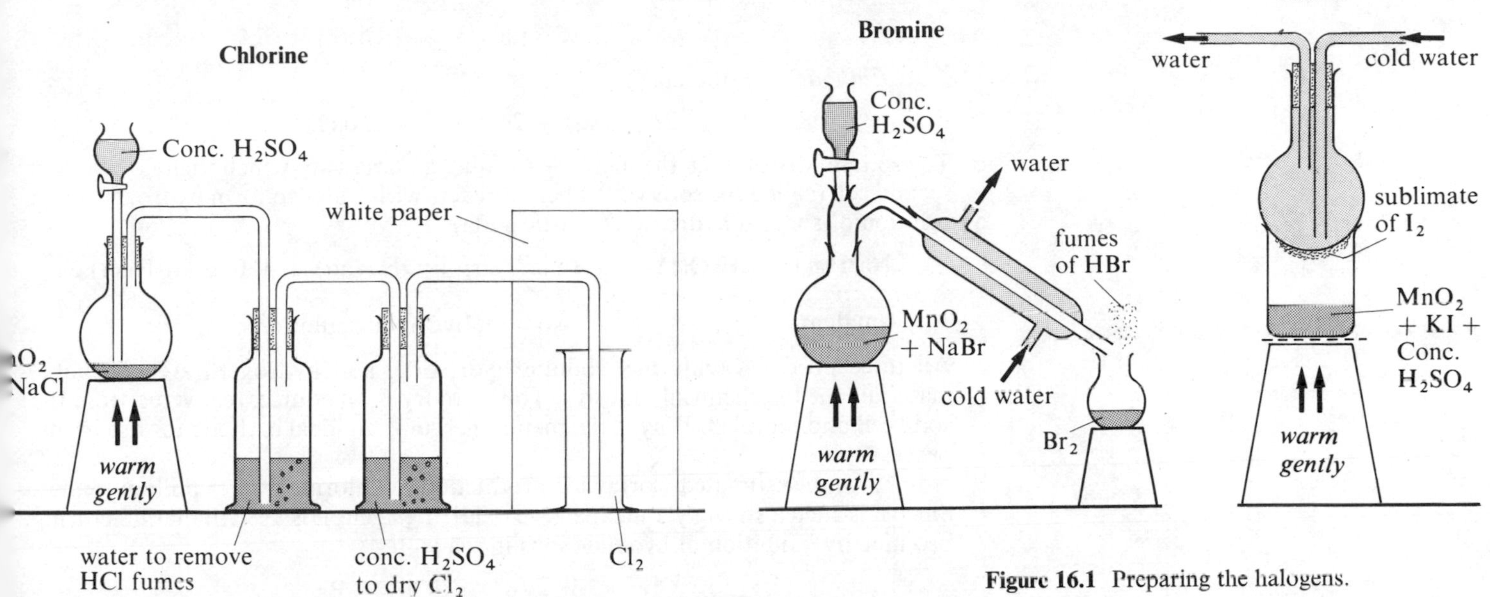


**Figure 16.1** Preparing the halogens.

Figure 16.1 shows how chlorine, bromine and iodine can be prepared on a small scale. These preparations should be carried out in a fume cupboard. All halogens are poisonous and should be handled with great care. Notice that each preparation is essentially the oxidation of halide ions by $MnO_2$ in the presence of conc.$H_2SO_4$, though the apparatus varies from one preparation to the next because chlorine is a gas, bromine is a liquid and iodine is a solid at room temperature.

$$2Hal^- \longrightarrow Hal_2 + 2e^-$$

$$MnO_2 + 4H^+ + 2e^- \longrightarrow Mn^{2+} + 2H_2O$$

- ○ Why is the acid necessary?
- ○ Why is conc.$H_2SO_4$ used rather than dilute $H_2SO_4$?

Some hydrogen halide is always evolved from the interaction of the halide and conc.$H_2SO_4$. The chlorine, for example, will be contaminated with hydrogen chloride fumes. Pure, dry chlorine can be obtained by passing the gas through water to remove the HCl fumes and then through conc.$H_2SO_4$ to dry it before collection by downward delivery.

- ○ Why is the white paper placed behind the gas jar in which $Cl_2$ collects?
- ○ The bromine which is collected will be contaminated with fumes of HBr. How could the liquid bromine be purified? (Hint: how are liquids normally purified?)
- ○ Why is cold water passed through the flask on which solid iodine is deposited?

Since chlorine is less reactive than fluorine it can be obtained industrially by the electrolysis of *aqueous* sodium chloride (brine) (figure 16.2).

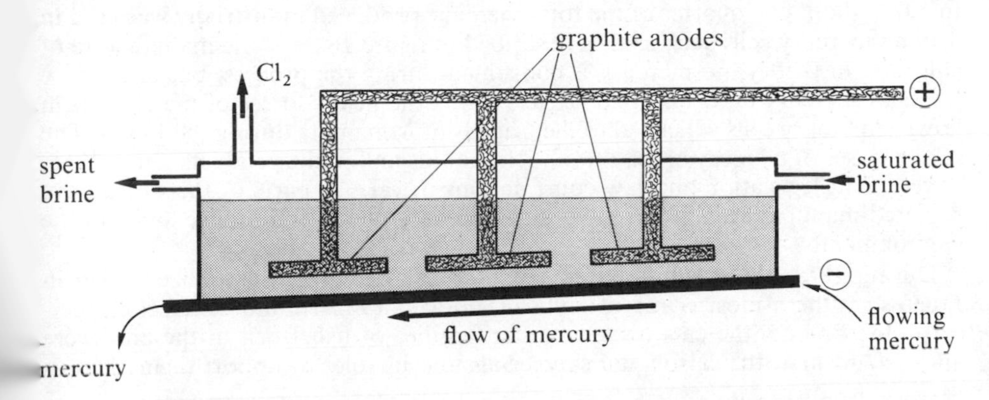


**Figure 16.2** A flowing mercury cell in which chlorine is manufactured by electrolysis of aqueous sodium chloride.

In the electrolytic process, chlorine is liberated at the graphite anodes whilst sodium is liberated at the mercury cathode.

*Anode* (+) (graphite)

$$2Cl^-(aq) \longrightarrow Cl_2(g) + 2e^-$$

*Cathode* (−) (mercury)

$$2Na^+(aq) + 2e^- \longrightarrow 2Na(l)$$

The sodium dissolves in the mercury forming an amalgam which then flows into a second cell (called the soda cell) where it reacts with water to form hydrogen and a solution of sodium hydroxide (caustic soda).

$$\underbrace{Na/Hg(l)}_{\text{amalgam}} + H_2O(l) \longrightarrow \underbrace{Na^+(aq) + OH^-(aq)}_{\text{sodium hydroxide solution}} + \tfrac{1}{2}H_2(g) + Hg(l)$$

All three products (chlorine, sodium hydroxide, and hydrogen) are invaluable materials for the chemical industry. The mercury is, of course, recovered from the soda cell and recycled. Why is the metal, mercury, an ideal cathode for use in this process?

In 1975, the estimated worldwide production of chlorine was 25 million tonnes. Since it is such a strong yet inexpensive oxidizing agent it is used in manufacturing bromine by oxidation of bromide ions in sea water.

$$Cl_2 + 2Br^- \longrightarrow 2Cl^- + Br_2$$

Unlike chlorine and bromine which normally exist in the −1 oxidation state, iodine occurs naturally in the +5 oxidation state as sodium iodate(V). Consequently, iodine is normally obtained by a reduction process using sodium hydrogensulphate(IV) as the reducing agent.

$$(IO_3^- + 6H^+ + 5e^- \longrightarrow \tfrac{1}{2}I_2 + 3H_2O) \times 2$$
$$(HSO_3^- + H_2O \longrightarrow HSO_4^- + 2H^+ + 2e^-) \times 5$$

A second, somewhat obsolete method of obtaining iodine is to extract the soluble iodides from dried seaweed with water and then displace the iodine from the solution by treatment with chlorine.

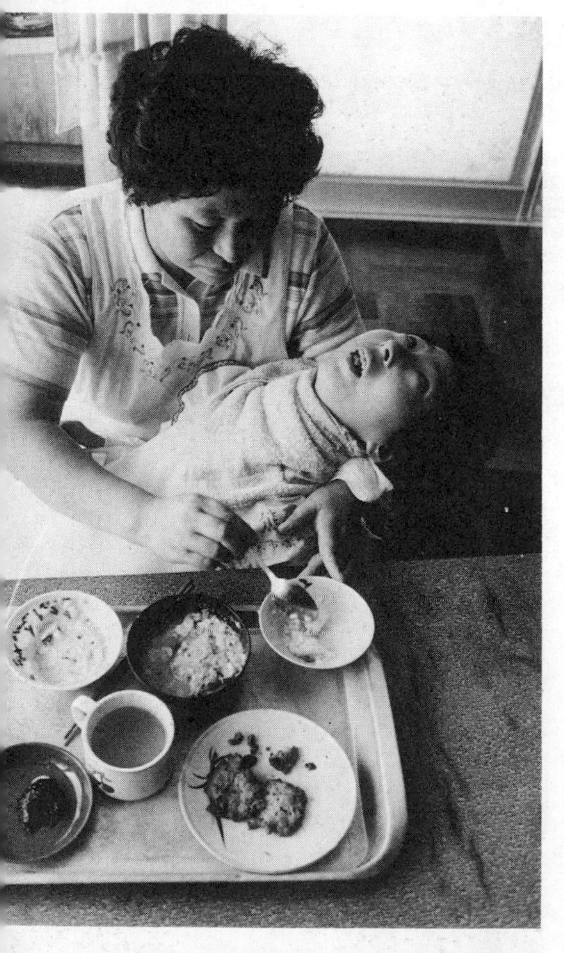

The appalling effects of mercury poisoning. Fifteen-year-old Tadsuko was deformed at birth because his mother ate fish from Minamata Bay during her pregnancy.

## 16.4 Mercury poisoning—the tragedy of Minamata

In 1970, about one quarter of the total mercury produced industrially was used in flowing mercury cells such as that described in figure 16.2 for the manufacture of chlorine. In theory, no mercury is consumed during the process, but in practice losses of mercury do occur. Sad to say, it was the heavy losses of mercury from flowing mercury cells which led to the tragedy at Minamata during the 1950's. The major losses of mercury are in the effluent liquids and in the products. Since 1965, government legislation and law court actions in various parts of the world have required manufacturers to improve their recovery plants and reduce losses to the environment.

During 1953, the inhabitants of Minamata, a heavily industrialized town in Japan's southern-most island of Kyushu, noticed a bizarre illness affecting their cats. One by one, the cats owned by the families of fishermen in the area were affected. At first, the cats would stagger around unable to support themselves as

*Left* In 1969, millions of fish were killed when the River Rhine became heavily polluted. The photograph shows fishermen removing dead fish from the river near Koblenz.

their muscles weakened. Later, they became paralysed, eventually going into a state of coma followed by death.

Unfortunately, the public health authorities were slow to realize the symptoms of heavy-metal poisoning in the animals owing to the widespread distribution of the fishing community.

In December of the same year, the first human case of this mysterious illness was recorded. During the next decade, 43 other humans died of the same disease and more than 60 others were permanently disabled.

Unable to solve the cause of this mysterious disease, the medical authorities eventually, in August 1956 sought the help of the medical department at Kumamoto University. By November 1956, the Kumamoto investigators had eliminated any naturally-occurring brain disease and had turned to the possibility of heavy-metal poisoning. By this time, other domestic animals, seabirds and the fish in Minamata Bay were dying of a similar disease to that in humans and cats.

Fish and shellfish were examined for toxic matter and found to contain 20 to 60 times as much mercury as would normally accumulate from the minute traces naturally present in seawater. All fishing in Minamata Bay was banned as it became obvious that fish, the main source of food in the area, were responsible for passing the toxic mercury on to man.

The search for the source of the mercury began immediately and was traced to a large industrial plant which produced plastics including poly(chloroethene) (PVC, polyvinyl chloride) and industrial organic chemicals.

Chlorine for making PVC was produced by the electrolysis of concentrated brine using a flowing mercury cathode, whilst several of the organic chemicals were prepared from ethanal (acetaldehyde, $CH_3CHO$) previously synthesized using a mercury(II) sulphate(VI) catalyst. A further significant fact was that large scale production of PVC commenced in 1952 shortly before the outbreak of the disease.

The effluents from these plants flowed into settling ponds and then via a canal into Minamata Bay. The mud from these settling ponds was found to contain as much as 700 parts per million (p.p.m.) wet weight of inorganic mercury compounds, mainly mercury(II) chloride and mercury(II) sulphide. Small concentrations of organic mercury compounds were present in the water effluent.

As a result of this the industrial plant started to build a new effluent treatment system to remove heavy metals, but a return to fishing resulted in a new outbreak of poisoning. Figure 16.3 shows a correlation of the production of chloroethene (vinyl chloride) at Minamata with the number of cases of Minamata disease.

*Above* Revolting effluent pollutes the ocean surface in Minamata Bay.

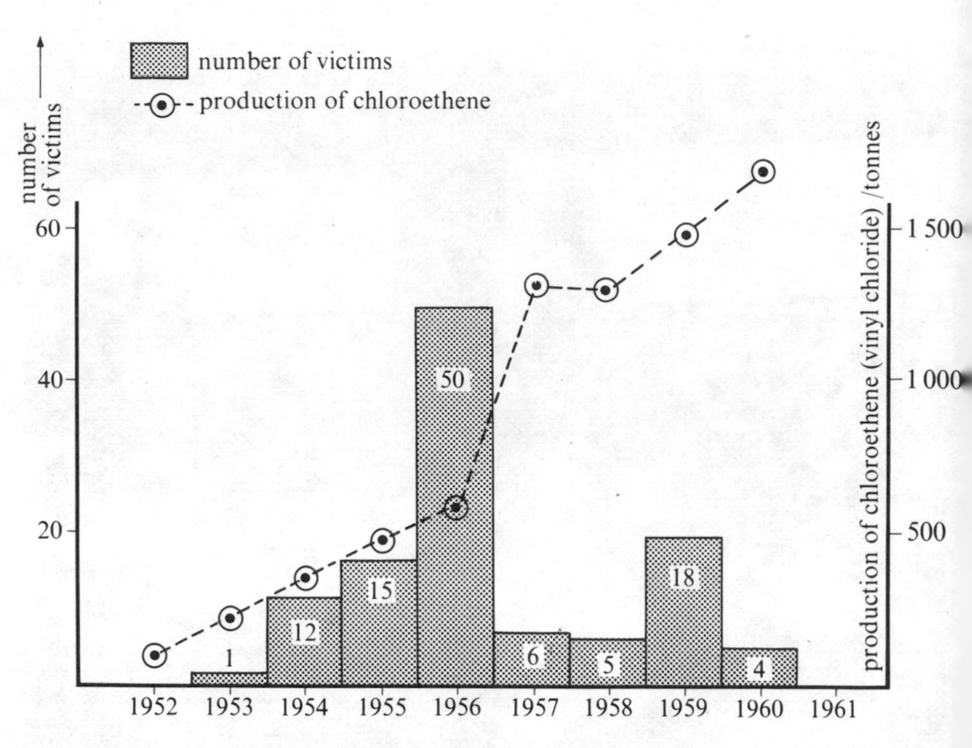


**Figure 16.3** The annual production of chloroethene and the number of cases of Minamata disease between 1952 and 1960.

Look at figure 16.3 and then answer the following questions.

- ○ Was the fishing ban in 1956 effective?
- ○ What happened after a partial return to fishing in 1958?
- ○ Did the new effluent system which came into use in early 1961 prove effective?

In 1960, methyl mercury ethanoate was identified in the effluent water from the industrial plant at Minamata and this compound was also identified in fish. The appalling tragedy had been solved, but this was not the end of Minamata disease. In their summary, the Kumamoto study group had stressed that the control of industrial contamination was so poor that further outbreaks of mercury poisoning were inevitable in the area. Less than four years later a similar outbreak of mercury poisoning occurred at Niigata in Japan. This time 28 people were critically ill and 5 of them died.

Napoleon in a nasty mood! Perhaps the sign of impending madness.

## 16.5 The effects of mercury and mercury compounds on the body

Mercury and its compounds are not the only poisons amongst metals. Barium, cadmium, lead and zinc and their compounds are also highly toxic to both animals and plants. They all act in a similar way to mercury. There is some evidence that Napoleon, Ivan the Terrible and Charles II may each have died from mercury poisoning.

Traces of mercury and its inorganic compounds containing $Hg^{2+}$ ions have always been present in nature. Therefore, all plants and animals will carry traces of mercury. Over millions of years, man and other living creatures have become tolerant to these small concentrations of mercury and it is more than likely that some forms of life may now be dependent on mercury and other poisonous metals as trace elements.

In fact, Mithridates VI Eupator, King of Pontus (132–63 B.C.) is reputed to have compounded 54 substances including mercury compounds as a universal antidote against poisoning. Mithridates was determined not to suffer a similar fate to other kings. Consequently, he protected himself against would-be poisoners by taking small, but steadily increasing doses of toxic materials.

By the nineteenth century, the toxic nature of mercury compounds had been noticed by practitioners in almost every trade in which mercury was used—felt-hatters, alchemists, physicians, etc. For example, it was the use of mercury(II) nitrate(V) in felt-making that led to the condition called 'hatters' shakes' or 'hatters' madness' in the felt trade. Perhaps mercury(II) nitrate(V) was responsible for the Hatter's madness in *Alice in Wonderland*.

*Above* The Mad Hatter's Tea Party from 'Alice in Wonderland'.

A BIOCHEMICAL INTERPRETATION OF MERCURY POISONING

Mercury has a strong affinity for sulphur and for the sulphhydryl group (—SH) in the cysteine units in proteins (figure 16.4).

```
       H                    H   H   O
       |                  / |   |   || \
  H2N—C—COOH           —(—N—C—C—)—
       |                  \     |     /
      CH2                      CH2
       |                        |
       S                        S
       |                        |
       H                        H

    cysteine            a cysteine unit
                          in a protein
```

**Figure 16.4** The structural formula of cysteine and the cysteine unit in a protein.

Alkyl-mercury compounds, such as methyl mercury ethanoate, are particularly dangerous because they can form strong covalent bonds to the sulphur atoms in these—SH groups (by replacing the hydrogen atoms) and maintain their destructive activity for long periods. The attachment of a mercury atom to the protein can seriously alter or hinder its properties. Proteins play an important part in cells as enzymes which catalyze biological processes and as constituents of cell membranes. Thus, the binding or attraction of mercury to sulphur in proteins can inhibit the activity of an enzyme and also interfere with the movement of materials across a cell membrane.

## 16.6 Structure and properties of the halogens

*All the halogens exist as diatomic molecules*; the two atoms being linked by a covalent bond (figure 16.5). These molecules persist in the gaseous, liquid and solid states. Table 16.2 shows some of the physical properties of the halogens.

They are all coloured—the depth of colour increasing with increase in atomic number. Fluorine is pale yellow, chlorine is pale green, bromine is red-brown and iodine is shiny black.

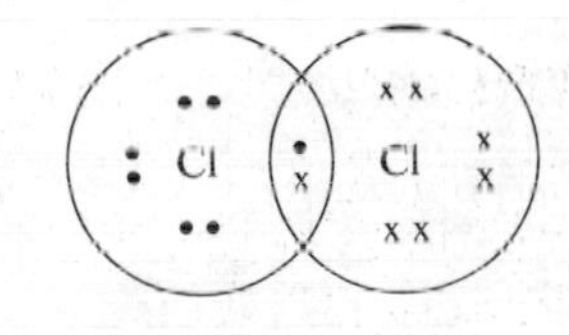


Electron structure 2,8,8 2,8,8

**Figure 16.5** Chlorine—a diatomic molecule in which two chlorine atoms are linked by a covalent bond.

**Table 16.2** Physical properties of the halogens.

| **Element** | Fluorine | Chlorine | Bromine | Iodine |
|---|---|---|---|---|
| **Atomic number** | 9 | 17 | 35 | 53 |
| **Electron configuration** | 2, 7 | 2, 8, 7 | 2, 8, 18, 7 | 2, 8, 18, 18, 7 |
| **Outer-shell electron configuration** | $2s^2\ 2p^5$ | $3s^2\ 3p^5$ | $4s^2\ 4p^5$ | $5s^2\ 5p^5$ |
| **Relative atomic mass** | 19.0 | 35.5 | 79.9 | 126.9 |
| **State at 20°C** | gas | gas | liquid | solid |
| **Colour** | pale-yellow | pale-green | red-brown | black |
| **Melting point/°C** | −220 | −101 | −7 | 113 |
| **Boiling point/°C** | −188 | −35 | 59 | 183 |
| **Molar heat of vaporization of liquid/kJ mole$^{-1}$** | 3.3 | 10.2 | 15 | 30 |
| **Solubility/g per 100g of water at 20°C** | reacts readily with water | 0.59 (reacts slightly) | 3.6 | 0.018 |

Notice also the change in volatility down the group; fluorine and chlorine are gases, bromine is a liquid and iodine is a solid. This decreasing volatility is, of course, related to the increasing strength of Van der Waals' forces with increasing relative molecular mass, and this results in increasing melting points, boiling points and molar heats of vaporization.

All the halogens except fluorine dissolve slightly in water and colour it. Fluorine is such a powerful oxidizing agent that it converts water to oxygen. The halogens, being non-polar simple molecules are, as we might expect, more soluble in organic solvents, the colour of the solution depending upon the particular halogen and solvent. In non-polar solvents such as tetrachloromethane and cyclohexane, chlorine is colourless, bromine is red and iodine is violet. In these solvents the elements exist as relatively free molecules as in the gas phase. In polar (electron donating) solvents such as water, ethanol and propanone (acetone), bromine, and particularly iodine tend to give brownish solutions. The change in colour is due to the formation of complexes involving charge-transfer (i.e. a donation of electrons) from the polar solvent molecules to the iodine molecule. It is thought that an electron normally located in an orbital of the solvent molecule can absorb energy from visible light and jump into an orbital of the iodine molecule. Thus, the complex removes visible light of a particular wavelength and appears coloured.

The blue compound which iodine forms with starch is also a complex. The iodine molecules are absorbed reversibly within the helical chains of glucose molecules which make up starch.

You will need to refer to table 16.2 in order to answer the following questions.

Astatine is the halogen element immediately below iodine in the periodic table. Predict the following properties of astatine.

○ Its colour
○ Its state at room temperature
○ Its relative atomic mass
○ Its melting point
○ Its molar heat of vaporization
○ The abundance of its compounds

## 16.7 Chemical properties of the halogens

All the halogens have one electron less than the noble gas which immediately follows them in the periodic table. As we might expect, their chemistry is dominated by a tendency to gain a completely filled outermost electron shell.

Consequently, *the halogens react with metals to form ionic compounds containing the* $Hal^-$ *ion. With non-metals and with some metals in high oxidation states, they tend to form simple molecular compounds in which the halogen is linked by a single covalent bond (—Hal) to the other element.*

### REACTIONS WITH METALS

The halogens are very reactive with metals, the vigour of reaction depending on

1 the position of the metal in the activity series and
2 the particular halogen which is reacting.

The reactivity of the halogen decreases with increase in atomic number. Fluorine is, in fact, the most reactive of all non-metals and because of its great reactivity it was not isolated until 1886. Fluorine combines readily and directly with all metals, whereas iodine reacts very slowly even at high temperatures with metals low in the activity series such as silver and gold. Gold and platinum are readily attacked by fluorine on heating but some metals, such as copper and nickel alloys, become coated with a superficial layer of fluoride which prevents further reaction. Hence, vessels of these alloys are used for preparing and storing the gas.

The standard heats of formation of each of the sodium halides (table 16.3) provide quantitative evidence for the relative reactivity of the halogens. The heat evolved is greatest for the reaction between sodium and fluorine and least for the reaction between sodium and iodine.

**Table 16.3** Properties of the halogens and their compounds.

| Element | Fluorine | Chlorine | Bromine | Iodine |
|---|---|---|---|---|
| Standard heat of formation of NaHal/kJ mole$^{-1}$ | −573 | −414 | −361 | −288 |
| Heat of atomization of $Hal_2$/kJ per mole of atoms formed | 79 | 121 | 112 | 107 |
| Bond energy $E$(Hal-Hal)/kJ per mole of bonds | 158 | 242 | 193 | 151 |
| Electrode potential $Hal_2$(aq), $2Hal^-$(aq)/Pt/Volts | 2.87 | 1.36 | 1.09 | 0.54 |

The most reactive halogen is fluorine; chlorine is the next most reactive, then bromine, then iodine. Thus, reactivity decreases as the relative atomic mass of the halogen increases in Group VII. A similar decrease in reactivity with increasing atomic mass is found in other groups of non-metals. This is the reverse of the trend in Group I and other groups of metals where reactivity increases as the relative atomic mass increases.

- List three factors which influence the reactivity of the halogen elements.
- How will each of these three factors influence the reactivity of fluorine compared to chlorine?
- Why is fluorine (F = 19.0) more reactive than chlorine (Cl = 35.5)?
- Why is sodium (Na = 23.0) less reactive than potassium (K = 39.1)?

### REACTIONS WITH NON-METALS

A similar pattern of reactivity emerges here also. Fluorine reacts directly with all non-metals except nitrogen, helium, neon and argon. It will even react with diamond and xenon on heating.

$$\text{C(diamond)} + 2F_2 \longrightarrow CF_4$$

$$Xe + 2F_2 \longrightarrow XeF_4$$

Fluorine will also attack glass, quartz and silica displacing oxygen.

$$SiO_2 + 2F_2 \longrightarrow SiF_4 + O_2$$

If these substances are carefully dried, the reaction is very slow. Thus, fluorine can be studied in dry glass equipment but any taps must be lubricated with 'fluorocarbon grease' containing fluorinated hydrocarbons rather than grease containing hydrocarbons since fluorine readily attacks organic compounds displacing hydrogen.

$$\text{—C—H} + F_2 \longrightarrow \text{—C—F} + HF$$

Chlorine and bromine are much less reactive than fluorine. They will not react directly with any of the noble gases, carbon, nitrogen or oxygen. Iodine does not combine with these elements nor with sulphur, but it reacts readily with phosphorus forming the triiodide.

$$\tfrac{1}{4}P_4 + \tfrac{3}{2}I_2 \longrightarrow PI_3$$

The relative reactivity of the halogens with non-metals is effectively illustrated by their reaction with hydrogen. Fluorine explodes with hydrogen even in the dark at −200°C. Chlorine and hydrogen explode in bright sunlight but react slowly in the dark. Bromine reacts with hydrogen only on heating and in the presence of a platinum catalyst, whilst iodine combines only partially and slowly with hydrogen even on heating.

$$H_2(g) + Cl_2(g) \longrightarrow 2HCl(g)$$

$$H_2(g) + I_2(g) \longrightarrow 2HI(g)$$

The great reactivity of fluorine in these reactions with non-metals is explained partly in terms of its low atomization energy (table 16.3) which means that the initial stage (in which the F—F bond breaks) requires little energy, and hence the activation energy for reactions involving fluorine is low. The low atomization energy of fluorine is reflected in a low F—F bond energy (table 16.3). Fluorine's reactivity is further explained by its ability to form very stable bonds with other elements. Look at table 11.3 and compare the bond energy of the C—F bond with the bond energies of other C—Hal bonds. How does the strength of the H—F bond compare with other H—Hal bonds?

## 16.8 The halogens as oxidizing agents

When the halogens combine with metals or non-metals they normally act as oxidizing agents and the elements with which they react have positive oxidation numbers in the resultant compounds.

When the halogens combine with metals to form ionic compounds, they gain electrons from the metals to form negative halide ions.

$$2Na(s) + Cl_2(g) \longrightarrow 2Na^+Cl^-(s)$$

The halogens accept electrons during these reactions and clearly act as oxidizing agents.

$$Hal_2 + 2e^- \longrightarrow 2Hal^-$$

Fluorine is the most reactive halogen and the most powerful oxidizing agent. *The order of decreasing power as oxidizing agents is* $F_2 > Cl_2 > Br_2 > I_2$. The electrode potentials ($Hal_2/Hal^-$) for the halogens shown in table 16.3 become less positive from fluorine to iodine reflecting the decreasing oxidizing power.

We have already noticed the different oxidizing powers of the halogens with metals, non-metals, and hydrogen. The following reactions further emphasize their relative oxidizing ability.

Fluorine, chlorine and bromine will all oxidize $Fe^{2+}(aq)$ to $Fe^{3+}(aq)$.

$$2Fe^{2+}(aq) \longrightarrow 2Fe^{3+}(aq) + 2e^-$$

$$Hal_2(aq) + 2e^- \longrightarrow 2Hal^-(aq)$$

Iodine, however, is such a weak oxidizing agent that it cannot remove electrons from iron(II) ions to form iron(III) ions. These observations are exactly as we might have predicted since $E^\ominus$ for $Fe^{3+}(aq), Fe^{2+}(aq)/Pt$ is 0.77 volts.

Fluorine and chlorine are such powerful oxidizing agents that they can oxidize various coloured dyes to colourless substances. Thus, indicators such as litmus and universal indicator are decolourized when exposed to these gases or to their aqueous solutions. Chlorine acts as an oxidizing agent when it is used for bleaching.

Chlorine acts as an oxidizing agent when it bleaches. The square of printed material on the left was bleached by immersion in chlorine water for a few hours.

When chlorine water ($Cl_2$(aq)) is added to aqueous KI, the solution becomes brown due to the formation of iodine. When bromine water ($Br_2$(aq)) is used in place of chlorine water, iodine is again liberated from KI(aq).

In these two reactions, iodide ions are oxidized to iodine.

$$2I^- \longrightarrow I_2 + 2e^-$$

The chlorine and bromine act as oxidizing agents by accepting electrons from iodide to form chloride and bromide respectively.

$$Cl_2 + 2e^- \longrightarrow 2Cl^-$$
$$Br_2 + 2e^- \longrightarrow 2Br^-$$

Chlorine and bromine can oxidize iodide to iodine, but iodine cannot oxidize chloride or bromide ions. This is because chlorine and bromine are more reactive than iodine, i.e. chloride will not release electrons to iodine.

Since chlorine is a stronger oxidizing agent than bromine, chlorine can oxidize bromide ions to bromine.

$$Cl_2 + 2Br^- \longrightarrow 2Cl^- + Br_2$$

Thus, when chlorine water is added to KBr(aq), yellow-orange bromine is produced.

- ○ Which halide ions can $F_2$ oxidize?
- ○ Which of the following species is the strongest reducing agent? $I^-$, $I_2$, $Cl^-$, $F_2$, $F^-$.
- ○ The standard electrode potential for the half reaction $MnO_4^-(aq) + 8H^+(aq) + 5e^- \longrightarrow Mn^{2+}(aq) + 4H_2O$ is 1.51 volts. Which halide ions could this half-cell oxidize under standard conditions? (Refer to the electrode potentials for halogens in table 16.3.)

### EXPLAINING THE RELATIVE OXIDIZING POWER OF THE HALOGENS

When halogens act as oxidizing agents they undergo the reaction:

$$\tfrac{1}{2}Hal_2 + e^- \longrightarrow Hal^-$$

We can picture the reaction in terms of three stages as shown in figure 16.6.

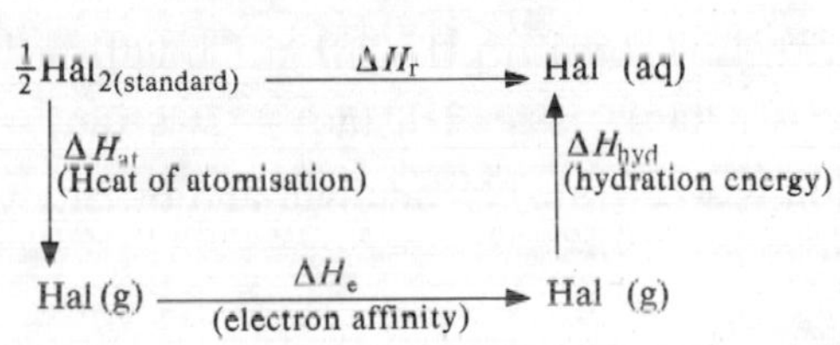

$$\begin{array}{ccc} \tfrac{1}{2}Hal_{2(standard)} & \xrightarrow{\Delta H_r} & Hal^-(aq) \\ \downarrow \Delta H_{at} \text{ (Heat of atomisation)} & & \uparrow \Delta H_{hyd} \text{ (hydration energy)} \\ Hal(g) & \xrightarrow[\text{(electron affinity)}]{\Delta H_e} & Hal^-(g) \end{array}$$

| | F | Cl | Br | I |
|---|---|---|---|---|
| $\Delta H_{at}$ /kJ mole$^{-1}$ | 79 | 121 | 112 | 107 |
| $\Delta H_e$ /kJ mole$^{-1}$ | −333 | −364 | −342 | −295 |
| $\Delta H_{hyd}$ /kJ mole$^{-1}$ | −457 | −381 | −351 | −307 |
| $\Delta H_r$ /kJ mole$^{-1}$ for $\tfrac{1}{2}Hal_2 \longrightarrow Hal^-(aq)$ | −711 | −624 | −581 | −495 |

**Figure 16.6** Stages in the conversion of diatomic halogen elements to aqueous halide ions.

The enthalpy change of each of these stages for F, Cl, Br, and I are also shown in figure 16.6. It would appear that the order of oxidizing power, $F_2 > Cl_2 > Br_2 > I_2$, is due mainly to the relative hydration energies and to a lesser degree to the small atomization energy of fluorine.

Since the lattice energies of ionic halides also get less exothermic from fluorides to iodides, the order of oxidizing power is the same for solid state reactions as for reactions in aqueous solution.

### ESTIMATING THE CONCENTRATION AND MOLAR AMOUNTS OF OXIDIZING AGENTS

Iodine is by far the weakest oxidizing agent of the four common halogens. Even mild oxidizing agents, such as $Fe^{3+}$(aq), can oxidize aqueous $I^-$ to $I_2$. Because of this iodine/thiosulphate(VI) titrations are used in the quantitative estimation of oxidizing agents.

The oxidizing agent to be estimated is first added to excess potassium iodide and iodine is liberated.

$$2I^-(aq) \longrightarrow I_2(aq) + 2e^-$$

The amount of iodine liberated is then determined by titration against a solution of sodium thiosulphate(VI) of known concentration.

$$I_2(aq) + 2e^- \longrightarrow 2I^-(aq)$$

$$\underset{\text{thiosulphate(VI)}}{2S_2O_3^{2-}(aq)} \longrightarrow \underset{\text{tetrathionate}}{S_4O_6^{2-}(aq)} + 2e^-$$

The end point of the titration can be detected by adding starch solution which gives a blue colour as long as any iodine is present. By calculating the amount of thiosulphate(VI) used in the titration, we can determine the amount of iodine liberated and hence the amount of oxidizing agent taken.

## 16.9 The reaction of halogens with water

The electrode potential for the oxidation of water to oxygen,

$$2H_2O(l) \longrightarrow O_2(g) + 4H^+(aq) + 4e^-$$

is $-1.23$ volts. The electrode potential data for the halogens in table 16.3 shows that iodine and bromine cannot oxidize water to oxygen but fluorine and chlorine are capable of doing so.

Fluorine reacts with water vapour to form oxygen and ozone.

$$2F_2(g) + 2H_2O(g) \longrightarrow 4HF(g) + O_2(g)$$
$$3F_2(g) + 3H_2O(g) \longrightarrow 6HF(g) + O_3(g)$$

The reaction of water with chlorine is very slow. In fact, chloric(I) acid (hypochlorous acid) is first formed which then decomposes to oxygen and hydrochloric acid.

$$Cl_2(g) + H_2O(l) \longrightarrow H^+(aq) + Cl^-(aq) + HClO(aq)$$
$$2HClO(aq) \longrightarrow 2H^+(aq) + 2Cl^-(aq) + O_2(g)$$

The second reaction is accelerated by sunlight and by catalysts such as platinum and metallic oxides.

## 16.10 The reactions of halogens with alkalis

Chlorine, bromine and iodine undergo very similar reactions with alkalis, the products depending upon the temperature at which the reaction occurs.

With cold, dilute alkali at 15°C, the halogen reacts to form a mixture of halide ($Hal^-$) and halate(I) (hypohalite, $HalO^-$).

$$Hal_2 + 2OH^-(aq) \longrightarrow Hal^-(aq) + HalO^-(aq) + H_2O(l)$$

The $HalO^-$ which is produced in the first reaction then decomposes to form halide and halate(V) ($HalO_3^-$).

$$3HalO^- \longrightarrow 2Hal^- + HalO_3^-$$

For chlorine, this second reaction is very slow at 15°C, but rapid at 70°C. Therefore, sodium chlorate(I) (sodium hypochlorite) can be obtained by passing chlorine into sodium hydroxide at 15°C, whilst sodium chlorate(V) ($NaClO_3$) is obtained by carrying out the same reaction at 70°C.

With bromine, both reactions are rapid at 15°C, but decomposition of $BrO^-$ is slow at 0°C.

With iodine, decomposition of $IO^-$ occurs rapidly even at 0°C so it is difficult to prepare $NaIO$ free from $NaIO_3$.

These two reactions of halogens with alkali involve **disproportionation**—*a change in which one particular molecule, atom or ion is simultaneously oxidized and reduced.*

When $Cl_2$ reacts with alkali to form $Cl^-$ and $ClO^-$, chlorine atoms are both oxidized and reduced. The oxidation number of one Cl atom in $Cl_2$ changes from 0 to $-1$ in $Cl^-$ (reduction) whilst that of the other changes from 0 to $+1$ in $ClO^-$ (oxidation). Oxidation numbers are shown above each atom in the following equation.

$$\overset{0}{Cl_2}(g) + 2\overset{-2}{O}\overset{+1}{H}{}^-(aq) \longrightarrow \overset{-1}{Cl}{}^-(aq) + \overset{+1}{Cl}\overset{-2}{O}{}^-(aq) + \overset{+1}{H_2}\overset{-2}{O}(l)$$

The decomposition of $ClO^-$ to $Cl^-$ and $ClO_3^-$ is also disproportionation.

- ○ Write an equation for this change.
- ○ What is the oxidation number of chlorine in each of $ClO^-$, $Cl^-$ and $ClO_3^-$?
- ○ What is the commonest oxidation number for chlorine in its compounds?
- ○ Write the formula of a compound containing chlorine in a different oxidation state to that in $Cl^-$, $ClO^-$ or $ClO_3^-$.

The halogens, other than fluorine, form compounds in which they have positive oxidation numbers up to $+7$. Figure 16.7 shows an oxidation number chart for chlorine. The exhibition of several oxidation numbers by each halogen is, of course, very different from the behaviour of metals in Groups I and II which exhibit only one oxidation number in their compounds. As we might expect, the most stable oxidation state for halogens is $-1$, but the relative stability of the $-1$ state decreases as the group is descended and the electronegativity of the halogen decreases.

Fluorine, the most electronegative element, never exhibits a positive oxidation number since it cannot form a compound in which it is the less electronegative element. In this respect, it is significant that fluorine reacts with alkalis differently to the other halogens, forming a mixture of fluoride and oxygen difluoride in both of which its oxidation number is $-1$.

$$2F_2(g) + 2OH^-(aq) \longrightarrow OF_2(g) + 2F^-(aq) + H_2O(l)$$

| **Oxidation number** | **Examples** |
|---|---|
| +7 | $Cl_2O_7$, $NaClO_4$ |
| +6 | $ClO_3$ |
| +5 | $NaClO_3$ |
| +4 | $ClO_2$ |
| +3 | $KClO_2$ |
| +2 | |
| +1 | $Cl_2O$, $NaClO$ |
| 0 | $Cl_2$ |
| −1 | $NaCl$ |

**Figure 16.7** The range of oxidation numbers for chlorine.

A further consequence of the decreasing electronegativity down Group VII is that the relative stability of the positive oxidation states increases with increasing relative atomic mass of the halogen.

Thus, the ease of decomposition of $HalO^-$ into $HalO_3^-$ (halogen in the $+5$ state) and $Hal^-$ is in the order $IO^- > BrO^- > ClO^-$ suggesting that $IO_3^-$ is relatively more stable than $BrO_3^-$ which in turn is more stable than $ClO_3^-$. In fact, the positive oxidation state has become sufficiently stable in iodine to permit the existence of iodine cations. Thus, iodine(I) nitrate(V) ($I^+NO_3^-$) and iodine(I) chlorate(VII) ($I^+ClO_4^-$) have been prepared in the presence of pyridine which forms a complex with the $I^+$ ion. Not surprisingly, these salts react with potassium iodide to give iodine,

$$I^+ + I^- \longrightarrow I_2$$

and when electrolysed they give iodine at the cathode. $I^+$ ions are probably also produced when iodine chloride (ICl) is dissolved in conc.$H_2SO_4$.

The existence of $I^+$ ions is evidence for the slight metallic character of iodine. Other features which show this are the metallic lustre of solid iodine and the small but definite conductivity of liquid iodine.

One sensible deduction that we might make from all this is that the halogens are likely to act as oxidizing agents in all their positive oxidation states (table 16.4).

**Table 16.4** The electrode potentials of the halogens and some of their oxyanions.

| **Electrode process** | $E^\ominus$**/Volts** |
|---|---|
| $F_2(g) + 2e^- \longrightarrow 2F^-(aq)$ | +2.87 |
| $2HClO(aq) + 2H^+(aq) + 2e^- \longrightarrow Cl_2(aq) + 2H_2O(l)$ | +1.59 |
| $2HBrO(aq) + 2H^+(aq) + 2e^- \longrightarrow Br_2(aq) + 2H_2O(l)$ | +1.57 |
| $2BrO_3^-(aq) + 12H^+(aq) + 10e^- \longrightarrow Br_2(aq) + 6H_2O(l)$ | +1.52 |
| $Cl_2(aq) + 2e^- \longrightarrow 2Cl^-(aq)$ | +1.36 |
| $2IO_3^-(aq) + 12H^+(aq) + 10e^- \longrightarrow I_2(aq) + 6H_2O(l)$ | +1.19 |
| $Br_2(aq) + 2e^- \longrightarrow 2Br^-(aq)$ | +1.09 |
| $HIO(aq) + H^+(aq) + 2e^- \longrightarrow I^-(aq) + H_2O(l)$ | +0.99 |
| $ClO^-(aq) + H_2O(l) + 2e^- \longrightarrow Cl^-(aq) + 2OH^-(aq)$ | +0.89 |
| $_2(aq) + 2e^- \longrightarrow 2I^-(aq)$ | +0.54 |
| $O^-(aq) + H_2O(l) + 2e^- \longrightarrow I^-(aq) + 2OH^-(aq)$ | +0.49 |

Notice two points from table 16.4.

1 The redox potentials for the various systems involving iodine are smaller than the corresponding systems involving bromine which in turn are less than those for chlorine. (i.e. $F_2 > Br_2 > Cl_2 > I_2$; $HClO > HBrO > HIO$; $BrO_3^- > IO_3^-$; $ClO^- > IO^-$.) This is further evidence for the increasing stability of positive oxidation states as Group VII is descended.

2 The $HalO_3^-$ ions are stable in alkali and require acid conditions before they become effective oxidizing agents—not surprising in view of the fact that each $HalO_3^-$ ion requires $6H^+$ ions during the reaction.

## 16.11 Reactions of halide ions

Some reactions of fluorides, chlorides, bromides and iodides in aqueous solution are summarized in table 16.5. All common halides are soluble except all lead halides, AgCl, AgBr and AgI. Notice in table 16.5 that precipitates of these halides are produced when aqueous solutions of halides are treated with either $Pb^{2+}$(aq) or $Ag^+$(aq).

$$Ag^+(aq) + Cl^-(aq) \longrightarrow AgCl(s)$$

$$Pb^{2+}(aq) + 2Cl^-(aq) \longrightarrow PbCl_2(s)$$

**Table 16.5** Reactions of aqueous halide ions.

| Solution added | $F^-$(aq) | $Cl^-$(aq) | $Br^-$(aq) | $I^-$(aq) |
|---|---|---|---|---|
| $Pb(NO_3)_2$(aq) | white ppte of $PbF_2$ | white ppte of $PbCl_2$ | yellow ppte of $PbBr_2$ | yellow ppte of $PbI_2$ |
| $AgNO_3$(aq) | no reaction | white ppte of AgCl | cream ppte of AgBr | yellow ppte of AgI |
| Solubility of AgHal in | | | | |
| (*a*) dil. $NH_3$(aq) | soluble | soluble | insoluble | insoluble |
| (*b*) conc. $NH_3$(aq) | soluble | soluble | soluble | insoluble |
| Effect of sunlight on AgHal | no effect | white AgCl turns purple-grey | cream AgBr turns green-yellow | no effect |

*The action of silver nitrate*(V) *solution followed by sunlight or ammonia solution can be used as a test for halide ions.*

$F^-$(aq) gives no precipitate of AgF.

$Cl^-$(aq) gives a white precipitate of AgCl which becomes purple-grey in sunlight and dissolves in conc.$NH_3$.

$Br^-$(aq) gives a cream precipitate of AgBr which becomes green-yellow in sunlight and dissolves in conc.$NH_3$.

$I^-$(aq) gives a yellow precipitate of AgI which is unaffected by sunlight and which is insoluble in conc.$NH_3$.

The colour changes which occur when AgCl and AgBr are exposed to sunlight results from the superficial conversion of these silver halides to silver and halogen.

$$Ag^+Br^-(s) \longrightarrow Ag(s) + \tfrac{1}{2}Br_2(g)$$

This photochemical change involving AgBr plays an essential part in black and white photography. Photographic plates and films contain silver bromide which decomposes to silver on exposure to light. During the development stage, the plate/film is treated with 'hypo' (sodium thiosulphate(VI) solution) which removes excess AgBr as a soluble complex ion, $[Ag(S_2O_3)_2]^{3-}$, and the silver remains on the plate/film as an opaque shadow.

The reactions of solid halides with concentrated sulphuric(VI) acid, with concentrated phosphoric(V) acid and with a mixture of manganese(IV) oxide plus concentrated sulphuric(VI) acid are summarized in table 16.6.

**Table 16.6** Reactions of solid halides.

| Reagent added | Fluoride | Chloride | Bromide | Iodide |
|---|---|---|---|---|
| Conc. $H_2SO_4$ | $HF(g)$ produced | $HCl(g)$ produced | $HBr(g)$ + a little $Br_2(g)$ produced | a little $HI(g)$ + $I_2(g)$ produced |
| Conc. $H_3PO_4$ | $HF(g)$ produced | $HCl(g)$ produced | $HBr(g)$ produced | $HI(g)$ produced |
| Conc. $H_2SO_4 + MnO_2$ | $HF(g)$ produced | $Cl_2(g)$ produced | $Br_2(g)$ produced | $I_2(g)$ produced |

When conc.$H_2SO_4$ is added to solid halides the first product is the hydrogen halide in each case. Being relatively volatile, these hydrogen halides are evolved as gases.

$$Hal^-(s) + H_2SO_4(l) \longrightarrow HHal(g) + HSO_4^-(s)$$

However, concentrated sulphuric(VI) acid is also an oxidizing agent and it is powerful enough to oxidize HBr and HI (but not HF and HCl) to $Br_2$ and $I_2$ respectively.

$$2HBr + H_2SO_4 \longrightarrow Br_2 + 2H_2O + SO_2$$

$$2HI + H_2SO_4 \longrightarrow I_2 + 2H_2O + SO_2$$

When concentrated $H_2SO_4$ is used in the presence of an even stronger oxidizing agent, such as $MnO_2$, the oxidizing conditions are sufficient to oxidize HCl to $Cl_2$, but HF is still not oxidized to $F_2$.

$$4HCl + MnO_2 \longrightarrow Cl_2 + MnCl_2 + 2H_2O$$

○ Write an equation for the reaction of concentrated phosphoric(V) acid ($H_3PO_4$) with NaBr(s).
○ Why does the conc.$H_3PO_4$ cause HBr(g) to be evolved during the reaction?
○ Why is no $Br_2$ produced in this reaction unlike that between conc.$H_2SO_4$ and NaBr(s)?

## 16.12 Uses of the halogens

Although the halogen elements have very few direct uses themselves, their compounds are used extensively in industry, in agriculture, in medicine and in the home. Chlorine is used as a cheap industrial oxidant in the manufacture of bromine (section 16.3), as a bleach and as a germicide in the treatment of household and swimming bath water supplies.

Iodine dissolved in alcohol, commonly known as 'tincture of iodine' is used as a mild antiseptic for cuts and scratches. Iodine is also mixed with the detergents used in cleaning dairy equipment.

Small quantities of fluorine are used in rocket propulsion, but much larger quantities are used to make uranium(VI) fluoride for the separation of $^{238}_{92}U$ and $^{235}_{92}U$.

$$UF_4(s) + F_2(g) \longrightarrow UF_6(s)$$

By allowing gaseous $UF_6$ to diffuse slowly it is possible to separate $^{238}_{92}UF_6$ from $^{235}_{92}UF_6$ and hence obtain the separate isotopes of uranium for use in atomic power stations.

Fluorine is also used to make a wide range of fluorocarbon compounds for use as refrigerants, aerosol propellants, anaesthetics and fire-extinguisher fluids. Many of these compounds are discussed in section 28.3. One of the most important fluorocarbons is poly(tetrafluoroethene)—PTFE frequently sold under the trade name 'Fluon' or 'Teflon' (figure 16.8).

$$\left( \begin{array}{c} F\;\;F \\ |\;\;\;| \\ -C-C- \\ |\;\;\;| \\ F\;\;F \end{array} \right)_n$$

**Figure 16.8** The structure of poly(tetrafluoroethene).

*Above* PTFE has an extremely low coefficient of friction and anti-stick properties. Because of this, thin layers of it are coated on the running surface of skis.

PTFE, like other fully-fluorinated hydrocarbons, is very unreactive and resists almost all corrosive chemicals. For this reason, it is used for valves, seals and gaskets in chemical plants and laboratories. It is also an excellent electrical insulator and is used for wire coverings. Furthermore, PTFE has an extremely low coefficient of friction and has anti-stick properties. Because of this, thin layers of it are coated on the cooking surface of non-stick sauce-pans and on the running surface of skis.

Chlorine is used in the manufacture of many familiar materials. Hydrogen chloride is produced by the reaction between chlorine and hydrogen, itself a by-product in the electrolytic manufacture of chlorine. Hydrogen chloride made in this way is dissolved in water to produce hydrochloric acid.

$$H_2(g) + Cl_2(g) \longrightarrow 2HCl(g) \xrightarrow{aq} 2H^+(aq) + 2Cl^-(aq)$$

Hydrochloric acid is the cheapest industrial acid. It is used industrially for removing (de-scaling) rust from steel sheets before galvanizing.

The other inevitable by-product of the electrolysis of brine is sodium hydroxide. The sodium hydroxide solution is treated with gaseous chlorine to obtain sodium chlorate(I) (sodium hypochlorite, NaClO) which is used as a bleach in laundries, as a disinfectant and in sewage treatment.

$$Cl_2(g) + 2NaOH(aq) \longrightarrow NaCl(aq) + NaClO(aq) + H_2O(l)$$

On warming, the NaClO decomposes to sodium chlorate(V) ($NaClO_3$) which is used as a weed killer.

$$3NaClO(aq) \xrightarrow{heat} 2NaCl(aq) + NaClO_3(aq)$$

In recent years, chlorine compounds have been developed for use as degreasing solvents such as tetrachloromethane and trichloroethene, as plastics such as PVC (section 26.7), as disinfectants and antiseptics such as 'dettol' and TCP (section 29.4) and as pesticides.

During the 1940's and 1950's a range of highly chlorinated aromatic compounds was developed and used as pesticides. Probably the best known of these compounds are DDT (dichlorodiphenyltrichloroethane), BHC (benzene hexachloride), aldrin, dieldrin and heptachlor (figure 16.9).

dichlorodiphenyltrichloroethane ('DDT')

benzene hexachloride ('BHC')

aldrin

dieldrin

**Figure 16.9** The structures of some chlorinated pesticides.

The spraying of large areas of land with these chlorine-containing pesticides has eliminated many insect-borne diseases, such as malaria, and led to enormous improvements in the quality and yield of crops. Unfortunately, however, these chlorinated compounds are so stable that they remain unchanged on the crops or accumulate in the soil. Furthermore, these compounds are fat soluble but not water soluble, which means that they concentrate, after ingestion, in the fatty tissues of birds and animals possibly reaching hazardous levels.

In the spring of 1956, large numbers of seed-eating birds were found dead in cereal-growing areas. Analysis showed that the corpses of these birds contained dieldrin and aldrin which had been sprayed on the spring-sown wheat. In 1964, the International Advisory Committee on Poisonous Substances used in Agriculture and Food Storage recommended that the use of dieldrin and aldrin should be reserved for the treatment of heavily-infected areas. At that time, no restrictions were placed on the use of DDT and BHC, but concern over their use continued to grow. It appears that insects can develop a tolerance to small, non-lethal doses of these chemicals. These insects are eaten by small carnivorous animals and birds which concentrate the DDT in their own fatty tissue. Larger predators who eat these small carnivores concentrate the DDT even further which may ultimately reach toxic levels in animals or birds several stages along the food chain. At one time, for instance, there was some concern that humans might be affected by drinking the milk of cows which had eaten grasses sprayed with DDT. The use of DDT was also thought to be responsible for the decreasing populations of birds of prey and the virtual disappearance of frogs, whose bodies were found to contain the chemical. As a result of concern over its use, the world consumption of DDT fell from 400 000 tonnes in 1963 to only about half that quantity in 1971.

Since 1972, both the British and American Governments have imposed restrictions on the use of DDT, but the need for an effective substitute is most urgent. Nevertheless, it is sensible to use less toxic insecticides for general purposes and reserve DDT for special control schemes.

A greenfinch killed by the excessive use of toxic pesticides.

## Summary

1 The halogens are a group of reactive non-metals, contrasting strongly with the alkali and alkaline-earth elements which are reactive metals.

2 Although the halogens show remarkable similarities in their properties and reactions, there is an obvious trend in their behaviour and reactivity with increase in atomic number. As atomic number increases, the halogens get less reactive.

3 The halogens are so reactive that they occur naturally only in compounds.

4 The halogens can be obtained by oxidation of halide ions.

5 The halogens exist as diatomic molecules. Their atoms have seven electrons in the outermost shell and, consequently, the chemistry of the halogens is dictated by a tendency to act as oxidizing agents in forming negative halide ions ($Hal^-$). The most stable oxidation number of the halogens is $-1$.

6 The order of oxidizing power for the halogens is $F_2 > Cl_2 > Br_2 > I_2$.

7 A disproportionation reaction is one in which a particular molecule, atom or ion is simultaneously oxidized and reduced. When halogens react with dilute alkali to form a mixture of $Hal^-$ and $HalO^-$, the halogen is both oxidized and reduced.

8 The action of $AgNO_3(aq)$ followed by sunlight or $NH_3(aq)$ can be used as a test for halide ions.

## Study questions

1 (a) Chlorine and sodium hydroxide are manufactured by the electrolysis of brine. Write equations to summarize what happens during the electrolysis when
   (i) precautions are taken to prevent the products mixing with each other,
   (ii) the products are deliberately mixed with one another.

(b) Iodine is obtained from iodate(V) by treatment with sodium hydrogensulphate(IV).
   (i) Write an equation (or half-equations) for the reaction of iodate(V) with hydrogensulphate(IV) ions in acid solution.
   (ii) Why must excess hydrogensulphate(IV) *not* be used?

(c) Fluorine can be obtained by the electrolysis of KF dissolved in liquid HF, but not by the electrolysis of KF dissolved in water. Why not?

(d) Astatine ($^{211}_{85}At$) has been made by bombarding bismuth with high-energy alpha particles.
   (i) Write a nuclear equation for this reaction.
   (ii) The name 'astatine' is derived from a Greek word meaning unstable. How do you think astatine might decompose? Write a nuclear equation for the decay process.

2 From your knowledge of the halogens, predict what happens in the following situations and write equations for any reactions which take place. (Ignore the radioactive nature of astatine.)
(a) Astatine vapour is mixed with hydrogen at 100°C.
(b) Astatine is added to aqueous sodium hydroxide.
(c) Concentrated sulphuric(VI) acid is added to solid sodium astatide.
(d) Aqueous silver nitrate(V) is added to aqueous sodium astatide.
(e) Astatine is added to sodium thiosulphate(VI) solution.

3 The halogens (F, Cl, Br and I) form a well-defined group of elements.
(a) Explain how the following support this statement:
(i) electron structure,
(ii) redox behaviour,
(iii) physical properties of the elements,
(iv) usual oxidation state.
(b) Describe four specific properties that show a regular gradation as the group is descended from F to I.
(c) Explain the meaning of the term 'electron affinity'. How does this vary among the halogens, and how is it related to their reactivity?
(d) Fluorine and fluorides show some properties not typical of the rest of the group. Mention three of these properties and suggest a reason or reasons for the difference.

4 Explain the following observations:
(a) A mass spectrograph of chlorine shows five particles with relative masses of 35, 37, 70, 72, and 74.
(b) As Group VII of the periodic table is descended, the halogens become weaker oxidizing agents.
(c) In its compounds, fluorine shows only one oxidation state, whereas chlorine shows several.
(d) Hydrogen fluoride has a higher boiling point than hydrogen chloride and hydrogen iodide.

5 The percentage of copper in a sample of brass was determined as follows. 2.0g of the brass was converted to $200cm^3$ of a solution of copper(II) nitrate(V) free from nitric(V) or nitric(III) (nitrous) acid and acidified with ethanoic (acetic) acid. $20.0cm^3$ of this solution liberated sufficient iodine from potassium iodide solution to react with $25.0cm^3$ of 0.1M sodium thiosulphate(VI) solution.
(Cu = 64, the reaction between $Cu^{2+}(aq)$ and $I^-(aq)$ can be represented as $2Cu^{2+}(aq) + 4I^-(aq) \rightarrow 2CuI(s) + I_2(aq)$.)
(a) Write an equation or half-equations for the reaction between iodine and sodium thiosulphate(VI).
(b) How many moles of $I_2$ were liberated by $20cm^3$ of aqueous $Cu^{2+}$ solution?
(c) How many moles of copper are there in $200cm^3$ of aqueous $Cu^{2+}$ solution?
(d) What is the percentage by weight of copper in the brass?

6 Read the second half of section 16.12 concerning pesticides.
(a) What is the correct systematic name for BHC?
(b) Why are chlorinated pesticides such as BHC and DDT soluble in fat, but not in water?
(c) Seed-eating birds are extremely vulnerable to chlorinated-pesticides sprayed on spring crops. Why is there much less danger to these birds when autumn-sown crops are treated with pesticides?
(d) Suggest an explanation for the pesticide-action of DDT.
(e) Why can insects develop a tolerance to small doses of chemicals such as DDT?
(f) Suppose you were given the task of synthesizing the ideal insecticide for spraying on spring-sown wheat. Make a list of the properties that you would look for in your product.

# Group IV – Carbon to Lead, Non – Metal to Metal

# 17

## 17.1 Introduction

*The similarity between elements in the same family which was so obvious in Groups I, II and VII is much less apparent in Group IV where there is a considerable change in the character of the elements as atomic number rises.* Carbon is unquestionably a non-metal, silicon and germanium are metalloids, whereas tin and lead show typical metallic properties.

Although carbon occurs naturally both as diamonds (Brazil and South Africa) and as graphite (Sri Lanka, Germany and U.S.A.), it occurs much more abundantly as carbon compounds in coal, oil and natural gas, in limestone and other carbonates and in living things.

The properties and uses of diamond and graphite were discussed in section 9.6. The high refractive index and dispersive power of diamonds led to their use as jewellery in the very earliest civilizations. The modern industrial uses of diamond almost all result from its hardness—drilling, cutting, grinding and for bearings in precision instruments such as watches.

**Table 17.1** The elements in Group IV.

| Group IV |
|---|
| C |
| Si |
| Ge |
| Sn |
| Pb |

*Above* A model of the structure of diamond.

*Right* A rough diamond being cut to form a gem. The cast iron surface below the diamond, which rotates 2500 turns per minute is coated with a paste of diamond powder in olive oil.

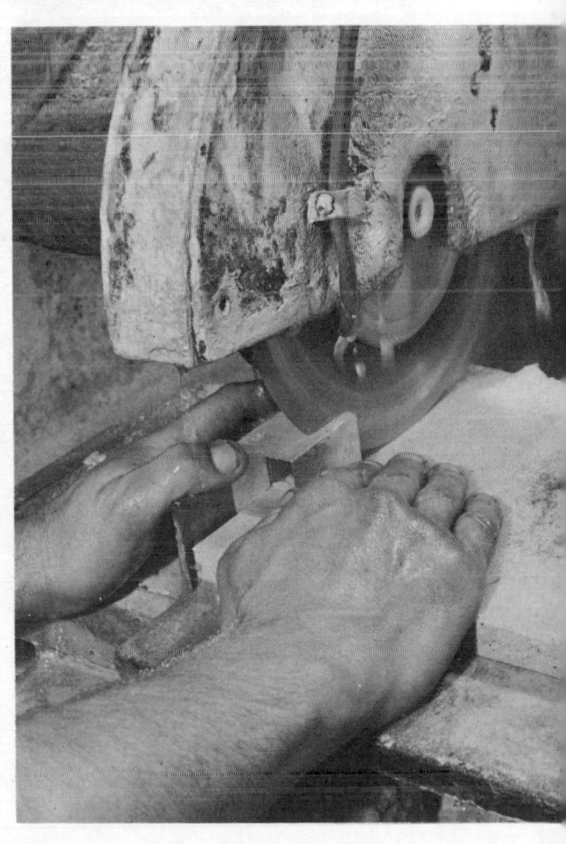

*Below* A diamond impregnated saw being used to cut glass.

The uses of graphite are related to its conducting and lubricating qualities. It is used in 'lead' pencils, as a high-temperature lubricant, as inert electrodes in various electrolytic processes and as a moderator for slowing down neutrons in nuclear reactors.

*Above* Small synthetic diamonds produced by the General Electric Company, U.S.A., in 1971. These were the first gem-quality diamonds ever created. No price could be put on their value if research and other intangible costs are included.

*Above* A model of the structure of graphite.

*Right* A carbon filament lamp.

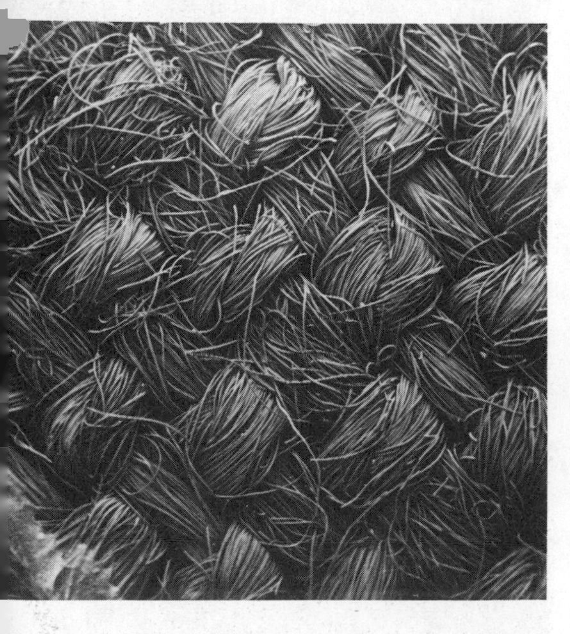

*Above* Carbon fibre cloth magnified 20 times.

*Right* The victorious Oxford crew of 1977 in their carbon-fibre-reinforced boat.

After oxygen, silicon is the most abundant element on the earth. It occurs as silicon(IV) oxide ($SiO_2$) in sand and sandstone and as many forms of silicate in rocks and clays. In comparison with carbon and silicon, germanium, tin and lead are rare elements. Traces of germanium are present in coal and accumulate in flue dust as germanium(IV) oxide, $GeO_2$. Silicon and germanium are used as semiconductors in transistorized electronic equipment. These two elements are obtained from their oxides, $SiO_2$ and $GeO_2$, by very similar processes.

The flow scheme in figure 17.1 shows how pure silicon is obtained industrially.

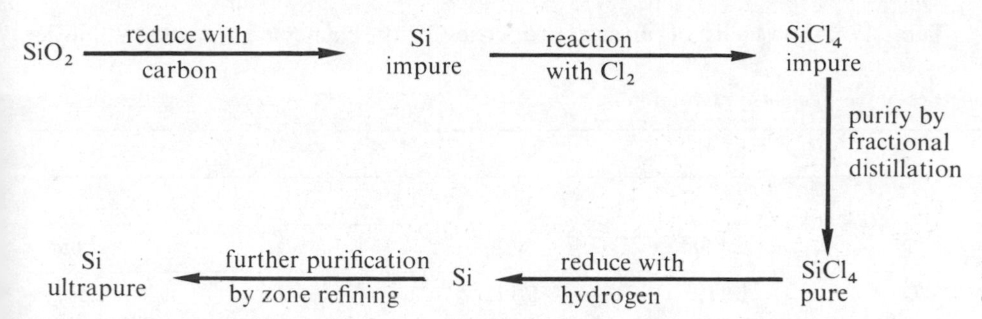


**Figure 17.1** A flow scheme showing the industrial manufacture of pure silicon.

The final zone refining process allows the production of ultra-pure silicon or germanium which are necessary for use in transistors. The material to be purified is packed in a cylindrical tube and suspended vertically (figure 17.2). At the top of the tube, a short length of the cylinder is surrounded by an electrical heating coil which melts a narrow band of material within the tube. During the zone refining operation, the tube is raised through the heating element and the zone of molten material moves slowly down the tube. As the sample moves away from the heating element, it begins to recrystallize, leaving impurities in the molten zone. In this way, the impurities collect in the molten phase and end up concentrated at one end of the tube.

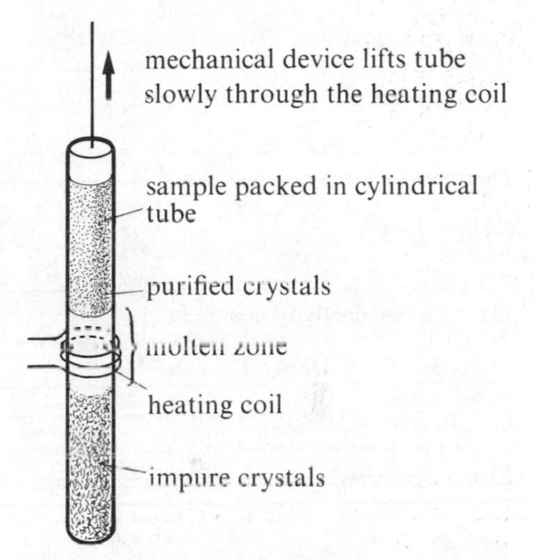


**Figure 17.2** The production of ultrapure silicon by zone refining.

○ Write equations for
  (i) the reduction of $SiO_2$ with carbon,
  (ii) the reaction of $Cl_2$ with Si.
○ Why is the impure silicon converted to $SiCl_4$?
○ Why can the impure silicon not be zone-refined?

Tin occurs as tin(IV) oxide ($SnO_2$) in cassiterite (tinstone) and lead occurs as the sulphide ore, galena (PbS). Lead is obtained by roasting the sulphide in air to obtain PbO, which is then reduced with carbon.

$$2PbS + 3O_2 \longrightarrow 2PbO + 2SO_2$$

$$PbO + C \longrightarrow Pb + CO$$

Tin can be obtained by direct reduction of $SnO_2$ with carbon.

$$SnO_2 + 2C \longrightarrow Sn + 2CO$$

Tin and lead have important uses as relatively inert metals. Tin is used to tin-plate steel which is then used to make the so called 'tins' for canning meats, soup, fruit, etc. Lead has been used as an inert material for gas and water pipes, for cable sheathing and for chemical vessels. Lead is also used for the plates of lead-acid accumulators (batteries) and as a screen from radioactivity.

Another important use of both tin and lead is in alloying. Indeed, there are several important alloys containing both of these metals, for example:
solder (50% Sn, 50% Pb),
pewter (80% Sn, 20% Pb) and
type metal (10% Sn, 75% Pb, 15% Sb).

*Above* The sixteenth-century leaded windows in St. Tysilio Church, Bryn Eglwys, Wales. Lead is an ideal material for holding the different-coloured pieces of glass in place on account of its malleability and inertness.

## 17.2 Variation in the physical properties of the elements

Table 17.2 shows a list of physical properties for the elements in Group IV. Notice

**Table 17.2** Various atomic and physical properties of the elements in Group IV.

| **Element** | C | Si | Ge | Sn | Pb |
|---|---|---|---|---|---|
| **Atomic number** | 6 | 14 | 32 | 50 | 82 |
| **Electron configuration (outer shell only)** | $2s^2\ 2p^2$ | $3s^2\ 3p^2$ | $4s^2\ 4p^2$ | $5s^2\ 5p^2$ | $6s^2\ 6p^2$ |
| **Atomic radius/nm** | 0.077 | 0.117 | 0.122 | 0.141 | 0.154 |
| **Melting point/°C** | 3730[d] | 1410 | 937 | 232 | 327 |
| **Boiling point/°C** | 4830[d] | 2680 | 2830 | 2270 | 1730 |
| **Density/g cm$^{-3}$** | 2.26[gr.] 3.51[d.] | 2.33 | 5.32 | 7.3 | 11.4 |
| **Thermal conductivity/J cm$^{-1}$ s$^{-1}$ K$^{-1}$** | 0.24[gr.] | 0.84 | 0.59 | 0.63 | 0.35 |
| **Conductivity** | fairly good[gr.] non-cond.[d.] | semi-cond. | semi-cond. | good | good |
| **Electrical conductivity/ohm$^{-1}$ m$^{-1}$** | | $1 \times 10^6$ | $2 \times 10^6$ | $8 \times 10^6$ | $5 \times 10^6$ |
| **Enthalpy of atomization/kJ mole$^{-1}$** | 716[gr] | 456 | 376 | 302 | 195 |
| **First ionization energy/kJ mole$^{-1}$** | 1086 | 787 | 760 | 707 | 715 |
| **Type of structure** | Giant molecular | Giant molecular similar to diamond | Giant molecular similar to diamond | Giant metallic | Giant metallic |

gr. = graphite  d. = diamond

how many of these properties vary more from one element to the next than with Group I or Group VII. These changes in property are, of course, related to the *increasing metallic (electropositive) character and the decreasing non-metallic (electronegative) character as atomic number rises.*

The structural changes from giant molecular lattices in carbon and silicon (see section 9.6) to giant metallic structures in tin and lead provide the key to any explanation of the changes in physical properties. Silicon and germanium crystallize in the same structure as diamond, whilst tin and lead have distorted close-packed metal structures. As the atoms get larger and the atomic radius increases, the interatomic bonding becomes weaker and the attraction of neighbouring nucleii for intervening electrons gets less.

The weaker interatomic forces result in a change in bonding from covalent to metallic down the Group and hence there is a decrease in melting point, boiling point, enthalpy of atomization and first ionization energy. At the same time, the increasing metallic character causes a general increase in density and conductivity.

The first ionization energy decreases considerably from carbon to silicon, but falls relatively little afterwards. The reason for this is that, after silicon, there is a larger increase in nuclear charge (associated with the filling of '*d*' and '*f*' sub-shells) to counterbalance the increase in atomic radius.

## 17.3 Variation in the chemical properties of the elements

Information concerning various chemical properties of the Group IV elements is given in table 17.3. *The group trends further emphasize the increase in metallic character down the group.* Notice how carbon, the first member of the group, is much more electronegative than the remainder.

**Table 17.3** Various atomic and chemical properties of the elements in Group IV.

| Element | C | Si | Ge | Sn | Pb |
|---|---|---|---|---|---|
| Atomic number | 6 | 14 | 32 | 50 | 82 |
| Electron configuration (outer shell only) | $2s^2\ 2p^2$ | $3s^2\ 3p^2$ | $4s^2\ 4p^2$ | $5s^2\ 5p^2$ | $6s^2\ 6p^2$ |
| Electronegativity | 2.5 | 1.8 | 1.8 | 1.8 | 1.8 |
| Electrode potential $M^{2+}(aq)/M(s)$/volts | | | 0.23 | −0.14 | −0.13 |
| $\Delta H_f^\ominus(XO_2)$/kJ mole$^{-1}$ | −394 | −910 | −551 | −581 | −277 |
| $\Delta H_f^\ominus(XO)$/kJ mole$^{-1}$ | −111 | | −212 | −286 | −217 |
| $\Delta H_f^\ominus(XH_4)$/kJ mole$^{-1}$ | −75 | +34 | +90 | +163 | |
| Bond energy, $E$(X—H)/kJ mole$^{-1}$ | 435 | 318 | 285 | 251 | |
| $\Delta H_f^\ominus(XCl_4)$/kJ mole$^{-1}$ | −136 | −640 | −544 | −511 | −320 |
| Bond energy, $E$(X—Cl)/kJ mole$^{-1}$ | 327 | 402 | 339 | 314 | 235 |

In general, chemical reactivity increases from carbon to lead. Electrode potentials show that only tin and lead are strong enough reducing agents to liberate hydrogen from dilute acids.

$$Pb(s) \longrightarrow Pb^{2+}(aq) + 2e^- \quad E^\ominus = 0.13 \text{ volts}$$

Lead will react very slowly with soft water containing dissolved oxygen to form $Pb(OH)_2$. It is this reaction which has been responsible for certain cases of lead poisoning in areas where householders have drunk hot water directly from lead pipes. The same problem does not arise in hard water areas where the lead piping develops a protective layer of insoluble lead sulphate(VI) or lead carbonate.

As the heats of formation would suggest, all the Group IV elements except lead react with oxygen on heating to form the dioxide.

$$C + O_2 \longrightarrow CO_2$$

$$Sn + O_2 \longrightarrow SnO_2$$

Lead, however, reacts to form PbO.

A study of the enthalpy of formation of the hydrides, $XH_4$, would suggest that only carbon could be expected to react directly with hydrogen. In practice, carbon does not react with hydrogen, even at very high temperatures, because of the large activation energy involved.

All the elements in Group IV react directly on heating with chlorine to form the tetrachloride, except lead which forms the dichloride (see $\Delta H_f^\ominus(XCl_4)$ values in table 17.3). However, the reaction between carbon and chlorine is so slow that $CCl_4$ (tetrachloromethane) is manufactured by the reaction between carbon disulphide and chlorine.

$$CS_2 + 3Cl_2 \longrightarrow CCl_4 + S_2Cl_2$$

## 17.4 General features of the compounds

The most striking feature of the compounds of these elements is the existence of two oxidation states, +2 and +4. The formation of compounds in which the elements show more than one stable oxidation number is typical of the *p*-block elements.

It is interesting to consider the relative stabilities of the +2 and +4 oxidation states for the different elements in Group IV. In carbon and silicon compounds, the +4 state is very stable relative to +2 whereas the +2 state is rare and easily oxidized to +4. Thus, CO reacts very exothermically to form $CO_2$ whilst SiO is too unstable to exist under normal conditions, although it has been obtained at 2 000°C. Germanium forms oxides in both +4 and +2 states. However, $GeO_2$

is rather more stable than GeO; $GeO_2$ does not act as an oxidizing agent whereas GeO is readily converted to $GeO_2$. In tin compounds, the +4 state is only slightly more stable than the +2 state. Thus, aqueous tin(II) ions are mild reducing agents, converting mercury(II) ions to mercury and iodine to iodide,

$$Sn^{2+} + Hg^{2+} \longrightarrow Sn^{4+} + Hg$$

$$Sn^{2+} + I_2 \longrightarrow Sn^{4+} + 2I^-$$

In lead compounds, however, the +2 state is unquestionably more stable and $PbO_2$ is a strong oxidizing agent, whilst PbO is relatively stable. Thus, $PbO_2$ can oxidize hydrochloric acid to chlorine and hydrogen sulphide to sulphur.

$$PbO_2 + 4HCl \longrightarrow PbCl_2 + Cl_2 + 2H_2O$$

*Notice the steady increase in the stability of the lower oxidation state relative to the higher oxidation state on moving down the group from carbon to lead* (figure 17.3).

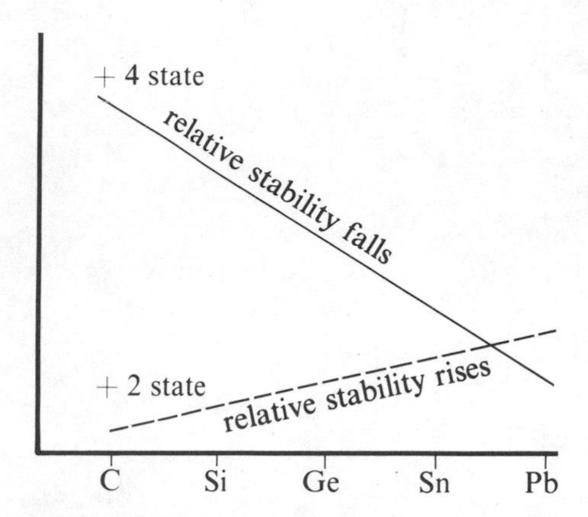


**Figure 17.3** Relative stabilities of the +4 and +2 oxidation states for the elements in Group IV.

The greater stability of the +2 oxidation state with respect to the +4 state as the atomic number rises is nicely illustrated by the standard electrode potentials of the $M^{4+}(aq)/M^{2+}(aq)$ systems for germanium, tin and lead.

$$Ge^{4+} + 2e^- \longrightarrow Ge^{2+} \qquad E^\ominus = -1.6 \text{ volts}$$

$$Sn^{4+} + 2e^- \longrightarrow Sn^{2+} \qquad E^\ominus = +0.15 \text{ volts}$$

$$Pb^{4+} + 2e^- \longrightarrow Pb^{2+} \qquad E^\ominus = +1.8 \text{ volts}$$

As the electrode potentials get more positive from $Ge^{4+}$ to $Pb^{4+}$, the oxidized form is more readily reduced to the +2 state.

All the Group IV elements have four electrons in their outermost shell ($ns^2np^2$) so it is not surprising that they show a well defined oxidation state of +4 (−4 in the hydrides). However, none of the elements forms an $M^{4+}$ cation in its solid compounds due to the high ionization energies involved in removing four successive electrons from an atom. Consequently, the bonding in the tetravalent compounds is predominantly covalent.

Compounds of tin and lead in which the Group IV element has an oxidation number of +2 (e.g. $PbF_2$, $PbCl_2$, PbO) are normally regarded as ionic. In these compounds, the $Sn^{2+}$ and $Pb^{2+}$ ions are believed to form by loss of the two '*p*' electrons in the outer shell, whilst the two '*s*' electrons remain relatively stable and unreactive in their filled sub-shell. This is sometimes referred to as the **'inert pair' effect**.

One of the most curious features regarding Group IV is the unique ability of carbon to form stable compounds containing long chains and rings of carbon atoms. This property, called **catenation**, results in carbon forming an enormous range of compounds. Indeed, there are many thousands of compounds containing only carbon and hydrogen of which methane ($CH_4$), ethane ($C_2H_6$), ethene ($C_2H_4$) and ethyne ($C_2H_2$) are four of the simplest.

**Table 17.4** Average energies for corresponding carbon and silicon bonds.

| Bond | Average bond energy $E$(X—Y)/kJ mole$^{-1}$ |
|---|---|
| C—C | 346 |
| C—O | 360 |
| Si—Si | 226 |
| Si—O | 464 |

This ability of carbon to catenate results from the fact that the C—C bond is almost as strong as the C—O bond (table 17.4). This makes the oxidation of carbon compounds to such products as carbon dioxide and water less energetically favourable than in the case of silicon, where the Si—Si bond is much weaker than the Si—O bond. Consequently, catenated compounds of silicon are energetically unstable with respect to $SiO_2$ and therefore do not occur naturally. Nevertheless, chemists have succeeded in synthesizing a whole series of silicon hydrides, called silanes, with as many as eleven silicon atoms linked together. In a similar fashion, three hydrides have been synthesized for germanium, but tin and lead form only one hydride each, $SnH_4$ and $PbH_4$. The ability of carbon to form long chains is considered further in section 24.1.

- ○ What structure and shape do you predict for the hydrides $CH_4$, $SiH_4$, $GeH_4$, $SnH_4$ and $PbH_4$?
- ○ How will the volatility of these hydrides vary from $CH_4$ to $PbH_4$?
- ○ The hydrides become less stable from $CH_4$ to $PbH_4$. Why is this?
- ○ Suggest a reason why there are no compounds of Si, Ge, Sn and Pb analogous to $C_2H_4$ and $C_2H_2$.

Some of the important properties of the hydrides, chlorides and oxides of the Group IV elements are summarized in tables 17.5, 17.6 and 17.7 respectively.

**Table 17.5** The preparation and properties of the simple hydrides of Group IV elements.

**Preparation**

1. Reaction of magnesium compounds (e.g. magnesium silicide) or magnesium alloy with dilute acid

$$Mg_2Si + 4H^+ \longrightarrow 2Mg^{2+} + SiH_4$$

2. Reduction of tetrachloride with lithium aluminium hydride at 0°C in ether

$$SnCl_4 + LiAlH_4 \longrightarrow SnH_4 + LiCl + AlCl_3$$

**Properties**

| | |
|---|---|
| Structure: | Simple molecular. |
| Molecular shape: | Tetrahedral (figure 17.4). |
| Volatility: | Low melting points and boiling points, all are gases at room temperature. |
| Thermal stability: | The X—H bonds become longer and weaker down the Group (table 17.3). Thus, the hydrides become less stable. $CH_4$, $SiH_4$ and $GeH_4$ are stable up to high temperatures, $SnH_4$ is unstable and decomposes slowly at room temperature whilst $PbH_4$ is so unstable that it cannot be isolated at room temperature. |
| Reducing powers: | As their ease of decomposition into the element + hydrogen increases, they become stronger reducing agents. Thermodynamically, the hydrides should all react with oxygen since their enthalpies of combustion are all negative. However, $CH_4$ is kinetically stable due to the high activation energy for its reaction with oxygen. |

**Figure 17.4** The molecular shape of the simple hydrides of Group IV elements.

**Table 17.6** The preparation and properties of the tetrachlorides of Group IV elements.

**Preparation**

1. Direct synthesis for $SiCl_4$, $GeCl_4$ and $SnCl_4$

$$Si + 2Cl_2 \longrightarrow SiCl_4$$

2. $CCl_4$ is prepared more easily by the action of $Cl_2$ on $CS_2$

$$CS_2 + 3Cl_2 \longrightarrow CCl_4 + S_2Cl_2$$

3. $PbCl_4$ is prepared by the reaction of cold conc. HCl with $PbO_2$.

**Properties**

| | |
|---|---|
| Structure: | Simple molecular. |
| Molecular shape: | Tetrahedral (similar to $XH_4$ hydrides). |
| Volatility: | Low melting points and boiling points. All are volatile liquids at room temperature. |
| Thermal stability: | As the X—Cl bonds become longer and weaker down the group, the tetrachlorides become less stable. Thus, $CCl_4$, $SiCl_4$ and $GeCl_4$ are stable even at high temperatures, $SnCl_4$ decomposes on heating to form $SnCl_2 + Cl_2$, whereas $PbCl_4$ decomposes readily to form $PbCl_2 + Cl_2$. |
| Hydrolysis: | All the chlorides (except $CCl_4$) are readily hydrolysed to form hydroxy-compounds + HCl $$SnCl_4 + 4H_2O \longrightarrow Sn(OH)_4 + 4HCl$$ |

Look closely for the following points in these tables:

1. The simple molecular structures of the hydrides and the tetrachlorides.
2. The decreasing thermal stability from C to Pb of the hydrides, tetrachlorides and dioxides.
3. The change in the nature of the dioxides down the group from acidic to amphoteric.
4. The change in the nature of the monoxides down the group from neutral to amphoteric.

**Table 17.7** The preparation and properties of the oxides of the Group IV elements.

**The dioxides** Group IV element in the +4 oxidation state

| **Oxide** | $CO_2$ | $SiO_2$ | $GeO_2$ | $SnO_2$ | $PbO_2$ |
|---|---|---|---|---|---|
| **Preparation** | Direct combination of element and oxygen | | | | Electrolytic oxidation of $Pb^{2+}$ in acid solution |
| **Properties** | | | | | |
| Boiling point/°C: | −78 | 2590 | 1200 | 1900 | decomposes on heating |
| Structure: | Simple molecular | Giant molecular | Intermediate between giant molecular and ionic | | |
| Nature: | Acidic<br>react with aq. alkalis giving $XO_3^{2-}$ salts<br>$CO_2 + 2OH^- \longrightarrow CO_3^{2-} + H_2O$<br>carbonate<br>$SiO_2 + 2OH^- \longrightarrow SiO_3^{2-} + H_2O$<br>silicate | | Amphoteric<br>(i) react with fused alkalis giving $XO_3^{2-}$ salts<br>$SnO_2 + 2OH^- \longrightarrow SnO_3^{2-} + H_2O$<br>stannate(IV)<br>$PbO_2 + 2OH^- \longrightarrow PbO_3^{2-} + H_2O$<br>plumbate(IV)<br>(ii) react with conc. acids forming +4 salts<br>$SnO_2 + 4H^+ \longrightarrow Sn^{4+}(aq) + 2H_2O$<br>$PbO_2 + 4HCl \longrightarrow PbCl_4 + 2H_2O$<br>conc. | | |
| Thermal stability: | stable even at high temps. | | | | Decomposes to PbO on warming<br>$PbO_2 \longrightarrow PbO + \frac{1}{2}O_2$ |

**The monoxides** Group IV elements in the +2 oxidation state.

| **Oxide** | CO | SiO | GeO | SnO | PbO |
|---|---|---|---|---|---|
| **Preparation** | Reduction of $CO_2$ with carbon<br>$CO_2 + C \rightarrow CO_2$ | Only exists at very high temps. | Reduction of $GeO_2$ with Ge | Heat appropriate hydroxide or nitrate(V)<br>$Pb(OH)_2 \longrightarrow PbO + H_2O$<br>$Pb(NO_3)_2 \longrightarrow PbO + 2NO_2 + \frac{1}{2}O_2$ | |
| **Properties** | | | | | |
| Boiling point/°C: | −191 | | | | 1470 |
| Structure: | Simple molecular | | Predominantly ionic | | |
| Nature: | Neutral oxides—<br>react with neither acids nor alkalis. | | Amphoteric oxides<br>(i) react with acids to form salts<br>$PbO + 2H^+ \longrightarrow Pb^{2+} + H_2O$<br>$SnO + 2H^+ \longrightarrow Sn^{2+} + H_2O$<br>(ii) react with alkalis to form salts<br>$PbO + OH^- + H_2O \longrightarrow Pb(OH)_3^-$<br>trihydroxyplumbate(II)<br>$SnO + OH^- + H_2O \longrightarrow Sn(OH)_3^-$<br>trihydroxystannate(II) | | |
| Thermal stability: | Readily oxidized to dioxide. (SiO, GeO and SnO revert to dioxide on standing in air.) | | | | Stable. |

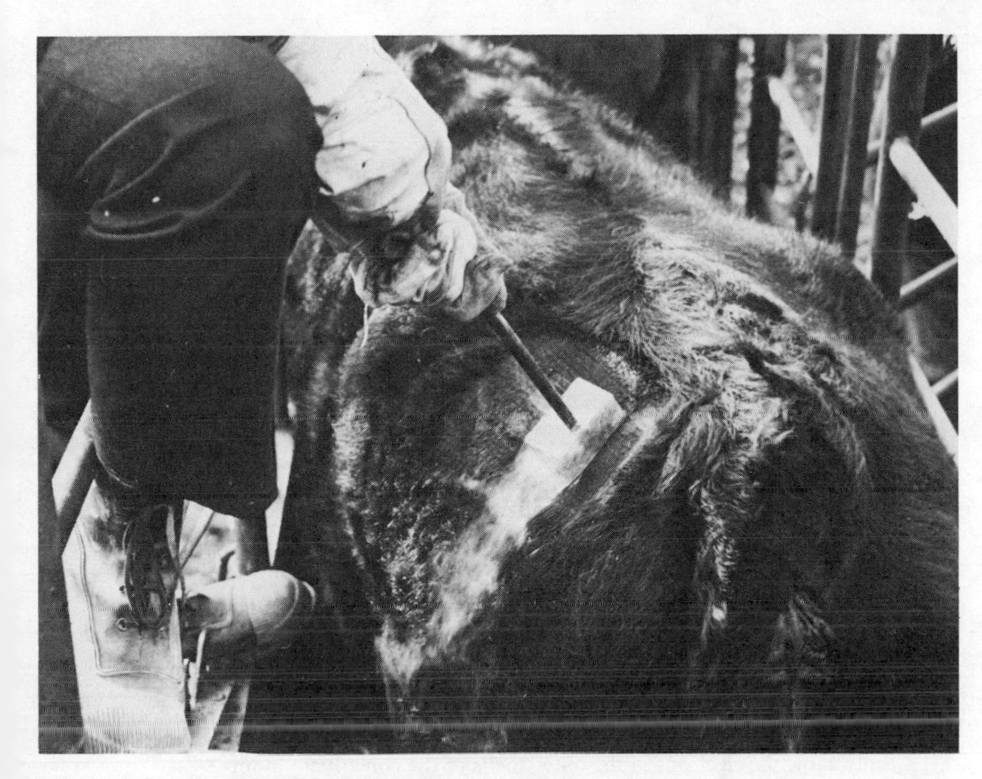

A bullock being freeze-branded using Drikold (solid carbon dioxide). The hide of the animal is marked by extreme cold in a similar way to extreme heat.

## Summary

1 The elements in Group IV emphasize the *differences* between one element and the next in a particular group of the periodic table.

2 Catenation is the ability of one element to form chains or rings in which its atoms are bonded to one another.

3 As we go down the group from carbon to lead:

(a) the elements change from non-metallic to metallic and this change in the elements is reflected in their properties and uses.

(b) the $+2$ oxidation state becomes more stable relative to the $+4$ oxidation state.

(c) the bonding in compounds changes from covalent to predominantly ionic. Covalent compounds become less stable down the group (e.g. hydrides and tetrachlorides) whilst ionic compounds become more stable (e.g. monoxides).

(d) the nature of the dioxides changes from acidic to amphoteric.

(e) the nature of the monoxides changes from neutral to amphoteric.

## Study questions

1 Discuss the following points with respect to the elements in Group IV of the periodic table, writing equations for any reactions which you mention.
  (a) The reactivity of the elements
  (b) The state, structure, and thermal stability of the hydrides
  (c) The state, structure, thermal stability and hydrolysis of the tetrachlorides
  (d) The acidic/basic character of the dioxides
  (e) The relative stabilities of the $+2$ and $+4$ oxidation states

2 (a) What is the principal ore of tin?
  (b) Explain briefly, with an equation, how tin is obtained from its principal ore.
  (c) What are the main uses of tin and its alloys?
  (d) What are the relative advantages and disadvantages of tin plating and galvanizing (zinc plating) iron in order to prevent corrosion?

3 (a) How and under what conditions does lead react with:
   (i) air (oxygen),
   (ii) water (soft and hard),
   (iii) hydrochloric acid,
   (iv) nitric(V) acid?
(b) Describe briefly the preparation of tin(II) oxide and tin(IV) oxide from tin.
(c) How (if at all) and under what conditions do tin(II) oxide and tin(IV) oxide react with:
   (i) oxygen,
   (ii) hydrochloric acid,
   (iii) sodium hydroxide?

4 The following passage describes the preparation of tin(IV) iodide.
Add 4.0g of powdered tin to a solution of 12.7g of iodine in 100cm$^3$ of tetrachloromethane. Reflux the mixture gently until the reaction is complete. Now filter the mixture through a pre-heated funnel and wash the residue with hot tetrachloromethane, adding the washings to the filtrate. Cool the filtrate in ice until orange crystals of tin(IV) iodide form. Filter and dry the crystals.
(a) Write an equation for the formation of tin(IV) iodide in the above preparation.
(b) Give two reasons for using $CCl_4$ as the solvent in this preparation.
(c) Calculate the maximum possible yield of tin(IV) iodide. Explain your calculation.
(d) How would you know when the reaction was complete?
(e) Why was the mixture filtered through a pre-heated funnel?
(f) Why was the residue washed with hot $CCl_4$?
(g) What would you predict for
   (i) the structure,
   (ii) the thermal stability of tin(IV) iodide?

5 (a) Draw electron dot/cross diagrams for carbon monoxide and carbon dioxide. (You need only show the electrons in the outer shells of the constituent atoms.)
(b) 'Carbon monoxide is iso-electronic with nitrogen'. Explain what is meant by this statement.
(c) Use the standard enthalpies of formation of $CO_2$ and CO in table 17.3 to find the enthalpy changes of the following reactions.

$$CO_2(g) + C(s) \longrightarrow 2CO(g)$$

$$CO(g) + \tfrac{1}{2}O_2(g) \longrightarrow CO_2(g)$$

Comment on the relative stabilities of $CO_2$ and CO.

6 (a) State four important similarities in the chemistry of tin and lead or their corresponding compounds.
(b) Explain the following in terms of atomic or electronic properties.
   (i) Tin(IV) compounds are more stable than tin(II) compounds.
   (ii) Lead(II) compounds are more stable than lead(IV) compounds.
   (iii) $PbCl_4$ has a simple molecular structure, whereas $PbCl_2$ is ionic.
(c) Tin and lead have variable oxidation states and form complex ions. Why then are they not classed as transition metals?
(d) Oil paintings containing lead(II) compounds as constituents of their pigments darken over many decades due to the reaction of traces of hydrogen sulphide in the atmosphere with the lead(II) salts.
   (i) Write an equation for the darkening process.
   (ii) Explain why dilute solutions of hydrogen peroxide can be used to restore the oil paintings.

# The Transition Metals 18

## 18.1 Introduction

The elements from scandium (atomic number 21) to zinc (atomic number 30) in the periodic table form what is generally regarded as the first sequence of transition elements. But, what exactly is a transition element and why is it that these elements have similarities to each other across their period, a feature which is not apparent in other parts of the periodic table?

The answer to these and other questions concerning the transition metals lies in their electronic structures shown in table 18.1. Indeed, *virtually all the properties of transition elements are related to their electronic structures and the relative energy levels of the orbitals available for their electrons.*

**Table 18.1** The electron structures of atoms and ions of the elements K to Zn. ((Ar) = electron structure of argon.)

| Element | Symbol | Electronic structure of atom | Common ion | Electronic structure of ion |
|---|---|---|---|---|
| potassium | K | $(Ar)4s^1$ | $K^+$ | (Ar) |
| calcium | Ca | $(Ar)4s^2$ | $Ca^{2+}$ | (Ar) |
| scandium | Sc | $(Ar)3d^14s^2$ | $Sc^{3+}$ | (Ar) |
| titanium | Ti | $(Ar)3d^24s^2$ | $Ti^{4+}$ | (Ar) |
| vanadium | V | $(Ar)3d^34s^2$ | $V^{3+}$ | $(Ar)3d^2$ |
| chromium | Cr | $(Ar)3d^54s^1$ | $Cr^{3+}$ | $(Ar)3d^3$ |
| manganese | Mn | $(Ar)3d^54s^2$ | $Mn^{2+}$ | $(Ar)3d^5$ |
| iron | Fe | $(Ar)3d^64s^2$ | $Fe^{2+}$ | $(Ar)3d^6$ |
| | | | $Fe^{3+}$ | $(Ar)3d^5$ |
| cobalt | Co | $(Ar)3d^74s^2$ | $Co^{2+}$ | $(Ar)3d^7$ |
| nickel | Ni | $(Ar)3d^84s^2$ | $Ni^{2+}$ | $(Ar)3d^8$ |
| copper | Cu | $(Ar)3d^{10}4s^1$ | $Cu^+$ | $(Ar)3d^{10}$ |
| | | | $Cu^{2+}$ | $(Ar)3d^9$ |
| zinc | Zn | $(Ar)3d^{10}4s^2$ | $Zn^{2+}$ | $(Ar)3d^{10}$ |

As the shells of electrons get further and further from the nucleus, successive shells become closer in energy. Thus, the difference in energy between the second and third shells is less than that between the first and second. By the time the fourth shell is reached, there is, in fact, an overlap between the third and fourth shells. In other words, from scandium onwards, the orbitals of highest energy in the third shell (i.e. the 3*d* orbitals) have higher energy than those of lowest energy in the fourth shell (the 4*s* orbital) (figure 18.1).

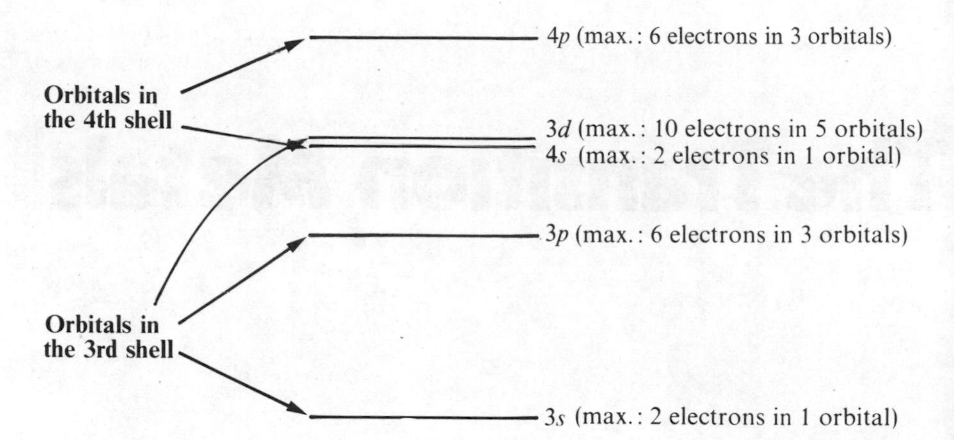


**Figure 18.1** Relative energy levels of the 3*s*, 3*p*, 3*d*, 4*s* and 4*p* orbitals.

The 3*d* sub-shell is 'on average' nearer the nucleus than the 4*s* sub-shell, but at a higher energy level. This means that once the 3*s* and 3*p* sub-shells are filled at argon, subsequent electrons enter the 4*s* sub-shell since it is at a lower energy level than the 3*d* sub-shell. Hence, potassium and calcium have respectively one and two electrons in the 4*s* sub-shell (figure 18.1). Once the 4*s* sub-shell is filled at calcium, electrons enter the 3*d* level. Hence scandium has the electron structure (Ar)$3d^1 4s^2$, titanium has the electron structure (Ar)$3d^2 4s^2$ and so on (figure 18.2).

| **Element** | | **Electronic structure** 3*d* | | | | | 4*s* |
|---|---|---|---|---|---|---|---|
| Scandium | (Ar) | ↑ | | | | | ↑↓ |
| Titanium | (Ar) | ↑ | ↑ | | | | ↑↓ |
| Chromium | (Ar) | ↑ | ↑ | ↑ | ↑ | ↑ | ↑ |
| Iron | (Ar) | ↑↓ | ↑ | ↑ | ↑ | ↑ | ↑↓ |
| Copper | (Ar) | ↑↓ | ↑↓ | ↑↓ | ↑↓ | ↑↓ | ↑ |

**Figure 18.2** The 'electrons-in-boxes' representation of the electronic structures of certain transition metals.

Notice the somewhat unexpected electron structures for chromium and copper. The arrangement of electrons in chromium is (Ar)$3d^5 4s^1$ (figure 18.2) not (Ar)$3d^4 4s^2$ which we might have expected. The explanation of this anomaly is that the electron structure (Ar)$3d^5 4s^1$ with half-filled 3*d* and 4*s* sub-shells has a lower energy level than (Ar)$3d^4 4s^2$. The extra stability of a half-filled sub-shell is thought to result from the occupation of each orbital by one electron and the consequent spreading and equal distribution of charge around an atom.

Copper atoms have an electron structure (Ar)$3d^{10} 4s^1$ (figure 18.2) rather than $3d^9 4s^2$. In this case, it appears that (Ar)$3d^{10} 4s^1$ with a filled 3*d* sub-shell and a half-filled 4*s* sub-shell is more stable than (Ar)$3d^9 4s^2$.

## 18.2 Ions of the transition metals

Look closely at table 18.1.

- ○ Which electrons does a calcium atom lose in forming a $Ca^{2+}$ ion?
- ○ Which electrons would you expect an iron atom to lose in forming an $Fe^{2+}$ ion? Which electrons are lost in practice by the Fe atom?
- ○ Why do you think $Fe^{3+}$ ions are more stable than $Fe^{2+}$ ions?

When transition metals form their ions electrons are lost first from the 4*s* sub-shell rather than the 3*d* sub-shell. Thus, $Fe^{2+}$ ions have the electron structure (Ar)$3d^6$

(figure 18.3) rather than (Ar)$3d^4 4s^2$ and $V^{3+}$ ions have the electron structure (Ar)$3d^2$ not (Ar)$4s^2$. At first sight, this would appear rather strange since, prior to occupation by electrons, the 4*s* level is energetically more stable than the 3*d* level. However, once the 3*d* level, which of course is closer to the nucleus, is occupied by electrons, these repel the 4*s* electrons even further from the nucleus and up to a higher energy level, higher in fact than the 3*d* level now occupied. Consequently, when transition metal atoms form ions, they lose electrons from the 4*s* level before the 3*d* level. This means that *all transition metals will have similar chemical properties* since these will be dictated by the behaviour of the electrons in the outermost shell.

This horizontal similarity contrasts sharply with the marked trends in traversing a period of the *s*- and *p*-block elements, say from sodium to argon. In the latter case, it is the differences between one element and the next which provide the most striking feature.

| **Atom/Ion** | | **Electronic structure** 3*d* | | | | | 4*s* |
|---|---|---|---|---|---|---|---|
| Fe | (Ar) | ↑↓ | ↑ | ↑ | ↑ | ↑ | ↑↓ |
| $Fe^{2+}$ | (Ar) | ↑↓ | ↑ | ↑ | ↑ | ↑ | |
| $Fe^{3+}$ | (Ar) | ↑ | ↑ | ↑ | ↑ | ↑ | |

**Figure 18.3** The 'electrons-in-boxes' representation of the electronic structures of Fe, $Fe^{2+}$ and $Fe^{3+}$.

## 18.3 What is a transition element?

The simplest answer to this question is to say that *transition elements are those in the 'd block' of the periodic table*. The neatness of this definition lies in the fact that it emphasizes the four blocks of elements in the periodic table. This division is certainly useful for the *s*-block elements in Groups I and II which are so alike in their properties, but much less satisfactory for the elements in the *p*-block which include metals such as aluminium, reactive non-metals within Group VII and the noble gases. Furthermore, the simple definition of transition metals as '*d*-block elements' is rather unsatisfactory since it would necessarily lead to the inclusion of scandium and zinc as transition elements. As you may have already realized, scandium and zinc show some fairly obvious differences to the elements in the sequence titanium to copper. They have only one oxidation state in their compounds (scandium 3+, zinc 2+), whereas the others have two or more. Their compounds are usually white, unlike the compounds of transition metals which are usually coloured, and they show little catalytic activity.

Owing to the divergence of scandium and zinc from typical transition metal properties, it would seem sensible to choose a more satisfactory definition for transition metals which excludes these elements, but includes the elements titanium to copper. In order to achieve this, we can define a transition metal as *one which forms at least one ion with a partially filled d sub-shell*. Look closely at table 18.1 and you will see that neither $Sc^{3+}$ nor $Zn^{2+}$ has a partly filled *d* sub-shell.

## 18.4 Trends across the period of transition metals

In building up the elements from sodium to argon, electrons are being added to the outer shell and the nuclear charge is increasing by the addition of protons. Since the added electrons shield each other only weakly from the extra nuclear charge, atomic radii decrease sharply from sodium to argon whilst the electronegativities and ionization energies steadily rise. The increasing number of outer-shell electrons also results in major differences in structure and chemical properties from one element to the next.

In traversing the series of metals from scandium to zinc, however, the nuclear charge is also increasing, but electrons are being added to an *inner d*-subshell. These inner *d*-electrons shield the outer 4*s* electrons from the increasing nuclear charge much more effectively than outer-shell electrons can shield each other and consequently the atomic radii decrease much less rapidly.

Similarly, electronegativities and ionization energies increase from scandium to zinc, but only marginally compared to the increase across period 3 from sodium to argon (table 18.2).

**Table 18.2** Electronegativities, ionization energies and electrode potentials for the elements Sc to Zn.

| | Sc | Ti | V | Cr | Mn | Fe | Co | Ni | Cu | Zn |
|---|---|---|---|---|---|---|---|---|---|---|
| **Metallic (atomic) radius/nm** | 0.16 | 0.15 | 0.14 | 0.13 | 0.14 | 0.13 | 0.13 | 0.13 | 0.13 | 0.13 |
| **Electronegativity** | 1.2 | 1.3 | 1.45 | 1.55 | 1.6 | 1.65 | 1.7 | 1.75 | 1.75 | 1.6 |
| **First ionization energy/kJ mole$^{-1}$** | 630 | 660 | 650 | 650 | 720 | 760 | 760 | 740 | 750 | 910 |
| **Second ionization energy/kJ mole$^{-1}$** | 1240 | 1310 | 1410 | 1590 | 1510 | 1560 | 1640 | 1750 | 1960 | 1700 |
| **Third ionization energy/kJ mole$^{-1}$** | 2390 | 2650 | 2870 | 2990 | 3260 | 2960 | 3230 | 3390 | 3560 | 3800 |
| **Electrode potential for $M^{2+}(aq) + 2e^- \rightarrow M(s)$/V** | | | −1.20 | −0.91 | −1.19 | −0.44 | −0.28 | −0.25 | +0.34 | −0.76 |
| **Electrode potential for $M^{3+}(aq) + 3e^- \rightarrow M(s)$/V** | −2.1 | −1.2 | −0.86 | −0.74 | −0.28 | −0.04 | +0.40 | | | |

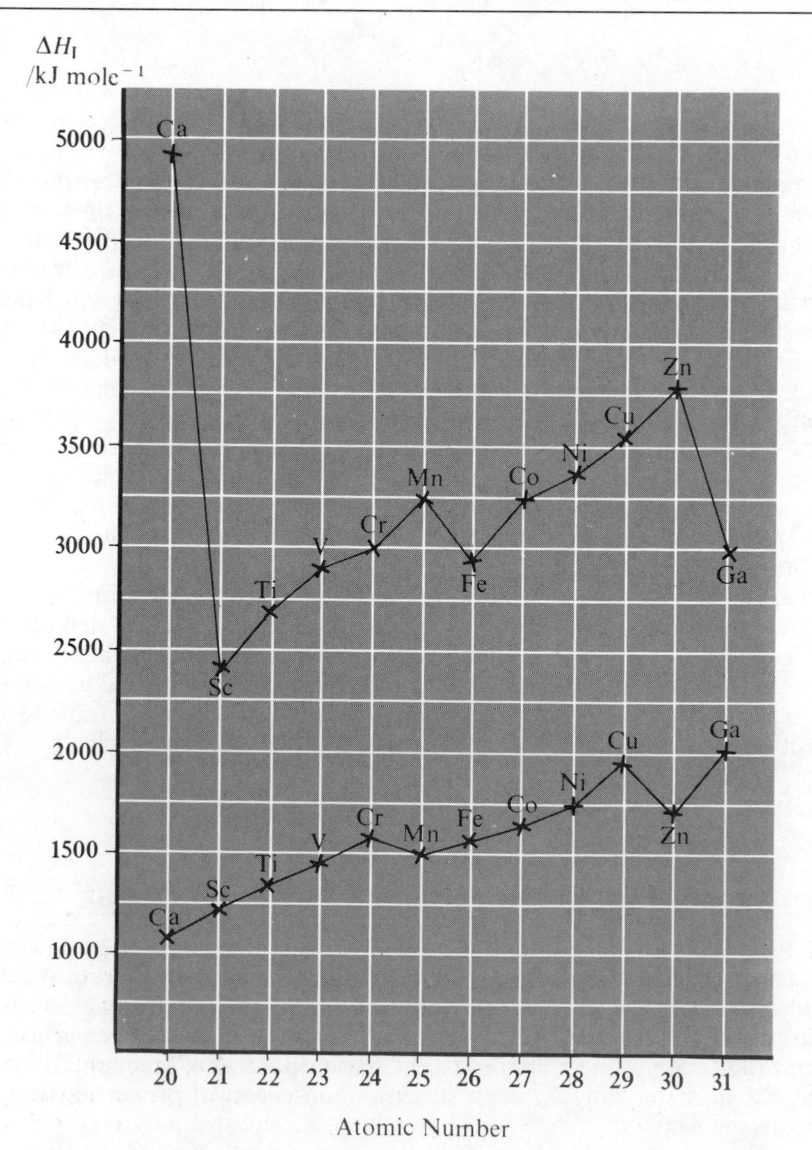


**Figure 18.4** Graphs of the second and third ionization energies of the elements from calcium to gallium.

The increasing electronegativity from scandium to copper means that the elements become slightly less metallic and this is reflected in the increasingly positive electrode potentials (sections 13.3 and 13.4) of their $M^{2+}$ and $M^{3+}$ ions. Look closely at figure 18.4 which shows graphs of the second and third ionization energies of the elements from calcium to gallium.

- ○ Write the electronic structure for the following:
  (a) Cu (b) $Cu^+$ (c) $Cu^{2+}$ (d) Cr (e) $Cr^+$ (f) $Mn^{2+}$ (g) $Zn^{2+}$.
  (Use (Ar) to represent the electron structure of argon as we have done before.)
- ○ By referring to electronic structures explain why:
  (a) the second ionization energies of both Cr and Cu are higher than those of the next element.
  (b) the third ionization energies of both Mn and Zn are higher than those of the next element.

Remember that the second ionization energy refers to the process

$$M^+(g) \longrightarrow M^{2+}(g) + e^-$$

and the third ionization to

$$M^{2+}(g) \longrightarrow M^{3+}(g) + e^-$$

## 18.5 General properties of the first transition series (Sc to Zn)

Most of the transition metals have a close-packed structure in which each atom has twelve nearest neighbours. Furthermore, transition metals have a relatively low atomic radius because the electrons being added to the 3*d* sub-shell as the series is traversed are nearer the nucleus than the electrons in the outermost 4*s* orbital. Consequently, the double effect of close packing and small atomic size results in strong metallic bonds between atoms. Hence, *the transition metals have higher melting points, higher boiling points, higher densities and higher heats of fusion and vaporization than metals such as potassium and calcium in the s-block of the periodic table* (table 18.3). The strong interatomic bonding in transition metals is also reflected in high tensile strengths and good mechanical properties generally.

**Table 18.3** Physical properties of the elements K to Cu.

| | s-block metals | | transition metals | | | | | | | | | |
|---|---|---|---|---|---|---|---|---|---|---|---|---|
| **Element** | K | Ca | Sc | Ti | V | Cr | Mn | Fe | Co | Ni | Cu | Zn |
| **Atomic radius/nm** | 0.24 | 0.20 | 0.16 | 0.15 | 0.14 | 0.13 | 0.14 | 0.13 | 0.13 | 0.13 | 0.13 | 0.13 |
| **Melting point/°C** | 64 | 850 | 1540 | 1680 | 1900 | 1890 | 1240 | 1540 | 1500 | 1450 | 1080 | 420 |
| **Boiling point/°C** | 770 | 1490 | 2730 | 3260 | 3400 | 2480 | 2100 | 3000 | 2900 | 2730 | 2600 | 910 |
| **Density/g cm$^{-3}$** | 0.86 | 1.54 | 3.0 | 4.5 | 6.1 | 7.2 | 7.4 | 7.9 | 8.9 | 8.9 | 8.9 | 7.1 |
| **Ionic radius/nm** | | | | | | | | | | | | |
| $M^+$ | 0.130 | | | | | | | | | | | |
| $M^{2+}$ | | 0.094 | | 0.090 | 0.088 | 0.084 | 0.080 | 0.076 | 0.074 | 0.072 | 0.070 | 0.074 |
| $M^{3+}$ | | | 0.081 | 0.076 | 0.074 | 0.069 | 0.066 | 0.064 | 0.063 | 0.062 | | |

Although the transition metals are less electropositive than the *s*-block metals, their electrode potentials indicate that all of them except copper should react with dilute solutions of strong acids, such as 1M HCl, to produce hydrogen and an aqueous solution of their ions. In practice, however, many of the metals react only slowly with dilute acids owing to protection of the metal from chemical attack by a thin impervious and unreactive layer of oxide. Chromium provides an excellent example of this, for despite its electrode potential, it can be used as a protective, non-oxidizing, non-corroding metal owing to the presence of an unreactive layer of $Cr_2O_3$. This can be compared with the protection of aluminium by a layer of $Al_2O_3$.

The ions of transition metals are smaller than those of the *s*-block metals in the same period (table 18.3). Owing to their smaller ionic radii and also to a larger charge in many of their ions, the charge/radius ratios for transition metals are greater than the values for *s*-block metals. These factors and the polarization of associated anions by small, highly charged cations result in the following properties of transition metal compounds compared to those of the *s*-block metals:
(a) their oxides and hydroxides in oxidation states 2+ and 3+ are less basic and less soluble;
(b) their salts are less ionic and less thermally stable;
(c) their salts and aqueous ions are more hydrated and more readily hydrolysed forming acidic solutions;
(d) their ions are more easily reduced.
It is interesting to compare these features of transition metal ions and their compounds with those of aluminium discussed in chapter 15.

Although the compounds of transition elements in oxidation states +2 and +3 are usually regarded as ionic, polarization of anions by small, highly charged cations is undoubtedly apparent and some of the oxides begin to show acidic features and the compounds begin to show covalent bonding. Thus, $Cr_2O_3$ and $Mn_2O_3$ are amphoteric and $Fe_2Cl_6$ is regarded as a molecular solid. Of course, these features begin to predominate in the higher oxidation states where the oxides become increasingly acidic and the compounds increasingly molecular. Hence, $CrO_3$ and $Mn_2O_7$ are regarded as simple molecular, acidic oxides.

Owing to the relatively small changes in ionic radii from scandium to copper, compounds of the simple hydrated +2 and +3 ions have very similar crystalline structures, hydration and solubility. Thus, all the $M^{3+}$ ions form an alum of the type $K_2SO_4.M_2(SO_4)_3.24H_2O$ and all the $M^{2+}$ ions form isomorphous double sulphates of formula $(NH_4)_2SO_4.MSO_4.6H_2O$.

## 18.6 Characteristic properties of transition metals and their compounds

VARIABLE OXIDATION STATES

Transition metals have electrons of similar energy in both the 3*d* and 4*s* levels. This means that one particular element can form ions of roughly the same stability by losing different numbers of electrons. Thus, all the transition metals from titanium to copper can exhibit two or more oxidation states in their compounds.

The formulae of the common oxides and chlorides of the elements scandium to zinc are shown in figure 18.5. Below these formulae, all the oxidation numbers of the elements are given, the more important oxidation states being emphasized by bold print. Notice that both scandium and zinc have only one common oxide, one common chloride, and one oxidation state in their compounds. Notice also, how closely the oxidation states of the elements in their common oxides and chlorides compare with the more important oxidation states of the elements.

The following generalizations emerge from a study of the oxidation states of transition metals.

(a) *the common oxidation states for each element include* +2 *or* +3 *or both.* +3 states are relatively more common at the beginning of the series whereas +2 states are more common towards the end.
(b) the highest oxidation states up to manganese correspond to the involvement of all the electrons outside the argon core; (4 for Ti, 5 for V, 6 for Cr and 7 for Mn). After this, the increasing nuclear charge binds the *d*-electrons more strongly and so one of the more important oxidation states is that which involves the weakly held electrons in the outer 4*s* shell only (2 for Fe, 2 for Co, 2 for Ni and 1 for Cu).
(c) the transition metals usually exhibit their highest oxidation states in compounds with oxygen or fluorine, two of the most electronegative elements.
(d) Ti, V, Cr and Mn never form simple ions in their highest oxidation state since this would result in ions of extremely high charge density. Hence the compounds of these elements in which they exhibit their highest oxidation state are either covalently bonded (e.g. $TiO_2$, $V_2O_5$, $CrO_3$, $Mn_2O_7$) or contain complex ions (e.g. $VO_3^-$, $CrO_4^{2-}$, $MnO_4^-$).

| | Sc | Ti | V | Cr | Mn | Fe | Co | Ni | Cu | Zn |
|---|---|---|---|---|---|---|---|---|---|---|
| **Common oxides** | $Sc_2O_3$ | $Ti_2O_3$ | $V_2O_3$ | $Cr_2O_3$ | $MnO$ | $FeO$ | $CoO$ | $NiO$ | $Cu_2O$ | $ZnO$ |
| | | $TiO_2$ | $V_2O_5$ | $CrO_3$ | $MnO_2$ | $Fe_2O_3$ | $Co_2O_3$ | | $CuO$ | |
| | | | | | $Mn_2O_7$ | | | | | |
| **Common chlorides** | $ScCl_3$ | $TiCl_3$ | $VCl_3$ | $CrCl_2$ | $MnCl_2$ | $FeCl_2$ | $CoCl_2$ | $NiCl_2$ | $CuCl$ | $ZnCl_2$ |
| | | $TiCl_4$ | | $CrCl_3$ | $MnCl_3$ | $Fe_2Cl_6$ | | | $CuCl_2$ | |

| **Oxidation numbers that occur in compounds** (Common oxidation numbers are in bold print) | Sc | Ti | V | Cr | Mn | Fe | Co | Ni | Cu | Zn |
|---|---|---|---|---|---|---|---|---|---|---|
| | | | | | **7** | | | | | |
| | | | | **6** | 6 | 6 | | | | |
| | | | **5** | 5 | 5 | 5 | 5 | | | |
| | | **4** | 4 | 4 | **4** | 4 | 4 | 4 | | |
| | **3** | **3** | **3** | **3** | 3 | **3** | **3** | 3 | 3 | |
| | | 2 | 2 | 2 | **2** | **2** | **2** | **2** | **2** | **2** |
| | | 1 | 1 | 1 | 1 | 1 | 1 | 1 | **1** | |

**Figure 18.5** Oxidation states of the elements Sc to Zn.

One of the most beautiful and effective demonstrations of the range of oxidation states of a transition metal can be shown by shaking a solution of vanadium(V) with zinc and dilute acid. The solution of vanadium(V) can be made by dissolving about 3g of ammonium vanadate(V) ($NH_4VO_3$) in 40cm³ of 2M NaOH and then adding 80cm³ of M $H_2SO_4$. This solution is a yellow colour owing to the presence of dioxovanadium(V) ions, $VO_2^+$, in acid solution.

$$VO_3^-(aq) + 2H^+(aq) \longrightarrow VO_2^+(aq) + H_2O(l)$$

When the yellow solution is shaken with granulated zinc or zinc amalgam, it changes gradually through green to blue oxovanadium(IV) ions, $VO^{2+}(aq)$, then to green vanadium(III) ions, $V^{3+}(aq)$, and finally to violet vanadium(II) ions, $V^{2+}(aq)$ (figure 18.6).

| **Oxidation state** | +5 | +4 | +3 | +2 |
|---|---|---|---|---|
| **Colour in aqueous solution** | yellow | blue | green | violet |
| **Ion** | $VO_2^+$ | $VO^{2+}$ | $V^{3+}$ | $V^{2+}$ |
| **Name** | dioxovanadium(V) ion | oxovanadium(IV) ion | vanadium(III) ion | vanadium(II) ion |

**Figure 18.6** The oxidation states of vanadium.

All the transition metals from titanium to copper exhibit oxidation numbers of +3 and +2 in their compounds. But what are the relative stabilities of the +3 and +2 states for the different elements? Why, for example, is manganese more stable in the +2 state than the +3 state, whilst the reverse is true for iron? Look closely at figure 18.7 which shows the electrode potentials of the $M^{3+}(aq)/M^{2+}(aq)$ systems for the elements from titanium to cobalt. There is a fairly steady rise in $E^\ominus$ values across the series interrupted by what appears to be an abnormally high value for manganese and an unusually low value for iron.

The more positive the value for $E^\ominus$, the more likely is the aqueous $M^{3+}$ ion to become reduced to $M^{2+}$.

$$M^{3+}(aq) + e^- \longrightarrow M^{2+}(aq)$$

Thus, low values of $E^\ominus$ for Ti, V, Cr and Fe indicate that the 3+ oxidation state is relatively more stable in these elements than in Mn and Co which have higher values of $E^\ominus$.

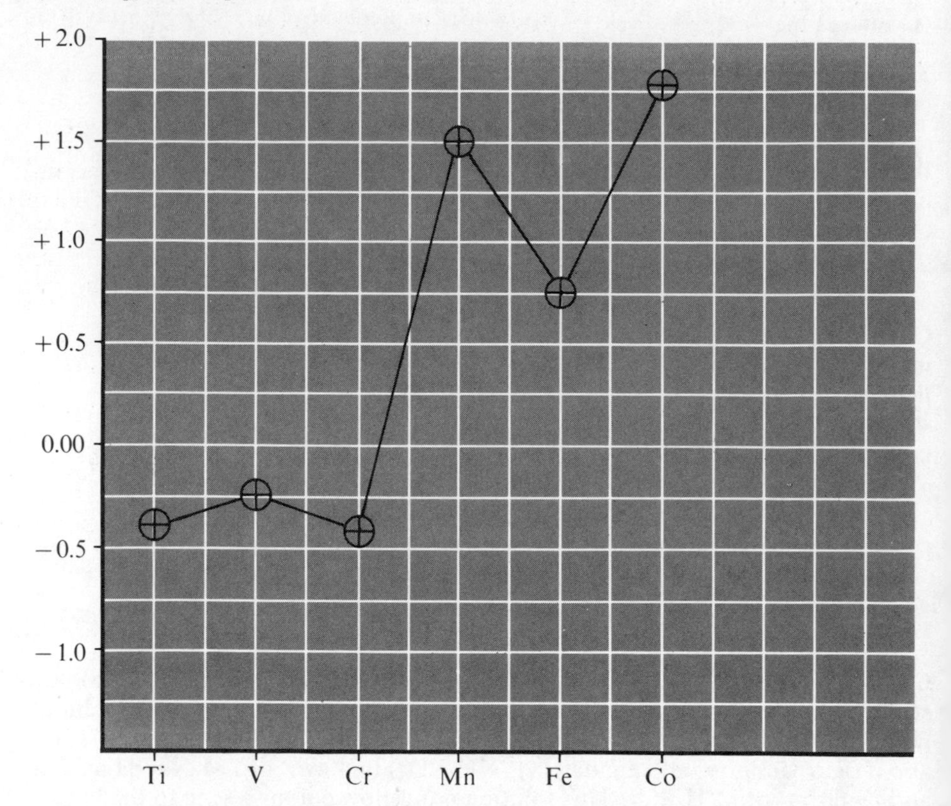


**Figure 18.7** Redox potentials of the $M^{3+}/M^{2+}$ systems for transition metals.

The relative stabilities of the +3 and +2 states in manganese and iron can be interpreted, to some extent, using an 'electrons-in-boxes' representation of their electronic structures (figure 18.8).

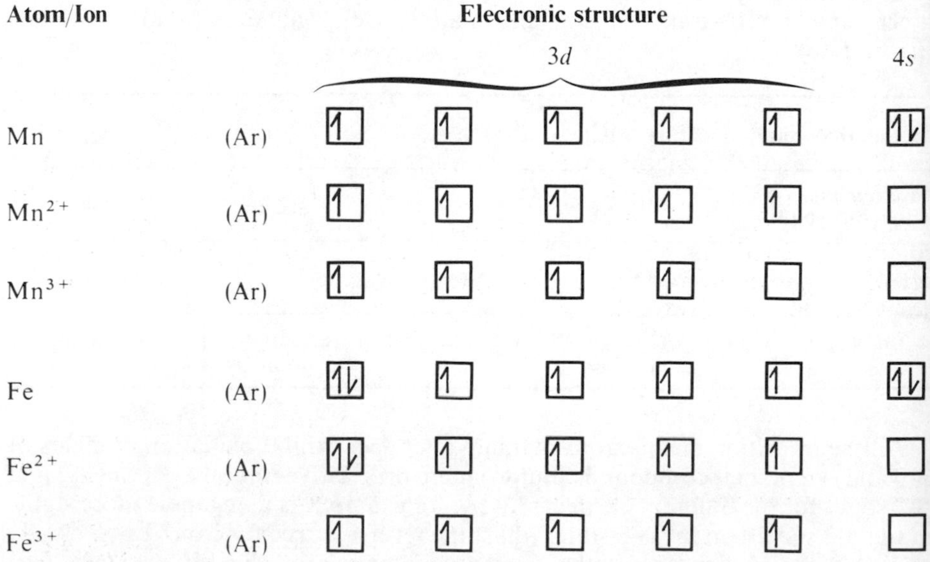

| **Atom/Ion** | | **Electronic structure** 3*d* | | | | | 4*s* |
|---|---|---|---|---|---|---|---|
| Mn | (Ar) | ↑ | ↑ | ↑ | ↑ | ↑ | ↑↓ |
| $Mn^{2+}$ | (Ar) | ↑ | ↑ | ↑ | ↑ | ↑ | |
| $Mn^{3+}$ | (Ar) | ↑ | ↑ | ↑ | ↑ | | |
| Fe | (Ar) | ↑↓ | ↑ | ↑ | ↑ | ↑ | ↑↓ |
| $Fe^{2+}$ | (Ar) | ↑↓ | ↑ | ↑ | ↑ | ↑ | |
| $Fe^{3+}$ | (Ar) | ↑ | ↑ | ↑ | ↑ | ↑ | |

**Figure 18.8** Electronic structures of manganese and iron and some of their respective ions.

$Mn^{2+}$ and $Fe^{3+}$ each have half-filled 3*d* orbitals which makes them more stable than $Mn^{3+}$ and $Fe^{2+}$ respectively. Hence, in manganese the +2 state is more stable than +3, whereas in iron the +3 state is more stable than +2.

- ○ Write the electronic structures of Cu, $Cu^{+}$ and $Cu^{2+}$ using the 'electrons-in-boxes' notation.
- ○ In which oxidation state would you expect copper to be more stable, + or +2?

○ From experience you will know that copper compounds usually exist in the +2 state. What factors might increase the relative stability of the +2 state over the +1 state for copper? We shall return to this anomalous stability of copper(II) over copper(I) in the next chapter.

FORMATION OF COMPLEX IONS

*(a) Investigating the stoichiometry of complex ions by colorimetry*

In chapter 13, we saw how the stoichiometry of a complex ion could be determined by titration. We now turn to another useful method of determining the formula of a complex—colorimetry. Of course, this method can only be used when the colour of the complex ion is quite different from the colours of the separate aqueous ions.

When $Fe^{3+}$ ions react with thiocyanate ions ($NCS^-$), a deep red colour is produced. We can write an equation for this reaction as

$$Fe^{3+}(aq) + xNCS^-(aq) \longrightarrow [Fe(NCS)_x]^{3-x}(aq)$$

In order to determine the stoichiometry of the complex, we must find the value of $x$. This can be done using the method of continuous variation. Mixtures of $Fe^{3+}$ and $NCS^-$ are made up containing $Fe^{3+}$:$NCS^-$ in molar proportions of 10:0, 9:1, 8:2, 7:3, 6:4, etc. In other words, the relative proportions of $Fe^{3+}$:$NCS^-$ are continuously varied from 10:0 to 0:10. Table 18.4 shows a typical set of results for the $Fe^{3+}/NCS^-$ investigation.

**Table 18.4** Results of the investigation of the stoichiometry of the complex formed between $Fe^{3+}$ and $NCS^-$.

| **Tube number** | 1 | 2 | 3 | 4 | 5 | 6 | 7 | 8 | 9 | 10 | 11 |
|---|---|---|---|---|---|---|---|---|---|---|---|
| **Volume of** $5 \times 10^{-3}$ M $Fe(NO_3)_3$/cm$^3$ | 10 | 9 | 8 | 7 | 6 | 5 | 4 | 3 | 2 | 1 | 0 |
| **Volume of** $5 \times 10^{-3}$ M KNCS/cm$^3$ | 0 | 1 | 2 | 3 | 4 | 5 | 6 | 7 | 8 | 9 | 10 |
| **Colorimeter reading** $\propto$ **Concentration of complex** | 8 | 20 | 32 | 40 | 48 | 52 | 45 | 36 | 24 | 16 | 0 |

The red colour in tube 3 is darker than that in tube 2 and that in tube 4 is even darker. In tube 2, there is a considerable excess of $Fe^{3+}$, in tube 3 a smaller excess of $Fe^{3+}$ and so on. At the other end of the series of tubes, the red colour fades again because $NCS^-$ is now in excess. Clearly, the tube containing the darkest coloured solution corresponds to molar proportions of $Fe^{3+}$ and $NCS^-$ which just react completely leaving neither in excess. In some cases, it might be possible to tell by eye which tube contains the darkest coloured solution, but the judgement can be done much more reliably and accurately using a colorimeter.

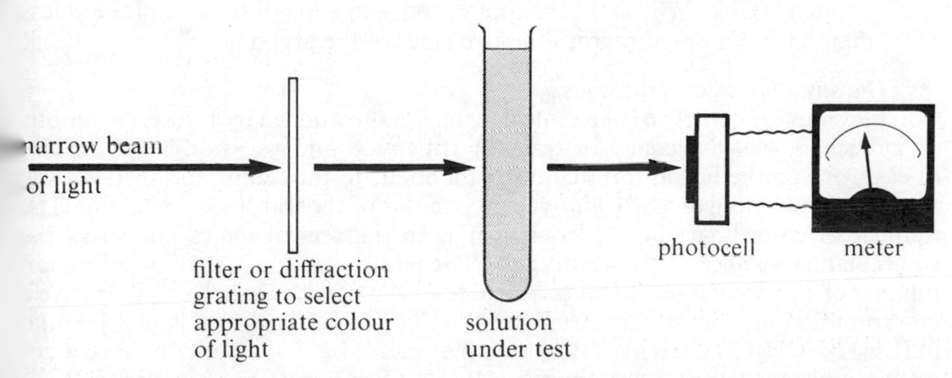


**Figure 18.9** A simplified diagram of a colorimeter.

In a colorimeter, a narrow beam of light passes through the solution under test owards a sensitive photocell (figure 18.9). In many colorimeters, it is possible to elect the most appropriate colour and wavelength of light by choosing a particular ilter or adjusting a diffraction grating. The current generated in the photocell is, f course, proportional to the amount of light transmitted by the solution which in turn depends upon the depth of colour of the complex ion. Thus, the current generated in the photocell will be greatest when the light transmitted by the solution is the greatest, i.e. when the complex colour is at its weakest. However, the meter is usually calibrated to show not the fraction of light transmitted but the fraction of light absorbed, since this will be proportional to the concentration of the complex in the test solution.

The results are represented graphically in figure 18.10.

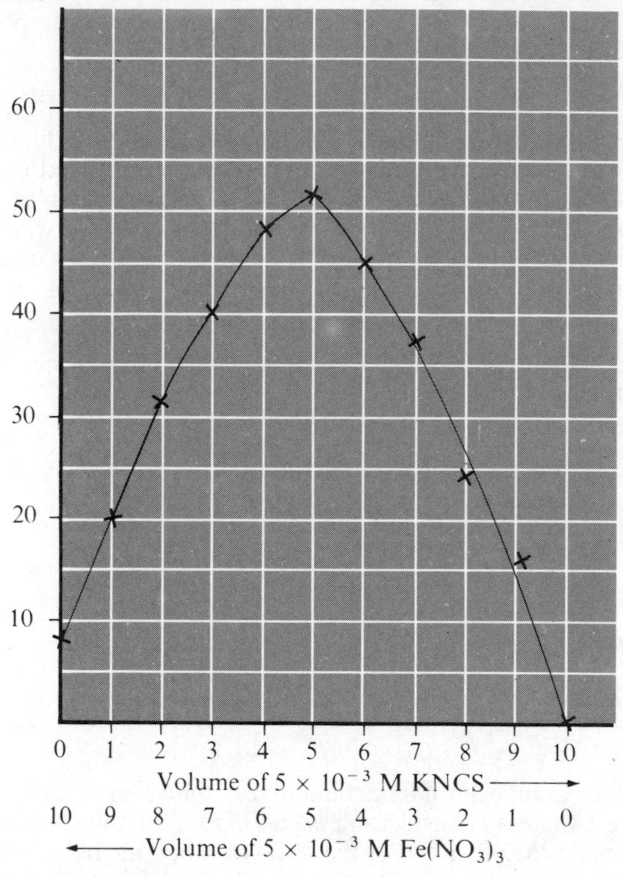


**Figure 18.10** An investigation of the stoichiometry of the complex formed between $Fe^{3+}$ and $NCS^-$.

- ○ In what molar proportions do $Fe^{3+}$ and $NCS^-$ react in forming the red complex ion?
- ○ Write an equation for the formation of the complex.
- ○ Why is the graph in figure 18.10 parabolic at its maximum rather than pointed? (Hint: Why is the maximum concentration of the complex ion less than one might predict from the two sides of the graph?)

*(b) The structure of complex ions*

Complex ions are composed of a central metal ion surrounded by a cluster of anions or molecules, called **ligands**. In transition metal complexes, non-bonded pairs of electrons on the ligand form co-ordinate bonds to the central ion by donating these unshared electron pairs into vacant orbitals of the transition metal ion. The number of co-ordinate bonds from ligands to the central ion is known as the **co-ordination number** of the central ion. Generally speaking, the co-ordination number of a particular ion is the same whatever the ligand. Thus, $Cu^{2+}$ ions have a co-ordination number of 4 in $[Cu(H_2O)_4]^{2+}$, in $[Cu(NH_3)_4]^{2+}$, in $[CuCl_4]^{2-}$ and in $[Cu(H_2NCH_2CH_2NH_2)_2]^{2+}$. On the other hand, $Fe^{3+}$ ions usually have a co-ordination number of 6 which is the case in $[Fe(H_2O)_6]^{3+}$, in $[FeF_6]^{3-}$, in $[Fe(CN)_6]^{3-}$ and in $[Fe(EDTA)]^-$, whilst $Ag^+$ ions normally have a co-ordination number of two ($[Ag(NH_3)_2]^+$, $[Ag(CN)_2]^-$).

In the case of the transition metals, the most common co-ordination number is six, but examples of four and two are not uncommon. In aqueous solution, transition metal ions exist as hydrated complexes with water molecules. Owing to th high charge density on the central metal ion, these hydrated complexes dissociate in a similar fashion to the hydrated $Al^{3+}$ ion (section 15.5).

$$[Fe(H_2O)_6]^{3+}(aq) \rightleftharpoons [Fe(H_2O)_5OH]^{2+}(aq) + H^+(aq)$$

Consequently, the aqueous solutions of most transition metal compounds such as $CuSO_4$(aq), $FeCl_3$(aq) and $Co(NO_3)_2$(aq) are acidic. In oxidation states higher than +3, the polarizing power of the central ion is so great that release of protons and loss of water molecules results in the formation of oxy-anions. For example, neither $[Cr(H_2O)_6]^{6+}$ nor $[Mn(H_2O)_6]^{7+}$ exist in aqueous solution. The loss of two water molecules and eight protons from each of these ions results in the formation of $CrO_4^{2-}$ and $MnO_4^-$ respectively.

$$[Cr(H_2O)_6]^{6+} \longrightarrow 2H_2O + 8H^+ + CrO_4^{2-}$$

$$[Mn(H_2O)_6]^{7+} \longrightarrow 2H_2O + 8H^+ + MnO_4^-$$

In complexes with a **co-ordination number of six**, the ligands usually occupy **octahedral positions** since the six electron pairs donated from the ligands towards the central metal ion in co-ordinate bonds are repelled as far as possible from each other (figure 18.11). In complexes with a **co-ordination number of four**, the ligands usually occupy **tetrahedral positions** although a few four-co-ordinated complexes (such as $[Cu(H_2O)_4]^{2+}$ and $[Cu(NH_3)_4]^{2+}$) have a square-planar structure (figure 18.11).

Complexes with a **co-ordination number of two** usually have a **linear arrangement** of ligands (figure 18.11).

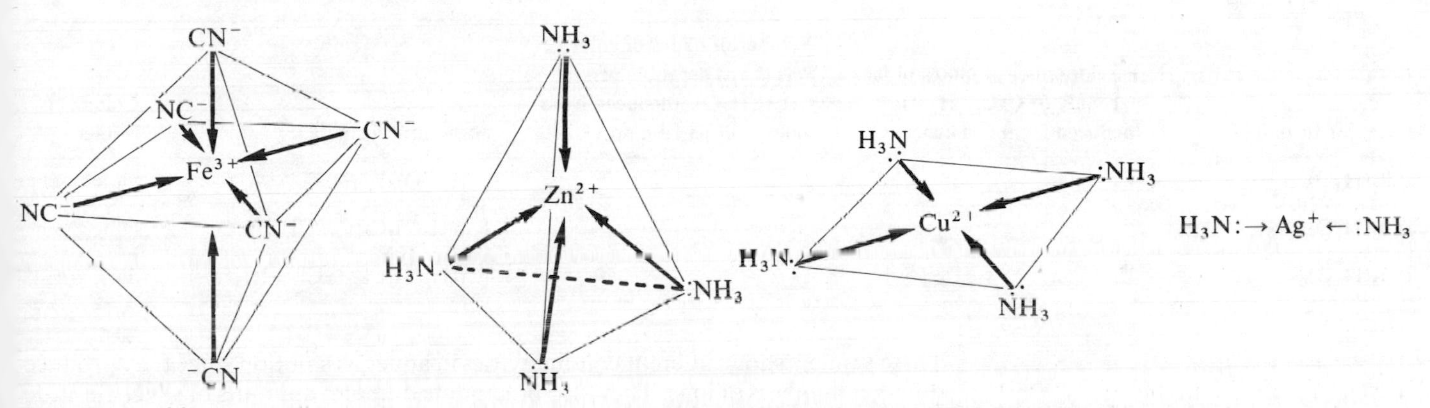


Complexes with a co-ordination number of 6 have an octahedral arrangement of ligands.

Complexes with a co-ordination number of 4 usually have a tetrahedral arrangement of ligands.

A few complexes with a co-ordination number of 4 have a square-planar arrangement of ligands.

Complexes with a co-ordination number of 2 have a linear arrangement of ligands.

**Figure 18.11** The stereochemistry of complex ions.

Owing to the precise stereochemical positions of the ligands in complex ions, isomerism can occur in four-co-ordinated and six-co-ordinated complexes. Isomers are compounds with the same molecular formula, but different arrangements of their constituent atoms. (See section 24.6)

The neutral complex $PtCl_2(NH_3)_2$ has two isomers.

- ○ Does $PtCl_2(NH_3)_2$ have a tetrahedral or a square-planar structure? Explain your answer.
- ○ The isomers of $PtCl_2(NH_3)_2$ are described as *cis* and *trans*. Draw these isomers, indicating which is the *cis* form and which is the *trans* form.

Octahedral complexes of the type $Ma_2b_4$ also exhibit *cis*/*trans* (geometric) isomerism. Figure 18.12 shows the two isomers for $Ma_2b_4$.

One of the most striking examples of this type of isomerism occurs with tetramminodichlorocobaltate(III) chloride ($[Co(NH_3)_4Cl_2]^+Cl^-$). In this compound, the two isomers have different-coloured crystals. The *cis* form is blue-violet in colour, whereas the *trans* form is green.

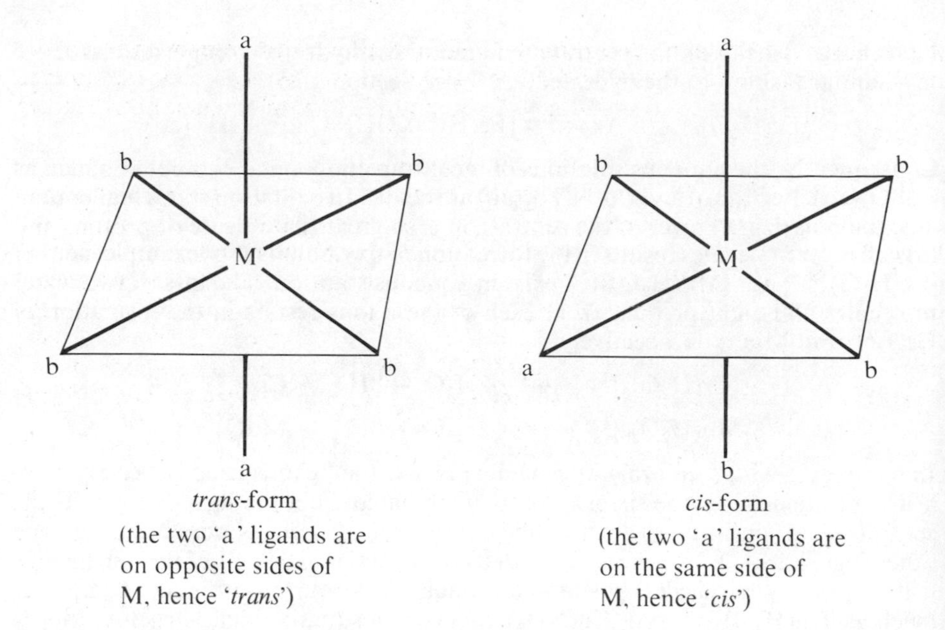


**Figure 18.12** Isomers of complexes of the type $Ma_2b_4$.

Another interesting situation arises with salts of the formula $CrCl_3(H_2O)_6$. Three differently coloured isomers have been isolated with this formula (table 18.5).

**Table 18.5** Isomers of $CrCl_3(H_2O)_6$

| Molecular formula | Total number of moles of ions per mole of $CrCl_3(H_2O)_6$ (deduced from conductivity) | Number of moles of chloride ions ($Cl^-$) per mole of $CrCl_3(H_2O)_6$ (deduced from amount of AgCl pptd.) | Structural formula | Colour of isomer |
|---|---|---|---|---|
| $CrCl_3(H_2O)_6$ | 4 | 3 | $[Cr(H_2O)_6]^{3+}.3Cl^-$ | violet |
| $CrCl_3(H_2O)_6$ | 3 | 2 | $[Cr(H_2O)_5Cl]^{2+}.2Cl^-.H_2O$ | light green |
| $CrCl_3(H_2O)_6$ | 2 | 1 | $[Cr(H_2O)_4Cl_2]^+.Cl^-.2H_2O$ | dark green |

These isomers show different conductivities in aqueous solution since they produce different numbers of ions. They also precipitate different amounts of silver chloride when treated with silver nitrate(v) solution since they contain different numbers of free $Cl^-$ ions.

### COLOURED COMPOUNDS

Most of the compounds of transition elements are coloured. The colour of these compounds can often be related to incompletely filled *d*-orbitals in the transition metal ion.

In general, when light hits a substance, part is absorbed, part is transmitted (provided the substance is transparent) and part may be reflected.

If all the incident radiation is absorbed then the substance looks '*black*'. If all the incident radiation is reflected, then the substance looks '*white*'. On the other hand, if only a very small proportion of the incident white light is absorbed and if all the radiations in the visible region of the spectrum are transmitted equally, then the substance will appear '*colourless*'.

However, many substances absorb preferentially the light photons of one or more regions of the visible spectrum, so that the transmitted or reflected light is relatively richer in the radiations of the remaining regions and the substance has a characteristic colour. For example, if a material absorbs all radiations in the yellow-orange-red region of the spectrum, it will appear blue in white light because only the radiations in the blue region, which are not absorbed, remain to be perceived.

When light energy is absorbed by a substance, an electron in the substance is promoted from an orbital of lower to one of higher energy. The atom or ion absorbing the radiation changes from what is described as its **ground state** (i.e. stable state) to an **excited state**. The different electron transitions involve the absorption of radiation of different frequencies (i.e. different quanta of energy) and if the absorbed frequencies are in the visible region of the spectrum, then the material appears coloured. Why are solutions of transition metal compounds usually coloured, whereas solutions of compounds of non-transition metals are usually colourless?

Consider first a solution of a non-transition metal compound such as aqueous sodium chloride. The sodium ion has the electronic structure $1s^2 2s^2 2p^6$ and the energy difference between the $2p$ orbitals and the next available orbital, $3s$, is very large. To undergo the electron transition $1s^2 2s^2 2p^6$ (ground state) $\rightarrow$ $1s^2 2s^2 2p^5 3s^1$ (excited state) a sodium ion must absorb a photon of high-energy radiation well beyond the range of visible light. Consequently, none of the photons of visible light has sufficient energy to promote electrons in $Na^+$ ions to even the lowest of their possible excited electronic states. Hence, $Na^+$(aq) does not absorb any photons of visible light and so solutions of sodium salts are usually colourless. The $Cl^-$(aq) ion is also colourless for the same reason.

Now let us consider an aqueous solution of titanium(III) chloride which contains the octahedral complex ions, $[Ti(H_2O)_6]^{3+}$, and $Cl^-$(aq) ions. A solution of titanium(III) chloride is violet and an absorption spectrum shows that the solution absorbs most effectively in the green-yellow region of the spectrum (figure 18.13).

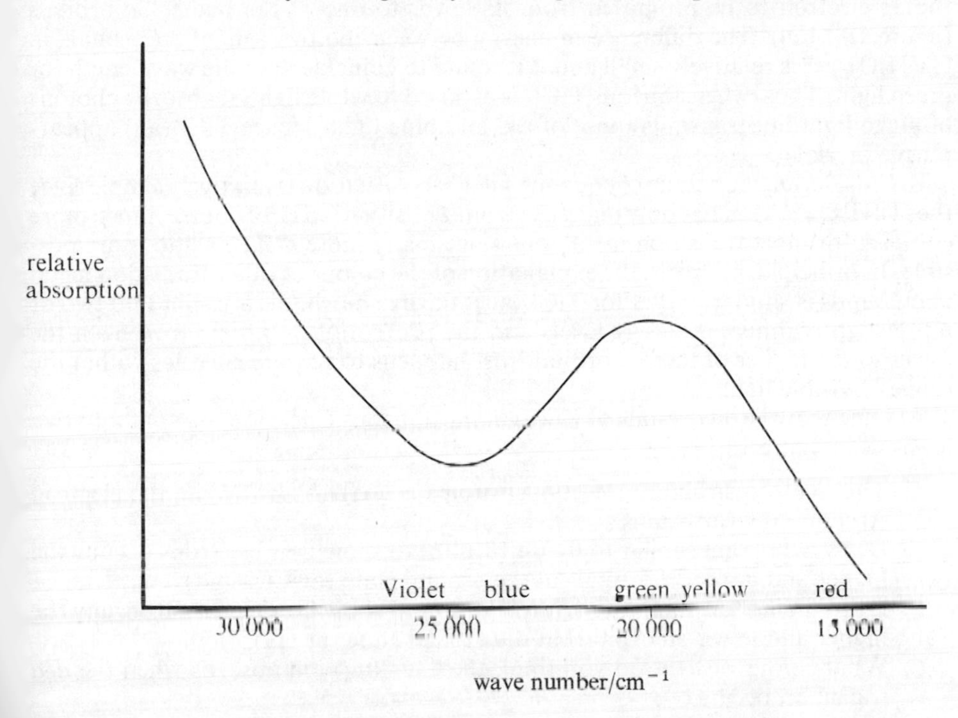


Figure 18.13 The absorption spectrum of $Ti(H_2O)_6^{3+}$.

Red, blue and violet radiations are absorbed less efficiently so the solution looks violet. Since both water and $Cl^-$(aq) are colourless, it must be the $Ti^{3+}$(aq) ion which is responsible for the colour of the solution.

The $Ti^{3+}$ ion has the electron structure (Ar)$3d^1$ and we might expect that, when a solution of titanium(III) chloride absorbs photons of visible light, electrons in the $Ti^{3+}$(aq) ion are promoted from (Ar)$3d^1$ (ground state) to (Ar)$4s^1$ (excited state). However, calculations show that such an excitation of $Ti^{3+}$, like that for $Na^+$, would require photons of radiation well beyond the visible region of the spectrum.

How then do we explain the absorption of green light by $Ti^{3+}$(aq)? Which electronic excitation is responsible for the absorption of green light? The answer lies in electronic transitions *within* the set of five $3d$ orbitals.

All five $3d$ orbitals have exactly the same energy level in the isolated gaseous $Ti^{3+}$ ion (figure 18.14(*a*)). However, when the $Ti^{3+}$ ion is surrounded by ligands, the $3d$ orbitals are no longer symmetrically arranged, those orbitals closer to the ligands are pushed to a slightly higher energy level than those orbitals further away. This splitting of the $3d$ orbitals in the octahedral $[Ti(H_2O)_6]^{3+}$ ion is shown in figure 18.14(*b*).

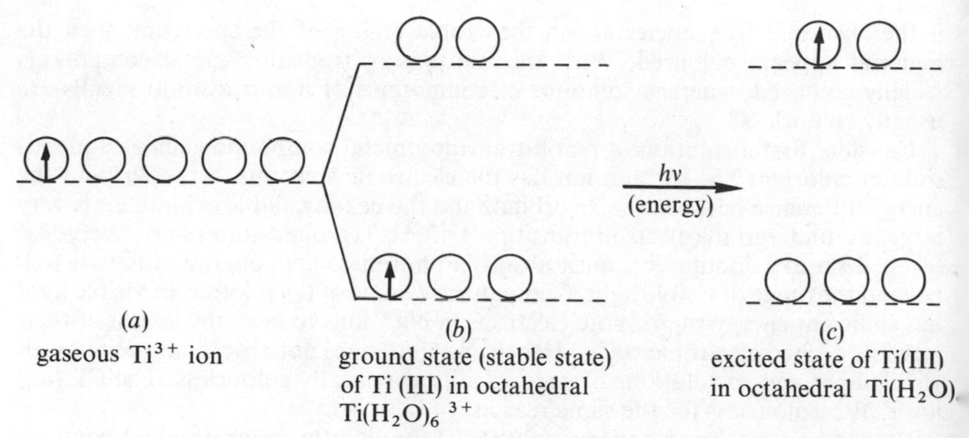


**Figure 18.14** Relative energy levels for the five 3*d* orbitals of the gaseous and hydrated $Ti^{3+}$ ion.

In the ground (stable) state, the single *d* electron will obviously occupy one of the *d* orbitals of slightly lower energy as shown in figure 18.14(*b*). As you will realize, it is now possible for the $Ti^{3+}$ ion in $[Ti(H_2O)_6]^{3+}$ to absorb sufficient energy for the 3*d* electron to be promoted from its lower to one of the higher 3*d* orbitals (figure 18.14(*c*)). The difference in energy between the two sets of 3*d* orbitals in $[Ti(H_2O)_6]^{3+}$ is relatively small and it is found to coincide with the wavelength for green light. Thus, when aqueous $TiCl_3$ is exposed to white light, it absorbs photons of green light but transmits those of red and blue light. Hence $TiCl_3$(aq) appears purple or violet.

In considering the colour of aqueous $TiCl_3$, the situation is relatively simple since the $[Ti(H_2O)_6]^{3+}$ ion has only one *d* electron. The situation is, of course, much more complex for most transition metal ions since many more *d–d* transitions are possible. In principle, however, the explanation of the colour of other transition metal compounds is similar to that for $TiCl_3$(aq): the five *d*-orbitals are split into two or more slightly different energy levels and the promotion of an electron from the lower to the higher of these *d*-orbitals just happens to require energies within the range of visible light.

Assume that the 3*d* orbitals of Cu(II) in the aqueous $Cu^{2+}$ ion are split into two levels as in figure 18.14.

- ○ The electron structure for copper atoms is $(Ar)3d^{10}4s^1$. Write the electron structure for $Cu^{2+}$ ions.
- ○ Draw a diagram similar to figure 18.14(*b*) to show how electrons occupy the higher and lower 3*d* orbitals in the ground state for $Cu^{2+}$(aq).
- ○ Draw a diagram similar to figure 18.14(*c*) to show how electrons occupy the higher and lower 3*d* orbitals in the excited state for $Cu^{2+}$(aq).
- ○ What colour of light do you think the $Cu^{2+}$(aq) ion absorbs when the *d–d* transition occurs?

The colour of a transition metal complex depends mainly on the nature of the central cation and is influenced somewhat less by the co-ordinating ligand. For example the colours of most $Cu^{2+}$ complexes are blue or violet as in the reaction scheme below, but those of $Cu^+$ are colourless.

$$\underset{\text{pale blue}}{[Cu(H_2O)_6]^{2+}} \xrightarrow{NH_3(aq)} \underset{\text{dark blue}}{[Cu(NH_3)_4]^{2+}} \xrightarrow{H_2NCH_2CH_2NH_2(aq)} \underset{\text{violet}}{[Cu(H_2NCH_2CH_2NH_2}$$

Since different oxidation states have different numbers of *d*-electrons, the colours of complexes undergo greater variation when the oxidation state of the transition metal varies. For example, when potassium manganate(VII) is heated with 50% KOH it is reduced through the following colour changes.

| substance present in alkaline soln. | $MnO_4^-$ | $MnO_4^{2-}$ | $MnO_3^-$ | $MnO_2$ | $Mn(OH)_3$ | $Mn(OH)_2$ |
|---|---|---|---|---|---|---|
| oxidation state | 7+ | 6+ | 5+ | 4+ | 3+ | 2+ |
| colour | purple | green | blue | dark brown | green | pink |

CATALYTIC PROPERTIES

Transition metals and their compounds are important **catalysts** in industry and in biological systems. Many of the transition metal ions are required by ourselves and other living things in minute, but definite quantities. Elements required in such small amounts are referred to as '**trace elements**'. Some of these trace elements including copper, manganese, iron, cobalt, nickel and chromium are essential for the effective catalytic activity of various enzymes (section 8.8). One of the most important enzymes containing copper is cytochrome oxidase. This enzyme is involved in the process whereby energy is obtained from the oxidation of food. In the absence of copper, cytochrome oxidase is completely inhibited and the animal or plant is unable to metabolize food effectively.

Numerous transition metals and their compounds are important industrial catalysts. Indeed, a large proportion of industrial catalysts are either transition metals or their compounds. Table 18.6 shows a list of some of the more important examples.

*Above* Virgin desert scrub side by side with reclaimed land in the background. The barren region known as the '90-mile desert' in South Australia was made productive simply by the application of minute quantities of zinc and copper 'trace elements' to the soil which made good mineral deficiencies that had prevented previous settlement of the area.

**Table 18.6** Transition metals and their compounds used as catalysts in industry.

| **Transition element** | **Substance used as catalyst** | **Reaction catalysed** |
|---|---|---|
| Ti | $TiCl_3/Al_2(C_2H_5)_6$ | $nC_2H_4 \longrightarrow \left(-\overset{\vert}{\underset{\vert}{C}}-\overset{\vert}{\underset{\vert}{C}}-\right)_n$ Polymerization of ethene ⟶ poly(ethene) |
| V | $V_2O_5$ or vanadate $(VO_3^-)$ | $2SO_2 + O_2 \longrightarrow 2SO_3$ (Contact process) |
| Fe | Fe or $Fe_2O_3$ | $N_2 + 3H_2 \longrightarrow 2NH_3$ (Haber process) |
| Ni | Ni | $RCH{=}CH_2 + H_2 \longrightarrow RCH_2CH_3$ Hardening of vegetable oils (e.g. manufacture of margarine) |
| Cu | Cu or CuO | $CH_3CH_2OH + \frac{1}{2}O_2 \longrightarrow CH_3CHO + H_2O$ (Oxidation of ethanol to ethanal (acetaldehyde)) |
| Pt | Pt | $2SO_2 + O_2 \longrightarrow 2SO_3$ (Contact process) |
| Pt | Pt | $4NH_3 + 5O_2 \longrightarrow 4NO + 6H_2O$ then $NO \longrightarrow NO_2 \longrightarrow HNO_3$ (Manufacture of nitric(V) acid from ammonia) |

Chemists believe that the catalytic activity of transition metals depends on their ability to exist in various states of oxidation or co-ordination. For example, in the **Contact Process**, vanadium compounds in the +5 state (either $V_2O_5$ or a vanadate) are used to oxidize sulphur dioxide to sulphur(VI) oxide.

$$SO_2 + \tfrac{1}{2}O_2 \xrightarrow{V_2O_5} SO_3$$

It is thought that the actual oxidation process involves two stages. In the first of these, $V^{5+}$ in the presence of oxide ions converts $SO_2$ to $SO_3$ and is simultaneously reduced to $V^{4+}$.

$$2V^{5+} + O^{2-} + SO_2 \longrightarrow 2V^{4+} + SO_3$$

In the second stage, $V^{5+}$ is regenerated from $V^{4+}$ by oxygen.

$$2V^{4+} + \tfrac{1}{2}O_2 \longrightarrow 2V^{5+} + O^{2-}$$

The overall process is, of course, the sum of these two stages.

$$\cancel{2V^{5+}} + \cancel{O^{2-}} + SO_2 \longrightarrow \cancel{2V^{4+}} + SO_3$$
$$\cancel{2V^{4+}} + \tfrac{1}{2}O_2 \longrightarrow \cancel{2V^{5+}} + \cancel{O^{2-}}$$

$$\text{Sum:}\quad SO_2 + \tfrac{1}{2}O_2 \longrightarrow SO_3$$

Transition metals and their compounds can catalyse reactions because they are able to introduce an entirely new reaction mechanism with a lower activation energy (section 23.9) than the uncatalysed reaction. Since the activation energy of the catalysed reaction is lower, the reaction rate is faster. This is shown diagrammatically in figure 18.15.

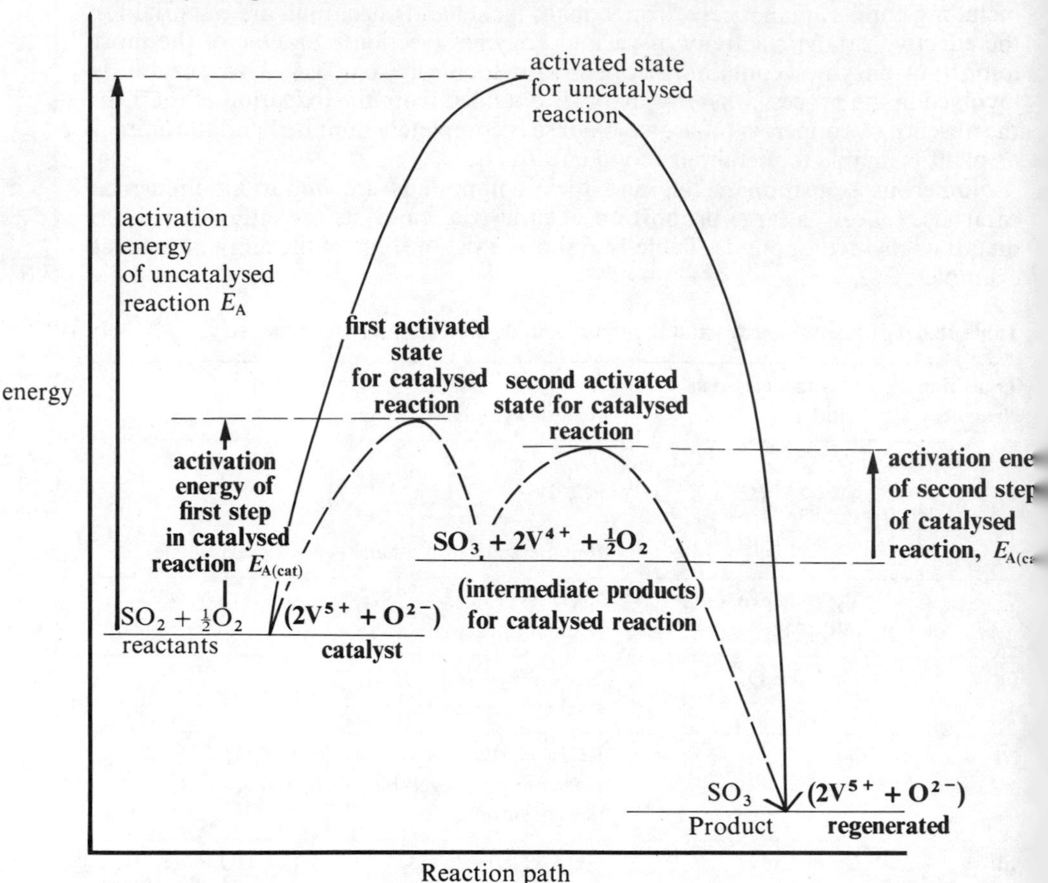


**Figure 18.15** The activation energies of catalysed and uncatalysed reactions.

During the course of this new reaction path, the catalyst undergoes changes in oxidation state but it is regenerated in its original form when the reactants form the products.

An excellent demonstration of the change in oxidation state of a transition metal during catalysis can be shown using cobalt(II) ions with a mixture of 2,3-dihydroxybutanedioate (tartrate) ions and hydrogen peroxide.

When aqueous solutions of 2,3-dihydroxybutanedioate (tartrate) ions and hydrogen peroxide are mixed there is little or no evolution of gas from the decomposition of hydrogen peroxide. If cobalt(II) chloride solution is added, the solution becomes pink and there is still little evolution of gas. After an induction period, the solution begins to darken, going brown and then green due to the presence of cobalt(III). At the same time, effervescence from the solution gradually increases and the maximum evolution of gas occurs when the solution is green. As the reaction subsides, the green colour of the cobalt(III)-complex fades to be replaced by the original pink of cobalt(II) ions.

## Summary

1 A transition metal can be defined as an element which forms at least one ion with a partially-filled *d* sub-shell.
2 Unlike those in other parts of the periodic table, the elements in a transition series have very similar properties to one another.
3 Many of the properties of transition elements can be attributed to their electronic structures and the relative energy levels of the orbitals available for their electrons.

4 Prior to occupation by electrons, the 4*s* level is energetically more stable than the 3*d* level. However, once the 3*d* level is occupied by electrons, the latter repel the 4*s* electrons further from the nucleus to an energy level higher than the 3*d* level now occupied. Thus, transition metals lose electrons from the 4*s* level before the 3*d* level in forming ions.
5 Transition metals have higher melting points, higher boiling points and higher densities than other metals.
6 Transition metals show the following characteristic properties:
 (a) variable oxidation states,
 (b) formation of complex ions,
 (c) coloured compounds,
 (d) catalytic properties.
7 The more important oxidation states for each transition element include +2 or +3 or both.
8 The stoichiometry of complex ions can be determined by titration or colorimetry.
9 Complex ions are composed of a central cation surrounded by a cluster of anions or molecules known as ligands. Non-bonded pairs of electrons on the ligand form co-ordinate bonds to the central metal cation.
10 The number of co-ordinate bonds from ligands to the central cation is known as the co-ordination number of the central ion. Generally speaking, the co-ordination number of a particular ion is the same whatever the ligand.
11 In complexes with a co-ordination number of six, the ligands usually occupy octahedral positions. In complexes with a co-ordination number of four the ligands occupy tetrahedral or square-planar positions. Complexes with a co-ordination number of two usually have a linear arrangement of ligands.
12 Transition metals and their compounds are important catalysts in industry and in biological systems.
13 Many enzymes rely on transition metals for their catalytic activity.

## Study questions

1 (a) By reference to copper or its compounds, give three properties that are characteristic of transition metals.
 (b) Explain the following:
  (i) Aqueous copper(II) sulphate(VI) turns blue litmus paper red.
  (ii) The addition of ammonia solution to aqueous copper(II) sulphate(VI) gives a pale blue precipitate at first and then a deep blue solution when more ammonia solution is added.
  (iii) Copper(I) compounds are sometimes white and rarely show catalytic properties whereas those of copper(II) are usually coloured and can act as catalysts for many reactions.

2 (a) Give the name and formula of:
  (i) one complex cation containing a transition metal,
  (ii) one complex anion containing a transition metal.
 (b) Describe the overall shape (octahedral, tetrahedral, etc.) of the two ions in part (a).
 (c) Outline the electronic structures of the ions that you chose in (a), restricting yourself to a consideration of the outer shell of each atom.

3 (a) How do the electronic structures of transition elements differ from those of the elements in the main groups of the periodic table?
 (b) Describe the electronic structure of either vanadium or chromium and discuss briefly how the more important oxidation states of the element are determined by this electronic structure.
 (c) Give three characteristic features of transition metals or their compounds (other than that of variable oxidation states). Illustrate your answer with reference to vanadium or chromium.

Complex ions can sometimes exhibit isomerism.
 (a) The compound $[Co(NH_3)_5Br]^{2+}SO_4^{2-}$ is isomeric with the compound $[Co(NH_3)_5SO_4]^+Br^-$.
  (i) What ions will these two isomers yield in solution?
  (ii) How would you confirm which isomer was which? (You are required to describe a simple positive test for each isomer.)

(iii) What is (*i*) the oxidation state,
(*ii*) the co-ordination number,
of cobalt in each complex ion?

(iv) Draw the structure of the $[Co(NH_3)_5Br]^{2+}$ ion indicating its shape and the co-ordinate bonds involved.

(b) The compound $NiCl_2(NH_3)_2$ has *cis-trans* isomers which have a complex non-ionic structure.

(i) Does $NiCl_2(NH_3)_2$ have a tetrahedral or a square-planar structure? Explain your answer.

(ii) Draw the *cis* and *trans* isomers for $NiCl_2(NH_3)_2$.

5 Ascorbate oxidase is a metallo-protein enzyme in plants. The enzyme consists of a protein associated with a copper(II) ion.

(i) The enzyme protein alone has no catalytic activity.

(ii) Copper(II) ions alone can act as a catalyst for the reaction, but much less efficiently than the metal-protein combination.

(iii) When egg-albumen is added to aqueous copper(II) ions, the mixture shows greater catalytic activity than copper(II) ions alone, but this activity is not so good as with the specific metal-protein in ascorbate oxidase.

(a) Write the electronic structure of copper(II) ions.

(b) Explain how copper(II) ions might act as catalysts.

(c) What explanation can you offer for observations (i), (ii) and (iii) above?

6 Early this century, the German scientist, Werner, succeeded in elucidating the problem concerning the five compounds of empirical formula $PtCl_4.6NH_3$, the properties of which are listed in the table below.

| **Compound** | **Empirical Formula** | **Total no. of ions formed from one $PtCl_4.6NH_3$** | **No. of moles of $Cl^-$ in one mole of $PtCl_4.6NH_3$** |
|---|---|---|---|
| A | $PtCl_4.6NH_3$ | 5 | 4 |
| B | $PtCl_4.6NH_3$ | 4 | 3 |
| C | $PtCl_4.6NH_3$ | 3 | 2 |
| D | $PtCl_4.6NH_3$ | 2 | 1 |
| E | $PtCl_4.6NH_3$ | 0 | 0 |

(a) What is the oxidation state of Pt in each of the compounds, A–E?

(b) The co-ordination number of Pt in each compound is six. Write a formula for each of the five compounds showing the complex ion and the other ions and/or molecules present.

(c) Each of the compounds forms an octahedral complex ion. Draw structures for the complex ions in A, B, C and D.

(d) Which of the complex ions in (c) have isomers?

(e) Draw structures to show the various structural isomers of A, B, C and D.

# Metals and the Activity Series 19

## 19.1 Iron: a typical transition metal

Iron typifies transition metals in many respects.
(a) It has **two** relatively stable **oxidation states** in its compounds (+2 and +3).
(b) It forms a variety of **complex ions** (such as $[Fe(H_2O)_6]^{2+}$, $[Fe(H_2O)_6]^{3+}$, $[Fe(CN)_6]^{4-}$ and $[Fe(CN)_6]^{3-}$).
(c) It has characteristically **coloured compounds**: pale green for iron(II) salts and yellow or brown for iron(III) salts.
(d) It has important **catalytic properties**. For example, it is used to catalyse the synthesis of ammonia in the Haber process.

The typical characteristics of transition metals were considered in detail in the last chapter.

How do transition metals compare and contrast with the typical metals of Group I and Group II such as sodium and calcium in terms of:

- melting points and boiling points,
- density,
- reactivity with water,
- reactivity with dilute HCl,
- number of oxidation states,
- colour of compounds,
- formation of complex ions?

## 19.2 The occurrence of iron

Apart from its presence in certain meteorites, iron does not occur native in the Earth's crust. Nevertheless, its compounds are abundant; the commonest being iron pyrites ($FeS_2$), haematite ($Fe_2O_3$), magnetite ($Fe_3O_4$) and limonite or brown iron ore ($Fe_2O_3 . H_2O$).

Iron is also present in silicates, sands and clays, and in all living matter, being essential for the production of haemoglobin in blood and chlorophyll in plants. Many pregnant women must take iron tablets (usually in the form of iron(II) sulphate(VI)) to enable them to produce sufficient haemoglobin for themselves and their growing babies without becoming anaemic.

As you will already realize, *iron is abundant in the Earth's crust. In addition, it is easily manufactured from its ores and it has desirable mechanical properties.* Consequently, iron has considerable economic and industrial importance.

*Above* A sample of 'kidney' iron ore.

## 19.3 The manufacture of iron

One hundred years ago, the total production of iron in the U.K. was only about 0 000 tonnes. Nowadays, its production in the U.K. exceeds 20 000 000 tonnes.

*The most important raw materials from which iron is obtained are haematite* ($Fe_2O_3$) *and brown iron ore* ($Fe_2O_3 . H_2O$). Other minerals containing iron, such as sulphides and carbonates must be roasted in air to obtain the oxide before reduction to the metal. In most parts of the world, there are still sufficiently large quantities of reasonably high quality $Fe_2O_3$ or $Fe_2O_3 . H_2O$ from which iron can be obtained.

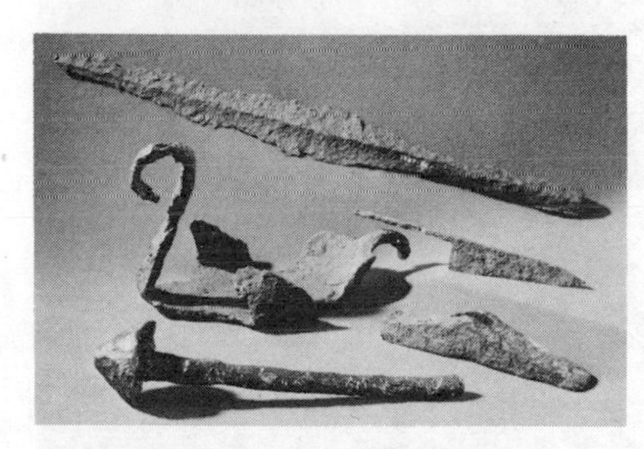

*Above* Roman iron objects. Successful methods for smelting and casting iron existed even in Roman times.

*The first stage in the manufacture of iron involves the preparation of the ore for reduction* in a blast furnace. This involves crushing the ore to produce lumps about the size of one's fist. These lumps are then pre-heated using hot gases from the furnace in order to drive off any water or other volatile impurities.

This prepared iron ore contains between 30 per cent and 95 per cent iron oxides, the main impurities being silicon(IV) oxide (silica, $SiO_2$) and aluminium oxide (alumina, $Al_2O_3$).

*The second stage in the manufacture of iron involves reduction of the iron oxides* with carbon (coke) in the presence of limestone (calcium carbonate) in a blast furnace (figure 19.1).

*Above* Mining iron ore: with the crushing plant and railway terminus in the foreground.

*Above right* When steel was first produced it was an expensive alloy. Thus, the first metal bridge which was erected at Coalbrookdale in 1799 was made of 400 tonnes of cast iron not steel.

*Above* Blast furnaces at Ravenscraig.

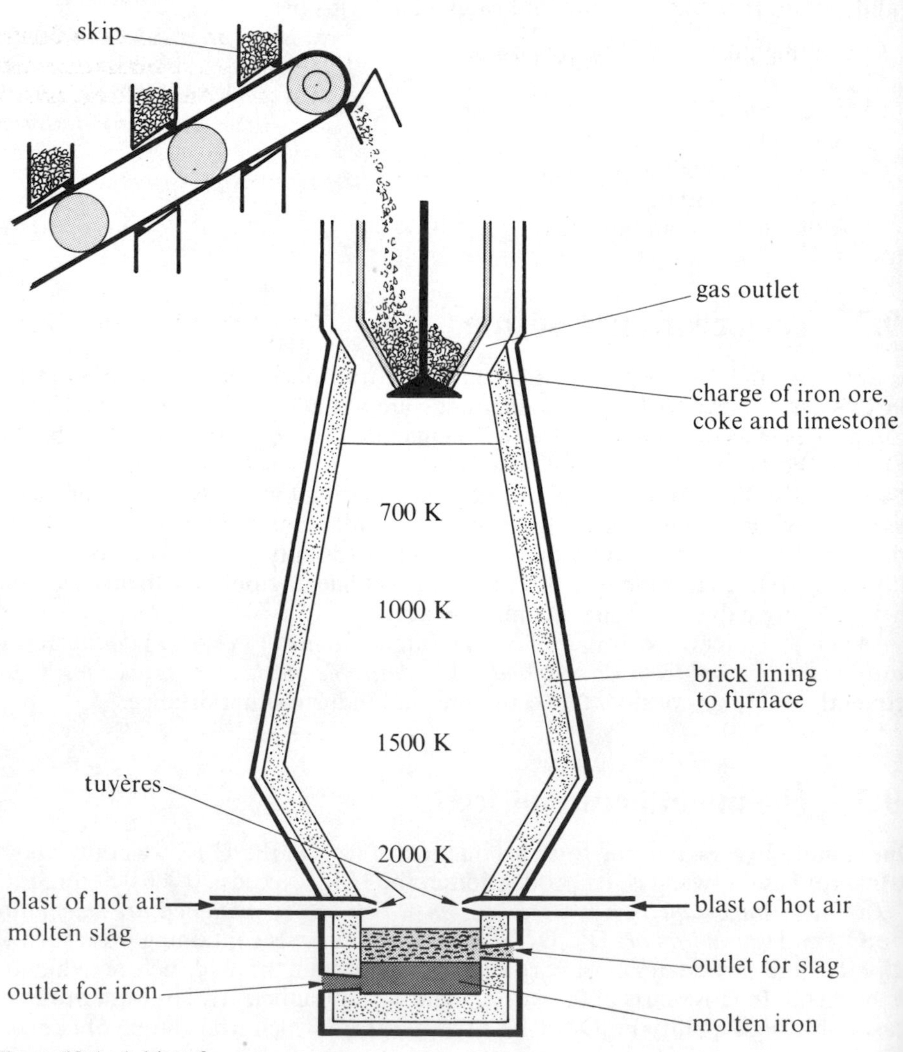


**Figure 19.1** A blast furnace.

The furnace itself is a tapered cylindrical tower about 30–40 metres in height. It is made of steel and lined with refractory bricks. The furnace is fed mechanically in such a way that no gas escapes during the charging. A mixture of iron ore, coke and limestone is added from the top of the furnace. At the same time, blasts of hot air at about 1 000 K (which give the furnace its name) are driven into the furnace through small holes or 'tuyères' near its base.

As the air enters the furnace, it reacts with coke in a highly exothermic reaction forming carbon dioxide.

$$C + O_2 \longrightarrow CO_2 \qquad \Delta H = -394\text{kJ}$$

This raises the temperature at the bottom of the furnace to about 2 000 K. As the $CO_2$ rises up the furnace, it reacts with more coke forming carbon monoxide.

$$CO_2 + C \longrightarrow 2CO \qquad \Delta H = +173\text{kJ}$$

The carbon monoxide then reduces the iron oxides to iron in the upper parts of the furnace, where the temperature is between 750 K and 1 000 K.

$$Fe_2O_3 + 3CO \longrightarrow 2Fe + 3CO_2$$

As the iron falls towards the base of the furnace near the tuyères, it eventually melts (m.pt. = 1 812 K) and flows to the hearth where it collects and is tapped off every few hours. A typical blast furnace produces upwards of 1 000 tonnes of iron every twenty-four hours.

So far, we have said nothing about the use of limestone during the blast furnace process. The limestone plays an essential part in the extraction of impurities from the iron.

At the high temperatures inside the furnace, the limestone decomposes into calcium oxide and $CO_2$.

$$CaCO_3 \longrightarrow CaO + CO_2$$

The calcium oxide then reacts with impurities in the iron, sand (silicon(IV) oxide) and alumina (aluminium oxide) forming calcium silicate(IV) and calcium aluminate(III) respectively.

$$CaO + SiO_2 \longrightarrow CaSiO_3$$

$$CaO + Al_2O_3 \longrightarrow CaAl_2O_4$$

The mixture of calcium silicate(IV) and calcium aluminate(III) which remains molten at the furnace temperature is known as '**slag**'. This flows to the bottom of the furnace where it floats on the molten iron and is tapped off at a different level, separately from the iron. The 'slag' is not wasted; it is used for road making materials, for cement manufacture and for lightweight building materials.

The iron obtained from the blast furnace is far from pure. It is known generally as '**pig iron**'. It is hard, but brittle and melts at about 1 500 K.

It is remelted and mixed with scrap steel and cooled in moulds to form '**cast iron**'. This has much the same properties as 'pig iron' and can be used for articles such as gates, pipes and lamp posts where cheapness is more important than strength.

*Above* Molten iron being tapped off from No 3 blast furnace at Llanwern, Newport

*Above* Workmen constructing a cast iron gate.

*Above* A large cast iron gun.

## 19.4 The conversion of iron into steel

Most of the pig iron manufactured nowadays is used for the production of **steel**, an alloy of iron with carbon and other elements such as manganese, nickel, chromium, tungsten, and vanadium.

The main impurities in pig iron and in mild steel are shown in table 19.1.

**Table 19.1** The main impurities in pig iron and in mild steel.

| **Impurity** | **% impurity in pig-iron** | **% impurity in mild steel** |
|---|---|---|
| Carbon | 3–5 | 0.15 |
| Silicon | 1–2 | 0.03 |
| Sulphur | 0.05–0.10 | 0.05 |
| Phosphorus | 0.05–1.5 | 0.05 |
| Manganese | 0.5–1.0 | 0.50 |

The carbon arises mostly from coke used during the reduction of iron ore. The silicon, sulphur and phosphorus are present as a result of the reduction of silicates, sulphates and phosphates by carbon and carbon monoxide during the blast furnace processes. The manganese arises from the reduction of manganese compounds present in the iron ore.

In order to obtain steel from pig iron, a large proportion of these impurities must be removed.

During the last fifty years, the techniques of **steel production** have undergone vast changes in scale and new processes have been developed to keep pace with the demands of quantity and quality. At present, there are only three major steel-making processes; **Open Hearth**, **Electric Arc**, and **Basic Oxygen**, the latter being a development of the old Bessemer process.

All of these processes share the same general principle which is to remove the impurities of carbon, silicon, sulphur and phosphorus from molten 'pig iron' by oxidation and then to add known quantities of other elements in order to obtain steel of the desired composition and properties.

During steel production, carbon and sulphur are oxidized to $CO_2$ and $SO_2$ which then escape as gases. On the other hand, silicon and phosphorus are oxidized to less volatile oxides, phosphorus(V) oxide and silicon(IV) oxide, which then combine with the lime added to the furnace to form 'slag'.

$$\left.\begin{array}{l} 6CaO + P_4O_{10} \longrightarrow 2Ca_3(PO_4)_2 \\ \qquad\qquad\qquad\quad \text{calcium phosphate(V)} \\ CaO + SiO_2 \longrightarrow CaSiO_3 \\ \qquad\qquad\qquad\quad \text{calcium silicate(IV)} \end{array}\right\} \text{'slag'}$$

THE BASIC OXYGEN PROCESS

In the Basic Oxygen Process, no external heating is required since the reactions which take place inside the furnace are very exothermic. Initially the furnace is charged with hot, molten pig iron and lime (figure 19.2).

Oxidation of the impurities in the iron is brought about by blowing high-purity oxygen on to the surface of the metal at great speed through water-cooled oxygen lances. The oxygen penetrates into the melt and oxidizes the impurities rapidly. The heat evolved as the impurities are oxidized maintains the contents of the furnace in a molten state, despite the rise in the melting point as impurities are removed.

At present, over 70% of the steel produced in the U.K. is obtained by the Basic Oxygen Process. However, this process is increasing in use because it is so fast; a large furnace is capable of converting 300 tonnes of pig iron into steel in only 40 minutes.

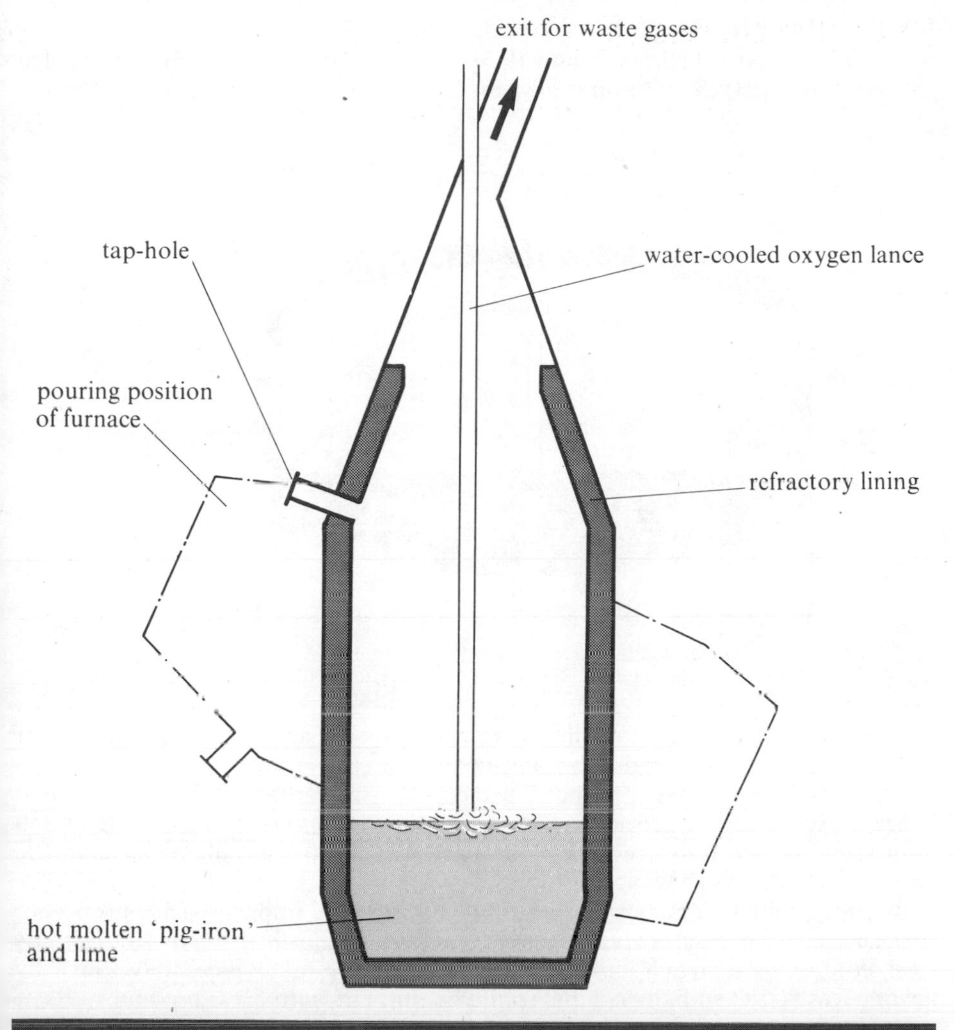


**Figure 19.2** The basic oxygen furnace.

Charging one of the basic oxygen furnaces with hot molten pig iron at the Frodingham Works, Scunthorpe.

THE OPEN HEARTH PROCESS

In the Open Hearth Process (figure 19.3), heat is provided by burning preheated gas or oil in a mixture of air and oxygen.

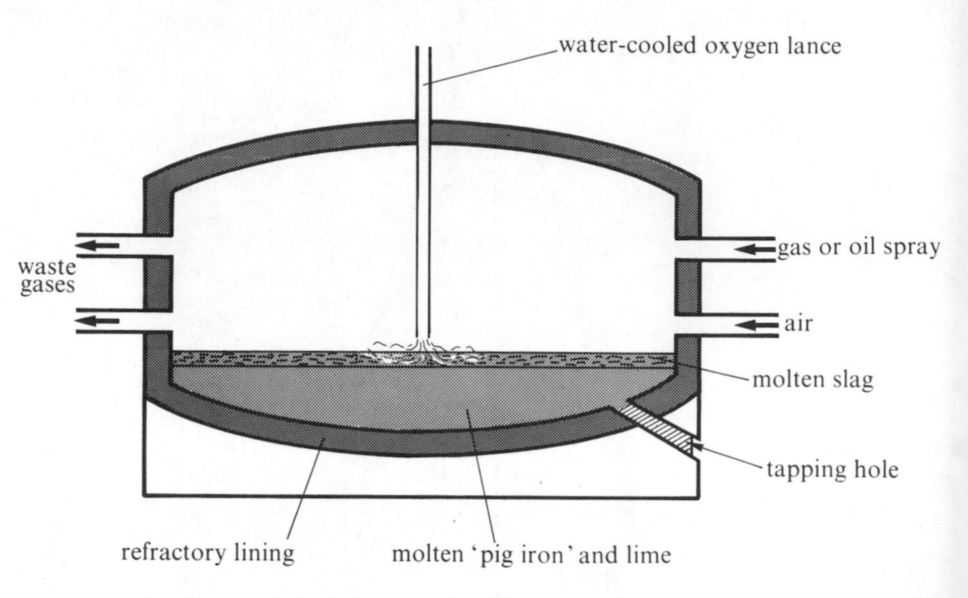


**Figure 19.3** The open hearth furnace.

Oxidation of the impurities in the iron is brought about by the injection of oxygen into the furnace through water-cooled lances.

About 10% of the steel produced in the U.K. is obtained by the Open Hearth Process. However, the process is slow (a large process may take a charge of 350 tonnes and convert it to steel in 10 hours) and it is gradually being replaced by the very much faster Basic Oxygen Process.

By adding different metals to the steel it is possible to impart specific properties to it, such as hardness, resistance to corrosion or high tensile strength. 'Manganese steel' containing as much as 13% manganese can be made incredibly tough by heating to 1 000°C and 'quenching' (rapid cooling) in water. It is used for parts of rock-breaking machinery and for railway cross-overs.

Stainless steel, containing about 20% chromium and 10% nickel is used for cutlery, surgical instruments and car bumpers.

'High speed' steels used as the cutting edges on lathes for metalwork, contain about 18% tungsten and 5% chromium. The tungsten hardens the steel and makes it less brittle. Using 'high-speed' steels, metal parts can be machined up to twenty times faster than ordinary steel.

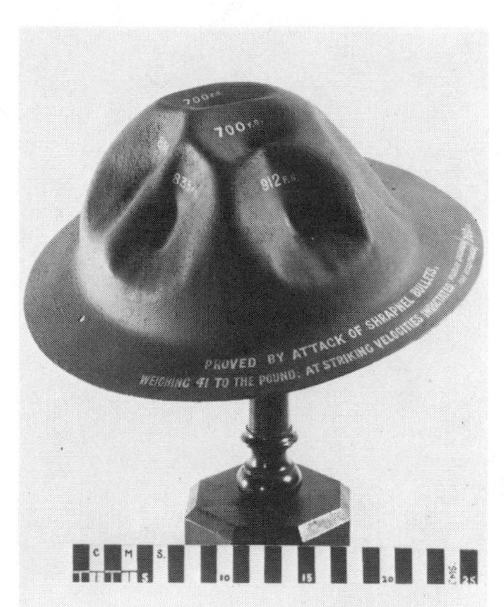


'Manganese steel' is used for military steel helmets. The dents in the steel are caused by bullets striking the helmet at the velocities shown. 912 F.S. means 912 feet per second.

## 19.5 The reactions of iron and its aqueous ions

Several reactions of iron and the aqueous $Fe^{2+}$ and $Fe^{3+}$ ions are summarized in figure 19.4.

Iron combines with most non-metals on heating, forming compounds such as $Fe_3O_4$, FeS, $FeCl_3$ and $FeI_2$. Naturally, those non-metals which are oxidizing agents, such as chlorine and oxygen, form products containing iron in the more oxidized +3 state.

The redox potentials involving iron and its +2 and +3 ions are given below.

**In 1.0M acid solution,** $Fe(s) \xrightarrow{+0.44} Fe^{2+}(aq) \xrightarrow{-0.77} Fe^{3+}(aq)$

**In 1.0M alkaline solution,** $Fe(s) \xrightarrow{+0.89} Fe(OH)_2(s) \xrightarrow{-0.56} Fe(OH)_3(s)$

These data show that the more stable oxidation state of iron in both acid and alkaline solution is +2.

However, *even mild oxidants such as oxygen are capable of oxidizing iron(II) to iron(III) and consequently, the more stable state in the presence of both air (oxygen) and water is +3.*

*Above* Wrought iron beetle. Wrought iron is very pure iron, it is much more malleable than steel or cast iron. Everything on this baroque style Volkswagen works.

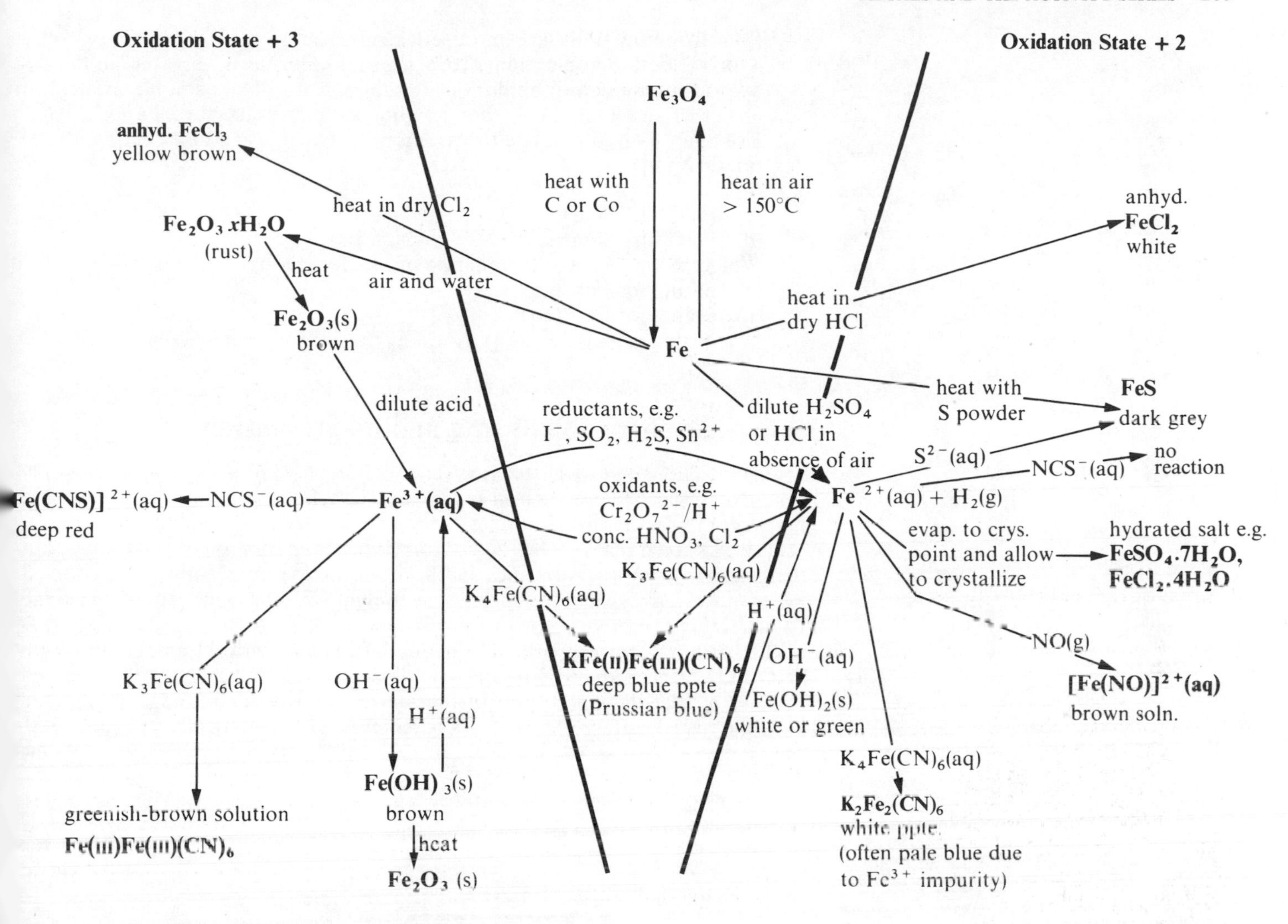


**Figure 19.4** The reactions of Fe, $Fe^{2+}(aq)$ and $Fe^{3+}(aq)$.

In acid solution, the conversion of $Fe^{2+}(aq)$ to $Fe^{3+}(aq)$ is slow (due to kinetic factors), but in alkaline or neutral solution the oxidation is rapid.

$$2Fe^{2+}(aq) \longrightarrow 2Fe^{3+}(aq) + 2e^-$$

$$\tfrac{1}{2}O_2 + H_2O + 2e^- \longrightarrow 2OH^-$$

Thus, aqueous solutions of $Fe^{2+}(aq)$ exposed to the air always contain $Fe^{3+}$ ions unless they are freshly prepared and acidic, whilst freshly prepared pale green $Fe(OH)_2$ rapidly turns dirty green and finally red-brown on contact with the air as it is oxidized to $Fe(OH)_3$.

The redox potentials quoted previously also confirm the fact that iron will react with dilute acids (in the absence of air) to yield solutions of iron(II) salts and hydrogen. However, dilute nitric(V) acid and concentrated sulphuric(VI) acid, both of which are oxidizing agents, yield a mixture of $Fe^{2+}$ and $Fe^{3+}$ ions together with the reduction products of the acids (mainly $NH_4^+$ and $SO_2$ respectively). On the other hand, concentrated nitric(V) acid and concentrated chromic(VI) acid have only a momentary effect on iron, after which the metal becomes 'passive'. It is thought that the initial attack of the acid coats the metal with an exceptionally tight, impervious and unreactive layer of oxide which prevents further attack. Not surprisingly, iron is sometimes treated in this way to prevent corrosion.

Try to answer these questions about the chemistry of iron.

- ○ Aqueous iron(II) salts are only slightly acidic whereas iron(III) salts are strongly acidic. Why is this? (Hint: your explanation should involve a consideration of the charge density on $Fe^{2+}$ and $Fe^{3+}$ and the possible hydrolysis of $[Fe(H_2O)_6]^{2+}$ and $[Fe(H_2O)_6]^{3+}$ in aqueous solution.)
- ○ Draw structural formulae to show the octahedral stereo-structures of the hexacyanoferrate(III) ion ($[Fe(CN)_6]^{3-}$) and the hexacyanoferrate(II) ion ($[Fe(CN)_6]^{4-}$).

- Iron(III) compounds are more covalent than iron(II) compounds. Why is this? (For example, anhydrous iron(III) chloride is very similar to anhydrous aluminium(III) chloride. It sublimes readily, is soluble in alcohol and ether and it has a co-ordinately-bonded dimer structure, $Fe_2Cl_6$.)
- Use figure 19.4 to tabulate the reactions of $Fe^{2+}$(aq) and $Fe^{3+}$(aq) with:
  (a) $OH^-$(aq)
  (b) $NCS^-$(aq)
  (c) $[Fe(CN)_6]^{4-}$(aq)
  (d) $[Fe(CN)_6]^{3-}$(aq)
  Which of the above ions would be most effective in:
  (i) testing for $Fe^{2+}$(aq)
  (ii) testing for $Fe^{3+}$(aq)
  (iii) distinguishing $Fe^{2+}$(aq) from $Fe^{3+}$(aq)?

## 19.6 The corrosion of iron and its prevention

Rusting costs our industrialized society millions of pounds each year, not only because of the need to protect iron and steel objects, but also because of the expense involved in replacing rusted articles.

It is well known that *rusting requires the presence of both oxygen and water*, but other factors (such as impurities in the iron surface, availability of dissolved oxygen and electrolytes in the solution in contact with the iron) can influence the rate of rusting considerably.

When rusting occurs a hydrated form of iron(III) oxide with variable composition ($Fe_2O_3.xH_2O$) is produced. This oxide is very permeable to both air and water and it cannot protect the metal from further corrosion, which continues unhindered below the rusted surface. The process is represented schematically in figure 19.5.

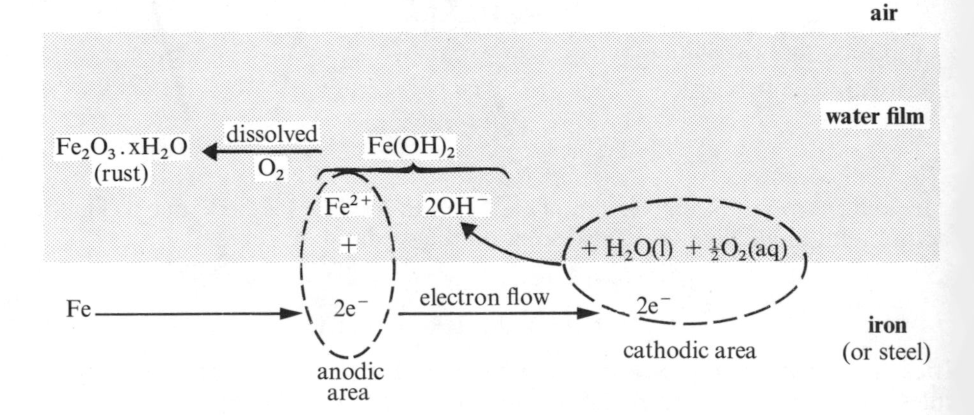


**Figure 19.5** The chemical reactions involved in rusting.

In our climate, any iron or steel object open to the atmosphere is often wet. The layer of water on the surface of the metal dissolves oxygen and carbon dioxide from the air. In addition to these, the water will also contain dissolved sulphur dioxide in any industrialized area and sodium chloride from salt spray near the sea.

Look closely at figure 19.5.
- Why do impurities in the iron surface hasten rusting?
- Why do electrolytes, particularly acids, accelerate rusting?
- Which substances, normally present in the atmosphere, can provide the $H^+$ ions to accelerate rusting?
- Why is rusting accelerated in polluted areas?

In the initial stages of rusting, iron(II) ions pass into solution at the anodic area which is pure iron, whilst reduction of atmospheric oxygen to hydroxide ions take place at the cathodic area where the concentration of dissolved oxygen is higher.

$$\textbf{Anode (+)}\quad Fe(s) \longrightarrow Fe^{2+}(aq) + 2e^-$$
$$\textbf{Cathode (−)}\quad H_2O(l) + \tfrac{1}{2}O_2(aq) + 2e^- \longrightarrow 2OH^-(aq)$$

The $Fe^{2+}$ and $OH^-$ ions diffuse away from the 'electrodes' and then precipitate as iron(II) hydroxide which is oxidized by dissolved oxygen to form rust.

Notice that the rust is formed by a secondary process *within the solution* as $Fe^{2+}$ and $OH^-$ ions diffuse away from the metal surface. Thus, the rust does not form as a protective layer in contact with the iron surface. However, if the solution contains a relatively high concentration of dissolved oxygen, $Fe^{2+}$ ions are converted into rust more rapidly, and a protective layer may be formed on the iron surface which retards further rusting.

Acid conditions (low pH) accelerate rusting by promoting the dissolution of iron. Electrolytes in the water and impurities in the iron also assist the process by increasing the conductivity of the solution and initiating cell action respectively.

Two general methods are used to protect iron and steel from corrosion: the **application of a protective layer** and the **application of a sacrificial metal.**

'Tin plate' being rolled into thin sheets at the British Steel Corporation's tin plate works in South Wales.

### APPLICATION OF A PROTECTIVE LAYER

Many different substances are used to protect iron and steel objects from rusting.

In the motor car industry, the chassis of vehicles are painted with phosphoric(V) acid which reacts with the iron to form an insoluble, tenacious film of iron(III) phosphate(V). Engine parts, like other machinery and tools are often protected by a film of oil or grease. However, the most common method of rust prevention is painting, often incorporating the use of dilead(II) lead(IV) oxide (red lead) or zinc chromate(VI) as protective priming coats.

Other methods are also used in the fight against corrosion. Iron can be alloyed with other metals, such as chromium, nickel and manganese in order to produce corrosion-resistant steels, or coated with non-rusting metals. Buckets and watering cans are coated with a layer of zinc ('galvanized') by dipping them in molten zinc at 450°C or by depositing the zinc by electrolysis. 'Tin-plate' is formed by immersing steel sheeting in molten tin whilst other articles, such as car bumpers, are chromium plated by making them the cathode in an electrolytic cell.

### APPLICATION OF A SACRIFICIAL METAL

In this case, the iron or steel must be connected to a more electropositive metal which is oxidized in preference to iron, thus preventing the formation of $Fe^{2+}$ ions and hence rust. 'Galvanizing' provides a good example of an electrochemical method of rust prevention. When the 'galvanized' surface is undamaged, zinc is protected from corrosion by a firmly adhering layer of zinc oxide. When the galvanized surface is scratched and the iron is exposed, zinc ions rather than iron(II) ions pass into solution at the anodic region because zinc is more electropositive. Electrode potentials confirm this fact.

| | $E^\ominus$/volts |
|---|---|
| $Fe(s) \longrightarrow Fe^{2+}(aq) + 2e^-$ | +0.44 |
| $Zn(s) \longrightarrow Zn^{2+}(aq) + 2e^-$ | +0.76 |

Thus, zinc is sacrificed in the protection of iron. 'Tin-plating', however, is not so effective as 'galvanizing'. Provided the coating is undamaged, the tin surface protects the iron below; but when the tin-plate is scratched rusting occurs, since iron, being more electropositive than tin, passes into solution as $Fe^{2+}$ ions and then forms rust.

| | $E^\ominus$/volts |
|---|---|
| $Fe(s) \longrightarrow Fe^{2+}(aq) + 2e^-$ | +0.44 |
| $Sn(s) \longrightarrow Sn^{2+}(aq) + 2e^-$ | +0.14 |

Cathodic protection is used to prevent corrosion in underground pipelines made of steel. Pieces of zinc or magnesium alloy are connected to the buried pipeline at intervals. In oil refineries, magnesium alloys are bolted inside distillation and cracking plant, so that these metals will corrode in preference to steel between regular maintenance inspections.

## 19.7 Copper: extraction and manufacture

Metallic copper does occur naturally in a few places, but the commonest minerals, from which 80% of the element is obtained, are the sulphides, copper pyrites ($CuFeS_2$) and copper glance ($Cu_2S$).

Most copper ores contain only a few per cent of the metal, so the first stage in obtaining copper is to crush the ore and concentrate it by froth flotation.

The second stage involves reduction of the sulphide ore to the metal. Initially, the concentrate is roasted in air to form copper(I) sulphide, iron(II) oxide and gaseous sulphur dioxide.

$$2CuFeS_2 + 4O_2 \longrightarrow Cu_2S + 2FeO + 3SO_2(g)$$

The product is then heated with silica in a closed furnace so that most of the iron(II) oxide reacts with the silica to form a molten slag which floats on the molten copper(I) sulphide and can be tapped off separately.

$$FeO + SiO_2 \longrightarrow \underset{\text{iron(II) silicate(IV) (slag)}}{FeSiO_3}$$

The impure copper(I) sulphide is then heated in air when part of it reacts, forming copper(I) oxide.

$$2Cu_2S + 3O_2 \longrightarrow 2Cu_2O + 2SO_2$$

The copper(I) oxide, mixed with unchanged copper(I) sulphide is now heated strongly in the absence of air, forming copper and sulphur dioxide.

$$2Cu_2O + Cu_2S \longrightarrow 6Cu + SO_2$$

The product is known as 'blister copper' because it releases bubbles of $SO_2$ as it solidifies and therefore gets a rather blistered appearance. It still contains two or three per cent impurities, mainly iron and sulphur.

The final stage in the manufacture of copper involves purification of the blister copper. This is usually achieved by electrolysis of copper(II) sulphate(VI) solution using a thin sheet of pure copper as the cathode and the impure blister copper as the anode. During the electrolytic purification, copper dissolves away from the anode, whilst a thickening deposit of pure copper appears on the cathode.

- ○ Write equations for the processes at the anode and the cathode during this electrolytic purification of blister copper.
- ○ If the impure blister copper contains metals above copper in the electrochemical series, such as iron, these metals will dissolve from the anode in preference to copper. Why then do these metals, which get into solution, not deposit on the pure copper cathode?
- ○ Suppose the impure blister copper contains metals below copper in the electrochemical series such as silver. Why do these metals not contaminate the copper which deposits on the cathode?

In practice, silver and gold can be obtained as valuable by-products which help to pay for the copper refining process.

The open cast copper mine at Chuquicamata, in the Atacama Desert, Chile, the largest copper producing mine in the world.

*Above left* Purifying crushed copper ore by froth flotation.

*Above right* Electrolytic refining of copper.

## 19.8 The properties and uses of copper

The many uses of copper result from its malleability and ductility, its high electrical and thermal conductivity, its resistance to corrosion and its ability to form alloys with other metals. The pure metal is used extensively as electrical conducting wires and cables, and also as pipes and radiators in central heating systems.

With other metals, copper provides a wide range of different alloys. Alloying copper with zinc produces various types of brass which are harder and stronger than pure copper. Alloying copper with tin produces bronze, an alloy which has greater tensile strength and is more readily cast into moulds than pure copper. Copper also alloys readily with other metals including nickel, aluminium and gold. For example, 75% copper and 25% nickel are used in our present 'silver coins', whilst 9 carat gold is about two-thirds copper and one-third gold

## 19.9 Copper: Dr Jekyll or Mr Hyde

Copper has a rather complex, two-faced character with two sets of very different compounds. *In the +2 state, copper shows typical transition metal properties, whilst in the +1 state its compounds are more like those of silver.*

The reactions of copper and the more important copper compounds are shown in figure 19.6 and various atomic properties of copper are tabulated in table 19.2.

**Table 19.2** Atomic properties of copper.

| | |
|---|---|
| Electron structure | $(Ar)3d^{10}4s^1$ |
| Heat of atomization/kJ mole$^{-1}$ | 339 |
| First ionization energy/kJ mole$^{-1}$ | 744 |
| Second ionization energy/kJ mole$^{-1}$ | 1956 |
| Hydration energy of $Cu^{2+}$(g)/kJ mole$^{-1}$ | −2244 |
| Hydration energy of $Cu^{+}$(g)/kJ mole$^{-1}$ | −481 |
| Standard electrode potential, Cu(s)/$Cu^{2+}$(aq)/Volts | −0.34 |
| Standard electrode potential, Cu(s)/$Cu^{+}$(aq)/Volts | −0.52 |
| Standard electrode potential, Pt/$Cu^{+}$(aq), $Cu^{2+}$(aq)/Volts | −0.15 |

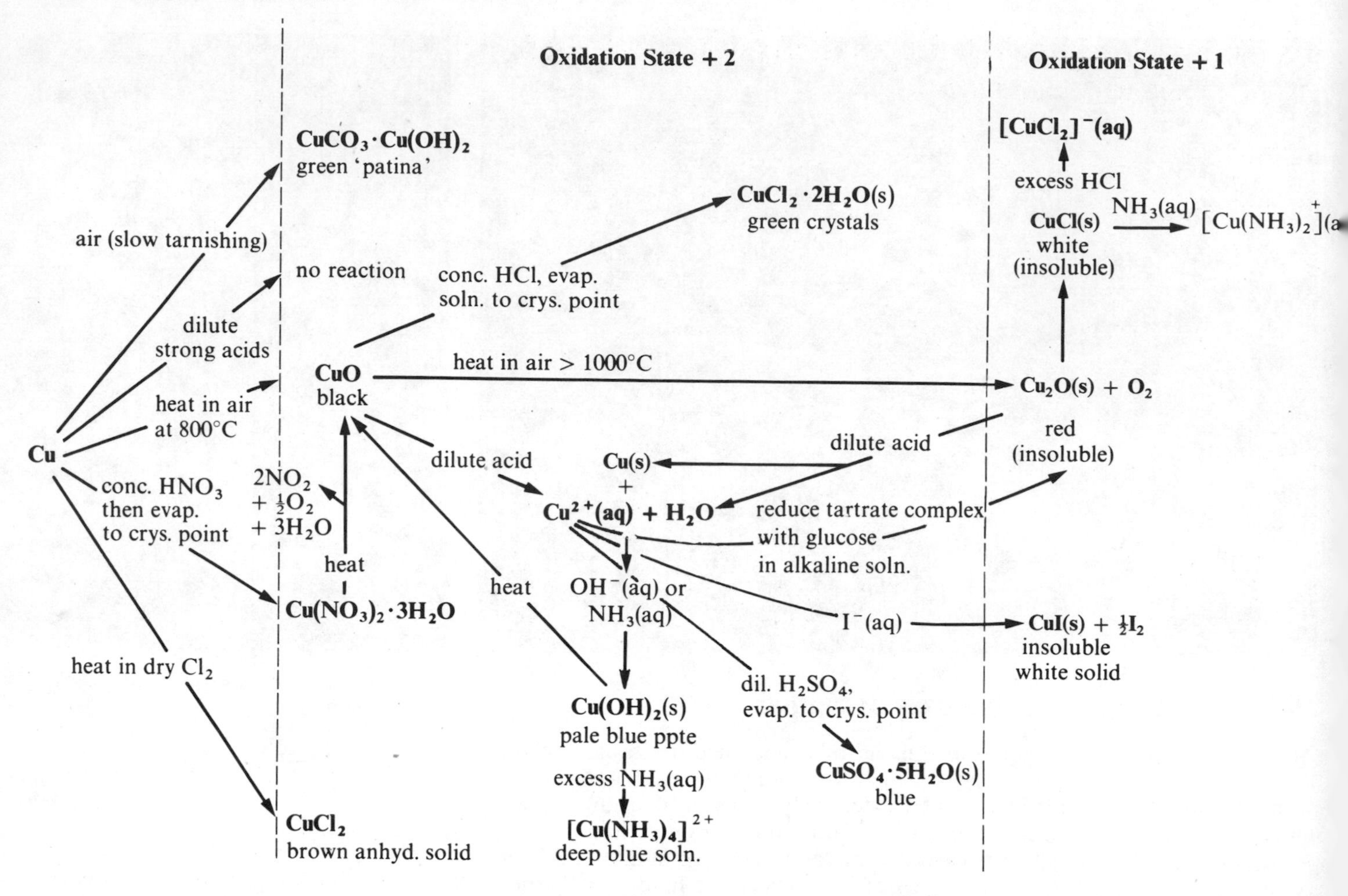


**Figure 19.6** The reactions of copper and copper compounds.

The chemical properties of copper can be related to its relatively high ionization energies, high heat of atomization, and relatively low heats of hydration which together result in its negative electrode potentials and generally low reactivity (see table 19.2).

The electrode potentials for both $Cu(s)/Cu^{2+}(aq)$ and $Cu(s)/Cu^{+}(aq)$ are negative. Consequently, copper will not react with dilute strong acids to produce hydrogen (figure 19.6).

$$2H^+(aq) + 2e^- \longrightarrow H_2(g) \qquad E^\ominus = 0.00 \text{ Volts}$$

However, copper can be oxidized very slowly to the copper(II) state ($CuCO_3 . Cu(OH)_2$) by oxygen and water vapour in the air, and much more rapidly by concentrated nitric(V) acid (figure 19.6). The following potentials further confirm this.

| | | $E^\ominus$/Volts |
|---|---|---|
| $Cu(s) \longrightarrow$ | $Cu^{2+}(aq) + 2e^-$ | $-0.34$ |
| $\frac{1}{2}O_2(g) + H_2O(g) + 2e^- \longrightarrow$ | $2OH^-(aq)$ | $+0.40$ |
| $NO_3^-(aq) + 2H^+(aq) + e^- \longrightarrow$ | $NO_2(g) + H_2O$ | $+0.80$ |

Furthermore, copper will react directly on heating with chlorine and oxygen (both strong oxidizing agents) forming copper(II) chloride and copper(II) oxide respectively (figure 19.6).

Notice that all these reactions of copper result in the formation of copper(II), rather than copper(I) compounds. This emphasizes yet another enigma concerning copper.

- ○ Write the electronic structure for Cu, $Cu^+$ and $Cu^{2+}$.
- ○ Which ion would you expect to be more stable, $Cu^+$ or $Cu^{2+}$? Explain your answer.

In terms of electron structure we would expect $Cu^+$ to be more stable than $Cu^{2+}$. Yet most common copper compounds are those of the element in the +2 state. Why is this? In the next section we must try to unravel this problem.

## 19.10 Exploring the relative stabilities of copper(I) and copper(II) compounds

Figure 19.7 shows two energy cycles relating to the formation of $Cu^+(aq)$ and $Cu^{2+}(aq)$ from copper. The cycles show that both $Cu^+(aq)$ and $Cu^{2+}(aq)$ are unstable relative to copper—not surprising, when one considers their negative electrode potentials. But, in this section, we are concerned not so much with the stability of these ions with respect to copper, but their stability with respect to each other.

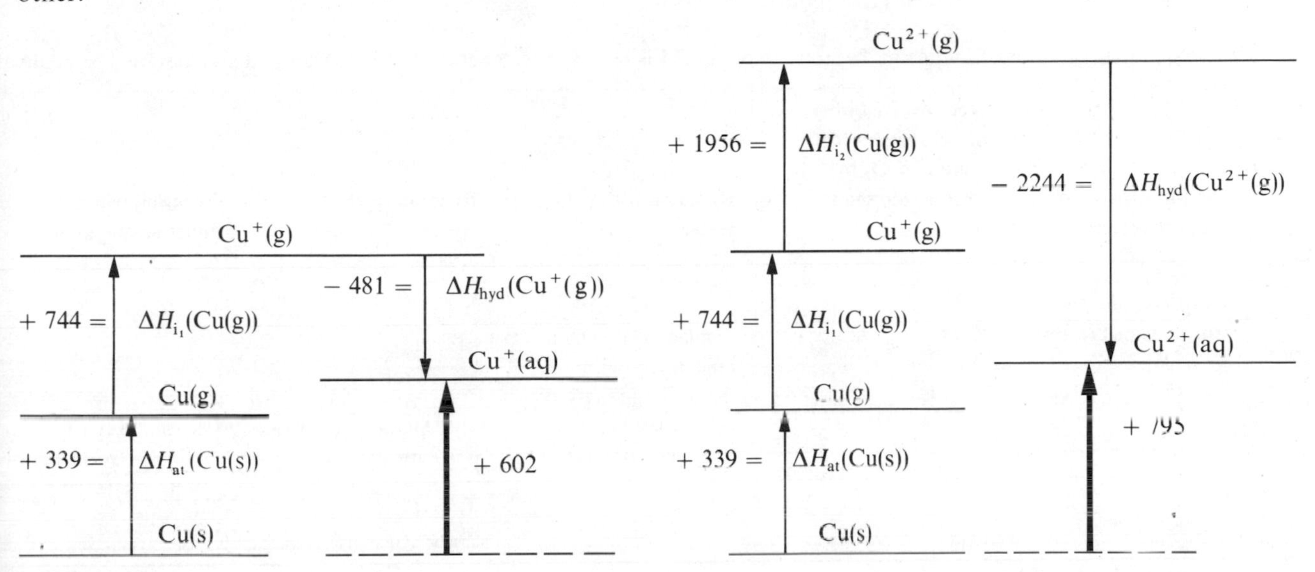


(*a*) The formation of $Cu^+(aq)$ from Cu

(*b*) The formation of $Cu^{2+}(aq)$ from Cu

**Figure 19.7** The formation of $Cu^+(aq)$ and $Cu^{2+}(aq)$ from Cu (all values in kJ mole$^{-1}$).

Using the values for the energy changes, $Cu(s) \rightarrow Cu^+(aq)$ and $Cu(s) \rightarrow Cu^{2+}(aq)$ in figure 19.7, we can now calculate the energy change for the process

$$2Cu^+(aq) \longrightarrow Cu(s) + Cu^{2+}(aq)$$

which involves the disproportionation of copper(I).

This is shown schematically in figure 19.8. The exothermic value for this change suggests that aqueous copper(I) ions are thermodynamically unstable with respect to aqueous copper(II) ions and copper and as we might expect, $Cu^+(aq)$ does not exist in aqueous solution.

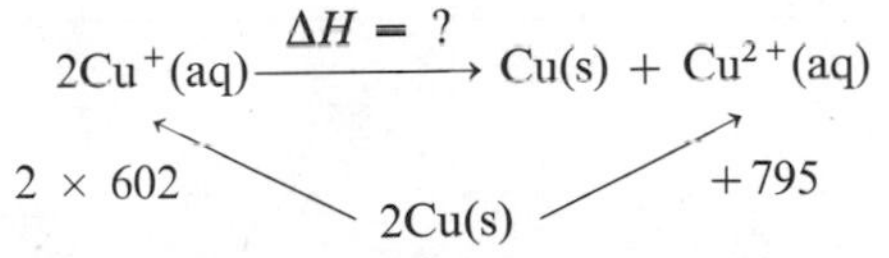

$$2Cu^+(aq) \xrightarrow{\Delta H = ?} Cu(s) + Cu^{2+}(aq)$$

2 × 602 (from 2Cu(s) to $2Cu^+(aq)$)    2Cu(s)    +795 (from 2Cu(s) to $Cu(s) + Cu^{2+}(aq)$)

By Hess's Law

$$2 \times 602 + \Delta H = 795$$

$$\therefore \Delta H = -409 \text{ kJ}$$

**Figure 19.8** Enthalpy change for the disproportionation of $Cu^+$

Thus, when solid copper(I) oxide ($Cu_2O$) is treated with dilute acids it forms a precipitate of copper and a solution of the copper(II) salt, not a solution of the copper(I) salt (figure 19.6).

$$Cu_2O(s) + H_2SO_4(aq) \longrightarrow Cu(s) + CuSO_4(aq) + H_2O$$

This means that the only copper(I) compounds which are stable in the presence of water must either be

(a) *insoluble* (e.g. $Cu_2O$, CuI and CuCl in figure 19.6) *or*

(b) *complexed with ligands other than water* (e.g. diamminocopper(I) ($[Cu(NH_3)_2]^+$) and dichlorocuprate(I) ($[CuCl_2]^-$) in figure 19.6.)

## 19.11 The reactions of metals and the activity series

Many of the reactions of metals can be summarized using either the periodic table or the activity series. The reactions of some common metals with oxygen, chlorine, water, steam and dilute acids are summarized in table 19.3 by writing the metals in order of their position in the activity (electrochemical) series.

The heat evolved when each metal reacts with one mole of oxygen and with one mole of chlorine are also included in the table as are the electrode potentials of the metals. Notice that there is a similar order of reactivity of these metals for each series of reactions.

**Table 19.3** Summary of the reactions of metals in the Activity Series (M is used as the general symbol for a metal and assumed to be divale

| Metal | Reaction with $O_2(g)$ on heating | Heat evolved when metal reacts with 1 mole of $O_2$ to form oxide shown /kJ | | Reaction with cold water | Reaction with steam | Reaction with dilute strong acids |
|---|---|---|---|---|---|---|
| K | form oxides (e.g. $Na_2O$) in limited supplies of $O_2$, but peroxides (e.g. $Na_2O_2$) with excess $O_2$ | $K_2O$ | 723 | displace $H_2(g)$ from cold water with decreasing reactivity (K, violently; Mg, very slowly) | displace $H_2(g)$ from steam with decreasing vigour, (K, very violently; Fe, very slowly) | displace $H_2(g)$ from dilute strong acids with decreasing vigour (K, explosively; Mg, very vigorously; Fe, steadily; Pb, very slowly) |
| Na | | $Na_2O$ | 832 | | | |
| Ca | burn with decreasing vigour to form oxides | CaO | 1272 | | | |
| Mg | | MgO | 1204 | | | |
| Al | | $Al_2O_3$ | 1114 | do *not* displace $H_2(g)$ from cold water | | |
| Zn | | ZnO | 697 | | | |
| Fe | | $Fe_2O_3$ | 548 | | | |
| Pb | do *not* burn, but only form a surface layer of oxide | PbO | 436 | | do *not* displace $H_2(g)$ from steam | |
| Cu | | CuO | 311 | | | do *not* displace $H_2(g)$ from dilute strong acids |
| Hg | | HgO | 182 | | | |
| Ag | do *not* burn or oxidize on surface | $Ag_2O$ | 61 | | | |
| Pt | | | | | | |
| Au | | $Au_2O_3$ | 54 | | | |
| General Eqtns | $2M + O_2 \rightarrow 2M^{2+}O^{2-}$ | | | $M + H_2O \rightarrow$ $\rightarrow M^{2+}O^{2-} + H_2$ | $M + H_2O \rightarrow$ $\rightarrow M^{2+}O^{2-} + H_2$ | |
| Half Eqtn | $2M \rightarrow 2M^{2+} + 4e^-$ $O_2 + 4e^- \rightarrow 2O^{2-}$ | | | $M \rightarrow M^{2+} + 2e^-$ $H_2O + 2e^- \rightarrow$ $\rightarrow H_2 + O^{2-}$ | $M \rightarrow M^{2+} + 2e^-$ $H_2O + 2e^- \rightarrow$ $\rightarrow H_2 + O^{2-}$ | |
| Comment | | | | the oxides which form react with water to form hydroxides | oxides of the top 4 metals react with water to form their hydroxides | |

Metals act as reducing agents in their reactions by donating electrons to a non-metal, water or $H^+$ ions.

$$\underset{\substack{\text{reducing}\\\text{agent}}}{M} \xrightarrow{\text{oxidation}} M^{2+} + 2e^-$$

Of the metals listed in table 19.3, potassium is the strongest reducing agent whereas gold is the weakest. Potassium donates its electrons most readily, gold least readily and, of course, the ease with which a metal loses electrons is directly linked to the value of its electrode potential and hence its position in the electrochemical series.

| Reaction with $Cl_2$ on heating | Heat evolved when metal reacts with 1 mole of $Cl_2(g)$ to form the chloride shown/kJ | | Strength as reducing agents | Reactions with other aqueous cations | Standard electrode potential, M(s) $/M^{n+}(aq)$/volts | Metal |
|---|---|---|---|---|---|---|
| | KCl | 873 | ↓ decreasing strength as reducing agents (i.e. electrons are donated less readily by the metal) | | 2.92 | K |
| | NaCl | 826 | | | 2.71 | Na |
| | $CaCl_2$ | 795 | | | 2.87 | Ca |
| | $MgCl_2$ | 642 | | | 2.37 | Mg |
| all the metals react with $Cl_2(g)$ on heating to form their chloride with decreasing vigour | $Al_2Cl_6$ | 464 | | any metal will displace from solution ions of a metal below it in the activity series | 1.66 | Al |
| | $ZnCl_2$ | 416 | | | 0.76 | Zn |
| | $FeCl_3$ | 270 | | | 0.44 ($Fe(s)/Fe^{2+}(aq)$) | Fe |
| | $PbCl_2$ | 359 | | | 0.13 | Pb |
| | $CuCl_2$ | 206 | | | −0.34 | Cu |
| | $HgCl_2$ | 230 | | | −0.79 ($Hg(l)/Hg_2^{2+}(aq)$) | Hg |
| | AgCl | 255 | | | −0.80 | Ag |
| | $PtCl_3$ | 142 | | | −1.20 | Pt |
| | $AuCl_3$ | 79 | | | | Au |
| $M + Cl_2 \rightarrow M^{2+}(Cl^-)_2$ | | | | | | General Eqtn |
| $M \rightarrow M^{2+} + 2e^-$<br>$Cl_2 + 2e^- \rightarrow 2Cl^-$ | | | | | $M(s) \rightarrow M^{2+}(aq) + 2e^-$ | Half Eqtn |
| metals with 2 oxidation states form the chloride of the metal in the highest state (e.g. $FeCl_3$ *not* $FeCl_2$) | | | | | | Comment |

## 19.12 The extraction of metals from their ores

Table 19.4 gives some information concerning the methods of extraction, the major sources, the annual productions and the uses of nine common metals.

**Table 19.4** The extraction and uses of metals.

| Metal | Date of discovery | Main ore from which metal is obtained | Method of extraction | Annual World Production /1000 tonnes (1976) | Price per tonne (1978) | Main uses | Reference to other sections in the book |
|---|---|---|---|---|---|---|---|
| Sodium | 1807 | Rock salt, NaCl | Electrolysis of molten NaCl | 400 | — | Manufacture of tetraethyllead(IV), sodium vapour lamps | 14.7–14.9, 19.14 |
| Magnesium | 1808 | Magnesite, $MgCO_3$ and $Mg^{2+}$ ions in sea water | Electrolysis of molten $MgCl_2$ | 200 | — | Low density alloys | 14.7–14.9, 19.14 |
| Aluminium | 1827 | Bauxite, $Al_2O_3.2H_2O$ | Electrolysis of $Al_2O_3$ in molten cryolite, ($Na_3AlF_6$) | 13 000 | £580 | Kitchen utensils, packaging foil, structural alloys in aircraft, automobiles | 15.4–15.6, 19.14 |
| Zinc | 1746 | Zinc blende, ZnS | Heat sulphide in air → oxide. Reduce oxide with carbon | 10 000 | £250 | Galvanizing iron, alloys (brass) | 19.14 |
| Iron | Ancient | Haematite, $Fe_2O_3$ | Reduce $Fe_2O_3$ with carbon monoxide | 890 000 | £100 | Most important structural metal (as steel), vehicles, engines, tools | 19.1–19.6, 19.14 |
| Tin | Ancient | Tinstone, $SnO_2$ | Reduce $SnO_2$ with carbon | 200 | £6600 | Tinplate (coating iron), alloys (e.g. solder, pewter, bronze, etc.) | 17.1–17.2 |
| Lead | Ancient | Galena, PbS | Similar to Zn or controlled heating with correct amount of air → Pb + $SO_2$ | 5000 | £350 | Roof and cable covering, water pipes, alloys (e.g. solder, pewter) | 17.1–17.2 |
| Copper | Ancient | Copper pyrites, $CuFeS_2$ (CuS + FeS) | Controlled heating with correct amount of air → Cu + $SO_2$ | 9000 | £700 | Electrical wires, cables, etc., alloys (e.g. brass, bronze) | 19.7–19.10, 19.14 |
| Mercury | Ancient | Cinnabar, HgS | Heat in air → Hg + $SO_2$ | 10 | — | Scientific equipment (e.g. barometers, thermometers, etc.), mercury vapour lamps | 19.14 |

Some of these extraction processes have already been considered in previous chapters. References to these earlier discussions are given in the final column of table 19.4.

- ○ Why is iron used in much larger quantities than any other metal?
- ○ Which compounds do metals most frequently exist as in their naturally occurring ores?
- ○ As a general rule, it has been said that the most electropositive metals occur in nature in combination with the most electronegative anions such as oxide and chloride, whereas the less electropositive metals are found in combination with the less electronegative anions such as sulphide. Do you think this is a good generalization? Illustrate your answer with several examples.

- ○ How do exceptionally poor metals such as gold and platinum occur in nature?
- ○ Why is there such a clear division between those metals discovered in ancient times and those discovered much later in the eighteenth and nineteenth centuries?

The extraction of all metals from their naturally occurring ores involves three stages.

(A) PURIFICATION AND CONCENTRATION OF THE ORE

Very few metal ores are sufficiently concentrated to be used without purification. The main impurities are usually earthy materials such as rocks, clays and sand. In many cases, the ore is separated from useless impurities before transportation in order to reduce costs.

Usually the material is first crushed and ground into small pieces. Magnetic material can then be separated by using strong electromagnets, or differences in density between the ore and its impurities may be sufficient to allow separation of crushed material on an agitated sloping table. Jets of water play across the line of inclination of the table so that the denser material accumulates at the lower end.

One of the most common methods of purifying metal ores is **froth flotation**. In this process, the finely ground ore is added to water which contains special oils known as 'frothing agents' such as pine oil, creosote and xanthates. Separation is possible because of the different densities of the materials and their different wetting characteristics with the frothing agent and water.

When air is blown through the mixture, air bubbles adhere to certain minerals, but not to others. The 'frothing agent' is carefully chosen so that particles of the ore become attached to air bubbles and float to the surface where they collect as a froth and are removed periodically while earthy particles sink to the bottom. The process of froth flotation is particularly suitable for concentrating such dense ores as galena (lead sulphide) and zinc blende (zinc sulphide).

An Egyptian worker smelting metal ore. Why is he using a blow pipe?

Egyptian workmen casting molten metal.

(B) REDUCTION OF THE PURIFIED ORE TO THE METAL

The various methods of reducing metal ores are discussed in section 19.14. The method used for a particular metal depends mainly on the position of the metal in the activity series.

(C) PURIFICATION OF THE METAL

The metal obtained by the initial reduction of the ore is usually contaminated with small amounts of the unchanged ore, with small amounts of other metals present in the ore and non-metals from the anions in the ore. For example, the 'blister copper' obtained by reduction of copper ores contains small amounts of unchanged copper sulphide, iron and sulphur. The purification of 'blister copper' is described in section 19.7.

Impure 'pig iron' obtained from the blast furnace contains about 8 % impurities which include carbon, silicon, sulphur, phosphorus and manganese. The removal of these impurities from 'pig iron' is described in section 19.4.

## 19.13 Factors influencing the choice of method used in reducing a metal ore

CHEMICAL FACTORS

The most important question to ask in this respect is 'How easily can the ore be reduced to the metal?' This, of course, will dictate the cost of the method used and, furthermore, it is directly related to the position of the metal in the activity series.

Other essential factors which must be considered are the accessibility of the ore, the ease with which the ore can be purified, the scale on which the metal is required and the most suitable type of furnace or electrolytic technique to be used.

The enormous quantities of iron required for the production of steels could not be extracted by any method other than carbon reduction, since this is the cheapest method available. In contrast, a much wider choice of techniques is possible for those metals which command higher prices but are required in smaller quantities. Thus, titanium can be obtained economically by reduction of titanium(IV) chloride using sodium in an inert atmosphere.

$$TiCl_4 + 4Na \longrightarrow Ti + 4NaCl$$

A derelict Cornish tin mine.

ECONOMIC FACTORS

The final choice of extraction method will almost invariably depend on cost. In this respect, quality of the ore is obviously important. Low-grade ores may not be economically worth exploitation; although it is interesting that Cornish tin mines, which closed down some decades ago owing to the poor quality of the ore, have now re-opened as the escalating prices of tin have made their operation economically profitable once more.

The demand for a particular metal will clearly affect its scale of production. Large-scale operations will necessitate a plentiful and preferably cheap reducing agent, such as coke, and possibly a cheap supply of electricity.

Consequently, almost all iron works are located near coalfields and aluminium smelters are usually situated near a source of cheap hydro-electric power.

Another crucial economic factor in the choice of extraction method relates to the value of any by-products. During the 1950's, a world surplus of zinc arose and prices sunk to an almost uneconomic level. Nevertheless, extraction of zinc continued because the process yielded sulphur dioxide for the manufacture of sulphuric(VI) acid which was in short supply at the time.

## 19.14 Metal extractions and the electrochemical series

ELECTROLYSIS OF FUSED COMPOUNDS FOR METALS AT THE TOP OF THE ELECTROCHEMICAL SERIES

This method is used to extract potassium, sodium, calcium, magnesium, and aluminium. Chemical reduction of the oxides of these metals by carbon or carbon monoxide is not feasible. The temperature required for reduction is too high to make the process economical or practical.

Furthermore, these metals cannot be obtained by electrolysis of their aqueous solutions since water in the solution would be discharged at the cathode in preference to the metal ions.

Since neither chemical reduction nor aqueous electrolysis is possible, the only viable method of extraction is electrolysis of their fused compounds, usually the chlorides.

In practice, the cathode at which the metal ions are discharged is usually made of steel, while chloride ions are discharged at a graphite anode.

$$\textit{Cathode}\ (-)\ \text{(steel)}\quad M^{2+}(l) + 2e^- \longrightarrow M(l)$$

$$\textit{Anode}\ (+)\ \text{(graphite)}\quad 2Cl^-(l) \longrightarrow Cl_2(g) + 2e^-$$

Large quantities of electricity are required not only to electrolyse the chloride but also to maintain the molten state. Suitable impurities are normally added to the chloride in order to reduce its melting point and thereby use less electrical energy in maintaining the molten state.

- ○ Why is the anode made of graphite rather than steel, in spite of the fact that steel is a better conductor than graphite?
- ○ Why must the metal liberated at the cathode be protected from the chlorine produced at the anode?
- ○ In the electrolytic manufacture of sodium, calcium chloride is added to sodium chloride to reduce the melting point of the latter. Will this result in any impurity in the sodium produced? Explain.

CHEMICAL REDUCTION OF COMPOUNDS FOR METALS IN THE MIDDLE OF THE ELECTROCHEMICAL SERIES

Metals in the middle of the activity series (such as zinc, iron, tin, lead and copper) can be extracted from their oxides and sulphides by chemical reduction. Very often, the sulphides are converted to oxides before reduction to the metal.

$$\underset{\text{zinc blende}}{2ZnS} + 3O_2 \longrightarrow 2ZnO + 2SO_2$$

*(a) Reduction of oxides by carbon*

The oxides are reduced by coke in a closed furnace forming the metal and carbon monoxide. This method is used to obtain zinc by heating a mixture of powdered zinc oxide and coke at 1 400°C. At this temperature, the zinc (b.pt. 907°C) distils from the furnace and condenses to a solid in the receivers.

$$ZnO + C \longrightarrow Zn + CO$$

*(b) Reduction of oxides by carbon monoxide*
Reduction of iron ore ($Fe_2O_3$) by carbon monoxide takes place in a blast furnace.

$$Fe_2O_3 + 3CO \longrightarrow 2Fe + 3CO_2$$

*(c) Self-reduction of sulphide ores*
Self-reduction plays an essential part in the extraction of both lead and copper from their sulphide ores.

Part of the sulphide is first converted to oxide by roasting in air.

$$2Cu_2S + 3O_2 \longrightarrow 2Cu_2O + 2SO_2$$

The supply of air is then cut off and the temperature is raised so that the rest of the sulphide reacts with oxide forming the metal and sulphur dioxide.

$$2Cu_2O + Cu_2S \longrightarrow 6Cu + SO_2$$

*(d) Reduction of compounds by more reactive metals*
This method is particularly useful for those metals such as chromium and titanium which are expensive and are required in only small quantities.

Thus, titanium is obtained by heating titanium(IV) chloride to 850°C with magnesium or sodium in an atmosphere of argon.

$$TiCl_4 + 2Mg \longrightarrow Ti + 2MgCl_2$$

$$TiCl_4 + 4Na \longrightarrow Ti + 4NaCl$$

- ○ Why is it necessary to use an atmosphere of argon?
- ○ Chromium is obtained from chromium(III) oxide by heating with aluminium powder. Write an equation for this process.

HEATING THE ORE ALONE OR DISPLACEMENT FROM AQUEOUS SOLUTION FOR METALS AT THE BOTTOM OF THE ELECTROCHEMICAL SERIES

*(a) Heating the ore alone*
The compounds of mercury and silver are so unstable that they decompose to the metal on heating. Consequently, mercury can be extracted from the ore, cinnabar (HgS), by heating in air.

$$HgS + O_2 \longrightarrow Hg + SO_2$$

The mercury distils over at the temperature of the furnace and is condensed in water-cooled receivers.

*(b) Displacement from aqueous solution*
In this case, the metal is precipitated from a solution of its ions by the addition of a metal higher in the electrochemical series.

Silver can be extracted from very low-grade ore by this method. The insoluble silver sulphide ore is first dissolved in a solution of cyanide ions with which it forms dicyanoargentate(I) ions.

$$Ag_2S(s) + 4CN^-(aq) \longrightarrow 2[Ag(CN)_2]^-(aq) + S^{2-}(aq)$$

Silver is then precipitated from the solution of dicyanoargentate(I) ions by the addition of powdered zinc dust.

$$2[Ag(CN)_2]^-(aq) + Zn(s) \longrightarrow [Zn(CN)_4]^{2-}(aq) + 2Ag(s)$$

## Summary

1 The world production of iron far exceeds that of any other metal. The reasons for this are that iron ores are abundant, rich, readily accessible and easily reduced to iron.

2 Iron can be alloyed easily with other elements to produce different steels with a great variety of properties and uses.

3 In many respects, iron is a typical transition metal. It produces coloured ions, it forms stable complexes, it shows variable oxidation state and the element and its compounds show catalytic activity.

4 Rusting requires the presence of both oxygen and water. During rusting, iron is first oxidized to $Fe^{2+}$ ions by dissolved oxygen in the water. Iron(II) hydroxide then precipitates and this is further oxidized by dissolved oxygen to rust (hydrated iron(III) oxide, $Fe_2O_3 . xH_2O$).

5 Two general methods are used to protect iron and steel from corrosion: the application of a protective layer or the application of a sacrificial metal.

6 The manufacture of copper from its ores involves three stages.
- (a) Conversion of the sulphide ore to $Cu_2O$ by roasting in air.
- (b) Reduction of the $Cu_2O$ to copper by heating with more sulphide ore in the absence of air.
- (c) Purification of the impure copper by electrolysis.

7 Although copper(II) compounds show typical transition metal properties, copper(I) compounds do not do so. Several of them are white, they show little or no catalytic activity and there are relatively few complexes containing $Cu^+$ ions.

8 $Cu^+(aq)$ is unstable in aqueous solution, disproportionating spontaneously into $Cu^{2+}(aq)$ and Cu. Thus, copper(I) compounds are stable in contact with water only if they are insoluble or when they are complexed with ligands other than water.

9 Metals at the top of the electrochemical series (e.g. K, Na, Ca, Mg, Al) are usually obtained by electrolysis of their fused compounds. Metals in the middle of the electrochemical series (e.g. Zn, Fe, Sn, Pb, Cu) are usually obtained by chemical reduction of their oxides with carbon or carbon monoxide or by self reduction of sulphide ores. Metals at the bottom of the electrochemical series (e.g. Cu, Hg, Ag, Au) are usually obtained by heating the ore alone or by displacement from aqueous solution.

## Study questions

1 (a) What do you understand by the term 'transition metal'?
(b) Which of the following do you regard as transition metals? Explain your answer.
(i) Scandium
(ii) Iron
(iii) Zinc
(c) Although the salts of transition elements are usually coloured, there are several copper(I) compounds which are white. Suggest an explanation for this.
(d) The densities of transition elements in the same period gradually increase with relative atomic mass (table 19.5). Why is this?

**Table 19.5** The densities of four metals in the first transition series.

| Metal | Relative atomic mass | Density/g $cm^{-3}$ |
|---|---|---|
| Vanadium | 50.9 | 6.1 |
| Manganese | 54.9 | 7.4 |
| Cobalt | 58.9 | 8.9 |
| Copper | 63.5 | 9.0 |

2 Comment upon and explain the following observations.
(a) An iron pipe, buried in the earth is protected from rusting by joining it to a bar of magnesium.
(b) A transition metal, M, forms only two isomers of formula $[M(NH_3)_4Cl_2]^+Cl^-$.
(c) Copper does not react appreciably with hydrochloric acid but will do so, evolving hydrogen steadily, in the presence of strong complexing agents.
(d) Iron immersed in copper(II) sulphate(VI) solution is rapidly coated with copper, but this does not occur when the iron has previously been dipped in concentrated nitric(V) acid.
(e) Copper(II) sulphate(VI) does not react with potassium chloride solution yet it reacts with potassium iodide solution to form a precipitate of copper(I) iodide and a solution of iodine.

3 Metals are usually isolated from their ores by reduction. Describe briefly the methods of reduction used for the following metals. Suggest reasons for the method chosen for each metal.
(a) Zinc from zinc blende ($ZnS$)
(b) Magnesium from magnesium chloride extracted from sea water
(c) Iron from haematite ($Fe_2O_3$)
(d) Silver from silver glance ($Ag_2S$)

4 10g of an impure iron(II) salt were dissolved in water and made up to $200cm^3$ of solution. $20cm^3$ of this solution acidified with dilute $H_2SO_4$ required $25cm^3$ of 0.03M $KMnO_4(aq)$ before a faint pink colour appeared.
(a) Write a balanced equation (or half-equations) for the reaction of acidified manganate(VII) ions with iron(II) ions.
(b) How many moles of iron(II) ions react with 1 mole of $MnO_4^-$ ions?
(c) How many moles of $Fe^{2+}$ react with $25cm^3$ of 0.03M $KMnO_4(aq)$?
(d) How many grams of $Fe^{2+}$ are there in the $200cm^3$ of original solution? ($Fe = 56$)
(e) What is the percentage by weight of iron in the impure iron(II) salt?

5 (a) Most metals exist naturally in a combined state because they are too reactive to occur native (uncombined). However, a few metals do occur native.
Give examples of:
(i) two metals which never occur native,
(ii) two metals which occur both native and combined,
(iii) two metals which almost always occur naturally in an uncombined form.
(b) Before extraction of the metal can begin, many ores must be purified and concentrated. Mention two different processes by which this is done.
(c) Why do metal extractions often produce slag?
(d) In many cases, the primary reduction of an ore does not produce metal of sufficient purity for the uses to which it will be put and the metal requires further purification. Outline two different processes by which further purification is carried out.

6 Economic considerations are important in deciding the method by which a metal is extracted from its ore.
How may economic considerations affect:
(a) the nature of the ore used,
(b) the nature of the reducing agent used,
(c) the location of the industrial plant at which extraction is carried out?
Give examples wherever possible to illustrate the point you make.

7 The method by which any metal is extracted is dependent upon the position of that metal in the electrochemical series.
(a) Highly reactive metals must be extracted by electrolysis.
(i) Give a brief outline of the extraction process for one such metal mentioning the electrolyte used, the materials of which the electrodes are made and the reactions taking place at each electrode.
(ii) Why are such reactive metals not obtained by chemical methods of reduction?
(b) A few metals are obtained by reduction of their purified ores with a second more reactive metal.
(i) Give a brief outline of the chemical principles of one such extraction process.
(ii) Why is this type of process fairly uncommon?
(c) Less reactive metals can be extracted by reduction of their ores with carbon.
(i) Give a brief outline of the chemical principles of the extraction of two different metals using carbon.
(ii) What are the disadvantages of reduction with carbon?

# 20 Equilibria

## 20.1 Introduction

Everyone has experienced the difficult position of having one's loyalty and responsibility pulled in two directions at the same time. More often than not, one has to settle for a compromise (or balance) between the alternatives.

In the same way, many physical and chemical processes also exist in a position of balance. In chapter 13, we studied the competition processes taking place in acid/base, redox and complexing processes. In these and other reactions, the starting materials or reactants are not always completely converted to the products, but sometimes reach an intermediate position or **equilibrium** in which both reactants and products are present.

In real life situations, any compromise in difficult circumstances is usually dictated by many factors—responsibility, loyalty, honesty, age, parents, finance, etc. Fortunately, the factors which dictate the balance position in chemical reactions are fewer and better understood than those in real life situations.

The tendency for materials to exist in a state of maximum disorder is a key factor in dictating the position of equilibrium. Disorder is, in fact, a more natural state than order. Even so, it may be rather difficult to convince your parents of this when you are next told to tidy your room.

## 20.2 Equilibria in physical processes

### LIQUID–VAPOUR EQUILIBRIUM: VAPOUR PRESSURE

When liquid bromine is shaken in a stoppered flask, some of the liquid evaporates forming an orange gas above the orange-red liquid. Gradually, the gas becomes thicker, but eventually the intensity of the brown gas does not change any more. However much we shake the flask, the colour remains constant and, provided we took more than a very small volume of liquid bromine, some liquid remains in the flask.

The constant colour of the gas within the flask and its constant vapour pressure suggest that a position of equilibrium has been reached between bromine liquid and bromine gas. This equilibrium can be summarized as:

$$Br_2(l) \rightleftharpoons Br_2(g)$$

What is happening to the molecules in the flask at equilibrium? Do all those in the gaseous phase remain as gas while all those in the liquid remain as liquid (**a static equilibrium**) or are some gas molecules becoming liquid molecules while an equal number of liquid molecules become gas molecules (**a dynamic equilibrium**)?

Since liquid and gas molecules are moving rapidly and randomly in all directions, it would seem likely that molecules in the flask are in a dynamic rather than a static equilibrium. Thus, the rate at which molecules leave the liquid surface and enter the vapour is just equal to the rate at which other molecules in the vapour return to the liquid. Random molecular activity occurs even after all the external signs of change have disappeared, but we cannot *see* the molecules moving, nor can we measure the rate at which they enter or leave the vapour.

The difference between a static and a dynamic equilibrium is illustrated in figure 20.1 using two familiar situations.

One of the key factors which dictates the position of equilibrium in any physical or chemical process is the tendency for materials to exist in the lowest energy state possible. Will equilibrium ever be achieved in the waterfall?

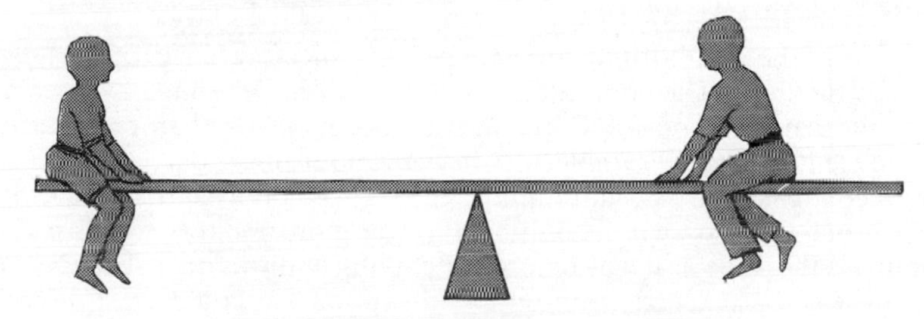

(*a*) Static equilibrium: children on a see-saw At the balance point (i.e. the equilibrium position) no movement of the children or the see-saw occurs.

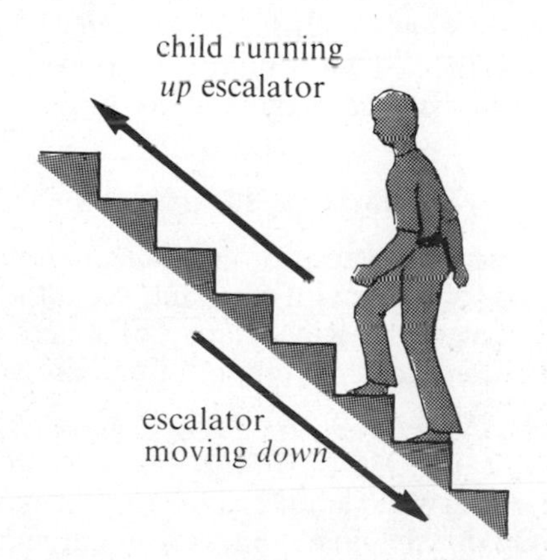


(*b*) Dynamic equilibrium: child ascending escalator at the same rate as the escalator descends. At the balance point (i.e. the equilibrium position) the child and the escalator are moving at the same rate in opposite directions. Thus, the equilibrium position is maintained.

**Figure 20.1** The difference between a static and a dynamic equilibrium.

Consider the equilibrium between bromine liquid and bromine vapour once again. Figure 20.2(*a*) shows the perfect balance between rates of evaporation and condensation at equilibrium. Suppose that some of the vapour is suddenly removed without affecting the system in any other way (figure 20.2(*b*)).

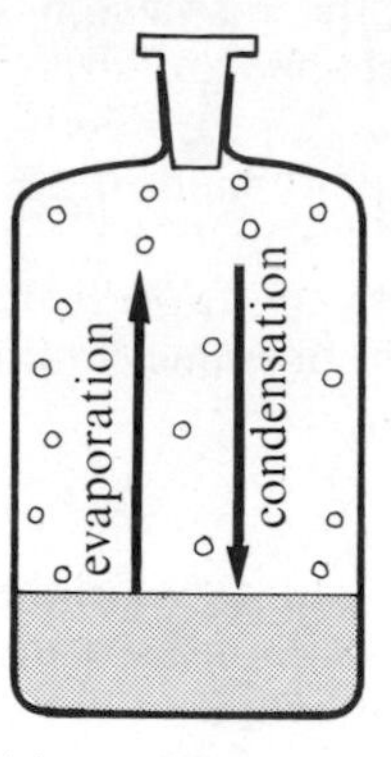


(*a*) **Equilibrium**
Rate of evaporation
= rate of condensation

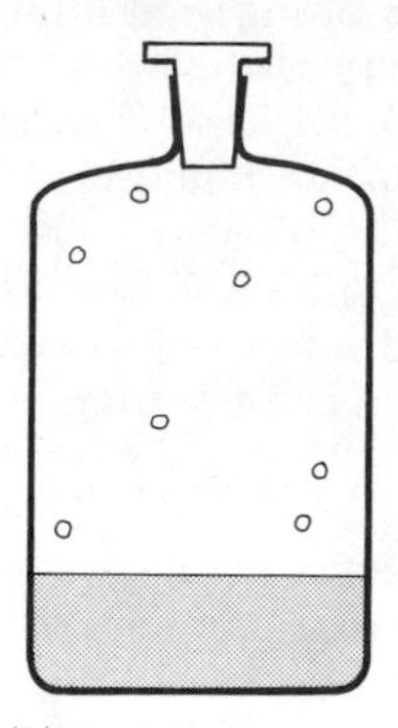

(*b*) **Imbalance**
Gas removed
from vapour phase

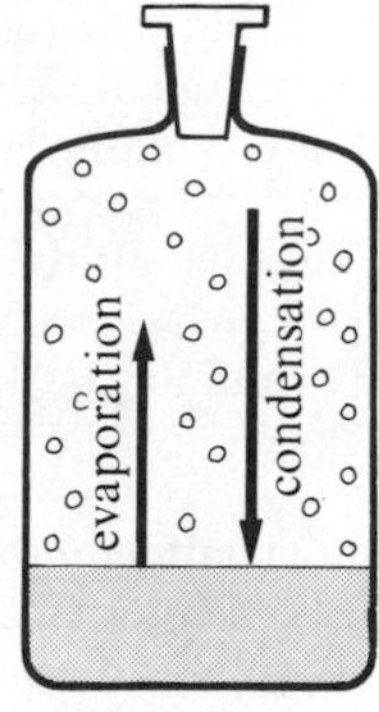


(*c*) **Imbalance**
Gas added to vapour phase
Rate of condensation
> rate of evaporation

**Figure 20.2** Liquid-vapour equilibria.

When the atmosphere is saturated with water vapour, water is in equilibrium with its vapour.

- ○ Is the rate of condensation of gas molecules still equal to the rate of evaporation of liquid molecules? Explain.
- ○ What happens to the concentration of molecules in the gas phase as time goes on? Explain.
- ○ How and when is equilibrium attained once more?

Suppose, now, the equilibrium is disturbed by injecting an excess of bromine vapour into the flask. The concentration of gas molecules suddenly rises and the rate of condensation increases. Condensation occurs faster than evaporation and thus the concentration of gas molecules in the vapour phase decreases. Gradually, the rate of condensation falls until finally the rate of condensation equals the rate of evaporation. The system is in equilibrium once more and the vapour pressure of the bromine is the same as it was before the equilibrium was disturbed. Notice that equilibrium can be reached from either direction. We could start simply with bromine liquid or with excess bromine vapour and *provided we use a closed container*, an equilibrium will eventually be obtained.

### SOLUTE–SOLUTION EQUILIBRIUM: SOLUBILITY

When one teaspoon of sugar is added to a cup of tea, all the sugar dissolves forming a solution. If more sugar is added, this also dissolves. If you are very sweet-toothed and continue to add sugar, a stage is eventually reached when no more will dissolve. The solution (tea) is now saturated with solute (sugar) at the temperature involved, and solute particles in the undissolved sugar are in equilibrium with solute particles in the solution.

$$\text{sugar(s)} \rightleftharpoons \text{sugar(aq)}$$

Provided the system is closed, no evaporation of solvent occurs and the amounts of dissolved and undissolved solute remain constant. As in the liquid/vapour system, equilibrium can be recognized by a constancy of macroscopic properties—the concentration of the solution and the amount of undissolved solute.

#### *INVESTIGATING THE NATURE OF THE EQUILIBRIUM BETWEEN A SOLUTE AND ITS SATURATED SOLUTION.*

What are the molecules doing when equilibrium is established between a solute and its saturated solution? Obviously, no changes are apparent at a macroscopic level but what is happening at a molecular (microscopic) level?

It would seem reasonable to predict that the equilibrium in this case is also dynamic with particles leaving the undissolved solute and entering the solution at the same rate as dissolved particles rejoin the crystalline solid.

In order to check these ideas about a dynamic equilibrium, we can 'label' some of the particles in either the saturated solution or in the undissolved solute using a radioactive material.

If solid radioactive $^{212}PbCl_2$ and saturated non-radioactive $PbCl_2(aq)$ are mixed together, no increase in the amount of dissolved lead chloride can occur. Nevertheless, if our ideas about a dynamic equilibrium are correct, there will be an interchange of $Pb^{2+}$ between the undissolved solid and the saturated solution. As a result, radioactive $^{212}PbCl_2$ will appear in the solution.

The background radiation of the saturated aqueous lead chloride is first measured. It is then shaken with radioactive solid $^{212}PbCl_2$ for ten minutes. After this, the saturated solution and undissolved solute are separated by centrifuging and the radioactivity of the saturated solution is measured a second time. The mixing and centrifuging are repeated at intervals and the radioactivity of the saturated solution is measured at each stage. Figure 20.3 shows how the radioactivity of the solution changes with time.

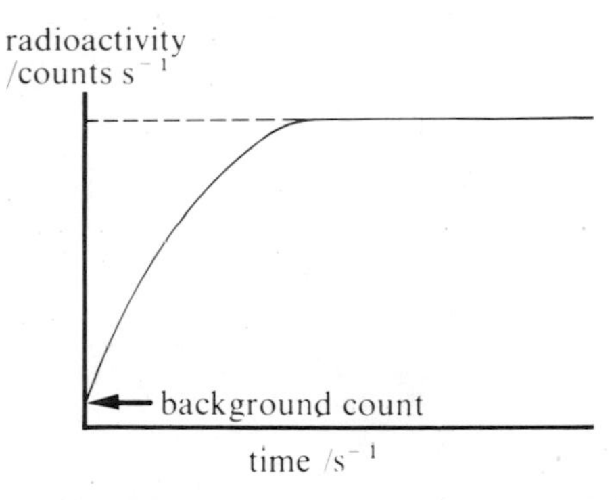


**Figure 20.3** Following the radioactivity of a saturated solution in contact with its radioactive solute.

The increasing radioactivity of the saturated solution suggests that solid radioactive $^{212}PbCl_2$ must have dissolved. But, the solution was saturated before $^{212}PbCl_2$ dissolved, so any radioactive particles dissolving from the solid have necessarily been replaced by non-radioactive particles crystallizing on to the solid from the solution. This means that a process of dynamic equilibrium is taking place between the undissolved solute and the saturated solution.

$$PbCl_2(s) \rightleftharpoons PbCl_2(aq)$$

Figure 20.3 shows that the radioactivity of the saturated solution eventually reaches a constant value. Why is this?

## 20.3 Characteristic features of a dynamic equilibrium

Considerations in the last section have highlighted four important features of dynamic equilibria.

1 *Equilibrium is characterized by constant macroscopic properties* under the given conditions of temperature, pressure and initial amounts of substances.
2 *At equilibrium, microscopic (molecular-scale) processes continue but these are balanced so that no overall macroscopic (large-scale) changes occur.* The particles participate in both forward and reverse processes, but the rate of the forward process is equal to the rate of the reverse process so that no net change results.
3 *The equilibrium can be attained from either direction* beginning with only the materials on one side of the change. Changes of this kind are described as **reversible** and the reactions which take place as equilibrium is approached are said to be reversible reactions.
4 *Equilibrium can only be achieved in a closed system.* An open system which allows matter to escape or to enter cannot reach equilibrium, unless there is absolutely no loss or gain of materials to or from the surroundings.

## 20.4 Equilibria in chemical reactions

THE DECOMPOSITION OF CALCIUM CARBONATE

When calcium carbonate is heated strongly, it decomposes forming calcium oxide and carbon dioxide.

$$CaCO_3(s) \xrightarrow{\text{heat}} CaO(s) + CO_2(g)$$

Normally, $CO_2$ escapes well away from the solid CaO produced and consequently their recombination to form $CaCO_3$ never occurs in an open container. The system can never reach equilibrium.

If, however, a few grams of $CaCO_3(s)$ are heated at 800°C in a *closed* container from which all the air has been removed, only part of the solid is decomposed no matter how long it is heated. What is more, the pressure of $CO_2$ inside the container rises, but then remains steady at a constant value. As long as the temperature is held constant at 800°C, the pressure remains constant at 25 kPa (0.25 atm.). The reaction appears to have reached an equilibrium since constant macroscopic properties have been achieved in a closed system.

In order to check that the system is indeed at equilibrium, we must show that it reaches the same macroscopic composition when approached from the opposite direction. Solid CaO is first placed in the reaction vessel which is then evacuated and refilled with $CO_2$ at a pressure well above 25 kPa. Finally, the container is closed and maintained at 800°C. The pressure begins to fall and becomes steady at 25 kPa (0.25 atm.). The same equilibrium pressure of $CO_2$ results whether we commence with $CaCO_3$ or with CaO and $CO_2$. This is further evidence that the materials are in an equilibrium which can be summarized as

$$CaCO_3(s) \rightleftharpoons CaO(s) + CO_2(g)$$

In this case, $CaCO_3$ is decomposing to form CaO + $CO_2$, at the same rate as CaO + $CO_2$ are reforming $CaCO_3$.

Although this particular system will come to equilibrium in a closed vessel, the industrial process is never allowed to do so. Millions of tonnes of calcium oxide are produced annually by heating limestone in large open kilns. The CaO is used in liming the soil and as a constituent of plaster and mortar.

A rotatory lime kiln at the Westbury Works, near Trowbridge, Wiltshire. The kiln is viewed from the lower heated end. Raw limestone slurry is pumped in at the higher end. As the slurry flows down the kiln, it is first dried and then decomposed to lime at the lower heated end. The kiln is inclined at 1 in 30 and makes one revolution per minute.

### THE IODINE CHLORIDE/IODINE TRICHLORIDE EQUILIBRIUM

When dry chlorine is passed over brown liquid iodine chloride (ICl), yellow crystals of iodine trichloride ($ICl_3$) form and the brown liquid disappears.

$$\underset{\text{pale green}}{Cl_2(g)} + \underset{\text{brown}}{ICl(l)} \longrightarrow \underset{\text{yellow}}{ICl_3(s)}$$

If the supply of chlorine is stopped and the $ICl_3$ is left in an open container, it slowly decomposes to a brown liquid (ICl) and a pale green gas ($Cl_2$) escapes.

$$\underset{\text{yellow}}{ICl_3(s)} \longrightarrow \underset{\text{brown}}{ICl(l)} + \underset{\text{pale green}}{Cl_2(g)}$$

This is another example of a reversible reaction. But what happens when a mixture of $Cl_2(g)$ and ICl(l) is left in a closed vessel from which the air has been removed?

Some yellow crystals of $ICl_3$ form, but not all the reactants are used up. As the reaction proceeds, the amounts of ICl and $Cl_2$ fall and the amount of $ICl_3$ rises (figure 20.4).

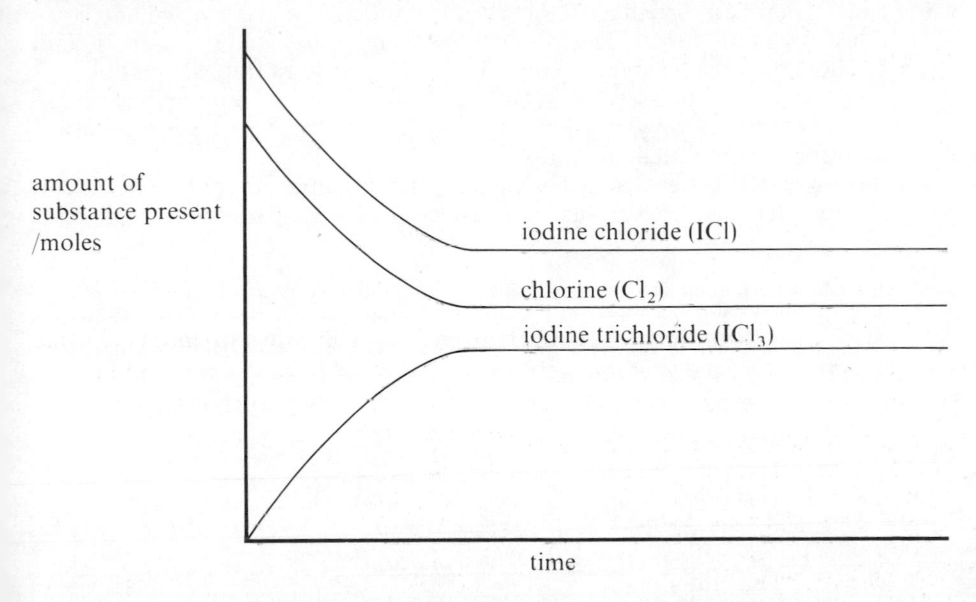


**Figure 20.4** Following the amount of substances present as equilibrium is approached in the reaction between ICl and $Cl_2$.

After some time, the composition of the mixture becomes constant and the amount of each reactant remains steady. The reactions appear to have attained an equilibrium since the macroscopic properties of the closed system have become constant.

○ What is happening to the molecules of $ICl_3$, ICl and $Cl_2$ inside the container at equilibrium?
○ What can be said about the rate of formation of $ICl_3$ and the rate of decomposition of $ICl_3$ at equilibrium?
○ How could you show experimentally that the equilibrium is a dynamic one? Look closely at figure 20.4.
○ What is the relationship between the decrease in amount of ICl, the decrease in amount of $Cl_2$ and the increase in amount of $ICl_3$?
○ Why is there always a constant difference between the amounts of ICl and $Cl_2$?
○ Why do the three curves become horizontal after the same time?

## 20.5 The equilibrium of a solute between two immiscible solvents: the Partition Coefficient

Try the following experiment for yourself. Dissolve a small crystal of iodine in 5cm$^3$ of aqueous potassium iodide. Notice that a brown solution forms and then carefully add 5cm$^3$ of tetrachloromethane to the mixture. The tetrachloromethane is clear at first, but when the mixture is shaken gently, iodine dissolves from the KI solution into the tetrachloromethane and a purple solution of iodine forms in the latter. Now shake vigorously. The density of the purple colour in the tetrachloromethane darkens whilst the brown colour of the KI(aq) becomes paler. Eventually, however, the colours of the two solutions remain constant. The iodine has distributed itself between the two solvents and an equilibrium has been attained. No matter how much the mixture is shaken, no further changes in colour intensity occur and the concentrations of iodine in the two solutions remain constant.

In the same way, other solutes which are significantly soluble in two immiscible solvents will distribute themselves between both solvents when the three substances are shaken together. These are further examples of equilibrium systems characterized by constant concentrations. But, how do the concentrations of the solute in the different solvents depend on the initial amount of solute taken? Is there any relationship between the equilibrium concentrations of the solute in the different solvents when different amounts of solute are used?

*INVESTIGATING THE PARTITION OF BUTANEDIOIC ACID (SUCCINIC ACID) BETWEEN WATER AND ETHOXYETHANE (DIETHYL ETHER).* One gram of butanedioic acid was shaken in a separating funnel with 25cm$^3$ of water and 25cm$^3$ of ethoxyethane (ether) until all the solid had dissolved and equilibrium had been established. The mixture was then left to stand for some time in order to allow the organic and aqueous layers to separate as fully as possible.

The two layers were then separated and the concentration of butanedioic acid in each layer was determined by titration against sodium hydroxide solution of known molarity using phenolphthalein indicator.

Further experiments were carried out using different initial amounts of butanedioic acid between 0.5 and 1.5 grams. The results obtained are shown in table 20.1.

**Table 20.1** Equilibrium concentrations of butanedioic acid in ether and in water.

| Experiment Number | Equilibrium concentration of butanedioic acid in ether layer/mole dm$^{-3}$ | Equilibrium concentration of butanedioic acid in water layer/mole dm$^{-3}$ |
|---|---|---|
| 1 | 0.023 | 0.152 |
| 2 | 0.028 | 0.182 |
| 3 | 0.036 | 0.242 |
| 4 | 0.044 | 0.300 |
| 5 | 0.052 | 0.358 |
| 6 | 0.055 | 0.381 |

When these equilibrium concentrations in table 20.1 are plotted graphically, a straight line is obtained (figure 20.5).

- ○ Would you have expected the graph to go through the origin? Explain.
- ○ What does the straight line graph suggest about the ratio of the equilibrium concentrations of butanedioic acid in the two solvents?
- ○ How does the ratio of concentrations depend upon the amount of butanedioic acid taken?

The results in figure 20.5 suggest that the ratio:

$$\frac{\text{concentration of butanedioic acid in ether}}{\text{concentration of butanedioic acid in water}} \text{ is constant}$$

Using square brackets, [ ], to denote the concentrations we can deduce from the graph in figure 20.5 that

$$\frac{[\text{butanedioic acid (ether)}]_{\text{eqm}}}{[\text{butanedioic acid (water)}]_{\text{eqm}}} = 0.15$$

This ratio is known as the **partition coefficient** (or the **distribution ratio**) for butanedioic acid between ether and water. *The partition coefficient is independent of the amount of solute taken and also independent of the volumes of the solvents used.* Thus, it is important to realize that it is *the ratio of concentrations and not the ratio of masses* of solute which matter. Other investigations of the ratio in which a dissolved substance distributes itself between two immiscible liquids at equilibrium also give a constant partition coefficient. In each case, the partition coefficient remains constant provided:

(i) the temperature is constant,
(ii) the solvents are immiscible and do not react with each other,
(iii) the solute does not react, associate or dissociate in the solvents.

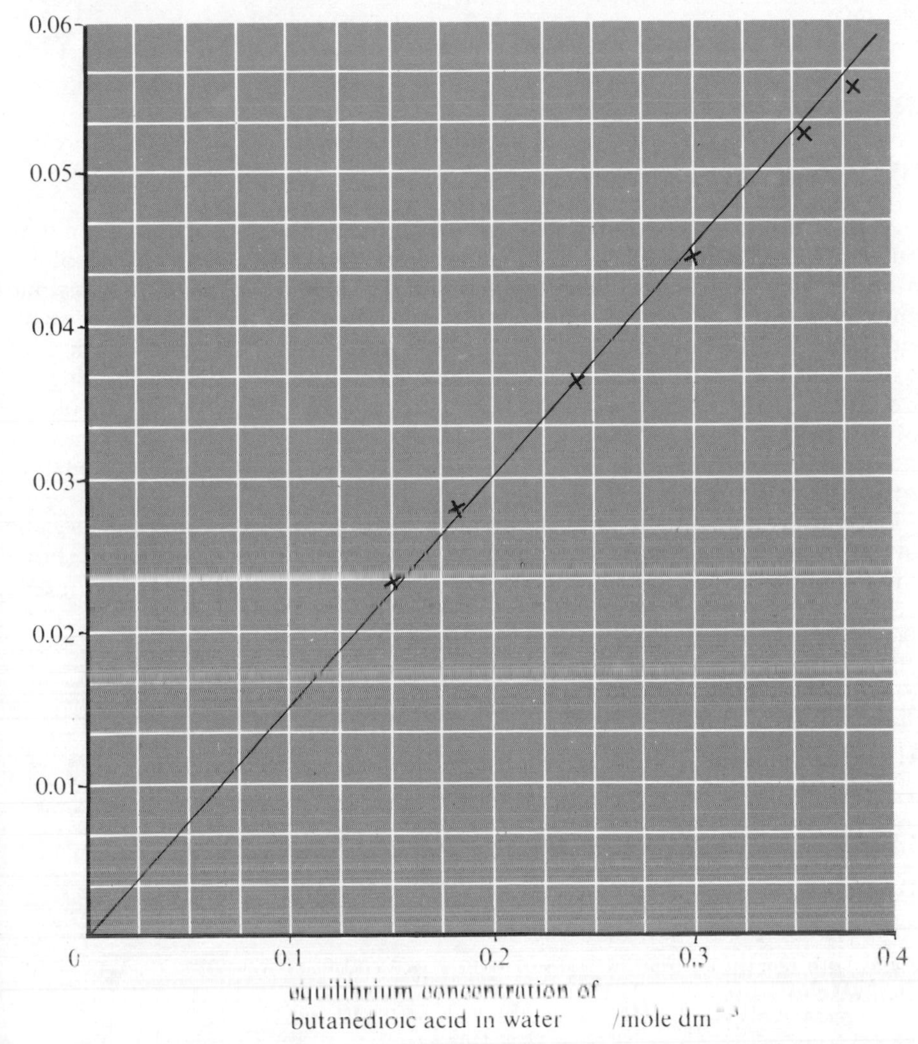


**Figure 20.5** A graph of the equilibrium concentration of butanedioic acid in ether against that in water.

## 20.6 The partition law and Henry's Law

The partition law is simply an application of a more generalized statement which says that *the concentrations of any single molecular species in two immiscible phases bear a constant ratio to each other at a fixed temperature*. In the case of the partition law, the immiscible phases are the two solvents and the single molecular species is the distributed solute.

The same general statement can also be applied to the distribution of a gas between its solution and vapour phases. As early as 1803, Henry had shown that *the mass of a gas dissolved in a given volume of liquid at a constant temperature is proportional to the pressure of the gas*; that is:

$$\text{mass of gas dissolved per unit volume} \propto \text{pressure of gas.}$$

This is usually referred to as **Henry's Law.**

Of course, the law is only true for those gases which do not react with their solvent. It cannot, for example, be applied to aqueous solutions of either ammonia or hydrogen chloride since both these gases react with water.

Henry's Law can be expressed in a second, more general form. The mass of gas dissolved per unit volume of the solvent is a measure of the concentration of gas in solution. Furthermore, the pressure of the gas is simply another way of expressing its concentration. Hence, the proportionality between the mass of gas dissolved and its pressure can be written as:

$$\text{Concentration of gas in solution} \propto \text{concentration of gas above solution}$$

$$\Rightarrow \frac{[\text{gas (in solution)}]_{\text{eqm}}}{[\text{gas (above solution)}]_{\text{eqm}}} = \text{constant}$$

## 20.7 Solvent extraction

The most important aspect of partition is its application in solvent extraction both industrially and in the laboratory. Organic compounds are generally more soluble in non-polar solvents such as methylbenzene (toluene), tetrachloromethane and ethoxyethane (ether) than in water, and these solvents are themselves immiscible with water. Thus, organic compounds can be extracted from aqueous solutions or suspensions by shaking with a non-polar organic solvent and then separating the two layers. The pure organic compound can then be obtained by distilling off the solvent. For example, penicillin is extracted from a dilute aqueous solution using trichloromethane whilst phenylamine (aniline) can be reclaimed from a mixture with water by using ether.

During solvent extraction, it is more efficient to use a given volume of organic solvent in small portions rather than all at once. The following calculation shows why this is so.

Suppose that 100cm$^3$ of ether is available for extracting the solute, X, from 100cm$^3$ of water, and that the partition coefficient of X between ether and water is 4.

FRACTION OF X EXTRACTED USING 100CM$^3$ OF ETHER ALL AT ONCE

After shaking the aqueous solution of X with ether, we know that:

$$\frac{\text{concentration of X in ether}}{\text{concentration of X in water}} = 4$$

$$\Rightarrow \frac{\text{mass of X in ether}}{\text{volume of ether}} \times \frac{\text{volume of water}}{\text{mass of X in water}} = 4$$

Let us suppose that $m_e$ is the mass of X in ether and $m_w$ is the mass of X in water. Then we can write:

$$\frac{m_e}{100} \cdot \frac{100}{m_w} = 4$$

$$\therefore \frac{m_e}{m_w} = \frac{4}{1}$$

i.e. four parts of X are in the ether and one part is in the water. Thus, using 100cm$^3$ of ether in one extraction, four-fifths of X are extracted into the organic solvent.

FRACTION OF X EXTRACTED USING TWO 50CM$^3$ PORTIONS OF ETHER

When 50cm$^3$ of ether is used to extract X from 100cm$^3$ of water we can write:

$$\frac{m_e}{50} \cdot \frac{100}{m_w} = 4$$

$$\therefore \frac{m_e}{m_w} = 4 \times \frac{50}{100} = \frac{2}{1}$$

i.e. two parts of X are in the ether and one part in the water. This means that two-thirds of X are extracted by the first 50cm$^3$ of ether leaving one-third in the water.

The second 50cm$^3$ portion of ether extracts two-thirds of the remaining one-third, i.e. $\frac{2}{3} \times \frac{1}{3} = \frac{2}{9}$ths of the original amount of X.

Thus, the total fraction of X extracted by two 50cm$^3$ portions of ether is $\frac{2}{3} + \frac{2}{9} = \frac{8}{9}$, which is more than the four-fifths extracted by one 100cm$^3$ portion of ether. The total fraction of X extracted would, of course, be even greater than eight-ninths if three portions of $33\frac{1}{3}$cm$^3$ were used, and so on.

## 20.8 The equilibrium constant

Earlier in this chapter we discovered that there was a constant value at one particular temperature for the ratio of concentrations of a solute between two immiscible solvents at equilibrium. Thus, it would seem sensible to investigate whether there is a similar relationship between the concentrations of reactants and products at equilibrium in chemical reactions.
Table 20.2 shows information concerning the following reaction at 731 K:

$$H_2(g) + I_2(g) \rightleftharpoons 2HI(g).$$

**Table 20.2** Initial and equilibrium concentrations of $H_2$, $I_2$ and HI.

| Experiment number | Initial concentrations/mole dm$^{-3}$ | | | Equilibrium concentrations/mole dm$^{-3}$ | | |
|---|---|---|---|---|---|---|
| | $[H_2(g)]$ | $[I_2(g)]$ | $[HI(g)]$ | $[H_2(g)]_{eqm}$ | $[I_2(g)]_{eqm}$ | $[HI(g)]_{eqm}$ |
| 1 | $2.40 \times 10^{-2}$ | $1.38 \times 10^{-2}$ | 0 | $1.14 \times 10^{-2}$ | $0.12 \times 10^{-2}$ | $2.52 \times 10^{-2}$ |
| 2 | $2.40 \times 10^{-2}$ | $1.68 \times 10^{-2}$ | 0 | $0.92 \times 10^{-2}$ | $0.20 \times 10^{-2}$ | $2.96 \times 10^{-2}$ |
| 3 | $2.44 \times 10^{-2}$ | $1.98 \times 10^{-2}$ | 0 | $0.77 \times 10^{-2}$ | $0.31 \times 10^{-2}$ | $3.34 \times 10^{-2}$ |
| 4 | $2.46 \times 10^{-2}$ | $1.76 \times 10^{-2}$ | 0 | $0.92 \times 10^{-2}$ | $0.22 \times 10^{-2}$ | $3.08 \times 10^{-2}$ |
| 5 | 0 | 0 | $3.04 \times 10^{-2}$ | $0.345 \times 10^{-2}$ | $0.345 \times 10^{-2}$ | $2.35 \times 10^{-2}$ |
| 6 | 0 | 0 | $7.58 \times 10^{-2}$ | $0.86 \times 10^{-2}$ | $0.86 \times 10^{-2}$ | $5.86 \times 10^{-2}$ |

In experiments 1, 2, 3 and 4 the sealed reaction vessel contains gaseous hydrogen and gaseous iodine initially. After a time, the composition of the mixture in the vessel, comprising unreacted hydrogen, unreacted iodine and hydrogen iodide, remains unchanged and equilibrium is attained. In experiments 5 and 6, equilibrium in the system is approached from the opposite direction and the reaction vessel contains only gaseous hydrogen iodide initially. Table 20.2 shows the initial and equilibrium concentrations of hydrogen, iodine and hydrogen iodide.

○ According to the equation

$$H_2(g) + I_2(g) \rightleftharpoons 2HI(g)$$

one mole of $H_2$ reacts with one mole of $I_2$ forming two moles of HI.
Check the data from experiments 1 and 2 to see whether:
no. of moles $H_2$ reacted = no. of moles $I_2$ reacted = $\frac{1}{2}$ no. of moles HI formed.

○ Why is $[H_2(g)]_{eqm} = [I_2(g)]_{eqm}$ in experiments 5 and 6?

○ How could you show that the system involving $H_2$, $I_2$ and HI is in dynamic equilibrium in these experiments?

By comparing the reaction $H_2(g) + I_2(g) \rightleftharpoons 2HI(g)$ with a partition equilibrium, it would seem reasonable to evaluate the ratio of product and reactant concentrations as:

$$\frac{[HI(g)]_{eqm}}{[H_2(g)]_{eqm}\,[I_2(g)]_{eqm}}$$

for each set of equilibrium concentrations to see if the expression is constant. These computed values are shown in the second column of table 20.3. The ratios are far from constant.

If, however, we evaluate the values for the ratio:

$$\frac{[HI(g)]^2_{eqm}}{[H_2(g)]_{eqm}\,[I_2(g)]_{eqm}}$$

we obtain the results in column three of table 20.3. The values obtained are constant, within experimental limits, and therefore we can write:

$$\frac{[HI(g)]^2_{eqm}}{[H_2(g)]_{eqm} \cdot [I_2(g)]_{eqm}} = \text{a constant} = 46.8 \text{ at } 731\text{K}$$

**Table 20.3** Equilibrium constants for the reaction $H_2 + I_2 \rightleftharpoons 2HI$.

| Experiment number | $\dfrac{[HI(g)]_{eqm}}{[H_2(g)]_{eqm}[I_2(g)]_{eqm}}$ | $\dfrac{[HI(g)]^2_{eqm}}{[H_2(g)]_{eqm}[I_2(g)]_{eqm}}$ |
|---|---|---|
| 1 | 1840 | 46.4 |
| 2 | 1610 | 47.6 |
| 3 | 1400 | 46.7 |
| 4 | 1520 | 46.9 |
| 5 | 1970 | 46.4 |
| 6 | 790 | 46.4 |

Notice that the ratio of concentrations which gives a constant value is related to the stoichiometry of the reaction.

$$H_2(g) + I_2(g) \rightleftharpoons 2HI(g)$$

The ratio uses the *second* power of the [HI] in the equilibrium expression and the first power for both $[H_2]$ and $[I_2]$. Thus, *the power to which we raise the concentration of each substance in the equilibrium expression is the same as its coefficient in the stoichiometric equation.*

The value of this ratio of concentrations at equilibrium is known as the **equilibrium constant** and represented by the symbol $K_c$.
Thus, for the reaction $H_2(g) + I_2(g) \rightleftharpoons 2HI(g)$

$$K_c = \frac{[HI(g)]^2_{eqm}}{[H_2(g)]_{eqm} \cdot [I_2(g)]_{eqm}}$$

Very often, the 'eqm' subscripts are omitted from the concentration terms since it is assumed that the concentrations involved in the expression for $K_c$ are necessarily equilibrium ones. Hence, the expression for $K_c$ is shortened to

$$K_c = \frac{[HI(g)]^2}{[H_2(g)].[I_2(g)]}$$

The subscript '$c$' indicates that $K_c$ is expressed in concentrations.

## 20.9 The equilibrium law

The equilibria in many other chemical reactions have also been studied and in each case an equilibrium constant has been obtained which relates to the stoichiometric equation in a similar manner to that for the hydrogen, iodine and hydrogen iodide system.

For example, the reaction:

$$Fe^{3+}(aq) + NCS^-(aq) \rightleftharpoons Fe(NCS)^{2+}(aq)$$

has been studied by colorimetry and shown to have an equilibrium constant in which:

$$K_c = \frac{[Fe(NCS)^{2+}(aq)]}{[Fe^{3+}(aq)]\,[NCS^-(aq)]}$$

These observations lead to the general statement known as the **Equilibrium Law** or the Law of Chemical Equilibrium.

If an equilibrium mixture contains substances A, B, C and D related by the equation

$$aA + bB \rightleftharpoons cC + dD$$

it is found experimentally that

$$\frac{[C]^c[D]^d}{[A]^a[B]^b} = K_c$$

where $K_c$, *the equilibrium constant, is constant at a given temperature.*

In writing expressions for the equilibrium constant of a reaction, it is conventional to put the concentrations of substances on the right hand side of the equation in the numerator and those of the substances on the left hand side in the denominator.

Thus, it is essential to relate any numerical value for an equilibrium constant to the particular equation concerned. For example, the equilibrium constant for the reaction $H_2 + I_2 \rightleftharpoons 2HI$:

$$K_c = \frac{[HI]^2}{[H_2][I_2]} = x,$$

but the equilibrium constant for the reaction $2HI \rightleftharpoons H_2 + I_2$ at the same temperature is:

$$K_c^{\prime} = \frac{[H_2][I_2]}{[HI]^2} = \frac{1}{x}$$

As you would expect, $K_c^{\prime} = \frac{1}{K_c}$

On the other hand, the equilibrium constant for the reaction:

$$\tfrac{1}{2}H_2 + \tfrac{1}{2}I_2 \rightleftharpoons HI \text{ is}$$

$$K_c^{\prime\prime} = \frac{[HI]}{[H_2]^{\frac{1}{2}}[I_2]^{\frac{1}{2}}} = \sqrt{x}$$

What is the relation between $K_c$ and $K_c^{\prime\prime}$?

$K_c$ has no units in reactions with equal numbers of particles on both sides of the equation because the concentration units cancel out in the expression for $K_c$. i.e. for the reaction:

$$H_2(g) + I_2(g) \rightleftharpoons 2HI(g)$$

$$K_c = \frac{[HI(g)]^2}{[H_2(g)][I(g)]} \qquad \text{units} = \frac{\cancel{(\text{mole dm}^{-3})^2}}{\cancel{(\text{mole dm}^{-3})}\cancel{(\text{mole dm}^{-3})}}$$

For reactions in which the numbers of reactant and product particles are not equal, $K_c$ will, of course have units.

What are the units of $K_c$ for the following reactions?

- $N_2(g) + 3H_2(g) \rightleftharpoons 2NH_3(g)$,
- $2NH_3(g) \rightleftharpoons N_2(g) + 3H_2(g)$,
- $2NO(g) + O_2(g) \rightleftharpoons 2NO_2(g)$,
- $NO(g) + \tfrac{1}{2}O_2(g) \rightleftharpoons NO_2(g)$.

### DETERMINATION OF EQUILIBRIUM CONSTANTS

The essential stages in determining an equilibrium constant are listed below.

(i) Write the stoichiometric equation.
(ii) Mix known molar amounts of either the reactants or the products.
(iii) Allow the mixture to reach equilibrium.
(iv) Determine the equilibrium concentration of at least one substance in the equilibrium mixture. This analysis might be carried out using colorimetry, pressure measurements, titration or some other method.
(v) Deduce the equilibrium concentrations of the other materials in the mixture from their initial concentrations using the stoichiometric equation.
(vi) Substitute the equilibrium concentrations in the expression for $K_c$.
(vii) Repeat the determination of $K_c$ using different initial concentrations.

In determining and using equilibrium constants it is important to realize the following points:

**1.** *The Equilibrium Law only applies to systems at equilibrium.*
**2.** *$K_c$ is constant only so long as the temperature remains constant.* If the temperature changes, the value of $K_c$ will change.
**3.** *The numerical value of $K_c$ is unaffected by any changes in concentration of either reactants or products.* Obviously, when more reactant is suddenly added to a system in equilibrium, more of the products will tend to form, but eventually the system will adjust itself at a new equilibrium position in which the concentrations of reactants and products give the same numerical value for $K_c$.

**4.** *The magnitude of $K_c$ provides a useful indication of the extent of a chemical reaction.* A large value for $K_c$ indicates a high relative proportion of products to reactants (i.e. an almost complete reaction) while a low value for $K_c$ indicates that only a small fraction of reactants has been converted to products.

**5.** *The equilibrium constant for a reaction indicates the extent of a reaction, but gives no information about the rate of reaction.* It tells us *how far*, but *not how fast* the reaction goes. In fact, the extent and the rate of a reaction are quite independent. For example, the conversion of sulphur dioxide and oxygen to sulphur(VI) oxide at 450°C occurs very slowly but almost completely, whereas the conversion of nitrogen oxide and oxygen to nitrogen dioxide at the same temperature occurs rapidly but only partially.

The following worked example will help you to understand the ideas we have covered so far and show you how equilibrium constants and equilibrium concentrations can be calculated.

When 1 mole of hydrogen iodide is allowed to dissociate in a 1.0dm³ vessel at 440°C, only 0.78 moles of HI are present at equilibrium. What is the equilibrium constant at this temperature for the reaction, $2HI \rightleftharpoons H_2 + I_2$?

| | $2HI$ | $\rightleftharpoons H_2$ | $+ \; I_2$ | |
|---|---|---|---|---|
| No. of moles initially | 1 | 0 | 0 | |
| No. of moles at equilibrium | 0.78 | 0.11 | 0.11 | Since 0.22 moles of HI have decomposed, 0.11 moles of $H_2$ and 0.11 moles of $I_2$ have formed at equilibrium. |
| Concentration at equilibrium /moles dm$^{-3}$ | $\frac{0.78}{1}$ | $\frac{0.11}{1}$ | $\frac{0.11}{1}$ | |

$$K_c = \frac{[H_2][I_2]}{[HI]^2} = \frac{0.11 \times 0.11}{(0.78)^2} = \frac{1}{50} = 0.02$$

$$\Rightarrow \underline{K_c \text{ for the reaction } = 0.02}$$

If 2 moles of hydrogen and 1 mole of iodine are mixed together in a 1.0dm³ vessel at 440°C, how many moles of HI, $H_2$ and $I_2$ will be present at equilibrium?

| | $2HI$ | $\rightleftharpoons H_2$ | $+ \; I_2$ | |
|---|---|---|---|---|
| No. of moles initially | 0 | 2 | 1 | |
| No. of moles at equilibrium | $2x$ | $2-x$ | $1-x$ | We have assumed that $2x$ moles of HI have formed at equilibrium. Hence, $x$ moles of $H_2$ and $x$ moles of $I_2$ must have disappeared. |
| Concentration at equilibrium /moles dm$^{-3}$ | $\frac{2x}{1}$ | $\frac{2-x}{1}$ | $\frac{1-x}{1}$ | |

$$\Rightarrow K_c = \frac{(2-x)(1-x)}{(2x)^2} = 0.02$$

$$\therefore \frac{2 - 3x + x^2}{4x^2} = 0.02$$

$$\Rightarrow 0.92x^2 - 3x + 2 = 0$$

Solving this quadratic equation using the formula,

$$x = \frac{-b \pm \sqrt{b^2 - 4ac}}{2a}, \text{ we get}$$

$$x = 0.935.$$

Hence, at equilibrium:
number of moles of HI $= 2x = 1.870$
number of moles of $H_2$ $= (2 - x) = 1.065$
number of moles of $I_2$ $= (1 - x) = 0.065$

See if you can now answer the following questions.

○ The equilibrium constant for the reaction

$$2NO_2(g) \rightleftharpoons N_2O_4(g)$$

at 298 K is 200 $mole^{-1}\ dm^3$.

(a) Write an expression for the equilibrium constant for the reaction.
(b) If the concentration of $N_2O_4(g)$ in the equilibrium mixture at 298 K is $2 \times 10^{-2}$ mole $dm^{-3}$, what is the concentration of $NO_2(g)$?
(c) Calculate the equilibrium constant at 298 K for the reaction

$$\tfrac{1}{2}N_2O_4(g) \rightleftharpoons NO_2(g)$$

○ The equilibrium constants for the synthesis of hydrogen chloride, hydrogen bromide and hydrogen iodide are given below.

| | $K_c$ |
|---|---|
| $H_2(g) + Cl_2(g) \rightleftharpoons 2HCl(g)$ | $10^{17}$ |
| $H_2(g) + Br_2(g) \rightleftharpoons 2HBr(g)$ | $10^9$ |
| $H_2(g) + I_2(g) \rightleftharpoons 2HI(g)$ | 10 |

(a) What do the values of $K_c$ tell you about the extent of each reaction?
(b) Which of these reactions would you regard as virtually complete conversions?

## 20.10 Equilibrium constants in gaseous systems

Equilibrium constants are normally expressed in terms of concentrations using the symbol, $K_c$. For reactions involving gases, however, it is usually more convenient to express the amount of gas present in terms of its partial pressure rather than its molar concentration.

Using the ideal gas equation, $pV = n\,RT$

$$\Rightarrow \quad p = \frac{n}{V}RT$$

where $p$ is the pressure in atmospheres, $n$ is the number of moles of gas, $V$ is the volume in cubic decimetres and $T$ is the temperature in Kelvins. In this case, $R$, the gas constant, has units of atm. $dm^3\ K^{-1}\ mole^{-1}$.

$$\therefore p = [\text{gas}]RT$$

where [gas] is the concentration of the gas in moles $dm^{-3}$. Thus, at a constant temperature, the pressure of a particular gas is proportional to its concentration; i.e.

$$p \propto [\text{gas}].$$

This means that for the equilibrium:

$$H_2(g) + I_2(g) \rightleftharpoons 2HI(g)$$

we can write either:

$$K_c = \frac{[HI(g)]^2}{[H_2(g)][I_2(g)]}$$

or

$$K_p = \frac{(P_{HI})^2}{(P_{H_2})(P_{I_2})}$$

Now, since

$$P_{HI} = [HI(g)]RT,$$

$$P_{H_2} = [H_2(g)]RT,$$

and

$$P_{I_2} = [I_2(g)]RT,$$

it follows that:

$$K_p = \frac{(P_{HI})^2}{(P_{H_2}).(P_{I_2})} = \frac{[HI(g)]^2(RT)^2}{[H_2(g)]RT\,[I_2(g)]RT} = \frac{[HI(g)]^2}{[H_2(g)][I_2(g)]} = K_c$$

In this particular example, $K_p = K_c$ and neither $K_p$ nor $K_c$ has any units, but this is not always the case. Just consider the reaction

$$N_2(g) + 3H_2(g) \rightleftharpoons 2NH_3(g).$$

$$K_p = \frac{(P_{NH_3})^2}{(P_{N_2}).(P_{H_2})^3} = \frac{[NH_3(g)]^2(RT)^2}{[N_2(g)]RT.[H_2(g)]^3(RT)^3}$$

$$= \frac{[NH_3(g)]^2}{[N_2(g)][H_2(g)]^3}.(RT)^{-2} = K_c(RT)^{-2}$$

In this case, $K_p = K_c(RT)^{-2}$

Can you see that the numerical value of $K_p$ is the same as that of $K_c$ only when there are the same number of moles on each side of the stoichiometric equation?

Although the S.I. unit of pressure is N $m^{-2}$, it is standard practice to use the atmosphere as the pressure unit in expressing $K_p$ values.

In general, $K_p = K_c(RT)^{\Delta n}$
where $\Delta n$ = number of moles on the right of the equation − number of moles on the left.

The equilibrium constant, $K_c$, for the reaction

$$N_2(g) + 3H_2(g) \rightleftharpoons 2NH_3(g)$$

at 620 K is 2 (mole $dm^{-3})^{-2}$.

- ○ What is the value of $K_p$ for this reaction at 620 K? (R = 0.082 atm $dm^3$ $K^{-1}$ $mole^{-1}$)
- ○ What are the units of $K_p$ for this reaction?

## 20.11 Heterogeneous equilibria

Most of the equilibria that we have discussed so far may be described as **homogeneous**. In these systems, all the reactants and products co-exist in the *same phase* at equilibrium: either they were all gases or they were all mixed together in aqueous solution.

In this section, we shall be exploring certain aspects of **heterogeneous equilibria** in more detail. In these systems, *two or more phases are present* at equilibrium; for example, a solid in equilibrium with its aqueous solution or a solid in equilibrium with a gas. We have already discussed three types of heterogeneous equilibria:

(i) liquid in equilibrium with its saturated vapour;
(ii) solid solute in equilibrium with its saturated solution;
(iii) solute partitioned between two immiscible solvents.

Data concerning the equilibrium of water with its vapour at various temperatures are given in table 20.4.

**Table 20.4** Saturation vapour pressures of water at various temperatures.

| **Mass of water taken/g** | **Pressure/N $m^{-2}$** | | |
|---|---|---|---|
| | **20°C** | **40°C** | **60°C** |
| 1 | $23.4 \times 10^2$ | $73.8 \times 10^2$ | $199 \times 10^2$ |
| 2 | $23.4 \times 10^2$ | $73.8 \times 10^2$ | $199 \times 10^2$ |
| 10 | $23.4 \times 10^2$ | $73.8 \times 10^2$ | $199 \times 10^2$ |
| 50 | $23.4 \times 10^2$ | $73.8 \times 10^2$ | $199 \times 10^2$ |

The important feature in these results is that the vapour pressure exerted by the water at a particular temperature is independent of the mass of water present. How can this be explained?

For equilibrium between water and its vapour:

$$H_2O(l) \rightleftharpoons H_2O(g)$$

we can write an equilibrium constant as:

$$K_c = \frac{[H_2O(g)]}{[H_2O(1)]} \quad \ldots\ldots\ldots\ldots\ldots\ldots\ldots (1)$$

Now, the value of $[H_2O(1)]$ is effectively constant, whatever the amount of water taken. We can, in fact, calculate its value from the density of water.

$$\text{Density of water} = 1\text{g cm}^{-3}$$

$$= 1\,000\text{g dm}^{-3}$$

$$\Rightarrow [H_2O(1)] = \frac{1\,000}{18} = 55.56 \text{ moles dm}^{-3}$$

Using equation (1) above we can write

$$K_c.[H_2O(1)] = K_c^{\text{!}} = [H_2O(g)]$$

where $K_c^{\text{!}}$ may be described as a modified equilibrium constant. Instead of writing, $K_c^{\text{!}} = [H_2O(g)]$, we can write:

$$K_p^{\text{!}} = P_{H_2O}$$

Thus, it is not surprising that the saturation vapour pressure of water is constant at a particular temperature.

Similar results are obtained when other liquids or solids are used in place of water. In each case, the results indicate that *the effective reacting concentration of a pure liquid or a pure solid is constant and is independent of the amount of liquid or solid present.*

In other words, [X(s)] and [X(l)] are constant whatever the amount of X taken, but [X(g)] and [X(aq)] will vary as the amount of X in a given volume varies.

An interesting example of heterogeneous chemical equilibrium involves the thermal dissociation of calcium carbonate.

$$CaCO_3(s) \rightleftharpoons CaO(s) + CO_2(g)$$

We can, of course, write $K_c = \dfrac{[CaO(s)][CO_2(g)]}{[CaCO_3(s)]}$

but $[CaCO_3(s)]$ and $[CaO(s)]$ are both constant, so the modified equilibrium constant for the system becomes:

$$K_c^{\text{!}} = [CO_2(g)]$$

or

$$K_p^{\text{!}} = P_{CO_2}$$

This suggests that at a particular temperature, there is a constant pressure (or concentration) of $CO_2$ in equilibrium with CaO(s) and $CaCO_3(s)$ no matter what masses of these two solids are present. This prediction is confirmed by experiment.

Thus, at 800°C, the pressure of $CO_2$ in equilibrium with $CaCO_3(s)$ and CaO(s) is 0.25 atm. Hence, the equilibrium constant at 800°C is:

$$K_p^{\text{!}} = P_{CO_2} = 0.25 \text{ atm.}$$

Solid $NH_4HS$ is allowed to dissociate in an evacuated vessel forming ammonia and hydrogen sulphide:

$$NH_4HS(s) \rightleftharpoons NH_3(g) + H_2S(g)$$

When excess $NH_4HS$ is used at 25°C, the total pressure at equilibrium is 0.6 atm.

- ○ Write an expression for the modified equilibrium constant for the system, $K_p^{\text{!}}$.
- ○ What are the partial pressures of ammonia and hydrogen sulphide at equilibrium?
- ○ What is the value of $K_p^{\text{!}}$ at 25°C?

## Summary

1 A dynamic equilibrium is characterized by the following features:
(a) Constant macroscopic properties.
(b) Continuing microscopic processes. Particles participate in both forward and reverse processes, but the rate of the forward process is equal to the rate of the reverse process so that no overall change occurs.
(c) Equilibrium attained from either direction.
(d) Equilibrium achieved only in a closed system.

2 When a dissolved solute, X distributes itself between two immiscible solvents A and B, at equilibrium:

$$\frac{\text{Concentration of X in solvent A}}{\text{Concentration of X in solvent B}} = \text{a constant}$$

This constant, called the distribution ratio or the partition coefficient, remains constant provided
(a) the temperature is constant,
(b) the solvents are immiscible and do not react with each other,
(c) the solute neither reacts, nor associates, nor dissociates in the solvents.

3 According to Henry's Law:
Mass of gas dissolved in a given volume of solvent $\propto$ pressure of gas above solvent at a constant temperature.

4 The most important aspect of partition is its application in solvent extraction both industrially and in the laboratory.

5 If an equilibrium mixture contains substances A, B, C and D related by the equation:

$$aA + bB \rightleftharpoons cC + dD$$

it is found experimentally that:

$$\frac{[C]^c[D]^d}{[A]^a[B]^b} = K_c$$

$K_c$, called the Equilibrium Constant, is constant at a given temperature. This experimental result is known as the Equilibrium Law.

6 Since the partial pressure of a gas is proportional to its concentration, we can express the equilibrium constant of a gaseous reaction either in terms of partial pressures or in terms of concentrations.

This means that for the equilibrium, $H_2(g) + I_2(g) \rightleftharpoons 2HI(g)$ we can write either:

$$K_c = \frac{[HI]^2}{[H_2][I_2]} \text{ or } K_p = \frac{(P_{HI})^2}{(P_{H_2}).(P_{I_2})}$$

7 Since the concentration of a solid or a pure liquid is constant, we can write modified equilibrium constants for heterogeneous equilibria excluding these constant concentrations. Thus, for the equilibrium,

$$AgCl(s) + 2NH_3(aq) \rightleftharpoons Ag(NH_3)_2^+(aq),$$

[AgCl(s)] is constant, so we can write a modified equilibrium constant:

$$K_c^{'} = \frac{[Ag(NH_3)_2^+(aq)]}{[NH_3(aq)]^2}$$

## Study questions

1 The partition coefficient of solute X between the solvent Y and the solvent Z is 9. Suppose you are given a solution containing 10g of X in 1 $dm^3$ of Z and asked to extract the X using 1 $dm^3$ of Y.
Calculate the % of X left in Z:
(a) using one extraction with the whole 1 $dm^3$ of Y,
(b) using two extractions each with 500$cm^3$ of Y,
(c) using three extractions each with $333\frac{1}{3}cm^3$ of Y.

2 The Mogul Oil Company is disturbed by the presence of the impurity, M in its four-star petrol. One $dm^3$ of petrol contains 5g of M. In an effort to reduce the concentration of M in the petrol, Mogul have discovered the secret solvent, S and the partition coefficient of M between petrol and S is 0.01.
(a) What is meant by the term 'partition coefficient'?
(b) Explain the principles of solvent extraction.
(c) Calculate the total mass of M removed from one $dm^3$ of petrol using
(i) one portion of $100cm^3$ of solvent, S,
(ii) two $50cm^3$ portions of solvent, S.

3 5 moles of ethanol, 6 moles of ethanoic acid, 6 moles of ethyl ethanoate and 4 moles of water were mixed together in a stoppered bottle at 15°C.
After equilibrium had been attained the bottle was found to contain only 4 moles of ethanoic acid.
(a) Write an equation for the reaction between ethanol and ethanoic acid to form ethyl ethanoate and water.
(b) Write an expression for the equilibrium constant, $K_c$, for this reaction.
(c) How many moles of ethanol, ethyl ethanoate and water are present in the equilibrium mixture?
(d) What is the value of $K_c$ for this reaction?
(e) Suppose 1 mole of ethanol, 1 mole of ethanoic acid, 3 moles of ethyl ethanoate and 3 moles of water are mixed together in a stoppered flask at 15°C.
How many moles of:
(i) ethanol, (ii) ethanoic acid, (iii) ethyl ethanoate, (iv) water are present at equilibrium?

4 At a certain temperature and a total pressure of 1 atm., $I_2$ vapour contains 40% by volume of I atoms:

$$I_2(g) \rightleftharpoons 2I(g)$$

(a) Calculate $K_p$ for the equilibrium.
(b) At what total pressure (without temperature change) would the percentage of I atoms be reduced to 20%?

5 (a) Deduce the relationship between $K_p$ and $K_c$ for the gaseous equilibria:
(i) $2NO(g) + O_2(g) \rightleftharpoons 2NO_2(g)$,
(ii) $NO(g) + \frac{1}{2}O_2(g) \rightleftharpoons NO_2(g)$.
(b) What are the units of $K_p$ and $K_c$ for the two equilibria referred to in (a)?

6 Consider the following reaction:

$$H_2(g) + I_2(g) \rightleftharpoons 2HI(g) \qquad \Delta H = -10 \text{ kJ.}$$

(a) Write an expression for the equilibrium constant in terms of partial pressures.
At a certain temperature, analysis of an equilibrium mixture of the gases yielded the following results:
$P_{H_2} = 2.5 \times 10^{-1}$ atm.,
$P_{I_2} = 1.6 \times 10^{-1}$ atm. and
$P_{HI} = 4.0 \times 10^{-1}$ atm.
(b) Calculate the equilibrium constant for the reaction. What are its units?
(c) In a second experiment at the same temperature, iodine and hydrogen iodide were mixed together, with each gas at a partial pressure of $3 \times 10^{-1}$ atm. What are the partial pressures of hydrogen, iodine and hydrogen iodide at equilibrium?
(d) In a third experiment at the same temperature, pure hydrogen iodide was injected into the flask at a pressure of $6 \times 10^{-1}$ atm. What are the partial pressures of hydrogen, iodine and hydrogen iodide at equilibrium?
(e) What effect, if any, will decreasing the temperature have on the value of $K_p$? Explain your answer.

# 21 Factors Affecting Equilibria

## 21.1 The effect of concentration changes on equilibria

When a system in equilibrium is suddenly disturbed, the system will respond in some way until equilibrium is eventually re-established.
Consider the equilibrium:

$$Fe^{3+}(aq) + NCS^{-}(aq) \rightleftharpoons FeNCS^{2+}(aq)$$
pale yellow colourless deep red

When $10^{-3}$ M iron(III) nitrate(V) solution is added to an equal volume of $10^{-3}$ M potassium thiocyanate, a red solution is produced owing to the formation of thiocyanatoiron(III) complex ions. The system forms an equilibrium mixture containing unreacted $Fe^{3+}$, unreacted $NCS^{-}$ and the product $FeNCS^{2+}$. But what happens to the equilibrium when one of the concentrations is suddenly changed? If a soluble iron(III) salt is added to an equilibrium solution containing $Fe^{3+}(aq)$, $NCS^{-}(aq)$ and $FeNCS^{2+}(aq)$, the colour of the solution becomes darker (figure 21.1).

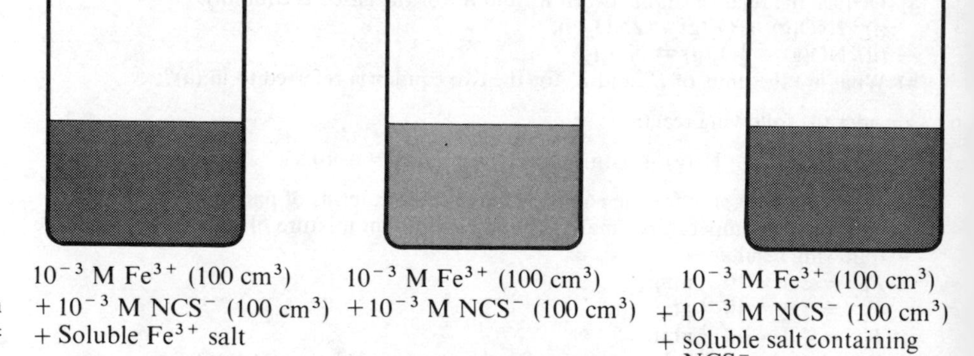


**Figure 21.1** The effect of $Fe^{3+}$ and $NCS^-$ on the equilibrium $Fe^{3+}(aq) + NCS^{-}(aq) \rightleftharpoons Fe(NCS)^{2+}(aq)$.

A new state of equilibrium is quickly attained in which the concentration of $FeNCS^{2+}(aq)$ is obviously greater than before the addition of $Fe^{3+}$. Increasing the concentration of $Fe^{3+}$ has increased the concentration of $FeNCS^{2+}(aq)$. In the same way, the concentration of $FeNCS^{2+}(aq)$ also rises when a soluble thiocyanate is added to the system. On the other hand, removal of $Fe^{3+}$ or $NCS^{-}$ from the equilibrium mixture causes the solution to become paler, suggesting that a decrease in the concentration of $Fe^{3+}$ or $NCS^{-}$ results in the conversion of some $FeNCS^{2+}$ into $Fe^{3+}$ and $NCS^{-}$ in an attempt to replace the substance removed.

The results of these experiments and others like them can be summarized by the following statement.

*If the concentration of one of the reacting substances in a reversible equilibrium is altered, the equilibrium will shift in such a way as to oppose the change in concentration.*

Thus, if a reactant is added to a system in equilibrium, that reaction will occur which uses up the added reactant. Conversely, if a reactant is removed, that reaction will occur which replenishes the removed reactant. The underlined statement above is a specific application of an important generalization known as **Le Chatelier's Principle**. The Frenchman, Henri Louis Le Chatelier, was one of the first chemist to investigate the effects of different factors, such as temperature, pressure and concentration on equilibria. After studying a considerable amount of data concerning equilibria, Le Chatelier proposed the following generalization.

*If a system in equilibrium is subjected to a change, processes occur which tend to counteract the change imposed.*

Although the concentration of individual substances in an equilibrium may vary over a wide range, it is important to remember that the equilibrium constant is always the same at constant temperature. This, of course, is the crucial point of the equilibrium law, which we can further emphasize by considering the effect of suddenly increasing the concentration of hydrogen in an equilibrium mixture of $H_2(g)$, $I_2(g)$ and HI(g). In the initial equilibrium mixture (figure 21.2),

$$[HI(g)] = 0.07\text{ M},$$

$$[H_2(g)] = 0.01\text{ M}$$

and

$$[I_2(g)] = 0.01\text{ M}.$$

$$\therefore K_c = \frac{[HI(g)]^2}{[H_2(g)][I_2(g)]} = \frac{(0.07)^2}{(0.01) \times (0.01)} = 49$$

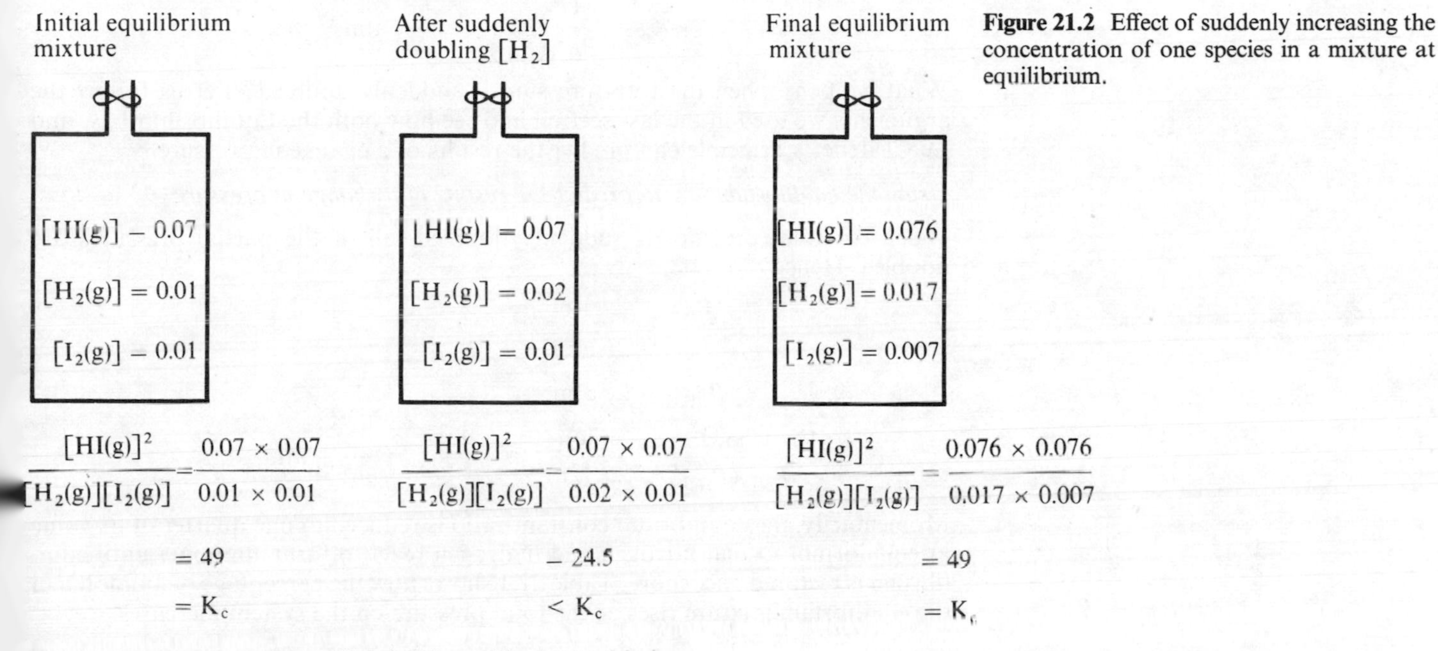


**Figure 21.2** Effect of suddenly increasing the concentration of one species in a mixture at equilibrium.

When the concentration of $[H_2(g)]$ is suddenly doubled,

$$\frac{[HI(g)]^2}{[H_2(g)][I_2(g)]} = \frac{(0.07)^2}{(0.02) \times (0.01)} = 24.5 < K_c$$

The system is no longer in equilibrium. In order to re-establish the equilibrium, the concentration of HI(g) must rise, whilst that of $H_2(g)$ and that of $I_2(g)$ must fall. This is achieved by an *overall* conversion of *some* of the hydrogen and iodine in the mixture to hydrogen iodide:

$$H_2(g) + I_2(g) \longrightarrow 2HI(g)$$

When equilibrium is restored once more (figure 21.2), we find that

$$[HI(g)] = 0.076\text{ M},$$

$$[H_2(g)] = 0.017\text{ M}$$

and

$$[I_2(g)] = 0.007\text{ M}$$

$$\Rightarrow \frac{[HI(g)]^2}{[H_2(g)][I_2(g)]} = \frac{(0.076)^2}{(0.017) \times (0.007)} = 49 = K_c$$

Notice that only part of the added hydrogen is used up in restoring equilibrium. The concentration of $[H_2(g)]$ was suddenly doubled from 0.01 M in the initial equilibrium to 0.02 M, yet when equilibrium is achieved once more the final concentration of hydrogen is not 0.01 M but 0.017 M. Obviously, [HI(g)] in the final equilibrium is greater than that in the initial equilibrium whilst $[I_2(g)]$ in the final equilibrium is less than that initially.

## 21.2 The effect of pressure changes on equilibria

If the partial pressure of *only one* of the gases in an equilibrium mixture is changed, the overall effect can be predicted in a similar fashion to those effects resulting from changes in concentration of one component as in the last section. But what happens when the *total* pressure of a gaseous system at equilibrium is suddenly increased or decreased? In this case, the partial pressures of *all* the gases increase or decrease. Consider, first, the reaction

$$N_2(g) + 3H_2(g) \rightleftharpoons 2NH_3(g)$$

for which

$$K_p = \frac{(P_{NH_3})^2}{P_{N_2}.(P_{H_2})^3}$$

Now, let us suppose that the equilibrium partial pressures of nitrogen, hydrogen, and ammonia are $a$, $b$, and $c$ atm. respectively.
Thus:

$$K_p = \frac{(P_{NH_3})^2}{P_{N_2}.(P_{H_2})^3} = \frac{c^2}{ab^3}\text{ atm}^{-2}$$

What happens when the total pressure is suddenly doubled? Let us follow the arguments we used in the last section and see how both the Equilibrium Law and Le Chatelier's Principle can predict the results of a change in pressure.

*Using the equilibrium law to predict the results of a change in pressure*

When the total pressure is suddenly doubled, all of the partial pressures are doubled. Hence

$$P_{N_2} = 2a \text{ atm},$$
$$P_{H_2} = 2b \text{ atm},$$
$$P_{NH_3} = 2c \text{ atm}.$$

$$\Rightarrow \frac{(P_{NH_3})^2}{P_{N_2}.(P_{H_2})^3} = \frac{(2c)^2}{2a.(2b)^3} = \frac{4c^2}{2a.8b^3} = \frac{c^2}{ab^3}.\frac{1}{4}\text{ atm}^{-2}$$

Momentarily, the equilibrium constant ratio is reduced to one quarter of its value at equilibrium so that nitrogen and hydrogen react to form ammonia until equilibrium is restored once more. Table 21.1 shows how the percentage of ammonia in the equilibrium mixture rises as the total pressure on the system increases.

**Table 21.1** The effect of pressure on the equilibrium percentage of ammonia in the system, $N_2 + 3H_2 \rightleftharpoons 2NH_3$.

| **Total pressure/atm** | 1 | 50 | 100 | 200 |
|---|---|---|---|---|
| **Equilibrium percentage of $NH_3$ at 723 K** | 0.24 | 9.5 | 16.2 | 25.3 |

*Using Le Chatelier's principle to predict the results of a change in pressure*

When the total pressure is suddenly increased, the molecules are crowded closer together. The additional pressure can be relieved if the molecules are able to react and reduce the number of molecules present.

In the reaction we are considering, one molecule of nitrogen reacts with three molecules of hydrogen to form two molecules of ammonia. In other words, four molecules of gas are reacting to form only two molecules of gas, and this reduction in the total number of gas molecules results in a reduction in the total pressure. Hence, any increase in pressure in the $N_2/H_2/NH_3$ system at equilibrium can be relieved by a conversion of nitrogen and hydrogen to ammonia. Conversely, a decrease in pressure will favour the formation of nitrogen and hydrogen since this results in an increase in the number of molecules present thereby counteracting the pressure reduction.

In general, for gaseous reactions, in which there is a change in the number of molecules, increase in pressure favours the reaction which produces fewer molecules and vice versa.

On the other hand, pressure has no effect on those gaseous reactions in which there is no change in the number of molecules. Consider the reaction:

$$H_2(g) + I_2(g) \rightleftharpoons 2HI(g)$$

for which the equilibrium partial pressures of hydrogen, iodine and hydrogen iodide may be given as $x$, $y$ and $z$ atm. respectively at a particular temperature.

- ○ Write an expression for the equilibrium constant, $K_p$, in terms of $x$, $y$, and $z$.
- ○ Suppose the overall pressure is halved. What are the partial pressures of hydrogen, iodine and hydrogen iodide now?
- ○ What is the value of the equilibrium constant expression

$$\frac{(P_{HI}(g))^2}{(P_{H_2}(g)).(P_{I_2}(g))}$$

when the overall pressure is suddenly halved?
- ○ Use the equilibrium law to explain why reducing the pressure has no effect on this system.
- ○ How does Le Chatelier's principle explain why pressure changes have no effect on the system?

## 21.3 The effect of catalysts on equilibria

The equilibrium constant expression includes *only* those substances shown in the overall stoichiometric equation. Catalysts do not appear in the overall equation for a reaction and, therefore, it is not surprising that they have no effect on the equilibrium position.

Experiments show that catalysts can increase the *rates* of both forward and backward reactions in an equilibrium and they enable equilibrium to be achieved much more rapidly, but they do not alter the concentrations of reacting substances at equilibrium.

## 21.4 The effect of temperature changes on equilibria

Although the equilibrium concentrations of reactants and products can vary over a wide range, the numerical value of the equilibrium constant remains constant at one particular temperature. This means that $K_c$ and $K_p$ are unaffected by catalysts or by changes in pressure and concentration. The equilibrium constant does, however, vary with temperature. Table 21.2 shows the values of $K_p$ at different temperatures for three important reactions together with the corresponding enthalpy changes for the complete conversion of reactants to products.

**Table 21.2** Values of $K_p$ for three different reactions at various temperatures.

| $N_2(g) + 3H_2(g) \rightleftharpoons 2NH_3(g)$ | | $N_2O_4(g) \rightleftharpoons 2NO_2(g)$ | | $2SO_2(g) + O_2(g) \rightleftharpoons 2SO_3(g)$ | |
|---|---|---|---|---|---|
| $\Delta H^\ominus = -92$ kJ | | $\Delta H^\ominus = +57$ kJ | | $\Delta H^\ominus = -197$ kJ | |
| $T/K$ | $K_p = \frac{(p_{NH_3})^2}{p_{N_2}\cdot(p_{H_2})^3}/atm^{-2}$ | $T/K$ | $K_p = \frac{(p_{NO_2})^2}{p_{N_2O_4}}/atm$ | $T/K$ | $K_p = \frac{(p_{SO_3})^2}{(p_{SO_2})^2\cdot p_{O_2}}$ |
| 400 | $1.0 \times 10^2$ | 200 | $1.9 \times 10^{-6}$ | 600 | $3.2 \times 10^3$ |
| 500 | $1.6 \times 10^{-1}$ | 300 | $1.7 \times 10^{-1}$ | 700 | $2.0 \times 10^2$ |
| 600 | $3.1 \times 10^{-3}$ | 400 | $5.1 \times 10$ | 800 | $3.2 \times 10$ |
| 700 | $6.3 \times 10^{-5}$ | 500 | $1.5 \times 10^3$ | 900 | 6.3 |
| 800 | $7.9 \times 10^{-6}$ | 600 | $1.4 \times 10^4$ | 1000 | 2.0 |

Notice that the two exothermic reactions have $K_p$ values which decrease with increase in temperature, whilst the endothermic reaction has $K_p$ values which increase as the temperature rises. Evidence from other investigations also fits this pattern of results and it is found that:

(i) equilibria in which the **forward reaction is exothermic (i.e. $\Delta H^\ominus$ negative)** have equilibrium constants that decrease as temperature rises;

(ii) equilibria in which the **forward reaction is endothermic (i.e. $\Delta H^\ominus$ positive)** have equilibrium constants that increase as temperature rises.

Look at the information in table 21.2.

- ○ How does the proportion of ammonia in the $N_2/H_2/NH_3$ system change as temperature increases?
- ○ What is the value of $\Delta H^\ominus$ for the reaction

$$2NH_3(g) \rightleftharpoons N_2(g) + 3H_2(g)?$$

What is the value of $K_p$ for this reaction at 400 K?

- ○ Predict the effect of increasing temperature on $K_p$ for the reaction

$$2NH_3(g) \rightleftharpoons N_2(g) + 3H_2(g)$$

Le Chatelier's principle can be used once more to predict the effect of temperature on chemical systems in equilibrium. Changing the temperature of a system in equilibrium naturally provides a constraint which the system will try to remove. Hence, increase in temperature favours the endothermic process which will absorb the additional heat, whilst decrease in temperature favours the exothermic process.

This may be summarized as:

*exothermic process favoured*
*be decrease in temperature*

A B C+D;

*endothermic process favoured*
*by increase in temperature*

$$\Delta H = -x\,\text{kJ}$$

An effective demonstration of the effect of temperature on the position of an equilibrium can be illustrated using three identical sealed tubes containing dark brown nitrogen dioxide ($NO_2$) in equilibrium with pale yellow dinitrogen tetraoxide ($N_2O_4$).

$$\underset{\text{pale yellow}}{N_2O_4(g)} \rightleftharpoons \underset{\text{dark brown}}{2NO_2(g)}$$

All three tubes contain the same amounts of $NO_2$ and $N_2O_4$ and initially they have the same brown appearance.

One tube is now placed in iced water, a second tube is left at room temperature and the third tube is placed in hot water (figure 21.3).

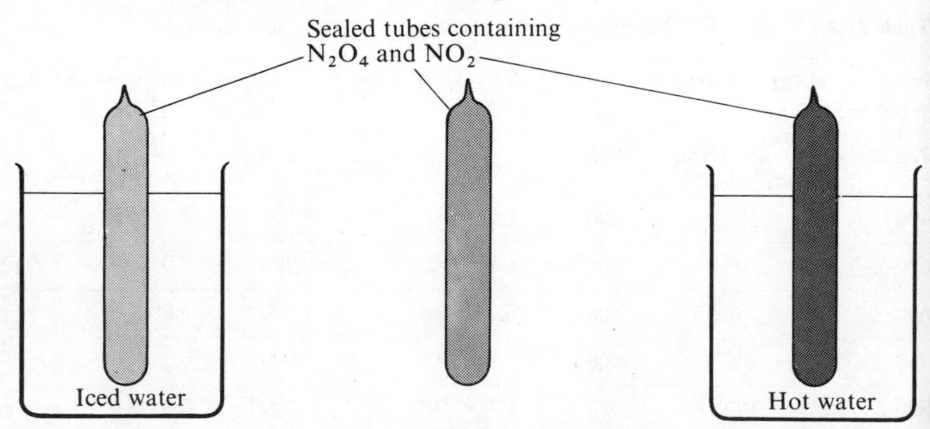


**Figure 21.3** The effect of temperature on the equilibrium $N_2O_4 \rightleftharpoons 2NO_2$.

The tube in cold water becomes much paler whilst that in hot water turns dark brown. This shows that the equilibrium in the reaction is displaced towards the formation of darker $NO_2$ at higher temperatures (i.e. in the endothermic direction), but towards the formation of paler $N_2O_4$ at lower temperatures.

## 21.5 Applying the principles of reaction kinetics and equilibria to industrial processes

The principles of reaction kinetics and chemical equilibria play an important part in the design and working conditions of industrial processes. The economic and commercial competitiveness of any process rests on the speed, efficiency and economy with which products can be obtained from starting materials. Clearly, many people (including managers, economists and engineers) will be involved in decisions about the methods and materials to be used in any industrial process, but the major problems confronting the chemist are to convert reactants into products:

(i) as quickly as possible,
(ii) as completely as possible.

The first of these problems is clearly a kinetic one; that of obtaining a maximum viable rate of reaction and rate of product formation.

The second problem is one of equilibrium and involves the choice of those conditions which favour an optimum proportion of the product in the equilibrium mixture.

The answer to each of these problems lies in the careful choice of reaction conditions such as temperature, pressure and choice of catalyst, whilst Le Chatelier's principle is invaluable in predicting those conditions for maximum yield of a particular product at equilibrium.

### THE MANUFACTURE OF SULPHURIC(VI) ACID: THE CONTACT PROCESS

The essential stages in the manufacture of sulphuric(VI) acid are shown diagrammatically in figure 21.4. Sulphur dioxide is first obtained by burning sulphur or by roasting sulphide ores in air. Before further oxidation to sulphur(VI) oxide, the $SO_2$ is mixed with excess air and thoroughly purified to prevent any impairment ('poisoning') of the catalyst by dust and other impurities. Finally, the $SO_3$ is combined with water to form sulphuric(VI) acid.

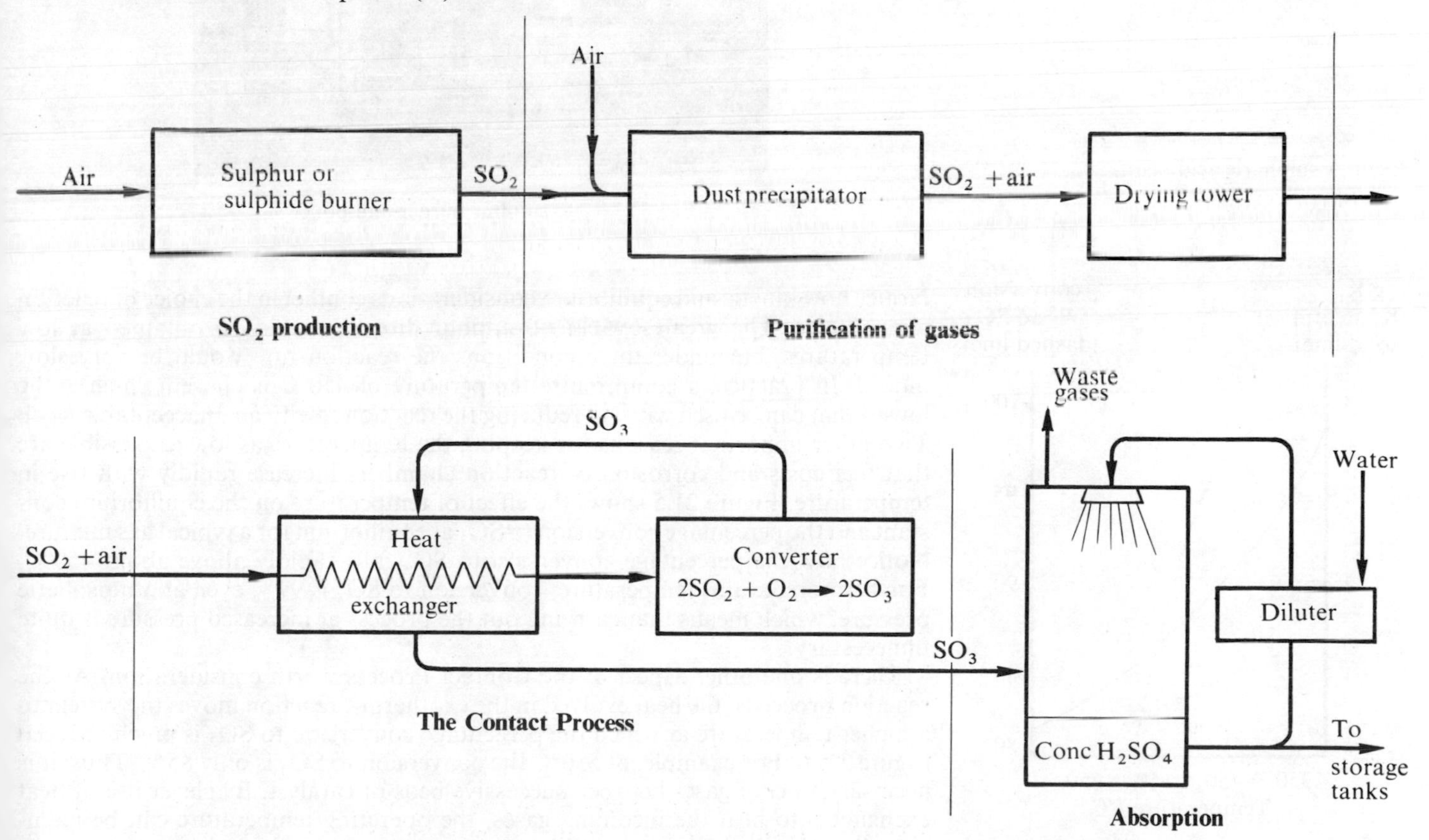


**Figure 21.4** Essential stages in the manufacture of sylphuric(VI) acid.

The bottleneck in the production of sulphuric(VI) acid is unquestionably the slow third stage: conversion of $SO_2$ to $SO_3$ in the Contact Process:

$$2SO_2(g) + O_2(g) \rightleftharpoons 2SO_3(g) \qquad \Delta H = -197 \text{ kJ}$$

How can the rate of production of $SO_3$ be increased?

(i) by increasing the concentration (pressure) of $O_2$ or $SO_2$,
(ii) by increasing the temperature,
(iii) by employing a catalyst.

Both vanadium compounds and platinum have been used as catalysts for the Contact Process. The vanadium catalysts (incorporating either $VO_3^-$ or $V_2O_5$) are less efficient than platinum, but cheaper and less susceptible to poisoning. Very few platinum catalyst plants have been built since 1945.

Since the reaction is exothermic and involves the conversion of three moles of reactants to two moles of products, Le Chatelier's principle predicts that the maximum yield of $SO_3$ at equilibrium will be obtained:

(i) at high pressure,
(ii) at low temperature.

*Right* A sulphuric acid plant.

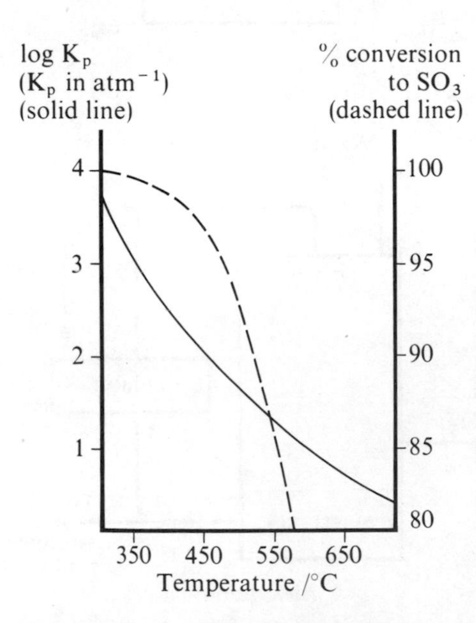


**Figure 21.5** The effect of temperature on the equilibrium constant and the percentage conversion to $SO_3$ at equilibrium for a typical mixture of $SO_2$ and $O_2$ at one atmosphere pressure.

Notice how kinetic and equilibrium considerations conflict in the choice of reaction temperature. The greatest yield of sulphur dioxide would be obtained at low temperatures, but under these conditions the reaction rate would be very slow indeed. In practice, a compromise temperature of 450°C is chosen which is the lowest that can be used without reducing the reaction rate to an unacceptable level. Two other important reasons for keeping the temperature as low as possible are that fuel costs and corrosion of reaction chambers increase rapidly with rise in temperature. Figure 21.5 shows the effect of temperature on the equilibrium constant and the percentage conversion to $SO_3$ at equilibrium for a typical gas mixture. Notice how the percentage conversion to $SO_3$ falls rapidly above about 450°C. Furthermore, at this temperature, conversion to $SO_3$ is 97% even at atmospheric pressure, which means that carrying out the process at increased pressure is quite unnecessary.

There is one other aspect of the Contact Process worth consideration. As the reaction proceeds, the heat evolved in the exothermic reaction moves the system to a higher temperature at which the percentage conversion to $SO_3$ is much reduced (figure 21.5). For example, at 550°C the conversion to $SO_3$ is only 85%. Thus, it is necessary to cool gases between successive beds of catalyst. By clever use of heat exchangers to heat the incoming gases, the operating temperature can be maintained at 450°C without external heating.

After passing through the heat exchange system, the product gases pass into an absorption tower where $SO_3$ dissolves in concentrated $H_2SO_4$. Direct absorption in water is unsatisfactory because the heat evolved vaporizes the $H_2SO_4$ which condenses as a fog of tiny droplets and is slow to settle out.

An unusual use for sulphur: an early advertisement for Frazer's sulphur tablets.

THE USES AND IMPORTANCE OF SULPHURIC(VI) ACID

Sulphuric(VI) acid is one of the most widely used chemicals. Nearly four million tonnes of sulphuric acid are manufactured each year in the U.K. alone. Figure 21.6 shows the main uses of sulphuric(VI) acid. The requirements for $H_2SO_4$ are so many and so diverse that its level of production can be used as a reliable guide to a country's industrial activity. What is more, problems of bulk storage of the acid mean that production must respond quickly to any changes in consumption. Hence, the economic and technical development of a country can be estimated from its sulphuric(VI) acid production.

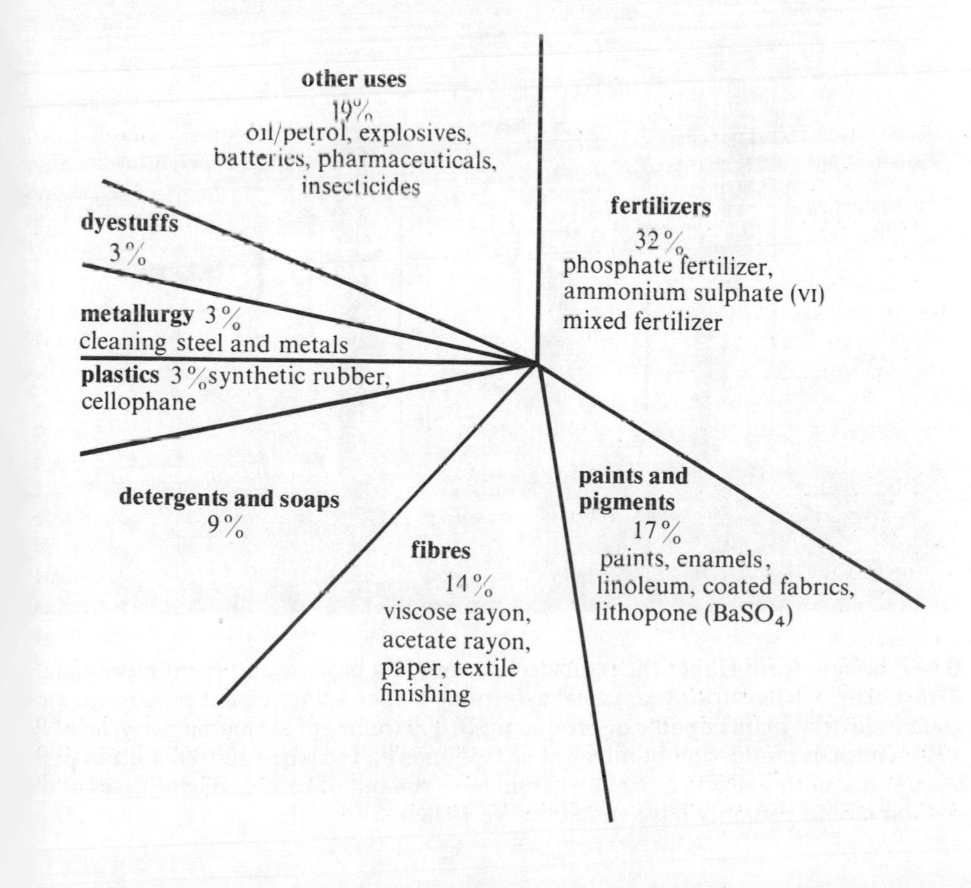


**Figure 21.6** The uses of sulphuric(VI) acid (%'s relate to 1971 values when the total consumption in the UK was 3 750 000 tonnes).

## THE MANUFACTURE OF AMMONIA: THE HABER PROCESS

During the last century, the agricultural industry began to require rapidly increasing quantities of nitrogenous fertilizers after it became known that these compounds could treble the yield of certain crops. Larger and larger supplies of food were being needed to feed growing populations, particularly in Europe and North America.

At the same time, the chemical industry also required increasing quantities of nitrogen compounds to make nitric(v) acid for dyes and explosives such as T.N.T. and dynamite.

Consequently, agriculture and industry were competing with each other for dwindling supplies of nitrogenous raw material. By 1900, Peruvian *guano*, a valuable fertilizer from the droppings of sea-birds, had already been worked out, and it was already clear that supplies of sodium nitrate(v) from Chile would soon become exhausted.

An alternative supply of nitrogen in the form of ammonia or nitrate(v) had to be found or the chemical industry would stagnate and the world's growing population would starve. Ironically, it was the war preparations in Germany between 1909 and 1914 which solved the problem. Military leaders in Germany realized that once war was declared, their country would be subjected to a strict blockade and importation of goods and raw materials from the rest of unoccupied Europe and America would cease. German industry, therefore, had to be capable of meeting its country's requirements for nitrogenous fertilizers, and the tremendous demand for explosives and hence for nitric(v) acid that a war would create.

In 1909, the leading German chemical company, Badische Anilin und Soda Fabric (BASF), turned its research expertise and financial resources towards the development of ammonia manufacture from atmospheric nitrogen. In the previous year, a young German research chemist, Fritz Haber, had discovered that at a temperature of 600°C and a pressure of 200 atmospheres, nitrogen and hydrogen formed an equilibrium mixture in the presence of a suitable catalyst.

$$N_2(g) + 3H_2(g) \rightleftharpoons 2NH_3(g); \quad \Delta H^\ominus = -92 \text{ kJ}$$

- ○ What conditions did Haber employ to increase the reaction rate?
- ○ Haber's experiments yielded an equilibrium mixture containing only 8% by volume of ammonia. What conditions of temperature and pressure does Le Chatelier's principle predict for maximum yield of ammonia at equilibrium?
- ○ Why do you think Haber employed a heat exchanger in his equipment?

*Above* Fritz Haber (1868–1934). Haber was the son of a merchant of Breslau. After studying chemistry, Haber obtained a post as lecturer at the technical college in Karlsruhe. It was here that he discovered his method of synthesizing ammonia, a discovery for which he was awarded the Nobel Prize for Chemistry in 1918.

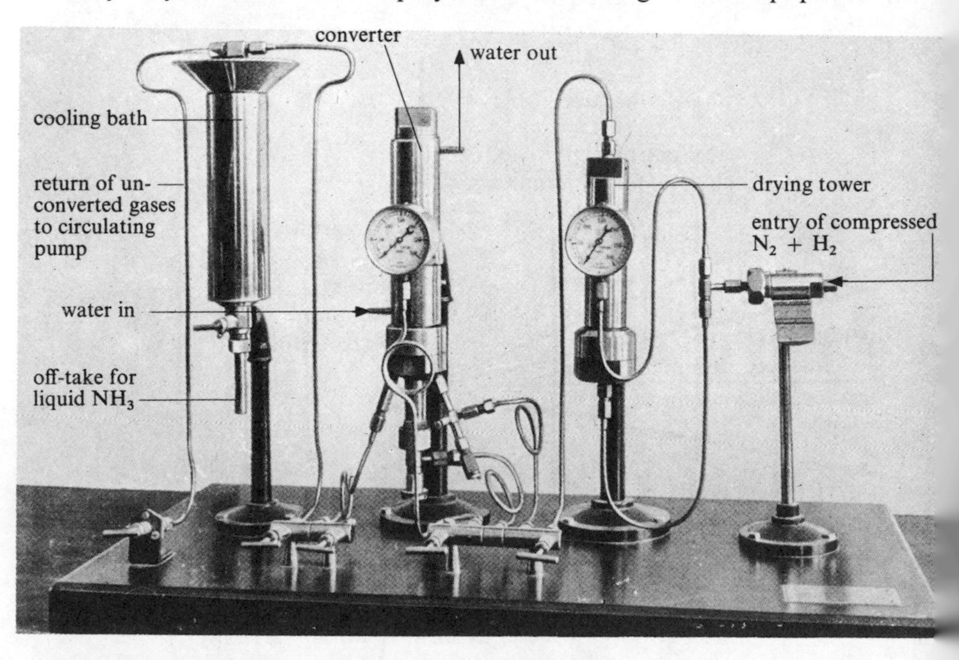


*Right* Haber's apparatus for the synthesis of ammonia.

BASF bought from Haber the rights to his ammonia process and spent more than £1m. during the next four years in transforming Haber's simple pilot process into a giant industrial plant capable of producing 10 000 tonnes of ammonia per year. By 1913, German production of nitrogen compounds had reached 120 000 tonnes per year. Without this effort, Germany would have run out of food and explosives and the war would certainly have ended before 1918.

*Above left* The world's first ammonia manufacturing plant built by BASF at Oppau in Germany and opened in 1912.

*Above* Carl Bosch (1874–1940). Bosch was the son of a plumber. After studying chemistry at university, Bosch joined BASF in 1899 and quickly gained a reputation as a brilliant chemical engineer. Bosch was responsible for developing an industrial plant to manufacture ammonia from Haber's laboratory process. Bosch was awarded the Nobel Prize for Chemistry in 1931 for his work on high-pressure reactions.

Figure 21.7 shows a flow diagram for the modern Haber process.

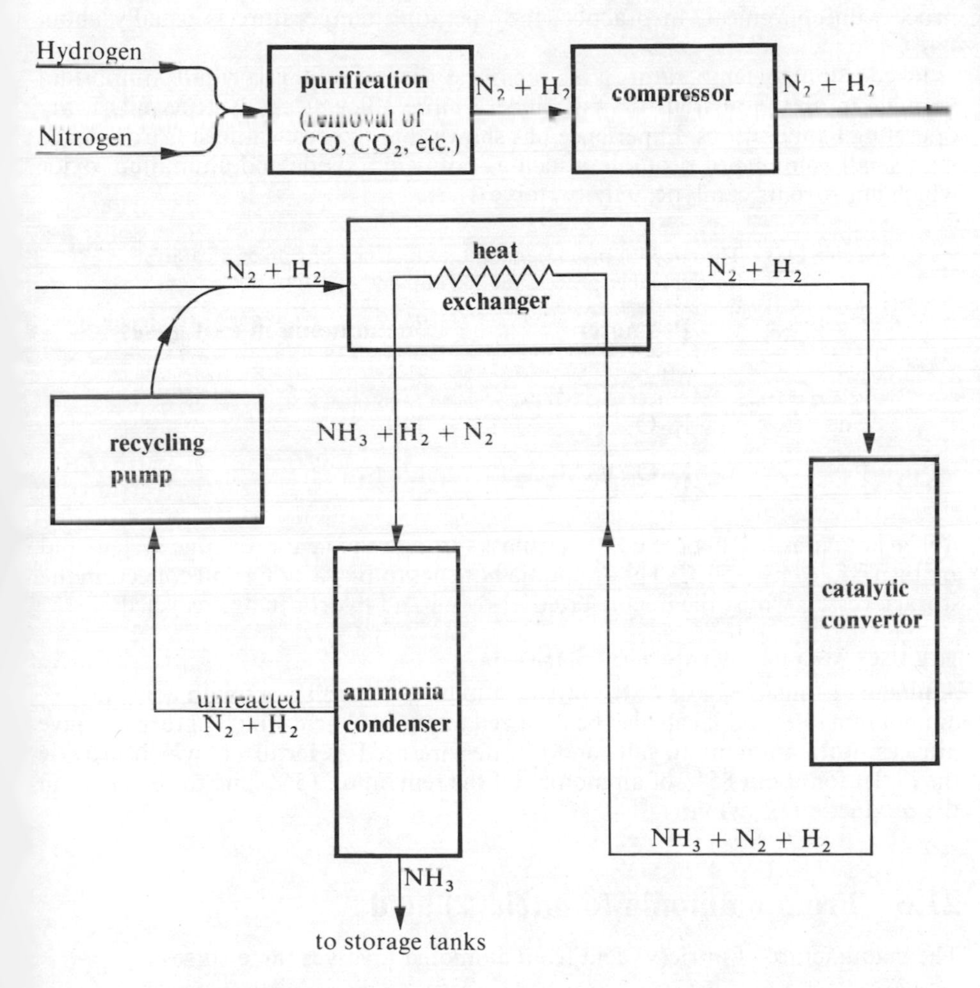


**Figure 21.7** A flow diagram for the Haber process.

Nitrogen can be obtained by the fractional distillation of liquid air, whilst hydrogen is obtained from naphtha (a mixture of hydrocarbons containing 5–9 carbon atoms) or natural gas. The production of hydrogen involves either catalysis with steam:

$$\underset{\text{in naphtha}}{C_6H_{14}(g)} + 6H_2O(g) \longrightarrow 6CO(g) + 13H_2(g)$$

$$\underset{\text{in natural gas}}{CH_4(g)} + H_2O(g) \longrightarrow CO(g) + 3H_2(g)$$

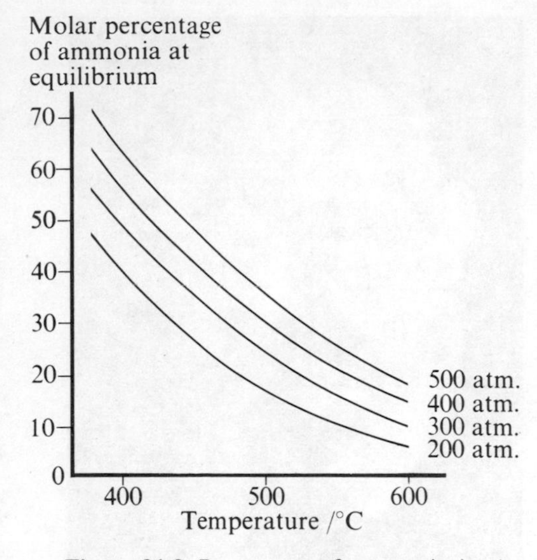


**Figure 21.8** Percentage of ammonia in the equilibrium mixture obtained from a 1:1 mixture of $N_2$ and $H_2$ at different temperatures and pressures.

or partial oxidation with oxygen:

$$\underset{\text{in naphtha}}{C_6H_{14}(g)} + 3O_2(g) \longrightarrow 6CO(g) + 7H_2(g)$$

$$\underset{\text{in natural gas}}{CH_4(g)} + \tfrac{1}{2}O_2(g) \longrightarrow CO(g) + 2H_2(g)$$

Thorough purification of both the nitrogen and hydrogen is necessary, not only to remove carbon monoxide from the hydrogen but also sulphur compounds, water vapour and carbon dioxide since these would poison the catalyst in the converter. Hot product gases are used to warm up the purified nitrogen and hydrogen in a heat exchanger before they enter the Haber process converter.

Le Chatelier's principle suggests that increase in pressure and decrease in temperature will increase the proportion of ammonia at equilibrium. These predictions are borne out by the results in figure 21.8.

High pressure obviously gives a higher yield of ammonia, but the higher the pressure the greater the cost and maintenance of equipment. Although pressures up to 600 atm have been used, the favoured pressure nowadays is 250 atm.

In contrast to pressure, the temperature must be low to give a high yield of ammonia, but at low temperature the rate of reaction is so slow that it makes the process uneconomical. In practice, the operating temperature is usually about 450°C.

In addition to temperature and pressure, the catalyst is a vitally important variable in any industrial process, since a more efficient catalyst permits lower operating temperatures. Experience has shown that the best catalyst is iron mixed with small amounts of promoters such as potassium oxide and aluminium oxide which improve its catalytic activity (table 21.3).

**Table 21.3** The effect of promoters on the efficiency of iron as a catalyst for the Haber process (at 200 atm and 400°C).

| Catalyst | Promoter | % ammonia in exit gases |
|---|---|---|
| Fe | nil | 3–5 |
| Fe | $K_2O$ | 8–9 |
| Fe | $K_2O + Al_2O_3$ | 13–14 |

The hot gases leaving the converter pass through the heat exchange system and are then cooled to −50°C. The ammonia liquefies (b.pt. −33°C) and collects in the storage vessels whilst the unconverted nitrogen and hydrogen are recycled.

### THE USES AND IMPORTANCE OF AMMONIA

Ammonia forms the basis of the nitrogen industry. It will react with acids to give ammonium salts and it can also be oxidized to nitric(V) acid which in turn can give nitrates. Both ammonium salts and nitrates are used as fertilizers which provide the outlet for about 85% of ammonia. Of the remaining 15%, one third is used in the production of nylon.

## 21.6 From ammonia to nitric(V) acid

The manufacture of nitric(V) acid from ammonia involves three stages.

1 Catalytic oxidation of ammonia to nitrogen oxide (NO).

$$4NH_3(g) + 5O_2(g) \rightleftharpoons 4NO(g) + 6H_2O(g) \qquad \Delta H^\ominus = -950\ \text{kJ}$$

2 Oxidation of nitrogen oxide (NO) to nitrogen dioxide ($NO_2$).

$$2NO(g) + O_2(g) \longrightarrow 2NO_2(g) \qquad \Delta H^\ominus = -114\ \text{kJ}$$

3 Reaction of nitrogen dioxide with water to form nitric(V) acid.

$$3NO_2(g) + H_2O(l) \longrightarrow 2HNO_3(aq) + NO(g) \qquad \Delta H^\ominus = -117\ \text{kJ}$$

Look closely at stage 1 above.

- ○ What conditions of temperature and pressure would give the maximum yield of nitrogen oxide (NO) at equilibrium?
- ○ In practice, the first stage is carried out by passing dry ammonia and air at 7 atm over a platinum gauze catalyst at 900°C. Explain why the conditions used in the process differ from those predicted in your answer to the last question.

The top sheet of platinum-alloy catalyst used in the manufacture of nitric acid from ammonia.

The product gases are cooled to 25°C and then mixed with more air so that nitrogen oxide is immediately oxidized to red-brown nitrogen dioxide.

$$2NO(g) + O_2(g) \longrightarrow 2NO_2(g)$$

In the third stage, nitrogen dioxide reacts with water in large absorption towers designed to ensure thorough mixing of the ascending gases and the descending solution. The final product contains about 60% nitric(V) acid. More concentrated acid can be obtained by distilling the 60% solution with concentrated sulphuric(VI) acid.

## 21.7 Fertilizers and explosives from nitric(V) acid

Nitric(V) acid plays an important part in the production of fertilizers and explosives which consume 75% and 15% of its production respectively.

### FERTILIZERS

Nitrogen is an essential element which is needed in large quantities for plant growth. It is required for the formation of proteins, chlorophyll and nucleic acids. Plants suffering from nitrogen deficiency become stunted with yellow leaves.

- ○ Why does a deficiency of nitrogen cause plants to become stunted?
- ○ Why does a deficiency of nitrogen cause yellowing of the leaves?

The intensive cropping of agricultural land means that the nitrogen removed from the soil must be replaced by the application of fertilizers if the soil is not to become barren and infertile. Nitrogenous fertilizers are usually nitrates or ammonium salts. Indeed, ammonium nitrate(V) ('Nitram') is the most widely used fertilizer in most countries, largely because of its very high percentage of nitrogen.

*Above* A range of ICI fertilizers. The numbers across the top of each bag represent the percentages of nitrogen(N), phosphorus oxides ($P_2O_5$ and $P_2O_3$) and potassium oxide $K_2O$) in each fertilizer. What do the terms 'Low-P', 'Hi-P', 'NP', 'Nitram', 'Nitro-chalk' and 'Kaynitro' stand for?

○ What is the percentage by mass of nitrogen in ammonium nitrate(V), $NH_4NO_3$? (N = 14, H = 1, O = 16)
○ How do you think ammonium nitrate(V) is obtained industrially?

'Nitro-chalk', which contains about 20 % nitrogen, consists of ammonium nitrate(V) crystals coated with chalk (calcium carbonate). It is used less frequently than 'Nitram', but it is non-deliquescent (unlike ammonium nitrate(V) crystals) and it is a convenient way of liming the soil at the same time.

*Above* 'Nitram' fertilizer.

*Above right* A huge silo in which the fertilizer, ammonium sulphate is stored.

EXPLOSIVES

Although 90 % of ammonium nitrate(V) is used in fertilizers, a large proportion of the remaining 10 % goes towards the production of explosives. Nowadays, virtually all explosives are manufactured by processes which use concentrated nitric(V) acid.

The formulae of some of these explosives are shown in figure 21.9.

cellulose nitrate
(nitrocellulose-gun cotton)

propane-1,2,3-triyl trinitrate
(nitroglycerine)

methyl-2,4,6-trinitrobenzene
(trinitrotoluene, T.N.T.)

2,4,6-trinitrophenol
(picric acid)

**Figure 21.9** Explosives from nitric(v) acid.

All of these substances undergo rapid chemical reactions evolving large amounts of gas and heat on explosion and the sudden development of great pressure. Their constituent carbon and hydrogen undergo combustion with oxygen from the nitro-groups forming carbon dioxide and water. For example, the explosion of 'nitroglycerine' made by the action of concentrated nitric(v) acid on glycerol (propane-1, 2, 3-triol), can be summarized by the equation:

$$4C_3H_5(NO_3)_3(l) \longrightarrow 12CO_2(g) + 10H_2O(g) + 6N_2(g) + O_2(g)$$

Notice that in this case, there is more than enough oxygen within the 'nitroglycerine' molecule for the carbon and hydrogen to undergo complete combustion to $CO_2$ and water. This is not the case with aromatic nitro compounds such as T.N.T. and picric acid which are frequently blended with compounds containing a high percentage of oxygen, such as chlorates and nitrates, to ensure complete oxidation during explosion.

The explosion of a French ammunitions depot in January 1871 during the Franco-Prussian War.

Dynamite is a general term for explosives which contain both nitroglycerine and nitrocellulose. Thus, cordite and gelignite are each different forms of dynamite.
The manufacture of explosives requires elaborate and very special safety precautions. You *should never* attempt to make any explosive yourself.

## Summary

1 If the concentration of one of the reacting substances in a reversible equilibrium is altered, the equilibrium will shift in such a way as to oppose the change in concentration.
2 The influence of different factors, such as temperature, pressure and concentration, on a system in equilibrium can be predicted using Le Chatelier's Principle: if a system in equilibrium is subjected to a change, processes occur which tend to counteract the change imposed.
3 The values of $K_c$ and $K_p$ are unaffected by changes in pressure or concentration. Temperature is the only factor which influences the values of $K_c$ and $K_p$. Changes in concentration or pressure may, however, result in changes in the concentration or partial pressure of substances in the equilibrium mixture. Catalysts affect neither the position of equilibrium nor the values of $K_c$ and $K_p$.
4 Equilibria for which the forward reaction is exothermic (i.e. $\Delta H^{\ominus}$ negative) have equilibrium constants that decrease as temperature rises.
Equilibria for which the forward reaction is endothermic (i.e. $\Delta H^{\ominus}$ positive) have equilibrium constants that increase as temperature rises.

## Study questions

1 Nowadays, hydrogen can be obtained from natural gas by partial oxidation with steam, which involves the endothermic reaction:

$$CH_4(g) + H_2O(g) \rightleftharpoons CO(g) + 3H_2(g)$$

(a) Write an expression for $K_p$ for this reaction.
(b) How will the value of $K_p$ be affected by
(i) increasing the pressure,
(ii) increasing the temperature,
(iii) using a catalyst?
(c) How will the composition of the equilibrium mixture be affected by:
(i) increasing the pressure,
(ii) increasing the temperature,
(iii) using a catalyst?

2 The first step in the manufacture of nitric(v) acid from ammonia involves the exothermic oxidation of ammonia to nitrogen oxide (NO) and steam.
(a) Write the equation for the reaction of ammonia with oxygen to form nitrogen oxide and steam.
(b) Predict, qualitatively, the conditions of temperature and pressure for maximum yield of nitrogen oxide in the equilibrium mixture.
(c) The industrial manufacture of nitrogen oxide from ammonia employs high temperature and a pressure of 7 atm. How and why are these industrial conditions different to those you predicted in (b) for maximum yield of nitrogen oxide at equilibrium?
(d) Describe, with equations, how nitrogen oxide produced by this process is converted to nitric(v) acid.

3 At 25°C, the value of $K_c$ for the following system is $10^{10}$:

$$Sn^{2+}(aq) + 2Fe^{3+}(aq) \longrightarrow Sn^{4+}(aq) + 2Fe^{2+}(aq)$$

(a) Write an expression for $K_c$ for this reaction.
(b) Explain why $K_c$ has no units.
(c) What is the value of $K_c$ for:
(i) $Sn^{4+}(aq) + 2Fe^{2+}(aq) \longrightarrow Sn^{2+}(aq) + 2Fe^{3+}(aq)$,
(ii) $2Sn^{2+}(aq) + 4Fe^{3+}(aq) \longrightarrow 2Sn^{4+}(aq) + 4Fe^{2+}(aq)$?

4 At 200°C, $K_c$ for the reaction

$$PCl_5(g) \rightleftharpoons PCl_3(g) + Cl_2(g) \qquad \Delta H^{\ominus} = +124 \text{ kJ}$$

has a numerical value of $8 \times 10^{-3}$.

(a) Write an expression for $K_c$ for this reaction.

(b) What are the units of $K_c$?

(c) What is the value of $K_c$ for the reverse reaction at 200°C and what are its units?

(d) How will the amounts of $PCl_5$, $PCl_3$ and $Cl_2$ in the equilibrium mixture change if (i) more $PCl_5$ is added, (ii) the pressure is increased, (iii) the temperature is increased?

(e) What would be the effect on $K_c$ if (i) more $PCl_5$ is added, (ii) the pressure is increased, (iii) the temperature is increased?

(f) A sample of pure $PCl_5$ was introduced into an evacuated vessel at 200°C. When equilibrium was obtained, the concentration of $PCl_5$ was $0.5 \times 10^{-1}$ mole dm$^{-3}$. What are the concentrations of $PCl_3$ and $Cl_2$ at equilibrium?

5 At 488 K the equilibrium constant for the reaction

$$COCl_2(g) \rightleftharpoons CO(g) + Cl_2(g)$$

is $2 \times 10^{-6}$ atm.

(a) Assuming that the total pressure at equilibrium is $P$ and the degree of dissociation of $COCl_2$ is $\alpha$, deduce a relationship between $K_p$, $\alpha$ and $P$. (Degree of dissociation ($\alpha$) is the fraction dissociated.)

(b) Calculate the degree of dissociation
(i) at 1 atm pressure assuming the temperature remains constant,
(ii) at 2 atm pressure assuming the temperature remains constant.
(Hint: when $\alpha$ is very small, $(1 - \alpha) \simeq 1$; $(1 + \alpha) \simeq 1$.)

6 State and explain what happens to the concentrations of hydrogen, carbon monoxide, and methanol which are in equilibrium according to the reaction

$$CO(g) + 2H_2(g) \rightleftharpoons CH_3OH(g) \qquad \Delta H = -92 \text{ kJ}$$

when

(a) the volume in which they are contained is suddenly reduced to half,
(b) the temperature is increased,
(c) the partial pressure of hydrogen is suddenly doubled,
(d) a catalyst is added,
(e) an inert gas is added to the system.

7 The densities of diamond and graphite are 3.5 and 2.3g cm$^{-3}$ respectively and the change from graphite to diamond is represented by the equation

$$C(\text{graphite}) \rightleftharpoons C(\text{diamond}) \qquad \Delta H = +2 \text{ kJ}$$

Is the formation of diamond from graphite favoured by
(a) high or low temperature,
(b) high or low pressure?
Explain your answers.

8 (a) Describe in outline the manufacture of *one* of the following industrial chemicals: (i) sulphuric(VI) acid, (ii) nitric(V) acid, (iii) ammonia.

(b) Discuss the chemical principles which determine the optimum operating conditions for the process which you describe.

(c) For the substance you have chosen, mention:
(i) two large scale uses,
(ii) two important features of its chemistry, giving reactions and equations to illustrate the points you make.

# 22 Ionic Equilibria in Aqueous Solution

## 22.1 Introduction

The equilibria between ions in aqueous solution are of particular interest because of their importance in industrial, analytical and biological processes. The principles and characteristics of these ionic equilibria are, of course, very similar to those used in considering other systems in chemical equilibrium.

In this chapter, we shall concentrate on two fundamental types of ionic equilibria: (i) the equilibrium between an undissolved solid solute and its dissolved species in solution, i.e. **solubility equilibria**; (ii) the equilibrium between a dissolved undissociated molecule and its dissociated ions, i.e. **dissociation equilibria**. In many cases, these dissociation equilibria will involve either acids or bases.

## 22.2 The solubility of sparingly soluble ionic solids in water

When increasing quantities of a sparingly soluble ionic solid are added to water, a saturated solution is eventually formed. Ions in the saturated solution are in equilibrium with the excess undissolved solute:

$$MX(s) \rightleftharpoons M^+(aq) + X^-(aq)$$

What is the relationship between the concentrations of the aqueous ions and the undissolved solute at equilibrium?

Table 22.1 shows the equilibrium concentrations of $Ag^+(aq)$ and $BrO_3^-(aq)$ ions in contact with undissolved $AgBrO_3$ when different initial volumes of 0.1M $AgNO_3$ and 0.1M $KBrO_3$ were added to 200cm³ of distilled water at 16°C.

**Table 22.1** Concentrations of $Ag^+(aq)$ and $BrO_3^-(aq)$ ions in contact with undissolved $AgBrO_3$ when different initial volumes of 0.1 M $AgNO_3$ and 0.1 M $KBrO_3$ are added to 200cm³ of distilled water.

| Initial volume of 0.1 M $AgNO_3$/cm³ | Initial volume of 0.1 M $KBrO_3$/cm³ | Concentration of $Ag^+(aq)$ at equilibrium /mole dm$^{-3}$ | Concentration of $BrO_3^-(aq)$ at equilibrium /mole dm$^{-3}$ | $[Ag^+]_{eqm} \times [BrO_3^-]_{eqm}$ /mole² dm$^{-6}$ |
|---|---|---|---|---|
| 40 | 10 | 0.0144 | 0.0024 | $3.45 \times 10^{-5}$ |
| 30 | 20 | 0.0081 | 0.0041 | $3.32 \times 10^{-5}$ |
| 25 | 25 | 0.0058 | 0.0058 | $3.36 \times 10^{-5}$ |
| 20 | 30 | 0.0042 | 0.0082 | $3.44 \times 10^{-5}$ |
| 10 | 40 | 0.0033 | 0.0102 | $3.37 \times 10^{-5}$ |

The concentrations of $BrO_3^-$(aq) were obtained by pipetting off a measured volume of the aqueous solution, adding acid followed by excess potassium iodide and then titrating the liberated iodine against sodium thiosulphate(VI) solution of known concentration.

Having determined the concentration of $BrO_3^-$(aq), the concentration of $Ag^+$(aq) can be calculated.

Notice that the products of the concentration of $Ag^+$(aq) and $BrO_3^-$(aq) are constant (column 5 in table 22.1). Thus, for this equilibrium we can write:

$$[Ag^+(aq)][BrO_3^-(aq)] = \text{a constant at a given temperature}$$
$$= 3.39 \times 10^{-5}\ \text{mole}^2\ \text{dm}^{-6}\ \text{at}\ 16°\text{C}.$$

In other words, the product of the concentrations of $Ag^+$ and $BrO_3^-$ is completely independent of the amount of $AgBrO_3$ present, provided there is some undissolved $AgBrO_3$ in contact with the solution. This situation is comparable to other heterogeneous systems involving solid/liquid or solid/gas equilibria (section 20.11).

How can we explain the constant value for the product $[Ag^+(aq)][BrO_3^-(aq)]$? When equilibrium between pure $AgBrO_3$ and its solution is reached, we have

$$AgBrO_3(s) \rightleftharpoons Ag^+(aq) + BrO_3^-(aq)$$

Hence we can write an equilibrium constant expression as

$$K_c = \frac{[Ag^+(aq)][BrO_3^-(aq)]}{[AgBrO_3(s)]}.$$

But, $[AgBrO_3(s)]$, which represents the concentration of a pure solid, is constant (see section 20.11) so,

$$[Ag^+(aq)][BrO_3^-(aq)] = K_c.[AgBrO_3(s)] = \text{a new constant.}$$

This new constant is known as the **solubility product** and is given the symbol, $K_{s.p.}$

Using the general formula $A_xB_y$ for a sparingly soluble salt, we can deduce a general expression for the solubility product as follows.

At equilibrium, let us suppose,

$$A_xB_y(s) \rightleftharpoons xA^{y+}(aq) + yB^{x-}(aq)$$

$$\text{Hence, } K_c = \frac{[A^{y+}(aq)]^x\,[B^{x-}(aq)]^y}{[A_xB_y(s)]}$$

But $[A_xB_y(s)]$ is constant, therefore

$$[A^{y+}(aq)]^x\,[B^{x-}(aq)]^y = K_{s.p.}, \text{ the solubility product of } A_xB_y$$

Write an expression for the solubility product of:

○ $Bi_2S_3$,
○ $AgCl$,
○ $PbI_2$.

The solubility products of some common compounds are given in table 22.2.

**Table 22.2** The solubility products of some common compounds at 25°C.

| Compound | Solubility product | Compound | Solubility product |
|---|---|---|---|
| Barium sulphate(VI) | $1.0 \times 10^{-10}$ | Lead(II) sulphate(VI) | $1.6 \times 10^{-8}$ |
| Calcium carbonate | $5.0 \times 10^{-9}$ | Lead(II) sulphide | $1.3 \times 10^{-28}$ |
| Calcium fluoride | $4.0 \times 10^{-11}$ | Nickel sulphide | $4.0 \times 10^{-21}$ |
| Calcium sulphate(VI) | $2.0 \times 10^{-5}$ | Silver bromide | $5.0 \times 10^{-13}$ |
| Copper(II) sulphide | $6.3 \times 10^{-36}$ | Silver chloride | $2.0 \times 10^{-10}$ |
| Lead(II) bromide | $3.9 \times 10^{-5}$ | Silver iodide | $8.0 \times 10^{-17}$ |
| Lead(II) chloride | $2.0 \times 10^{-5}$ | Zinc sulphide | $1.6 \times 10^{-24}$ |
| Lead(II) iodide | $7.1 \times 10^{-9}$ | | |

The solubility product of a salt can also be obtained from its solubility. The following example shows how this is done.

A saturated solution of silver chloride contains $1.46 \times 10^{-3}$g dm$^{-3}$ at 18°C. What is the solubility product of silver chloride at this temperature?

The solubility of silver chloride at 18°C $= 1.46 \times 10^{-3}$g dm$^{-3}$

$$= \frac{1.46 \times 10^{-3}}{143.5} \text{ mole dm}^{-3}$$

$$= 1 \times 10^{-5} \text{ mole dm}^{-3}$$

Now at equilibrium,

$$AgCl(s) \rightleftharpoons Ag^+(aq) + Cl^-(aq)$$

$$\therefore [Ag^+(aq)] = 1 \times 10^{-5}\text{M and}$$

$$[Cl^-(aq)] = 1 \times 10^{-5}\text{ M}$$

$$\Rightarrow K_{s.p.}(AgCl) = [Ag^+(aq)][Cl^-(aq)] = 1 \times 10^{-5} \times 1 \times 10^{-5}$$

$$= 10^{-10} \text{ mole}^2 \text{ dm}^{-6}$$

so, the solubility product of silver chloride at 18°C is $10^{-10}$ mole$^2$ dm$^{-6}$.

Conversely, the solubility of a salt can be obtained from its solubility product as the following calculation shows.

The solubility product of silver carbonate at 20°C is $8 \times 10^{-12}$ mole$^3$ dm$^{-9}$. What is its solubility at this temperature?

Let us suppose the solubility is $s$ mole dm$^{-3}$.

At equilibrium,

$$Ag_2CO_3(s) \rightleftharpoons 2Ag^+(aq) + CO_3^{2-}(aq)$$

$\therefore$ if the solubility of $Ag_2CO_3$ is $s$ mole dm$^{-3}$

$$[Ag^+(aq)] = 2s \text{ and } [CO_3^{2-}(aq)] = s$$

$$\therefore K_{s.p.}(Ag_2CO_3) = [Ag^+(aq)]^2[CO_3^{2-}(aq)] = 8 \times 10^{-12} \text{ mole}^3 \text{ dm}^{-9}$$

$$= (2s)^2 s = 8 \times 10^{-12}$$

$$\therefore 4s^3 = 8 \times 10^{-12}$$

$$\Rightarrow s^3 = 2 \times 10^{-12}$$

$$\therefore s = 1.25 \times 10^{-4} \text{ mole dm}^{-3}$$

i.e. the solubility of silver carbonate at 20°C is $1.25 \times 10^{-4}$ mole dm$^{-3}$.

Notice that the concentration of $Ag^+$ is both doubled and squared in the solubility product expression relative to the solubility of $Ag_2CO_3$. This is because the stoichiometry of the equation means that the concentration of $Ag^+$ will be double the solubility (i.e. $2s$ not $s$) and the expression for $K_{s.p.}$ requires this concentration of $Ag^+$ to be squared (i.e. $(2s)^2$ not $2s$).

## 22.3 Limitations to the solubility product concept

The solubility product concept is valid only for saturated solutions in which the total concentration of ions is no more than about 0.01M. For concentrations greater than this the value of $K_{s.p.}$ is no longer constant. This means that it is quite inappropriate to use the solubility product concept for soluble compounds such as NaCl, $CuSO_4$ and $AgNO_3$. As a consequence of this, the numerical value of solubility products is always very small, rarely exceeding $10^{-4}$. For substances of extremely low solubility, $K_{s.p.}$ may be less than $10^{-40}$.

The solubility product of a sparingly soluble salt is essentially a modified equilibrium constant and, like other equilibrium constants, its value will change with temperature. Consequently, the temperature at which a solubility product is measured should always be specified unless it relates to the selected standard temperature of 298 K.

## 22.4 Using the solubility product concept

THE COMMON ION EFFECT

Although the solubility product of a particular salt is constant at constant temperature, the concentrations of the individual ions may vary over a very wide range. When a saturated solution is obtained by dissolving the pure salt in water the concentrations of the ions produced are in the ratio determined by the stoichiometry of the compound. For example, the concentrations of $Ca^{2+}$ and $F^-$ ions in pure saturated calcium fluoride solution must be in the ratio 1:2. However, when a saturated solution is obtained by mixing two solutions containing a common ion, there may be a big difference in the concentration of the ions of any sparingly soluble electrolyte. In these cases, solubility products can be used to determine the concentration of ions remaining in solution. We can illustrate this by considering the solubility of $BaSO_4$, first in water and then in 0.1M sodium sulphate(VI) solution.

$$(K_{s.p.}(BaSO_4) = 1 \times 10^{-10}\ \text{mole}^2\ \text{dm}^{-6}).$$

*Solubility of $BaSO_4$ in water.*
Suppose the solubility of $BaSO_4$ in water $= s$ mole dm$^{-3}$

$BaSO_4(s) \rightleftharpoons Ba^{2+}(aq) + SO_4^{2-}(aq)$

$\therefore\ K_{s.p.}(BaSO_4) = [Ba^{2+}][SO_4^{2-}]$

$\Rightarrow 1 \times 10^{-10} = s \times s = s^2$

$\Rightarrow s = 10^{-5}$ mole dm$^{-3}$

$\therefore$ solubility of $BaSO_4$ in water $= 10^{-5}$mole dm$^{-3}$.

*Solubility of $BaSO_4$ in 0.1M $Na_2SO_4$.*
Suppose the solubility of $BaSO_4$ in 0.1M $Na_2SO_4$ $= s'$ mole dm$^{-3}$

$$BaSO_4(s) \rightleftharpoons Ba^{2+}(aq) + SO_4^{2-}(aq)$$

$$Na_2SO_4(s) \longrightarrow 2Na^+(aq) + SO_4^{2-}(aq)$$

In this case, $[Ba^{2+}(aq)] = s'$ mole dm$^{-3}$

but, $[SO_4^{2-}(aq)] = (s' + 0.1)$ mole dm$^{-3}$

$\Rightarrow K_{s.p.} = [Ba^{2+}][SO_4^{2-}] = s'(s' + 0.1)$

$\Rightarrow s'(s' + 0.1) = 1 \times 10^{-10}$

Now since $s' \ll 0.1$; $(s' + 0.1) \simeq 0.1$

$\Rightarrow s' \times 0.1 = 1 \times 10^{-10}$

$\Rightarrow s' = 10^{-9}$ mole dm$^{-3}$

$\therefore$ solubility of $BaSO_4$ in 0.1M $Na_2SO_4$ $= 10^{-9}$ mole dm$^{-3}$

This calculation illustrates the important generalization known as the **common ion effect**, i.e. *in the presence of either $A^+$ or $B^-$ from a second source, the solubility of the salt AB is reduced.*

Calculate the solubility of silver chloride
- in water,
- in 0.1M NaCl.

$(K_{s.p.}(AgCl) = 2.0 \times 10^{-10}\ \text{mole}^2\ \text{dm}^{-6})$

PREDICTING PRECIPITATION

Another important application of solubility products is that they enable chemists to predict the maximum concentrations of ions in a solution at a given temperature and hence whether or not precipitation will occur.

Suppose we mix a $10^{-3}$M solution of $Ca^{2+}$ ions with an equal volume of a $10^{-3}$M solution of $SO_4^{2-}$ ions at 25°C. Will a precipitate of $CaSO_4$ form?

The solubility product for calcium sulphate(VI) is $2 \times 10^{-5}$ mole$^2$ dm$^{-6}$ at 25°C; that is,

$$K_{s.p.}(CaSO_4) = [Ca^{2+}][SO_4^{2-}] = 2 \times 10^{-5}\ \text{mole}^2\ \text{dm}^{-6}$$

Immediately after mixing equal volumes of the two solutions and before any precipitation has occurred,

$$[Ca^{2+}] = [SO_4^{2-}] = 5 \times 10^{-4}\text{M}$$

The formation of stalagmites and stalactites in caves results from precipitation of calcium carbonate from saturated solution.

(The concentration of each ion is halved since each solution is diluted by mixing with the other.) Hence, the ionic product for $CaSO_4$ immediately after mixing,

$$[Ca^{2+}][SO_4^{2-}] = 5 \times 10^{-4} \times 5 \times 10^{-4}$$
$$= 25 \times 10^{-8} = 2.5 \times 10^{-7}\ \text{mole}^2\ \text{dm}^{-6}$$

This ionic product is less than the value of $K_{s.p.}$ for $CaSO_4$ and so no precipitate will form.

Let us now suppose that we mix equal volumes of $10^{-2}$M solutions. Immediately after mixing,

$$[Ca^{2+}] = [SO_4^{2-}] = 5 \times 10^{-3}\text{M}$$

and the ionic product $= [Ca^{2+}][SO_4^{2-}]$

$$= 5 \times 10^{-3} \times 5 \times 10^{-3}$$
$$= 25 \times 10^{-6}$$
$$= 2.5 \times 10^{-5}\ \text{mole}^2\ \text{dm}^{-6}$$

Stag's horn coral growing off the shore of the Seychelle Islands in the Indian Ocean.

In this case, the ionic product is greater than the solubility product and therefore precipitation of $CaSO_4$ occurs. The concentrations of aqueous $Ca^{2+}$ and $SO_4^{2-}$ ions are lowered by the reaction

$$Ca^{2+}(aq) + SO_4^{2-}(aq) \longrightarrow CaSO_4(s),$$

until the product $[Ca^{2+}][SO_4^{2-}]$ is reduced from $2.5 \times 10^{-5}$ to $2.0 \times 10^{-5}$.

The precipitation of solids from aqueous solution is of great importance in nature and industry. Stalagmites and stalactites precipitate slowly from water in which the concentrations of $Ca^{2+}(aq)$ and $CO_3^{2-}(aq)$ have an ionic product greater than the solubility product of calcium carbonate. Coral reefs grow in a similar fashion. In this case, the concentration of $Ca^{2+}$ and $CO_3^{2-}$ ions in the immediate vicinity of the coral must be large enough to precipitate calcium carbonate from the surrounding sea water.

SELECTIVE PRECIPITATION

In both qualitative and quantitative analysis, the differing solubilities of salts can be used as a means of separating different substances from each other by carefully selected precipitation reactions. Just suppose that we have a solution containing magnesium chloride, calcium chloride and barium chloride. How can a separation of the metal ions be achieved? Since both magnesium and calcium chromates(VI) are soluble, whereas barium chromate(VI) is insoluble, addition of a solution of $K_2CrO_4(aq)$ will precipitate $BaCrO_4(s)$, which can then be removed by filtration.

The remaining solution now contains $Mg^{2+}$ and $Ca^{2+}$ ions. However, $MgSO_4$ is soluble, while $CaSO_4$ is insoluble. Thus, addition of $Na_2SO_4(aq)$ to the mixture will precipitate $CaSO_4$ and leave $Mg^{2+}(aq)$ in solution. Finally, the $Mg^{2+}(aq)$ can be removed as solid $MgCO_3$ by adding $Na_2CO_3$ solution.

Notice that the order in which reagents are added is important. If we had added $Na_2SO_4$ solution before adding $K_2CrO_4$ solution, a mixture of $BaSO_4$ and $CaSO_4$ would have been precipitated. On the other hand, if $Na_2CO_3$ solution had been added to the mixture of the three cations, $MgCO_3$, $CaCO_3$ and $BaCO_3$ would all have been precipitated. Thus, both the precipitating reagents and their order of addition must be selected with great care.

The results in table 22.3 show how three similar cations ($Cu^{2+}$, $Zn^{2+}$ and $Ni^{2+}$) can be separated by careful, selective precipitation.

**Table 22.3** Selective precipitation of CuS, ZnS, and NiS.

| Cation under test | $H_2S$ bubbled into acidic solution of cation | $H_2S$ bubbled into neutral solution of cation | $H_2S$ bubbled into alkaline solution of cation |
|---|---|---|---|
| $Cu^{2+}$ | Black ppte of CuS<br>$Cu^{2+} + S^{2-} \longrightarrow CuS(s)$ | Black ppte of CuS | Black ppte of CuS |
| $Zn^{2+}$ | No ppte | White ppte of ZnS<br>$Zn^{2+} + S^{2-} \longrightarrow ZnS(s)$ | White ppte of ZnS |
| $Ni^{2+}$ | No ppte | No ppte | Black ppte of NiS<br>$Ni^{2+} + S^{2-} \longrightarrow NiS(s)$ |

When hydrogen sulphide is bubbled through a solution containing these cations, the following equilibrium is established.

$$H_2S(aq) \rightleftharpoons 2H^+(aq) + S^{2-}(aq)$$

In acid solution, $H^+$ ions will displace the equilibrium to the left and reduce the concentration of $S^{2-}(aq)$ to very low values. Hence, only those metal sulphides with very low solubility products (such as CuS, $K_{s.p.}(CuS) = 6.3 \times 10^{-36}$) will be precipitated. i.e. in acid solution;

$$[Cu^{2+}][S^{2-}] > K_{s.p.}(CuS) = 6.3 \times 10^{-36}$$

$$\text{but } [Zn^{2+}][S^{2-}] < K_{s.p.}(ZnS) = 1.6 \times 10^{-24}$$

$$\text{and } [Ni^{2+}][S^{2-}] < K_{s.p.}(NiS) = 4 \times 10^{-21}$$

Thus CuS is precipitated, but ZnS and NiS do not precipitate.

In neutral solution, the concentration of $S^{2-}(aq)$ from $H_2S$ will be higher and this enables ZnS with a higher solubility product ($K_{s.p.}(ZnS) = 1.6 \times 10^{-24}$) to precipitate as well as CuS.

When the solution is made slightly alkaline with ammonia, the $H_2S$ equilibrium is displaced further to the right and the concentration of $S^{2-}(aq)$ is even higher. Under these conditions NiS is precipitated as well as CuS and ZnS.

Thus, by careful adjustment of the pH before passing in $H_2S$, it is possible to separate a mixture of $Cu^{2+}$, $Zn^{2+}$ and $Ni^{2+}$.

The oyster, in growing its shell, must adjust conditions so that the concentration of carbonate ions and calcium ions is large enough to precipitate calcium carbonate from sea water.

## 22.5 The strengths of acids and bases

The strengths of different acids and bases can be compared quickly, though only very approximately, using conductivity measurements.

Strong electrolytes, such as hydrochloric acid and sodium hydroxide, are virtually completely dissociated into ions in aqueous solutions and are therefore better conductors than weak electrolytes, such as ethanoic (acetic) acid, which are only partially dissociated:

$$HCl(aq) \rightleftharpoons H^+(aq) + Cl^-(aq) \quad \text{almost complete dissociation}$$

$$CH_3COOH(aq) \rightleftharpoons H^+(aq) + CH_3COO^-(aq) \quad \text{only partial dissociation}$$

The simple descriptive terms 'strong' and 'weak' are much too limited and inaccurate as a method of comparing the strengths of electrolytes and therefore chemists looked for a more accurate and quantitative comparison. In the case of acids, relative strengths can be compared by measuring the concentration of $H^+$ ions or by measuring their **pH**. The 'p' in pH comes from the German word 'potenz' meaning power and the 'H' from $[H^+]$.

*The pH of a solution is the negative logarithm to base ten of the molar hydrogen ion concentration,* i.e.

$$pH = -\lg [H^+(aq)]^*$$

Since the hydrogen ion concentrations in aqueous solution range from about $10^{-15}$ to 10 mole $dm^{-3}$, it is particularly convenient to have a scale that is both negative and logarithmic. The negative sign produces pH values which are positive for almost all solutions encountered in practice, whilst the logarithmic scale reduces the extremely wide variation in $[H^+(aq)]$ to a narrow range of pH from about 15 to $-1$.

The following examples show how pH can be obtained from $[H^+(aq)]$ and vice-versa.

(a) What is the pH of $10^{-1}$M HCl?
Since HCl is fully dissociated and monobasic,

$$[H^+(aq)] \text{ in } 10^{-1}\text{M HCl} = 10^{-1}\text{M}$$

$$\therefore pH = -\lg [H^+(aq)] = -\lg [10^{-1}] = -1(-1) = +1$$

* Notice that the accepted abbreviation for logarithm to base 10 is lg not $\log_{10}$.

(b) What is the pH of $10^{-3}$M $H_2SO_4$?
Since $H_2SO_4$ is fully dissociated and dibasic,

$[H^+(aq)]$ in $10^{-3}$M $H_2SO_4$ = $2 \times 10^{-3}$M

$$\therefore pH = -\lg [H^+(aq)] = -\lg (2 \times 10^{-3})$$
$$= -(+0.30 - 3.00)$$
$$= -(-2.70)$$
$$= +2.70$$

(c) The pH of pure water at 25°C is 7. What is its hydrogen ion concentration?

$$pH = -\lg [H^+(aq)]$$
$$\Rightarrow 7 = -\lg [H^+(aq)]$$
$$\therefore \lg [H^+(aq)] = -7$$
$$\therefore [H^+(aq)] = 10^{-7}$$

What are the pH values of the following solutions:
- ○ $10^{-3}$M HCl,
- ○ 1.0M HCl,
- ○ 3M HX which is only 50% dissociated?

## 22.6 The dissociation of water

When water is purified by repeated distillation, its conductance falls to a constant, low but definite value. Even the purest water has a tiny electrical conductivity. This is further evidence that water forms ions as a result of its own dissociation, i.e.

$$H_2O(l) \rightleftharpoons H^+(aq) + OH^-(aq)$$

Obviously, the concentration of ions is very small, as the pH of pure water shows, and the equilibrium in this reaction lies far to the left. Nevertheless, we can write an equilibrium constant for the dissociation of water as

$$K_c = \frac{[H^+][OH^-]}{[H_2O]}$$

Now, since only a minute trace of the water is ionized, each cubic decimetre of water will contain virtually 1 000g of $H_2O$, and as 1 mole of water weighs 18g, we can say,

$$[H_2O] \text{ in water} = \frac{1\,000}{18} = 55.55\text{M which is constant}$$

Thus, we can incorporate this constant for $[H_2O]$ in the value of $K_c$, just as we did with the concentration of undissolved solute when considering the solubility product of a sparingly soluble salt.

i.e. $K_c[H_2O] = [H^+][OH^-]$ = a new constant, $K_w$

(constant) (constant)

This overall constant, $K_w$, is called the **ionic product for water**.

At 25°C, $[H^+] = [OH^-] = 10^{-7}$M,

hence $K_w = [H^+][OH^-] = 10^{-7} \times 10^{-7} = 10^{-14}$ mole² dm⁻⁶.

- ○ Why is $[H^+] = [OH^-]$ in pure water?
- ○ At 25°C, $K_w = 10^{-14}$. What is the value of $K_c$ for the reaction
$$H_2O(l) \rightleftharpoons H^+(aq) + OH^-(aq) \text{ at } 25°C?$$
- ○ Look at the information in table 22.4. How does the value of $K_w$ change with increase in temperature?
- ○ Explain the effect of temperature on $K_w$ using Le Chatelier's principle.
$$H_2O(l) \rightleftharpoons H^+(aq) + OH^-(aq); \quad \Delta H = +58\text{ kJ}$$

**Table 22.4** Values of $K_w$ at various temperatures.

| Temperature/°C | $K_w$/mole² dm$^{-6}$ |
|---|---|
| 0 | $0.11 \times 10^{-14}$ |
| 10 | $0.30 \times 10^{-14}$ |
| 20 | $0.68 \times 10^{-14}$ |
| 25 | $1.00 \times 10^{-14}$ |
| 50 | $5.47 \times 10^{-14}$ |
| 100 | $51.3 \times 10^{-14}$ |

## 22.7 The pH scale

In pure water and in neutral solutions such as sodium chloride, $H^+$ and $OH^-$ ions arise only from the ionization of water.

Hence in water and in neutral solutions,

$$[H^+] = [OH^-]$$

and at 25°C, $[H^+] = [OH^-] = 10^{-7}$, ∴ pH = 7.
However, in acidic and alkaline solutions $H^+$ and $OH^-$ ions may arise from sources other than water. Nevertheless, the system

$$H_2O(l) \rightleftharpoons H^+(aq) + OH^-(aq)$$

is still in equilibrium and so the product $[H^+][OH^-]$ remains constant for all solutions at the same temperature. Thus, it is possible to determine both the $[H^+]$ and the $[OH^-]$ in any solution.

For example, in $10^{-2}$M HCl,

$$[H^+] = 10^{-2} \text{ and } \therefore \text{pH} = 2,$$

but since $[H^+][OH^-] = 10^{-14}$ for this solution,

$$[OH^-] \text{ in } 10^{-2}\text{M HCl} = 10^{-12}.$$

Likewise, in $10^{-1}$M NaOH,

$$[OH^-] = 10^{-1},$$

but since $[H^+][OH^-] = 10^{-14}$ for this solution,

$[H^+]$ in $10^{-1}$M NaOH $= 10^{-13}$ and ∴ pH = 13.

These results will help you to appreciate the following important generalizations.
*For neutral solutions,*
$[H^+] = [OH^-] = 10^{-7}$ and pH = 7 at 25°C.
*For acidic solutions,*
$[H^+] > [OH^-]$ and pH < 7.
*For basic (alkaline) solutions,*
$[H^+] < [OH^-]$ and pH > 7.
Figure 22.1 relates the pH scale to the hydrogen ion concentration and to changing acidity and alkalinity.

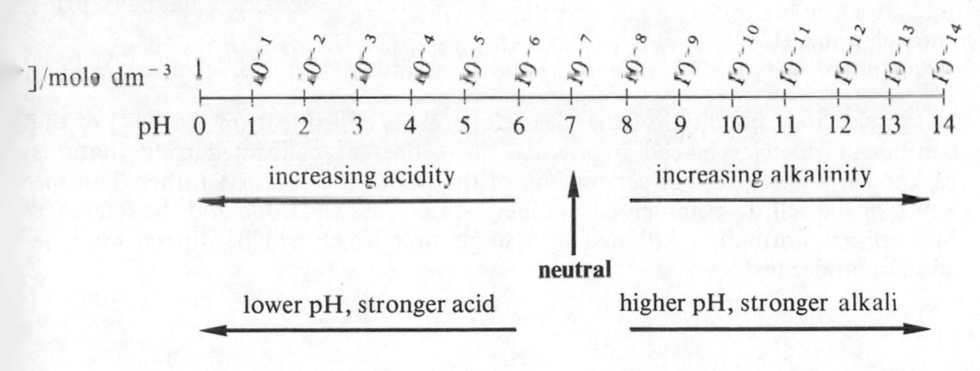


**Figure 22.1** The pH scale.

## 22.8 The measurement of hydrogen ion concentration and pH

The most obvious method of measuring the hydrogen ion concentration of a solution is by using the hydrogen half-cell (hydrogen electrode) described in section 13.3. Under standard conditions, using a 1.0M solution of $H^+$ ions, hydrogen gas at 1 atm and a temperature of 298 K, the hydrogen half-cell is assigned an electrode potential of 0.00 volts. Thus, when the standard hydrogen electrode is combined with another electrode to form a complete cell, the numerical value of the overall e.m.f. will be dictated by the second electrode. However, if the concentration of $H^+$ ions in the hydrogen half-cell changes, then the electrode potential of this cell is no longer 0.00 volts and so the overall e.m.f. will also change. By constructing a calibration curve showing the overall e.m.f. against the $[H^+]$ in the hydrogen half-cell using various solutions of known $[H^+)$, it is possible to place the hydrogen electrode in any solution and obtain its pH from the resulting e.m.f., by using the predetermined calibration curve.

Unfortunately, the hydrogen half-cell is awkward and inconvenient to use. It requires the use of a bulky hydrogen cylinder, it is difficult to adjust the gas pressure so as to provide a steady and satisfactory flow rate, it takes some time to reach equilibrium and the platinum electrode is easily 'poisoned' by impurities in the gas or in the solution.

Consequently, alternative and more convenient electrodes were sought, of which the so called '**glass electrode**' is the most widely used.

The glass electrode consists of a platinum electrode in a solution of fixed acidic pH. The electrode is placed inside a thin glass membrane permeable to $H^+$ ions. This half-cell is combined with a reference half-cell which is usually either a calomel electrode (mercury in contact with mercury(I) chloride in saturated potassium chloride) or a silver/silver chloride electrode (silver wire coated with silver chloride in saturated potassium chloride). This arrangement is attached to a sensitive voltmeter to form a **pH meter** (figure 22.2).

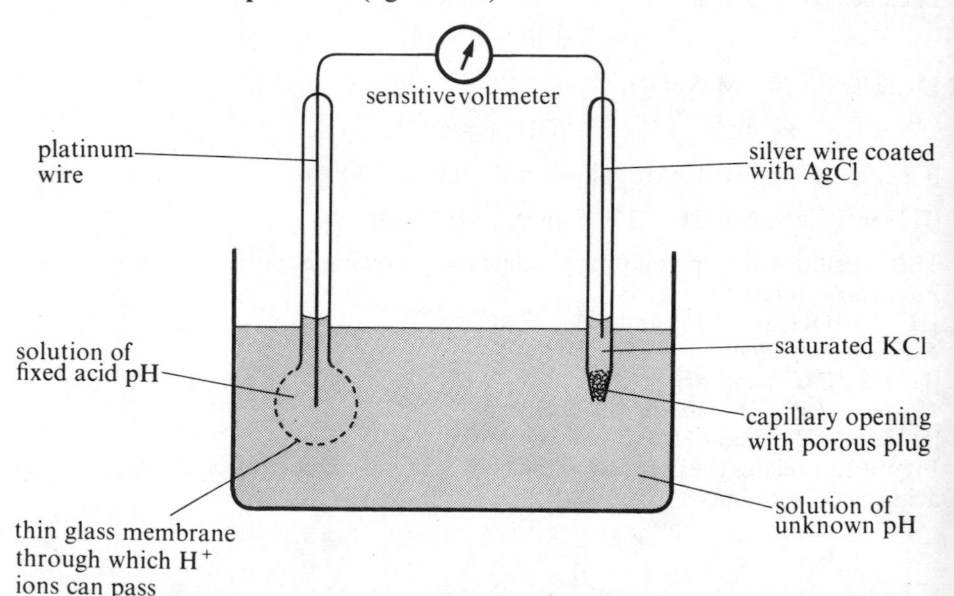


**Figure 22.2** A pH meter consisting of a glass electrode and a reference silver/silver chloride half-cell.

The electrode potential of the glass electrode is dependent on the $[H^+]$ of the solution in which it is placed. In practice, the voltmeter is calibrated using solutions of known pH to give a direct reading of the pH of the solution rather than the e.m.f. of the cell. In commercial pH meters, the glass electrode and the reference half-cell are normally combined in a single unit which can be dipped into the solution under test.

## 22.9 Dissociation constants of acids and bases

Although the pH of a solution provides some measure of the strength of a constituent acid or base, the use of pH is very limited in this context since its value will change as the concentration changes. Consequently, chemists looked for a more useful, yet quantitative, means of representing the strengths of acids and bases. They found this by considering the dissociation equilibria of these substances in aqueous solution.

When the weak acid HA is dissolved in water, we can write

$$HA(aq) \rightleftharpoons H^+(aq) + A^-(aq)$$

Hence,
$$K_c = \frac{[H^+(aq)][A^-(aq)]}{[HA(aq)]}$$

In dealing with acids, $K_c$ is usually replaced by the symbol, $K_a$, which is known as the **dissociation constant** of the acid. By eliminating the (aq) state symbols, we can simplify the last equation to

$$K_a = \frac{[H^+][A^-]}{[HA]}$$

Since dissociation constants are effectively equilibrium constants, they are unaffected by concentration changes and are influenced only by changes in temperature. Hence, the numerical value of $K_a$ provides an accurate measure of the extent to which an acid is dissociated (i.e. the strength of the acid).

The greater the extent of dissociation, the greater are $[H^+]$ and $[A^-]$, the larger is $K_a$ and the stronger is the acid. The values of $K_a$ for some acids are given in table 22.5.

**Table 22.5** The values of $K_a$ for some acids.

| **Acid** | **Equilibrium in aqueous solution** | **$K_a$ at 25°C /mole dm$^{-3}$** |
|---|---|---|
| Sulphuric(VI) acid | $H_2SO_4 \rightleftharpoons H^+ + HSO_4^-$ | very large |
| Nitric(V) acid | $HNO_3 \rightleftharpoons H^+ + NO_3^-$ | 40 |
| Trichloroethanoic acid | $CCl_3COOH \rightleftharpoons H^+ + CCl_3COO^-$ | $2.3 \times 10^{-1}$ |
| Dichloroethanoic acid | $CHCl_2COOH \rightleftharpoons H^+ + CHCl_2COO^-$ | $5.0 \times 10^{-2}$ |
| Sulphuric(IV) acid | $H_2SO_3 \rightleftharpoons H^+ + HSO_3^-$ | $1.6 \times 10^{-2}$ |
| Chloroethanoic acid | $CH_2ClCOOH \rightleftharpoons H^+ + CH_2ClCOO$ | $1.3 \times 10^{-3}$ |
| Nitric(III) acid | $HNO_2 \rightleftharpoons H^+ + NO_2^-$ | $4.7 \times 10^{-4}$ |
| Methanoic acid | $HCOOH \rightleftharpoons H^+ + HCOO^-$ | $1.6 \times 10^{-4}$ |
| Benzoic acid | $C_6H_5COOH \rightleftharpoons H^+ + C_6H_5COO^-$ | $6.4 \times 10^{-5}$ |
| Ethanoic acid | $CH_3COOH \rightleftharpoons H^+ + CH_3COO^-$ | $1.7 \times 10^{-5}$ |
| Hydrated aluminium ion | $[Al(H_2O)_6]^{3+} \rightleftharpoons H^+ + [Al(H_2O)_5OH]^{2+}$ | $1.0 \times 10^{-5}$ |
| Carbonic acid | $H_2CO_3 \rightleftharpoons H^+ + HCO_3^-$ | $4.5 \times 10^{-7}$ |
| Hydrogen sulphide | $H_2S \rightleftharpoons H^+ + HS^-$ | $8.9 \times 10^{-8}$ |
| Boric acid | $H_3BO_3 \rightleftharpoons H^+ + H_2BO_3^-$ | $5.8 \times 10^{-10}$ |
| Hydrogen peroxide | $H_2O_2 \rightleftharpoons H^+ + HO_2^-$ | $2.4 \times 10^{-12}$ |
| Water | $H_2O \rightleftharpoons H^+ + OH^-$ | $1.0 \times 10^{-14}$ |

The value of $K_a$ for an acid can be calculated using the expression above if we know the concentration of the acid and its pH. The following example shows how this is done.

The pH of 0.01M ethanoic(acetic) acid ($CH_3COOH$) is 3.40 at 25°C. What is the dissociation constant of ethanoic acid at this temperature?

$$CH_3COOH(aq) \rightleftharpoons H^+(aq) + CH_3COO^-(aq)$$

$$\Rightarrow K_a = \frac{[H^+][CH_3COO^-]}{[CH_3COOH]}$$

Since the concentration of $H^+$ ions arising from the water is much smaller than the concentration of those from the acid, we can say that

$[H^+] \simeq [CH_3COO^-]$

and $[CH_3COOH] = 0.01 - [H^+] \simeq 0.01$

if $[H^+] \ll 0.01$, which it will be since ethanoic acid is a weak acid.

$pH = -\lg [H^+] = 3.40$

Hence, $\lg [H^+] = -3.40 = (-4.00 + 0.60)$

$\therefore [H^+] = 4.0 \times 10^{-4}$ mole dm$^{-3}$

and $[CH_3COO^-] = 4.0 \times 10^{-4}$ mole dm$^{-3}$

Now, assuming $[CH_3COOH] = 0.01$ mole dm$^{-3}$

$$K_a = \frac{[H^+][CH_3COO^-]}{[CH_3COOH]} = \frac{4 \times 10^{-4} \times 4 \times 10^{-4}}{0.01}$$

$$K_a = 1.60 \times 10^{-5} \text{ mole dm}^{-3}$$

By reversing this calculation, it is possible to predict the pH of a solution of a weak acid if we know its molarity and its dissociation constant. The following questions will help you to calculate the pH of 1.0M benzoic acid ($C_6H_5COOH$).

$$K_a(C_6H_5COOH) = 6.4 \times 10^{-5} \text{ mole dm}^{-3}$$

- O Write an equation for the dissociation of $C_6H_5COOH$.
- O Using this equation, write an expression for the dissociation constant of benzoic acid.
- O Assuming $[H^+] = [C_6H_5COO^-]$ and

  $[C_6H_5COOH] \gg [H^+]$

  calculate $[H^+]$ for 1.0M benzoic acid.
- O What is the pH of 1.0M benzoic acid?

Just as acid dissociation constants can be used to compare the strengths of different acids, we can also use base dissociation constants for bases.

If the base BOH is in equilibrium with water as

$$BOH \rightleftharpoons B^+ + OH^-,$$

$$\text{then } K_b = \frac{[B^+][OH^-]}{[BOH]}$$

where $K_b$ is known as the dissociation constant of the base.

## 22.10 Acid–base indicators

*Acid–base indicators* such as methyl orange, phenolphthalein and bromothymol blue, *are substances which change colour according to the hydrogen ion concentration of the solution or liquid to which they are added* (figure 22.3). Consequently, they are used to test for acidity and alkalinity and to detect the end-point in acid–base titrations.

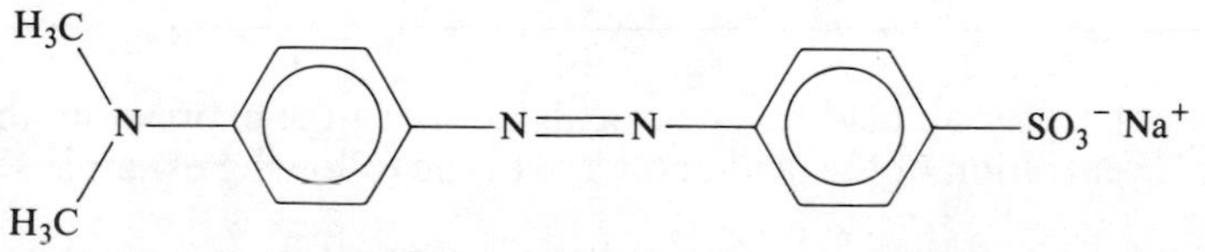

$H_3C$, $H_3C$ N—(ring)—N=N—(ring)—$SO_3^-$ $Na^+$

Methyl orange (yellow form)

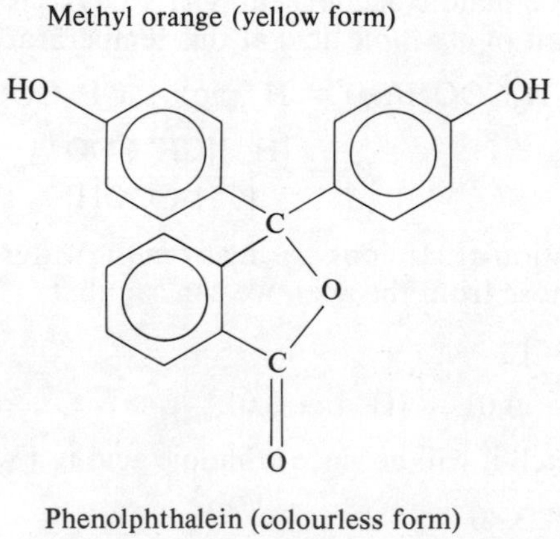

HO, OH, C, O, C, O

Phenolphthalein (colourless form)

**Figure 22.3** The formulae of two common indicators.

Most indicators can be regarded as weak acids of which either the undissociated molecule or the dissociated anion, or both, are coloured. If we take methyl orange as our example and write the undissociated molecule as HMe,

$$\underset{\text{red}}{HMe} \rightleftharpoons \underset{\text{colourless}}{H^+} + \underset{\text{yellow}}{Me^-}$$

Addition of acid (i.e. $H^+$ ions) displaces this equilibrium to the left, so that $[HMe] \gg [Me^-]$ and the solution becomes red.

On the other hand, when alkali (containing $OH^-$ ions) is added to methyl orange it removes $H^+$ ions forming water and the equilibrium in the above system moves to the right in order to replace some of the $H^+$ ions. In this case, $[Me^-] \gg [HMe]$, and the methyl orange becomes yellow.

The dissociation of phenolphthalein in aqueous solution can be represented as

$$HPh \rightleftharpoons H^+ + Ph^-$$

○ What colour is phenolphthalein in
(a) strongly acid solution,
(b) strongly alkaline solution?

○ What are the colours of
(a) HPh (b) $Ph^-$ for phenolphthalein?

Indicators can be regarded as weak acids and so it is possible to determine their dissociation constants. Using HIn for the undissociated form of the indicator, we can write,

$$HIn(aq) \rightleftharpoons H^+(aq) + In^-(aq)$$

$$\Rightarrow K_a(HIn) = \frac{[H^+][In^-]}{[HIn]}$$

The numerical values of these dissociation constants can, of course, be obtained by measuring the pH of a solution of known molarity for each indicator (see section 22.8). The dissociation constants for some indicators are shown in table 22.6.

**Table 22.6** The values of $K_a$ for some indicators.

| **Indicator** | **$K_a$ at 25°C /mole dm$^{-3}$** |
|---|---|
| Phenolphthalein | $7 \times 10^{-10}$ |
| Bromothymol blue | $1 \times 10^{-7}$ |
| Litmus | $3 \times 10^{-7}$ |
| Methyl orange | $2 \times 10^{-4}$ |

## THE END-POINT OF AN INDICATOR

The aim of any titration is to determine the volumes of two solutions which just react with each other. Thus, the **end point**, which is the point at which the titration is stopped, must coincide with the **equivalence point** for the two reacting solutions. In order to achieve this, the indicator should change colour sharply at the equivalence point on adding a single drop of either acid or alkali. At the exact end-point of the titration, the colour of the indicator will be mid-way between the acid colour of HIn and the alkaline colour of $In^-$, since

$$[HIn] = [In^-].$$

Now,
$$K_a(HIn) = \frac{[H^+][In^-]}{[HIn]}$$

But, at the end-point, $[HIn] = [In^-]$

$\therefore$ At the end-point, $K_a(HIn) = [H^+]$

$\Rightarrow$ pH at end-point $= -\lg [H^+] = -\lg K_a(HIn)$.

The pHs at the end-point for the four indicators previously mentioned are shown in table 22.7.

**Table 22.7** The pH at the end point for some indicators.

| **Indicator** | **pH at end point $= -\log K_a(HIn)$** |
|---|---|
| Phenolphthalein | 9.1 |
| Bromothymol blue | 7.0 |
| Litmus | 6.5 |
| Methyl orange | 3.7 |

THE RANGE OF AN INDICATOR

The colour change of an indicator is due to the change from one coloured form to another. Near the end-point, both coloured forms will be present in appreciable quantities and it is not possible to say precisely when the two forms are at equal concentrations. Thus, the eye cannot judge the exact end-point, and indicators effectively change colour over a range of about 2 pH units.

*The range of an indicator is the pH range over which it changes colour.* The values in tables 22.6 and 22.7 show that the dissociation constants of indicators differ widely; consequently, they change colour over widely differing ranges of pH. This last point is illustrated more clearly in figure 22.4.

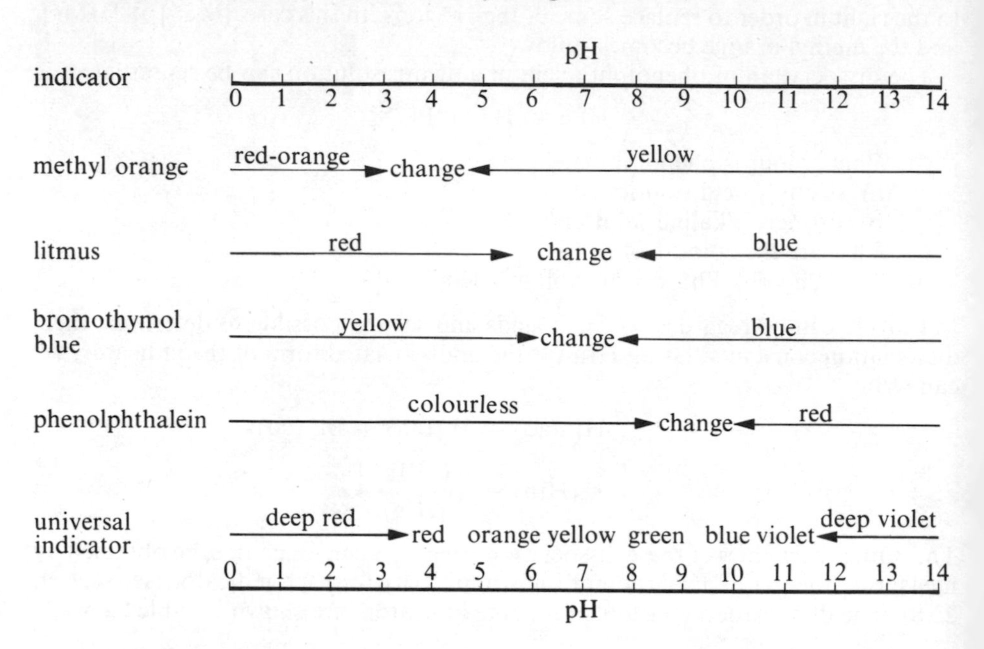


**Figure 22.4** The pH ranges of some indicators.

Look closely at figure 22.4.

- ○ What is the pH of pure water?
- ○ What colours will the following indicators give in pure water:
  - (a) methyl orange,
  - (b) phenolphthalein,
  - (c) bromothymol blue?
- ○ When too much phenolphthalein is added to water, it forms a cloudy solution. Why is this? (*Hint*: Phenolphthalein is an organic compound of high relative molecular mass.)

Many indicators, like phenolphthalein, are only slightly soluble in water, and are therefore prepared for use in alcohol or in a mixture of alcohol and water. Because of this, the colour of an indicator may be confusing before it is added to an aqueous solution.

Notice that the end-point of each indicator is in the centre of its pH range. Figure 22.4 also shows that many indicators will change colour over a range of pH well away from 7. This is an important point as many students misunderstand the use of indicators and expect them all to change colour at pH 7. In fact, it is because different indicators change colour at different pHs that acid–base titrations have such a wide application in industry and in the laboratory.

## 22.11 pH changes during titrations

During a titration there is a change in pH as alkali is added to acid or vice versa. At the equivalence point, the pH must change sharply by several units for it to be identifiable using an indicator.

The change in pH during the course of a titration depends largely upon the strength of the acid and alkali used.

### TITRATING STRONG ACID AGAINST STRONG ALKALI

The graph in figure 22.5 shows how the pH changes during the titration of 50cm³ of 0.1M HCl with 0.1M NaOH. As the alkali is added, the pH changes slowly at

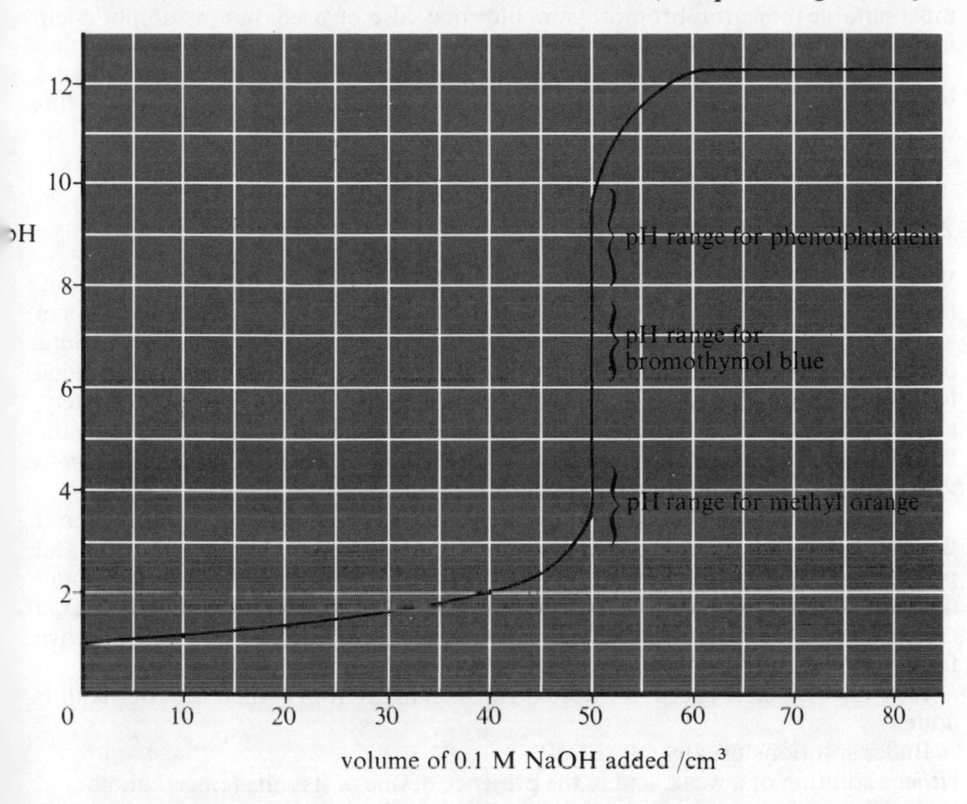


**Figure 22.5** pH changes during the titration of 50 cm³ of 0.1M HCl with 0.1M NaOH.

first, and then very rapidly from about 3.5 to 9.5 at the equivalence point. Thus, any indicator which changes colour between pH 3.5 and 9.5 will identify the equivalence point—any one of methyl orange, bromothymol blue or phenolphthalein can be used in this case.

### TITRATING STRONG ACID AGAINST WEAK ALKALI

The graph in figure 22.6 shows how the pH changes when 0.1M $NH_3$ solution (weak alkali) is added to 50cm³ of 0.1M HCl. As before, there is little variation in pH

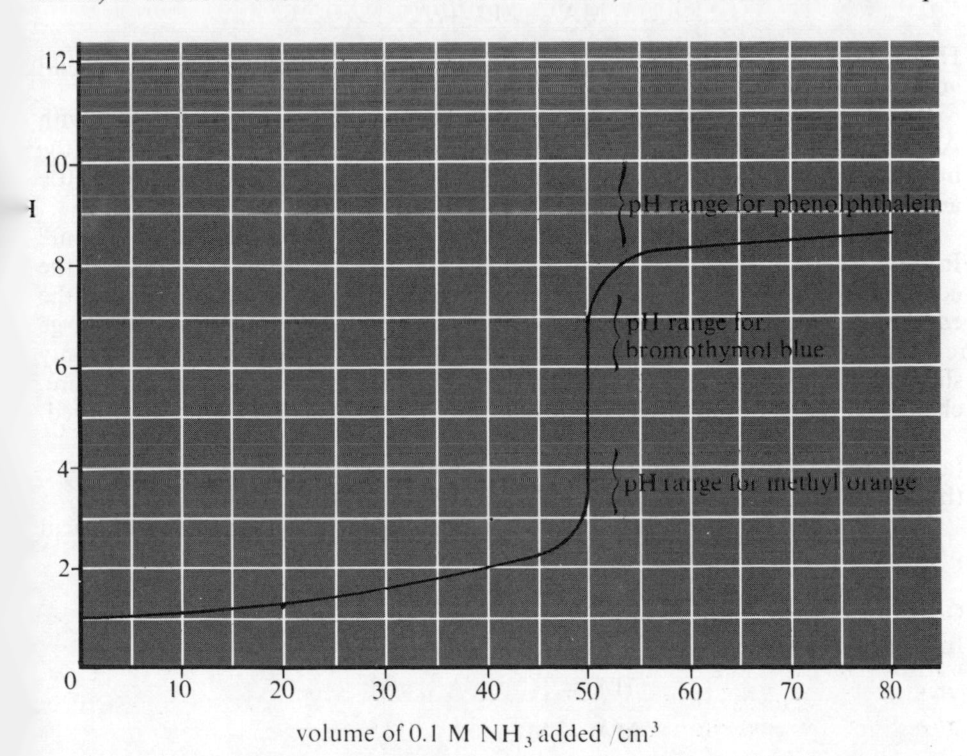


**Figure 22.6** pH changes during the titration of 50cm³ of 0.1M HCl with 0.1M $NH_3$.

when the alkali is first added, but at the equivalence point the pH changes rapidly from about 3.5 to 7.0. Thus, any indicator which changes colour between 3.5 and 7.0 will identify the equivalence point accurately. In this case, methyl orange is a most suitable indicator, bromothymol blue may also be used, but phenolphthalein is useless since it does not begin to change colour until about pH 8.

The pH changes during the titration of a weak acid with a strong alkali and a weak acid with a weak alkali are discussed in study question 10 at the end of this chapter.

## 22.12 Buffer solutions

When $0.1 cm^3$ of 1.0M HCl is added to $1 dm^3$ of water or sodium chloride solution, the pH changes sharply from 7.0 to 4.0 (i.e. by 3 units of pH). Clearly, the pH's of water and sodium chloride solution are extremely sensitive to even small additions of acid or alkali. If this happened when small amounts of acid or alkali were added to biological systems any living organisms would be killed instantly. Fortunately, animals and plants are protected against sharp changes in pH, resulting from the addition of small amounts of acid or alkali, by the presence of **buffers**—*solutions which resist changes in pH on addition of acid or alkali.*

Buffers are also important in many industrial processes where the pH must not deviate very much from an optimum value. Furthermore, several synthetic and processed foods must be prepared in a buffered form so that they may be eaten and digested in our bodies without undue change in pH. Under most circumstances, a change of only 0.5 units in the pH of blood, which is normally 7.4, is likely to prove fatal.

How does a buffer act? How does it resist changes in pH when acid or alkali is added?

Buffer solutions usually consist of:
*either* a solution of a weak acid in the presence of one of its salts (e.g. ethanoic acid and sodium ethanoate, carbonic acid and sodium hydrogencarbonate);
*or* a solution of a weak base in the presence of one of its salts (e.g. ammonia solution and ammonium chloride).

In order to understand how a buffer works, we can consider the hypothetical weak acid, HA, in a solution with its salt MA. In this solution, HA will be slightly dissociated whilst MA is fully dissociated into ions.

$$HA \rightleftharpoons H^+ + A^-$$
$$MA \rightarrow M^+ + A^-$$

Hence the mixture contains a relatively high concentration of un-ionized HA (an acid) and a relatively high concentration of $A^-$ (a base).

If acid is suddenly added to this system, the $H^+$ ions in the acid will combine with $A^-$ ions to form un-ionized HA. Provided there is a large reservoir of $A^-$ ions in the buffer, nearly all the added $H^+$ ions are removed. Thus, $[H^+]$ changes very little and the pH is only slightly altered.

What happens when alkali, such as sodium hydroxide, is added to the system? In this case, the $OH^-$ ions combine with $H^+$ ions to form water. This reduces the concentration of $H^+$ ions in the buffer, but more HA dissociates to restore the equilibrium and the $[H^+]$ rises almost to its original value. Provided there is a large reservoir of HA in the buffer, the $[H^+]$ changes very little and again the pH is only slightly altered. By having these reserves of both HA and $A^-$ in the buffer mixture, changes in the pH resulting from the addition of acid or alkali can be minimized.

Essentially, the stable pH of the buffer is due to
(a) a high $[A^-]$ which traps added $H^+$ ions and
(b) a high [HA] which can supply $H^+$ ions to trap added $OH^-$ ions.

Solutions of this kind which provide a 'buffer' against the effects of adding acid or alkali are therefore known as buffers.

### CALCULATING THE pH OF BUFFER SOLUTIONS

In a buffer composed of the weak acid HA and its salt MA

$$HA \rightleftharpoons H^+ + A^-, \text{ and}$$
$$MA \rightarrow M^+ + A^-.$$

Hence, we can write an expression for the dissociation constant of HA as

$$K_a = \frac{[H^+][A^-]}{[HA]}$$

$$\therefore [H^+] = K_a \cdot \frac{[HA]}{[A^-]}$$

Now, in the buffer mixture, [HA] is effectively the concentration of acid taken ([acid]) since the acid will be only very slightly dissociated in the presence of its salt, and $[A^-]$ is effectively the concentration of salt taken ([salt]) since the salt is fully dissociated into ions.
Thus we can write,

$$[H^+] \simeq K_a \times \frac{[acid]}{[salt]}$$

This last equation explains why the $[H^+]$, and therefore the pH of a buffer, is affected very little by dilution since the ratio [acid]/[salt] will remain constant on dilution.

Using the last equation, notice the special case which applies when [acid] = [salt], i.e.

$$[H^+] \simeq K_a$$

The following example will help you to understand the points which have just been discussed.
A buffer solution was made by adding 3.28g of sodium ethanoate to $1dm^3$ of 0.01M ethanoic acid. What is the pH of the resulting buffer?

$$(K_a(CH_3COOH)) = 1.7 \times 10^{-5} \text{ mole dm}^{-3})$$

$$K_a = \frac{[H^+][CH_3COO^-]}{[CH_3COOH]}$$

$$\therefore [H^+] = K_a \frac{[CH_3COOH]}{[CH_3COO^-]}$$

But, $[CH_3COOH] = [acid] = 0.01M$

and $[CH_3COO^-] = [salt] = \frac{3.28}{82} = 0.04M$

$$\therefore [H^+] = 1.7 \times 10^{-5} \times \frac{0.01}{0.04} = 4.25 \times 10^{-6}M$$

$$\therefore \text{pH of the buffer} = -\lg[H^+] = -\lg(4.25 \times 10^{-6})$$

$$= -(+0.63 - 6.00) = -(-5.37)$$

$$\Rightarrow \text{pH of the buffer} = 5.37$$

In order to appreciate the action of a buffer let us first consider the change in pH when $1cm^3$ of M NaOH is added to $1dm^3$ of the buffer in the last example, and then the change in pH when $1cm^3$ of M NaOH is added to $1dm^3$ of 0.01M ethanoic acid.

*1 The change in pH when $1cm^3$ of M NaOH is added to $1dm^3$ of buffer in the last example*

pH of buffer initially = 5.37.

When NaOH is added to the buffer it reacts with $CH_3COOH$ and forms $CH_3COONa$ (i.e. $[CH_3COOH]$ falls and $[CH_3COO^-]$ rises). $1cm^3$ of 1.0M NaOH is 0.001 moles of $OH^-$. This removes 0.001 moles of $CH_3COOH$ and forms 0.001 moles of $CH_3COONa$.
∴ After adding 0.001 moles of NaOH,

$$[CH_3COOH] = 0.010 - 0.001 = 0.009$$

$$[CH_3COO^-] = 0.040 + 0.001 = 0.041$$

$$\Rightarrow [H^+] = K_a \times \frac{[acid]}{[salt]} = 1.7 \times 10^{-5} \times \frac{0.009}{0.041}$$

$$[H^+] = 3.73 \times 10^{-6}$$

$$\therefore \text{pH after adding NaOH} = 5.43$$

*Therefore, the pH of the buffer changes by only 0.06 units.*

*2 The change in pH when 1cm³ of M NaOH is added to 1dm³ of 0.01M ethanoic acid.*
Before the addition of alkali,

$$[H^+] = [CH_3COO^-] \ll [CH_3COOH]$$

$$\therefore \text{ using } K_a = \frac{[H^+][CH_3COO^-]}{[CH_3COOH]}$$

$$[H^+]^2 = K_a \times [CH_3COOH] = 1.7 \times 10^{-5} \times 10^{-2} = 17 \times 10^{-8}$$

$$\therefore [H^+] = 4.12 \times 10^{-4}$$

$$\therefore \text{pH of 0.01M ethanoic acid} = 3.39$$

On adding 0.001 moles of NaOH,

$$[CH_3COOH] = 0.010 - 0.001 = 0.009$$

and $[CH_3COO^-] = 0.001$

$$\therefore [H^+] = K_a \times \frac{[\text{acid}]}{[\text{salt}]} = 1.7 \times 10^{-5} \times \frac{0.009}{0.001} = 1.53 \times 10^{-4}$$

$$\Rightarrow \text{pH after adding NaOH} = 3.82$$

*Therefore, the pH of the 0.01M ethanoic acid changes by 0.43 units on adding alkali.* This is more than seven times the pH change of the buffer and it illustrates the pH-stabilizing nature of the buffer very nicely.

The main use of buffers in the laboratory is in preparing solutions of known and constant pH. These solutions cannot be made by preparing acid or alkaline solutions of a given concentration, because the pH of such solutions will vary slightly as gases from the atmosphere, such as $CO_2$, dissolve in them, or as traces of alkali dissolve from the glass vessel.

Buffer solutions are also important in medicine and in agriculture since the pH's of living systems must be maintained at certain critical values. Because of this, intravenous injections must be carefully buffered so as not to change the pH of the blood from its normal value of 7.4. In the same way, most fermentation processes must be buffered as relatively small changes in the pH would cause the death of the fermenting organisms. In living systems, the buffering action is usually provided by $H_2CO_3$ and $HCO_3^-$, by $H_2PO_4^-$ and $HPO_4^{2-}$ and by various proteins which can both accept and donate $H^+$ ions.

## Summary

1 For the sparingly soluble strong electrolyte, $A_xB_y$, in contact with excess undissolved solute,

$$[A^{y+}]^x[B^{x-}]^y = K_{s.p.}$$

$K_{s.p.}$, known as the solubility product of $A_xB_y$, is constant at constant temperature.

2 The solubility product concept is valid only for saturated solutions in which the total concentration of ions is no more than about 0.01M. It is therefore inappropriate for soluble salts such as NaCl, $KNO_3$ and $CuSO_4$.

3 In the presence of either $A^{y+}$ or $B^{x-}$ from a second source, the solubility of the salt $A_xB_y$ is reduced. This generalization is known as the common ion effect.

4 Precipitation of an insoluble salt occurs when

$$\text{ionic product} > \text{solubility product.}$$

In this case, precipitation occurs until

$$\text{ionic product} = \text{solubility product.}$$

5 The pH of a solution provides a quantitative measure of its acidity or alkalinity.

$$\text{pH} = -\lg [H^+]$$

6 At 25°C, the product, $[H^+][OH^-] = 10^{-14}$ for all aqueous solutions.
For neutral solutions, $[H^+] = [OH^-] = 10^{-7}$ and pH = 7.
For acidic solutions, $[H^+] > [OH^-]$ and pH $< 7$.
For alkaline solutions, $[H^+] < [OH^-]$ and pH $> 7$.
7 For the weak acid, HA, dissolved in water

$$K_a = \frac{[H^+][A^-]}{[HA]}$$

$K_a$, known as the dissociation constant of the acid, is constant at constant temperature.
The numerical value of $K_a$ provides a quantitative indication of the extent to which an acid is dissociated (i.e., the strength of the acid). The greater the extent of dissociation, the greater are $[H^+]$ and $[A^-]$, the larger is $K_a$ and the stronger the acid.
8 $K_a$, like $K_{s.p.}$, is a modified equilibrium constant.
9 Acid–base indicators change colour according to the $[H^+]$ of the solution to which they are added. They are used to test for acidity and alkalinity and to detect the end-point in acid–base titrations.
10 The range of an indicator is the pH range over which it changes colour.
11 Buffers are solutions which resist changes in pH on dilution or on addition of acid or alkali. Animals and plants are protected against changes in pH by the presence of buffers.

## Study questions

1 The solubility of silver bromide in water is $7 \times 10^{-7}$ mole dm$^{-3}$ at 25°C. Calculate its solubility product.

2 (a) Calculate the solubility of silver ethanedioate (oxalate) ($Ag_2C_2O_4$) in water. ($K_{s.p.}$ ($Ag_2C_2O_4$) $= 5 \times 10^{-12}$ mole$^3$ dm$^{-9}$).
(b) How would you expect the value of a solubility product to vary with temperature? Explain your answer.
(c) Explain why $Ag_2C_2O_4$ is very soluble in dilute nitric(V) acid, but only sparingly soluble in water.

3 The solubility product of lead(II) sulphate(VI), $PbSO_4$, in water, is $1.6 \times 10^{-8}$ mole$^2$ dm$^{-6}$.
(a) Calculate the solubility of lead(II) sulphate(VI) in
(i) pure water,
(ii) 0.1M $Pb(NO_3)_2$ solution,
(iii) 0.01M $Na_2SO_4$ solution.
(b) Why is lead(II) sulphate(VI) more soluble in water than in any solution containing either $Pb^{2+}$ or $SO_4^{2-}$ ions?
(c) Use your understanding of solubility product to explain the so-called 'common ion' effect. Illustrate your answer with an example.

4 Suppose the acid, HX, is a weak electrolyte.
(a) What happens to $[H^+]$ in a solution of HX if:
(i) water is added,
(ii) gaseous HCl is added,
(iii) solid NaX is added?
(b) Assume that $K_a(HA) = 10^{-6}$, $K_a(HB) = 10^{-8}$ and $K_a(HC) = 10^{-10}$ mole dm$^{-3}$ respectively.
(i) Which solution has the highest $[H^+]$, 1.0M HA, 1.0M HB or 1.0M HC?
(ii) What is the $[H^+]$ in the solution with the highest value of $[H^+]$?

5 Chloric(I) (hypochlorous) acid, HClO, is a weak acid.
$K_a(HClO) = 3.2 \times 10^{-8}$ mole dm$^{-3}$
(a) Calculate the $[H^+]$ and $[OH^-]$ in $1.25 \times 10^{-2}$ M HClO.
(b) What is the pH of $1.25 \times 10^{-2}$ M HClO?

6 Explain the following observations:
(a) When 1.0M hydrochloric acid is diluted, the pH rises, but eventually reaches a static value and does not change on further dilution.
(b) A solution of iron(III) chloride is acid to litmus.
(c) Phenolphthalein can be used to determine the equivalence point in the titration of ethanoic acid with sodium hydroxide, but methyl orange is no use in this case.
(d) Zinc sulphide dissolves in dilute hydrochloric acid, but not in ethanoic acid or water.
(e) The pH of $10^{-8}$ M HCl is not 8.

7 Benzoic acid, $C_6H_5COOH$, is a weak monobasic acid ($K_a = 6.4 \times 10^{-5}$ mole dm$^{-3}$).
(a) Explain how a mixture of benzoic acid and sodium benzoate can act as a buffer on the addition of small amounts of either HCl(aq) or NaOH(aq).
(b) What is the $[H^+]$ in 0.02M benzoic acid?
(c) What is the pH of 0.02M benzoic acid?
(d) What is the pH of a solution containing 7.2g of sodium benzoate in 1 dm$^3$ of 0.02M benzoic acid?
(e) By how much will the pH change if 1cm$^3$ of 1.0M NaOH is added to the buffer in part (d)?

8 Assuming that the pH of blood is maintained at 7.4 by the acid, $H_2PO_4^-$, and its salt, $HPO_4^{2-}$, calculate the ratio of the concentration of $H_2PO_4^-$ to that of $HPO_4^{2-}$ in blood. ($K_a(H_2PO_4^-) = 6.4 \times 10^{-8}$ mole dm$^{-3}$).

9 Calculate the pH of
(a) $10^{-4}$ M HCl,
(b) $10^{-4}$ M $Ba(OH)_2$,
(c) 1.0M $H_2X$ which is only 50% dissociated,
(d) 0.01M propanoic acid ($K_a = 1.45 \times 10^{-5}$ mole dm$^{-3}$),
(e) 1.0M $NH_4OH$ ($K_b = 1.7 \times 10^{-5}$ mole dm$^{-3}$).

10 Figure 22.7 shows how the pH changes when 0.1M $CH_3COOH$ is titrated against 0.1M NaOH (curve a) and when 0.1M $CH_3COOH$ is titrated against 0.1M $NH_3$ (curve b).

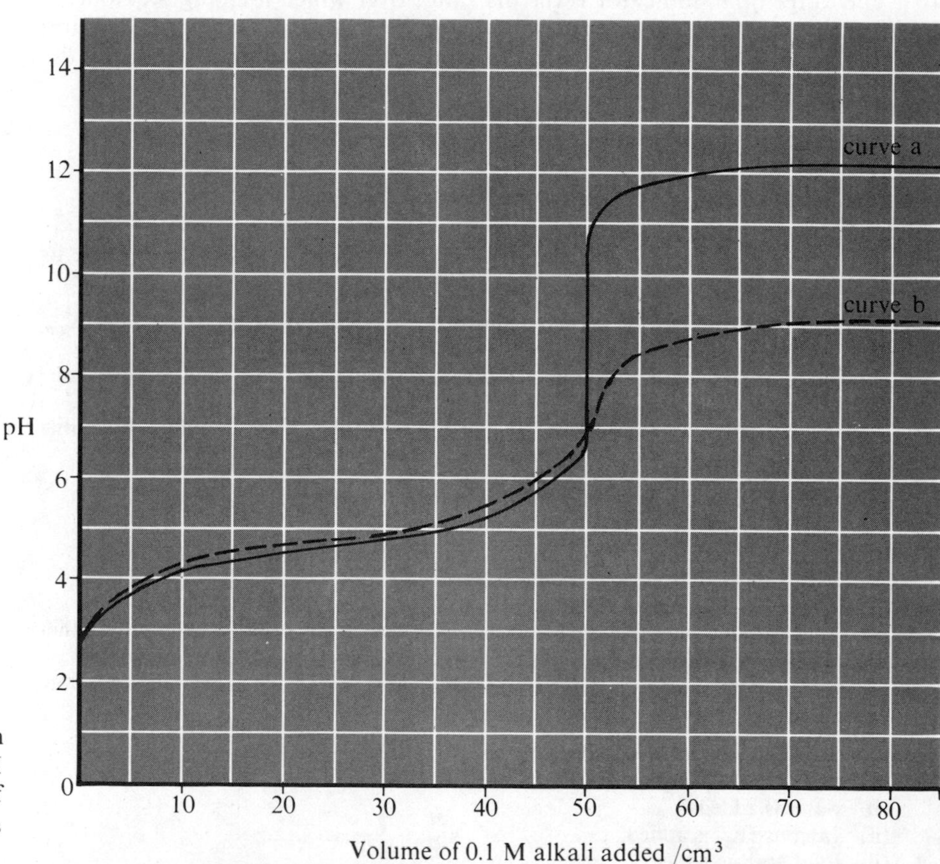


**Figure 22.7** pH changes during the titration of 50cm$^3$ of 0.1M $CH_3COOH$ with 0.1M NaOH (curve a) and during the titration of 50cm$^3$ of 0.1M $CH_3COOH$ with 0.1M $NH_3$ (curve b).

(a) The pH of 0.1M HCl is 1.0. Why is the pH of 0.1M $CH_3COOH$ about 2.8 rather than 1.0?
(b) Is methyl orange a suitable indicator to use when titrating 0.1M $CH_3COOH$ against 0.1M NaOH? Explain.
(c) Is phenolphthalein a suitable indicator to use when titrating 0.1M $CH_3COOH$ against 0.1M NaOH? Explain.
(d) No indicator will detect the equivalence point with accuracy when titrating 0.1M $CH_3COOH$ against 0.1M $NH_3$. Why is this?
(e) In view of the statement in (d), how could you determine the equivalence point in titrating a weak acid against a weak alkali?

# Reaction Rates 23

## 23.1 Introduction

The rates at which chemical reactions occur are just as important to you as they are to the industrialist and the chemical engineer.

At home you might be interested in the rate at which you can boil an egg or bake a cake. Out-of-doors you might be interested in the rate at which the car is rusting, the rate at which the lettuces are growing and possibly the rate at which the stonework of buildings is being weathered by acidic gases in the atmosphere.

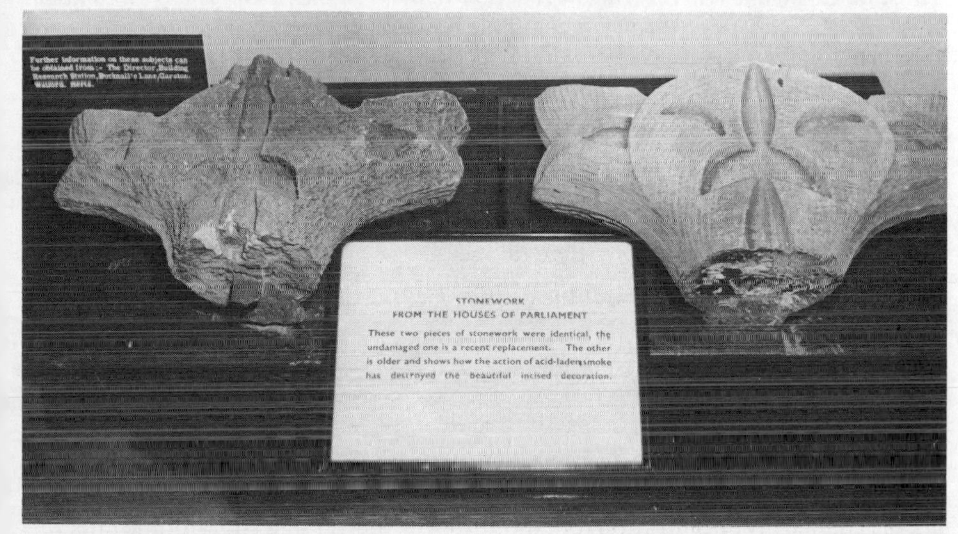


Stonework from the Houses of Parliament. The left hand specimen has been slowly weathered by acidic gases in the atmosphere.

In industry, engineers and other workers will be closely concerned with the rates of chemical reactions in industrial processes and in constructional engineering. These might include the rate at which ammonia can be obtained from nitrogen and hydrogen, the rate at which concrete sets or the rate of growth of a particular fruit or vegetable crop.

Industrialists and chemical engineers are not satisfied with merely turning one substance into another. In most cases, they want to perform reactions and obtain products rapidly, easily and as cheaply as possible. Time and money are important in industry, and it is often necessary to accelerate reactions so that they are economically worthwhile.

At normal temperatures and pressures and in the absence of a catalyst, ammonia cannot be obtained from nitrogen and hydrogen. Fortunately, chemical engineers have found that a viable and economic reaction rate results when the process is carried out at 250 atmospheres pressure and 450°C in the presence of an iron catalyst (section 21.5).

$$N_2(g) + 3H_2(g) \longrightarrow 2NH_3(g)$$

Besides industrial processes, virtually all biological reactions rely on the presence of catalysts. Almost every chemical reaction in your body is controlled by one or more catalysts, whether that reaction is the relatively simple hydrolysis of starch to sugars or the highly complex replication of DNA which forms the genes in the nuclei of your cells. And of course, the same can be said for all other animals and plants. These biological catalysts, called **enzymes**, are usually proteins.

Reaction rates are also of archaeological importance. Archaeologists can estimate the age of rocks, fossils or prehistoric remains by a process known as radioactive dating, in which they measure the concentration of a decaying radioactive isotope such as $^{14}_{6}C$ in the object under scrutiny.

- ○ Why does a pressure cooker enable vegetables to be cooked more rapidly?
- ○ What conditions or processes are used to slow down the rate at which perishable foods deteriorate?
- ○ How do gardeners accelerate the growth of their crops?

## 23.2 The concept of reaction rate

During a chemical reaction, reactants are being converted to products and the reaction rate tells us how fast the reaction is taking place by indicating how much of a reactant is consumed or how much of a product forms in a given time. Hence,

$$\text{Reaction rate} = \frac{\text{change in amount (or concentration) of a substance}}{\text{time taken}}$$

Thus, *we can define reaction rate or reaction velocity as the rate of change of amount or concentration of a particular reactant or product.*

When acidified hydrogen peroxide is added to a solution of potassium iodide, iodine is formed.

$$H_2O_2 + 2I^- + 2H^+ \longrightarrow 2H_2O + I_2$$

The concentration of iodine rises from 0 to $10^{-5}$ mole dm$^{-3}$ in 10 seconds.

$$\therefore \text{Reaction rate} = \frac{\text{change in concentration of iodine}}{\text{time taken}}$$

Using the symbol Δ to represent the change in a particular quantity, we can write,

$$\frac{\Delta[I_2]}{\Delta t} = \frac{10^{-5}\text{ mole dm}^{-3}}{10\text{s}}$$

$$= 10^{-6}\text{ mole dm}^{-3}\text{ s}^{-1}$$

Strictly speaking, this result gives the average reaction rate over the ten seconds that it took for the concentration of iodine to become $10^{-5}$ mole dm$^{-3}$. By measuring the change in concentration (or amount) over shorter and shorter time intervals we obtain an increasingly accurate estimate of the reaction rate at any moment. The disadvantages of 'clock' techniques such as this one, in which the rate is obtained as the inverse of the time for a certain proportion of the reaction to occur, are illustrated more effectively in figure 23.1. Provided the reaction has gone only a little way towards completion, very little error is introduced, but serious errors result if the 'end point' is, say half-way to completion.

Ideally, we should make the time interval almost zero and then we obtain, what is effectively, the reaction rate at a particular instant, i.e.

$$\frac{\Delta[I_2]}{\Delta t}_{\Delta t \to 0} = \frac{d[I_2]}{dt}$$

In practice, one usually plots a graph of the concentration or amount of a particular substance against time and then the reaction rate can be obtained at particular times by drawing tangents to the resulting curve. This technique is illustrated in section 23.5.

Normally it is convenient to express reaction rates in mole dm$^{-3}$ s$^{-1}$ or in mole s$^{-1}$, but occasionally it is more convenient to use minutes or even hours as the unit of time.

In the reaction of acidified hydrogen peroxide with potassium iodide,

$$H_2O_2 + 2I^- + 2H^+ \longrightarrow 2H_2O + I_2,$$

we measured the reaction rate in terms of the rate of formation of iodine.

Rate of formation of iodine $= 10^{-6}$ mole $dm^{-3}$ $s^{-1}$

○ What is the rate of consumption of $H_2O_2$ in this reaction?
○ What is the rate of consumption of $I^-$ in this reaction?

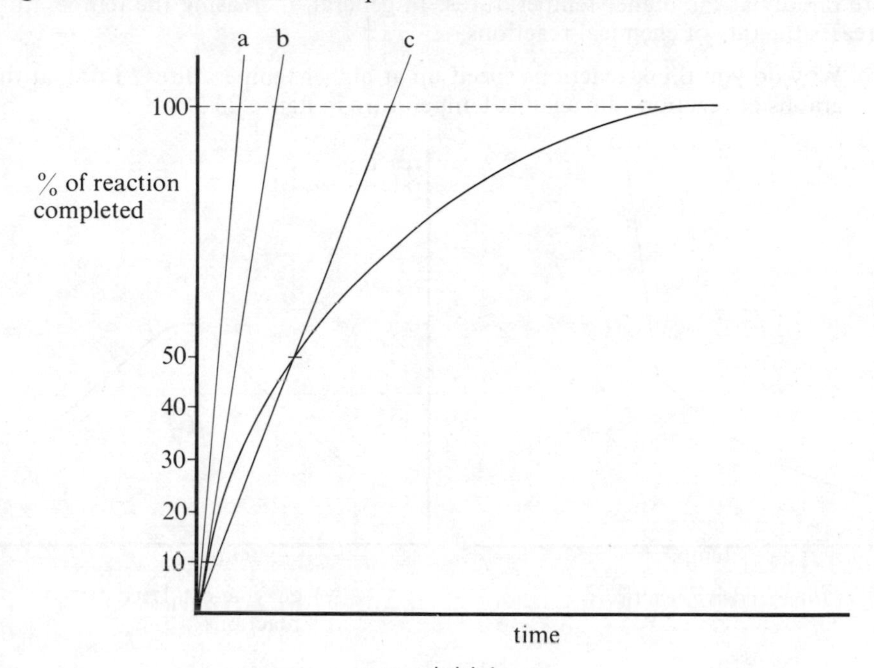


a = true initial rate
b = average reaction rate for 10% completion
c = average reaction rate for 50% completion

**Figure 23.1** Errors in the 'clock technique' for measuring reaction rates.

## 23.3 Factors affecting the rate of a reaction

Our studies have already indicated several factors which can influence the rate of a reaction.

THE AVAILABILITY OF REACTANTS AND THEIR SURFACE AREA

Anyone who has camped knows that it is easier to start a fire using thin sticks rather than tree trunks. Similarly, magnesium powder will react much more rapidly than magnesium ribbon with dilute sulphuric(VI) acid. In general, the smaller the size of reacting particles, the greater is the total surface area exposed for reaction and consequently, the faster the reaction. In the case of heterogeneous systems, in which the reactants are in different states, the area of contact between the reacting substances will influence the reaction rate considerably. In homogeneous systems, reacting substances normally occur in their *maximum* state of subdivision as individual particles in the gaseous or aqueous phase. In this case, the idea of surface area becomes meaningless and it obviously does not affect the reaction rate in any way.

THE CONCENTRATION OF REACTANTS

Increasing the concentration of a reactant normally causes an increase in the rate of a reaction, but this is not always the case. Furthermore, the different reactants can affect the rate of a particular reaction in different ways. For example, when nitrogen oxide reacts with oxygen,

$$2NO(g) + O_2(g) \longrightarrow 2NO_2(g),$$

the reaction rate doubles when the oxygen concentration doubles, but doubling the concentration of NO quadruples the rate of reaction. The effect of concentration on reaction rates is considered further in sections 23.5 and 23.6.

THE TEMPERATURE OF THE REACTANTS

Milk and other perishable foods 'go bad' much more rapidly in summer than in winter. In summer, the chemical reactions in the deterioration processes occur more rapidly at the higher temperatures. In general, increasing the temperature increases the rate of chemical reactions.

○ Why do you think reactions speed up at higher temperatures? Look at the graphs of reaction rate against temperature in figure 23.2.

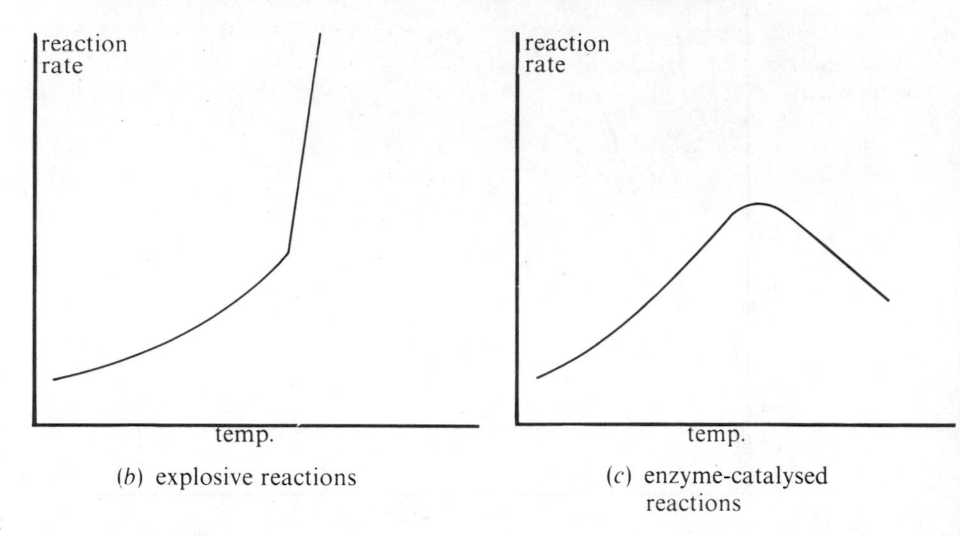


**Figure 23.2** Graphs of reaction rate against temperature.

○ Why is there a sudden change in the rate of an explosive reaction at one particular temperature?

○ The rate of an enzyme-catalysed reaction increases at first, but then decreases as the temperature increases. Why is this? (How are enzymes susceptible to increase in temperature?)

The effect of temperature on reaction rates is considered further in sections 23.8 and 23.9.

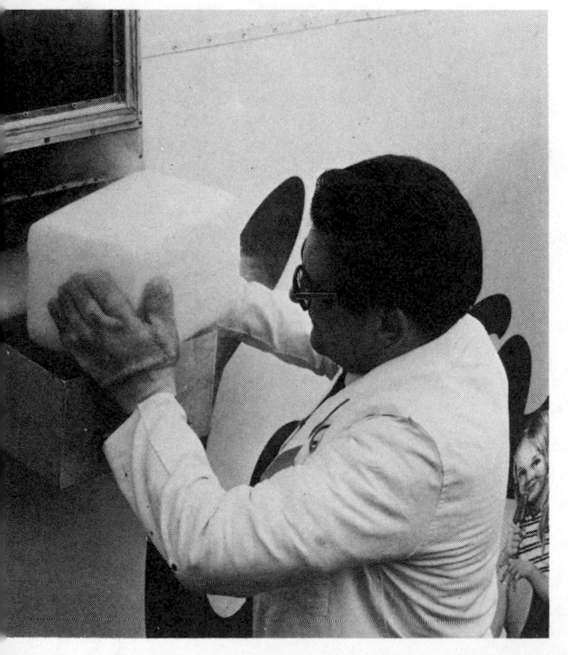

The last of six blocks of solid $CO_2$ (Drikold) being loaded into the refrigeration system of a Walls' ice cream van.

CATALYSTS

Catalysts are substances which alter the rate of chemical reactions without undergoing any overall chemical change themselves. Although catalysts are not consumed in reactions, they are believed to participate by forming intermediate compounds in the conversion of reactants to products.

Normally, catalysts are used to accelerate reactions. Certain catalysts can, however, be used to slow reactions down. For example, propane-1,2,3-triol (glycerine) is sometimes added to hydrogen peroxide as a negative catalyst in order to slow down its rate of decomposition.

Catalysts play an important part in biological and industrial processes by enabling reactions to take place which would never occur in their absence. Many large-scale industrial processes including the manufacture of ammonia, sulphuric(VI) acid, nitric(V) acid, ethene, poly(ethene) (polythene), and poly(phenylethene) (polystyrene) rely heavily on the use of catalysts. Almost every chemical reaction in every animal, plant and micro-organism requires its own specific enzyme (catalyst).

LIGHT

Photosynthesis and photography both involve light-sensitive reactions. The leaves of plants contain chlorophyll, a green pigment, which can absorb radiation in the visible region of the electromagnetic spectrum and use this energy to synthesize chemicals and provide food for the plant.

During photosynthesis, plants transform carbon dioxide and water into oxygen and sugars such as glucose.

$$6CO_2(g) + 6H_2O(1) \xrightarrow{hv} C_6H_{12}O_6(aq) + 6O_2(g) \qquad \Delta H = +2\,820\text{ kJ}$$

In the absence of sunlight, energy is no longer provided and photosynthesis ceases.

White silver chloride turns purple and finally dark grey when it is exposed to sunlight which provides the energy required to decompose the silver chloride.

$$AgCl(s) \longrightarrow Ag(s) + \tfrac{1}{2}Cl_2(g)$$

The use of silver salts in photography depends on photosensitivity of this kind.

The reactions of halogens with hydrogen and with alkanes are other examples of photochemical reactions.

Thus, chlorine reacts slowly with hydrogen or methane in diffused daylight, but explosively when exposed to intense ultraviolet radiation. The effect of sunlight is believed to result from its ability to split chlorine molecules into highly reactive single atoms, known as **radicals**, which contain an unpaired electron (figure 23.3). See also section 25.5.

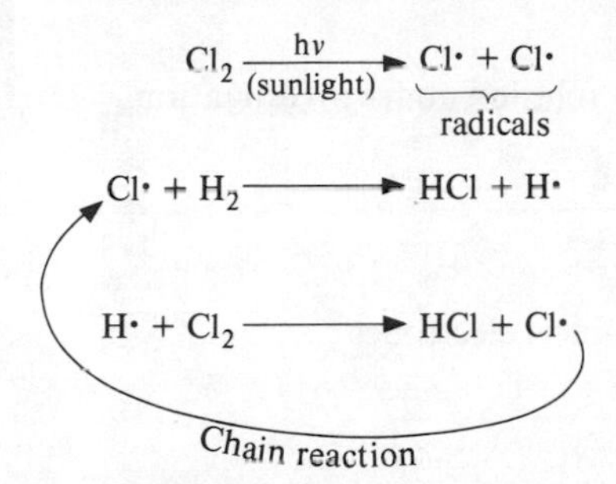


**Figure 23.3** The reaction of chlorine with hydrogen.

## 23.4 Measuring reaction rates

The rate of a chemical reaction can be obtained by following some property which alters with the extent of the reaction. By analysing the reaction mixture at suitable intervals, it is possible to determine the concentration of both reactants and products at different times and hence obtain a measure of the reaction rate (i.e. the rate at which the concentration of a particular substance changes with time).

The choice of analytical method depends on the reaction under consideration. The following techniques illustrate four possible approaches.

TITRIMETRIC ANALYSIS

This is particularly suitable for reactions in solution such as that between iodine and propanone catalysed by acid:

$$CH_3.CO.CH_3(aq) + I_2(aq) \xrightarrow{H^+} CH_2I.CO.CH_3(aq) + H^+(aq) + I^-(aq)$$

or the hydrolysis of an ester (such as methyl methanoate) by alkali:

$$HCOOCH_3 + OH^- \longrightarrow HCOO^- + CH_3OH$$

In these cases, the reaction can be followed by removing and analysing small portions of the reaction mixture at intervals. Very often, the removed portion must be added to some reagent which will stop the reaction (i.e. 'quench' it), thereby preventing further changes in concentration before the analysis is carried out. For example, in studying the reaction between iodine and propanone, portions of the reaction mixture can be pipetted into sodium hydrogencarbonate solution which 'quenches' the reaction by neutralising the acid catalyst. The quenched mixture can then be analysed by titration of the unreacted iodine against a standard solution of sodium thiosulphate(VI).

COLORIMETRIC ANALYSIS

This method is especially convenient for those systems in which one of the substances is coloured (e.g. the reaction of iodine with propanone or the reaction of bromine with methanoic (formic) acid).

The intensity of colour can be followed during the reaction using a photoelectric colorimeter (figure 18.9), and from these measurements the concentration of the coloured species can be obtained at different times.

CONDUCTIMETRIC ANALYSIS

Many reactions in aqueous solution involve ions and changes in the number of ions present as the reaction proceeds. Consequently the electrical conductivity of the solution will change during the reaction and this can be used to determine the changing concentrations of reactants and products with time. Essentially, this method involves immersing two inert electrodes in the reaction mixture and then following the change in electrical conductivity of the solution with time (figure 23.4).

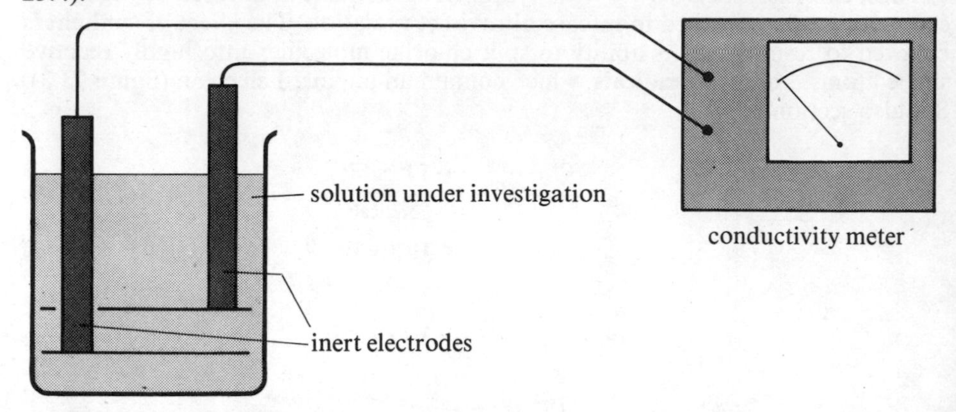


**Figure 23.4** Following the change in electrical conductivity of a solution with time.

PRESSURE MEASUREMENTS

This technique is particularly suitable for reactions in the gas phase which involve changes in pressure when the system is kept in a vessel of constant volume. For example, the gaseous decomposition of 2-methyl-2-iodopropane at constant volume can be followed conveniently by measuring the pressure at suitable time intervals.

$$H_3C{-}C(CH_3)(I){-}CH_3(g) \longrightarrow (H_3C)_2C{=}CH_2(g) + HI(g)$$

The last three methods have one great advantage over titrimetric analysis in that samples need not be removed from the reacting mixture. In these three cases, the extent of the reaction is determined at intervals of time by an external method without disturbing the reaction mixture.

It is important to realize that measurements on the reacting system do not give the rate of reaction directly; they simply give the concentration of a particular reactant or product at a given time. By plotting a graph of the concentration against time, it is possible to determine the reaction rate, (i.e. the change in concentration with time, $d[X]/dt$) from the gradient of the tangent at a given point (figure 23.5).

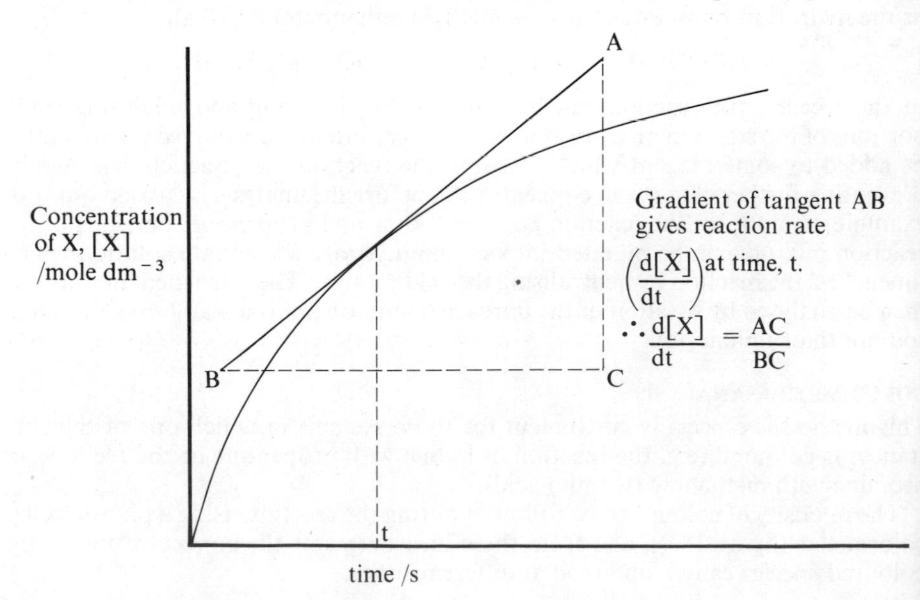


**Figure 23.5** Obtaining the rate of reaction at a given time from a graph of concentration against time.

## 23.5 Investigating the effect of concentration on the rate of a reaction

We must now consider reactions in more detail in order to discover how reaction rates are influenced by the concentration of reactants. A convenient reaction to study is that between bromine and methanoic acid (formic acid) in aqueous solution.

The reaction is catalysed by acid.

$$Br_2(aq) + HCOOH(aq) \xrightarrow{H^+} 2Br^-(aq) + 2H^+(aq) + CO_2(g)$$

The reaction can be followed colorimetrically by measuring the intensity of the red-brown bromine at suitable time intervals. By plotting a calibration curve of known bromine concentrations against colorimeter readings, it is possible to deduce the concentrations of bromine corresponding to the colorimeter readings obtained in the experiment. Some typical results are shown in table 23.1. The concentration of methanoic acid was kept constant throughout this experiment by having it present in large excess.

The concentrations of bromine in table 23.1 are plotted graphically against time in figure 23.6.

**Table 23.1** Results for the kinetic study of the reaction between bromine and methanoic acid.

| Time/s | $[Br_2]$ /mole dm$^{-3}$ |
|---|---|
| 0 | 0.0100 |
| 30 | 0.0090 |
| 60 | 0.0081 |
| 90 | 0.0073 |
| 120 | 0.0066 |
| 180 | 0.0053 |
| 240 | 0.0044 |
| 360 | 0.0028 |
| 480 | 0.0020 |
| 600 | 0.0013 |

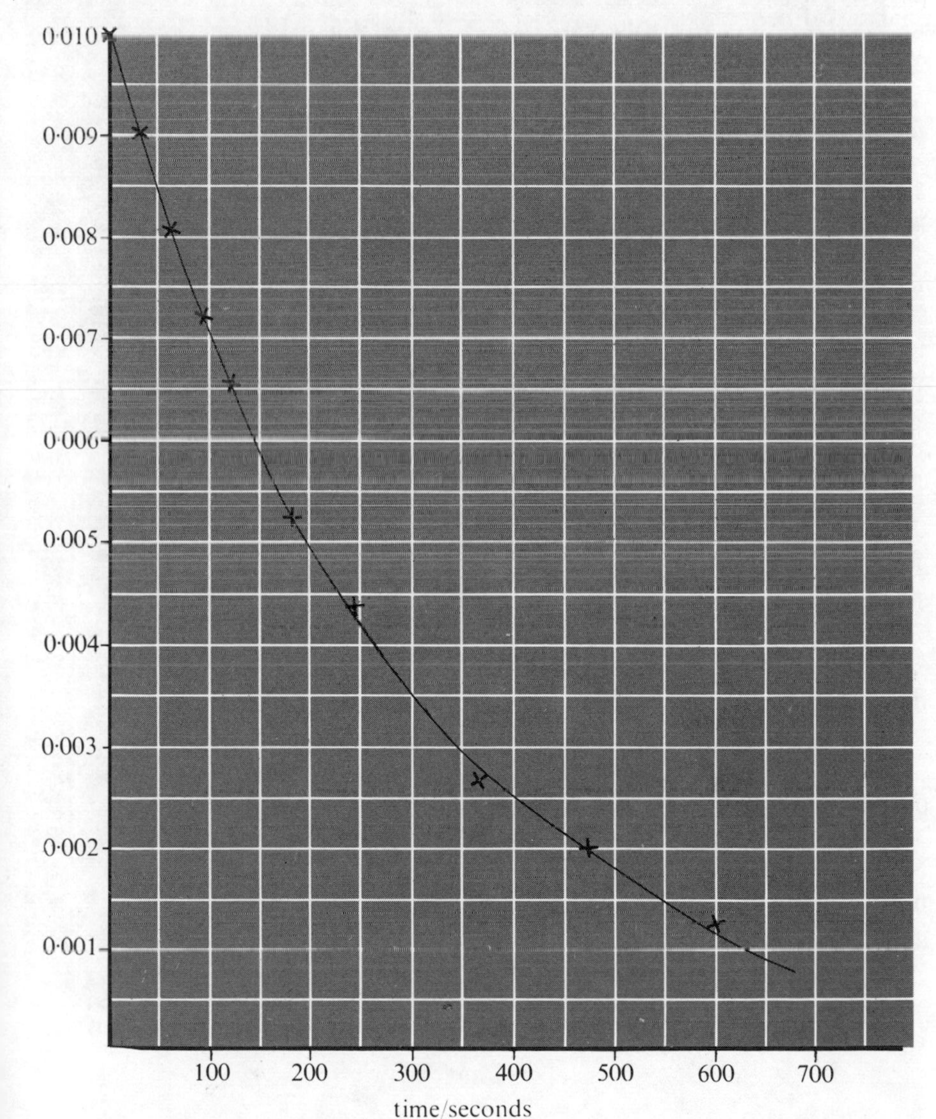


**Figure 23.6** The variation of bromine concentration with time in the reaction between methanoic acid and bromine.

The concentration of bromine ($[Br_2]$) falls during the course of the reaction and the rate of the reaction can be expressed in terms of the rate at which the bromine concentration changes:

$$\text{Reaction rate} = -\text{rate of change of concentration of bromine}$$
$$= -\frac{d[Br_2]}{dt}$$

Notice the negative sign in the last expression; it is necessary because the reaction rate must be positive, whereas the rate of change of $[Br_2]$ is negative since the bromine is disappearing.

In order to obtain the reaction rate at any given time, we must draw a tangent to the curve at this particular time and measure its gradient. Various values of the reaction rate corresponding to different bromine concentrations at different times are shown in table 23.2. These values of reaction rate are plotted vertically against bromine concentration in figure 23.7.

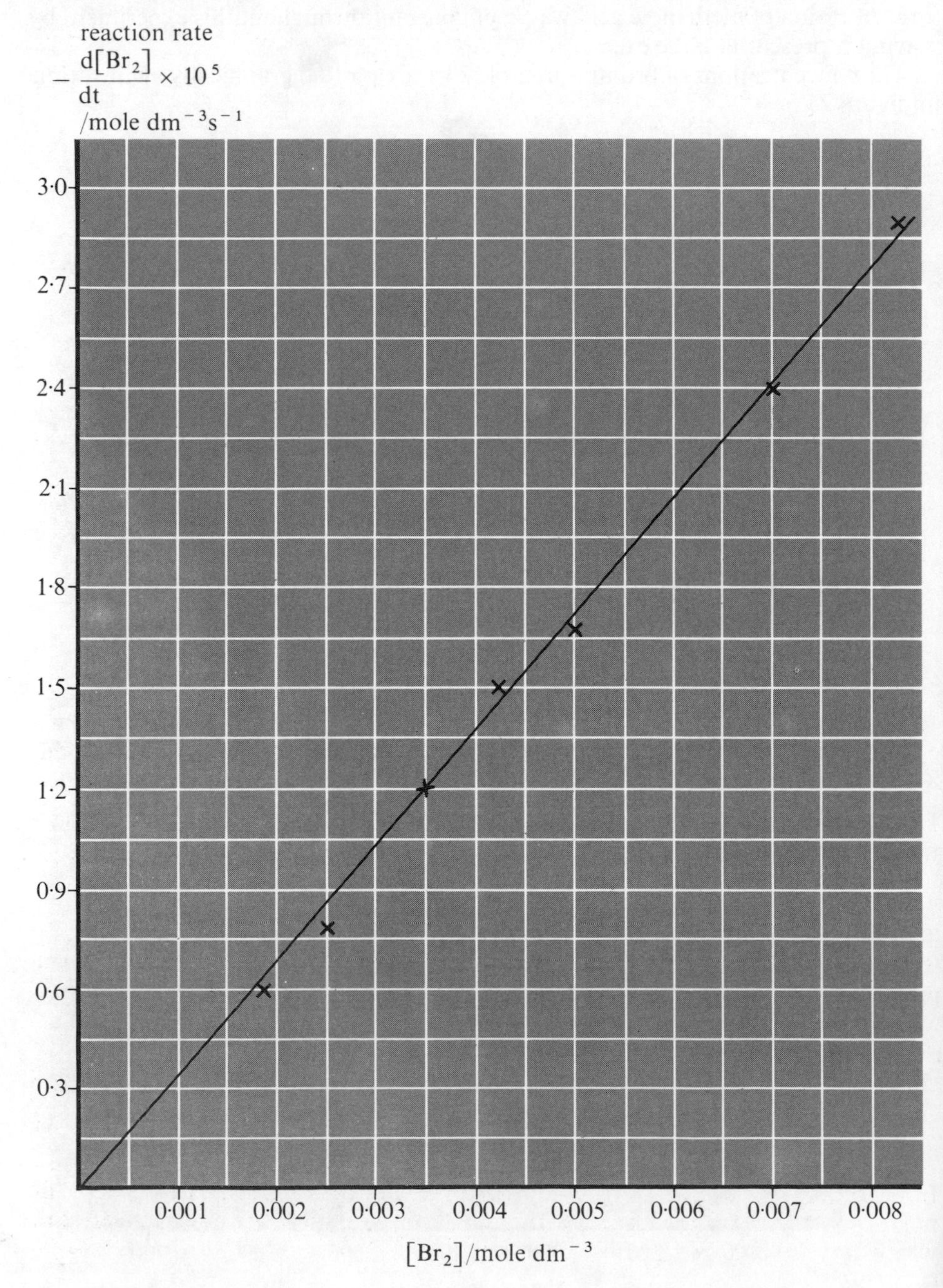


**Figure 23.7** Variation of reaction rate with bromine concentration.

**Table 23.2** Various values of the reaction rate corresponding to different bromine concentrations.

| Time/s | $[Br_2]$ obtained from figure 23.6 /mole $dm^{-3}$ | Reaction rate $(-d[Br_2]/dt)$ obtained from gradients in figure 23.6/mole $dm^{-3}$ $s^{-1}$ |
|---|---|---|
| 50 | 0.0083 | $2.9 \times 10^{-5}$ |
| 100 | 0.0070 | $2.4 \times 10^{-5}$ |
| 200 | 0.0050 | $1.7 \times 10^{-5}$ |
| 250 | 0.0042 | $1.5 \times 10^{-5}$ |
| 300 | 0.0035 | $1.2 \times 10^{-5}$ |
| 400 | 0.0025 | $0.8 \times 10^{-5}$ |
| 500 | 0.00185 | $0.6 \times 10^{-5}$ |

- ○ How does the bromine concentration change with time?
- ○ How does the reaction rate change with time?
- ○ Is the rate of reaction affected by the bromine concentration?
- ○ How does the rate of reaction depend on the bromine concentration?
- ○ Write a mathematical expression relating reaction rate to bromine concentration.

The graph in figure 23.7 shows that the reaction rate is directly proportional to the bromine concentration, i.e.

$$\text{Reaction rate} \propto [Br_2]$$

$$\Rightarrow \text{Reaction rate} = k[Br_2]$$

where $k$ is a constant, known as the **rate constant** or the **velocity constant** for the reaction.

## 23.6 Order of reaction and rate equations

Experiments show that the rates of most reactions can be related to the concentrations of individual reactants by an equation of the form,

$$\text{Rate} = k[X]^n$$

This expression, in which X is the reactant under consideration and $n$ is usually 0, 1 or 2, is known as a **rate equation** or **rate law**. The value of $n$ gives the **order of the reaction.**

When $n = 0$, the reaction rate is said to be **zero order** with respect to X, i.e.

$$\text{Rate} = k[X]^0$$

$$\text{but, since } [X]^0 = 1,$$

$$\text{Reaction rate} = k.$$

In other words, the reaction rate is independent of the concentration of X. This means that changing the concentration of X will not affect the rate of a reaction which is zero order with respect to X (figure 23.8).

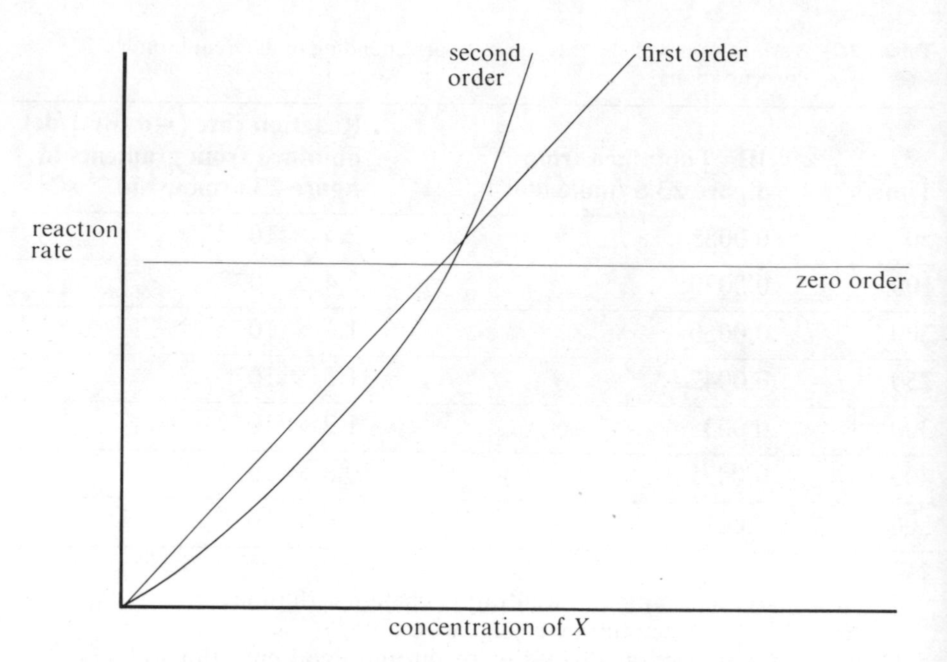


**Figure 23.8** Variation of reaction rate with concentration for reactions which are zero, first, and second order.

When $n = 1$, the reaction rate is proportional to $[X]^1$ and the reaction is said to be **first order** with respect to X (figure 23.8). When $n = 2$, the reaction rate is proportional to $[X]^2$ and the reaction is said to be **second order** with respect to X (figure 23.8).

What is the order with respect to $Br_2$ for the reaction we studied in the last section?

Look at the information in table 23.3 which relates to the reaction between hydrogen and nitrogen oxide at 800°C.

$$2H_2(g) + 2NO(g) \longrightarrow 2H_2O(g) + N_2(g)$$

**Table 23.3** Information concerning the rate of reaction between hydrogen and nitrogen oxide at 800°C.

| **Experiment number** | **Initial concentration of nitrogen oxide /mole dm$^{-3}$** | **Initial concentration of hydrogen /mole dm$^{-3}$** | **Initial rate of production of nitrogen /mole dm$^{-3}$ s$^{-1}$** |
|---|---|---|---|
| 1 | $6 \times 10^{-3}$ | $1 \times 10^{-3}$ | $3 \times 10^{-3}$ |
| 2 | $6 \times 10^{-3}$ | $2 \times 10^{-3}$ | $6 \times 10^{-3}$ |
| 3 | $6 \times 10^{-3}$ | $3 \times 10^{-3}$ | $9 \times 10^{-3}$ |
| 4 | $1 \times 10^{-3}$ | $6 \times 10^{-3}$ | $0.5 \times 10^{-3}$ |
| 5 | $2 \times 10^{-3}$ | $6 \times 10^{-3}$ | $2.0 \times 10^{-3}$ |
| 6 | $3 \times 10^{-3}$ | $6 \times 10^{-3}$ | $4.5 \times 10^{-3}$ |

In experiments 1, 2 and 3 the initial concentration of nitrogen oxide is the same.

- ○ What happens to the initial rate, when the initial concentration of hydrogen is doubled in experiment 2 compared to experiment 1?
- ○ What happens to the initial rate, when the initial concentration of hydrogen is trebled in experiment 3 compared to experiment 1?
- ○ How does the reaction rate depend on $[H_2]$?
- ○ What is the order of the reaction with respect to hydrogen?

In experiments 3, 4 and 5 the initial concentration of hydrogen is the same. When [NO] is doubled in experiment 5 compared to experiment 4, the reaction rate is *not* doubled, *but* quadrupled (i.e. $2^2$).

Similarly, when [NO] is trebled (experiment 6 compared to experiment 4), the reaction rate is *not* trebled, *but* increases ninefold (i.e. $3^2$). These results show that

$$\text{Reaction rate} \propto [NO]^2 \text{ and}$$

$$\text{Reaction rate} \propto [H_2]$$

$\therefore$ The order of reaction with respect to NO = 2
and order of reaction with respect to $H_2$ = 1
We can combine these results in a single rate equation as

$$\text{Reaction rate} = k[NO]^2[H_2]$$

*k, the rate constant, is constant for a given reaction at a particular temperature.*
It is important to realize that *the rate equation can only be obtained experimentally; it cannot be deduced either theoretically or from the stoichiometric equation.*

- ○ Using the concentrations of NO and $H_2$ and the initial rate of reaction in experiment 1, determine the value of $k$ from the rate equation, Rate = $k[NO]^2[H_2]$.
- ○ What are the units of $k$ in this case?

In experiment 2, $[NO] = 6 \times 10^{-3}$ mole dm$^{-3}$,
$[H_2] = 2 \times 10^{-3}$ mole dm$^{-3}$ and
rate = $6 \times 10^{-3}$ mole dm$^{-3}$ s$^{-1}$.
Substituting these values in Rate = $k[NO]^2[H_2]$, we get

$$6 \times 10^{-3} = k(6 \times 10^{-3})^2 \times 2 \times 10^{-3}$$

$$\Rightarrow k = \frac{6 \times 10^{-3}}{(6 \times 10^{-3})^2} \times \frac{1}{2 \times 10^{-3}} = \frac{10^6}{12} = 8.33 \times 10^4$$

By substituting the units for reaction rate and concentrations in the rate equation, we can determine the units of $k$. For example, in the NO/$H_2$ reaction:

$$k = \left(\frac{\text{rate}}{[NO]^2[H_2]}\right) \frac{\cancel{\text{mole dm}^{-3}}\text{s}^{-1}}{(\text{mole dm}^{-3})^2.\cancel{(\text{mole dm}^{-3})}}$$

$$= \left(\frac{\text{rate}}{[NO]^2[H_2]}\right) \text{mole}^{-2}\text{ dm}^6\text{ s}^{-1}$$

Thus, since the rate constant is constant at a fixed temperature, our results show that $k = 8.33 \times 10^4$ mole$^{-2}$ dm$^6$ s$^{-1}$ at 800°C.

Suppose we take the general case of a reaction,

$$xA + yB \longrightarrow \text{products},$$

with a rate equation which can be expressed as

$$\text{Rate} = k[A]^m[B]^n$$

The indices $m$ and $n$ are known as the **orders of the reaction with respect to A and B** respectively. The **overall order** of the reaction is described as $(m + n)$.

It is therefore important to be clear whether we are discussing the overall order of the reaction or the order with respect to an individual reactant.

*The order of a reaction with respect to a given reactant is the power of that reactant's concentration in the experimentally determined rate equation.*
*The overall order of the reaction is the sum of the powers of the concentration terms in the rate equation.*

The reaction between propanone and iodine in aqueous solution,

$$CH_3COCH_3(aq) + I_2(aq) \longrightarrow CH_2ICOCH_3(aq) + H^+(aq) + I^-(aq)$$

is catalysed by $H^+$ ions.
Experiments show that the rate law can be expressed as:

$$\text{Rate} = k[CH_3COCH_3][H^+]$$

- ○ What is the order of the reaction with respect to
  - (a) propanone,
  - (b) iodine,
  - (c) $H^+$ ions?
- ○ What is the overall order of the reaction?

This particular reaction shows very convincingly that the stoichiometric equation tells us nothing about the rate equation, since $I_2$, which appears in the stoichiometric equation, does not feature in the rate equation. Furthermore, $H^+$ ions which feature in the rate equation do not appear in the stoichiometric equation.

## 23.7 The half-life for a first order reaction and for radioactive decay

Look closely at figure 23.6 once more.

- O How many seconds does it take for the concentration of bromine to fall from 0.010M to 0.005M?
- O How many seconds does it take for the concentration of bromine to fall from 0.005M to 0.0025M?
- O How many seconds does it take for the concentration of bromine to fall from 0.0025M to 0.00125M?

The results in figure 23.6 show that at any point during the reaction it takes 200 seconds for the concentration of bromine to halve, (i.e. to fall from a concentration of say $x$M to $\frac{x}{2}$M).

*Other first order reactions also have constant half-lives.*

Thus, the kinetics of a first order reaction with its constant half-life are similar to the decay of a radioactive isotope which also has a constant half-life and obeys a first order rate equation. (See figure 23.9 and section 6.4.)

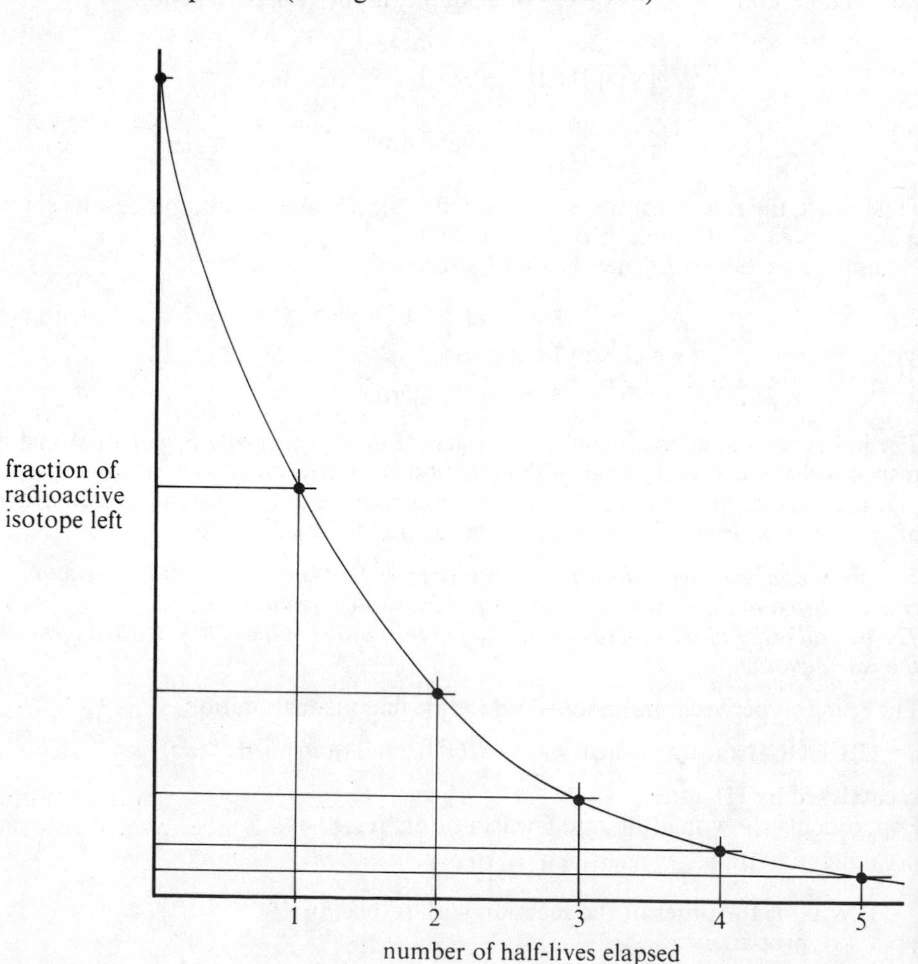


**Figure 23.9** The fraction of a radioactive isotope remaining after 1, 2, 3, 4 and 5 half-lives.

Indeed, the constancy of the half-lives of radioactive isotopes forms the basis of archaeological dating, whether such dating applies to organic remains containing radioactive $^{14}_{6}C$ or mineral remains containing radioactive $^{238}_{92}U$.

Half-lives range from exceedingly small fractions of a second (such as $10^{-21}$ seconds for $^{5}_{2}He$) to thousands of millions of years (such as $4.5 \times 10^{9}$ years for $^{238}_{92}U$). The half-life of a radioactive isotope is a fundamental property unaffected by chemical and physical changes.

### DATING GEOLOGICAL REMAINS

By knowing the half-life of a radioactive isotope, we can deduce the time it would take for a certain proportion of the substance to decay. Thus, the decay of $^{238}_{92}U$ to $^{206}_{82}Pb$ provides a method of dating rocks in the earth's crust. If we assume that the uranium-bearing rocks in the earth's crust originally contained $^{238}_{92}U$ but no $^{206}_{82}Pb$, then the present ratio of $^{238}_{92}U : ^{206}_{82}Pb$ in the rocks can be used to calculate the time which has elapsed since the rocks formed. Using this technique, it is found that the ages of different rocks vary from forty million to four thousand million years. Geologists often take the larger of these two values to be the age of the earth.

### DATING THE REMAINS OF LIVING THINGS USING CARBON-14

The basis of carbon dating is the simultaneous production and disintegration of radioactive $^{14}_{6}C$. As high-energy cosmic radiation passes through the earth's atmosphere, it strikes atoms splitting them into sub-atomic fragments including neutrons. These neutrons (which are produced mainly in the upper atmosphere) react with nitrogen atoms producing $^{14}_{6}C$.

$$^{1}_{0}n + ^{14}_{7}N \longrightarrow ^{14}_{6}C + ^{1}_{1}H$$

At the same time as $^{14}_{6}C$ is being formed, the radioactive isotope is also decaying to $^{14}_{7}N$ by $\beta$-decay (section 6.3).

$$^{14}_{6}C \longrightarrow ^{14}_{7}N + ^{0}_{-1}e$$

As a result of this simultaneous formation and decay of $^{14}_{6}C$, the atmosphere contains a constant concentration of $^{14}_{6}C$, present as $^{14}_{6}CO_2$. This $^{14}_{6}CO_2$ enters plants via photosynthesis and eventually vegetable foods containing $^{14}_{6}C$ are eaten and metabolized by animals. Thus, all living things have a constant proportion of their carbon in the form of $^{14}_{6}C$. However, when the animal or plant dies, replacement of $^{14}_{6}C$ ceases, but decay of $^{14}_{6}C$ continues.

Suppose that carbon-14 makes up $x$% of the carbon in living things and that the half-life of $^{14}_{6}C$ is 5 800 years.

- ○ What percentage of $^{14}_{6}C$ will the remains of a plant contain 5 800 years after it has died?
- ○ How long will it take for the percentage of $^{14}_{6}C$ in a dead plant to fall from $x/2$% to $x/4$%?
- ○ Approximately how old is an object which has $x/8$% of $^{14}_{6}C$?

By comparing the $^{14}_{6}C$ content of archaeological specimens with that of similar materials living at the present time, it is possible to estimate the age of the specimen. In this way, radiocarbon dating has been widely used to establish Egyptian chronology and to check the authenticity of ancient remains such as the Dead Sea scrolls.

## 23.8 Investigating the effect of temperature on the rate of a reaction

When dilute hydrochloric acid is added to sodium thiosulphate(VI) solution ($Na_2S_2O_3(aq)$), the solution becomes cloudy as sulphur is precipitated. As the precipitate gradually thickens its yellow colour becomes apparent.

$$S_2O_3{}^{2-}(aq) + 2H^+(aq) \longrightarrow S(s) + SO_2(g) + H_2O(l)$$

How can we measure the reaction rate in this case?

Add $10 cm^3$ of 1.0M HCl to $50 cm^3$ of 0.05M $Na_2S_2O_3(aq)$, with both solutions at the same temperature, and mix the contents thoroughly. Place the flask above an ink cross on white paper and measure the interval between the addition of HCl and the obscuring of the ink cross as the precipitate thickens.

**Table 23.4** Results from the investigation of the effect of temperature on the reaction between HCl(aq) and $Na_2S_2O_3$(aq).

| Temperature/K | Time for ink-cross to disappear/s |
|---|---|
| 296 | 135 |
| 298 | 119 |
| 302 | 90 |
| 307 | 62 |
| 311 | 45 |
| 317 | 37 |
| 320 | 34 |
| 326 | 24 |
| 332 | 20 |

The results in table 23.4 were obtained when the experiment was carried out at different temperatures.

Notice that a rise in temperature of about 10 K roughly doubles the reaction rate. For example, the cross disappears in 119 seconds at 298 K, but in about half the time (62 secs) at 307 K; the cross disappears in 90 seconds at 302 K and in half the time (45 secs) at 311 K.

Why is it that such a small rise in temperature can cause such a large percentage increase in the reaction rate?

## 23.9 Explaining the increase in reaction rate with temperature: the collision theory

As the temperature rises, the average speeds of the reacting particles increase. Consequently, there are more collisions per second and this results in an increase in the rate of reaction.

The increase in collision frequency seems to explain the increase in reaction rate as temperature rises, but does it also explain how *rapidly* the rate increases—the rate of many gaseous and aqueous reactions often doubles for a temperature rise of only 10 K?

Using the kinetic theory, we can predict the relative increase in number of collisions when the temperature rises by 10 K. The kinetic energy of a particle is proportional to its absolute temperature:

$$\tfrac{1}{2}mV^2 \propto T,$$

but the mass of a given particle remains constant

$$\Rightarrow V^2 \propto T$$

$$\therefore \frac{V_1^2}{V_2^2} = \frac{T_1}{T_2} \qquad \text{(equation 1)}$$

where $V_1$ is the velocity at temperature $T_1$, and $V_2$ is the velocity at temperature $T_2$.

Now, suppose that the average speed of a particle is $V$ at 300 K, what will its average speed be at 310 K?

Substituting in equation 1,

$$\frac{V_1^2}{V^2} = \frac{310}{300}$$

$$\therefore V_1 = \sqrt{\frac{310}{300}}\,V = \sqrt{1.033}\,V = 1.016V$$

∴ The average speed at 310 K is only 1.016 times greater than at 300 K, i.e. it has increased by only 1.6 %.

Since the frequency of collisions depends on the average speed of the particles, we might expect the rate of collisions and hence the reaction rate to be 1.6 % greater at 310 K than at 300 K. In practice, the reaction rate roughly doubles between 300 and 310 K, i.e. it increases by approximately 100 %.

Clearly, the *simple* collision theory cannot account adequately for the increase in reaction rate as temperature rises. How then are we to explain in molecular terms the relatively large increase in reaction rate with temperature?

During a chemical reaction, bonds are first broken and then others are formed. Consequently, energy is required to break bonds and start this process, whether the overall reaction is exothermic or endothermic. Therefore, it is reasonable to assume that *particles do not always react when they collide* since they may not have sufficient energy for the necessary bonds to be broken.

A reaction occurs only as a result of those collisions between particles which possess more than a certain minimum amount of energy known as the **activation energy**. This minimum amount of energy is required to enable chemical bonds to stretch and break and rearrangements of atoms, ions and electrons to occur as the reaction proceeds.

The reaction can be imagined to proceed as shown in figure 23.10.

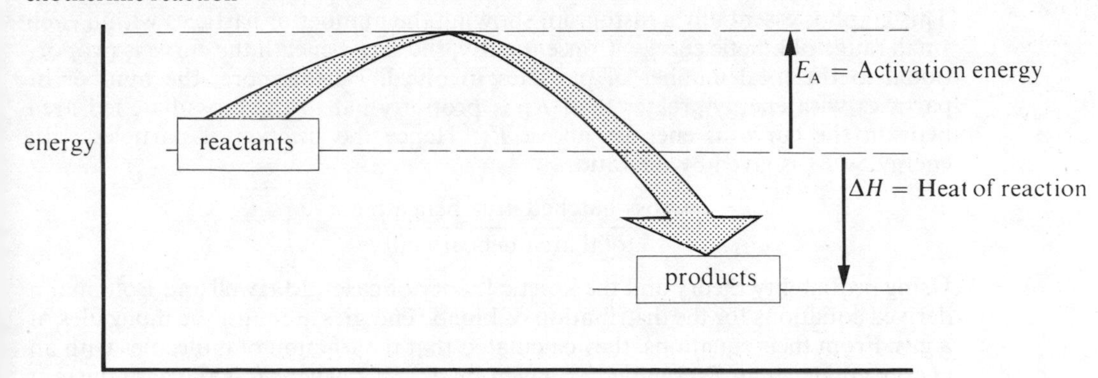


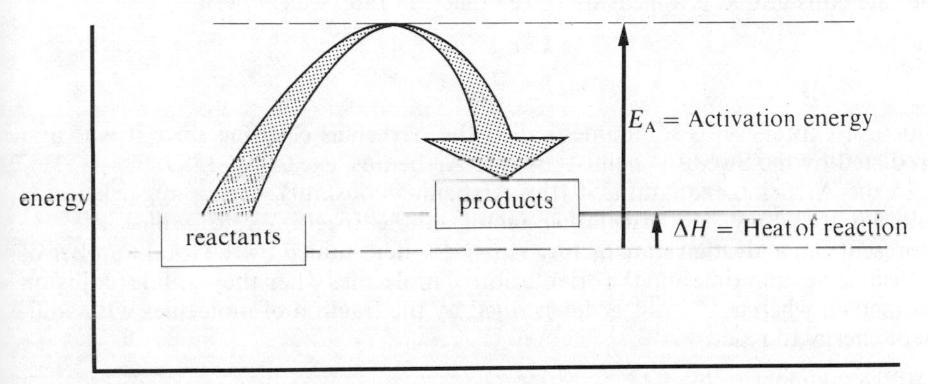


**Figure 23.10** The progress of an exothermic and an endothermic reaction according to the collision theory.

The diagrams show the relationship between $E_A$ and $\Delta H$ for an exothermic and an endothermic reaction. $E_A$ is related to the rate constant of a reaction (see later in this section). It gives information concerning *how fast* the reaction occurs. If the activation energy is very large, only a small proportion of molecules have enough energy to react, so the reaction proceeds very slowly. If, however, the activation energy is very small, most of the molecules have sufficient energy to react and the reaction proceeds very fast.

On the other hand, $\Delta H$ is related to the equilibrium constant of a reaction (section 20.8) and it gives information concerning *how far* a reaction goes towards completion.

The fact that a certain minimum energy is needed initially for reactions to occur is well illustrated by fuels and explosives which usually require a small input of energy to initiate their exceedingly exothermic reactions.

If this development to the collision theory is correct, it would be useful to know what fraction of the particles have more than the minimum energy required for reaction. In the case of a gas, the energy of particles is largely kinetic energy ($\frac{1}{2}mv^2$) and for particles of a given mass this is determined by their velocity. As we discovered in section 10.9, the distribution of velocities (and hence energies) can be determined by an apparatus similar to that used by Zartmann (figure 10.3).

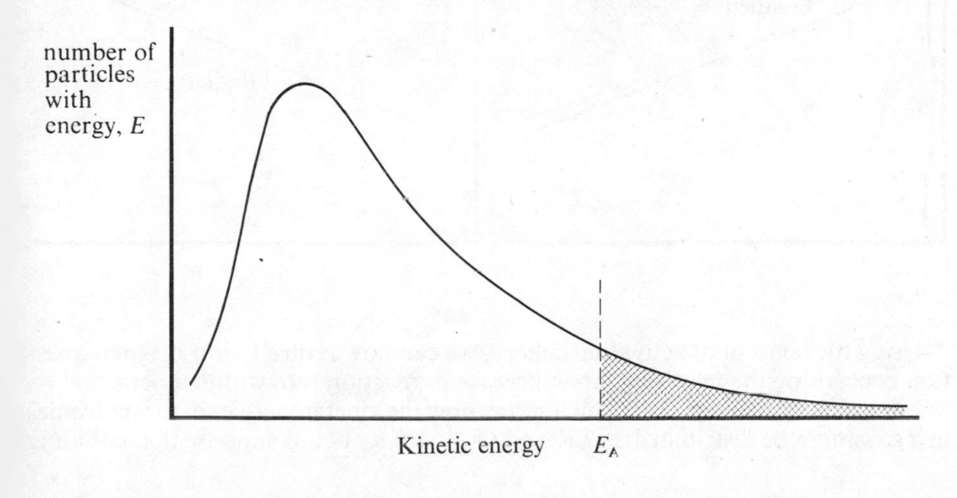


**Figure 23.11** Distribution of kinetic energies of particles in the gas phase.

The graph in figure 23.11 shows how the energies of particles are distributed. This graph is essentially a histogram showing the number of particles within each small range of kinetic energy. Consequently, the area beneath the curve is proportional to the total number of particles involved. Furthermore, the number of particles with energy greater than $E_A$ is proportional to the cross-hatched area beneath the curve at energies above $E_A$. Hence the *fraction* of particles with energy > $E_A$ is given by the ratio,

$$\left(\frac{\text{Cross-hatched area beneath curve}}{\text{total area beneath curve}}\right).$$

Using probability theory and the kinetic theory of gases, Maxwell and Boltzmann derived equations for the distribution of kinetic energies amongst the molecules of a gas. From their equations, they calculated that the fraction of molecules with an energy greater than $E_A$ J mole$^{-1}$ is given by $e^{-E_A/RT}$, where $R$ is the gas constant (8.3 J K$^{-1}$ mole$^{-1}$) and $T$ is the absolute temperature.

This suggests that at a given temperature $T$ reaction rate $\propto e^{-E_A/RT}$. Now, since the rate constant, $k$, is a measure of the reaction rate, we can write

$$k \propto e^{-E_A/RT},$$

$$\Rightarrow k = Ae^{-E_A/RT}.$$

This last expression is sometimes called the **Arrhenius equation** since it was first predicted by the Swedish chemist, Svante Arrhenius as early as 1889.

In the Arrhenius equation, $A$ (the Arrhenius constant), can be regarded as a **collision frequency** and **orientation factor** in the reaction rate, whilst $e^{-E_A/RT}$ represents an **activation state factor**. Thus, $A$ is determined by the total number of collisions per unit time and the orientation of molecules when they collide (collision geometry), whereas $e^{-E_A/RT}$ is determined by the fraction of molecules with sufficient energy to react.

If we take logs in $k = Ae^{-E_A/RT}$

$$\ln k = \ln A + \ln e^{-E_A/RT}$$

$$\Rightarrow \ln k = \ln A - \frac{E_A}{RT}$$

Comparing the last equation with

$$y = c + mx,$$

we can see that a graph of ln $k$ against $1/T$ should be a straight line with gradient $-E_A/R$ (figure 23.12). $E_A$ is the activation energy in J mole$^{-1}$ and $R$ is the gas constant (8.3 J K$^{-1}$ mole$^{-1}$). Thus, by plotting a graph of the logarithm to base $e$ of velocity constants against the reciprocal of the absolute temperature, we can measure the gradient ($-E_A/R$) and hence calculate the activation energy ($E_A$).

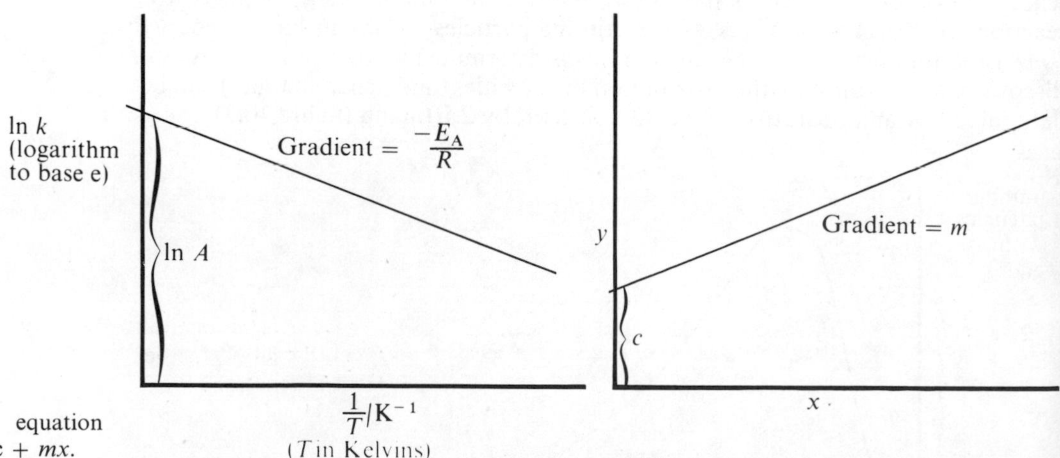


**Figure 23.12** Comparing the equation ln $k$ = ln $A$ − $E_A/RT$ with $y$ = c + $mx$.

Using the concept of activation energy, we can now return to our original question concerning the relatively large increase in reaction rate with temperature.

Look closely at figure 23.13 which shows how the kinetic energies of the molecules in a gas might be distributed at $T$ K and ($T$ + 10) K. Let us suppose that colliding

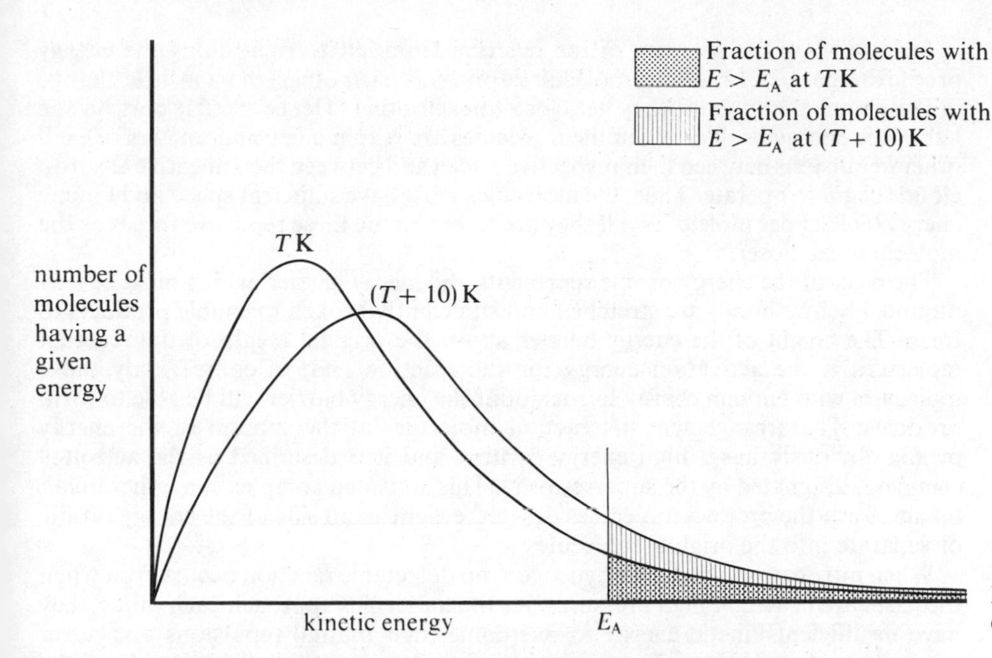


**Figure 23.13** Distribution of the kinetic energies of gas molecules at $T$ and $(T + 10)$ K.

molecules must have a kinetic energy of $E_A$ before a reaction takes place. Notice in figure 23.13 that only a small fraction of molecules (indicated by the dotted area) have sufficient energy to react. However, when the temperature rises by 10 K, the fraction of molecules with sufficient energy to react (indicated by the vertically-lined area) roughly doubles and so the reaction rate also doubles.

## 23.10 Catalysis

In a chemical reaction, existing bonds must be broken and new bonds must form as reactant molecules are converted to products. In order to do this, energy is needed, usually in the form of heat. The 'energy barrier' (as we might call the activation energy) for the reaction between $N_2$ and $H_2$ to produce $NH_3$ is shown in figure 23.14. The horizontal axis of the diagram shows the progress of the reaction from initial reactants to final products. The curve showing the energy of the materials throughout the reaction is usually referred to as an **energy profile**.

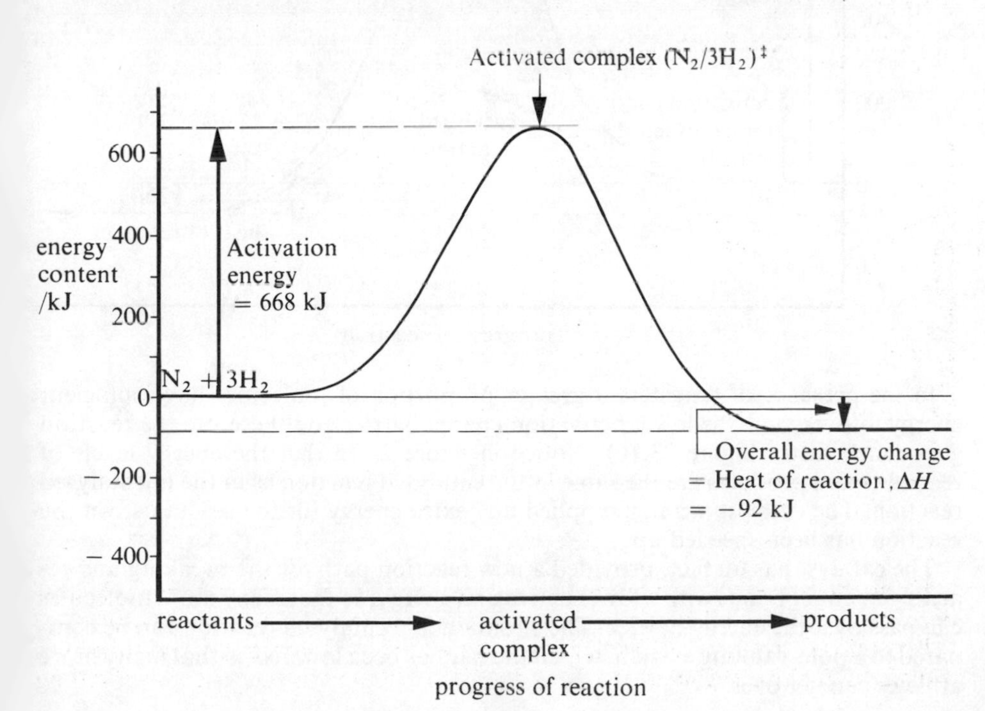


**Figure 23.14** An energy profile for the reaction, $N_2 + 3H_2 \rightarrow 2NH_3$.

Let us follow the progress of the reaction from left to right along the energy profile in figure 23.14. As the molecules approach each other, there is little change in their total energy until they get close to each other. Hence the flat part on the left of the energy profile. When the molecules are within a few nanometres of each other, repulsions between their respective nuclei and between their negative electron clouds begin to operate. Thus, the molecules must have sufficient speed and kinetic energy (668 kJ per mole of $N_2$) if they are to overcome these repulsive forces as the molecules get closer.

The peak of the energy profile represents an 'energy barrier' which must be surmounted before bonds are stretched and sufficiently broken to enable products to form. The height of the energy barrier above the original height of the reactant molecules is the activation energy for the reaction and, of course, only those molecules with enough energy to surmount the energy barrier will be able to form products. The arrangement of reactant molecules at the summit of the energy profile obviously has a high energy content and it is described as the **activated complex** (designated by the superscript ‡). This activated complex can either break up and form the product molecules down the right-hand side of the energy profile or separate into the original molecules.

When nitrogen is mixed with hydrogen, no detectable reaction occurs even when the gases are heated at high pressures. As the molecules approach each other, they have insufficient kinetic energy to overcome their mutual repulsions and never reach the activated state. They rise part of the way up the left-hand side of the energy profile, repel one another and separate again.

We can, however, speed up the reaction by using a **catalyst** (sections 18.6 and 21.3). *A catalyst can be defined as a substance which alters the rate of a reaction without itself undergoing any permanent chemical change.* Thus, a small amount of catalyst is capable of catalysing an infinite amount of reaction and often at a much faster rate than the uncatalysed reaction.

When the reaction,

$$N_2(g) + 3H_2(g) \longrightarrow 2NH_3(g),$$

is catalysed by tungsten, the activation energy is much lower than in the uncatalysed reaction (figure 23.15).

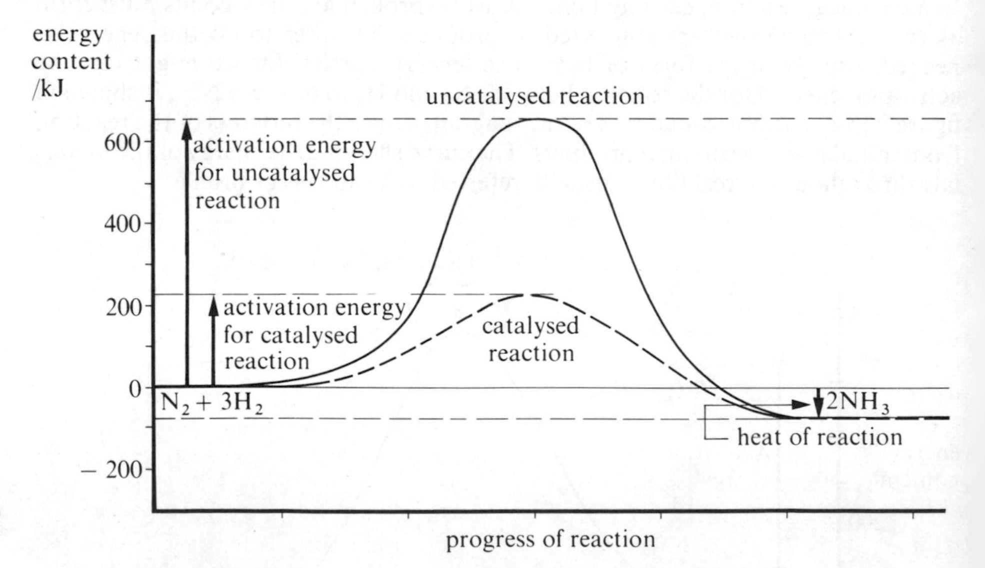


**Figure 23.15** Energy profiles for the reaction, $N_2 + 3H_2 \rightarrow 2NH_3$, (*a*) uncatalysed and (*b*) catalysed by tungsten.

In the presence of tungsten, a greater proportion of molecules have sufficient energy to overcome the lower activation energy barrier and therefore the reaction goes much faster (figure 23.16). Notice in figure 23.15 that the energy levels of reactants and products are the same in the catalysed reaction as in the uncatalysed reaction. The catalyst has not supplied any extra energy for the reactants, yet the reaction has been speeded up.

The catalyst has, in fact, provided a new reaction path for the breaking and rearrangement of bonds with a lower activation energy, so that many more molecules can pass over the energy barrier. The situation in a catalysed reaction can be compared to a pole-vaulting event in which the bar has been lowered so that many more athletes can get over.

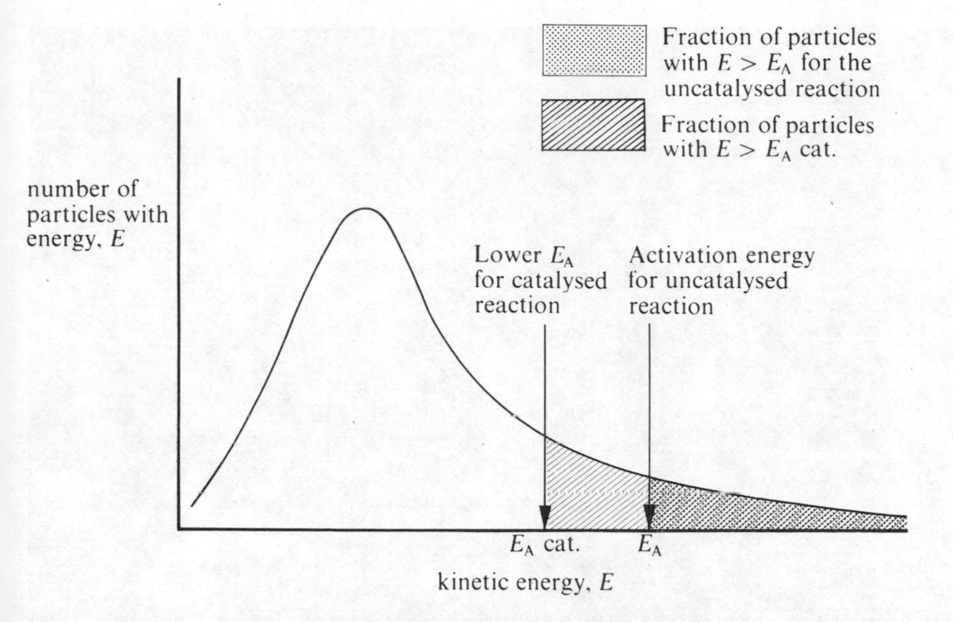


**Figure 23.16** Distribution of the kinetic energies of reacting particles and the activation energies for catalysed and uncatalysed reactions. Notice the greater proportion of molecules which have an energy > the activation energy for the catalysed reaction.

The reaction between nitrogen and hydrogen involves gases passing over a solid catalyst. Reactions of this kind in which the reactants are in a different physical state to the catalyst are said to involve **heterogeneous catalysis.**

In contrast to this, catalysed reactions in which reactants and catalyst are mixed together in the same state are said to involve **homogeneous catalysis**.

Many of the important catalysed processes in industry involve heterogeneous catalysis, whilst the action of enzymes in biological systems usually involves homogeneous catalysis.

○ Name the catalyst used in each of the following industrial processes;
(a) the Haber Process,
(b) the Contact Process,
(c) the oxidation of ammonia to nitrogen oxide in the manufacture of nitric(V) acid.

○ Do these processes involve heterogeneous or homogeneous catalysis?

Many chemists believe that some form of heterogeneous catalysis could help to reduce the atmospheric pollution from car exhaust fumes. The toxic chemicals in exhaust fumes, such as oxides of nitrogen, carbon monoxide and unburnt petrol are normally blown away and do not cause any problems. Under certain conditions, however, and particularly in traffic-dense city streets, these fumes cause a great deal of atmospheric pollution. Consequently, industrial chemists have been looking for a solid catalyst which could be put into the exhaust systems of all vehicles and which would then catalyse the conversion of these polluting chemicals to fully-oxidized, harmless exhaust products. For further discussion see page 399.

Another interesting development in the field of catalysis has been the increasing use of enzymes in industrial processes. These include the use of papain (a protein-hydrolysing enzyme) to remove the haze in beer and to tenderize beef, and the manufacture of fruit juices, beer, vitamins and pharmaceutical products using enzymes extracted from animal tissues, plants, yeasts and fungi. The so-called 'biological' washing powders contain enzymes which attack animal and plant tissues and so are especially useful for removing the stains caused by biological materials such as food and blood.

A strong pole-vaulter has sufficient height (potential energy) to clear the cross bar (activation energy barrier).

Many catalysts are highly specific and the details of their operation are still not fully understood. The final choice of catalyst for many industrial processes may often be a combination of scientific deduction, trial and error or simply inspired guess-work. Indeed, the precise composition of industrial catalysts is often a closely guarded secret.

Spear thistle (*Cirsium arvense*) before and after treatment with 2,4-D. 2,4-D and similar substances, act as plant hormones by controlling the rate of metabolic processes and hence the rate of growth in plants. They do this by controlling the synthesis and activity of enzymes. 2,4-D is such a powerful growth promotor in broad leaved plants that it can make them grow so rapidly that they exhaust their food supply and die, which has happened to the thistle in the lower photograph. Thin leaved plants, such as grass surrounding the thistle are unaffected. Hence 2,4-D can be used as a selective herbicide. Its full name is 2,4-dichlorophenoxyethanoic acid and its formula is $OCH_2COOH$

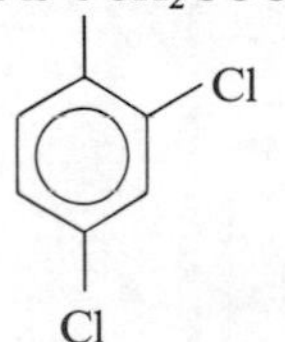


## 23.11 The uses of reaction rate studies

The most obvious use of reaction rate studies is the information which they give us about the rates of chemical reactions. This is important not only for industrial processes in which time and the efficient use of resources are crucial, but also for biological processes, archaeological dating and many other reactions, such as rusting and burning, which affect our everyday lives. Nevertheless, the most exhaustive studies of reaction rates have usually been done on industrial processes in which chemists, engineers and economists endeavour to obtain maximum product from the minimum amount of raw material, using minimum fuel in the minimum possible time. When viewed in this light, it is not surprising that catalysts have such a widespread use in industry.

Another important aspect of rate studies is the way in which the order of a reaction can be used to interpret the reaction on a molecular level. By considering the order of a reaction with respect to the different reactants, chemists can speculate about the sequence in which bonds break and atoms rearrange during the reaction and hence suggest what is usually described as a **reaction mechanism**.

As an example, let us consider the reaction between oxygen and hydrogen bromide at 700 K.

$$4HBr(g) + O_2(g) \longrightarrow 2H_2O(g) + 2Br_2(g)$$

This equation indicates that four HBr molecules react with one $O_2$ molecule. If the reaction were to take place in a single step, these five molecules would need to collide with each other simultaneously—an extremely improbable event. As this reaction occurs quite rapidly at 700 K, it is likely that it proceeds by a sequence of steps rather than by a single step involving the simultaneous collision of five molecules. In fact, *most chemical reactions which proceed at a measurable rate are believed to take place in a series of simple steps. This series of simple reactions is known as the reaction mechanism.*

Quantitative studies of the reaction between HBr and $O_2$ show that the reaction rate is proportional to both the concentration of HBr and the concentration of $O_2$, i.e. the reaction is first order with respect to both HBr and $O_2$. We can summarize this information in the rate equation,

$$\text{Rate} = k[HBr][O_2]$$

The stoichiometric equation involves HBr and $O_2$ in molar proportions of 4:1, yet the rate equation shows proportions of 1:1. How are we to explain this?

The overall reaction must take place in a series of simpler steps which satisfy both the rate equation and the stoichiometric equation.

The following mechanism has been proposed for the reaction:

| | | | | |
|---|---|---|---|---|
| $HBr + O_2$ | $\longrightarrow$ | $HBrOO$ | Step 1 | Slow |
| $HBrOO + HBr$ | $\longrightarrow$ | $2HBrO$ | Step 2 | Fast |
| $HBrO + HBr$ | $\longrightarrow$ | $H_2O + Br_2$ | Step 3 | Fast |
| $HBrO + HBr$ | $\longrightarrow$ | $H_2O + Br_2$ | Step 4 | Fast |
| Overall $4HBr + O_2$ | $\longrightarrow$ | $2H_2O + 2Br_2$ | | |

Notice that each step in the reaction mechanism involves the collision of only two molecules, a much more likely event than the simultaneous collision of five molecules. Notice also the suggestion that the first step in the mechanism is slow whilst the others are fast. This explains why the reaction rate is proportional to both [HBr] and $[O_2]$.

The first step producing HBrOO is very much a 'bottleneck' in the oxidation of hydrogen bromide. Thus, HBrOO forms slowly, but is immediately consumed in the fast second step by reaction with HBr. No matter how rapid the second, third, and fourth steps are, they can produce water and $Br_2$ only as fast as the slowest stage in the sequence. Hence, those factors that determine the rate of formation of HBrOO determine the overall rate of reaction. The formation of HBrOO is the one step which dictates the rate because it is the slowest stage in the reaction mechanism. Hence, the slowest stage in a mechanism is called the **rate-determining step**.

The reaction of iodine with propanone in acid solution,

$$I_2 + CH_3COCH_3 \longrightarrow CH_2ICOCH_3 + H^+ + I^-$$

is found to be first order with respect to $CH_3COCH_3$, zero order with respect to $I_2$ and first order with respect to $H^+$.

- O Write a rate equation for the reaction.
- O What is the overall order of reaction?
- O Which substances are probably involved in the slow, rate-determining step of the reaction?
- O Try to write a possible mechanism for the reaction.

It would seem that the rate-determining step in the reaction involves propanone and $H^+$, but not iodine. Hence the suggested mechanism is:

$$H\text{–}CH_2\text{–}C(=O)\text{–}CH_3 + H^+ \longrightarrow H\text{–}CH_2\text{–}\overset{+}{C}(OH)\text{–}CH_3 \qquad \text{Step 1 Slow}$$

$$H\text{–}CH_2\text{–}\overset{+}{C}(OH)\text{–}CH_3 \longrightarrow H\text{–}CH{=}C(OH)\text{–}CH_3 + H^+ \qquad \text{Step 2 Fast}$$

$$H\text{–}CH{=}C(OH)\text{–}CH_3 + I_2 \longrightarrow H\text{–}CHI\text{–}CI(OH)\text{–}CH_3 \qquad \text{Step 3 Fast}$$

$$H\text{–}CHI\text{–}CI(OH)\text{–}CH_3 \longrightarrow H\text{–}CHI\text{–}C(=O)\text{–}CH_3 + H^+ + I^- \qquad \text{Step 4 Fast}$$

The reaction rate is dictated by the first, slow stage which requires only the participation of $CH_3COCH_3$ and $H^+$. Once this first step is completed, the remaining steps take place rapidly and so the reaction rate is independent of the concentration of iodine and $[I_2]$ does not feature in the rate equation. If this mechanism is correct, then the reaction of $Br_2$ with propanone should take place at a similar rate to iodination. This is, in fact, found to be so.

Both of the reaction mechanisms that we have considered so far involve an initial slow step, but this is not always the case. In order to illustrate this point, we can consider the reaction between bromide and bromate(v) ions in acid solution.

$$5Br^- + BrO_3^- + 6H^+ \longrightarrow 3Br_2 + 3H_2O$$

Kinetic studies show that the reaction is fourth order overall; first order with respect to bromide, first order with respect to bromate(v) and second order with respect to $H^+$ ions.

$$\text{Rate} = k[Br^-][BrO_3^-][H^+]^2$$

The immediate deduction to make from this is that the rate-determining step involves one $Br^-$, one $BrO_3^-$ and two $H^+$ ions, but the simultaneous collision of four ions is most improbable. A more likely explanation is that the slow rate-determining step is preceded by faster reactions. Hence, the suggested mechanism involves the initial formation of HBr and $HBrO_3$ in two fast reactions, followed by a reaction between these two substances which is the rate-determining step.

$$H^+ + Br^- \longrightarrow HBr \qquad \text{Step 1 Fast}$$

$$H^+ + BrO_3^- \longrightarrow HBrO_3 \qquad \text{Step 2 Fast}$$

$$HBr + HBrO_3 \xrightarrow{\text{rate-determining step}} HBrO + HBrO_2 \qquad \text{Step 3 Slow}$$

The HBrO and $HBrO_2$, produced in the slow step, now react rapidly with more HBr, forming bromine and water.

$$HBrO_2 + HBr \longrightarrow 2HBrO \qquad \text{Step 4 Fast}$$

$$HBrO + HBr \longrightarrow H_2O + Br_2 \qquad \text{Step 5 Fast}$$

## Summary

1

$$\text{Reaction rate} = \frac{\text{change in amount (or concentration) of a substance}}{\text{time taken}}$$

2 The rate of a reaction can be affected by:
(a) the availability of reactants and their surface area,
(b) concentration (partial pressure for gases),
(c) temperature,
(d) catalysts,
(e) light.

3 For the hypothetical reaction

$$aA + bB + cC \longrightarrow \text{products},$$

it is found experimentally that the reaction rate can be expressed as

$$\text{Reaction rate} = k[A]^{\alpha}[B]^{\beta}[C]^{\gamma},$$

which is known as the rate equation for the reaction. $k$ is known as the rate constant and its value is constant for a given reaction at a particular temperature. When all reactant concentrations are 1.0M; $k$, the rate constant, is numerically equal to the reaction rate.

4 The rate equation can only be obtained experimentally. It cannot be deduced from the stoichiometric equation.

5 The order of a reaction with respect to a given reactant is the power of that reactant's concentration in the rate equation.

For the hypothetical reaction just mentioned,
order of reaction with respect to A = $\alpha$,
order of reaction with respect to B = $\beta$,
order of reaction with respect to C = $\gamma$.

The overall order of a reaction is the sum of the powers of the concentration terms in the rate equation.
Hence, overall order of the hypothetical reaction = $\alpha + \beta + \gamma$.

6 The half-life of a first order reaction is constant. The kinetics of a first order reaction with its constant half-life are similar to the decay of a radioactive isotope, which also has a constant half-life.

7 A reaction occurs only as a result of those collisions between particles which possess more than a certain minimum amount of energy known as the activation energy.

8 A catalyst is a substance which alters the rate of a reaction without itself undergoing any permanent chemical change. Those catalysts which speed up a reaction do so by introducing an entirely different reaction mechanism with a much lower activation energy than in the uncatalysed reaction.

9 Most chemical reactions proceed by a sequence of simple steps, each involving only one or two particles. This series of simple steps is known as the reaction mechanism.
The slowest step in the reaction mechanism dictates the overall reaction rate and is usually known as the rate-determining step.

10 The mechanism for a reaction is related to the rate equation. It cannot be deduced from the stoichiometric equation.

## Study questions

1 (a) Give one example in each case of a reaction which takes place:
(i) instantaneously,
(ii) at a moderate rate,
(iii) rapidly at a high temperature, but not at all at room temperature.
(b) Explain why the reactions you chose in (a) (i) and (a) (iii) behave as they do.

2 (a) Why is it difficult to hard boil an egg on the top of Mount Everest?
(b) Is it difficult to fry an egg there? Explain.

3 For the gaseous reaction,

$A(g) + B(g) \longrightarrow C(g) + D(g)$ it is found that,

Reaction rate $= k[A]^2[B]$

How many times does the rate increase or decrease if:
(a) the partial pressures of both A and B are doubled,
(b) the partial pressure of A doubles, but that of B remains constant,
(c) the volume of the reacting vessel is doubled,
(d) inert gas is added, which doubles the overall pressure whilst the partial pressures of A and B remain constant,
(e) the temperature rises by 30°C?

4 (a) Draw a sketch graph of the percentage reactant remaining against time, for a zero order and a first order reaction. (Assume, in each case, that it takes 10 minutes for the amount of reactant to fall from 100% to 50%.)
(b) The half-life of radioactive $^{238}_{92}U$ is $4.5 \times 10^9$ years. It takes about $4.5 \times 10^9$ years for half of a given amount of uranium to disintegrate by radioactive emission and turn into lead. When uranium and lead are found together in rocks, the age of the rock may be deduced. In a particular sample, uranium and lead are found in molar proportions of 1:3.
Estimate the age of the rock, stating any assumptions which you make.

5 The isotope $^{24}_{11}Na$ decays by $\beta$-particle emission to a stable nuclide. The half-life of $^{24}_{11}Na$ is 15 hours.
(a) Write a nuclear equation for the decay process (balanced for mass and atomic number).
(b) The rate of emission of $\beta$-particles from a freshly prepared sample of $^{24}_{11}Na$ is observed for 60 hours by means of a suitable method. Draw a sketch graph showing the rate of emission of $\beta$-particles (vertically) against time, during the first sixty hours.
(c) 2.0g of $^{24}_{11}Na$ is allowed to decay. What mass of $^{24}_{11}Na$ will be left after sixty hours?

6 (a) Explain the following terms:
(i) order of reaction,
(ii) rate constant,
(iii) half-life,
(iv) activation energy,
(v) activated state.
(b) The rate constants ($k$) for the decomposition of hydrogen iodide at different absolute temperatures are given in table 23.5.
(i) Plot a graph of $\ln k$ ($\log_e k$) against $1/T$.
(ii) Use this graph to obtain a value for the activation energy for the decomposition of hydrogen iodide.
(The gas constant, $R = 8.3\ JK^{-1}\ mole^{-1}$.)

**Table 23.5** The rate constants for the decomposition of hydrogen iodide at different temperatures.

| Rate constant, $k$ /mole$^{-1}$ dm$^3$ s$^{-1}$ | Temperature/K |
|---|---|
| $3.75 \times 10^{-9}$ | 500 |
| $6.65 \times 10^{-6}$ | 600 |
| $1.15 \times 10^{-3}$ | 700 |
| $7.75 \times 10^{-2}$ | 800 |

7 The rate of the reaction,

$$H_2O_2(aq) + 2I^-(aq) + 2H^+(aq) \longrightarrow I_2(aq) + 2H_2O(l)$$

may be calculated by measuring the time for the first appearance of $I_2$ in the solution, i.e. the time required for the concentration of $I_2$ to reach $10^{-5}$ mole dm$^{-3}$.
(a) For a particular experiment in which initially $[H_2O_2] = 0.010M$, $[I^-] = 0.010M$ and $[H^+] = 0.10M$;
calculate the reaction rate if $I_2$ first appears after 6 seconds.
(b) In a second experiment in which initially $[H_2O_2] = 0.005M$, $[I^-] = 0.010M$ and $[H^+] = 0.10M$;
calculate the reaction rate if $I_2$ first appears after 12 seconds.
(c) From these calculations show that the reaction is first order with respect to $H_2O_2$.
(d) Given the further information that the rate law is,
Reaction rate $= k[H_2O_2][H^+][I^-]$, calculate the rate constant, $k$.
(e) What are the units of $k$?
(f) Predict the rate of reaction when
$[H_2O_2] = 0.05M$, $[H^+] = 0.10M$ and $[I^-] = 0.02M$.

8 In an experiment to study the acid-catalysed reaction of propanone ($CH_3COCH_3$) with iodine, 50cm³ of 0.02M $I_2$ were mixed with 50cm³ of acidified 0.25M propanone solution. 10cm³ portions of the reaction mixture were removed at 5 minute intervals and added rapidly to excess $NaHCO_3$(aq). The remaining iodine was then titrated against $Na_2S_2O_3$(aq). The graph in figure 23.17 shows the volume of $Na_2S_2O_3$(aq) required to react with the remaining iodine at different times from the start of the reaction.

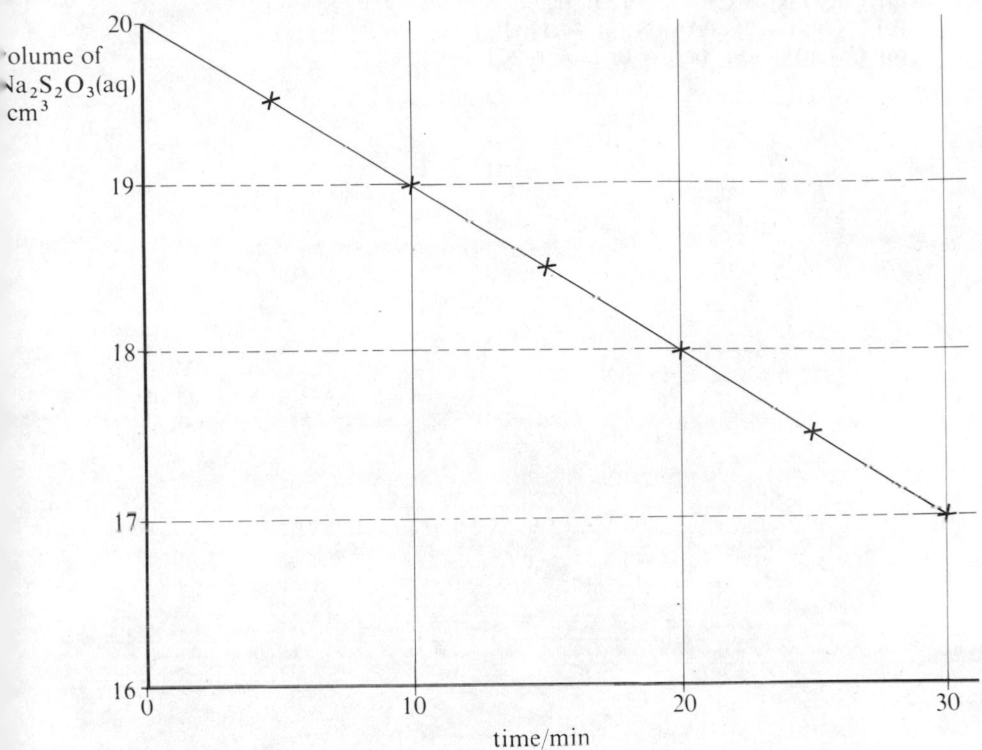


**Figure 23.17** Volume of $Na_2S_2O_3$(aq) required to react with the remaining iodine at different times during the reaction of iodine with propanone.

(a) Why are the 10cm³ portions of the reaction mixture added rapidly to excess $NaHCO_3$(aq) before titration with $Na_2S_2O_3$(aq)?
(b) What is the rate of reaction in terms of cm³ of $Na_2S_2O_3$(aq) $min^{-1}$?
(c) How does the *rate of* change of iodine concentration vary during the experiment?
(d) Is the reaction rate dependent on the concentration of iodine?
(e) What is the order of reaction with respect to iodine?
(f) Write an equation for the reaction between $S_2O_3^{2-}$ and $I_2$.
(g) What is the molarity of $I_2$ in the 100cm³ of reaction mixture at time = 0 min?
(h) Use the graph to predict the volume of $Na_2S_2O_3$(aq) which reacts with 10cm³ of the reaction mixture at time = 0 min.
(i) What is the molarity of the $Na_2S_2O_3$(aq) used in the titrations?
(j) Suppose the reaction is first order with respect to propanone. What would be the rate of reaction (in cm³ $Na_2S_2O_3$(aq) $min^{-1}$) if 0.50M propanone were used in place of 0.25M?

9 Two gases, X and Y, react according to the equation

$$X(g) + 2Y(g) \longrightarrow XY_2(g)$$

Experiments were performed at 400 K in order to determine the order of this reaction and the following results were obtained.

| Experiment number | Initial concentration of X /mole $dm^{-3}$ | Initial concentration of Y /mole $dm^{-3}$ | Initial rate of formation of $XY_2$ /mole $dm^{-3}$ $s^{-1}$ |
|---|---|---|---|
| 1 | 0.10 | 0.10 | 0.0001 |
| 2 | 0.10 | 0.20 | 0.0004 |
| 3 | 0.10 | 0.30 | 0.0009 |
| 4 | 0.20 | 0.10 | 0.0001 |
| 5 | 0.30 | 0.10 | 0.0001 |

(a) What is the order of this reaction with respect to (i) X, (ii) Y?
(b) Write a rate equation for the reaction of X with Y.
(c) Using the rate equation, predict a possible mechanism for this reaction.
(d) Using the results from experiment 1, calculate the numerical value of the rate constant, *k*.

(e) What are the units of $k$?
(f) What further experiments would you carry out in order to find the activation energy of the reaction between X and Y?
(g) Why are chemists interested in obtaining orders of reaction and rate equations?

10 Suggest experimental means by which the rates of the following reactions could be followed.
(a) $CaCO_3(s) \rightarrow CaO(s) + CO_2(g)$
(b) $2NO(g) + 2H_2(g) \rightarrow N_2(g) + 2H_2O(g)$
(c) $Cl_2(aq) + 2Br^-(aq) \rightarrow Br_2(aq) + 2Cl^-(aq)$

# Introduction to Carbon Chemistry 24

## 24.1 The uniqueness of carbon

The number of compounds containing carbon and hydrogen whose formulae are known to chemists was recently estimated to be about two and a half million. This is far more than the number of compounds of all the other elements put together. Why does carbon have this unique ability to form an enormous number of compounds? And how can we hope to begin to study more than a tiny fraction of these compounds? There are three important properties of carbon that enable it to form so many stable compounds.

*(a) Carbon has a fully shared octet of electrons in its compounds*

For example in methane the outer shell electrons are shared as shown in figure 24.1. This means that the carbon atoms have no lone pairs or empty orbitals in their outer shells and are thus unable to participate in co-ordinate bonding. The inability of carbon to bond in this way once it has an octet of electrons is responsible for the kinetic stability of its compounds (see below).

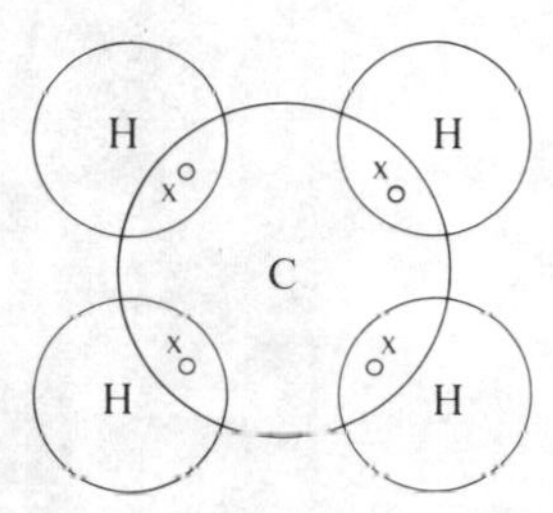


**Figure 24.1** Electron sharing in methane.

*(b) Carbon can form strong single, double and triple bonds to itself*

The stability of the single C–C bond can be seen by comparing the bond energies in table 24.1.

It is worth comparing carbon and silicon here since silicon might be expected to show similarities to carbon, being the next member of group IV. Note the strength of the C–C bond compared with that of the Si–Si bond, and also the high strength of the C–H bond; all but a handful of the vast range of carbon compounds also contain hydrogen. Of course, it is insufficient to consider simply the strength of bonds between carbon atoms and hydrogen atoms. If carbon compounds are to be stable they must be stable under normal conditions, and that means in the presence of air. In fact, compounds containing carbon and hydrogen are not stable relative to their oxidation products, carbon dioxide and water. We would therefore expect them to react with oxygen exothermically. This, of course, they do. For example, methane:

$$CH_4(g) + 2O_2(g) \longrightarrow CO_2(g) + 2H_2O(g) \qquad \Delta H = -890 \text{ kJ}$$

**Table 24.1** Some average bond energies.

| Bond | Bond energy/kJ mole$^{-1}$ |
|---|---|
| C—C | 346 |
| C=C | 610 |
| C≡C | 835 |
| Si—Si | 226 |
| Si=Si | 318 (estimated) |
| S—S | 272 |
| C—O | 360 |
| Si—O | 464 |
| C—H | 413 |
| Si—H | 318 |

Most people are familiar with this reaction: it occurs in most gas-fired appliances in Britain, since North Sea gas is largely methane. It is a familiar fact that although methane is energetically unstable relative to its combustion products, it does not react with air until quite high temperatures, that is, it needs lighting before it burns. This is because the reaction between methane and oxygen has a high activation energy which must be supplied before the reaction will proceed. Thus methane, like most compounds containing carbon and hydrogen, is energetically unstable in the presence of air, but kinetically stable (see chapter 13). Compounds containing silicon and hydrogen are also energetically unstable relative to their combustion products:

$$\underset{\text{silane}}{SiH_4(g)} + 2O_2(g) \longrightarrow SiO_2(s) + 2H_2O(g) \qquad \Delta H = -1\,428 \text{ kJ}$$

This reaction is much more exothermic than the corresponding one for methane, largely due to the very high energy of the Si–O bond. Unlike methane, silane is not kinetically stable in the presence of oxygen; the activation energy of the above reaction is quite low, and silane bursts into flame spontaneously in air. Other compounds containing silicon and hydrogen behave similarly, so there is no huge range of silicon compounds to compare with that of carbon. However, the high strength of the Si–O bond relative to the Si–Si and Si–H bonds means that silicon exists naturally as highly stable silicon(IV)oxide (sand) and silicates.

The ability of carbon to form strong bonds to itself means that it can **catenate**, i.e. form chains and rings of varying size which are the basis of its many stable compounds. The kinetic stability of hydrocarbons in air is very important to society for it means they can be stored and, barring accidents, the energy of their oxidation released when it is required. This makes hydrocarbons, of which oil is our major source, the most important modern fuels.

In the presence of air, hydrocarbons are energetically unstable, but kinetically stable. They can, therefore, be safely stored for long periods at room temperature . . .

. . . but at high temperatures they undergo rapid combustion.

*(c) Carbon can form four covalent bonds*

The bond energies given in table 24.1 suggest that sulphur should be able to form reasonably stable bonds to itself, which it does. However, sulphur forms only two bonds, so a chain of sulphur atoms cannot have side-groups attached to it. In comparison, the ability of carbon to form four bonds enables a chain of carbon atoms to have many different groups attached and this leads to a wide diversity of compounds.

## 24.2 Organic Chemistry

The diversity of carbon chemistry is responsible for the diversity of life itself. The ability to form a virtually unlimited range of compounds has led to an almost unlimited range of living organisms constructed out of molecules containing carbon. You yourself are a unique individual because you contain unique proteins; only carbon could form the basis of a range of compounds diverse enough to provide a different one for every individual.

Because the major source of compounds containing carbon and hydrogen is living or once-living material (animals, plants, coal, oil), it was originally thought that only living organisms could produce these compounds. This has since been shown to be untrue, but the name **organic** has continued to be applied to that branch of chemistry concerned with the study of compounds containing C–H bonds. This includes the vast majority of carbon compounds, but traditionally compounds such as CO, $CO_2$ and carbonates have been considered to belong to the field of inorganic chemistry.

The position of carbon as the basis of the molecules of life means that the study of organic chemistry is of central importance in understanding the chemistry, and therefore the biology, of living systems (the chemical study of living systems is called **biochemistry**). A knowledge of organic chemistry enables chemists to develop and manufacture drugs, agricultural chemicals, anaesthetics and other chemicals

Carbon compounds surround us. Paper, paint, plastic, plants, our clothes and we ourselves all contain carbon.

whose interactions with life processes are important to man. Many other organic chemicals, less directly related to biological compounds, are of prime importance to modern society, for example the many polymers (polythene, nylon) whose properties of flexibility and elasticity are a direct consequence of carbon's unique ability to form chains.

There can be little doubt of the importance of an understanding of organic chemistry, but with such a multiplicity of compounds to consider, we need some means of simplifying and systematizing our study.

## 24.3 Functional groups

The ability of carbon to form strong bonds to itself and to hydrogen leads to the formation of stable compounds, called hydrocarbons, containing only carbon and hydrogen. The simplest hydrocarbons, containing only single bonds, are the **alkanes**, an example of which is butane:

$$\begin{array}{ccccccccc} & & H & & H & & H & & H & \\ & & | & & | & & | & & | & \\ H & - & C & - & C & - & C & - & C & - & H \\ & & | & & | & & | & & | & \\ & & H & & H & & H & & H & \end{array}$$

Consider however the compound butan-1-ol:

$$\begin{array}{ccccccccc} & & H & & H & & H & & H & \\ & & | & & | & & | & & | & \\ H & - & C & - & C & - & C & - & C & - & OH \\ & & | & & | & & | & & | & \\ & & H & & H & & H & & H & \end{array}$$

Butane and butan-1-ol, despite their structural similarities, have very different properties: butane is a gas, butan-1-ol is a liquid; butane has no effect on sodium, butan-1-ol reacts with the evolution of hydrogen. Clearly, the OH group in butan-1-ol has a profound effect in modifying the properties of the unreactive butane skeleton to which it is attached. The OH group in butan-1-ol is an example of a **functional group**. Experiments have shown that a given functional group, such as OH, has much the same effect whatever the size and shape of the hydrocarbon skeleton to which it is attached. This greatly simplifies the study of organic compounds because all molecules containing the same functional group can be considered as members of a family, with similar properties. Of course, as the hydrocarbon chain gets bigger it increasingly dominates the properties of the compound, so that members of the family show a steady gradation of physical and chemical properties as the size of the hydrocarbon portion increases.

A family of compounds containing the same functional group is called a **homologous series**. Butan-1-ol is a member of the homologous series of **alcohols**, all of which contain the OH group. The first two members of the series of alcohols are methanol and ethanol:

```
   H              H  H
   |              |  |
H—C—OH        H—C—C—OH
   |              |  |
   H              H  H
methanol        ethanol
```

For any homologous series it is possible to write a general formula for the series in terms of the number of carbon atoms it contains. For example, the general formula of the alcohols is $C_nH_{2n+1}OH$.

The main functional groups considered in this book and their corresponding homologous series are shown in table 24.2. The methods of naming the different compounds will be explained as they are encountered.

The idea of functional groups can also be applied to compounds containing more than one group. Thus, the properties of the molecule as a whole can be predicted by considering the modifying effect of each functional group.

Refer to table 24.2 to answer these questions:

- To what homologous series does propanal, $CH_3CH_2CHO$ belong?
- To what homologous series does ethoxyethane, $CH_3CH_2OCH_2CH_3$ belong?

**Table 24.2** Common functional groups.

| Functional group | Name of homologous series | Example |
|---|---|---|
| —OH | alcohols | $CH_3OH$ methanol |
| $—NH_2$ | amines | $CH_3NH_2$ methylamine |
| —C(=O)OH | carboxylic acids | $CH_3COOH$ ethanoic acid |
| >C=C< | alkenes | $H_2C{=}CH_2$ ethene |
| —C≡C— | alkynes | HC≡CH ethyne |
| —Halogen | halogeno compounds | $CH_3Cl$ chloromethane |
| —C(=O)H | aldehydes | $CH_3CHO$ ethanal |
| >C=O | ketones | $CH_3COCH_3$ propanone |
| —O— | ethers | $CH_3OCH_3$ methoxymethane |

## 24.4 Finding the formula of organic compounds

If the properties of a compound are to be predicted from a knowledge of its functional groups, it is necessary to know the structural formula of the compound, showing the position and nature of each group.

### DETERMINATION OF EMPIRICAL FORMULAE

Chapter 1 explains how the empirical formula of a compound may be found from its percentage composition. The composition of organic compounds is normally found by **combustion analysis**. A known mass of a compound is burned and the carbon dioxide and water formed are collected and measured. Other elements that may be present, such as nitrogen and halogens, can also be estimated. From the masses of the combustion products the empirical formula can be calculated. In modern laboratories this composition analysis is performed automatically by machines.

*Example*
A compound **X** containing only carbon, hydrogen and oxygen was subjected to combustion analysis. 0.1g of the compound on complete combustion gave 0.228g of carbon dioxide and 0.0931g of water. Calculate the empirical formula of the compound.

First, calculate the mass of C and H in 0.1g of the compound.

$$44\text{g } CO_2 \text{ contains } 12\text{g C} \Rightarrow \text{mass of C in 0.1g } \mathbf{X} = \frac{12}{44} \times 0.228 = 0.0621\text{g}$$

$$18\text{g } H_2O \text{ contains } 2\text{g H} \Rightarrow \text{mass of H in 0.1g } \mathbf{X} = \frac{2}{18} \times 0.0931 = 0.0103\text{g}$$

$$\text{Mass of C + H in 0.1g } \mathbf{X} = 0.0621 + 0.0103 = 0.0724\text{g}$$

$$\Rightarrow \text{Mass of O in 0.1g} = 0.1 - 0.0724 = 0.0276\text{g}$$

$$\therefore \text{Ratio by mass C : H : O is } 0.0621 : 0.0103 : 0.0276$$

$$\Rightarrow \text{Ratio by moles C : H : O is } \frac{0.0621}{12} : \frac{0.0103}{1} : \frac{0.0276}{16}$$

$$= 0.00518 : 0.0103 : 0.00173$$

$$= 3 : 6 : 1$$

$\therefore$ Empirical formula of **X** is $C_3H_6O$.

## DETERMINATION OF MOLECULAR FORMULAE

Once the empirical formula has been found, in order to arrive at the molecular formula of the compound it is necessary to know its relative molecular mass. Several methods are available for doing this; traditional methods include measurement of gas density (chapter 10) and measurement of osmotic pressure. A quicker and more accurate modern method for finding relative molecular masses is by mass spectrometry, described in chapter 1. The mass spectrum obtained for the compound **X** in the example above is shown in figure 24.2.

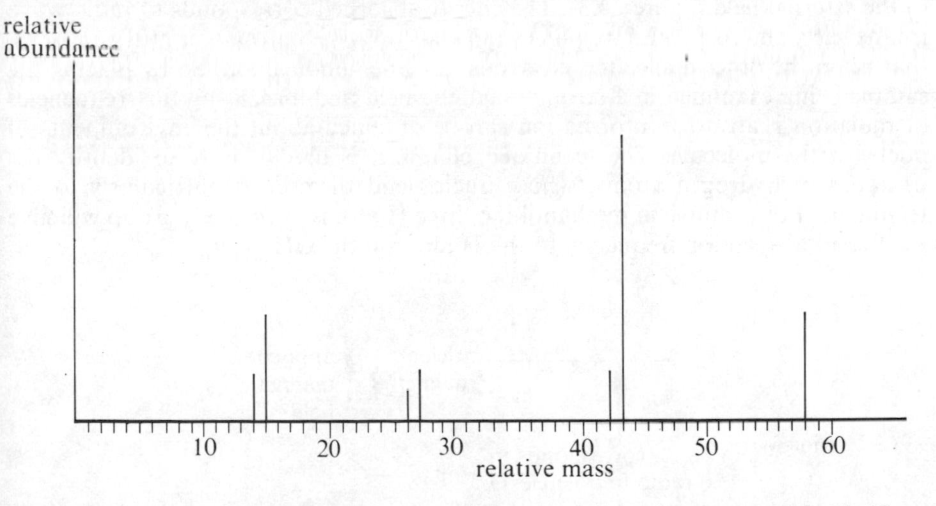


**Figure 24.2** Mass spectrum for compound **X**, empirical formula $C_3H_6O$.

The relative mass of the heaviest particle recorded in the spectrum is 58. If we assume that this corresponds to the intact molecule with a single positive charge, i.e. $X^+$, the relative molecular mass of X must be 58. With an empirical formula of $C_3H_6O$, X could have molecular formula $C_3H_6O$, $C_6H_{12}O_2$, $C_9H_{18}O_3$ and so on, but since its relative molecular mass is known to be 58, the only possible molecular formula is $C_3H_6O$.

## DETERMINATION OF STRUCTURAL FORMULAE

The molecular formula of a compound gives the number of atoms of the different elements in one molecule of the compound, but it gives no information about the way the atoms are arranged. A compound with molecular formula $C_2H_6O$, for example, could have one of two structural formulae:

| $CH_3OCH_3$ | $CH_3CH_2OH$ |
|---|---|
| methoxymethane | ethanol |

The structural formula of ethanoic acid is given in table 24.2.

- ○ What is its empirical formula?
- ○ What is its molecular formula?
- ○ Write one other structure with the same molecular formula.

In order to find the exact structural formula of a compound, more information is needed in order to decide which functional groups it contains. This information may be obtained from various studies some of which are indicated below:

*(a) The mass spectrum of the compound*

When an organic compound passes through the ionization chamber of a mass spectrometer, many of its molecules become fragmented. Each fragment gives a line in the mass spectrum; from the position of the line the relative mass of the fragment can be found and the nature of the fragment deduced. It is then possible, by piecing together the fragments, to deduce the structure of the parent molecule. For example, if we look again at figure 24.2, we see that the mass spectrum of the compound **X** (molecular formula $C_3H_6O$) has prominent peaks at 15 and 43. These probably relate to $CH_3{}^+$ and $CH_3CO^+$ respectively, which in turn suggest the structural formula $CH_3COCH_3$ (propanone). The other possible structure for X, $CH_3CH_2CHO$, propanal, is ruled out because this would give a very strong peak at mass 29, corresponding to both $CH_3CH_2{}^+$ and $CHO^+$. Try to identify the other peaks in the mass spectrum of **X**.

Much of the work of investigating the structure of organic compounds is performed in modern laboratories by complex machines. This is a nuclear magnetic resonance (nmr) instrument at Brunel University.

*(b) Other spectroscopic methods*

The two modern techniques of nuclear magnetic resonance and infrared spectroscopy, like mass spectrometry, require sophisticated and expensive automated machinery not normally possessed by schools. The details of these methods are beyond the scope of this book and the briefest outline only is given below.

(i) *Nuclear magnetic resonance (n.m.r.)* The nuclei of atoms possess spin, just as their electrons do. Nuclei also possess charge, and the rotation of this charge about its axis of spin is equivalent to a circulating electric current. Any circulating current has an associated magnetic field, so the nuclei behave like tiny magnets, lining themselves up with an external magnetic field. If energy is supplied to the nucleus, it can be made to change the orientation of its magnetic field so it is opposed to the external field (figure 24.3). The energy absorbed corresponds to radio wave frequencies, and its precise frequency depends on the environment of the nucleus, that is, on the other nuclei and electrons in its neighbourhood. So by placing the sample being examined in a strong magnetic field and measuring the frequencies of radiation it absorbs, information can be obtained about the environments of nuclei in the molecule. The technique of n.m.r. is used mainly to identify the positions of hydrogen atoms, whose nuclei lend themselves particularly to the technique. For example in methanol the three H atoms in the $CH_3$ group will give a different absorption frequency to the H atom in the OH group.

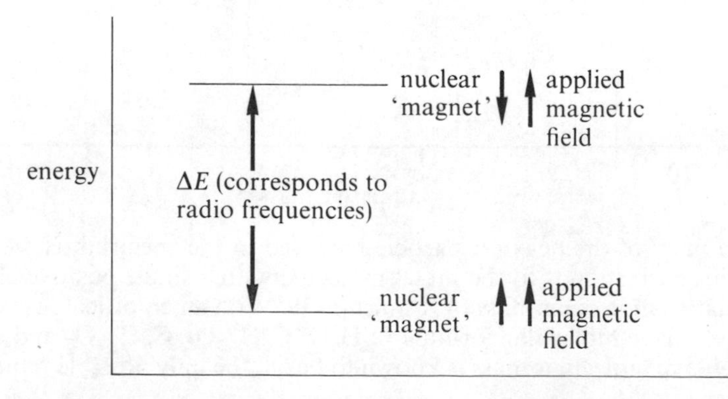


**Figure 24.3** Two orientations of the nuclear 'magnet' in an external magnetic field. The energy difference between the two orientations is the basis of the technique of n.m.r.

(ii) *Infrared spectroscopy* Infrared radiation has a longer wavelength than visible light, occupying the wavelength range from about 2 500 nm to about 25 000 nm. When infrared radiation containing a broad spectrum of wavelengths is shone through a sample of a compound, radiation of a specific frequency, and therefore wavelength, is absorbed in causing the molecule to rotate and in making bonds in the molecule vibrate. The wavelength of radiation absorbed by the bonds in a particular functional group is characteristic of that group, so measurement of the wavelengths at which a compound absorbs infrared radiation provides information about its structural formula. Table 24.3 gives the characteristic absorption wavelengths of some bonds commonly encountered in organic chemistry.

*(c) Physical properties of the compound, such as boiling point*
Physical properties are dependent on structure and provide information that can help elucidate the structural formula. For example methoxymethane is a gas at room temperature, its boiling point being 248K. In contrast, ethanol, with the same molecular formula, is a liquid with boiling point 341K. All members of the homologous series of alcohols tend to have high boiling points relative to other compounds of comparable relative molecular mass.

*(d) Chemical properties of the compound*
Each functional group has certain chemical characteristics. For example alcohols such as ethanol react with sodium to liberate hydrogen while ethers such as methoxymethane do not. Thus a knowledge of the chemical properties of a compound can provide indications of the functional groups it contains.

**Table 24.3** Characteristic absorption wavelengths in the infrared region for vibration of some bonds commonly encountered in organic compounds.

| Bond | Approximate maximum absorption wavelength/$10^{-6}$m |
|---|---|
| C—H | 3.4 |
| C═C | 6.1 |
| C≡C | 4.5 |
| C—Cl | 14 |
| C—Br | 16 |
| C—I | 20 |
| C═O | 5.8 |

## 24.5 Writing structural formulae

The structural formula of a compound shows which atoms are joined to each other. This information is crucial in determining the properties of the compound. The actual shape of the molecule, too, affects its properties, and it is helpful if the structural formula can give some indication of the *stereochemistry* of the molecule.

The methane molecule, as was shown in section 7.7, is tetrahedral in shape. This tetrahedral arrangement of bonds is common to all saturated carbon atoms (that is, carbon atoms bonded to four other groups). Unfortunately the tetrahedral shape creates a problem in representing structural formulae, because it is difficult to show a three-dimensional shape on two-dimensional paper. Figure 24.4 shows eight ways of representing the structural formula of pentane.

(a)

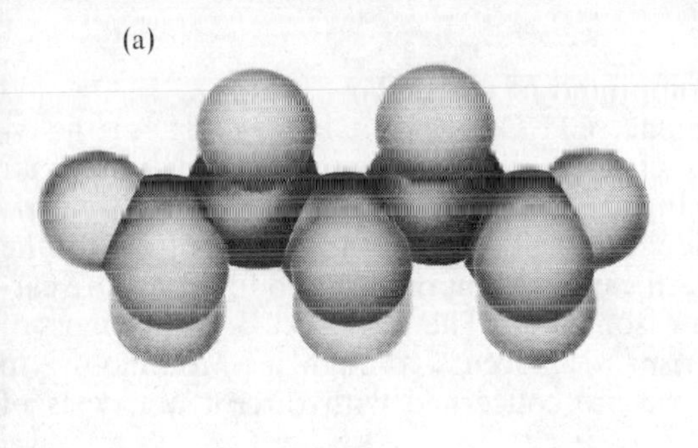

(b)

```
     H H    H H
     \ /    \ /
      C      C
H   /   \  /   \    H
  C       C      C
 / \     / \    / \
H   H   H   H  H   H
```

(c)

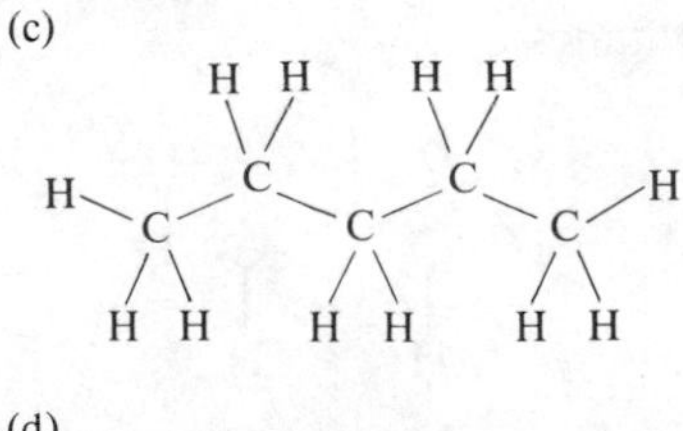

(d)

```
 /\/\
```

(e)

```
   H   H   H   H   H
   |   |   |   |   |
H— C — C — C — C — C —H
   |   |   |   |   |
   H   H   H   H   H
```

(f)

```
         H  H
          \/
          C
H      /    \
   C          C—H
  / \  H      \H
 H   H  \    /
          C      H
       H/   \C /
            /   \ H
          H
```

(g)

```
   H   H   H
   |   |   |
H— C — C — C —H
   |   |   |
   H   H   |
           |
       H — C — H
           |
       H — C — H
           |
           H
```

(h) $CH_3CH_2CH_2CH_2CH_3$

**Figure 24.4** Representations of the structural formula of pentane.

Possibly the most accurate representation is the **space-filling model** (figure 24.4(*a*)) and it would be useful if you could try building this model yourself. Space-filling models show the extent of the electron cloud of each atom accurately, but they are not appropriate to the printed page. Figure 24.4(*b*) attempts to represent the tetrahedral carbon atom by showing the bonds in the plane of the paper as single lines and those projecting in front or behind the paper as thicker, tapering lines. The idea is to give perspective by making the bond appear wider at the end nearer the observer. This system is of limited use, and the tapering lines are often omitted (figure 24.4(*c*)). Occasionally a **skeletal structure** (figure 24.4(*d*)) is used, in which only carbon–carbon bonds and functional groups are shown. A common and easy way of writing formulae is shown in figure 24.4(*e*), but this has drawbacks because it represents the bond angles at each carbon atom as 90° instead of 109°. This type of diagram shows its limitations when an attempt is made to represent rotation about a single bond: compare figure 24.4(*g*) with figure 24.4(*f*). It is easy to see that (*f*) can be obtained from (*c*) by rotating about the bond between the second and third carbon atoms, but (*g*) appears to be a different compound to (*e*), though both in fact represent pentane. Finally, figure 24.4(*h*) shows an abbreviated way of writing a structural formula.

These different ways of representing structural formulae will be used as appropriate in different parts of this book.

○ Write a full structural formula for $CH_3CH_2CHOHCH_2CH_3$
○ Write a skeletal formula for

```
            H
            |
          H-C-H
    H       |  H   H   H
    |       |  |   |   |
  H-C-------C--C---C---C-H
    |       |  |   |   |
    H       H  H   H   H
```

## 24.6 Isomerism

It was seen in section 24.4 that a compound of molecular formula $C_2H_6O$ could have two possible structural formulae, $CH_3OCH_3$ or $CH_3CH_2OH$, methoxymethane or ethanol. Compounds such as these, possessing the same molecular formula but with their atoms arranged in different ways, are called **isomers**. Isomerism is very common in organic chemistry and is another of the reasons for the multiplicity of carbon compounds. All carbon compounds with four or more carbon atoms, and many with less, show isomerism. There are $4.11 \times 10^9$ isomers of molecular formula $C_{30}H_{62}$. Isomerism also occurs, though less commonly, in inorganic chemistry. In this book we are concerned with three main types of isomerism.

### STRUCTURAL ISOMERISM

Structural isomers differ in which atom is attached to which. Methoxymethane and ethanol are structural isomers: in the former the oxygen atom is attached to two carbon atoms while in the latter it is attached to a carbon and a hydrogen. Structural isomers can be members of different homologous series, like the pair just mentioned, or members of the same series, like the following two structural isomers of molecular formula $C_4H_{10}$, which are both alkanes:

```
    H   H   H   H                  H   H   H
    |   |   |   |                  |   |   |
  H-C---C---C---C-H     and      H-C---C---C-H
    |   |   |   |                  |   |   |
    H   H   H   H                  H   |   H
                                       |
                                     H-C-H
                                       |
                                       H

       butane                    methylpropane
```

```
               H  H  H
               |  |  |
Note that H—C—C—C—H is not a third isomer of C4H10, having been formed
               |  |  |
               H  H
                H—C—H
                  |
                  H
```

from butane simply by rotating about a single bond. It is possible to rotate a structure freely about any single carbon–carbon bond, but structures that can be interconverted by rotating about a bond in this way are not isomeric. This third structure appears at first sight to be a separate isomer, but this is due to the shortcomings of representing the carbon bond angles as 90° instead of 109°. A little experimenting with a molecular model will make this point clear.

- ○ How many isomers are there of molecular formula $C_5H_{12}$?
- ○ Write the structural formulae of all the alcohols (i.e. the compounds with the OH functional group) of molecular formula $C_4H_{10}O$.

Structural isomers usually show considerable differences in physical and chemical properties, even if they are members of the same homologous series. Thus, the boiling point of butane is 273 K while that of methylpropane is 261 K. It is not really surprising that structural isomers differ, since any functional group will be influenced by its environment and will therefore have its properties modified by the atoms to which it is attached.

## OPTICAL ISOMERISM

Light is a form of electromagnetic radiation and can be regarded as consisting of waves. A ray of normal light consists of waves which vibrate in many directions at right angles to the direction of travel of the ray (figure 24.5(*a*)).

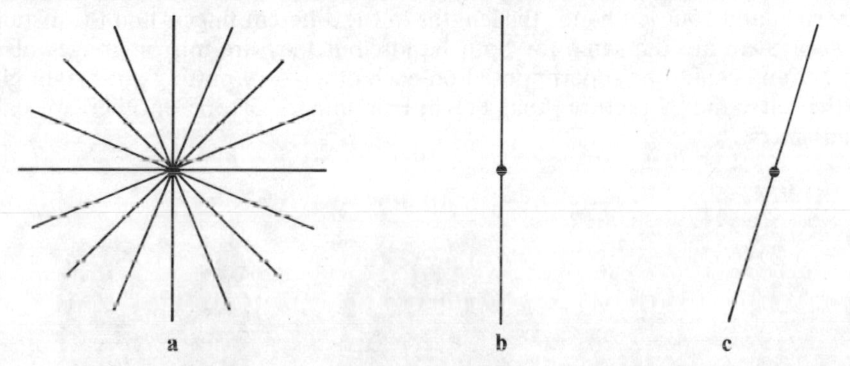


**Figure 24.5** Polarized light. **a** Normal light ray travelling towards the observer: each line represents a wave seen 'end-on'. **b** Plane-polarized light: only waves in a single plane are present. **c** Plane-polarized light from **b** after passing through an optically active solution.

Certain materials have the ability to remove from normal light all waves except those vibrating in a single plane (see figure 24.5(*b*)). The light is then said to be **plane polarized**. It is rather like the light being combed as it passes through the polarizer. A well known polarizer is polaroid, which is used in the lenses of some sunglasses.

It has been known for a long time that certain chemical substances have the ability to rotate the plane of polarized light. That is, if polarized light is shone through a solution of the substance in a suitable solvent, the plane in which the polarized light vibrates will be rotated to the right or left (figure 24.5(*c*)). These substances are said to be **optically active**: an example is 2-hydroxypropanoic acid (lactic acid, $CH_3CHOHCOOH$), the substance responsible for the sour taste in sour milk. Chemists have isolated two forms of lactic acid: one of these rotates polarized light to the right, the other rotates it to the left. For a given concentration of solution the two forms rotate light to exactly the same extent, but in opposite directions. In addition, crystals of the two forms are found to be mirror-images of one another. Apart from these differences, the forms are physically and chemically identical. Since the two forms of lactic acid are chemically identical, they must contain exactly the same groups attached to each other in the same way. Their difference can only lie in the way the groups are arranged relative to each other in space. Furthermore they must be exact opposites in this respect since their effect on polarized light is equal and opposite.

A consideration of the molecular structures of different optically active substances shows they have one thing in common: they all possess asymmetric molecules, that is to say, their molecules have no centre, axis or plane of symmetry and they are different from their own mirror images. Optical activity is shown by all substances with asymmetric molecules: such substances can have two isomeric forms which are mirror-images of one another and which rotate polarized light in opposite directions.

The simplest type of asymmetric molecule is one in which four different groups are attached to the same carbon atom—lactic acid is such a molecule. Figure 24.6

OH, C, HOOC, $CH_3$, H — (a) D-lactic acid.

imaginary mirror

OH, C, H, $CH_3$, COOH — (b) L-lactic acid

**Figure 24.6** The enantiomeric forms of lactic acid. The two isomers can be seen to be images of one another reflected in the imaginary mirror.

shows the two forms, called D-lactic acid and L-lactic acid, but you should try making models of each of the forms to satisfy yourself that they are indeed different molecules and that they are mirror images. They can be shown to be different by trying to superimpose the two molecules on one another: it is impossible to arrange them so that all the groups correspond in position. Note, though, that in both D- and L-lactic acid all the groups are attached together in the same way, and that the spacings of the various groups are the same in each isomer. This is why the two forms of lactic acid have identical chemical properties. The relationship between optical isomers such as D- and L-lactic acid is like the relation between your right and your left hand: the lengths of the different fingers and the distances between them are the same for both hands, but they are mirror images of one another and cannot be superimposed on each other—try putting your right glove on the left hand. Structures that are mirror images of one another are called **enantiomers**.

Much of the early work on optical isomerism was done by the great French scientist Louis Pasteur. These models, constructed for Pasteur, represent crystals of the two forms of ammonium 2-hydroxybutanedioate. Like the molecules themselves, the two types of crystal are mirror-images of one another. The model on the left is of one of the optically active forms of potassium, 2,3-dihydroxybutanedioate.

Many organic compounds and some inorganic ones have asymmetric molecules and therefore show optical isomerism. Optically active compounds are very common in nature: almost all amino acids (the building units of proteins) are optically active, as are all the sugars. The interesting thing is that most naturally occurring optically active compounds occur as one isomer only: thus all naturally occurring glucose is D-glucose, and all naturally occurring amino acids are L-isomers. This is because the enzymes that produce and break down these substances are so specific that they can accommodate only one type of molecular configuration. Presumably, if we went through the looking-glass like Alice all the glucose in our bodies would become L-glucose.

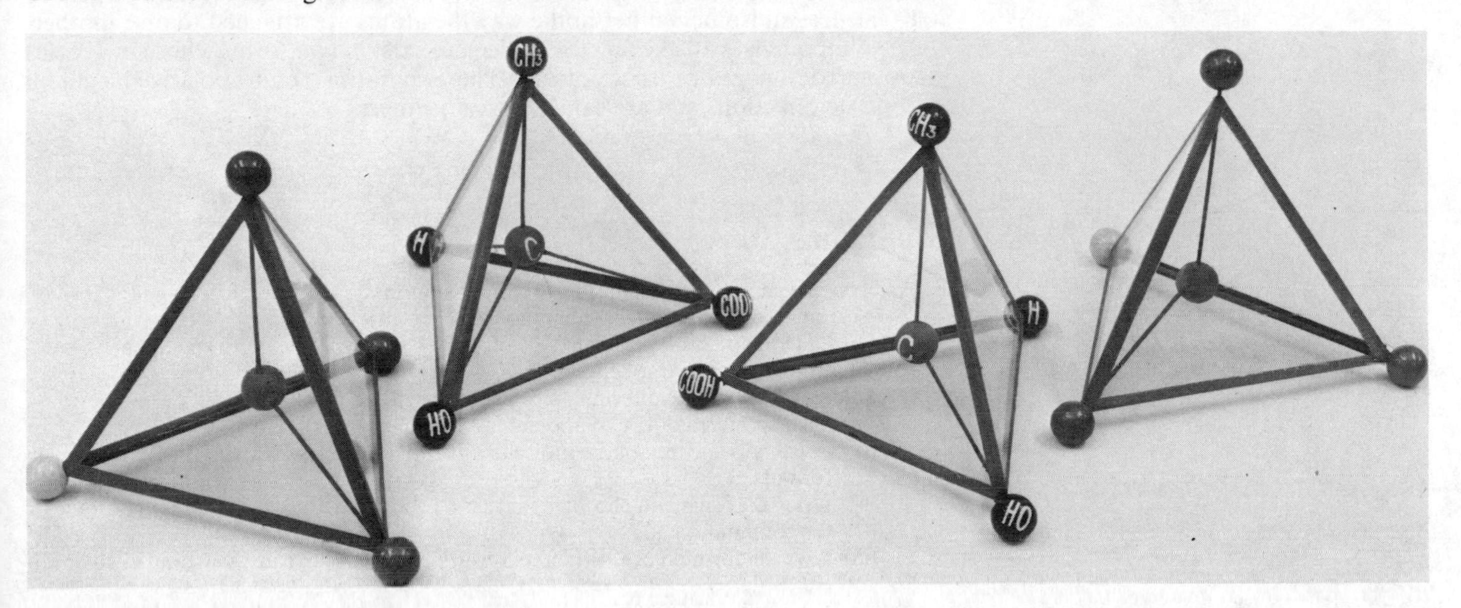


Early models constructed to illustrate optical isomerism. The two central models are of (left) L-lactic acid and (right) D-lactic acid.

Many optically active compounds have complex structures and it is difficult to tell quickly whether or not their molecules are asymmetric. For simpler molecules it is safe to say that if the molecule contains a carbon atom to which four different groups are attached, it will show optical isomerism.

- ○ Would you expect (*a*) $CH_2ClF$ (*b*) CHClBrF to show optical isomerism?
- ○ Which of the isomeric alcohols whose structures you wrote earlier in this section would show optical isomerism? Write the structures of the enantiomers.
- ○ Write the structure of the first alkane to show optical isomerism.

Lactic acid can be extracted from natural sources such as sour milk or muscle tissue and the acid from these sources shows optical activity. It is also possible to prepare lactic acid in the laboratory from fairly simple starting materials, but the acid prepared in this way shows no optical activity at all. This is because most laboratory methods for preparing lactic acid produce the D- and L-isomers in equal amounts: the two isomers of course cancel each other out in their effect on polarized light. This kind of mixture of optical isomers containing equal molar quantities of D- and L-isomers is called a **racemic mixture**. Because of the identical properties of the isomers it is very difficult to separate them from a racemic mixture, though methods do exist for doing so. The process of separation is called **resolution**.

### GEOMETRIC ISOMERISM

This type of isomerism is described in section 26.4.

## Summary

1 The ability of carbon atoms to form strong bonds to four other atoms, including other carbon atoms, enables it to form a huge range of organic compounds.
2 Organic compounds, though kinetically stable, are energetically unstable relative to their combustion products.
3 Organic compounds can be regarded as having a basic hydrocarbon skeleton with functional groups attached. Compounds containing the same functional group form a family with similar properties called a homologous series.
4 The empirical formula of an organic compound can be found by analysis of its combustion products.
5 The molecular formula can be found from the empirical formula once the relative molecular mass is known.

6 The structural formula shows the precise arrangement of atoms and can be found from a knowledge of some of the properties of the compound.
7 Compounds with the same molecular formula but with their atoms arranged in different ways are called isomers.
8 Structural isomers differ in the way the atoms are attached to one another.
9 Compounds with asymmetric molecules exist in two forms whose molecules are mirror images or enantiomers. The two forms rotate polarized light in opposite directions and are called optical isomers.

## Study questions

1 An organic compound was subjected to combustion analysis. 1.0g of the compound formed 1.37g carbon dioxide, 1.12g water and no other products.
(a) Calculate the percentage by mass of carbon and of hydrogen in the compound.
(b) What other element must be present?
(c) Calculate the empirical formula of the compound.
(d) The mass spectrum of the compound is shown below.
(i) Use this to find the relative molecular mass of the compound, and thus its molecular formula.
(ii) Using the fragments shown in the mass spectrum, deduce the structural formula of the compound.
(iii) Give the formula of the fragment to which each peak can be attributed.

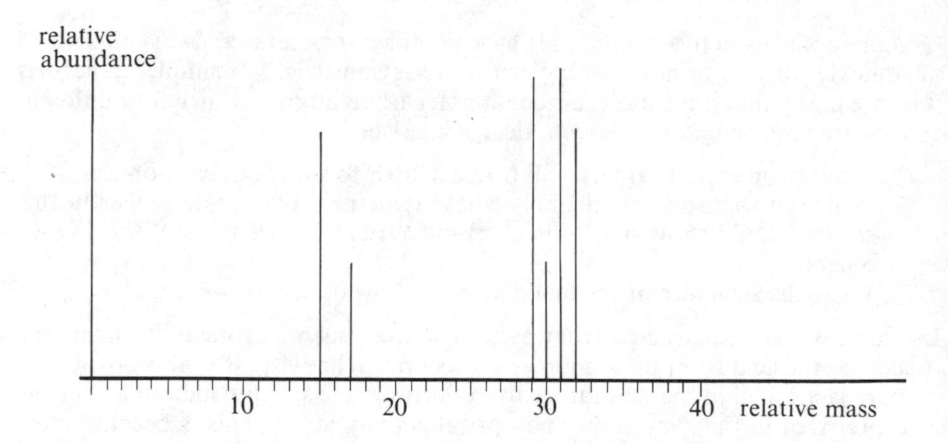


2 **A**, **B** and **C** are isomeric compounds of molecular formula $C_3H_8O$. Two of the compounds are members of the same homologous series. The table gives some data about **A**, **B** and **C**.

| | **A** | **B** | **C** |
|---|---|---|---|
| Boiling point /K | 370 | 284 | 356 |
| Density/g cm$^{-3}$ | 0.80 | 0.72 | 0.79 |

(a) Which two compounds are members of the same homologous series?
(b) Write the structural formulae of all three possible isomers of $C_3H_8O$.
(c) Use table 24.2 to decide to which homologous series each of the isomers you have drawn belongs.
(d) Is it possible from the data given to say which of the isomers you have drawn is **A**, which is **B** and which is **C**? Explain your answer.

3 Use table 24.2 to decide to which homologous series the following compounds belong.
(a) $CH_3CH_2CH_2OH$
(b) $CH_3CH_2COCH_3$
(c) $CH_3CH_2Cl$
(d) $CH_3CH_2COOH$
(e) $CH_3CH{=}CHCH_3$
(f) $H_2NCH_2COOH$
(g) $CH_3CH_2OCH_2CH_2CH_3$

4 Write the full structural formulae of all the isomers of the following, stating which type of isomerism is involved:
(a) $C_3H_7Cl$
(b) $C_6H_{14}$
(c) $C_2H_3Cl_2Br$

5 Silicones are unreactive polymers with the structure

```
   CH3     CH3     CH3     CH3     CH3
    |       |       |       |       |
  —Si—O—Si—O—Si—O—Si—O—Si—
    |       |       |       |       |
   CH3     CH3     CH3     CH3     CH3
```

Look at the bond energies given in table 24.1, and suggest a reason why silicones are inert and unreactive while silanes (compounds analagous to alkanes but containing silicon atoms instead of carbon atoms) are unstable and ignite spontaneously in air.

6 Fluorocarbons are compounds analogous to alkanes but containing fluorine instead of hydrogen, e.g. $CF_4$, $CF_3CF_3$ etc. They are extremely unreactive, being quite stable in air even at high temperatures. Long-chain fluorocarbons such as 'Teflon' (—$CF_2$—$CF_2$—$CF_2$—$CF_2$—$CF_2$—) are used as unreactive corrosion-resistant materials for gaskets and protective coatings.

Use the bond energies given below to explain why fluorocarbons are stable in air while hydrocarbons are energetically unstable.

| **Bond** | **Bond energy/kJ mole$^{-1}$** |
|---|---|
| C—H | 413 |
| C—F | 485 |
| H—O | 463 |
| F—O | 234 |

7 A compound containing carbon, hydrogen and nitrogen only was analysed. 0.1g of the compound on combustion gave 0.228g of carbon dioxide and 0.124g of water. On reduction of the compound, all the nitrogen in it was converted to ammonia and it was found that 0.1g of the compound gave ammonia equivalent to 17.2cm$^3$ of 0.1M hydrochloric acid when titrated.

(a) How many moles of carbon, hydrogen and nitrogen are there in 0.1g of the compound?

(b) What is the empirical formula of the compound?

(c) The relative molecular mass of the compound was found to be 116. What is its molecular formula?

(d) The infrared spectrum of the compound indicated the presence of the —$NH_2$ group. Suggest one possible structure for the compound.

8 The following structural formulae represent only three different substances—some of the formulae are equivalent to others. Which formulae are equivalent to which?

```
A     H  H  H  H
      |  |  |  |
    H—C—C—C—C—H
      |  |  |
      H  H  H
         H—C—H
           |
         H—C—H
           |
           H
```

B (skeletal formula: zigzag chain)

C $CH_3CH_2CH_2CH_2CH_2CH_3$

```
D             H
              |
            H—C—H
              |
      H  H  H  |  H  H
      |  |  |  |  |  |
    H—C—C—C—C—C—C—H
      |  |  |  |  |  |
      H  H  H  H  H  H
```

E (skeletal formula: branched zigzag chain)

F $CH_3(CH_2)_5CH_3$

G $CH_3CH_2CH_2CHCH_2CH_3$ with $CH_3$ on the CH

```
H        H        H        H
         |        |        |
       H—C————————C————————C—H
         |        |        |
       H—C—H      H      H—C—H
         |                 |
       H—C—H             H—C—H
         |                 |
         H                 H
```

9 The earth's crust contains only 0.036% by mass of carbon, compared with 49% oxygen and 26% silicon. Despite its relatively low abundance on earth, carbon is the element that forms the basis of all living things. What are the properties of carbon that make it so suitable for this rôle?

# 25 Petroleum and Alkanes

## 25.1 Crude oil

In 1973 the Arab oil producers of the Middle East, who supply a large proportion of the crude oil used in the Western world, decided to cut back production of crude oil and at the same time to raise its price. Overnight, a cheap and abundant raw material became expensive and insecure. The effects of this move were felt in every area of life in Britain, so dependent is it on crude oil. Motor fuels, lubricating oils and heating oils naturally increased in price, as did electricity, much of which is generated in oil-fired power stations. But many other commodities from plastics and detergents to paints and antifreeze quickly became far more expensive than they had been in the days when oil was cheap. For as well as supplying the majority of our energy needs, crude oil is the source of 70% of Britain's organic chemicals. It is the most important of all modern raw materials.

Crude oil was formed from the remains of small marine animals and plants that were buried in the beds of the seas millions of years ago. The decay of these remains under the layers of overlaying rock formed the liquid known as crude oil or petroleum (from Greek words meaning 'rock oil'). Similar conditions led to the formation of the natural gas that is often found associated with crude oil as well as in deposits on its own, such as those in the North Sea.

Crude oil is a complex mixture of hydrocarbons. It has no uses in its raw form: if it is to provide useful products its components must be partly separated and if necessary modified. Once crude oil has been located and extracted, it must therefore be transported to a refinery where it is processed. The fundamental process is primary distillation, the details of which will be familiar to most readers and are simply summarized here in diagrammatic form in figure 25.1.

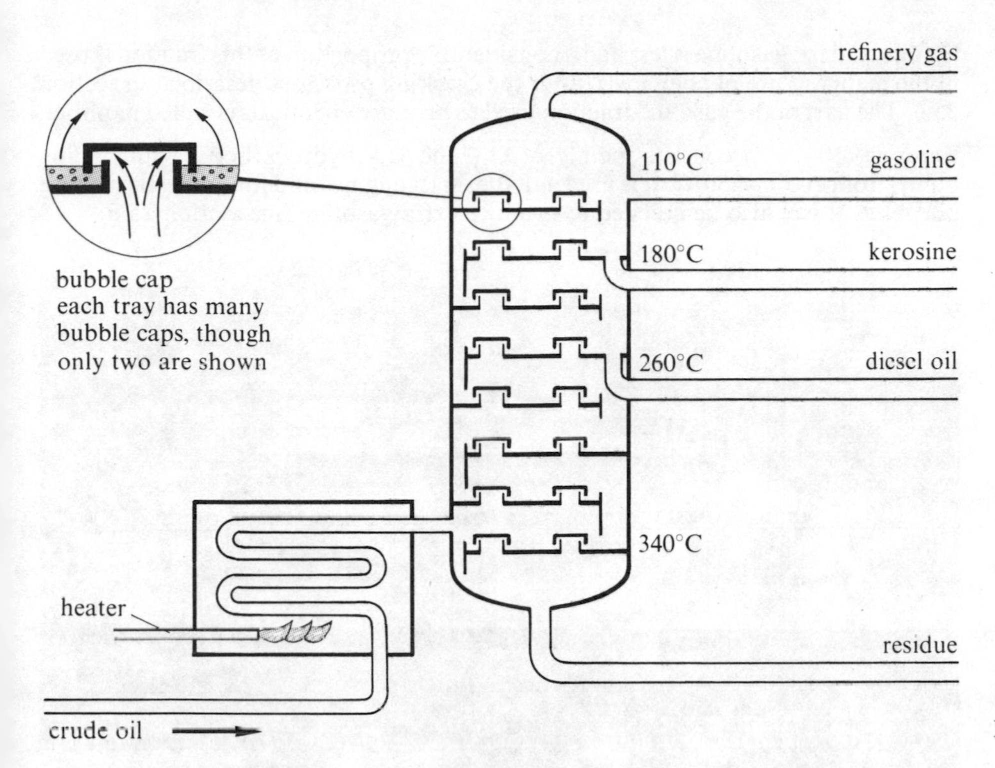


**Figure 25.1** Primary distillation of crude oil.

The different fractions are used as follows.

**Refinery gas** (1–2 % of crude oil) is similar in composition to natural gas, containing those hydrocarbons that are gases at normal temperatures. This comprises the alkanes with one to four carbon atoms in their molecules, with methane as the major component. The main use of refinery gas is as a gaseous fuel, but, like natural gas, it can also be used as a starting point for making petrochemicals, since most organic chemicals are built up from small molecules containing one, two or three carbon atoms.

**Gasoline** (15–30 %) is a complex liquid mixture of hydrocarbons containing mainly $C_5$ to $C_{10}$ compounds whose boiling points range from 40°C to 180°C. The major use of gasoline is of course as a fuel in internal combustion engines (see section 25.6). In the U.S.A., where the motor car is the predominant method of transport, fuel taxes are low and cars are built to consume large quantities of fuel. Gasoline is therefore in great demand and American refineries use the entire gasoline fraction for blending in motor fuels. In Western Europe and most of the rest of the world,

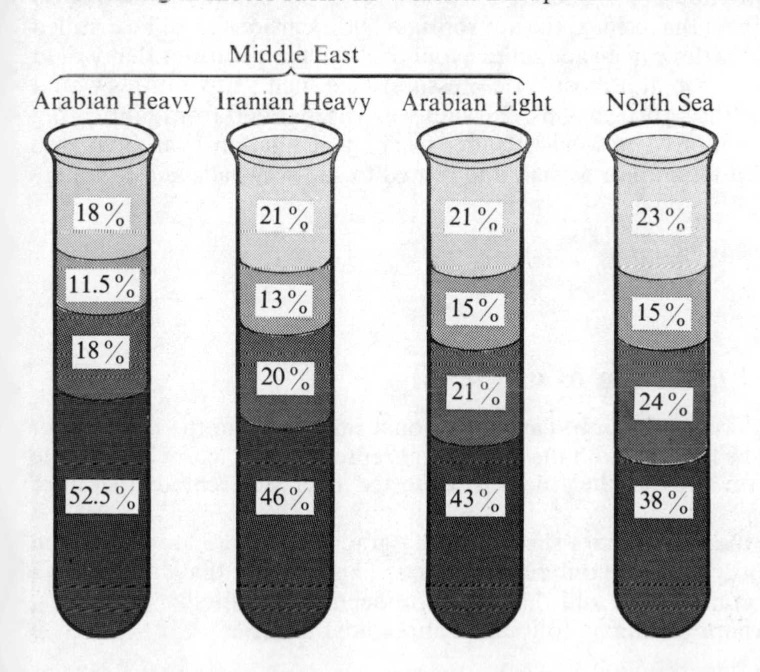


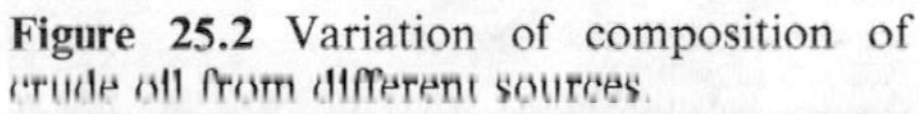

**Figure 25.2** Variation of composition of crude oil from different sources.

Different crudes contain varying proportions of the different fractions. The output of the refinery can be arranged to suit market demands by blending different crudes and by converting heavy fractions into lighter ones by cracking.

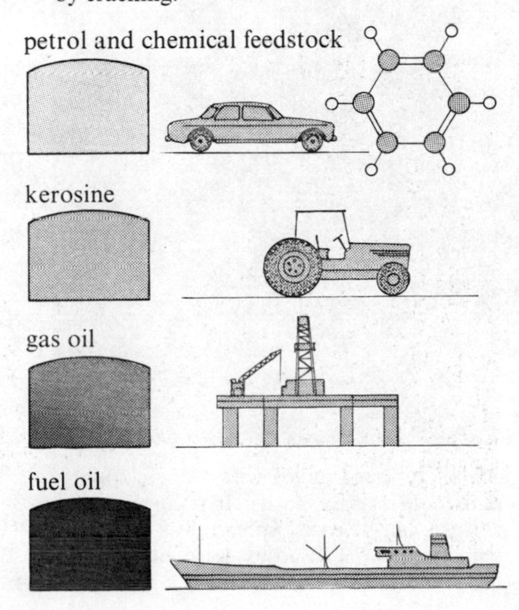

the demand for gasoline is less and a considerable proportion of this fraction is used in the manufacture of chemicals, after the cracking processes described in section 25.6. The part of the gasoline fraction used to produce chemicals is called naphtha.

**Kerosine** (10–15 %) consists mainly of $C_{11}$ and $C_{12}$ hydrocarbons, with boiling points from 160°C–250°C. It is used as a fuel in jet engines and for domestic heating purposes. It can also be cracked to produce extra gasoline (see section 25.6).

*Right* The major use of kerosine is as jet fuel.

**Diesel oil** or **gas oil** (15–20 %) containing $C_{13}$–$C_{25}$ compounds, boils in the range 220°C–350°C. Its use in diesel engines, where the fuel is ignited by compression instead of by a spark, is well known. Diesel oil is also used in furnaces for industrial heating purposes, and like kerosine can be cracked to produce extra gasoline.

**Residue** (40–50 %) The residual oil from the primary distillation boils above 350°C and is a highly complex mixture of involatile hydrocarbons. Most of it is used as fuel oil in large furnaces such as those in power stations or big ships. A proportion of it, however, is used to make **lubricating oils** and **waxes**. Both these materials contain $C_{26}$–$C_{28}$ hydrocarbons; when pure these hydrocarbons are solid, but lubricating oil contains a complex mixture, each member of which depresses the melting point of the others so that the mixture is a liquid. In paraffin wax the components are similar enough in structure to form a solid. To obtain lubricating oil and paraffin wax from the residue, the appropriate hydrocarbons must be distilled off. The distillation is done in a vacuum to avoid the high temperatures that would be needed for distillation at atmospheric pressure since such temperatures would tend to decompose the hydrocarbons. Paraffin wax is separated from lubricating oil by solvent extraction. The solid left after vacuum distillation is an involatile tarry material called **bitumen** or **asphalt** and is used to surface roads and to waterproof materials.

*Above* Bitumen mixed with stone chippings is used to surface roads. It is applied hot, when it can be easily spread, but at normal temperatures its viscosity is so high that it hardly flows at all.

## 25.2 The composition of crude oil

The distillation of crude oil can be carried out on a small scale in the laboratory: you will probably be familiar with the experiment represented in figure 25.3. Table 25.1 gives the properties of the fractions collected over different temperature ranges.

Notice that all these properties show a steady gradation: there are no sudden changes in the properties of the different fractions. This suggests that crude oil is a mixture of many components and that the components have similar properties, many of them perhaps belonging to the same homologous series.

**Table 25.1** Properties of fractions obtained in laboratory distillation of crude oil.

| Property | Room temp. to 70°C | 70°C–120°C | 120°C–170°C | 170°C–220°C |
|---|---|---|---|---|
| **Colour** | pale yellow | yellow | dark yellow | brown |
| **Viscosity** | runny | fairly runny | fairly viscous | viscous |
| **Behaviour when ignited** | burns readily: clean yellow flame | quite easily ignited: yellow flame, some smoke | harder to ignite: quite smoky flame | hard to ignite: smoky flame |

**Figure 25.3** The laboratory distillation of crude oil. Fractions are collected over several different boiling ranges.

Hydrocarbons are the main and the most important components of crude oil. Three different homologous series of hydrocarbons are present: **alkanes**, which will be examined in more detail later in this chapter, **cycloalkanes**, considered in section 25.7, and **aromatics** (compounds containing benzene-type rings), considered in chapter 27. The actual proportions of the different classes present depend on the source of the oil, but alkanes and cycloalkanes form the large majority, with aromatics making only about 10% of the hydrocarbon total.

Hydrocarbons make up the more volatile fractions of crude oil, the increasing boiling ranges of the fractions corresponding to the increasing size of the molecules, as indicated in the last section. The involatile tarry residue left after vacuum distillation consists of compounds with large molecules called resins and asphalts. These contain mainly carbon and hydrogen, with some oxygen and sulphur.

*Left* The pouring characteristics of different petroleum fractions. From left to right: gasoline, diesel oil, fuel oil.

Look again at table 25.1.

- ○ Why are the first fractions easiest to ignite?
- ○ Why do the higher fractions burn with the smokiest flames?
- ○ Suggest a reason why the fractions become increasingly viscous.

We will now consider in detail the largest group of compounds present in crude oil, the alkanes.

## 25.3 Naming alkanes

The alkanes are **saturated hydrocarbons**. 'Saturated' means that they contain the maximum amount of hydrogen possible, with no double or triple bonds between carbon atoms. The general formula of the alkanes is $\mathbf{C_nH_{2n+2}}$. As we have seen in chapter 24, it is possible to have alkanes with straight or with branched chains, for example

```
CH3—CH2—CH2—CH2—CH3        straight-chain
CH3—CH2—CH—CH3             branched-chain
        |
        CH3
```

**Table 25.2** Straight-chain alkanes.

| Formula | Name |
|---|---|
| $CH_4$ | methane |
| $CH_3CH_3$ | ethane |
| $CH_3CH_2CH_3$ | propane |
| $CH_3CH_2CH_2CH_3$ | butane |
| $CH_3(CH_2)_3CH_3$ | pentane |
| $CH_3(CH_2)_4CH_3$ | hexane |
| $CH_3(CH_2)_8CH_3$ | decane |
| $CH_3(CH_2)_{18}CH_3$ | eicosane |

Table 25.2 gives the names of some simple straight-chain alkanes.

It will be seen that after the first four alkanes the name is formed by adding the suffix **-ane** to the Greek root (e.g. *pent-* five, *hex-* six) indicating the number of carbon atoms in the molecule. The first part of the name indicates the number of carbon atoms, the ending -ane indicates that it is an alkane. All compounds based on alkane skeletons are named by this method, with a stem to indicate the number of carbon atoms and a suffix to indicate the functional group.

Branched-chain alkanes are named by considering them as straight-chain alkanes with side groups attached. For example:

```
CH3—CH2—CH—CH3
        |
        CH3
```

is regarded as butane with $CH_3$— attached to the *second* atom. It is therefore called 2-methylbutane. The name methyl indicates the $CH_3$— group, which is just methane with a hydrogen atom removed so it can be attached to another atom. Similarly $CH_3CH_2$— is called ethyl, $CH_3CH_2CH_2$— propyl, and so on. Side groups of this kind are called **alkyl groups**; their general formula is $\mathbf{C_nH_{2n+1}}$. The symbol R is often used to represent a general alkyl group. Another example of a branched chain alkane is:

```
             CH3
             |
CH3—CH2—CH—CH—CH3
        |
        CH2—CH3
```

This molecule can be regarded as a five carbon chain with a methyl and an ethyl group attached to the second and third carbon atoms respectively. It is therefore called 3-ethyl-2-methylpentane. Note that if the carbon atoms were numbered from the left, the name would be 3-ethyl-4-methylpentane. This name is not used because the convention is to use the name that includes the lowest numbers. Note also that for each side group the name includes a number showing the carbon atom to which the group is attached. Thus

```
    CH3
    |
CH3—C—CH2—CH3
    |
    CH3
```

is called 2,2-dimethylbutane, and

```
    CH3
    |
CH3—C—CH—CH2—CH2—CH2—CH3
    | |
  CH3 CH2—CH3
```

is called 2,2-dimethyl-3-ethylheptane.

So the procedure for naming an alkane, given its structural formula, is:

1 Look for the longest unbranched chain in the molecule.
2 Look for the side groups attached to the main chain and the numbers of the carbon atoms to which they are attached.
3 The name then consists of the name of the longest unbranched chain, pre-

fixed by the names of the side groups and the numbers of the carbon atoms to which they are attached.

Name the following alkanes:

$$CH_3{-}\underset{CH_3}{CH}{-}\underset{CH_3}{CH}{-}\overset{CH_3}{\underset{CH_3}{C}}{-}CH_3 \qquad CH_3{-}\overset{CH_3}{\underset{CH_3}{C}}{-}CH_3 \qquad CH_3(CH_2)_6CH_3$$

Write the formulae of the following alkanes:

2,2,4-trimethylhexane, methylpropane, 5-ethyldecane.

The system of nomenclature described above is called the IUPAC system (IUPAC stands for International Union of Pure and Applied Chemistry) and can be extended to apply to all organic compounds. Using this system, it is possible to write the structural formula of any compound from its name. The IUPAC system is gradually replacing the older, less systematic way of naming organic compounds. IUPAC names will be used wherever possible in this book. The only disadvantage of the system is that for complicated molecules the name becomes very cumbersome; in such cases systematic nomenclature is sensibly abandoned and a less informative but more easily spoken name is used. Thus the name 2,3,4,5,6-pentahydroxyhexanal is dropped in favour of 'glucose'.

## 25.4 Physical properties of alkanes

The alkanes form a homologous series and as with all homologous series the members show a gradual change in physical properties as the number of carbon atoms in their molecules increases. This has already been seen in the distillation of crude oil, which is after all largely composed of alkanes. Successive fractions from crude oil have successively higher boiling ranges and steadily increasing viscosity. The fractions range from gas (refinery gas), through liquids (gasoline, kerosine) to solids (paraffin wax, bitumen).

Some properties of individual straight-chain alkanes are shown in table 25.3.

**Table 25.3** Physical properties of straight-chain alkanes.

| Number of carbon atoms | Formula | Name | State (at 298 K) | Boiling point/K | Melting point/K | Density /g cm$^{-3}$ |
|---|---|---|---|---|---|---|
| 1 | $CH_4$ | methane | g | 112 | 90 | 0.424 |
| 2 | $C_2H_6$ | ethane | g | 184 | 101 | 0.546 |
| 3 | $C_3H_8$ | propane | g | 231 | 85 | 0.501 |
| 4 | $C_4H_{10}$ | butane | g | 273 | 138 | 0.579 |
| 5 | $C_5H_{12}$ | pentane | l | 309 | 143 | 0.626 |
| 6 | $C_6H_{14}$ | hexane | l | 342 | 178 | 0.657 |
| 7 | $C_7H_{16}$ | heptane | l | 371 | 182 | 0.684 |
| 8 | $C_8H_{18}$ | octane | l | 399 | 216 | 0.703 |
| 9 | $C_9H_{20}$ | nonane | l | 424 | 219 | 0.718 |
| 10 | $C_{10}H_{22}$ | decane | l | 447 | 243 | 0.730 |
| 11 | $C_{11}H_{24}$ | undecane | l | 469 | 247 | 0.740 |
| 12 | $C_{12}H_{26}$ | dodecane | l | 489 | 263 | 0.749 |
| 15 | $C_{15}H_{32}$ | pentadecane | l | 544 | 283 | 0.769 |
| 20 | $C_{20}H_{42}$ | eicosane | s | 617 | 310 | 0.785 |

Notice that these properties show a gradual steady change as the number of carbon atoms in the molecules increases. Figure 25.4 shows graphically the smooth increase of boiling point with increasing number of carbon atoms.

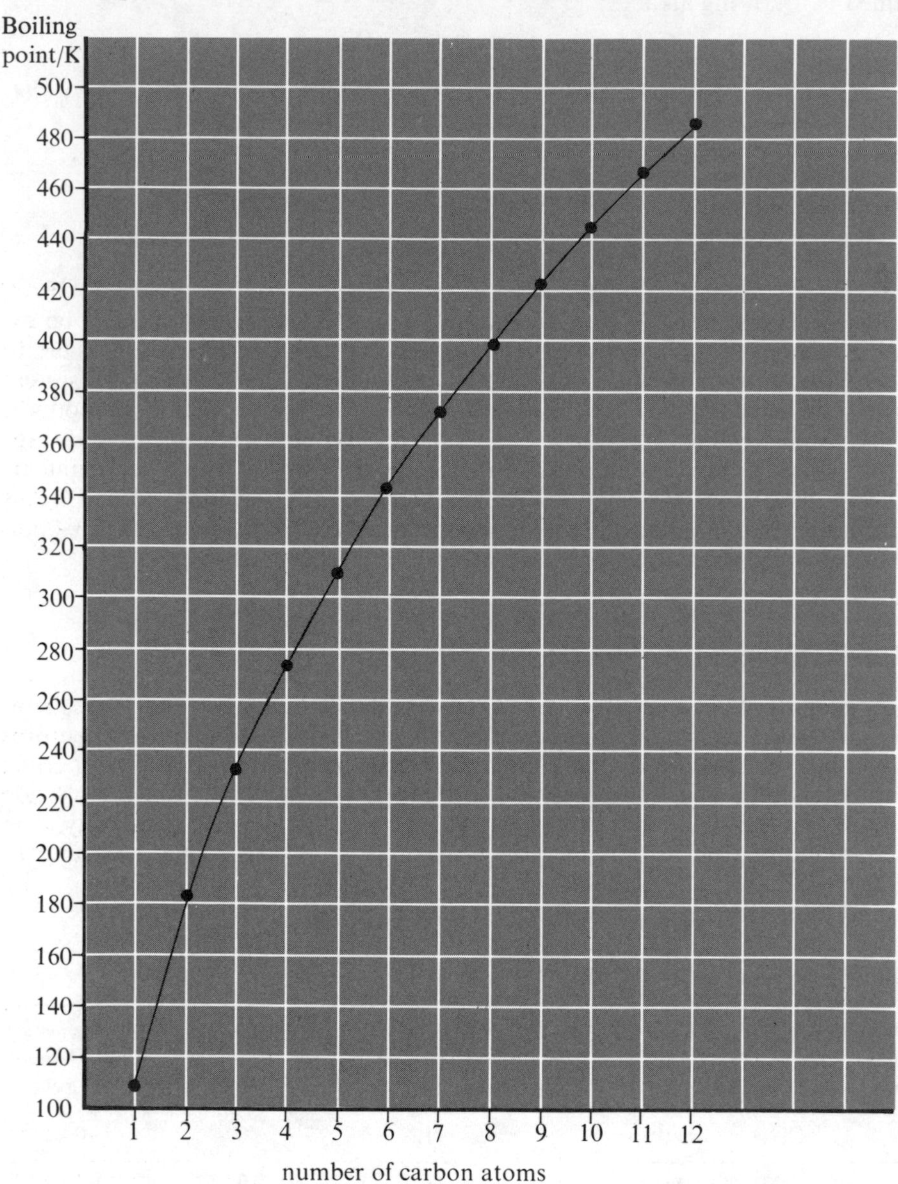


**Figure 25.4** Variation of boiling point of straight-chain alkanes with number of carbon atoms.

A steady variation in physical properties is characteristic of all homologous series, and makes it possible to predict the properties of a compound from the properties of its homologues.

Use the data in table 25.3 to predict:

- ○ The density of tridecane ($C_{13}H_{28}$),
- ○ The boiling point of tetradecane ($C_{14}H_{30}$),
- ○ The formula of the first alkane to be solid at 20°C.

As table 25.3 shows, the first four straight-chain alkanes ($C_1$–$C_4$) are gases. The next twelve ($C_5$–$C_{16}$) are liquids, and the remainder are solids. Solid alkanes of high molecular mass are similar in nature to polythene, which is effectively composed of long-chain alkane molecules. Alkanes are all less dense than water and therefore float on it. Alkanes are colourless when pure. The viscosity of liquid alkanes increases with increasing molecular mass.

Grebes soaked in oil following a blow out at an offshore oil rig. Hydrocarbons are less dense than water and therefore float on it: hence the danger to beaches and wildlife from leaking tankers and oil wells.

The steady change in physical properties of alkanes is directly related to the effect of steadily increasing molecular size. The increase in boiling point can be attributed to the increasing forces of attraction between molecules of increasing size (see section 8.4). The higher viscosity of the higher alkanes is due to the tendency of the long molecules to become 'tangled up' with one another.

The trends in physical properties considered so far have been those among the straight-chain alkanes. Branched-chain alkanes do not show the same steady gradation of properties as straight-chain alkanes: the variation in molecular structure is too great for trends to be clear. On the whole, the effect of branching is to increase volatility and reduce density. Thus the boiling point of pentane is 309 K while that of its highly branched isomer 2,2-dimethylpropane is 283 K.

## 25.5 Reactivity of alkanes

The effects of various reagents on hexane, a typical alkane, are shown in table 25.4.

**Table 25.4** The effect of common reagents on hexane.

| Reagent | Effect |
|---|---|
| air | no effect cold. Burns when heated |
| sodium hydroxide solution | no effect hot or cold |
| concentrated sulphuric(VI) acid | no effect hot or cold |
| potassium manganate(VII) (potassium permanganate) solution | no effect hot or cold |
| bromine | no effect in dark. Bromine slowly decolorized in sunlight |

Clearly hexane is unreactive, being unaffected by acids, alkalis, dehydrating agents or aqueous oxidizing agents. A close look at the substances in the table with which hexane *does* react, namely bromine and oxygen, shows two things. First, neither of these reagents has any centre of electrical charge in their molecules: they are non-polar. This is in contrast with the other substances in the table, which are all polar reagents. Second, before reaction can occur, energy must be supplied: heat in the case of oxygen, light in the case of bromine.

The failure of ions or polar molecules to have any effect on alkane molecules is due to the non-polarity of the C–H bond. Carbon and hydrogen are very close in electronegativity, so the electron pair in the covalent bond between carbon and hydrogen is fairly evenly shared: consequently there is little polarity in the C–H bond. The electron pair in the C–C bond is of course also evenly shared, so this bond is non-polar too. Thus there are no polar bonds in alkane molecules, and so no centres of electrical charge to attract normally reactive species like $H^+$, $OH^-$ and $MnO_4^-$.

Alkanes, then, are unreactive towards polar or ionic reagents, but will react with non-polar substances like oxygen or bromine. The question is how. Some mechanism other than the simple attraction between two polar groups must be involved.

### FREE RADICAL REACTIONS

We have seen that hexane and bromine react in sunlight. A similar reaction occurs between methane and chlorine and this one will be considered as it affords a simple example. Methane and chlorine do not react at all in the dark, but in sunlight an explosive reaction occurs, forming chloromethane and hydrogen chloride:

$$CH_4(g) + Cl_2(g) \xrightarrow{\text{light}} CH_3Cl(g) + HCl(g) \qquad \Delta H = -98\ \text{kJ}$$

This sort of reaction, in which an atom or group of atoms in a molecule is replaced, is called a **substitution** reaction. Like all reactions between covalent molecules, it involves breaking some bonds, for which energy must be supplied, and making new bonds, when energy is released.

Breakage, or fission, of covalent bonds can occur in two ways. Both the electrons in the bonding pair can be gained by the same atom; this atom will then possess a net negative charge, while the other atom in the bond will acquire a positive charge:

$$X:Y \longrightarrow [X:]^- + Y^+$$

This type of bond fission, producing two oppositely charged ions, is called **heterolytic fission**. Alternatively, the pair of electrons in the covalent bond may split up, one electron going to each of the atoms involved in the bond:

$$X:Y \longrightarrow X\cdot + Y\cdot$$

$X\cdot$ and $Y\cdot$ possess unpaired electrons; they are therefore very reactive, tending to attract an electron from another atom or molecule, reforming an electron pair. An atom or group of atoms possessing an unpaired electron is called a **free radical**. This kind of bond fission, in which each atom retains one electron from the pair, is called **homolytic fission**.

- ○ Which type of bond fission would you expect for (*a*) a bond between identical atoms, (*b*) a bond between atoms whose electronegativities differ widely, (*c*) a bond between atoms whose electronegativities are similar?
- ○ Which type of bond fission would you expect in (*a*) the Cl–Cl bond, (*b*) the C–H bond?

The reaction between methane and chlorine does not proceed in the dark at room temperature because there is insufficient energy available to start the reaction by breaking the necessary bonds. Splitting $Cl_2$ into two chlorine atoms requires 242 kJ mole$^{-1}$, while the process

$$CH_4 \longrightarrow CH_3\cdot + H\cdot$$

requires 435 kJ mole$^{-1}$.

The Cl–Cl bond is thus the weaker of the two, and when light is shone on the reaction mixture, chlorine molecules are supplied with the energy necessary to split them into atoms. This stage is called **initiation**. These chlorine atoms, being free radicals are highly reactive, and when they collide with a methane molecule they combine with one of its hydrogen atoms, forming a new free radical:

$$CH_4 + Cl\cdot \longrightarrow CH_3\cdot + HCl \qquad \Delta H = +4 \text{ kJ}$$

The $CH_3\cdot$ free radical then reacts with another chlorine molecule:

$$CH_3\cdot + Cl_2 \longrightarrow CH_3Cl + Cl\cdot \qquad \Delta H = -97 \text{ kJ}$$

and so the process continues. These two reactions enable a chain reaction to occur: they are **propagation** steps. Note that each propagation step involves the breakage and the formation of a bond: the net energy change is therefore relatively small. The reaction chain ends when two free radicals collide and combine; this is called **termination** and is highly exothermic:

$$Cl\cdot + Cl\cdot \longrightarrow Cl_2; \qquad \Delta H = -242 \text{ kJ}$$

$$CH_3\cdot + Cl\cdot \longrightarrow CH_3Cl; \qquad \Delta H = -339 \text{ kJ}$$

$$CH_3\cdot + CH_3\cdot \longrightarrow C_2H_6; \qquad \Delta H = -346 \text{ kJ}$$

Each chain may go through 100 to 10 000 cycles before termination occurs. The processes are extremely rapid, hence the explosive nature of the reaction. The net result of such reactions is the formation of large amounts of $CH_3Cl$ and HCl and small amounts of $C_2H_6$. (Further substitution may occur, forming $CH_2Cl_2$, $CHCl_3$ and $CCl_4$, see the next section.) The overall energy change of 93 kJ mole$^{-1}$ evolved represents the difference between the energy released in forming C–Cl and H–Cl bonds and the energy absorbed in breaking C–H and Cl–Cl bonds. The activation energy is supplied by the light which initiates the reaction.

Practically all the reactions of alkanes proceed by free-radical mechanisms, characterized by high activation energies and a tendency to proceed rapidly in the gas phase. Some of these reactions will now be considered.

## 25.6 Important reactions of alkanes

Alkanes show little reactivity towards all the common polar and ionic reagents. Indeed the alkanes were once known as the **paraffins**, from the Latin words *parum* (little) and *affinitas* (affinity). Hence there are only a few reactions of alkanes, but these are of great importance.

BURNING

Although alkanes are kinetically stable in the presence of oxygen, they are energetically unstable with respect to their oxidation products. Therefore when the necessary activation energy is supplied, as when they are ignited, combustion occurs. The reaction, which has a free-radical mechanism, occurs rapidly in the gas phase. Because it is a gas-phase reaction, liquid and solid alkanes must first be vaporized and hence less volatile alkanes burn less readily. The combustion products are carbon dioxide and water if the oxygen supply is plentiful, for example:

$$C_7H_{16}(g) + 11O_2(g) \longrightarrow 7CO_2(g) + 8H_2O(l) \qquad \Delta H = -485 \text{ kJ}$$

If the oxygen supply is limited, the products may include carbon monoxide and carbon.

- ○ Write an equation for the complete combustion of octane, $C_8H_{18}$.
- ○ Estimate the heat of combustion of octane given that the heats of combustion of hexane and heptane are $-4195$ kJ mole$^{-1}$ and $-4854$ kJ mole$^{-1}$ respectively.
- ○ Write an equation for the combustion of octane to form carbon monoxide and water.

The combustion of alkanes is a gas-phase reaction. The wick of a candle provides a surface from which the molten wax vaporizes.

The combustion of hydrocarbons in general and alkanes in particular is of immense importance, for it occurs in power stations, furnaces, domestic heaters, candles, gas heaters, internal combustion engines and many other devices essential to a technological society. An understanding of the nature of the combustion process is indispensable to the design of such devices.

An important example is the **petrol engine**. In the cylinder of a motor-car engine a mixture of gasoline vapour (which contains mostly $C_5$ to $C_{10}$ alkanes) and air is ignited by an electric spark, producing an explosive reaction which drives the piston down. The rate at which this reaction occurs and the ease with which it is initiated are very important for the efficiency of the engine. If the explosion occurs too rapidly, heat will be dissipated instead of being converted to useful kinetic energy. If the explosion starts too early the pistons are subjected to harmful jarring and 'knocking' or 'pinking' of the engine occurs. This is more likely to occur in the engines of high-performance cars, where high compression-ratios tend to cause premature ignition. Gasoline mixtures that are rich in straight-chain alkanes such as heptane ignite very readily and explode rapidly, causing 'knocking' and inefficient combustion. The combustion of branched-chain alkanes such as 2,2,4-trimethylpentane (iso-octane) is much smoother and more controlled, so gasoline mixtures rich in branched-chain alkanes are more efficient fuels and less likely to cause knocking. The **octane rating** of 2,2,4-trimethylpentane is set at 100, and that of heptane at 0. Thus gasoline mixtures rich in branched-chain alkanes burn smoothly and efficiently in high-performance engines and have high octane numbers.

There are two ways of meeting the demand of modern high-performance engines for fuels with high octane numbers. One is to produce artificial gasoline mixtures that are rich in branched-chain alkanes (see below). The other is to add an anti-knock compound to gasoline. The anti-knock compound normally used is tetraethyllead(IV), $Pb(CH_3CH_2)_4$. When burned, this compound produces small particles of lead oxide which tend to combine with free radicals produced in the chain reaction of combustion, thus slowing it down and making it smoother. To prevent lead accumulating in the engine, 1,2-dibromoethane ($CH_2BrCH_2Br$) is also added to the gasoline. This results in the formation of lead bromide, which is volatile and is swept away in the car exhaust. The unfortunate effect of this is to add lead to the atmosphere, whence it can be inhaled. At present there is considerable concern that leaded petrol may be increasing the concentration of lead in the blood of city-dwellers to near the point where symptoms of toxicity such as brain damage may occur.

Lead is not the only pollutant produced by petrol engines. Table 25.5 gives the composition of typical petrol engine exhaust.

*Above* Car exhaust is a major source of pollution in modern cities.

**Table 25.5** Gases in petrol engine exhaust.

| Gas | Percentage by volume |
|---|---|
| carbon dioxide | 9 |
| oxygen | 4 |
| hydrogen | 2 |
| carbon monoxide | 4–9 |
| hydrocarbons | up to 0.2 |
| aldehydes | 0.004 |
| nitrogen oxides | 0.05–0.4 |
| sulphur dioxide | 0.006 |
| ammonia | 0.0006 |
| [lead (as solids) | 4 mg m$^{-3}$] |

*Above* A catalytic converter for fitting to car exhaust systems.

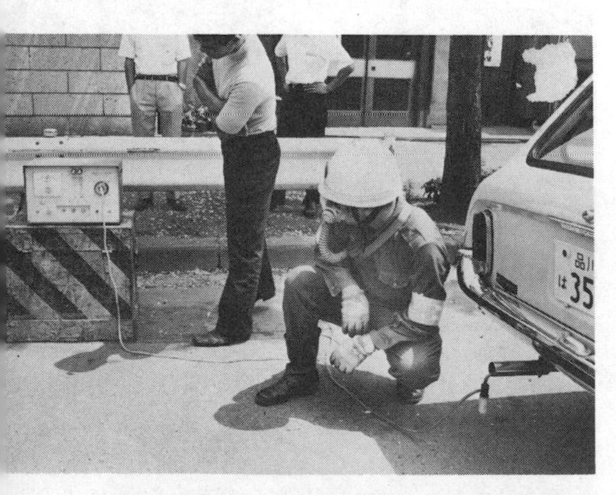

*Above* Tokyo police carry out spot checks on car exhausts in an attempt to control pollution. Owners of cars whose exhausts contain more than 5.5% carbon monoxide are ordered to tune their engines.

- ○ What single gas omitted from table 25.5 makes up the remainder of car exhaust?
- ○ Why does exhaust contain carbon monoxide?
- ○ Why does exhaust contain oxides of nitrogen?
- ○ Why does exhaust contain hydrocarbons?

The most dangerous pollutants in car exhaust, apart from lead, are probably carbon monoxide, hydrocarbons and oxides of nitrogen. Carbon monoxide is very toxic because it forms a stable compound with haemoglobin in the blood, making the haemoglobin unable to transport oxygen. In a confined space the carbon monoxide in exhaust gases can be fatal; in the open the danger is less, though the carbon monoxide concentration in a busy street is undesirably high. It is worth noting, though, that the carbon monoxide level in a railway compartment full of smokers is twice as high as that in a busy London street.

Oxides of nitrogen are very toxic, but their concentration in car exhausts is probably not high enough to make them as hazardous as the carbon monoxide. The worst effect of oxides of nitrogen and of unburnt hydrocarbons in exhaust is not seen in Britain. It is limited mainly to the city of Los Angeles, where these substances take part in the complex sequence of chemical reactions that lead to the formation of the notorious photochemical smog. The seven million inhabitants of Los Angeles own four million cars which between them consume thirty million dm$^3$ of petrol per day. The exhaust gases produced in this automobile-dominated city combine with particular geographical conditions to produce a unique smog of especially unpleasant character. It is not surprising that the Californian authorities have introduced legislation designed to reduce the level of pollutants in exhaust gases. One way of doing this is to install a catalyst unit in the exhaust system which converts the carbon monoxide, nitrogen oxides and hydrocarbons to harmless products. A problem associated with the use of such catalytic converters is that the catalysts are poisoned by lead.

The pollution caused by petrol engines is considerable even in Britain, where smog has tended to be caused by the burning of coal rather than petrol. Elsewhere oil is responsible for many other forms of pollution, such as the smoke from diesel engines and spillages from oil tankers. As far as air pollution is concerned, though, in Britain coal is still at least as much of a nuisance as oil. And as far as danger to health is concerned, neither of these sources can compare with the hazards caused by tobacco smoke.

### REACTION WITH HALOGENS

The reaction of methane with chlorine has already been considered in some detail. This reaction produces several products:

$$CH_4 + Cl_2 \longrightarrow \underset{\text{chloromethane}}{CH_3Cl} + HCl$$

$$CH_3Cl + Cl_2 \longrightarrow \underset{\text{dichloromethane}}{CH_2Cl_2} + HCl$$

$$CH_2Cl_2 + Cl_2 \longrightarrow \underset{\text{trichloromethane (chloroform)}}{CHCl_3} + HCl$$

$$CHCl_3 + Cl_2 \longrightarrow \underset{\text{tetrachloromethane (carbon tetrachloride)}}{CCl_4} + HCl$$

The proportions of different products formed depend on the proportions of chlorine and methane used. All four products are useful in industry, but their separation from one another involves distillation. Chlorination of alkanes is therefore of limited use for preparing chloroalkanes, particularly in the case of the higher alkanes where the number of possible products is very large.

Similar reactions occur between other alkanes and fluorine, chlorine and bromine. The reaction with fluorine occurs in the absence of sunlight and is very vigorous.

### CRACKING

When alkanes are heated to high temperatures their molecules vibrate strongly enough to break bonds and form smaller molecules, one of which is an alkene. For example:

$$\underset{\text{undecane}}{C_{11}H_{24}} \longrightarrow \underset{\text{nonane}}{C_9H_{20}} + \underset{\text{ethene}}{H_2C{=}CH_2}$$

Such a reaction is known as **cracking**.

- ○ In which petroleum fraction would $C_{11}H_{24}$ be found?
- ○ In which fraction would $C_9H_{20}$ be found?
- ○ Why must one of the products of cracking always be an alkene?

By using a catalyst, cracking can be made to occur at fairly low temperatures: this is known as catalytic cracking.

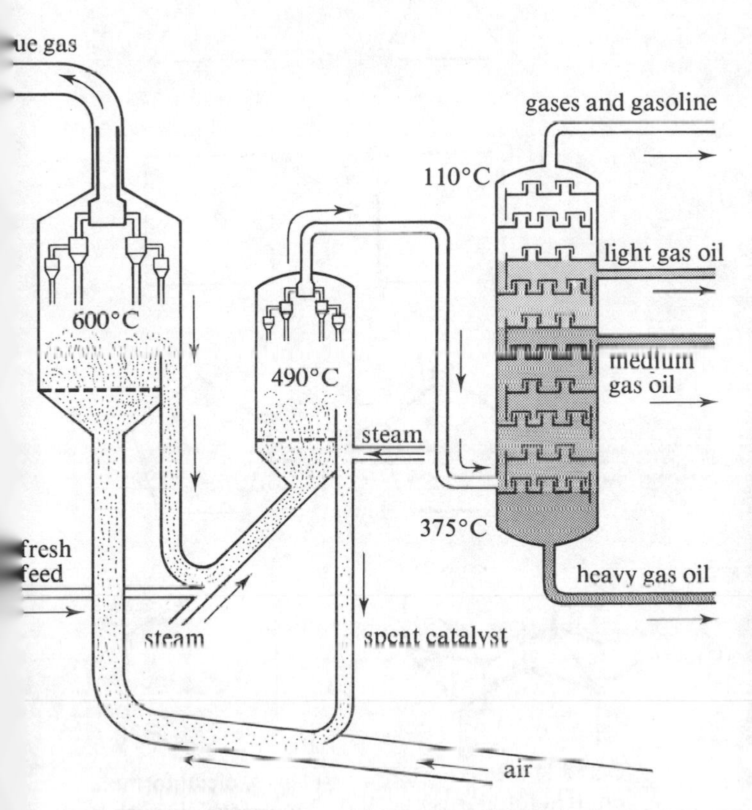

*Above* A catalytic cracking unit. The workings of the plant shown in the photograph are explained in the corresponding diagram. The reaction actually occurs in the central unit: the catalyst is regenerated in the left hand unit and the products are separated in the distillation tower on the right.

Cracking is very important in the petroleum industry. It is used:

*(a) To provide extra gasoline*

The example above shows how undecane, a member of the kerosine fraction, can be cracked to produce nonane, a component of gasoline. Thus heavier fractions can be cracked to produce extra gasoline. Furthermore, cracking tends to produce branched-chain rather than straight-chain alkanes, and therefore gasoline of a high octane rating. A process similar to cracking, in which straight-chain molecules are effectively broken up and then reassembled as branched-chain molecules and aromatic rings, can be used to convert low-grade gasoline to high-grade fuel. This process is called **reforming**.

*(b) As a source of alkenes*

Because they are so unreactive, alkanes are not a good starting point from which to make the many organic chemicals derived from crude oil. Alkenes, with their reactive double bonds, are a more suitable starting point. The petrochemical industry uses vast quantities of ethene and propene as units for building larger organic molecules. We have seen that cracking reactions always produce alkenes, and under the right conditions large yields of ethene and propene can be obtained. Alkenes are thus produced as by-products from cracking heavy fractions (such as $C_{11}H_{24}$) and by cracking some of the gasoline fraction (which is called naphtha when used as a source of alkenes). In countries such as the U.S.A. where gasoline is in high demand, alkenes are made by cracking $C_2$ to $C_4$ gases.

The reactions of alkenes and the products obtained from them are discussed in the next chapter.

Which pump dispenses a higher proportion of straight chain alkanes?

## 25.7 Cycloalkanes

The alkanes considered so far have all had open-chain molecules, that is molecules which come to an end at some point. It is also possible for alkane molecules to form rings: such compounds are named by using the prefix **cyclo-**. Some cycloalkanes are shown in figure 25.5.

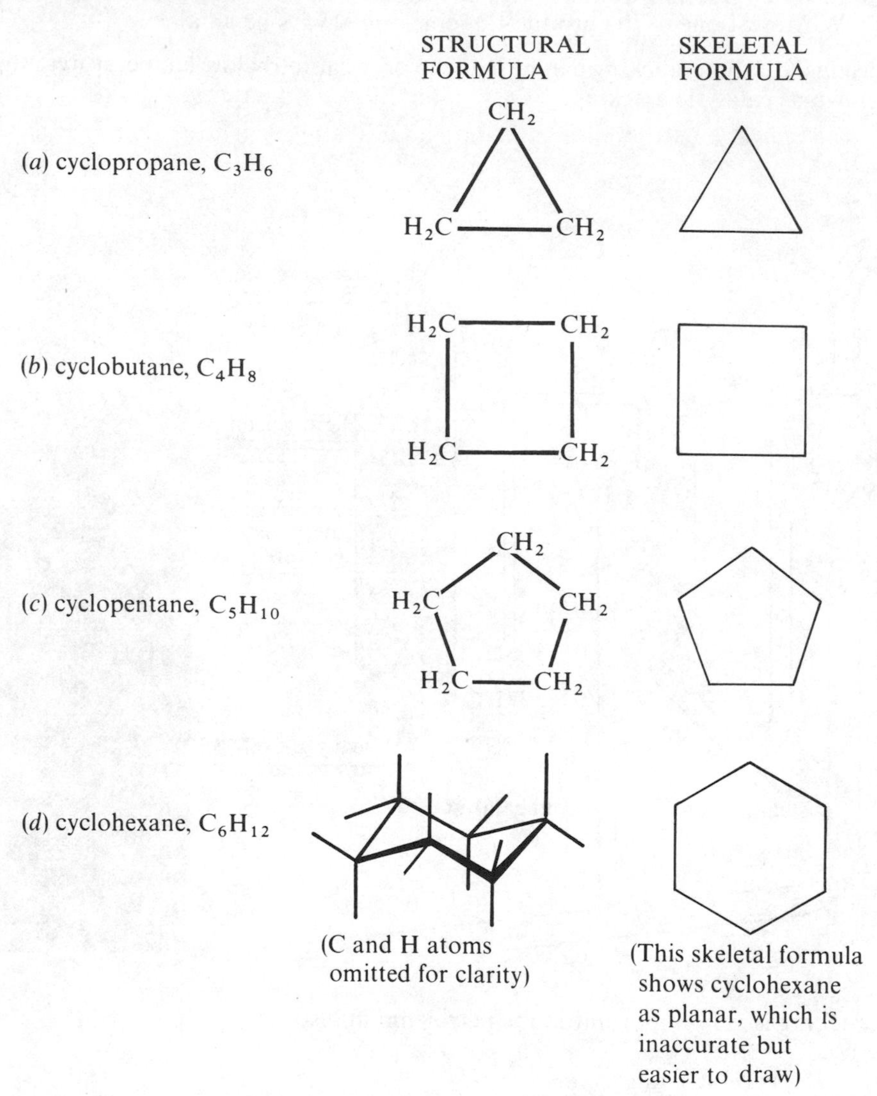


**Figure 25.5** Some cycloalkanes.

Cyclopropane, cyclobutane and cyclopentane all have planar molecules. The natural bond angles around a saturated carbon atom are 109° (the angle at the centre of a tetrahedron). In cyclopropane the bond angle is only 60°, since the carbon atoms within the molecule form an equilateral triangle. Hence there is considerable bond strain in this molecule, so cyclopropane is unstable and reactive, tending to break open its ring. The bond angle in cyclobutane is 90°, so the bond strain is less, but nevertheless considerable: cyclobutane is more reactive than open-chain alkanes. Cyclopentane, with a bond angle of 108°, has little bond strain and is therefore stable. Cyclohexane and the higher cycloalkanes can relieve bond strain by puckering, so that their rings are no longer planar: these cycloalkanes are therefore stable and very similar in properties to open-chain alkanes. Figure 25.5(*d*) shows the puckered structure of cyclohexane, in which all the carbon atoms have an unstrained bond angle of 109°. You should try building this molecule with models to see exactly how bond strain is eliminated.

The general formula of cycloalkanes is $\mathbf{C_nH_{2n}}$, compared with $C_nH_{2n+2}$ for open-chain alkanes. Although cyclopentane, cyclohexane and higher cycloalkanes behave similarly to open-chain alkanes their melting and boiling points tend to be somewhat higher. Cycloalkanes are present in considerable quantities in crude oil. Industrially the most important is cyclohexane, used in the manufacture of nylon (section 32.9).

# Summary

1 Crude oil is a complex mixture of hydrocarbons consisting of alkanes, cycloalkanes and aromatic compounds. It is separated into useful fractions by fractional distillation.
2 Alkanes are saturated hydrocarbons with general formula $C_nH_{2n+2}$.
3 Straight-chain alkanes are named by adding the suffix *-ane* to a prefix indicating the number of carbon atoms in the molecule.
4 Branched-chain alkanes are named by considering their molecules as straight-chain alkanes with side-groups attached.
5 Alkanes show a steady gradation in physical properties with increasing molecular size.
6 Alkanes are chemically unreactive toward polar or ionic reagents, but they react with reagents such as oxygen or halogens by free-radical mechanisms.
7 Alkanes burn in a plentiful supply of oxygen to form carbon dioxide and water. They undergo substitution reactions with halogens in sunlight.
8 Alkane molecules break down to smaller molecules at high temperatures: this process is called cracking and is important in the petroleum industry.
9 Cycloalkanes are saturated hydrocarbon rings with general formula $C_nH_{2n}$.

# Study questions

1 Name the following alkanes

(a) $CH_3—CH_2—CH_2—CH_2—CH_2—CH_3$

(b) 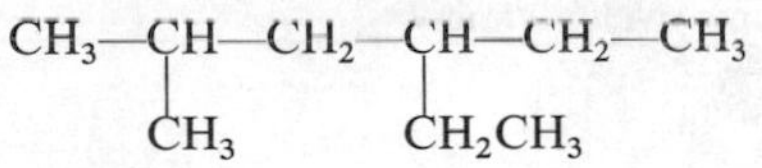 $CH_3—CH(CH_3)—CH_2—CH(CH_2CH_3)—CH_2—CH_3$

(c) 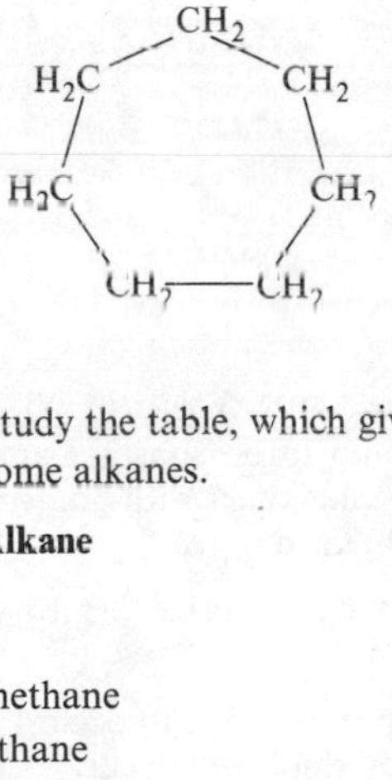 ring: $CH_2$, $H_2C$, $CH_2$, $H_2C$, $CH_2$, $CH_2—CH_2$

(d) 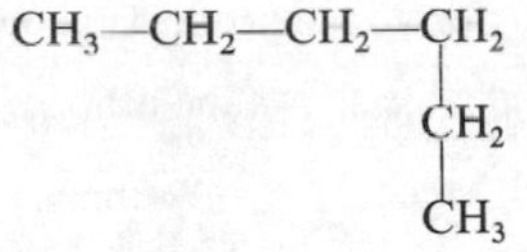 $CH_3—CH_2—CH_2—CH_2—CH_2—CH_3$ (drawn with the chain bent: $CH_3—CH_2—CH_2—CH_2$ / $CH_2$ / $CH_3$)

(e) 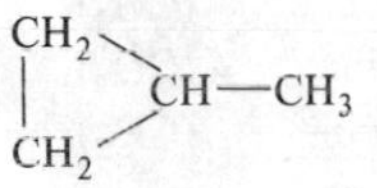 ring: $CH_2$, $CH_2$, $CH—CH_3$

(f)

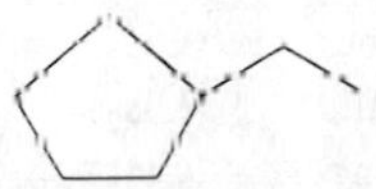

2 Study the table, which gives the heats of combustion and the relative molecular masses of some alkanes.

| **Alkane** | **No. of carbon atoms, *n*** | **Relative molecular mass** | **Heat of combustion $\Delta H_c$/kJ mole$^{-1}$** |
|---|---|---|---|
| methane | 1 | 16 | −890 |
| ethane | 2 | 30 | −1560 |
| propane | 3 | 44 | −2220 |
| butane | 4 | 58 | −2877 |
| pentane | 5 | 72 | −3509 |
| methylbutane | 5 | 72 | −3503 |
| 2,2-dimethylpropane | 5 | 72 | −3517 |
| hexane | 6 | 86 | −4195 |

(a) Plot a graph of $\Delta H_c$ against $n$ for the first six straight-chain alkanes.
(b) Does $\Delta H_c$ increase by the same amount for each extra carbon atom in an alkane chain?
(c) If so, what is the average increase in $\Delta H_c$ per carbon atom?
(d) Your answer to (c) represents the heat of combustion of which structural group?
(e) Where does your graph intercept the $\Delta H_c$ axis? What is the physical significance of this intercept?
(f) Compare the $\Delta H_c$ values for the three isomeric alkanes with five carbon atoms. Why are they so similar?
(g) Work out the heat of combustion per gram of butane, pentane and hexane respectively. Comment on your result.

3 Write (a) full structural formulae; (b) molecular formulae for the following alkanes:
(i) ethylcyclohexane,
(ii) 1,2-dimethylcyclopentane,
(iii) 2,2,3-trimethylbutane,
(iv) 2,2-dimethyl-3,4-diethylheptane.

4 For the purposes of this question, assume that petrol has a density of 0.75g $cm^{-3}$ and is a mixture of hydrocarbons all of formula $C_8H_{18}$. Assume 0.5$cm^3$ of tetraethyllead(IV) are added to every $dm^3$ (litre) of petrol. Tetraethyllead(IV), $Pb(C_2H_5)_4$, is a liquid of density 1.6g $cm^{-3}$.
(a) Why is tetraethyllead(IV) added to petrol?
(b) Write an equation for the complete combustion of petrol in excess air.
(c) How many moles of $C_8H_{18}$ are there in 1$dm^3$ of petrol?
(d) Calculate the volume of air needed to burn completely 1$dm^3$ of petrol (assume air is 20% oxygen by volume and that 1 mole of a gas occupies 24 $dm^3$ under ordinary conditions).
(e) Calculate the volume of nitrogen emitted as exhaust when 1$dm^3$ of petrol burns. (Assume all the nitrogen in the air intake passes into the exhaust.)
(f) Calculate the mass of nitrogen emitted as exhaust when 1$dm^3$ of petrol burns.
(g) Calculate the mass of carbon dioxide produced when 1$dm^3$ of petrol burns.
(h) Assuming that all the lead in the tetraethyllead(IV) passes out in the exhaust, calculate the mass of lead passed out in the exhaust when 1$dm^3$ of petrol burns.

5 This question is about octane ($C_8H_{18}$) and cyclooctane($C_8H_{16}$).
(a) Draw the structural formula of octane.
(b) Draw the structural formula of cyclooctane, showing the exact shape you would expect for the molecule.
(c) Suppose octane and cyclooctane were reacted in turn with chlorine so that only one hydrogen atom in each molecule was substituted by chlorine. Draw structural formulae for all the products you would expect in each case.

6 The table shows the heats of combustion of some cycloalkanes.

| **Name** | **Formula** | **Heat of combustion, $\Delta H_c$/kJ mole$^{-1}$** |
|---|---|---|
| cyclopropane | $(CH_2)_3$ | −2090 |
| cyclobutane | $(CH_2)_4$ | −2740 |
| cyclopentane | $(CH_2)_5$ | −3320 |
| cyclohexane | $(CH_2)_6$ | −3948 |
| cycloheptane | $(CH_2)_7$ | −4635 |
| cyclooctane | $(CH_2)_8$ | −5310 |
| cyclononane | $(CH_2)_9$ | −5980 |
| cyclodecane | $(CH_2)_{10}$ | −6630 |

(a) For each compound, calculate the heat of combustion per $CH_2$ group.
(b) Would you expect the heat of combustion per $CH_2$ group to be constant among: (i) open-chain alkanes; (ii) unstrained cycloalkanes; (iii) cycloalkanes with ring-strain?
(c) Explain your answers and comment on the figures you obtained in (a).

7 Write the structural formulae of all the products you would expect to be formed when ethane reacts with excess chlorine in sunlight.

8 For each of the following, state whether or not you would expect reaction to occur to any significant extent. Write equations for any reactions which you think will occur.
(a) Chlorine is bubbled through hexane in sunlight.
(b) Sodium metal is added to warm hexane.
(c) Hexane is boiled with acidified potassium dichromate(VI) solution.
(d) Chlorine and hexane vapour are heated in the dark.
(e) Hexane and hydrogen are heated in sunlight.

9 Methane has been suggested and used as an alternative to gasoline as a fuel for motor cars. What modifications would be necessary to an ordinary car to enable it to run on methane? What would be the advantages of using methane instead of gasoline?

10 Natural gas from the Texas Panhandle has the percentage composition by volume shown in the table.

| Component | Percentage by volume |
|---|---|
| Methane | 80.9 |
| Ethane | 6.8 |
| Propane | 2.7 |
| Butane and higher alkanes | 1.6 |
| Nitrogen | 7.9 |
| Carbon dioxide | 0.1 |

The natural gas is processed in the following way:
Carbon dioxide is first removed, then ethane, propane, butane and higher alkanes are separated off and mainly used to make alkenes, especially ethene.

(a) How might the carbon dioxide be removed from the gas?
(b) How might ethane and the higher alkanes be separated off?
(c) What would the remaining natural gas be used for?
(d) Why is it not necessary to remove the nitrogen from the gas?
(e) Why is ethene an important industrial chemical?
(f) Ethene is produced from ethane by a cracking reaction that eliminates hydrogen. Write an equation for this reaction. What conditions would be needed to make the reaction occur?
(g) In Europe ethene is manufactured from naphtha (part of the gasoline fraction from petroleum distillation) rather than ethane. Suggest a reason for this difference.

# 26 Unsaturated Hydrocarbons

## 26.1 Importance of alkenes

What do PVC raincoats and antifreeze have in common? Or polythene bottles and adhesives? Like many other things in everyday use, they are made from ethene, the simplest alkene and the most versatile organic compound in use today. In the UK, about a million tonnes of ethene are produced and consumed every year. Ethene, $CH_2{=}CH_2$, with its reactive double bond, can be used as a building block to prepare complex organic molecules. Propene, $CH_3CH{=}CH_2$, is used in a similar way, though on a smaller scale. The necessary large quantities of these alkenes are manufactured by cracking processes described in the last chapter.

## 26.2 Nomenclature of alkenes

Ethene and propene are the first two members of the homologous series of **alkenes**. All members of this series contain a double carbon–carbon bond. They therefore have two atoms of hydrogen less than the corresponding alkane and their general formula is $\mathbf{C_nH_{2n}}$. Because they contain less than the maximum amount of hydrogen they are said to be **unsaturated**.

Alkenes are named using the same general rules as those described for alkanes in the last chapter, but adding the suffix **-ene** instead of -ane together with a number indicating the position of the double bond in the chain. Thus the molecule $CH_3CH{=}CHCH_3$ is named but-2-ene. Note that although the double bond joins carbon atoms 2 and 3, the number 2 is used because this is the lower. Table 26.1 gives the formulae and names of some more alkenes.

**Table 26.1** Nomenclature of alkenes.

| Formula | Name |
|---|---|
| $CH_3CH_2CH_2CH{=}CH_2$ | pent-1-ene |
| $CH_3CH_2CH{=}CHCH_3$ | pent-2-ene |
| $CH_3C(CH_3){=}CHCH_3$ | 2-methylbut-2-ene |
| $CH_2{=}CHCH_2CH{=}CH_2$ | penta-1,4-diene |
| (cyclohexene ring) | cyclohexene |

○ Name these alkenes:

(a) $CH_2{=}CHCH_2CH_3$

(b) $CH_2{=}C(CH_3){-}CH_3$

(c) 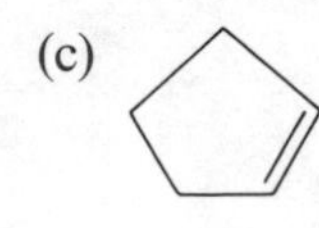

○ Write structural formulae for these alkenes:
(a) hex-2-ene
(b) buta-1,3-diene
(c) 2,3-dimethylbut-2-ene.

## 26.3 The nature of the double bond

When ethene is bubbled through bromine, the bromine is decolorized. No sunlight is needed. A colourless liquid, immiscible with water, is formed, but no hydrogen bromide is produced. This reaction is clearly different in nature to the reaction of bromine with alkanes such as ethane, in which sunlight is needed to make the reaction occur and hydrogen bromide is produced. The reaction between ethene and bromine is an **addition reaction**; one of the two carbon–carbon bonds breaks, enabling bonds to be formed to bromine:

$$H_2C=CH_2 + Br-Br \longrightarrow H-\underset{Br}{\overset{H}{C}}-\underset{Br}{\overset{H}{C}}-H$$

1,2-dibromoethane
(a colourless liquid)

Addition reactions are characteristic of all alkenes and alkenes are much more reactive than alkanes. This reactivity is perhaps a little surprising at first: we might expect a double bond to be stronger than a single one, and therefore more stable.

Double bonds are indeed stronger than single bonds, as table 26.2 shows, but the energy of a C=C bond is not twice that of a C–C bond.

**Table 26.2** Bond energies and bond lengths of C—C, C=C, and C≡C.

| | **C—C (in ethane)** | **C=C (in ethene)** | **C≡C (in ethyne)** |
|---|---|---|---|
| bond energy/kJ mole$^{-1}$ | 346 | 598 | 813 |
| bond length/nm | 0.154 | 0.134 | 0.121 |

This suggests that the two bonds in C=C may not be identical. In fact, two kinds of covalent bond are thought to be involved.

(a) A bond situated symmetrically between the two carbon atoms, formed by the overlap of two orbitals. This is called a **sigma- (σ-) bond** (see figure 26.1(*a*)).
(b) A bond formed by the 'sideways' overlap of two 2*p* orbitals. Because each *p*-orbital has two lobes, this bond has two regions, one above and one below the plane of the molecule. It is called a **pi- (π-) bond** (see figure 26.1(*b*)).

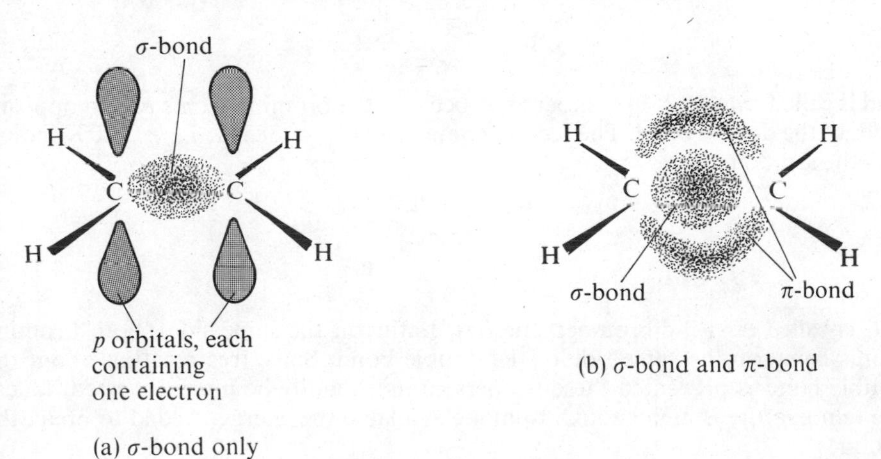


**Figure 26.1** σ- and π-bonds in ethene. Note that the plane of the molecule is perpendicular to the page.

Since the two electrons of the $\pi$-bond are not situated axially between the carbon atoms, they are not 'on average' as close to the nuclei of the atoms as the electron pair in the $\sigma$-bond. Therefore they do not attract the nuclei so strongly, so the $\pi$-bond is not as strong as the $\sigma$-bond. Nevertheless, the $\pi$- and $\sigma$-bonds together are stronger than the single $\sigma$-bond that links the carbon atoms in ethane. Consequently the carbon atoms in ethene are held together more strongly than those in ethane and the C=C bond is shorter than the C–C bond.

If double bonds are stronger than single bonds, why are alkenes more reactive than alkanes? When bromine adds across the double bond in ethene, the $\pi$-bond is broken: this requires energy. The energy used in breaking the one $\pi$-bond, however, is more than repaid by the energy released when two new bonds are made to bromine atoms. Hence alkenes are *energetically* unstable with respect to their products in an addition reaction. They are also *kinetically* unstable, because the high electron density in the double bond tends to attract electron-deficient groups (called **electrophiles**), thus initiating addition reactions. Before considering the mechanism of these reactions in detail, there is another important consequence of the double bond to consider.

## 26.4 Cis-trans isomerism

X-ray diffraction evidence shows that the ethene molecule is planar and this fits in with our ideas on shapes of molecules (section 7.7) for ethene has no lone pairs. The three groups round each carbon atom are arranged trigonally, at approximately 120° to one another, so the ethene molecule is drawn thus:

```
H       H
 \     /
  C = C
 /     \
H       H
```

Unlike ethane the ethene molecule cannot be rotated about the bond between the carbon atoms. It is possible to rotate about a $\sigma$-bond, because this does not affect the orbital overlap. With a double bond, though, rotation would involve breaking the $\pi$-bond and this requires more energy than is available at normal temperatures.

Now consider the compound 1,2-dibromoethene, BrCH=CHBr. This can be made by addition of bromine to ethyne (HC≡CH). The product is a colourless non-polar liquid, boiling point 381K, melting point 266K and immiscible with water. However, another compound is known with the same formula BrCH=CHBr. This second substance is also a colourless liquid, immiscible with water, but its boiling point is 383 K, its melting point is 220 K and its molecule is appreciably polar. These two compounds exhibit a kind of isomerism called ***cis-trans*** or **geometric** isomerism, which arises from the lack of free rotation about a double bond. The first compound ($T_b$ = 381 K, $T_m$ = 266 K, non-polar) has the structure

```
 H       Br
  \     /
   C = C
  /     \
Br       H
```

and is called *trans*-1,2-dibromoethene, because the bromine atoms are on opposite sides of the double bond. The second compound ($T_b$ = 383 K, $T_m$ = 220 K, polar) has the structure

```
 H       H
  \     /
   C = C
  /     \
Br       Br
```

and is called *cis*-1,2-dibromoethene (*cis*, Latin: on the same side), both bromine atoms being on the same side of the double bond. Since free rotation about the double bond is prevented these isomers cannot readily be interconverted, unless the temperature is high enough to make available the energy needed to break the $\pi$-bond.

*Cis-trans* isomerism is common in compounds containing double bonds. The isomers normally have similar chemical properties but often their physical properties are markedly different. *Cis-trans* isomerism is not limited to compounds with C=C double bonds: it can arise wherever rotation about a bond is restricted, for example in ring compounds.

- ○ Why is *cis*-1,2-dibromoethene polar while its *trans* isomer is not?
- ○ Suggest a reason for the large difference in melting point between the two isomers.
- ○ Would you expect but-2-ene, $CH_3CH{=}CHCH_3$, to have *cis* and *trans* isomers? If so, write their structures.
- ○ Would you expect 1,1-dibromoethene, $Br_2C{=}CH_2$, to have *cis* and *trans* isomers? If so, write their structures.
- ○ Why are H–C(H)(Br)–C(Br)(H)–H and H–C(H)(Br)–C(H)(Br)–H not considered to be isomers?

## 26.5 Mechanism of addition to a double bond

Ethene and bromine undergo an addition reaction to form 1,2-dibromoethane. The reaction occurs in the dark at room temperature, which suggests that a free-radical mechanism like that described in section 25.5 is not involved. The mechanism is believed to involve heterolytic rather than homolytic fission.

Bromine molecules and ethene molecules are both symmetrical and non-polar. As a bromine molecule approaches an ethene molecule, however, it becomes polarized by the π-electrons of the double bond, which constitute a region of high negative charge density:

$$H_2C{=}CH_2 \;\cdots\; Br{-}Br \longrightarrow H_2C{=}CH_2 \;\cdots\; Br^{\delta+}{-}Br^{\delta-}$$

A loose association is formed between the ethene and bromine molecules. Negative charge moves from the double bond towards the positively charged bromine atom. At the same time, electrons in the Br–Br bond are repelled towards the negatively charged Br atom. The net result is the formation of a C–Br bond and the production of two ions:

$$H_2C{=}CH_2 \;\cdots\; Br^{\delta+}{-}Br^{\delta-} \longrightarrow H_2\overset{\oplus}{C}{-}CH_2Br \;+\; Br^{\ominus}$$

The positively charged ion, called a **carbonium ion**, is very unstable and quickly combines with the $Br^-$ to form 1,2-dibromoethane:

$$H_2\overset{\oplus}{C}{-}CH_2Br + Br^- \longrightarrow H_2BrC{-}CBrH_2$$

This mechanism is accepted by most chemists, though the exact structure of the intermediate carbonium ion is not altogether clear. Evidence for this mechanism comes from observation of the reaction in the presence of $Cl^-$ ions, when it is found that 1-chloro-2-bromoethane, $ClCH_2—CH_2Br$, is formed as well as 1,2-dibromoethane. This suggests that the intermediate carbonium ion has indeed been formed and has reacted with $Cl^-$ as well as $Br^-$ ions.

- ○ What product would you expect from an addition reaction between ethene and hydrogen chloride molecules? Write its formula.
- ○ In the reaction between ethene and HCl, which end of the HCl molecule would you expect to attack the double bond initially?
- ○ What would be the structure of the intermediate carbonium ion in this reaction?

## 26.6. Important reactions of alkenes

Nearly all the important reactions of alkenes are addition reactions, many having mechanisms similar to that for the reaction with bromine. The product of an addition reaction can often be predicted from a knowledge of the position at which the molecule being added to the alkene will cleave into a positive and a negative portion. Predict the structure of the products of addition reactions between ethene and:

- ○ $H_2O$,
- ○ HOCl,
- ○ $H_2SO_4$

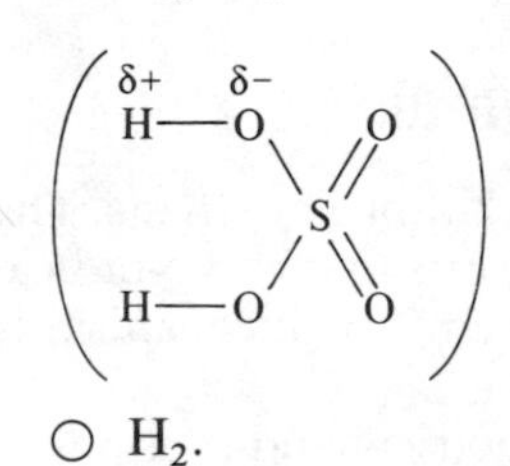


- ○ $H_2$.

You can check your answers below.

The reactions given below all involve ethene, which is industrially the most important alkene. The reactions of ethene are in any case typical of alkenes in general, for the alkenes are a well-graded homologous series. Physically, alkenes are similar to alkanes, with boiling points generally a little lower (e.g. ethane 185 K, ethene 169 K; propane 231 K, propene 225 K).

Many of these reactions can be carried out in the laboratory, though some require special conditions normally only available on an industrial scale. Industrially, of course, ethene is produced by cracking light hydrocarbon fractions; in the laboratory it can be conveniently produced by the dehydration of ethanol (see section 29.6).

### REACTION WITH HALOGENS

We have already seen that bromine reacts with ethene under normal conditions to form 1,2-dibromoethane. This compound is added to gasoline as a scavenger for lead (see section 25.6). As expected, chlorine also reacts with ethene under normal conditions:

$$CH_2{=}CH_2 + Cl_2 \longrightarrow CH_2Cl{-}CH_2Cl$$

The product, 1,2-dichloroethane, is used to manufacture chloroethene ('vinyl chloride'), from which PVC is made. Fluorine reacts explosively with ethene, but the reaction with iodine is rather slow.

Decolorization of bromine water is a useful test-tube reaction to detect a double bond.

### REACTION WITH HYDROGEN

Hydrogen and ethene do not react under normal conditions, but in the presence of a finely divided metal catalyst, usually nickel at about 140°C, ethane is produced:

$$H_2C{=}CH_2 + H_2 \xrightarrow{Ni} CH_3CH_3$$

This reaction is a useful one analytically, because by measuring the number of moles of hydrogen absorbed by one mole of a hydrocarbon, the number of double (or triple) bonds in its molecule can be established. Hydrogenation of double bonds is used to convert edible oils into margarine. Vegetable oils such as palm oil consist of esters of long chain carboxylic acids (see section 31.5) which contain double bonds. Treatment of these oils with hydrogen in the presence of a nickel catalyst saturates the carbon chains, raising the melting point of the oil so that it becomes solid that is, a fat, at room temperature. Margarine has the advantage over butter that it contains less of the fats such as cholesterol which are hazardous to health. Being made from a vegetable rather than an animal source, it is cheaper than butter. By controlling the amount of hydrogenation, the margarine can be made as soft or hard as required.

Which contains more unsaturated molecules?

- ○ Predict the formula of the product of the reaction between propene and chlorine.
- ○ Predict the formula of the product of the reaction between but-2-ene and hydrogen.

REACTION WITH HYDROGEN HALIDES

Ethene reacts with concentrated aqueous solutions of hydrogen halides in the cold:

$$CH_2{=}CH_2 + HX \longrightarrow CH_3CH_2X$$

Initial attack on the double bond is by hydrogen ions, forming the intermediate carbonium ion

$$H_2\overset{\oplus}{C}{-}CH_3$$

which then reacts with halide ions to form the product.

The reaction of hydrogen chloride with ethene produces chloroethane, which when reacted with a sodium–lead alloy produces tetraethyllead(IV), the petrol anti-knock additive.

When propene reacts with a hydrogen halide such as HCl, there are two possible products:

$$CH_3CH{=}CH_2 \xrightarrow{HCl} CH_3CH_2CH_2Cl \ \textbf{(A)}\ \text{1-chloropropane}$$

$$CH_3CH{=}CH_2 \xrightarrow{HCl} CH_3CHClCH_3 \ \textbf{(B)}\ \text{2-chloropropane}$$

**A** and **B** result from intermediate carbonium ions **C** and **D** respectively:

$$CH_3CH_2\overset{\oplus}{C}H_2 \ \textbf{(C)} \qquad CH_3\overset{\oplus}{C}HCH_3 \ \textbf{(D)}$$

When this reaction is actually carried out, it is found that much more **B** is formed than **A**. The explanation of this can be found by comparing the stability of the two carbonium ions **C** and **D**. **D** is the more stable of these two, because of a very important property of an alkyl group: *an alkyl group tends to donate electrons slightly to any carbon atom to which it is attached.*

In ion **D** there are two methyl groups donating electrons to the positively-charged carbon atom; in **C** there is only one ethyl group doing so. So the positive charge is stabilized slightly more in **D** than in **C**, because the donated electrons tend to cancel out the charge. **D** is therefore the more stable of the two possible intermediates (though still very unstable), so it tends to persist longer, making it more likely to combine with $Cl^-$ to form the product **B**. This means that **B** is the major product, though a certain amount of **A** is also formed.

- ○ When 2-methylpropene ($CH_3—\underset{\displaystyle CH_3}{\underset{|}{C}}{=}CH_2$) reacts with HCl, what are the structures of the two possible intermediate carbonium ions?
- ○ Which of these two ions is the more stable?
- ○ What will be the major product of the reaction between 2-methylpropene and HCl?

In general, *when an acidic molecule HA adds to an asymmetric alkene, the major product is the one in which the hydrogen atom attaches itself to the carbon atom already carrying the larger number of hydrogen atoms.* This rule is known as **Markovnikoff's Rule**. Thus in the example above,

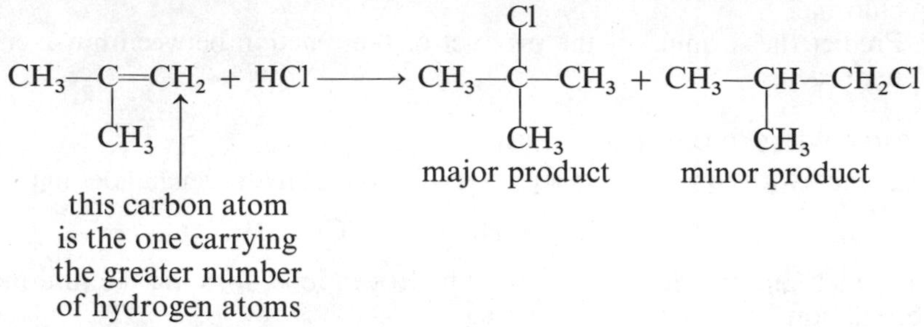

$$CH_3—\underset{\displaystyle CH_3}{\underset{|}{C}}{=}\underset{\uparrow}{CH_2} + HCl \longrightarrow \underset{\text{major product}}{CH_3—\underset{\displaystyle CH_3}{\underset{|}{\overset{\displaystyle Cl}{\overset{|}{C}}}}—CH_3} + \underset{\text{minor product}}{CH_3—\underset{\displaystyle CH_3}{\underset{|}{CH}}—CH_2Cl}$$

this carbon atom is the one carrying the greater number of hydrogen atoms

Predict the major products of reactions between

- ○ $CH_3CH_2CH{=}CH_2$ and HBr
- ○ $CH_3\underset{\displaystyle CH_3}{\underset{|}{C}}{=}CHCH_3$ and HI

REACTION WITH SULPHURIC(VI) ACID: HYDRATION

Sulphuric(VI) acid is another strong acid with which ethene undergoes addition; concentrated sulphuric(VI) acid reacts with ethene in the cold:

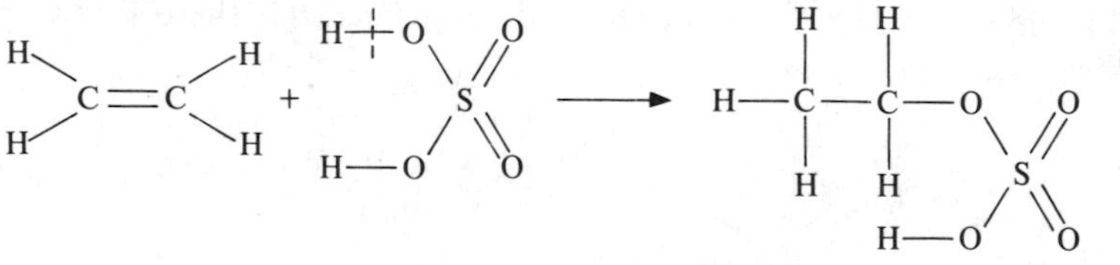

$$\begin{matrix} H & & H \\ & C{=}C & \\ H & & H \end{matrix} + \begin{matrix} H{+}O & & O \\ & S & \\ H—O & & O \end{matrix} \longrightarrow H—\underset{\displaystyle H}{\underset{|}{\overset{\displaystyle H}{\overset{|}{C}}}}—\underset{\displaystyle H}{\underset{|}{\overset{\displaystyle H}{\overset{|}{C}}}}—O—\underset{\displaystyle H—O\quad O}{\overset{\displaystyle O}{S}}$$

ethyl hydrogensulphate(VI)

When added to water and warmed, ethyl hydrogensulphate(VI) is readily hydrolyzed to ethanol:

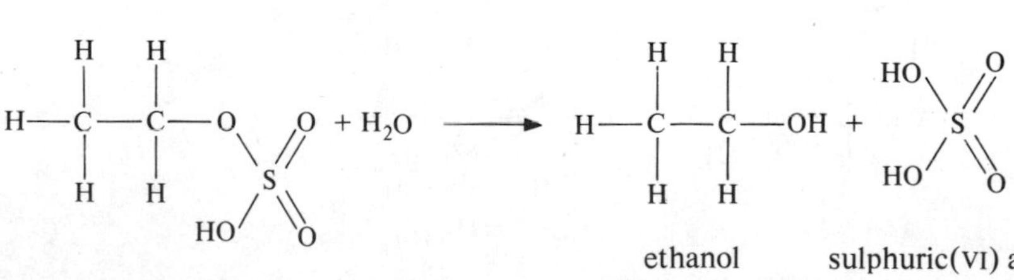

$$H—\underset{\displaystyle H}{\underset{|}{\overset{\displaystyle H}{\overset{|}{C}}}}—\underset{\displaystyle H}{\underset{|}{\overset{\displaystyle H}{\overset{|}{C}}}}—O—\underset{\displaystyle HO\quad O}{\overset{\displaystyle O}{S}} + H_2O \longrightarrow \underset{\text{ethanol}}{H—\underset{\displaystyle H}{\underset{|}{\overset{\displaystyle H}{\overset{|}{C}}}}—\underset{\displaystyle H}{\underset{|}{\overset{\displaystyle H}{\overset{|}{C}}}}—OH} + \underset{\text{sulphuric(VI) acid}}{\begin{matrix} HO & & O \\ & S & \\ HO & & O \end{matrix}}$$

The overall effect of these two reactions is to combine ethene with water to form ethanol, sulphuric(VI) acid being regenerated. At one time this was the most important method of manufacturing ethanol from ethene, but nowadays most ethanol is manufactured by the direct catalytic hydration of ethene in the vapour phase:

$$CH_2{=}CH_2(g) + H_2O(g) \xrightarrow[\text{on celite}\ 330°C,\ 60\ \text{atm.}]{H_3PO_4} CH_3CH_2OH(g)$$

The reaction with sulphuric acid is still used to produce propan-2-ol from propene.

REACTION WITH POTASSIUM MANGANATE(VII)

Ethene will decolorize acidified potassium manganate(VII) (potassium permanganate). This reaction is a useful test-tube diagnosis for a double bond. The reaction with manganate(VII) is a complicated oxidation, producing ethane-1,2-diol, $HOCH_2CH_2OH$, which is itself further oxidized if excess manganate(VII) is present. The equation for this reaction, like all redox reactions in organic chemistry, can be written using the half-equation method described in chapter 2:

$$CH_2{=}CH_2 + 2H_2O \longrightarrow HOCH_2CH_2OH + 2H^+ + 2e^- \quad \text{(i)}$$

$$MnO_4^- + 8H^+ + 5e^- \longrightarrow Mn^{2+} + 4H_2O \quad \text{(ii)}$$

Multiplying (i) by 5 and (ii) by 2 and adding, we get

$$5CH_2{=}CH_2 + 2H_2O + 2MnO_4^- + 6H^+ \longrightarrow 5HOCH_2CH_2OH + 2Mn^{2+}$$

Potassium manganate(VII) is much too expensive for this process to be useful for making ethane-1,2-diol industrially: this important compound is made instead from epoxyethane (see below).

REACTION WITH OXYGEN

Like all hydrocarbons, ethene burns in air. This is an unimportant reaction, a waste of such a useful compound. In the presence of a finely divided silver catalyst at 180°C, ethene can be made to combine with oxygen much more fruitfully forming epoxyethane:

$$2CH_2{=}CH_2 + O_2 \xrightarrow[180°C]{Ag} 2\ CH_2{-}CH_2 \text{ (bridged by O)}$$

Epoxyethane, with its strained three-membered ring, is unstable and can be readily converted to a number of useful products. In particular it reacts with water to form ethane-1,2-diol (ethylene glycol):

$$CH_2{-}CH_2 \text{ (bridged by O)} + H_2O \longrightarrow HOCH_2CH_2OH$$

This is an important organic chemical, used mainly as antifreeze but also for manufacturing polyester fibres.

Epoxyethane can also be made to join to itself, forming a polymer of structure $-OCH_2CH_2OCH_2CH_2O-$. This polymer, when added to the water used in fire hoses in the proportion 30 parts per million, reduces the friction between the water and the hose walls so successfully that the range of the hose is doubled.

A polymer made from epoxyethane can be used to reduce the friction between water and the walls of fire hoses.

REACTION WITH ITSELF

Perhaps the most important industrial reaction of ethene is with itself, to form poly(ethene). This is described in the next section.

All the above reactions have referred to ethene, but they are applicable to alkenes in general. Predict the outcome of the following reactions:

- $CH_3CH{=}CH_2 + H_2$ over Ni catalyst
- $CH_3CH{=}CHCH_3 + H_2SO_4$ (concentrated)
- excess $CH_3CH{=}CH_2 + KMnO_4$ in acid
- $CH_3CH{=}CHCH_3 + Br_2$

## 26.7 Addition polymerization

POLYMERS

About 80% of the world's output of organic chemicals is used to make polymers. The use of polymers by man is not new: wood, cotton, wool and rubber are all naturally occurring polymeric materials. However, the manufacture and use of *synthetic* polymers has only really got under way since the second world war, but their use has increased very rapidly and they are steadily replacing traditional natural materials. Synthetic polymers are usually cheaper than natural materials and are often better suited to their particular function since chemists are now able to produce a polymer to suit most specifications. One drawback of the increased use of synthetic polymers is the problem of their disposal. Synthetic polymer molecules, not being of natural origin, cannot be broken down by the enzymes in bacteria—they are said to be non-biodegradable. Furthermore, as polymers cannot be economically recycled (unlike metals) their disposal presents a considerable problem: plastic containers are among the worst forms of modern litter.

Synthetic polymers present a disposal problem, since they are not biodegradable.

**Polymers** are long-chain molecules made by joining together many small molecules. The small molecules from which the polymer is built are called **monomers**. Polymers vary widely in physical properties such as strength, flexibility and softening temperature. Some, such as polyesters, are very strong and not readily stretched and are therefore suitable for use as fibres. Others, such as polythene, are more easily deformed and are classed as plastics.

The properties of a polymer are, of course, dependent on the properties of the chains it contains. Four factors are particularly important.

*(a) Chain length*. The characteristic polymer properties of flexibility and tensile strength arise from the chain structure of polymer molecules. Strength and melting point increase with chain length. Typical polymer properties become apparent for average chain lengths of 50 monomer units and upwards; mechanical strength increases up to about 500 units, after which it changes only slightly. It is necessary to talk in terms of average chain lengths where polymers are concerned, since individual chains vary in length even in a pure sample of polymer.

*(b) Intermolecular forces*. If the intermolecular forces between chains are high, the polymer will tend to be strong and difficult to melt.

*(c) Branching*. Highly branched polymer chains cannot pack together so well as straight chains: highly branched polymers therefore tend to have lower tensile strength and to melt more easily (see figure 26.2).

*(d) Cross-linking*. Some polymers have extensive cross-linking between chains. This forms a rigid network, making a hard material (figure 26.2). Such polymers are called *thermosetting* (see section 29.7).

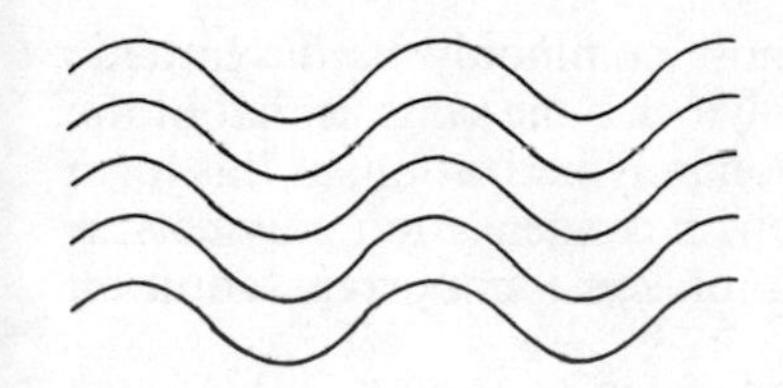

(*a*) Polymer with few branched chains, e.g. high density polythene. Chains regularly oriented and well packed with extensive crystalline regions. High tensile strength, high melting point.

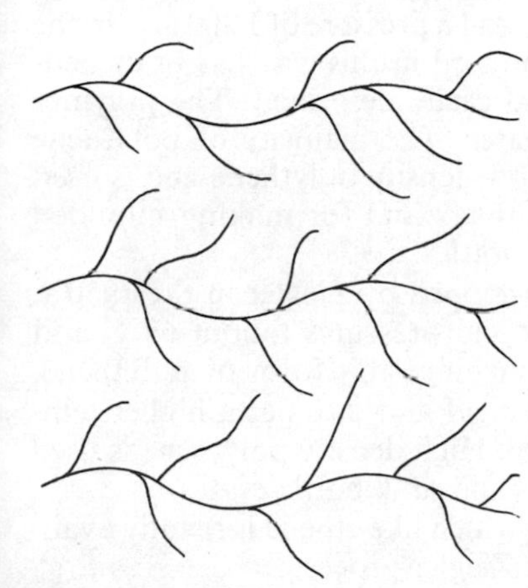

(*b*) Polymer with many branched chains, e.g. low density polythene. Chains irregularly packed: largely amorphous with few crystalline regions. Lower tensile strength, lower melting point.

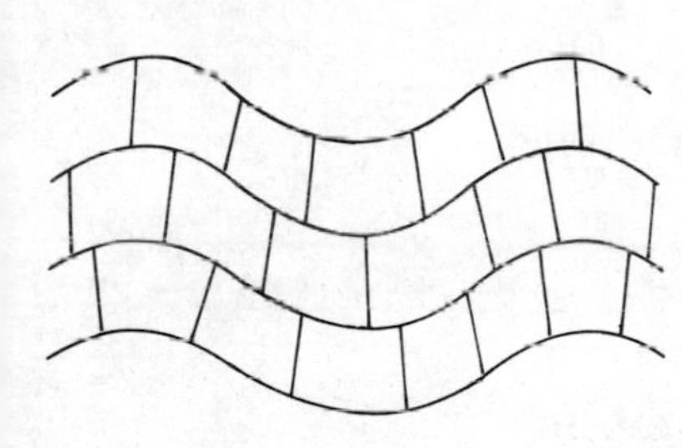

(*c*) Polymer with much cross-linking, e.g. bakelite. Rigid and hard, fairly brittle.

**Figure 26.2** Physical properties of polymers.

Forces between polymer molecules are considered further in section 8.4.

ADDITION POLYMERIZATION

Molecules containing double bonds are particularly useful monomers as they can usually be made to undergo addition reactions among themselves. The double bond in each molecule can be thought of as breaking open, enabling the free bonds to link with one another forming a chain.

Thus with ethene:

$$H_2C{=}CH_2 \qquad H_2C{=}CH_2 \qquad H_2C{=}CH_2$$

$$\downarrow$$

$$\left[ -CH_2-CH_2- \qquad -CH_2-CH_2- \qquad -CH_2-CH_2- \right]$$

$$\downarrow$$

$$-CH_2-CH_2-CH_2-CH_2-CH_2-CH_2-$$

The product, **poly(ethene)** or **polythene**, is the most commonly used synthetic polymer. Note that the empirical formula of the polymer is the same as that of the monomer: $CH_2$. This is always the case with addition polymerization, as this form of polymerization is called, but this is not the case with condensation polymerization, which is described in section 31.5. Details of some important addition polymers are now given.

*(a) Polythene.* Polythene was discovered by ICI in 1933. The method they used to make it involved heating ethene at about 200°C and a pressure of 1 200 atm in the presence of traces of oxygen. The polythene produced in this way has branched-chains and is therefore fairly readily melted and easily deformed. The polymer melts at about 105°C and softens in boiling water. The majority of polythene manufactured today is of this form: it is called low density polythene and is used for making film and sheeting for bags and wrappers and for making moulded articles such as washing-up bowls and 'squeezy' bottles.

An insulated fish box made from high density polythene.

Another method of making polythene was developed by Ziegler in the 1950's. This process uses catalysts at low temperatures and pressures (about 60°C and 1 atm). The molecules produced have little branching: this form of polythene, called high density polythene, is therefore more rigid and melts at a higher temperature (about 135°C) than the low density form. High density polythene is used for moulding rigid articles such as bleach bottles and milk bottle crates.

*(b) Poly(propene) (polypropylene).* Propene, which like ethene is readily available from petroleum, can be polymerized by the Ziegler process:

$$\begin{array}{c} \mathrm{(CH_3)(H)C{=}C(H)(H)} \quad \mathrm{(CH_3)(H)C{=}C(H)(H)} \quad \mathrm{(CH_3)(H)C{=}C(H)(H)} \\ \downarrow \\ -\mathrm{C(CH_3)(H)}-\mathrm{C(H)(H)}-\mathrm{C(CH_3)(H)}-\mathrm{C(H)(H)}-\mathrm{C(CH_3)(H)}-\mathrm{C(H)(H)}- \end{array}$$

These two bottles, one made from high-density polythene and the other from low-density polythene, were heated in an oven. Which is which?

Poly(propene) produced in this way has the $CH_3$— side groups arranged in a highly regular fashion so that its chains pack together closely, producing a material similar to high density polythene. It is used in mouldings and film and can be made into a fibre. It is increasingly used to make ropes, an example of a synthetic polymer replacing a natural one; polypropene has the advantage that being non-biodegradable it does not rot.

*(c) Poly(chloroethene) (PVC).* Chloroethene (vinyl chloride) polymerizes to form poly(chloroethene) or polyvinyl chloride (PVC):

$$\begin{array}{c} \mathrm{(Cl)(H)C{=}C(H)(H)} \quad \mathrm{(Cl)(H)C{=}C(H)(H)} \quad \mathrm{(Cl)(H)C{=}C(H)(H)} \\ \downarrow \\ -\mathrm{C(Cl)(H)}-\mathrm{C(H)(H)}-\mathrm{C(Cl)(H)}-\mathrm{C(H)(H)}-\mathrm{C(Cl)(H)}-\mathrm{C(H)(H)}- \end{array}$$

Drink crates made from poly(propene).

The polar C–Cl bond results in considerable intermolecular attraction between the polymer chains, making PVC a fairly strong material. Its best feature is its versatility: by incorporating additives it can be adapted to suit many uses. It is used among

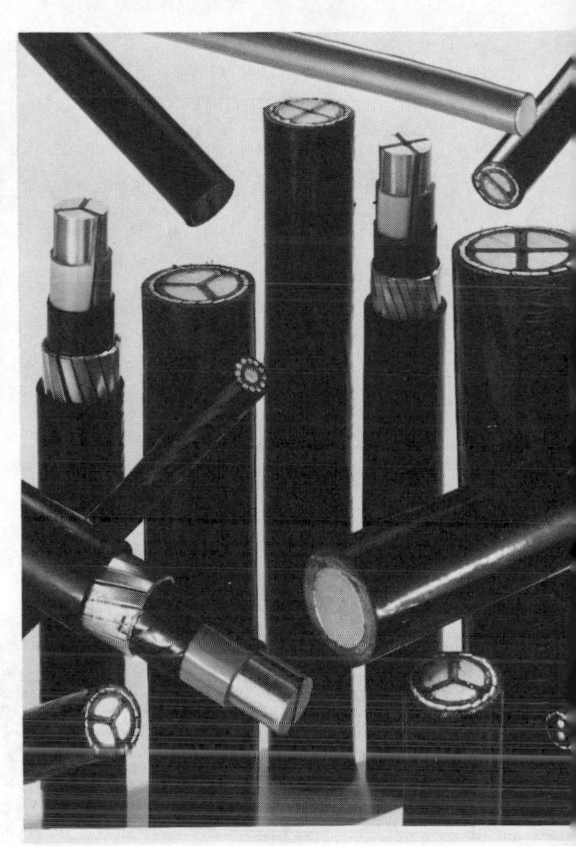

*Right* PVC-insulated cables.
*Above* Illuminated signs made from coloured Perspex.
*Left* Non-stick frying pan and electric iron coated in PTFE.

other things for coating fabrics, for covering wires and cables and for making gramophone records.

*(d) Other addition polymers.* Table 26.3 shows some other addition polymers, their monomers and their uses.

**Table 26.3** Addition polymers.

| | | | | |
|---|---|---|---|---|
| **Polymer:** systematic name | poly(phenylethene) | poly(tetrafluoroethene) | poly(methyl 2-methyl-propenoate) | poly(propenonitrile) |
| **Polymer:** common name | polystyrene | PTFE, 'Teflon', etc. | 'Perspex' | 'Acrilan', etc. |
| **Monomer:** systematic name | phenylethene | tetrafluoroethene | methyl 2-methyl-propenoate | propenonitrile |
| **Monomer:** formula | $C_6H_5CH{=}CH_2$ | $CF_2{=}CF_2$ | $(CH_3)(CH_3)C{-}COOCH_3$ | $CH_2{=}CHCN$ |
| **Monomer:** common name | styrene | tetrafluoroethylene | methyl methacrylate | acrylonitrile |
| **Properties** | brittle (but cheap) | very stable. Low friction, anti-stick properties | transparent | strong, fibre properties |
| **Uses** | expanded polystyrene for insulation. Plastic toys, etc. | non-stick coatings on pans. Insulators. | as a substitute for glass | making textiles (wool substitute) |

Use table 26.3 to answer these questions.

- ○ Draw a section of the chain structures of poly(phenylethene) and poly(propenonitrile).
- ○ Suggest a reason why poly(phenylethene) is brittle.
- ○ Suggest a reason why poly(propenonitrile) is a strong material, suitable for making fibres.

## 26.8 Alkynes

**Alkynes** are hydrocarbons containing triple bonds. The simplest is **ethyne**, commonly called acetylene, H—C≡C—H. Alkynes are named in a similar manner to alkenes, using the suffix **-yne** instead of -ene.

- ○ Write the structural formula of but-1-yne.
- ○ Name the compound with formula $CH_3CH_2C{\equiv}CCH_3$.
- ○ What is the general formula of the alkynes?

The triple bond in alkynes can be thought of as consisting of a σ-bond and two π-bonds (see section 26.3). As in ethene, the σ-bond is formed by the axial overlap of two orbitals; the π-bonds are each formed by the overlap of two 2*p* orbitals. Because the 2*p* orbitals are at right angles to one another and have two lobes each, the two π-bonds comprise four regions of negative charge around the C–C bond. These regions tend to coalesce and form a cylinder of charge (see figure 26.3). Since ethyne has no lone pairs, its molecule is linear.

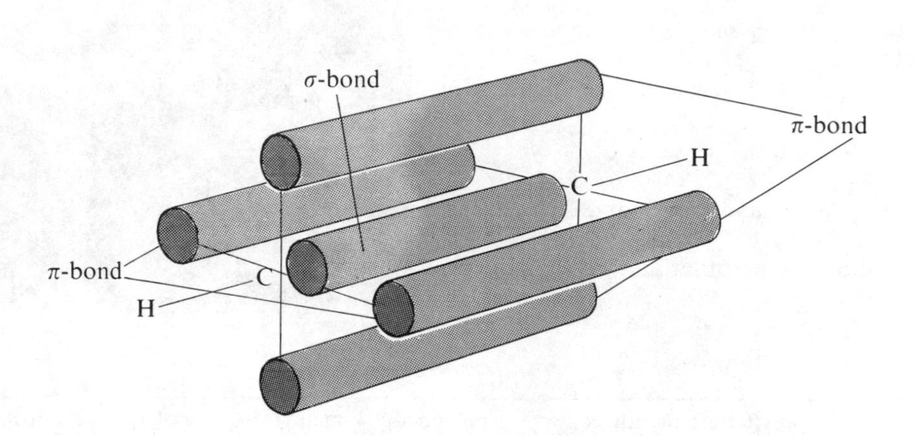


(*a*) Three-dimensional representation. Limited regions of bonding orbitals shown.

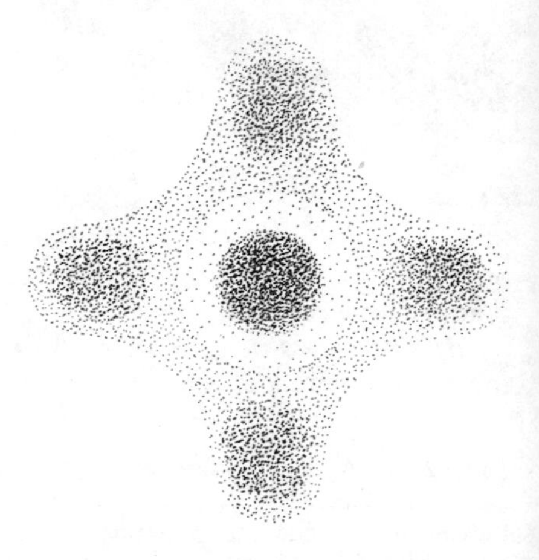

(*b*) End-on view. Depth of stippling represents electron density.

**Figure 26.3** Bonding in ethyne.

With a high density of electrons in the triple bond, ethyne may be expected to attract electrophiles and undergo addition reactions like those of ethene. This is indeed the case: ethyne reacts with most of the substances that undergo addition with ethene. With the two π-bonds, ethyne might be expected to be more reactive toward electrophiles than ethene. In fact, although it is *energetically* less stable with respect to its addition products, it reacts rather more slowly—that is, it is *kinetically* slightly more stable. Thus while ethene decolorizes bromine water almost instantaneously, with ethyne the colour takes a few minutes to disappear.

Because it has two π-bonds to add across, ethyne can give two different addition products, depending on the number of moles reacting. Thus, with hydrogen in the presence of a nickel catalyst:

$$H{-}C{\equiv}C{-}H + H_2 \xrightarrow[150°]{Ni} H_2C{=}CH_2 \quad \text{(one π-bond broken)}$$

$$\text{or} \quad H{-}C{\equiv}C{-}H + 2H_2 \xrightarrow[150°]{Ni} H_3C{-}CH_3 \quad \text{(both π-bonds broken)}$$

In practice, when ethyne reacts with hydrogen in the presence of a nickel catalyst, a mixture of ethane and ethene is formed. By carefully adjusting the quantities and conditions, one of these can be obtained in preference to the other. Thus substitution of a palladium/barium sulphate(VI) catalyst in place of nickel gives almost entirely ethene.

OTHER ADDITION REACTIONS OF ETHYNE

*(a) With hydrogen halides.* For example:

$$H{-}C{\equiv}C{-}H \xrightarrow{HCl} \underset{\text{chloroethene}}{H_2C{=}CHCl} \xrightarrow{HCl} \underset{\text{1,1-dichloroethane}}{H{-}CH_2{-}CHCl{-}Cl}$$

The second stage of this addition accords with Markovnikoff's Rule (see section 26.6), the second molecule of HCl adding onto chloroethene so that the hydrogen atom bonds to the carbon atom already carrying the larger number of hydrogen atoms.

*(b) With halogens.* Ethyne and chlorine explode spontaneously when mixed:

$$H{-}C{\equiv}C{-}H + Cl_2 \longrightarrow 2C + 2HCl$$

If, however, the reactants are mixed in the presence of an inert material such as kieselguhr (almost pure silicon(IV) oxide) the free radicals responsible for this explosive reaction are absorbed and expected addition reactions can take place:

$$H{-}C{\equiv}C{-}H \xrightarrow{Cl_2} \underset{\text{1,2-dichloroethene}}{HClC{=}CClH} \xrightarrow{Cl_2} \underset{\text{1,1,2,2-tetrachloroethane}}{Cl{-}CHCl{-}CHCl{-}Cl}$$

Predict the formulae of both possible products of the following reactions:

- ○ $CH_3C{\equiv}CH + H_2$ over nickel catalyst
- ○ $CH_3C{\equiv}CCH_3 + Br_2$
- ○ $CH_3CH_2C{\equiv}CH + HI$

FORMATION OF SALTS FROM ETHYNE

Ethyne has one characteristic that is quite distinct from ethene. The hydrogen atoms in ethyne are slightly acidic and can be replaced by metal ions to form salts called **dicarbides**. Ethyne is an extremely weak acid: with a $K_a$ of about $10^{-20}$ mol $dm^{-3}$ it is much weaker than water, but it is about $10^{10}$ times stronger than ethene. For example, if ethyne is bubbled through ammoniacal silver nitrate(V), a yellow precipitate of silver dicarbide is produced:

$$HC{\equiv}CH(g) + 2Ag^+(aq) \longrightarrow \underset{\text{silver dicarbide}}{Ag{-}C{\equiv}C{-}Ag(s)} + 2H^+(aq)$$

This reaction is shown by all alk-1-ynes (alkynes with the triple bond at the end of the chain), and it provides a useful test for such compounds.

An important dicarbide is that of calcium, $CaC_2$, often simply called calcium carbide. It is made by heating calcium oxide and coke at 2 000°C in an electric furnace. Calcium dicarbide, unlike silver dicarbide, is ionic, containing the $(C{\equiv}C)^{2-}$ ion. It reacts readily with water to form ethyne:

$$(C{\equiv}C)^{2-} + 2H_2O \longrightarrow HC{\equiv}CH + 2OH^-$$

At one time this was a highly important reaction for manufacturing ethyne. With the decline of the coal industry and the high price of electricity, it has become too expensive and nowadays most ethyne is made from petroleum by cracking reactions. Ethyne used to be a very important material for the manufacture of organic chemicals, but it has lost this position to ethene, which is a great deal cheaper to produce. If oil becomes very expensive it is possible that the carbide route to ethyne may again become economic and ethyne could regain its importance.

Ethyne is still used to manufacture some chloroethene (for making PVC) and 1,1,2,2-tetrachloroethane (a solvent). It is important as a fuel in 'oxy-acetylene' torches for cutting and welding metal: a flame temperature of about 2 800°C can

Acetylene (ethyne) lamps like this one were once common on bicycles. The upper compartment contains water which drips onto calcium dicarbide contained in the lower compartment. The ethyne produced burns at the jet.

Ethyne burning in pure oxygen gives a flame temperature high enough to melt most metals.

be obtained by the combustion of ethyne in pure oxygen. Storage of ethyne under pressure presents a problem. The heat of formation of ethyne is +227 kJ mole$^{-1}$ which means that it is energetically unstable with respect to carbon and hydrogen. It is therefore liable to decompose explosively when compressed on its own, so it is stored dissolved in propanone under pressure.

## Summary

1 Alkenes are unsaturated hydrocarbons containing a carbon-carbon double bond in their molecules. Their general formula is $C_nH_{2n}$ and they are named using the suffix *-ene* together with a number to indicate the position of the double bond.
2 Double bonds consist of a $\sigma$-bond situated axially between the carbon atoms and a weaker non-axial $\pi$-bond.
3 *Cis* and *trans* isomers differ in the way their substituents are arranged about a double bond.
4 The characteristic reaction of alkenes is electrophilic addition. In general:

$$>C=C< \;+\; X{-}Y \longrightarrow -\overset{X}{\underset{|}{C}}-\overset{Y}{\underset{|}{C}}-$$

5 The mechanism of electrophilic addition involves initial attack on the $\pi$-electrons, often leading to the formation of an intermediate carbonium ion.
6 Some important reactions of a double bond are shown in figure 26.4.

$>C=C<$ —polymerize→ $-C-C-C-C-$
$>C=C<$ —$O_2$, Ag catalyst→ $-C-C-$ (with O bridging the two carbons)
$>C=C<$ —$H_2$, Ni catalyst→ $-C(H)-C(H)-$
$>C=C<$ —$Br_2$→ $-C(Br)-C(Br)-$
$>C=C<$ —HBr→ $-C(Br)-C(H)-$
$>C=C<$ —$KMnO_4$→ $-C(OH)-C(OH)-$
$>C=C<$ —$H_2SO_4$→ $-C(H)-C(HSO_4)-$ —$H_2O$→ $-C(H)-C(OH)-$

**Figure 26.4** Reactions of a double bond.

7 Many compounds containing double bonds readily undergo addition polymerization. The most important case is ethene, which polymerizes to polythene.
8 Alkynes are hydrocarbons containing a triple bond: they are named using the suffix *-yne*.
9 Alkynes undergo two-stage addition:

$$-C\equiv C- \;+\; X{-}Y \longrightarrow >C(X)=C(Y)<$$

$$>C(X)=C(Y)< \;+\; X{-}Y \longrightarrow X-C(X)-C(Y)-Y$$

10 The terminal H atom in the alk-1-ynes is acidic and can be replaced by metal ions to form dicarbides.

## Study questions

1 (a) Name the following compounds:

(i) $CH_3CH_2CH_2CH_2CH{=}CH_2$

(ii)

(iii)

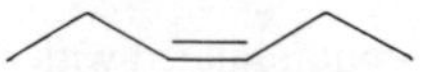

(iv) $CH_3C{\equiv}CCH_3$

(v) $CH_3C{\equiv}CCHCH_3$ (with $CH_3$ on the CH)

(b) Write structural formulae for these compounds:

(i) propyne

(ii) *cis*-pent-2-ene

(iii) cycloocta-1,3,5,7-tetraene

(iv) 2-methylpent-2-ene.

2 1,1-Dichloroethene, $Cl_2C{=}CH_2$ readily undergoes addition polymerization.

(a) What is the systematic name of the polymer?

(b) Write the structure of a section of the polymer chain.

(c) Would you expect the polymer to have a higher or a lower melting point than poly-(chloroethene) (PVC)? Explain your answer.

3 (a) For each of the following compounds, say whether you would expect them to show geometric (*cis-trans*) isomerism. For those compounds that show geometric isomerism, draw the structures of all the possible isomers.

(i) 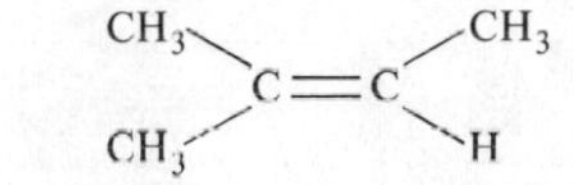 $(CH_3)_2C{=}C(CH_3)H$

(ii) 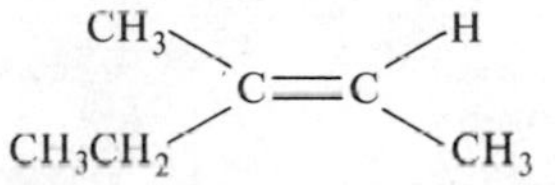 $(CH_3)(CH_3CH_2)C{=}C(H)(CH_3)$

(iii) 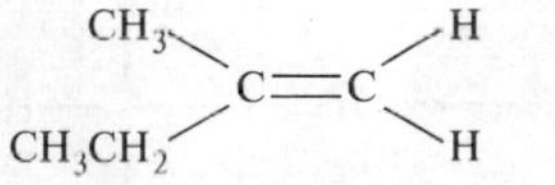 $(CH_3)(CH_3CH_2)C{=}CH_2$

(iv) 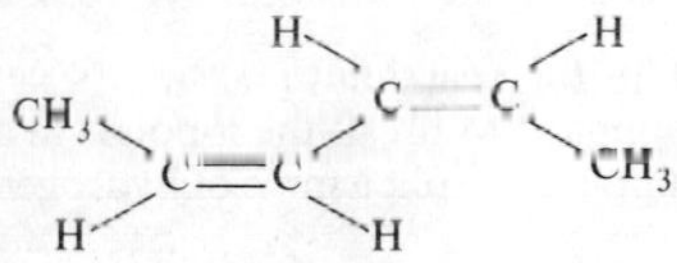 $CH_3CH{=}CH{-}CH{=}CHCH_3$

(b) 1,2-Dichlorocyclopropane, $CH_2$ / $ClCH{-}CHCl$ (ring), has *cis* and *trans* isomers. Explain why this is so, and draw the structures of the two isomers.

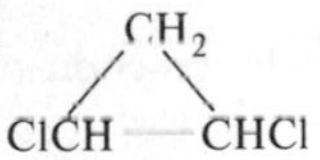

4 A hydrocarbon **A** contains 87.8% carbon and 12.2% hydrogen by mass. Its relative molecular mass is 82. **A** decolorizes bromine water and in the presence of a nickel catalyst reacts with hydrogen to form **B**. 0.1g of **A** was found to absorb 27.3cm$^3$ of hydrogen (measured at s.t.p.). **B** does not decolorize bromine water.

(a) What is the empirical formula of **A**?

(b) What is the molecular formula of **A**?

(c) How many moles of hydrogen react with 1 mole of **A**?

(d) How many double bonds does **A** have in its molecule?

(e) What is the molecular formula of **B**?

(f) Suggest structural formulae for **A** and **B**.

5 Propan-2-ol is an important alcohol used as a solvent, as a de-icing fluid and for manufacturing propanone (acetone). Propan-2-ol is manufactured as follows:

Impure propene, containing traces of ethene, is passed up an absorption tower down which 85% sulphuric(VI) acid trickles. The propene reacts to form 2-propyl hydrogensulphate(VI):

$$CH_3CH{=}CH_2 + H_2SO_4 \longrightarrow CH_3CH(HSO_4)CH_3$$

Virtually no 1-propyl hydrogensulphate(VI) is produced, and hardly any of the ethene present as impurity reacts. The 2-propyl hydrogensulphate(VI) produced in this way is added to water, when a highly exothermic reaction occurs, 2-propyl hydrogensulphate(VI) being hydrolyzed to propan-2-ol:

$$CH_3-\underset{\displaystyle HSO_4}{\underset{|}{CH}}-CH_3 + H_2O \longrightarrow CH_3-\underset{\displaystyle OH}{\underset{|}{CH}}-CH_3 + H_2SO_4$$

(a) How is propene obtained industrially? Why is it contaminated with ethene?
(b) When propene reacts with sulphuric(VI) acid, initial attack is by $H^+$ on the $\pi$-electrons of the double bond. Write the structures of the two possible carbonium ions formed as a result of this attack.
(c) Which of the two carbonium ions formed in (b) is the more stable?
(d) Use your answer to (c) to explain why virtually no 1-propyl hydrogensulphate(VI) is formed in the reaction of propene with sulphuric(VI) acid.
(e) Write the structure of the carbonium ion formed when ethene is attacked by $H^+$.
(f) Is the carbonium ion in (e) more or less stable than the more stable carbonium ion formed by propene (refer back to your answer to (c))?
(g) Use your answer to (f) to explain why hardly any of the ethene contaminating the propene reacts with sulphuric(VI) acid, whereas the propene reacts readily.
(h) Why is it important that 2-propyl hydrogensulphate(VI) is virtually the only product of the reaction of the gases with sulphuric(VI) acid?

6 How would you distinguish between the compounds in each of the following pairs, using simple test-tube reactions?
(a) $CH_3CH_2C{\equiv}CH$ and $CH_2{=}CHCH{=}CH_2$
(b) hex-1-ene and cyclohexane
(c) $CH_3CH_2C{\equiv}CH$ and $CH_3C{\equiv}CCH_3$.

7 Some bond energies are given below.
C=C (in ethene) 598 kJ $mole^{-1}$
C—C (in ethane) 346 kJ $mole^{-1}$
C—H (general) 413 kJ $mole^{-1}$
H—H 436 kJ $mole^{-1}$
Consider the reaction of ethene with hydrogen:

$$\begin{matrix} H & & & & H \\ & \diagdown & & \diagup & \\ & & C{=}C & & \\ & \diagup & & \diagdown & \\ H & & & & H \end{matrix} + H_2 \longrightarrow H-\underset{\displaystyle H}{\overset{\displaystyle H}{\underset{|}{\overset{|}{C}}}}-\underset{\displaystyle H}{\overset{\displaystyle H}{\underset{|}{\overset{|}{C}}}}-H$$

(a) What conditions are needed for this reaction to occur?
(b) How much energy must be supplied to break the $\pi$-bonds in a mole of ethene?
(c) How much energy must be supplied to split a mole of hydrogen molecules into hydrogen atoms?
(d) How much energy is released when two new C–H bonds are formed by a mole of ethene?
(e) Calculate the energy change for the reaction of a mole of ethene with a mole of hydrogen.

8 Alkenes such as ethene and propene have been described as the building blocks of the organic chemical industry. Discuss this statement, giving examples. What particular features of the chemistry of alkenes make them suitable for this rôle and why are alkanes unsuitable?

9 Predict the structures of the products, if any, of the following reactions.

(a) [cyclohexene] + $KMnO_4$ (aq)

(b) [1-methylcyclohexene] + HI

(c) $CH_3C{\equiv}CCH_3$ + ammoniacal silver nitrate(V)
(d) $CH_3CH_2C{\equiv}CH$ + excess HBr
(e) $CH_3CH{=}CHCH_3$ reacts with itself, i.e. polymerizes.

10 Consider the following compounds.

**A** hex-2-ene $CH_3CH_2CH_2CH{=}CHCH_3$
**B** hex-1-yne $CH_3CH_2CH_2CH_2C{\equiv}CH$
**C** hex-2-yne $CH_3CH_2CH_2C{\equiv}CCH_3$
**D** hexane $CH_3CH_2CH_2CH_2CH_2CH_3$

**E** cyclohexane

```
        CH2
      /     \
  H2C         CH2
   |           |
  H2C         CH2
      \     /
        CH2
```

The answers to the following questions may be one or more than one of the above compounds.

(a) Which would decolorize bromine in the absence of sunlight?
(b) Which would give a yellow precipitate with an ammoniacal solution of silver nitrate(v)?
(c) Which would react with chlorine only when heated or exposed to light?
(d) Which would absorb 1 mole of hydrogen per mole in the presence of a nickel catalyst?
(e) Which has *cis* and *trans* isomers?
(f) Which are unsaturated?
(g) In which are there three or more carbon atoms arranged in a straight line?

# 27 Aromatic Hydrocarbons

## 27.1 Aromatic hydrocarbons

In the last chapter we considered unsaturated compounds with double and triple bonds. There is another important class of unsaturated compounds that needs to be considered separately because the properties here are so different from the alkenes and alkynes. This class of compounds is called the **aromatic hydrocarbons** and its simplest and most important member is **benzene, $C_6H_6$**. The name 'aromatic' was originally used because some derivatives of these hydrocarbons have pleasant smells. It is now known that just as many of them smell unpleasant, and in any case many of the aromatic vapours are toxic, so it is unwise to smell them. The name aromatic has been retained to indicate certain chemical characteristics rather than odorous properties.

A. Kekulé, 1829–1896

## 27.2 The structure of benzene

The molecular formula of benzene has been known since 1834 to be $C_6H_6$. The exact structural formula, however, posed a problem for many years. With such a high C:H ratio it must clearly be highly unsaturated. A possible structure might be hexatetraene, e.g. $CH_2{=}C{=}CH{-}CH{=}C{=}CH_2$. Such a structure would be expected to have two isomeric monosubstituted chlorocompounds, $C_6H_5Cl$:

$$ClCH{=}C{=}CH{-}CH{=}C{=}CH_2 \text{ and } CH_2{=}C{=}CCl{-}CH{=}C{=}CH_2.$$

Only one form of chlorobenzene has ever been isolated. The structure of benzene must therefore be one in which all six hydrogen atoms occupy equivalent positions.

The problem was solved by **Kekulé** in 1865. He proposed a ring structure in which alternate carbon atoms were joined by double bonds:

H C C C C C C H H H H H or

Here is Kekulé's description of how this structure occurred to him.

> 'I turned my chair to the fire and dozed. Again the atoms were gambolling before my eyes. This time the smaller groups kept modestly in the background. My mental eye, rendered more acute by repeated visions of this kind, could now distinguish larger structures, of manifold conformation; long rows, sometimes more closely fitted together; all twining and twisting in snakelike motion. But look. What was that? One of the snakes had seized hold of its own tail, and the form whirled mockingly before my eyes. As if by a flash of lightning I awoke.'

This structure, called the Kekulé or cyclohexatriene structure, explains many of the properties of benzene and was accepted for a long time. It still leaves unanswered, though, some problems concerned with bond length and thermochemistry.

Theoretische Betrachtungen.

schlossene Kette (einen symmetrischen Ring), die noch sechs freie Verwandtschaftseinheiten enthält.

offene Kette. geschlossene Kette.

Diese Ansicht über die Constitution der aus sechs Kohlenstoffatomen bestehenden, geschlossenen Kette wird vielleicht noch deutlicher wiedergegeben durch folgende graphische Formel, in welcher die Kohlenstoffatome rund und die vier Verwandtschaftseinheiten jedes Atomes durch vier von ihm auslaufende Linien dargestellt sind:

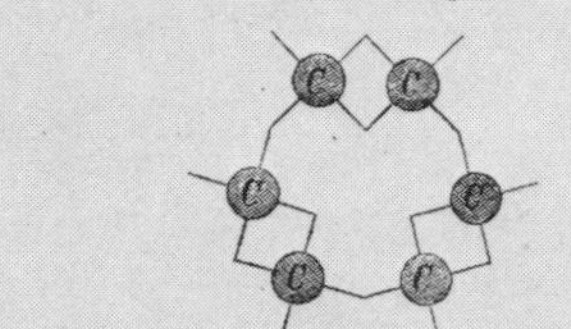


Von dieser geschlossenen Kette leiten sich nun, wie gleich ausführlicher gezeigt werden wird, alle die Verbindungen ab, die man gewöhnlich als aromatische Substanzen bezeichnet. Die offene Kette ist vielleicht im Chinon, im Chloranil und den wenigen Körpern anzunehmen, die zu beiden in näherer Beziehung stehen. Auch diese Körper können indess auf die geschlossene Kette bezogen und von ihr abgeleitet

Part of a German paper written by Kekulé describing his proposed structure for benzene.

## BOND LENGTHS

X-ray diffraction studies show that benzene is planar: this is to be expected from the Kekulé structure. X-ray diffraction also shows that all the C–C bonds in benzene are the same length:

carbon-carbon bond length in all bonds in benzene, 0.139nm
carbon-carbon single bond length in cyclohexane, 0.154nm
carbon-carbon double bond length in cyclohexene, 0.133nm

The Kekulé model would suggest unequal carbon-carbon lengths, alternating between double and single bond values. In fact we find a constant bond length, somewhere between the value for a single and a double bond.

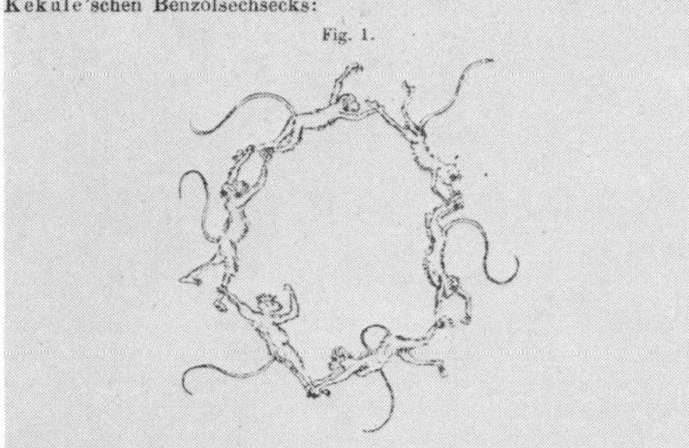
reichen, so erhält man ein höchst vollkommenes Analogon des Kekulé'schen Benzolsechsecks:

Fig. 1.

Nun aber besitzt der genannte Macacus cynocephalus ausser seinen eigentlichen vier Händen noch ein fünftes Greifwerkzeug in Form eines caudalen Appendix. Zieht man diesen mit in Betracht, dann gelingt es, die 6 Individuen des gezeichneten Ringes auch noch in anderer Weise mit einander zu verbinden. So entsteht das nachfolgende Bild:

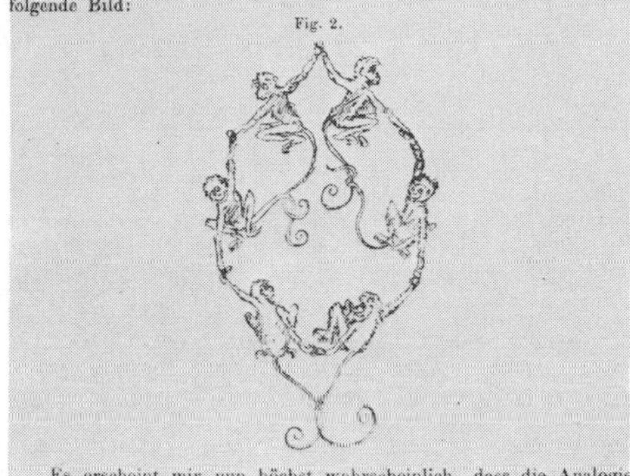
Fig. 2.

Es erscheint mir nun höchst wahrscheinlich, dass die Analogie

An article lampooning Kekulé's structure by proposing a ring made up of six monkeys. The monkeys join hands to form single bonds and link tails to make double bonds.

## THERMOCHEMISTRY OF BENZENE

It is interesting and instructive to work out a theoretical value for the heat of formation of benzene on the basis of the Kekulé model, and to compare it with the experimental value obtained from the heat of combustion.

The heat of formation of gaseous benzene is the heat change when a mole of gaseous benzene is formed from its elements:

$$6C(s) + 3H_2(g) \longrightarrow C_6H_6(g)$$

Relevant data are:

heat of atomization of C(s) 715 kJ (mole of C atoms)$^{-1}$
heat of atomization of $H_2$(g) 218 kJ (mole of H atoms)$^{-1}$
bond energy of C=C (average) 610 kJ mole$^{-1}$
bond energy of C—C (average) 346 kJ mole$^{-1}$
bond energy of C—H (average) 413 kJ mole$^{-1}$

Work out the heat of formation of benzene by the following stages.

1 Calculate the energy needed to produce
 (a) six moles of gaseous carbon atoms from C(s)
 (b) six moles of gaseous hydrogen atoms from $H_2$(g)
2 Calculate the energy released when
 (a) three moles of C–C bonds are formed from gaseous atoms
 (b) three moles of C=C bonds are formed from gaseous atoms
 (c) six moles of C–H bonds are formed from gaseous atoms.
3 Use your answers to 1 and 2 to calculate the total energy change when a mole of gaseous benzene is formed from its elements.
4 Compare your answer with the experimental value of +82 kJ mole$^{-1}$.
5 Do your results suggest that benzene is more or less stable than the Kekulé structure?

The theoretical heat of formation of gaseous benzene based on the Kekulé structure is +252 kJ mole$^{-1}$, about 170 kJ mole$^{-1}$ greater (i.e. more endothermic) than the experimental value of +82 kJ mole$^{-1}$. This implies that the actual structure of benzene is considerably more stable than the Kekulé structure and agrees reasonably closely with the stabilization energy of benzene obtained using heats of hydrogenation in section 11.11.

## ELECTRON DELOCALIZATION IN BENZENE

The extra stability of benzene and the fact that its C–C bonds are all of equal length can be explained using the following model, which is currently accepted by most chemists.

The carbon atoms in the ring are bonded to one another and to their hydrogen atoms by $\sigma$-bonds. This leaves one unused $p$ orbital on each carbon, each containing a single electron. These $p$ orbitals are perpendicular to the plane of the ring, with one lobe above and one below this plane (figure 27.1(*a*)). Each $p$ orbital overlaps sideways with the two neighbouring orbitals to form a single $\pi$-bond that extends as a ring of charge above and below the plane of the molecule (figure 27.1(*b*)).

The electrons in the $\pi$-bond cannot be said to 'belong to' any particular carbon atom. Each electron is free to move throughout the entire $\pi$ system, so the electrons are said to be **delocalized**. It is this delocalization that gives benzene its extra stability: any system in which electron delocalization can occur is stabilized. The reason for this is not hard to see: electrons tend to repel one another, so a system in which they are as far apart from one another as possible will enjoy minimum repulsion and will therefore be stabilized.

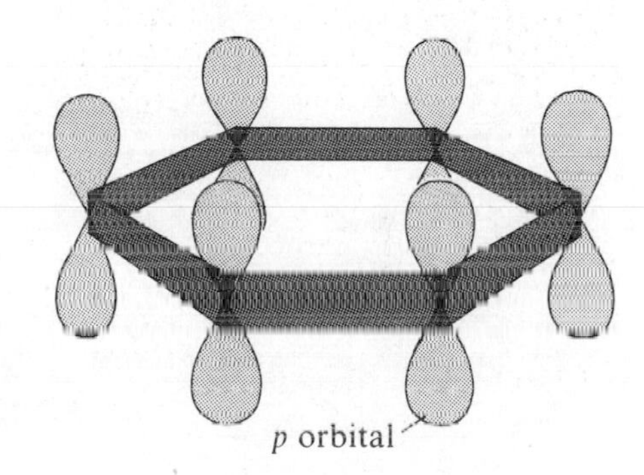


(a) $\sigma$-bonded skeleton (C and H atoms omitted for clarity)

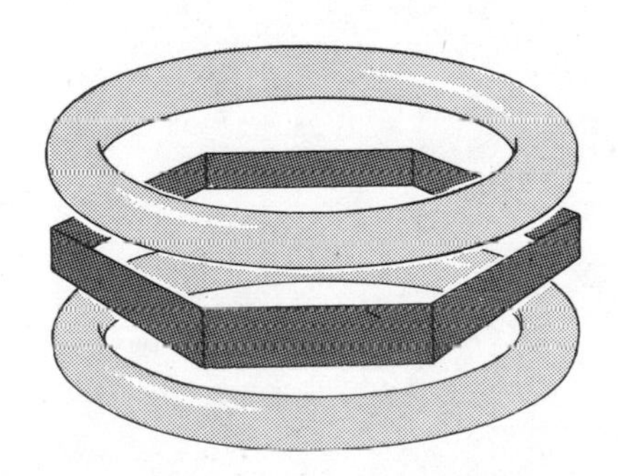

(b) $\sigma$-skeleton with $\pi$-bonds

**Figure 27.1** Bonding in benzene. Note that the plane of the molecule is perpendicular to the paper.

To conform with this model the structural formula of benzene is nowadays usually written as rather than . As we shall see in section

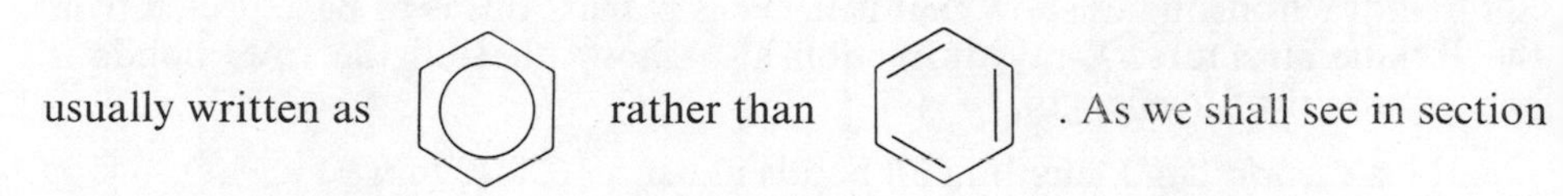

27.5, the delocalization of π-electrons has a profound effect on the chemical properties of benzene.

The word **aromatic** is used to describe any system that is stabilized by a ring of delocalized π-electrons. Non-aromatic compounds (such as alkanes and alkenes) are called **aliphatic**.

The structure described above means that the benzene molecule is planar, symmetrical and non-polar. Its lack of polarity results in benzene being a liquid at room temperature and immiscible with water. Its boiling point is 80°C and its melting point 6°C. The surprisingly high melting point is due to the ease with which highly symmetrical benzene rings can pack into a crystal lattice: compare it with that of the structurally similar but less symmetrical methylbenzene, $C_6H_5CH_3$, which melts at −95°C.

## 27.3 Nomenclature of aromatic compounds

Benzene is not the only aromatic hydrocarbon: many others exist, either substituted forms of benzene or compounds containing different ring systems. Some other aromatic hydrocarbons are discussed in section 27.9. Aromatic hydrocarbons are sometimes known as **arenes**. The name 'benzene' comes from *gum benzoin*, a natural product containing benzene derivatives. The name *pheno*, derived from the Greek 'I bear light' (benzene was originally isolated from illuminating gas by Michael Faraday), was at one time suggested as an alternative to 'benzene'. It was not adopted, but it survives in the word **phenyl** used for the $C_6H_5$— group. The phenyl group is an example of an **aryl** group.

The hydrogen atoms in a benzene ring can be substituted by other atoms and groups, as we shall see in section 27.5. Some examples are given in table 27.1.

**Table 27.1** Some derivatives of benzene.

| Substituent group | Systematic name | Other name |
|---|---|---|
| methyl, —$CH_3$ | methylbenzene | toluene |
| chloro, —Cl | chlorobenzene | — |
| nitro, —$NO_2$ | nitrobenzene | — |
| hydroxy, —OH | phenol | — |
| amino, —$NH_2$ | phenylamine | aniline |
| carboxylic acid, —COOH | benzoic acid | — |

Where more than one hydrogen atom is substituted, numbers are used to indicate which of the six possible ring positions are concerned.

Thus

$CH_3$ Cl 1 2 3 4 5 6 — methyl-2-chlorobenzene

$CH_3$ Cl — methyl-3-chlorobenzene

$CH_3$ Cl — methyl-4-chlorobenzene

The methyl group is regarded as occupying the 1 position and the ring is numbered clockwise as shown in the first formula. Two more examples:

$NH_2$ $NO_2$
3-nitrophenylamine

$CH_3$ $CH_3$
1,2-dimethylbenzene

Note that other numbers could be used (e.g. 3,4-dimethylbenzene instead of 1,2-dimethylbenzene), but the numbers actually employed are the lowest ones possible.

Use these rules and table 27.1 to write the structural formulae of the following compounds:

- ethylbenzene
- 2-methylphenol
- 1,3-dinitrobenzene.

Name the following:

- Br
- COOH, $CH_3$
- $CH_3$, $O_2N$, $NO_2$, $NO_2$

## 27.4 The importance of benzene

Detergents, polystyrene, nylon and insecticides can all be made from benzene, which is industrially as well as chemically the most important arene. Most benzene is normally manufactured from oil by **catalytic reforming**. In the presence of a catalyst, $C_6$–$C_8$ hydrocarbons from the gasoline fraction rearrange their molecules, producing a variety of aromatic hydrocarbons including benzene. In the past, benzene was produced as a by-product of the destructive distillation of coal: this can be carried out on a small scale in the laboratory experiment illustrated in figure 27.2.

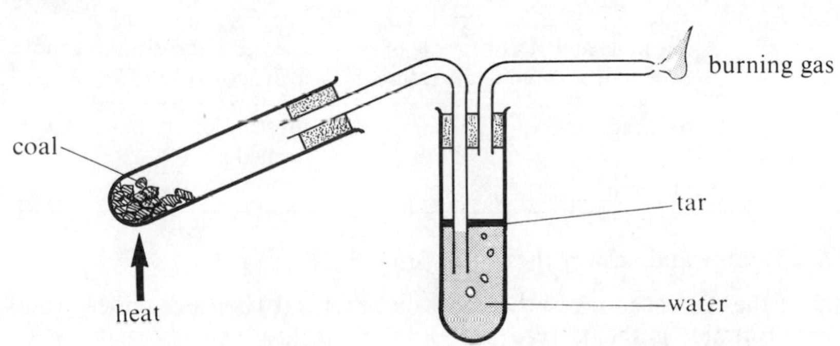


Figure 27.2 The destructive distillation of coal.

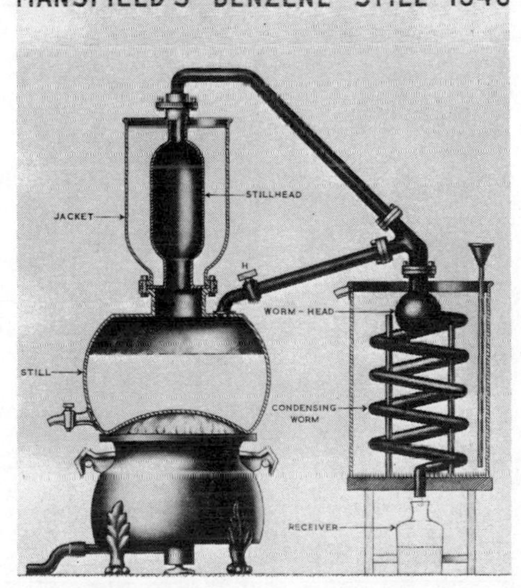


Early equipment for distilling benzene from coal tar.

*Right* A catalytic reforming plant in an oil refinery, used to convert gasoline fractions to a mixture of aromatic hydrocarbons.
*Above* A coal tar distillery, used to produce aromatic hydrocarbons. This rather messy method of production has been largely replaced by catalytic reforming of oil fractions.

Coal is heated in the absence of air: the gases produced are bubbled through water. Coal tar condenses and floats on the water, leaving a gaseous fuel called coal gas. The solid residue is coke. On an industrial scale this process was once the basis of the gas industry in Britain, also producing coke for boilers and for steel-making. The tar obtained contains a wide range of aromatic compounds which can be separated by distillation. At one time all aromatic hydrocarbons were produced from coal tar, but with the development of catalytic reforming, the decline of coal and the rise of North Sea gas, most aromatic hydrocarbons now come from oil.

## 27.5 Chemical characteristics of benzene

Table 27.2 compares some reactions of cyclohexane, cyclohexene and benzene.

**Table 27.2** Some reactions of cyclohexane, cyclohexene, and benzene.

| **Reagent** | **cyclohexane** | **cyclohexene** | **benzene** |
|---|---|---|---|
| bromine (in dark) | no reaction | bromine decolorized, no HBr evolved | no reaction with bromine alone, in presence of iron filings, bromine decolorized and HBr fumes evolved |
| acidified potassium manganate(VII) (potassium permanganate) | no reaction | manganate(VII) decolorized | no reaction |
| hydrogen over very finely divided nickel catalyst | no reaction | one mole absorbs one mole of hydrogen at room temperature | one mole absorbs three moles of hydrogen at 150°C |
| mixture of concentrated nitric(V) and concentrated sulphuric(VI) acids | no reaction | oxidized | substitution reaction: yellow oil formed |

Look at the table and answer these questions.

- Does the evidence suggest that (a) cyclohexene, (b) benzene undergo addition with bromine in the dark?

○ Does benzene undergo addition or substitution with bromine in the presence of iron filings?
○ Does benzene undergo catalytic hydrogenation as readily as cyclohexene?

As table 27.2 shows, benzene undergoes addition reactions far less readily than we might expect for so unsaturated a compound. Indeed, as the reactions with bromine in the presence of iron filings and with concentrated nitric(V) acid in the presence of concentrated sulphuric(VI) acid suggest, substitution reactions are more characteristic of benzene than addition reactions. If we remember the delocalization of $\pi$-electrons in benzene, it is quite easy to see why addition reactions are difficult. For example, if a molecule of benzene underwent addition with a molecule of bromine, the ring of delocalized $\pi$-electrons would be broken:

$$C_6H_6 + Br_2 \longrightarrow C_6H_6Br_2$$

This would require the input of considerably more energy than is needed to break the one double bond in cyclohexene. In its reactions, therefore, benzene has a tendency to maintain its $\pi$-electron ring intact and to undergo **substitution** rather than addition (see study question 10).

## 27.6 Mechanism of substitution reactions of benzene

Consider as an example the nitration of benzene. Benzene reacts with a mixture of concentrated nitric(V) acid and concentrated sulphuric(VI) acid (called a **nitrating mixture**) at 50°C to form nitrobenzene:

$$C_6H_6 + HNO_3 \xrightarrow[50°C]{c.\ H_2SO_4} C_6H_5NO_2 + H_2O$$

nitrobenzene
(a yellow oil)

This is a substitution reaction, hydrogen having been substituted by a nitro group, $NO_2$.

The reaction of benzene with concentrated nitric(V) acid alone is slow, whilst pure sulphuric(VI) acid at 50°C has practically no effect on benzene. This suggests that the sulphuric(VI) acid must somehow react with the nitric(V) acid, producing a species that then reacts with benzene. There is good evidence that this species is $NO_2^+$, the nitryl cation, formed by the removal of $OH^-$ from $HNO_3$ by sulphuric(VI) acid:

$$HNO_3 + 2H_2SO_4 \longrightarrow NO_2^+ + 2HSO_4^- + H_3O^+$$

Here nitric(V) acid is acting as a base in the presence of the stronger sulphuric(VI) acid. The $NO_2^+$ ion is a strong *electrophile*. It therefore tends to attack the negative $\pi$-electron system in benzene. First a loose association is formed:

$$C_6H_6 + NO_2^+ \longrightarrow C_6H_6\text{---}NO_2^+$$

The $NO_2^+$ then attacks one of the carbon atoms of the ring, forming a bond to it and disrupting the delocalized $\pi$ system:

$$C_6H_6\text{---}NO_2^+ \longrightarrow [C_6H_6NO_2]^+$$

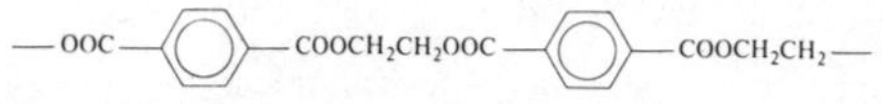

Polyester (section 31.5)

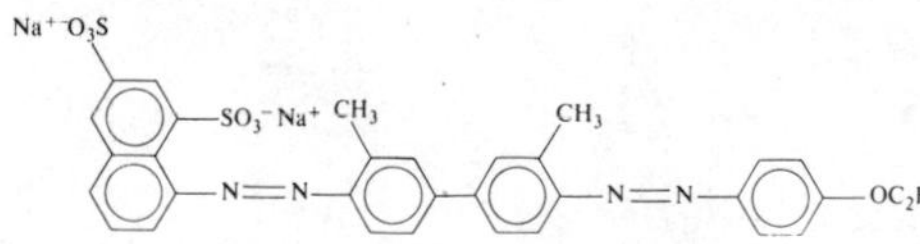

'Direct Red 39' – a red dye (section 32.7)

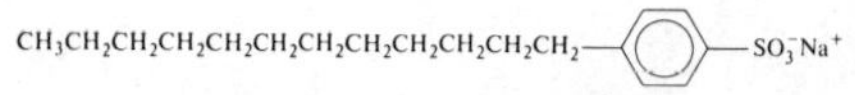

A synthetic detergent (section 31.3)

**Figure 27.3** Important aromatic compounds: polyester to make clothes, dyes to colour them, detergents to wash them.

The formation of this intermediate requires the input of considerable energy to break the delocalized $\pi$ system: the reaction therefore has a fairly high activation energy. The intermediate cation then breaks down, either reforming benzene or producing nitrobenzene and the delocalized $\pi$ system reforms, releasing energy:

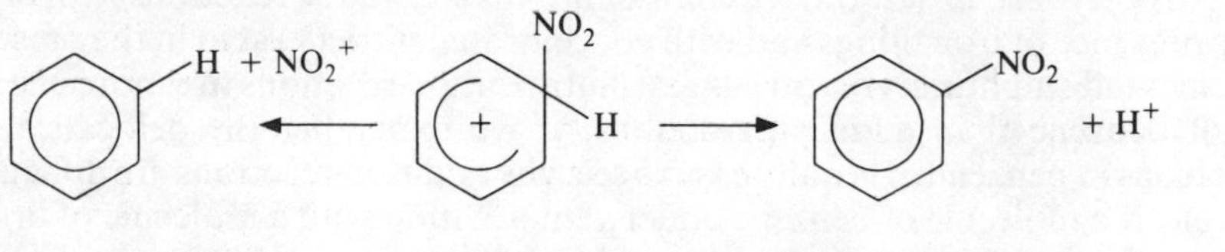

Notice that the first two stages of this mechanism, up to the formation of the intermediate cation, are similar to the stages in addition to an alkene double bond. The difference is that the aromatic cation loses its charge by loss of $H^+$, thus regaining aromatic character, rather than by combining with an anion to form an addition product, as is the case with alkenes where there is no delocalization energy to be lost or gained.

This kind of mechanism, in which an electron-deficient group (electrophile) attacks the $\pi$-electron system of the benzene ring to form an intermediate which breaks down to give substitution products, is called **electrophilic substitution**. It is the characteristic reaction of benzene and other aromatic hydrocarbons.

## 27.7 Important electrophilic substitution reactions of benzene

NITRATION

The detailed mechanism of this reaction has already been considered in the last section. It is used industrially to manufacture nitrobenzene, from which phenylamine (aniline), $C_6H_5NH_2$, is produced by reduction. Phenylamine is used to manufacture dyes (see chapter 32).

SULPHONATION

If benzene and concentrated sulphuric(VI) acid are refluxed together for several hours, benzenesulphonic acid is formed:

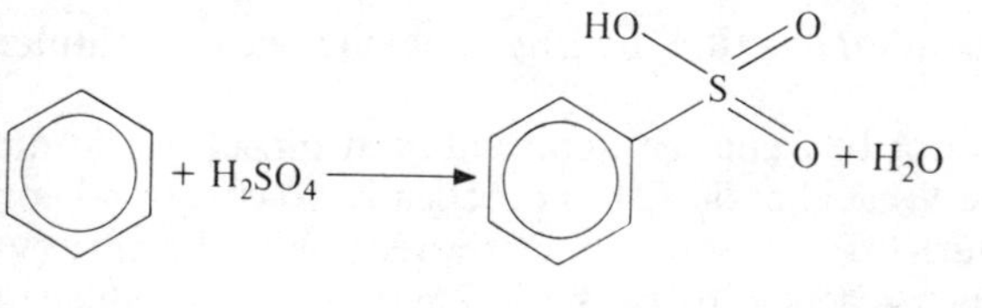

$$C_6H_6 + H_2SO_4 \longrightarrow C_6H_5SO_2OH + H_2O$$

This reaction is known as **sulphonation**. The electrophile that initially attacks the benzene ring is thought to be $SO_3$, which carries a large partial positive charge on the sulphur atom.

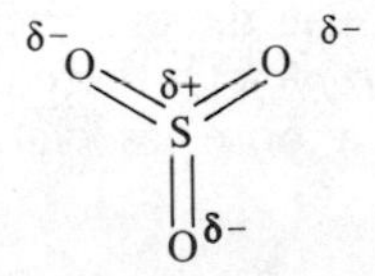

This theory is borne out by the fact that benzene is sulphonated in the cold by 'fuming sulphuric acid', which is a solution of $SO_3$ in concentrated sulphuric(VI) acid, but not by concentrated sulphuric(VI) acid alone at room temperature.

Benzenesulphonic acid is important industrially as an intermediate in the manufacture of phenol, used in the manufacture of plastics (section 29.7). Phenol, $C_6H_5OH$, is a simple derivative of benzene, but it cannot be made by the direct reaction of $^-OH$ with benzene because the high electron density in the $\pi$ system makes the ring unreactive toward negatively charged groups. When the sulphonate group is attached to a benzene ring, however, the electron-withdrawing effect of the group gives rise to a positive charge on the carbon atom to which it is attached.

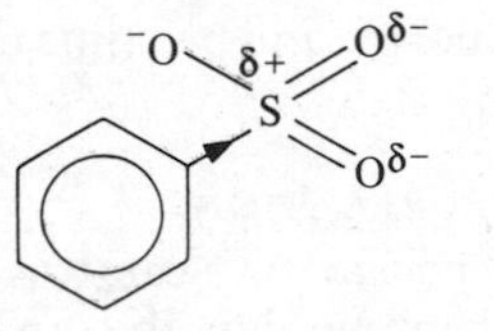

This makes the carbon atom susceptible to attack by negatively charged groups such as $^{-}OH$, which can displace the $SO_3^{2-}$ group. Consequently when benzenesulphonic acid is heated with molten sodium hydroxide phenol is formed:

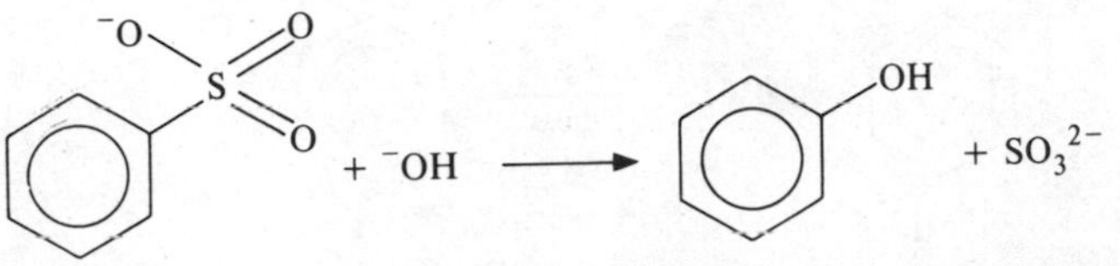

$$C_6H_5SO_3^{-} + {}^{-}OH \longrightarrow C_6H_5OH + SO_3^{2-}$$

This is one of several methods used today to manufacture phenol. What other group might have a similar effect to the sulphonate group in making the carbon atom it is attached to susceptible to attack by negatively charged groups?

### HALOGENATION

Benzene does not react with chlorine, bromine or iodine on their own in the dark. This is because the non-polar halogen molecule has no centre of positive charge to initiate electrophilic attack. However, in the presence of a catalyst such as iron filings, iron(III) bromide or aluminium chloride, benzene is substituted by chlorine or bromine:

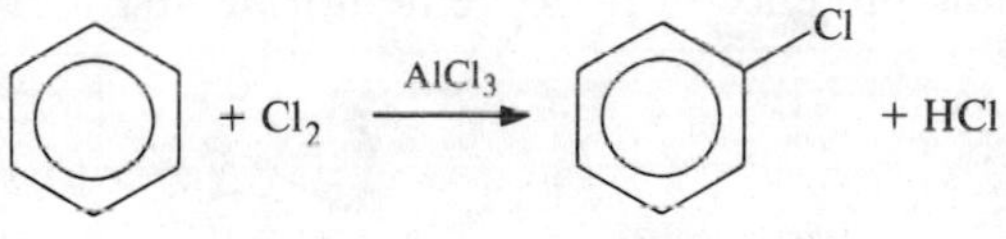

$$C_6H_6 + Cl_2 \xrightarrow{AlCl_3} C_6H_5Cl + HCl$$

chlorobenzene

The catalyst (called a **halogen-carrier**) is thought to work by inducing polarization in the halogen molecule by accepting a lone pair from it:

$$\overset{\delta+}{Cl}—\overset{\delta-}{Cl}: \longrightarrow AlCl_3$$

The positively charged end of the halogen molecule is now electrophilic and attacks the benzene ring.

- ○ What would you expect to be the product of a substitution reaction between benzene and bromine(I) chloride, BrCl?
- ○ Iodine is too unreactive to substitute benzene even in the presence of a halogen carrier. Quite good yields of iodobenzene can however be obtained by reacting benzene with iodine(I) chloride, ICl. Explain why.

The electron-withdrawing tendency of the Cl group in chlorobenzene has a similar effect to that of the $—SO_3H$ group in benzenesulphonic acid, making the carbon atom to which it is attached susceptible to nucleophilic attack. For example, when boiled with hot, aqueous sodium hydroxide under pressure, chlorobenzene undergoes nucleophilic substitution to form phenol:

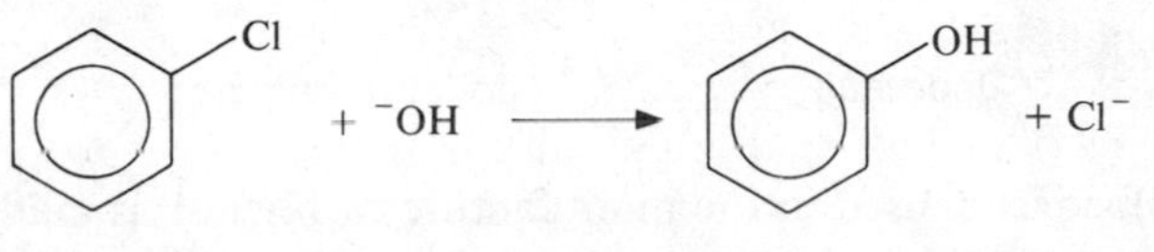

$$C_6H_5Cl + {}^{-}OH \longrightarrow C_6H_5OH + Cl^{-}$$

This reaction provides another method for the manufacture of phenol from benzene. Other reactions of chlorobenzene are described in chapter 28.

Interhalogen compounds such as BrCl will substitute benzene without a halogen-carrier, since they are already polarized. BrCl reacts with benzene to form bromobenzene.

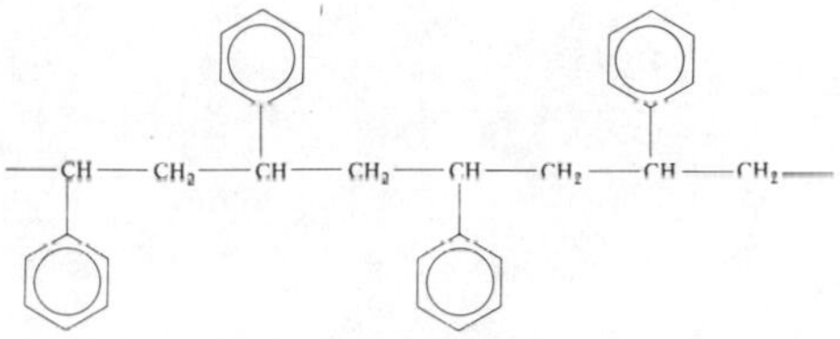

Polystyrene (section 26.7)
Made from phenylethene (styrene) (section 27.7)

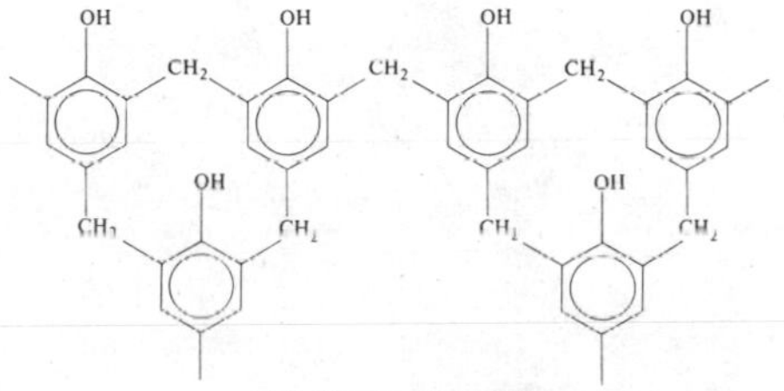

Bakelite (section 29.7)
Made from phenol and methanal

**Figure 27.4** Plastics manufactured from aromatic compounds.

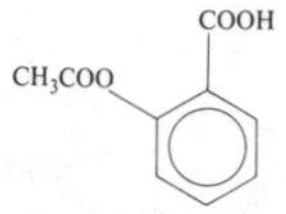

COOH
$CH_3COO$

Aspirin (section 31.6)

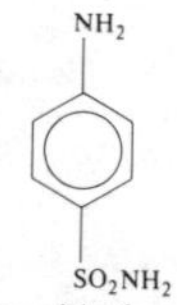

$NH_2$

$SO_2NH_2$

A sulphonamide (section 32.6)
Sulphonamides were among the earliest antibacterial drugs.

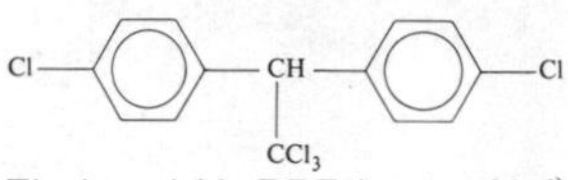

Cl — CH — Cl
$CCl_3$

The insecticide DDT (section 30.6)

**Figure 27.5** Some aromatic compounds of biological importance.

In the presence of sunlight, chlorine and bromine undergo *addition* reactions with benzene (see section 27.8).

ALKYLATION: FRIEDEL–CRAFTS REACTION

Just as aluminium chloride can be used as a catalyst to polarize halogen molecules and cause them to substitute a benzene ring, the same catalyst can be used to bring about the substitution of a benzene ring by halogenoalkanes. For example, if benzene is warmed with chloromethane and aluminium chloride under anhydrous conditions, a substitution reaction occurs and methylbenzene is formed:

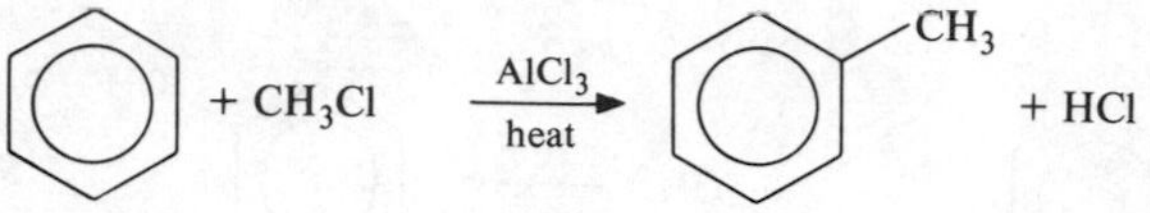

$$C_6H_6 + CH_3Cl \xrightarrow[\text{heat}]{AlCl_3} C_6H_5CH_3 + HCl$$

As before, the aluminium chloride accepts an electron pair from the chlorine atom, polarizing the chloromethane molecule:

$$\overset{\delta+}{CH_3}\text{—}\overset{\delta-}{Cl}: \longrightarrow AlCl_3$$

Polarization may reach the extreme where a carbonium ion is actually formed. The positively charged methyl group attacks the benzene ring and electrophilic substitution occurs. This is an example of a **Friedel-Crafts** reaction. Such reactions occur between aromatic hydrocarbons and any combination of reagents that can give rise to a positively charged carbon atom: the latter includes alkenes and alcohols as well as halogenoalkanes. For example, ethene and benzene undergo a Friedel-Crafts reaction in the presence of hydrogen chloride and aluminium chloride, to form ethylbenzene:

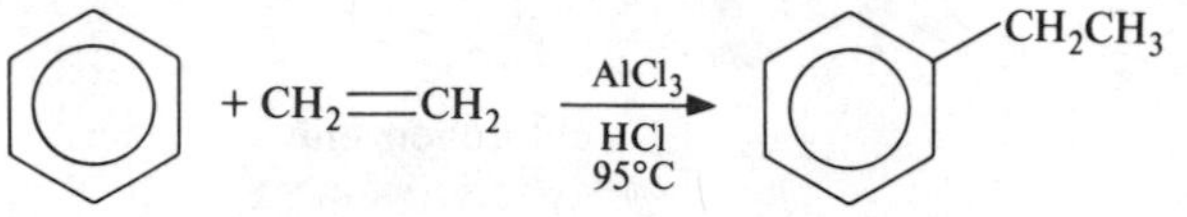

$$C_6H_6 + CH_2{=}CH_2 \xrightarrow[\text{HCl, 95°C}]{AlCl_3} C_6H_5CH_2CH_3$$

ethylbenzene

This reaction is used industrially to manufacture ethylbenzene from which phenylethene (styrene) is made by catalytic dehydrogenation:

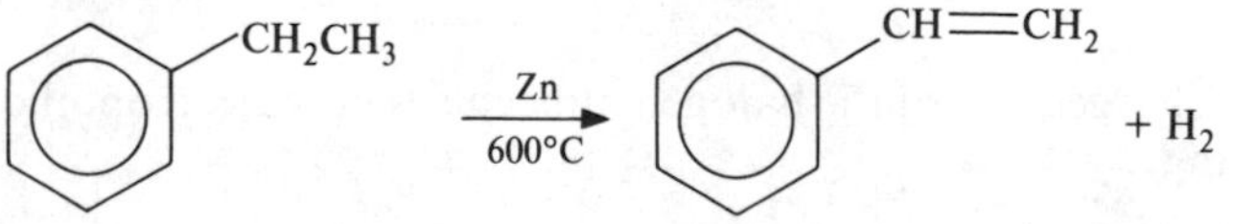

$$C_6H_5CH_2CH_3 \xrightarrow[600°C]{Zn} C_6H_5CH{=}CH_2 + H_2$$

phenylethene

- ○ What important plastic is manufactured from phenylethene?
- ○ Ethylbenzene could be manufactured by a Friedel-Crafts reaction between chloroethane and benzene. Why is ethene used in preference to chloroethane?

Dodecylbenzene, important in the manufacture of detergents (see section 31.3) is made by a Friedel-Crafts reaction between benzene and dodecene:

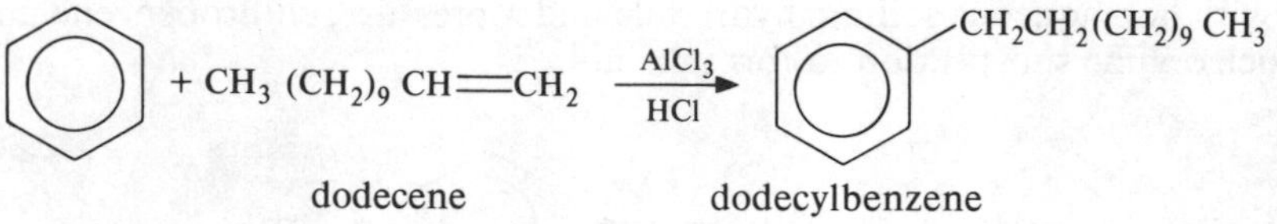

$$C_6H_6 + CH_3(CH_2)_9CH{=}CH_2 \xrightarrow[\text{HCl}]{AlCl_3} C_6H_5CH_2CH_2(CH_2)_9CH_3$$

dodecene — dodecylbenzene

(1-Methylethyl)benzene, used in the manufacture of phenol, is made by a similar reaction between benzene and propene (see study question 7).

# 27.8 Other important reactions of benzene

The characteristic reactions of benzene and other aromatic hydrocarbons involve electrophilic substitution, because this type of reaction retains the delocalized $\pi$-electron system. Reactions involving the disruption of this system do occur, however.

ADDITION REACTIONS

1 *With hydrogen.* Alkenes undergo addition with hydrogen in the presence of a nickel catalyst. Benzene also gives this reaction, but considerably higher temperatures are required than for aliphatic compounds. The higher temperature is needed because extra energy must be supplied to break up the delocalized $\pi$-electron system.

$$3H_2 + \text{C}_6\text{H}_6 \xrightarrow[150°\text{C}]{\text{Raney nickel}} \text{C}_6\text{H}_{12}$$

cyclohexane

The catalyst, known as Raney nickel, is a form of nickel with an extremely high surface area, and is very active.

If one mole of benzene and one mole of hydrogen are reacted in this way, one third of the benzene is converted into cyclohexane and the remainder is left unreacted. We might expect some cyclohexadiene, and cyclohexene, to be formed, but this does not in fact happen. Can you suggest a reason why?

The catalytic hydrogenation of benzene is important industrially in the manufacture of cyclohexane, from which nylon is made (section 32.9).

2 *With chlorine.* Benzene can undergo addition as well as substitution reactions with chlorine. In ultraviolet light, chlorine adds to benzene to form 1,2,3,4,5,6-hexachlorocyclohexane. The need for light suggests a free-radical mechanism (section 25.5).

$$\text{C}_6\text{H}_6 + 3Cl_2 \xrightarrow{\text{light}} \text{C}_6\text{H}_6\text{Cl}_6$$

There are eight geometric isomers for 1,2,3,4,5,6-hexachlorocyclohexane due to the restricted rotation about the C–C bonds. Try drawing them or making models of them. One of these eight forms is a very effective insecticide, known commercially as Gammexane or B.H.C.

BURNING

Benzene burns in air with a sooty, smoky flame. This sort of flame is characteristic of all hydrocarbons containing a high percentage of carbon.

The remains of the Nypro plant at Flixborough after the 1974 explosion. As the result of a leak, cyclohexane, made from benzene and used in the manufacture of nylon, mixed with air and exploded.

## 27.9 Other arenes

Some examples of arenes other than benzene are shown in table 27.3.

**Table 27.3** Some arenes.

| Systematic name | Other name | Molecular formula | Structural formula |
|---|---|---|---|
| methylbenzene | toluene | $C_7H_8$ | $CH_3$ |
| 1,3-dimethylbenzene | para-xylene | $C_8H_{10}$ | $CH_3$, $H_3C$ |
| naphthalene | — | $C_{10}H_8$ | |
| anthracene | — | $C_{14}H_{10}$ | |

METHYLBENZENE

The reforming reactions used to manufacture benzene from petroleum (see section 27.4) also produce considerable quantities of **methylbenzene** (commonly called toluene). This substance is used in the manufacture of plastics and explosives (see below), but much more is produced than can be consumed in this way. The majority of the methylbenzene produced is added to motor fuel to increase its octane rating: it does not have the pollution disadvantage of tetraethyllead(IV). Some methylbenzene is converted to benzene, which is more useful.

REACTIONS OF METHYLBENZENE

We have seen that the properties of aromatic compounds are very different from those of aliphatic ones. Methylbenzene has an aromatic portion (the benzene ring) and an aliphatic portion (the —$CH_3$ group). These two portions will make different contributions to the properties of methylbenzene and will have a modifying effect on one another.

1 *The $CH_3$ group*. The $CH_3$ group shows some reactions we would expect of an alkyl group, for example it can be substituted by chlorine. This reaction occurs when chlorine is bubbled into boiling methylbenzene in sunlight:

$$C_6H_5CH_3 + Cl_2 \xrightarrow{\text{sunlight}} C_6H_5CH_2Cl + HCl$$

○ Write the structure of two other compounds that might form when the $CH_3$ group is substituted by chlorine.

○ Do the reaction conditions suggest that homolytic fission or heterolytic fission of the chlorine molecule is involved?

○ Under certain conditions chlorine will substitute the *ring* in methylbenzene instead of the $CH_3$ group. What will these conditions be?

The $CH_3$ group in methylbenzene does not behave as a typical alkyl group in all its reactions. The benzene ring with its regions of high electron density has a modifying effect on any group that is attached to it: we shall see later that a given functional group behaves differently depending on whether it is attached to an aliphatic or to an aromatic molecule. One example of the way the ring modifies the properties of the $CH_3$ group in methylbenzene is the reaction with potassium manganate(VII). Alkanes, as we have seen, are inert to oxidation, but the alkyl group in methylbenzene can be oxidized by alkaline manganate(VII) to give benzoic acid. The manganate(VII) is reduced to green manganate(VI).

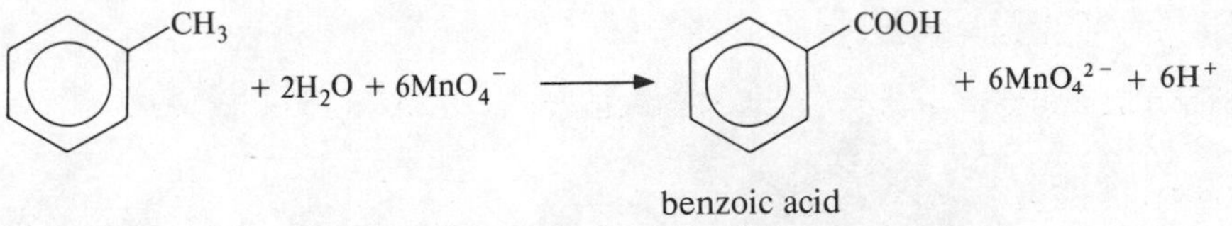

$$C_6H_5CH_3 + 2H_2O + 6MnO_4^- \longrightarrow C_6H_5COOH + 6MnO_4^{2-} + 6H^+$$

benzoic acid

Note that the ring is not affected, another indication of its stability.

2 *The aromatic ring*. If chlorine is bubbled through methylbenzene in the absence of sunlight and in the presence of a halogen carrier such as $AlCl_3$, the ring is substituted instead of the $CH_3$ side-chain. This reaction proceeds by the electrophilic substitution mechanism described in section 27.6, rather than by the free radical mechanism involved in side-chain chlorination. A mixture of two isomers is obtained:

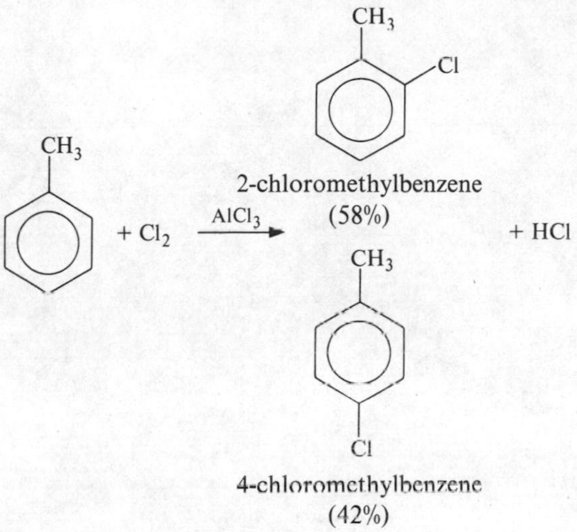

$$C_6H_5CH_3 + Cl_2 \xrightarrow{AlCl_3} \text{2-chloromethylbenzene (58\%)} + \text{4-chloromethylbenzene (42\%)} + HCl$$

Virtually none of the other possible isomer, 3-chloromethylbenzene, is produced. Furthermore, the reaction proceeds at a considerably higher rate than the corresponding reaction of chlorine with benzene. Clearly the $CH_3$ group has influenced the aromatic ring, making it more susceptible to electrophilic substitution and dictating the positions in which it is substituted. This effect is discussed further in the next section.

An important substitution reaction of methylbenzene is nitration. When heated with a mixture of concentrated nitric(V) acid and concentrated sulphuric(VI) acid, methylbenzene is substituted by one, two or three $NO_2$ groups, depending on the conditions. The main products are

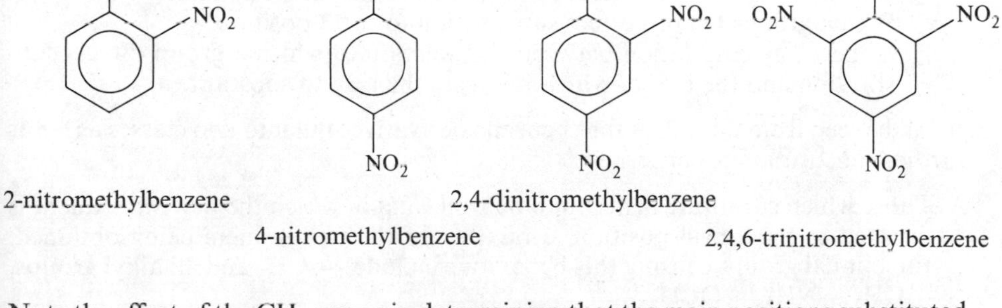

2-nitromethylbenzene

4-nitromethylbenzene

2,4-dinitromethylbenzene

2,4,6-trinitromethylbenzene

Note the effect of the $CH_3$ group in determining that the main positions substituted are 2,4 and 6 rather than 3 or 5. 2,4,6-Trinitromethylbenzene (trinitrotoluene or TNT) is an important high explosive. It is fairly resistant to shock and so can be used in shells without risk of explosion under the shock of firing from a gun. When detonated it decomposes forming large volumes of CO, $H_2O$ and $N_2$ at high temperature. The sudden expansion of these gases is responsible for the explosive force of TNT. TNT has the advantage that it is a solid which melts below 100°C, so it can be melted with steam and poured into its container.

Blowing up a large building. High explosives such as TNT owe their explosive force to the production of large volumes of gas at high temperature.

## 27.10 Position of substitution in benzene derivatives

We have already seen that the methyl group in methylbenzene activates the ring toward electrophilic substitution and favours substitution in positions 2,4 and 6 rather than 3 or 5. In fact, *any substituent group* attached to a benzene ring affects the rate and the position at which further substitution occurs. Table 27.4 shows the main products of mononitration (i.e. substitution by one nitro group) of different benzene derivatives, and whether they are nitrated faster or slower than benzene.

Look at the table and answer these questions.

- ○ Which groups tend to direct substitution to the 2 or 4 position?
- ○ Which groups tend to direct substitution to the 3 position?
- ○ Is there any correlation between the position to which a group directs substitution and the rate at which it causes the ring to substitute?

It can be seen from table 27.4 that benzene derivatives fall into two classes as far as further substitution is concerned:

**A** Those which substitute faster than benzene and in which the new substituent is directed to the 2 or 4 position, a mixture of the two isomers being obtained. Functional groups causing this behaviour include: —$CH_3$ and all alkyl groups, —OH, —$NH_2$, —$OCH_3$.

**B** Those which substitute slower than benzene and in which the new substituent is directed to the 3 position. Functional groups causing this behaviour include —COOH, —$SO_3H$, —$NO_2$.

In practice, a mixture of all possible isomers is obtained, but these rules give the main products. Note that the rules apply whatever the nature of the new substituent, not just to nitration. Note too, that as far as monosubstitution is concerned, the 6 position is equivalent to the 2 position and the 5 to the 3.

**Table 27.4** Mononitration products of benzene derivatives.

| Compound | Main products of mononitration | Rate of nitration relative to benzene |
|---|---|---|
| methylbenzene ($CH_3$) | $CH_3$, $NO_2$ (2-nitro); $CH_3$, $NO_2$ (4-nitro) | Faster |
| phenol (OH) | OH, $NO_2$ (2-nitro); OH, $NO_2$ (4-nitro) | Faster |
| nitrobenzene ($NO_2$) | $NO_2$, $NO_2$ (3-nitro) | Slower |
| phenylamine ($NH_2$) | $NH_2$, $NO_2$ (2-nitro); $NH_2$, $NO_2$ (4-nitro) | Faster |
| benzoic acid (COOH) | COOH, $NO_2$ (3-nitro) | Slower |

○ Write structural formulae for the main products you would expect from the following substitution reactions. Assume monosubstitution occurs in each case.

(a) $CH_3$-benzene $\xrightarrow[AlCl_3]{CH_3Cl}$

(b) $NO_2$-benzene $\xrightarrow[AlCl_3]{Cl_2}$

(c) $CH_2CH_2CH_3$-benzene $\xrightarrow[c.H_2SO_4]{c.HNO_3}$

(d) COOH-benzene $\xrightarrow{c.H_2SO_4}$

## Summary

1 Aromatic hydrocarbons (arenes) are ring compounds stabilized by electron delocalization.
2 Benzene, $C_6H_6$, has a symmetrical planar ring of six carbon atoms with a ring of delocalized $\pi$-electrons above and below the plane of the molecule.
3 Benzene is manufactured by catalytic reforming of petroleum fractions.
4 Disubstituted benzene derivatives are named by referring to numbered positions on the ring.
5 The $C_6H_5$— group is called phenyl.
6 Benzene undergoes addition considerably less readily than alkenes.
7 The characteristic reaction type of benzene is electrophilic substitution. Some important substitution reactions are shown on the following diagram.

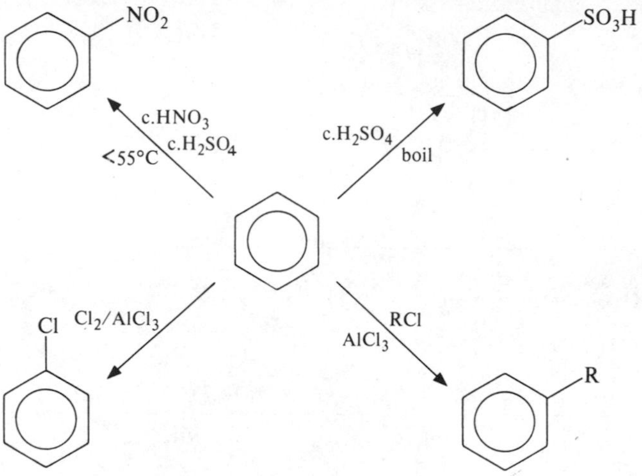

$NO_2$ c.$HNO_3$ c.$H_2SO_4$ <55°C

$SO_3H$ c.$H_2SO_4$ boil

Cl $Cl_2/AlCl_3$

R RCl $AlCl_3$

8 A benzene ring considerably influences the properties of functional groups attached to it. A functional group influences the position and rate of substitution of the benzene ring to which it is attached.
9 Methylbenzene shows some properties of an alkane and some of an arene, but the properties of the $CH_3$ group are substantially modified by the ring.

## Study questions

1 Name the following benzene derivatives:

(a) Br, Br

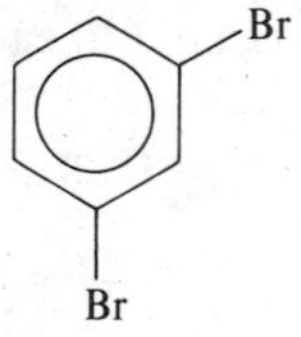

(b) $CH_3$, $CH_2CH_3$

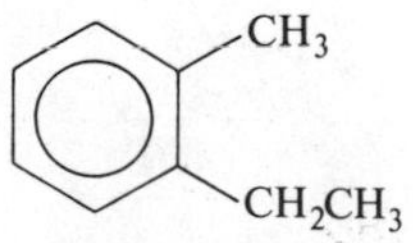

(c) $CH_3$, $CH_3$, $CH_3$

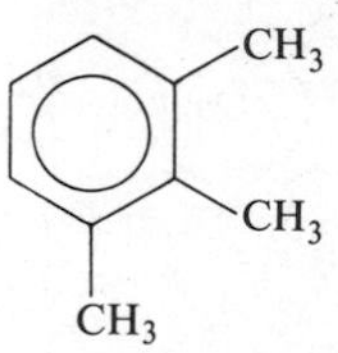

(d) Br, $NO_2$, Cl

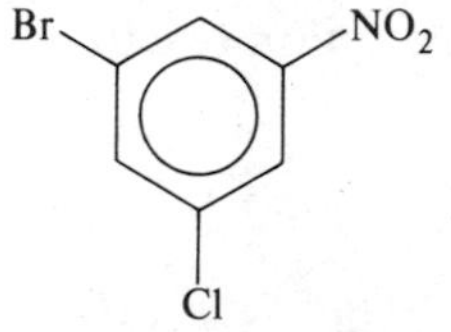

(e) COOH, $NO_2$

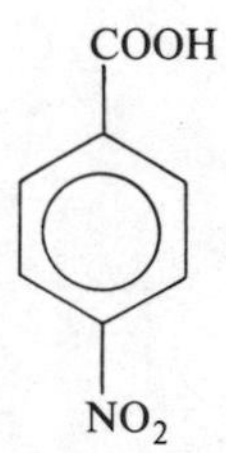

(f) OH, Cl

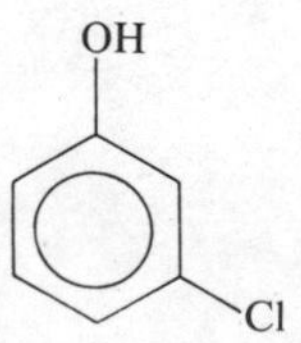

2 Write structural formulae for the following benzene derivatives:
(a) 2,4,6-trinitrophenol
(b) 1,4-dichlorobenzene
(c) 4-nitrophenylamine
(d) 2-methylbenzenesulphonic acid
(e) 2-hydroxybenzoic acid
(f) 2-chlorophenylamine.

3 Consider the catalytic hydrogenation of cyclohexene:

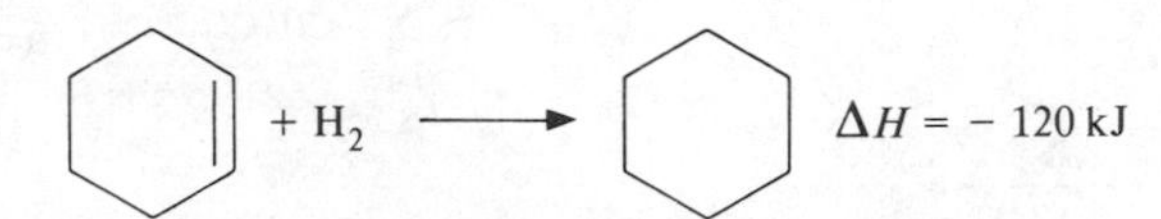


(a) Suggest a suitable catalyst for this reaction.
(b) Assuming benzene has the cyclohexatriene (Kekulé) structure 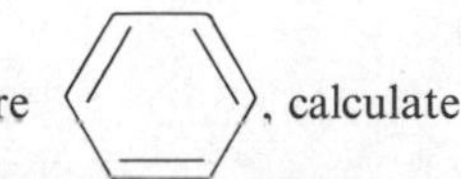, calculate the expected value of its heat of hydrogenation to cyclohexane, using the data above.
(c) Compare your answer in part (b) with the experimental value:

$+ 3H_2 \longrightarrow \quad \Delta H = -208\ \text{kJ}$

(d) Suggest why the two values differ.

4 If deuterium chloride, DCl, is dissolved in methylbenzene, no reaction occurs. If, however, anhydrous aluminium chloride is added to the solution, hydrogen atoms on the aromatic ring are rapidly substituted by deuterium atoms. If excess DCl is present, all five ring hydrogens are substituted, forming:

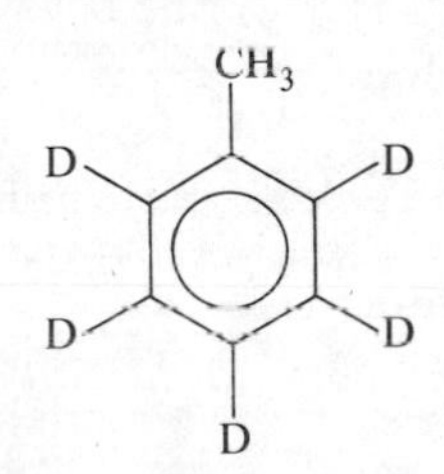


(a) Write the structure of the complex formed between $AlCl_3$ and DCl.
(b) What is the effect of the $AlCl_3$ on the polarization of the DCl molecule?
(c) Why is $AlCl_3$ effective in causing DCl to substitute the aromatic ring?
(d) Why are none of the hydrogen atoms on the $CH_3$ side-chain substituted, even in the presence of excess DCl?

5 How would you distinguish between the members of the following pairs of compounds, using simple chemical tests?

(a)
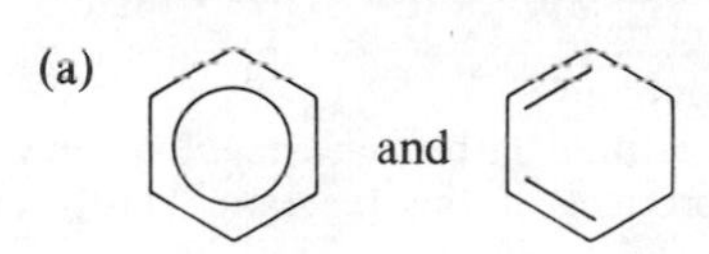


(b)
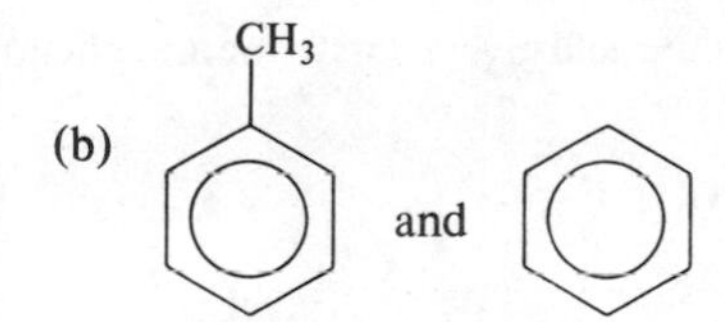


(c)
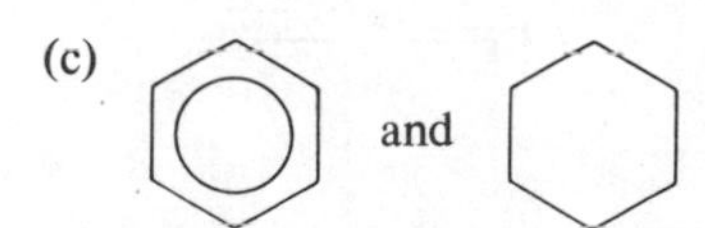


(d) and $CH_2{=}CH{-}CH{=}CH{-}CH{=}CH_2$

6 Predict the major products of the following reactions.

(a) Methylbenzene ($CH_3$ on benzene ring) $\xrightarrow[300°C]{H_2/Ni}$

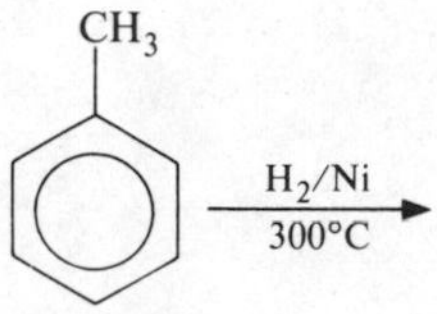

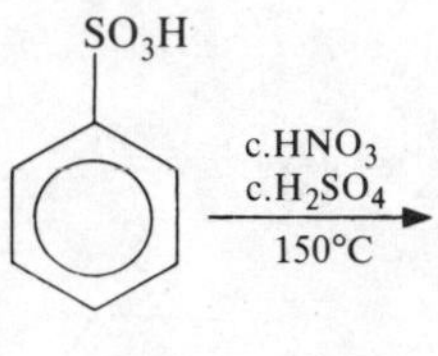

(b) Benzenesulfonic acid ($SO_3H$ on benzene ring) $\xrightarrow[150°C]{c.HNO_3,\ c.H_2SO_4}$

(c) 1,4-dimethylbenzene ($CH_3$ and $CH_3$ on benzene ring) $\xrightarrow[warm]{alkaline\ KMnO_4(aq)}$

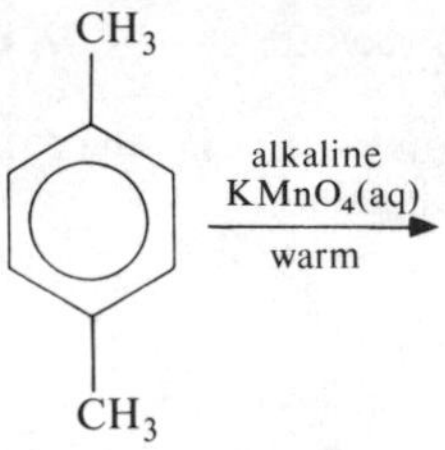

(d) 1,3-dinitrobenzene ($NO_2$ and $NO_2$ on benzene ring) $\xrightarrow[120°C]{c.HNO_3,\ c.H_2SO_4}$

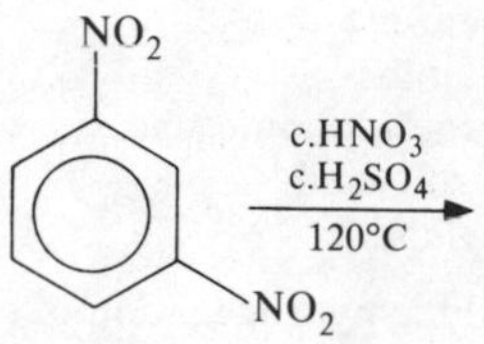

(e) Benzene $\xrightarrow[AlCl_3,\ warm]{CH_3CHCl\ CH_3}$ (monosubstituted product only)

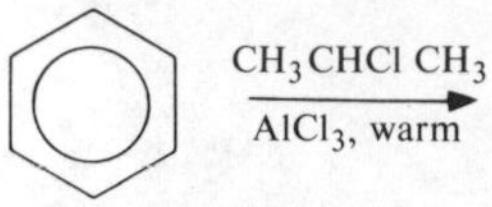

(f) Methylbenzene ($CH_3$ on benzene ring) $\xrightarrow[cold,\ in\ dark]{Br_2/FeBr_3}$

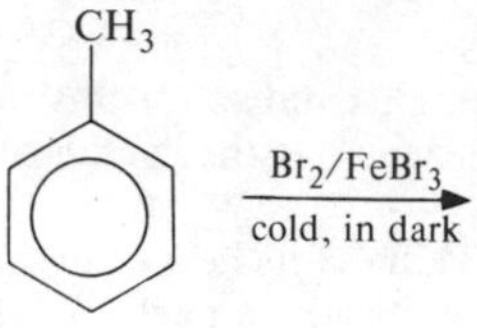

7 (1-methylethyl)benzene, commonly called cumene, has the structure

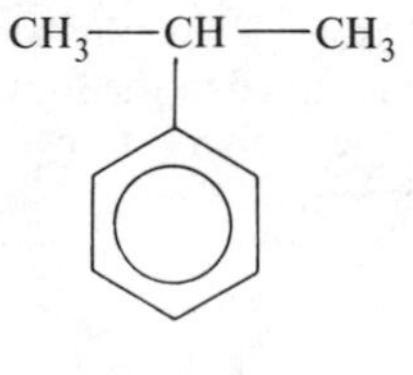

$CH_3—CH—CH_3$ (with benzene ring attached to CH)

and is an important intermediate in a process used to manufacture phenol and propanone (acetone). It is manufactured by a Friedel-Crafts-type reaction between propene and benzene in the presence of an acid catalyst:

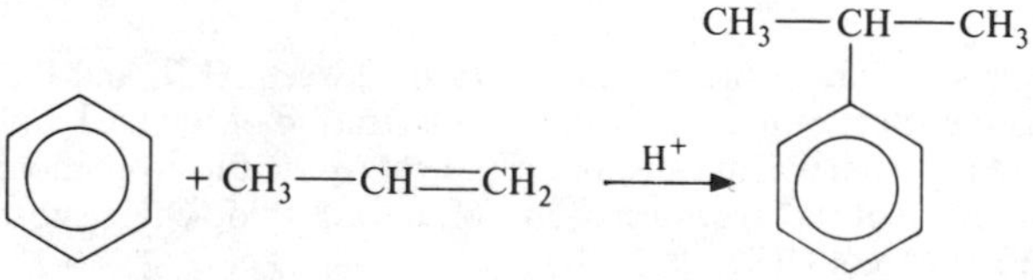

$$C_6H_6 + CH_3—CH{=}CH_2 \xrightarrow{H^+} CH_3—CH(C_6H_5)—CH_3$$

The reaction is believed to proceed via a carbonium ion intermediate.
(a) Write the structures of the two possible carbonium ions formed by the attack of $H^+$ on the double bond of propene.
(b) Which is the more stable of these two carbonium ions?
(c) Show how this carbonium ion can attack and substitute the benzene ring. Explain why (l-methylethyl)benzene is virtually the only product of this reaction, hardly any propylbenzene being produced.

Treatment of the product with air followed by dilute acid gives propanone and phenol:

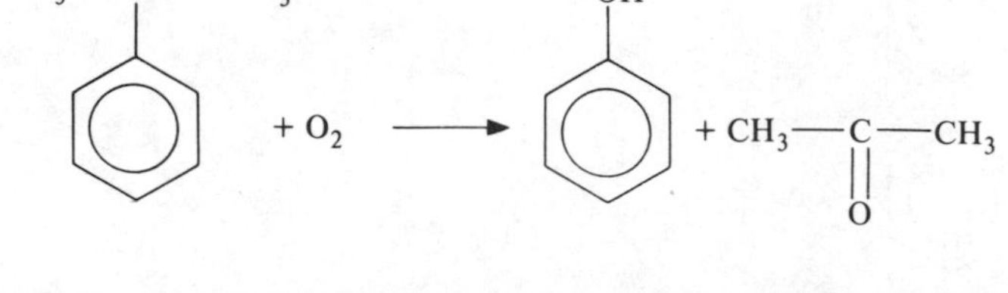

$$CH_3—CH(C_6H_5)—CH_3 + O_2 \longrightarrow C_6H_5OH + CH_3—\underset{\underset{O}{\|}}{C}—CH_3$$

(d) Why is this the most economic of the several methods available for manufacturing phenol?

8 (a) Write structural formulae for all the compounds of molecular formula $C_8H_{10}$ containing one benzene ring.
(b) For each of these compounds, write the formulae of all the possible mononitration products (not just the ones you would expect for the major product).
(c) For one of the compounds, state which of the mononitration products you would expect to be produced in the majority.

9 Consider the following compounds.

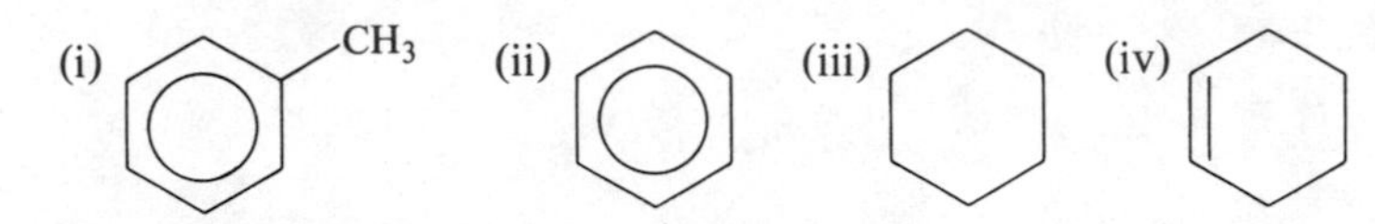


(v) $CH_3CH_2CH{=}CHCH_2CH_3$

(a) Which are aromatic hydrocarbons?
(b) Which are cyclic compounds?
(c) Which are unsaturated hydrocarbons?
(d) Which have a planar ring in their molecule?
(e) Which would decolorize bromine water in the dark?
(f) Which would evolve fumes of HBr when treated with bromine and iron filings?
(g) Which would react with alkaline potassium manganate(VII) solution?

10 Consider two possible reactions of chlorine with benzene.

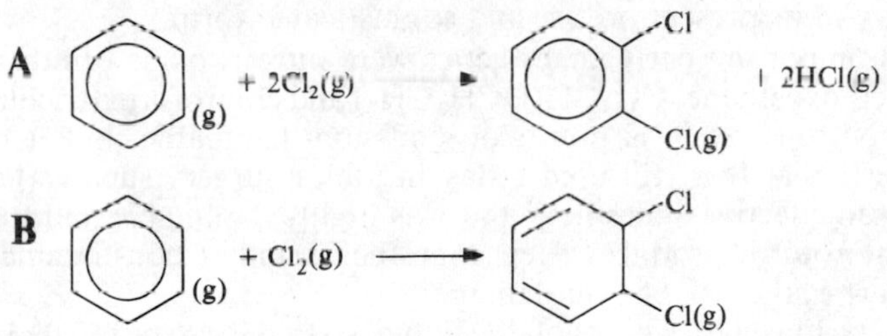

**A** (g) + $2Cl_2(g)$ → Cl, Cl(g) + 2HCl(g)

**B** (g) + $Cl_2(g)$ → Cl, Cl(g)

Some relevant bond energies (in kJ $mole^{-1}$) are: Cl–Cl 242; C–Cl (general value) 339; C–H (in benzene) 430; H–Cl 431.
(a) Which of the two reactions is addition and which is substitution?
(b) Calculate the energy change in reaction **A** by the following stages.
  (i) How much energy is needed to split two chlorine molecules into atoms?
  (ii) How much energy is needed to break two C–H bonds in benzene?
  (iii) How much energy is released when two new C–Cl bonds form?
  (iv) How much energy is released when two HCl molecules are formed?
  (v) What is the total energy change for reaction **A**?
(c) Calculate the energy change in reaction **B** by the following stages.
  (i) The energy needed to break one C–C bond in benzene, i.e. convert

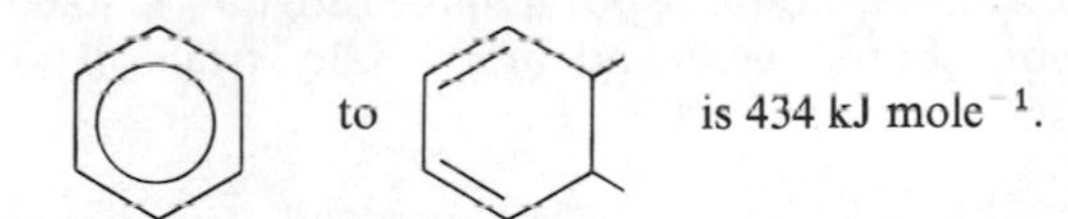


is 434 kJ $mole^{-1}$.

  (ii) How much energy is needed to split one chlorine molecule into atoms?
  (iii) How much energy is released when two C–Cl bonds form?
  (iv) What is the total energy change in reaction **B**?
(d) Which reaction is more likely to occur between chlorine and benzene, **A** or **B**? Give reasons for your answer.

# 28 Organic Halogen Compounds

Early surgery was primitive and performed without anaesthetics.

## 28.1 Anaesthetics

Before the advent of anaesthetics, surgery was a savage and primitive affair. It was agony for the patient, and surgeons were therefore only prepared to operate if it was absolutely essential, for example in wartime the amputation of a damaged limb that would otherwise become gangrenous. Anaesthetics enabled surgery to develop from crude carpentry to its present precise and sophisticated form.

Three of the most important early anaesthetics were nitrous oxide (dinitrogen oxide, $N_2O$), ether (ethoxyethane, $CH_3CH_2OCH_2CH_3$) and chloroform (trichloromethane, $CHCl_3$). Nitrous oxide is non-toxic and non-flammable, but it only produces light anaesthesia. It is still used today in minor surgery such as tooth extraction. Ether is an effective anaesthetic but it is highly flammable and therefore dangerous. Chloroform produces deep anaesthesia and is non-flammable, but it is toxic and carries the risk of liver damage.

The ideal anaesthetic must be a gas or volatile liquid, so that it can be inhaled and absorbed via the lungs. It must be non-flammable. It must produce deep anaesthesia but must be non-toxic. In 1951, ICI began the search for the ideal anaesthetic. They decided to look at **halogenoalkanes**. It was known that the substitution of chlorine atoms into an alkane chain gave it anaesthetic properties, but also made it toxic. Thus dichloromethane, $CH_2Cl_2$, is a fairly weak anaesthetic with little toxicity, trichloromethane (chloroform), $CHCl_3$, is stronger and more toxic, and tetrachloromethane is a very strong anaesthetic and also very toxic. The introduction of halogen atoms into an alkane skeleton also tends to make it non-flammable. Fluorine is useful in this respect as the C–F bond is very stable, so fluoroalkanes are unreactive, non-flammable and non-toxic. ICI therefore looked for a short chain halogenoalkane containing fluorine for inertness and chlorine for anaesthetic properties, with a suitable boiling point in the range 40°C to 60°C. They concentrated on two-carbon compounds, and after many trials produced the compound

```
F\       /H
F—C —— C—Cl
F/       \Br
```

The bromine atom was introduced to produce a substance with a sufficiently high boiling point to be conveniently stored. This substance, 2-bromo-2-chloro-1,1,1-trifluoroethane, was given the name Halothane, and since 1956 has been in widespread use in hospitals.

Halothane illustrates several of the important properties of halogenoalkanes: the increasing reactivity of the C–Hal bond as we go from F to I; the decreasing volatility of R–Hal as we move in the same direction and the effect of halogen atoms in reducing the flammability of a hydrocarbon chain.

Although organic halogen compounds are uncommon in nature a study of their properties is important to chemists as they have many uses in industry and in the laboratory.

## 28.2 Nomenclature

Organic halogen compounds are named using the prefixes **fluoro-**, **chloro-**, **bromo-**, and **iodo-**, numbered if necessary to indicate the position of the halogen atom in the molecule. Thus,

$CH_3CH_2Cl$ chloroethane $CH_3CHBrCH_3$ 2-bromopropane.

If the molecule contains more than one halogen atom of the same kind, the prefixes **di-**, **tri-**, etc. are used. Thus,

$CH_2ClCH_2Cl$ 1,2-dichloroethane,

$CHCl_2CHClCH_3$ 1,1,2-trichloropropane.

Name these compounds.
- ○ $CH_3CH_2CHICH_3$
- ○ $CH_3CHCl_2$

Write formulae for these compounds.
- ○ 1,3,5-tribromobenzene
- ○ 1,2-dibromo-3-chloropropane

## 28.3 The nature of the carbon-halogen bond

UNREACTIVE HALOGENOALKANES: THE C–F BOND

The C–F bond is very strong: compare its bond energy of 485 kJ mole$^{-1}$ with 435 for C–H and 327 for C–Cl. The C–F bond is thus very unreactive, as we have already seen in our discussion of Halothane. Consequently, fluorocarbons, compounds containing fluorine and carbon only, are extremely inert (see study question 6, chapter 24). The C–Cl bond is more reactive than C–F but nevertheless highly chlorinated compounds such as $CCl_4$ and $CHCl_3$ are fairly inert and in particular are non-flammable.

An important group of compounds is the chlorofluorocarbons, often known by the trade name 'Freons'.
Some examples are:
$CCl_3F$ trichlorofluoromethane,
$CCl_2F_2$ dichlorodifluoromethane,
$CCl_2F—CClF_2$ 1,1,2-trichloro-1,2,2-trifluoroethane.

These compounds have some useful properties:
(*a*) They are very stable, and in particular are non-flammable.
(*b*) They have low toxicity.
(*c*) They are odourless.
(*d*) They are volatile, with a wide range of boiling points.

Some examples of their uses are:
(i) *Refrigerants.* Dichlorodifluoromethane, $CCl_2F_2$, has a boiling point of $-30°C$, which makes it a useful refrigerant fluid. In the refrigerator the fluid is liquefied by compression then vaporized by sudden expansion, which gives a cooling effect. $CCl_2F_2$ is particularly useful for this purpose because it is unreactive so it does not corrode the machinery, and it is non-toxic so there is no danger from possible leaks.
(ii) *Aerosol propellants.* Dichlorodifluoromethane is also suitable for use as an aerosol propellant, either alone or mixed with other chlorofluorocarbons. Under pressure in the aerosol can it is a liquid, but when the valve is opened some of it vaporizes, carrying with it the active components: insecticide, paint, hair lacquer, or whatever. The stability of $CCl_2F_2$ means that it does not interact with the active component, and its lack of toxicity is of course vital. In 1976, 495 million aerosol packs were sold in the United Kingdom. Between them these pumped large quantities of chlorofluorocarbons into the atmosphere. Here the inertness of these compounds may have a disadvantage, for they tend to persist in the atmosphere for long periods, and there is evidence that they may interfere with the formation of the ozone layer that is vital in screening the earth from excess ultraviolet radiation. More research needs to be done before the case against chlorofluorocarbons is proven, but if it is the use of aerosol packs may have to be limited.

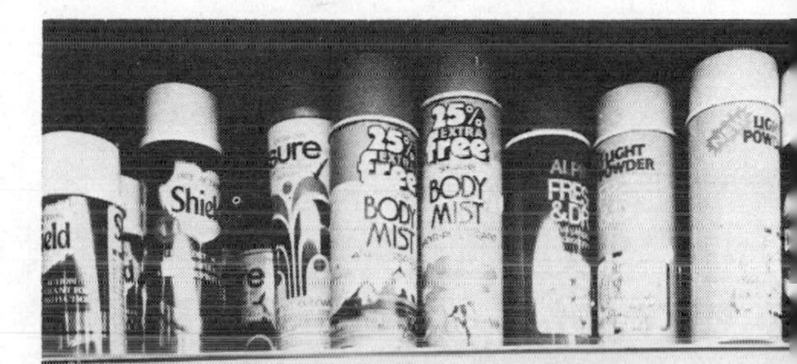

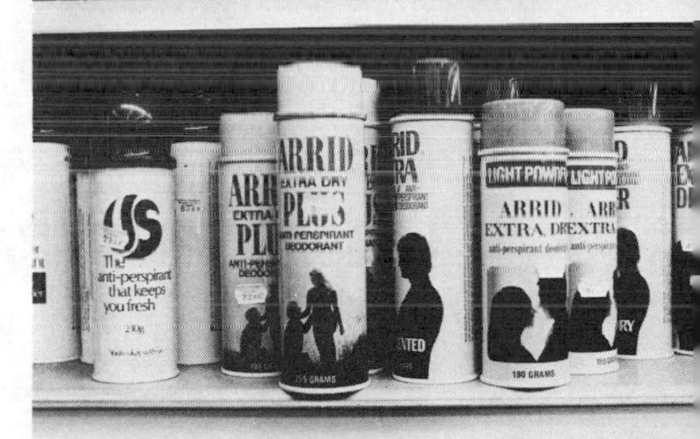

The stability and lack of toxicity of chlorofluorocarbons like $CCl_2F_2$ make them useful aerosol propellants.

Degreasing machinery using a halogeno-alkane solvent.

(iii) *Solvents*. Chlorinated hydrocarbons tend to be very good solvents for non-polar materials. For example, dichloromethane will partially dissolve old paint and is therefore used as a paint-stripper. Fluorocarbons tend to be less effective than chlorocarbons as solvents. Fluorochlorocarbons have intermediate solvent properties and have many applications where a moderate solvent is needed, for example in cleaning and degreasing machinery and electronic circuits. Here a powerful solvent might damage plastic components in the machinery or in the circuit: fluorochlorocarbons will dissolve grease but are not powerful enough solvents to affect components. Particularly useful for these purposes is 1,1,2-trichloro-1,2,2-trifluoroethane, a liquid that boils at 48°C; it is also used for dry-cleaning delicate materials such as suede or fur.

### REACTIVE HALOGENOALKANES: THE C–BR BOND

The C–F bond is not typical of the halogenoalkanes as a whole, which tend to be much more reactive than fluoroalkanes. To get some idea of the nature of the C–Hal bond, we will consider some bromine-containing compounds.

Table 28.1 compares the effect of adding silver nitrate(V) solution to sodium bromide, 1-bromobutane and bromobenzene. In the case of the two organic liquids, vigorous shaking is necessary since they do not mix appreciably with the aqueous phase.

**Table 28.1** Effect of aqueous silver nitrate(V) on bromine-containing compounds at room temperature.

| Sodium bromide | 1-Bromobutane | Bromobenzene |
|---|---|---|
| pale yellow precipitate appears immediately | no reaction at first: faint precipitate appears after several minutes | no reaction even after several hours |

Look at table 28.1 and answer these questions.

- ○ What is the pale yellow precipitate produced in the reaction between silver nitrate(V) and sodium bromide?
- ○ Write an ionic equation for this reaction.
- ○ Why does 1-bromobutane produce no immediate precipitate with silver nitrate(V), even though it contains bromine?
- ○ Suggest a reason why a precipitate appears after several minutes.
- ○ Compare the reactivity of bromobenzene with that of 1-bromobutane.

The C–Br bond in 1-bromobutane is covalent: 1-bromobutane therefore contains no $Br^-$ ions, so it does not produce a precipitate of silver bromide with silver nitrate(V). The slow appearance of a precipitate of silver bromide suggests that $Br^-$ ions are slowly being produced. Why?

Bromine is more electronegative than carbon and the C–Br bond is thus polar:

$$\overset{\delta+}{\text{C}}\text{—}\overset{\delta-}{\text{Br}}$$

The partial positive charge on the carbon atom tends to attract groups, such as $NH_3$, $^-OH$ and $H_2O$, carrying lone pairs of electrons. Such groups are called **nucleophiles**. Water molecules from the aqueous silver nitrate(V) can act as nucleophiles since their oxygen atoms carry two lone pairs and a partial negative charge:

$$\overset{\delta+}{\text{H}_2}\overset{\delta-}{\ddot{\text{O}}}\text{:}$$

This negative charge is attracted to the partially positive carbon atom in 1-bromobutane and a substitution reaction takes place, releasing bromide ions:

$$CH_3CH_2CH_2\text{—}\overset{\delta+}{C}H_2\text{—}\overset{\delta-}{Br} + :\ddot{O}H_2 \longrightarrow CH_3CH_2CH_2\text{—}CH_2\text{—}\overset{+}{O}H_2 + Br^- \longrightarrow CH_3CH_2CH_2\text{—}CH_2\text{—}OH + H^+$$

butan-1-ol

This is an example of a **nucleophilic substitution** reaction. Such reactions are typical of halogenoalkanes: their mechanism is discussed further in the next section. The failure of bromobenzene to react under these conditions suggests that the C–Br bond is stronger in aromatic than in aliphatic compounds.

The C–Hal bond is polar, but not polar enough to have an appreciable effect on the physical properties of organic halogen compounds. They are all immiscible with water and as table 28.2 shows, their volatility is determined more by the size and number of halogen atoms they contain than by the polarity of the bonds. Thus iodomethane is a liquid at room temperature while chloromethane is a gas, even though the latter has a more polar molecule.

**Table 28.2** Boiling points of some halogen compounds.

| Compound | State at 298 K | Boiling point/K |
|---|---|---|
| $CH_3F$ | g | 195 |
| $CH_3Cl$ | g | 249 |
| $CH_3Br$ | g | 277 |
| $CH_3I$ | l | 316 |
| $CH_2Cl_2$ | l | 313 |
| $CHCl_3$ | l | 335 |
| $CCl_4$ | l | 350 |
| $C_6H_5Cl$ | l | 405 |

## 28.4 Nucleophilic substitution

We have seen that water can act as a nucleophile toward halogenoalkanes. In general

$$RHal + H_2O \longrightarrow ROH + HHal$$

This reaction is slow, as the slow appearance of the AgBr precipitate in the last section showed. If $^-OH$ is used as a nucleophile instead of $H_2O$ the reaction is quicker: $^-OH$ with its full negative charge is a better nucleophile. Thus bromoethane forms ethanol quite rapidly when heated under reflux with aqueous sodium hydroxide (see figure 28.1 for diagram of reflux apparatus).

$$CH_3CH_2Br + {}^-OH \longrightarrow CH_3CH_2OH + Br^-.$$

The mechanism of reactions between halogenoalkanes and hydroxide ion has been extensively studied by chemists. As suggested in the last section, it involves attack on the partial positive charge on the carbon atom, e.g.

$$H-\overset{H}{\underset{H}{C}}-\overset{H}{\underset{H}{C}}{}^{\delta+}-Br^{\delta-} \quad {}^-OH \longrightarrow H-\overset{H}{\underset{H}{C}}-\overset{H}{\underset{H}{C}}-OH + Br^-$$

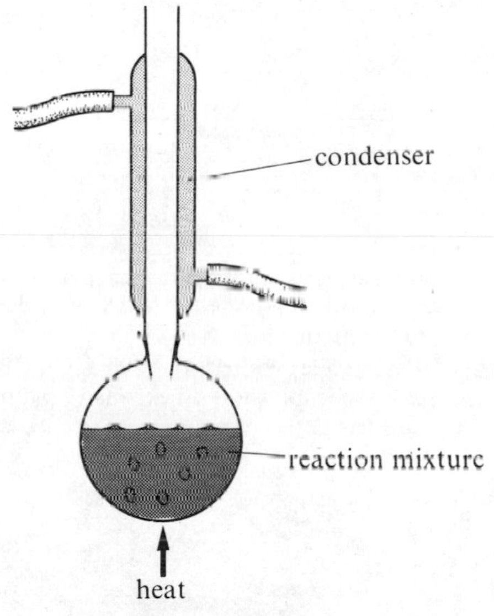

**Figure 28.1** Reflux apparatus used for boiling an organic reaction mixture. The condenser prevents escape of volatile reagents.

In some cases, the halide ion may leave before the $^-OH$ actually attacks, so that a carbonium ion (section 26.5) is formed as an intermediate. In other cases, the C–Hal bond may break at the same time as the C–OH bond forms.

The substitution reaction outlined above between $CH_3CH_2Br$ and $^-OH$ is general for all halogenoalkanes and can be used as a method for preparing alcohols:

$$RHal + {}^-OH \longrightarrow ROH + Hal^-$$

For a given alkyl group R, the iodo compound reacts most readily, the bromo compound less so and the chloro compound reacts least readily. This is because the C–Hal bond becomes progressively stronger as we pass from I to Cl.

As we saw in the last section, if the halogen is attached to a benzene ring, substitution is more difficult. It is thought that lone pairs of electrons on the halogen atom interact with the delocalized $\pi$-electron system of the benzene ring, strengthening the C–Hal bond. In addition, the high electron density on the aromatic ring tends to repel the approaching negatively charged $^{-}OH$ ion. Consequently, chlorobenzene reacts with aqueous sodium hydroxide at a reasonable rate only at 300°C under a pressure of 200 atmospheres. (This reaction affords one way of manufacturing phenol—see section 27.7.) Compare this with the behaviour of chlorobutane, which is readily converted to butanol by refluxing with aqueous sodium hydroxide under ordinary conditions.

Predict the organic product of the reactions when aqueous sodium hydroxide is boiled with

- 2-chloropropane
- 1,2-dibromoethane
- 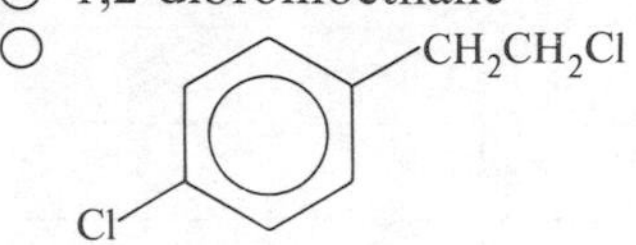



## 28.5 Important reactions of halogenoalkanes

Organic halogen compounds have important uses. Several of these have already been mentioned, and their importance in insecticides is considered in section 16.12. Other uses are discussed in chapter 25. Organic halogen compounds occur hardly at all in nature so they must be synthesized. They are usually manufactured from alkenes or alkanes: chloroethane, for example, used in the manufacture of tetraethyllead(IV) (section 25.6), is made by the addition reaction between ethene and hydrogen chloride. Tetrachloromethane is made by the substitution reaction between chlorine and methane. In the laboratory, halogenoalkanes are usually prepared from alcohols (see section 29.6).

As well as being used in their own right, halogenoalkanes are important intermediates in synthesis. That is, they can be used to introduce a reactive site into a hydrocarbon molecule which can then be substituted by another group which could not be introduced directly. Examples of groups that can be introduced in this way are discussed later in this section. This kind of synthesis is important in small-scale preparations such as those carried out in the laboratory or in the manufacture of pharmaceuticals. Many drugs have complicated organic molecules and their synthesis involves building up a complex molecule from a simple starting compound. Such synthesis often involves many steps. For example, the synthesis of phenobarbitone (a sleep-inducing drug) from methylbenzene involves eight steps (see figure 28.2). Note in this example how a chlorine atom is used to provide an active site in the relatively unreactive methylbenzene molecule so that other groups can then be introduced.

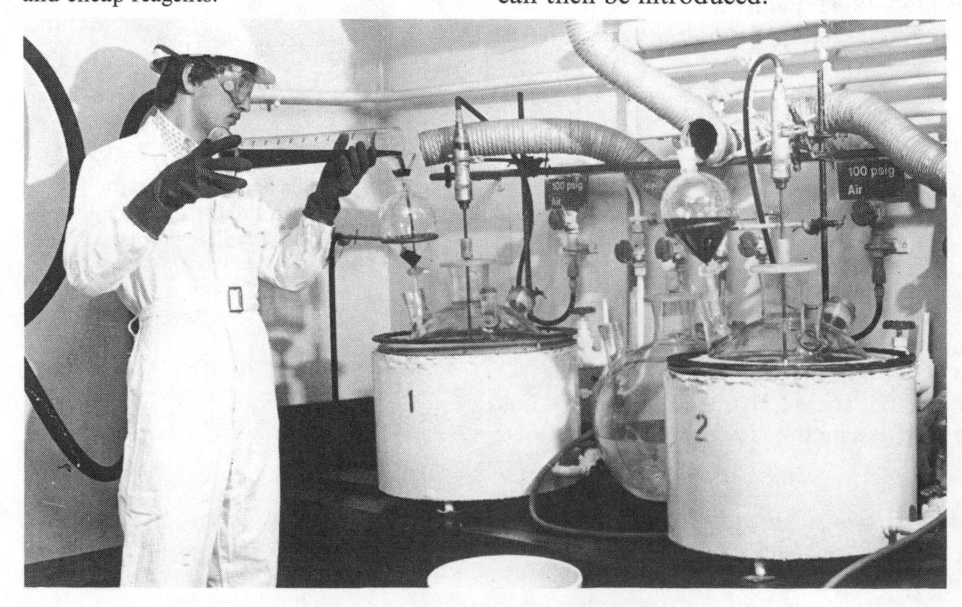

*Left* Small scale synthesis of a pharmaceutical. Laboratory methods are used and the reagents are often expensive.
*Right* Large scale synthesis of ethene. On the large scale the emphasis is on expensive plant and cheap reagents.

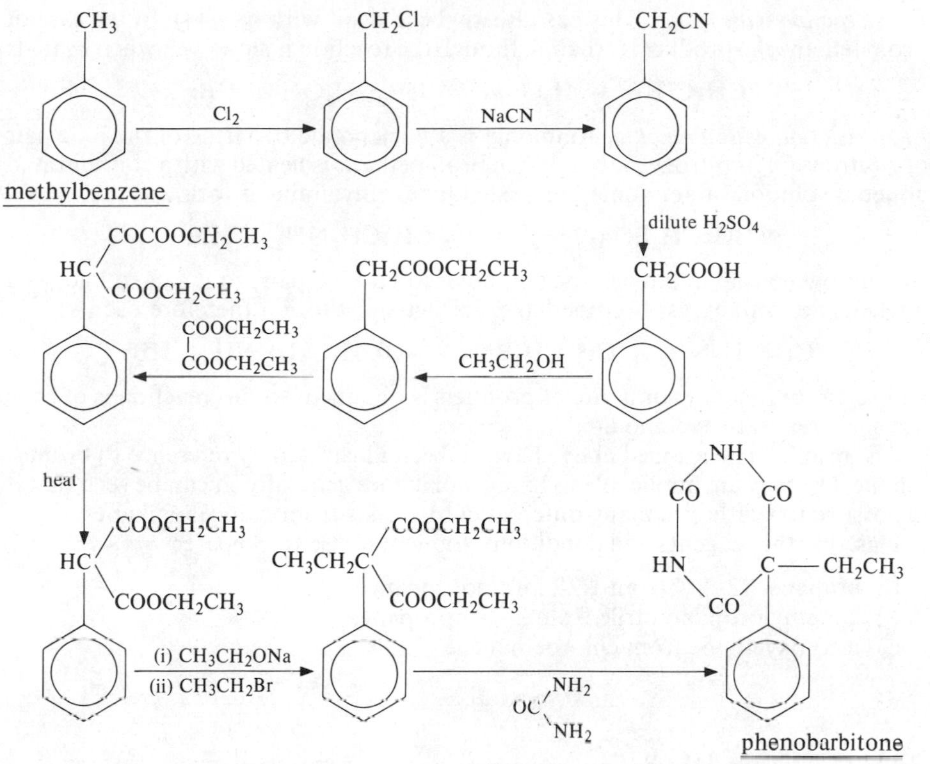

$CH_3$ (methylbenzene) $\xrightarrow{Cl_2}$ $CH_2Cl$ $\xrightarrow{NaCN}$ $CH_2CN$ $\xrightarrow{\text{dilute } H_2SO_4}$ $CH_2COOH$ $\xrightarrow{CH_3CH_2OH}$ $CH_2COOCH_2CH_3$ $\xrightarrow{COOCH_2CH_3 / COOCH_2CH_3}$ $HC(COCOOCH_2CH_3)(COOCH_2CH_3)$ $\xrightarrow{\text{heat}}$ $HC(COOCH_2CH_3)_2$ $\xrightarrow{\text{(i) } CH_3CH_2ONa \text{ (ii) } CH_3CH_2Br}$ $CH_3CH_2C(COOCH_2CH_3)_2$ $\xrightarrow{OC(NH_2)_2}$ phenobarbitone

methylbenzene

phenobarbitone

**Figure 28.2** Synthesis of phenobarbitone. Phenobarbitone is a hypnotic, used in sleeping pills.

The synthetic methods described in this chapter are used for small-scale operations, but they are relatively expensive and therefore not used for large-scale, high-tonnage industrial operations. On this scale high pressure, high temperature catalytic processes, very different from those used in the laboratory, tend to be employed as they are more economic.

Halogenoalkanes undergo substitution reactions with a wide range of nucleophiles. Species that can act as nucleophiles include not only those carrying a full negative charge ($^-OH$, $^-CN$, $CH_3COO^-$, $CH_3CH_2O^-$) but also neutral molecules carrying an unshared pair of electrons ($H_2O$, $NH_3$).

Predict the structure of the compound formed by the reaction of bromoethane with each of the nucleophiles mentioned above. The answers are given below, together with the conditions needed for the reaction; do not look at these until you have predicted the structures yourself.

*(a) Reaction with hydroxide ion.* This has already been dealt with (p. 445).

*(b) Reaction with cyanide ion.* When bromoethane is heated under reflux with a solution of potassium cyanide in ethanol, propanonitrile is formed:

$$CH_3CH_2Br + {}^-CN \longrightarrow CH_3CH_2CN + Br^-$$

This reaction is useful in synthesis as a means of increasing the length of a carbon chain.

*(c) Reaction with ethanoate ion.* If bromoethane is warmed with dry silver ethanoate, ethyl ethanoate, an ester, is formed:

$$CH_3COO^-Ag^+ + CH_3CH_2Br \longrightarrow CH_3COOCH_2CH_3 + AgBr$$

Ethanoate ion is only a weak nucleophile: the reaction is encouraged by using the silver salt which removes bromide ion as insoluble silver bromide.

*(d) Reaction with ethoxide ion.* Bromoethane reacts with an ethanolic solution of sodium ethoxide, $Na^{+-}OCH_2CH_3$ (see section 29.5) to form an ether, ethoxyethane:

$$CH_3CH_2Br + CH_3CH_2O^- \longrightarrow CH_3CH_2OCH_2CH_3 + Br^-$$

The ethoxide ion can also act as a base under these conditions, in which case a different product, ethene, is formed (see next section).

Just as both water and hydroxide ion can act as nucleophiles, the latter more effectively because of its full negative charge, so both ethanol and ethoxide ion give nucleophilic substitution reactions with halogenoalkanes. Ethanol reacts considerably slower than ethoxide ion as we would expect, but the product is the same:

$$CH_3CH_2Br + CH_3CH_2OH \longrightarrow CH_3CH_2OCH_2CH_3 + HBr$$

*(e) Reaction with water.* This has already been dealt with (p. 444). In the case of bromoethane the product is ethanol, though the reaction is slow even when heated:

$$CH_3CH_2Br + H_2O \longrightarrow CH_3CH_2OH + HBr$$

*(f) Reaction with ammonia.* Ammonia is a nucleophile by virtue of the lone pair of electrons on its nitrogen atom. When bromoethane is heated with a concentrated aqueous solution of ammonia in a sealed tube, ethylamine is formed:

$$CH_3CH_2Br + NH_3 \longrightarrow CH_3CH_2NH_2 + HBr$$

The nitrogen atom in ethylamine still possesses a lone pair, so it can still act as a nucleophile, with excess bromoethane, further substitution therefore occurs:

$$CH_3CH_2NH_2 + CH_3CH_2Br \longrightarrow (CH_3CH_2)_2NH + HBr$$

and so on. In practice a mixture of products is obtained, so this reaction is of little use as a preparative method.

The reactions mentioned above have all been illustrated by reference to bromoethane, but they are applicable to halogenoalkanes generally. It can be seen that it is possible to synthesize many different compounds from halogenoalkanes.

Describe the reagents and conditions you would use to prepare:

- propane-1,2-diol from 1,2-dibromopropane,
- 2-methylpropanonitrile from 2-iodopropane,
- methoxyethane from chloroethane.

## 28.6 Elimination

In the reactions we have considered so far, $^-OH$ has acted as a nucleophile in its reactions with halogenoalkanes. For example, when 2-bromopropane is refluxed with aqueous sodium hydroxide, the hydroxide ion acts as a nucleophile and substitution occurs:

$$CH_3CHBrCH_3 + {}^-OH \longrightarrow CH_3CHOHCH_3 + Br^-$$

Under certain conditions, $^-OH$ can act as a base instead of a nucleophile, removing $H^+$ from a halogenoalkane. In this case, the C–Br bond breaks at the same time as $^-OH$ removes $H^+$ from the neighbouring carbon atom and an alkene is formed. For example, when 2-bromopropane is refluxed with a solution of sodium hydroxide in ethanol instead of in water, propene is formed:

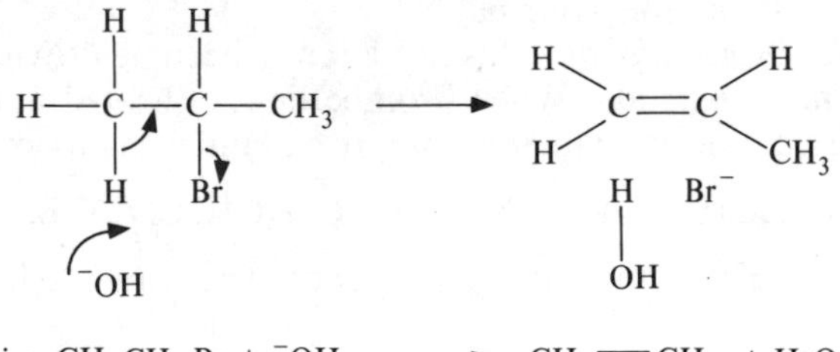

$$H-\underset{\underset{H}{|}}{\overset{\overset{H}{|}}{C}}-\underset{\underset{Br}{|}}{\overset{\overset{H}{|}}{C}}-CH_3 \quad {}^-OH \longrightarrow \begin{matrix} H \\ H \end{matrix}\!>C{=}C<\!\begin{matrix} H \\ CH_3 \end{matrix} \quad H{-}OH \quad Br^-$$

$$\text{i.e. } CH_3CH_2Br + {}^-OH \longrightarrow CH_2{=}CH_2 + H_2O + Br^-$$

In the first equation, the curved arrows indicate the movement of a pair of electrons (a bonding pair or a lone pair). This is a conventional method of indicating movement of an electron pair and will be used where appropriate in this and the following chapters.

The overall effect of this reaction is to eliminate HBr from the molecule of 2-bromopropane. This is an example of an **elimination reaction**. Notice that the reagent, sodium hydroxide, is the same as for the substitution reaction outlined at the beginning of this section. In the substitution reaction, $^-OH$ acted as a nucleophile; in the elimination reaction it acts as a base. By altering the conditions, we can alter the manner in which it acts: in aqueous solution it behaves as a nucleophile, in ethanolic solution it behaves as a base. Under each set of conditions, both substitution and elimination will occur, but by controlling the conditions we can ensure that one particular reaction occurs to a greater extent than the other.

Try to predict the effect of an elimination reaction in a halogenoalkane that contains *two* halogen atoms: for example, what would be the product if 1,2-dibromoethane were heated under reflux with a solution of sodium hydroxide in ethanol?

## 28.7 Acyl halides

So far we have considered molecules in which a halogen atom is attached directly to an alkyl or aryl group. An important group of halogenated compounds is the acyl halides, in which a halogen atom is attached to a $>C{=}O$ group.

An example is ethanoyl chloride, $CH_3{-}\underset{\underset{O}{\|}}{C}{-}Cl$, usually written $CH_3COCl$.

- ○ Would you expect the C=O bond to be polar? If so, which atom will carry a positive charge and which a negative charge?
- ○ Would you expect the C atom in $-COCl$ to carry a larger or smaller positive charge than the C atom in $-CH_2Cl$?
- ○ Would you expect $CH_3COCl$ to react more or less readily with nucleophiles than $CH_3CH_2Cl$?
- ○ Write the structure of the product you would expect from a reaction between $CH_3COCl$ and water. Would you expect this reaction to be more or less vigorous than the reaction between $CH_3CH_2Cl$ and water?

**Acyl halides** are compounds with the general formula **RCOHal**, where R is an alkyl or aryl group. They are named using the suffix **-oyl halide** after a stem indicating the number of carbon atoms in the molecule. Thus $CH_3CH_2COBr$ is called propanoyl bromide and $CH_3CH_2CH_2COCl$ is called butanoyl chloride. The presence of a polar $^{\delta+}C{=}O^{\delta-}$ bond in the molecule means that the carbon atom in RCOCl carries a considerably larger positive charge than in halogenoalkanes. This attracts nucleophiles more strongly, so acyl halides undergo nucleophilic substitution considerably more readily than the halogenoalkanes. For example, when water is added to ethanoyl chloride (which is a liquid), a violent reaction occurs: the liquid boils and clouds of hydrogen chloride fumes are evolved:

$$CH_3COCl + H_2O \longrightarrow \underset{\text{ethanoic acid}}{CH_3COOH} + HCl$$

Acyl chlorides react with all the nucleophiles that will substitute halogenoalkanes. Predict the structure of the product of the reaction between ethanoyl chloride, $CH_3COCl$, and each of the nucleophiles $NH_3$ and $CH_3CH_2OH$. The answers appear below.

Like halogenoalkanes, acyl halides are important in synthesis. Just as RHal is used to attach an alkyl group R to a nucleophile, RCOHal is used to attach an acyl group RCO. Because of their greater reactivity, acyl halides will react rapidly with those nucleophiles, such as water and alcohols, that react only slowly with halogenoalkanes.

Some reactions of ethanoyl chloride:

*(a) With ammonia.* Ethanoyl chloride reacts violently with an aqueous solution of ammonia at room temperature. The product is ethanamide, an amide. Amides have the general formula $RCONH_2$ and should not be confused with amines, $RNH_2$.

$$CH_3COCl + NH_3 \longrightarrow \underset{\text{ethanamide}}{CH_3CONH_2} + HCl$$

The C=O group in ethanamide has a strong electron-withdrawing effect. It tends to withdraw electrons from the $-NH_2$ group, making the nitrogen atom much less nucleophilic than in $NH_3$ or in amines such as $CH_3CH_2NH_2$. Ethanamide is therefore not a strong nucleophile, despite carrying a lone pair on its nitrogen atom, so further substitution does not occur and ethanamide is the only product. Compare this with the reaction of chloroethane with ammonia, which gives several products.

*(b) With alcohols.* Ethanoyl chloride reacts vigorously with alcohols at room temperature to form esters. For example, with ethanol:

$$CH_3CH_2OH + CH_3COCl \longrightarrow \underset{\text{ethyl ethanoate}}{CH_3COOCH_2CH_3} + HCl$$

As expected, the reaction occurs much more readily than the corresponding reaction of chloroethane with ethanol.

The reactions of ethanoyl chloride described above are of course quite general and apply to all acyl chlorides.

## Summary

1 Halogenoalkanes are named using a prefix (*chloro*-, etc.) to indicate the nature of the halogen atom and a number to indicate its position on the alkane chain.
2 The C–F bond is very strong. Fluorocarbons are very stable.
3 The C–Hal bond is polar and tends to attract nucleophiles to the positively-charged carbon atom. Nucleophilic substitution is the characteristic reaction of halogenoalkanes. Some examples are shown in this diagram.

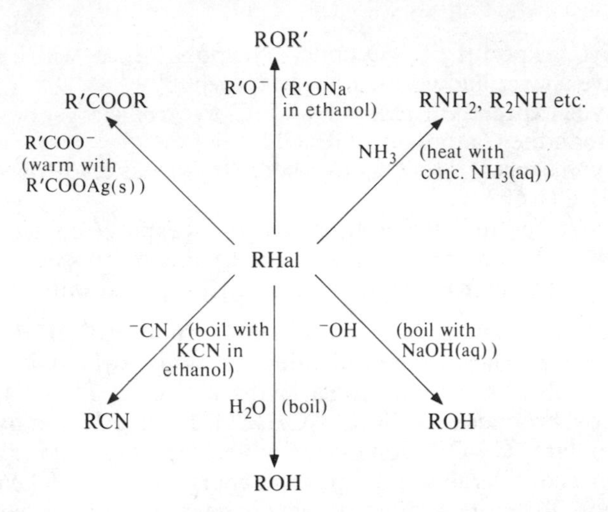


4 The Ar–Hal (Ar = aryl group) bond is considerably less reactive than the R–Hal bond.
5 In the presence of a strong base, halogenoalkanes can undergo elimination reactions, forming alkenes.
6 Acyl halides contain the —COCl group. They undergo nucleophilic substitution considerably more readily than halogenoalkanes.

## Study questions

1 The table below gives the formulae of some halogenoalkanes and their boiling points.

| Formula | Boiling point/°C |
|---|---|
| $CCl_3F$ | 24 |
| $F_2C—CF_2$<br>$F_2C—CF_2$ | −6 |
| $CCl_2F—CClF_2$ | 48 |
| $CBrF_3$ | 149 |

(a) Name each compound.
(b) Which would be suitable for use as an aerosol propellant?
(c) Which would be suitable as a degreasing solvent?
(d) Which would be a suitable fire-extinguisher?

2 Consider the following compounds.

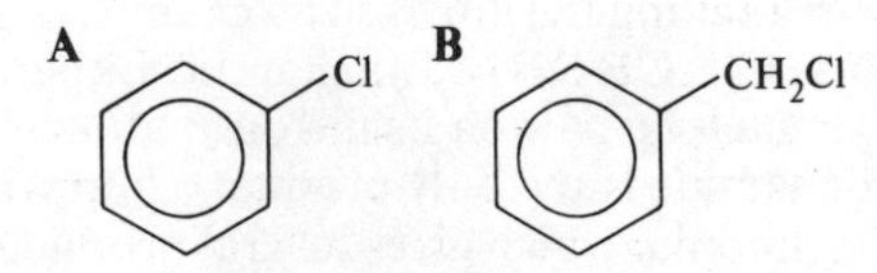


**C** $CH_3CH_2Cl$ **D** $CH_3CH_2COCl$
**E** $CH_3CHBrCH_3$

(a) Which are halogenoalkanes?
(b) Which is an acyl halide?
(c) Which would react most readily with cold water?
(d) Which would react least readily with aqueous sodium hydroxide?
(e) Which would form an amide when reacted with aqueous ammonia?
(f) Which would have the lowest boiling point?

3 This question concerns the hydrolysis of three different halogenoalkanes, 1-chlorobutane, 1-bromobutane and 1-iodobutane.
Four drops of each halogenoalkane are added separately to three separate tubes, each containing 1cm$^3$ of 0.1M silver nitrate(v) solution and standing in a water-bath at 60°C. The results are as follows.
1-Chlorobutane—slight cloudiness after three minutes. Still only slightly cloudy after fifteen minutes.
1-Bromobutane—slightly cloudy after one minute, opaque after three minutes, coagulation and precipitation after six minutes.
1-Iodobutane—immediately opaque, yellow precipitate within first minute.
(a) What is the precipitate formed in each case?
(b) Why does the precipitate form?
(c) In each case a substitution reaction is occurring between the halogenoalkane and a nucleophile. What is the nucleophile involved?
(d) Which halogenoalkane undergoes substitution most readily and which least readily?
(e) Explain your answer to (d), given the following bond energies in kJ mole$^{-1}$: C–Cl (in chloroethane) 339; C–Br (in bromoethane) 284; C–I (in iodoethane) 218.

4 Predict the products of reactions between the following pairs of substances.
(a) $CH_3CH_2CH_2Br$ and $NH_3$
(b) $CH_2ICH_2CH_2CH_2I$ and $CH_3CH_2CH_2COO^-Ag^+$
(c) $C_6H_5$–$CH_2Br$ and KCN (ethanolic)
(d) cyclohexyl–Br and NaOH(aq)
(e) cyclohexyl–Br and NaOH (ethanolic)
(f)

$$CH_3-\overset{\displaystyle Br}{\underset{\displaystyle Br}{\overset{|}{\underset{|}{C}}}}-CH_3$$

and $CH_3O^-Na^+$
(g) $C_6H_5$–COCl and $CH_3CH_2OH$

5 Use your knowledge of nucleophilic substitution reactions to predict the structural formulae of the products of reactions between bromoethane and the following.
(a) sodium hydrogensulphide ($Na^{+-}SH$)
(b) sodium ethynide ($HC{\equiv}C^-Na^+$)
(c) potassium nitrate(III) ($K^+NO_2^-$)
(d) potassium chloride
(e) lithium tetrahydridoaluminate, $LiAlH_4$, (nucleophile = $H^-$)
(f) sodamide ($Na^{+-}NH_2$).

6 An organic liquid **A** is immiscible with water. When 1.00g of **A** was refluxed for several hours with excess aqueous sodium hydroxide it slowly dissolved in the aqueous layer. The resultant solution was neutralized with nitric(v) acid and excess silver nitrate(v) solution was added, when a precipitate of silver bromide was formed. The precipitate was filtered off and dried and its mass was found to be 2.00g. The filtrate was distilled and a high-boiling liquid **B** was obtained. **B** contained 39% C, 9.5% H and 51.5% O by mass. Suggest structural formulae for **A** and **B** and explain the above reactions.

(C = 12, H = 1, O = 16, Br = 80, Ag = 108)

7 Two compounds **X** and **Y** have the same molecular formula, $C_3H_5OCl$. **X** reacts vigorously with cold water to give a strongly acidic solution, but **Y** has no visible reaction with cold water. Suggest structural formulae for **X** and **Y**.

8 The reaction of 3-chloro-3-ethylpentane with aqueous sodium carbonate at room temperature yields two products, one an alcohol and one an alkene.
(a) Write the formula of 3-chloro-3-ethylpentane.
(b) Consider the reaction that produces the alcohol.
 (i) What type of reaction is this?
 (ii) Write the structural formula of the alcohol.
 (iii) Write an equation for the reaction.
(c) Consider the reaction that produces the alkene.
 (i) What type of reaction is this?
 (ii) Suggest a possible structural formula for the alkene.
 (iii) Write an equation for the reaction.
(d) What changes in reaction conditions might favour the formation of the alkene?

9 Give reagents, reaction conditions and equations to show how you would convert
(a) $CH_2{=}CH_2$ to $CH_3CH_2CN$ (2 steps)
(b) $CH_2{=}CH_2$ to $CH_3OCH_2CH_2OCH_3$ (2 steps)

(c) $C_6H_5COCl$ to $C_6H_5CONH_2$ (1 step)

(d) $C_6H_5CH_3$ to $C_6H_5CH_2OH$ (2 steps)

# Alcohols, Phenols and Ethers 29

## 29.1 Fermentation

The yeast on a vat of fermenting beer. The yeast has multiplied tenfold and the best is retained for the next brewing. The remainder is sold to the manufacturers of yeast extract.

When Noah left the Ark he promptly planted a vineyard and was soon drinking its produce. Man has been fermenting grape-juice for at least ten thousand years, and probably fermenting honey for even longer. The reason for doing this is, of course, that fermentation of sugar by yeast produces ethanol, the well-known intoxicant. Ethanol is a member of the homologous series of alcohols and is commonly just called 'alcohol'.

$$\underset{\text{a sugar}}{C_6H_{12}O_6} \longrightarrow 2\underset{\text{ethanol}}{CH_3CH_2OH} + 2CO_2$$

This is an exothermic reaction that provides the yeast with energy.

Alcoholic fermentation occurs naturally wherever sugar-containing materials such as fruit are allowed to decay: in autumn it is quite common to see drunken (and therefore dangerous) wasps which have been feeding off fermented fruit. Fermentation can also be carried out under more controlled conditions, as in the manufacture of alcoholic drinks. The source of sugar varies: it may be grapes (for wine), honey (for mead), malted grain (for beer), apples (for cider) or indeed any sugar-containing fruit or plant. From the point of view of yeast, the ethanol produced in fermentation is a toxic waste product, which kills the yeast at concentrations greater than about 15% by volume. It is therefore impossible to produce alcoholic drinks containing more than 15% alcohol by fermentation alone. There are reasons for producing beverages with greater alcoholic content than this: for one thing they can be stored longer, because alcohol in high concentration is toxic to bacteria. Another reason is simply that some people like drinks containing alcohol in high concentrations. To satisfy these requirements, fermented liquids are distilled to increase their alcoholic concentration and produce spirits. For example, distillation of wine produces brandy. A typical spirit contains about 40% ethanol. In Britain alcoholic strength is measured in degrees proof: 'proof spirit' is an ancient standard, defined as the weakest mixture of pure alcohol and water which when poured over gunpowder and ignited will allow the powder to burn. 100% proof spirit corresponds to about 57% ethanol: thus typical spirits are about 70% proof.

All the members of the homologous series of alcohols are toxic to a greater or lesser extent. The first member of the series, methanol, is much more toxic than ethanol and is added to industrial alcohol (on which no excise duty is charged) to make it unpalatable. It is then called methylated spirits. Unfortunately a few people drink it nevertheless, the eventual result being blindness or death. The higher alcohols are moderately toxic, unpleasant-tasting compounds. A mixture of these compounds is produced in small amounts during fermentation: the mixture is called *fusel oil* (*fusel* is German for 'bad liquor'). Ethanol itself is intoxicating in small amounts but toxic in large amounts. In small amounts it has the effect of making people relaxed and for this reason it is undoubtedly socially useful. In large amounts it has a serious effect on mental and physical performance and its use can be very dangerous, particularly to drivers of motor cars. It is addictive if regularly taken in large quantities. Ethanol is firmly established as an accepted social drug with many advantages, but there is little doubt that if it were newly introduced today and its properties were known, it would be banned as a dangerous drug.

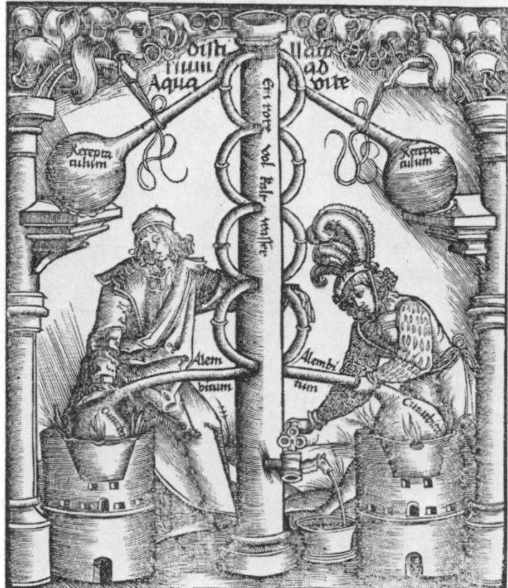


An early method for the distillation of brandy.

Fermentation is used throughout the world to produce alcoholic drinks. At one time it was also a major source of organic chemicals. Ethanol, produced by the fermentation of cheap carbohydrate such as grain or molasses, was once a major feedstock in the production of aliphatic compounds. Its place as a feedstock has now been taken by the much cheaper ethene; indeed most industrial ethanol is now produced from ethéne. Nevertheless, ethanol still has important uses as a solvent, and other alcohols are of industrial importance too.

## 29.2 Nomenclature of alcohols

In this chapter we shall be considering organic **hydroxy compounds**, containing the —OH group. A study of the properties of the OH group is important to the chemist because of the industrial importance of compounds containing this functional group and because of its wide occurrence in biological molecules. Aliphatic and aromatic hydroxy compounds, as we shall see, differ considerably in their properties. They are, therefore, regarded as two distinct groups of compounds: aliphatic hydroxy compounds are called **alcohols** and aromatic ones are called **phenols**.

Alcohols are named using the suffix **-ol**, preceded if necessary by a number to indicate its position in the carbon skeleton. Thus,

$CH_3OH$ is methanol, $CH_3CH_2CH_2OH$ is propan-1-ol,

$CH_3CHOHCH_3$ is propan-2-ol and $CH_3{-}C(CH_3)(OH){-}CH_3$ is 2-methylpropan-2-ol.

Alcohols with structures of the form $RCH_2OH$, where R is an alkyl or aryl group or hydrogen, are called **primary** alcohols (e.g. propan-1-ol above). Those with the structure $R_1R_2CHOH$ are called **secondary** alcohols (e.g. propan-2-ol), and those with the structure $R_1R_2R_3COH$ are called **tertiary** alcohols (e.g. 2-methylpropan-2-ol).

Primary, secondary and tertiary alcohols have some important differences in chemical reactivity, as we shall see.

Some alcohols, particularly biologically occurring ones, contain more than one OH group in their molecule. They are known as **polyhydric** alcohols. They are named using the suffixes **-diol**, **-triol**, etc., depending on how many —OH groups they contain. Thus,

$CH_2OHCH_2OH$ ethane-1,2-diol,

$CH_2OHCHOHCH_2OH$ propane-1,2,3-triol.

Compounds in which the —OH group is attached to an aromatic ring are called phenols, the simplest and most important is phenol itself.

Phenols have the OH group attached *directly* to the ring. One of the two compounds shown in figure 29.1 is a phenol. Which one?

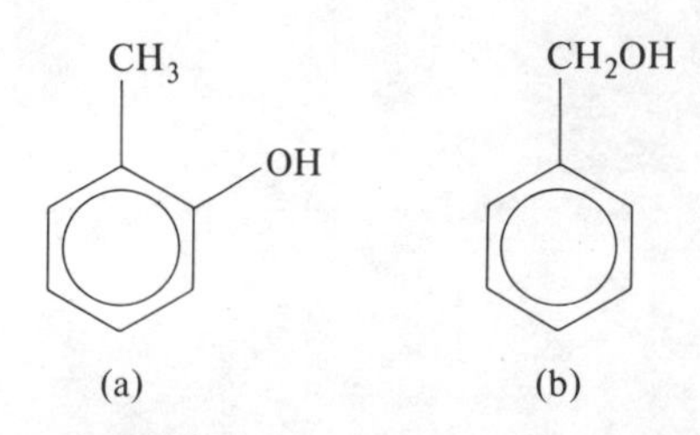


**Figure 29.1** Which of these two compounds would be classed as a phenol?

Name the following:

(a) $CH_3CHOHCH_2CH_3$

(b) H—C(H)(OH)—C(CH$_3$)(OH)—$CH_3$

- Write formulae for the following:
  (a) butane-1,2,4-triol
  (b) 2-methylpentan-2-ol
- From the previous questions, classify those alcohols containing a single OH group as primary, secondary or tertiary.

## 29.3 Alcohols as a homologous series

Reference has already been made to the industrial and biological importance of compounds containing the OH group. Industrially, alcohols are manufactured by the hydration of alkenes (see section 26.6). In the laboratory they can be made by nucleophilic substitution reactions between halogenoalkanes and $^-OH$ as described in section 28.4.

The alcohols illustrate very well the steady change in physical properties that occurs when a homologous series is ascended. The OH group has a profound effect on the physical properties of any molecule of which it is a part: alcohols show marked physical differences from the organic compounds we have encountered so far. Consider three compounds of similar relative molecular mass: propane ($M_r = 44$), chloromethane ($M_r = 50.5$) and ethanol ($M_r = 46$). Ethanol is far less volatile than the other two: it is a liquid at room temperature while the others are gases. (Boiling points 231K, 249K and 351K respectively.) Furthermore, ethanol is soluble in water in all proportions, while the other two are practically insoluble. The cause of ethanol's anomalous properties is hydrogen bonding (see section 8.5). Hydrogen bonds form between O–H groups of adjacent ethanol molecules, giving ethanol relatively high intermolecular forces and therefore relatively low volatility (see figure 29.2). Hydrogen bonding between ethanol molecules and water molecules explains why these two liquids are miscible in all proportions.

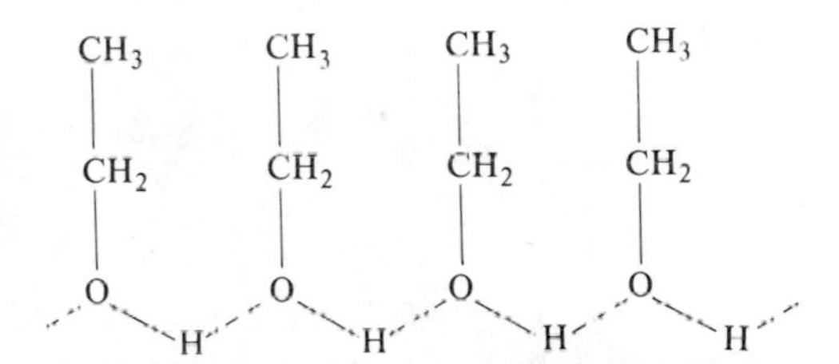


**Figure 29.2** Hydrogen bonding between ethanol molecules. Hydrogen bonds are represented by dotted lines.

Pure ethanol has a strong affinity for water and tends to absorb it from the atmosphere. Furthermore, pure ethanol cannot be obtained from its aqueous solution by distillation alone: the two liquids form a constant-boiling mixture containing 95.6% ethanol which distils over unchanged. The remaining 4.4% water must be removed by a chemical drying agent such as calcium oxide in order to produce 100% ethanol (called **absolute alcohol**).

Figure 29.3 (overleaf) shows the boiling points of the first seven straight-chain primary alcohols and the first seven straight-chain alkanes. Look at the graph and answer the questions following it.

As the homologous series of alcohols is ascended, the influence of the OH group becomes less and less important compared to that of the increasingly large hydrocarbon portion, so the properties of the higher alcohols tend more and more towards those of the corresponding alkane. This trend can be seen in solubility as well as volatility, as table 29.1 shows.

**Table 29.1** Solubility of alcohols in water.

| Name | Formula | Solubility/g per 100g of water |
|---|---|---|
| methanol | $CH_3OH$ | infinite (miscible in all proportions) |
| ethanol | $CH_3CH_2OH$ | infinite |
| propan-1-ol | $CH_3CH_2CH_2OH$ | infinite |
| butan-1-ol | $CH_3CH_2CH_2CH_2OH$ | 8.0 |
| pentan-1-ol | $CH_3CH_2CH_2CH_2CH_2OH$ | 2.7 |
| hexan-1-ol | $CH_3CH_2CH_2CH_2CH_2CH_2OH$ | 0.6 |

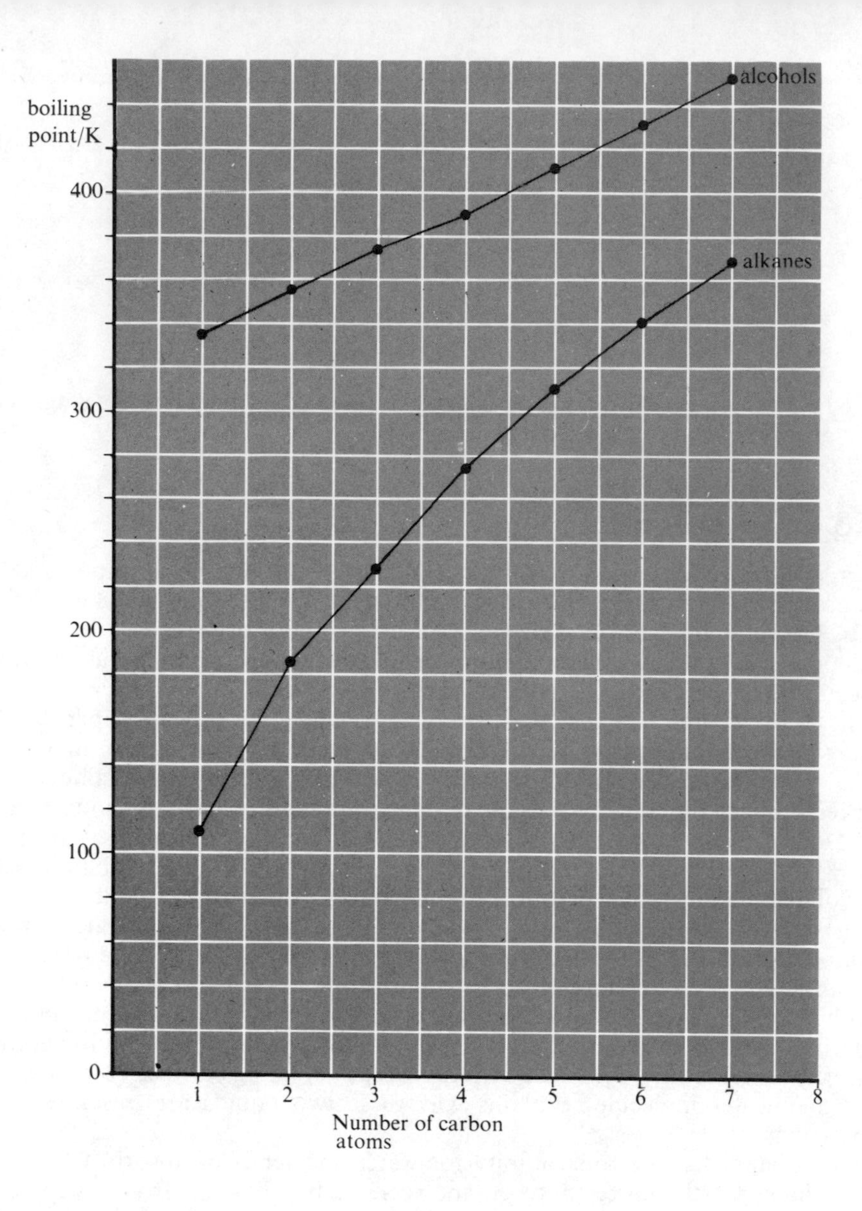


**Figure 29.3** Boiling points of the first seven straight-chain alkanes and straight-chain primary alcohols.

- ○ Why are the boiling points of alcohols higher than those of the corresponding alkanes?
- ○ Why do the differences in boiling points between corresponding alcohols and alkanes get less as the number of carbon atoms increases?
- ○ Where would the two graphs intersect, and what is the physical significance of the point of intersection?

Containing both a highly polar OH group and a non-polar hydrocarbon portion, the lower alcohols such as ethanol tend to be good solvents for polar as well as non-polar solutes. For example, both sodium hydroxide (ionic) and hexane (molecular) dissolve well in ethanol. This property makes ethanol, and to a lesser extent methanol and propanol, valuable solvents in the laboratory and in industry. One example is the use of ethanol as a base for after-shave lotions: the ethanol dissolves both the water-insoluble oils which provide the aroma and the water which makes up the bulk of the preparation. In Germany propan-2-ol is commonly used for this purpose instead of ethanol.

Hydrogen bonding between OH groups also has an effect on the viscosity of alcohols, particularly those with more than one OH group in their molecule. Thus ethanol has a viscosity of $1.06 \times 10^{-3}$ Pa s at 298K, about the same as water, but propane-1,2,3-triol (commonly called glycerine) is very thick and sticky, with a viscosity of $942 \times 10^{-3}$ at the same temperature, because of extensive interaction between its molecules which carry three OH groups each.

In phenol, the large, non-polar benzene ring to some extent dominates the OH group. Thus phenol is only partially soluble in water (9.3g in 100g water at 20°C). It is a crystalline solid, melting point 43°C.

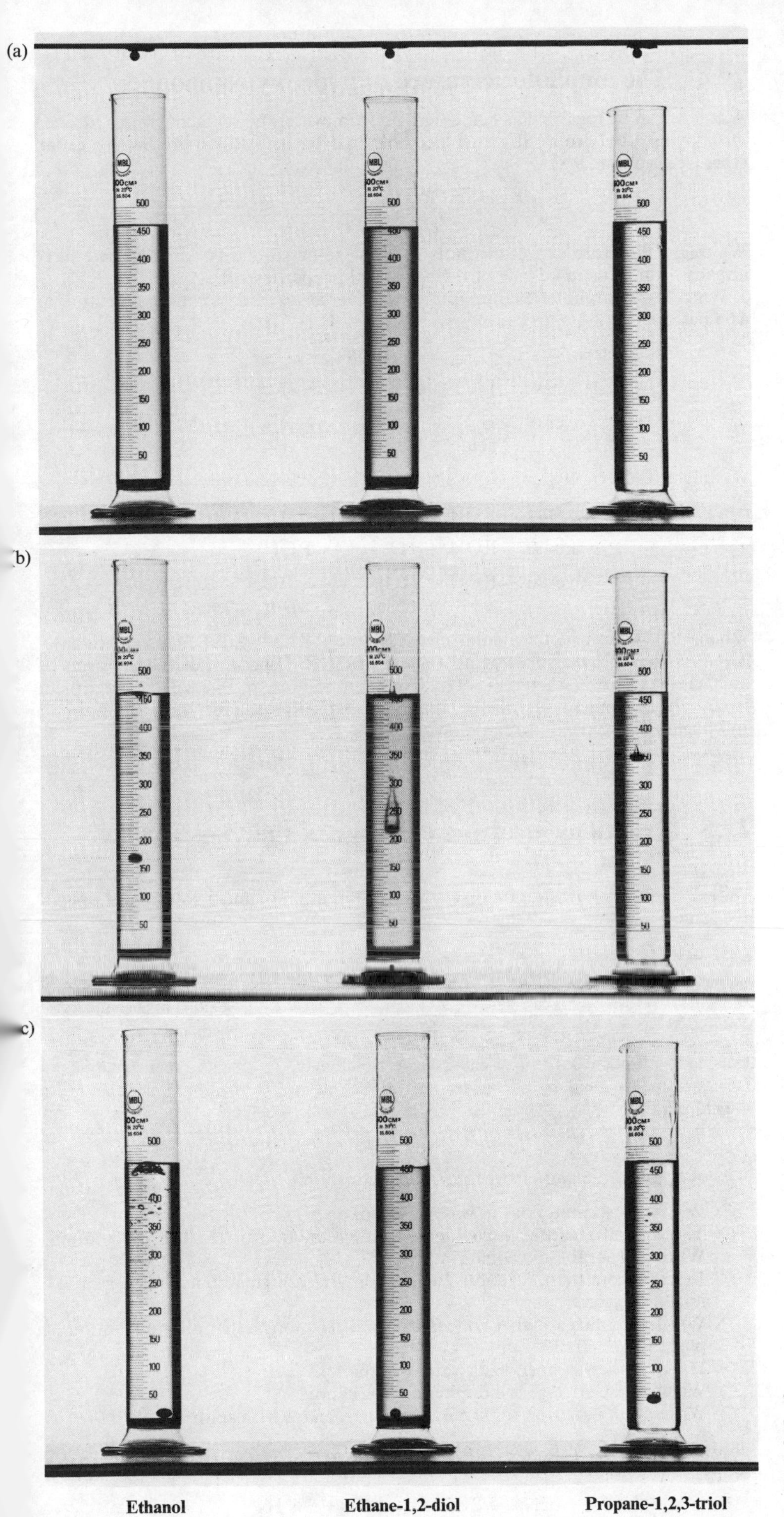


The effect of hydrogen bonding on viscosity. (a) Identical ball bearings are held magnetically above (from left to right) ethanol, ethane-1,2-diol, propane-1,2,3-triol. (b) and (c) the ball bearings are released simultaneously and fall through the liquids.

## 29.4 The amphoteric nature of hydroxy compounds

Alcohols can be regarded as being derived from water, by replacing one hydrogen atom by an alkyl group. If we replace both hydrogens by alkyl groups, we get an **ether** (see section 29.8).

| H—O—H | R—O—H | R—O—R |
|---|---|---|
| water | an alcohol | an ether |

We might therefore expect alcohols to show some similarity to water: we have already found this to be true of their physical properties.

Water is an amphoteric compound: it can act as an acid, donating a proton, or as a base, accepting a proton:

$$\text{as an acid:}\quad H_2O \longrightarrow {}^-OH + H^+$$

$$\text{as a base:}\quad H_2O + H^+ \longrightarrow H_3O^+$$

$$\text{overall:}\quad \underset{\text{acid}}{H_2O} + \underset{\text{base}}{H_2O} \longrightarrow {}^-OH + H_3O^+$$

We might also expect alcohols to show amphoteric behaviour:

$$\text{as an acid:}\quad ROH \longrightarrow RO^- + H^+$$

$$\text{as a base:}\quad ROH + H^+ \longrightarrow ROH_2{}^+$$

$$\text{overall:}\quad \underset{\text{acid}}{ROH} + \underset{\text{base}}{ROH} \longrightarrow RO^- + ROH_2{}^+$$

When it acts as an acid, the alcohol cleaves at the O–H bond. When it acts as a base, it can, as we shall see, subsequently cleave at the R–O bond. Both these forms of bond cleavage are characteristic of hydroxy compounds, and we will consider them separately. In doing so we will take ethanol as a typical aliphatic hydroxy compound and phenol as a typical aromatic one.

## 29.5 Reactions involving cleavage of the O—H bond

ACIDITY

Table 29.2 shows how sodium reacts with water, ethanol and a solution of phenol in ethanol.

**Table 29.2** Reaction of a small piece of sodium with different hydroxy compounds.

| **Water** | **Ethanol** | **Solution of phenol in ethanol** |
|---|---|---|
| floats, melts, rushes about on surface, rapid evolution of hydrogen | sinks, steady evolution of hydrogen, does not melt | sinks, rapid evolution of hydrogen, does not melt |

Look at the table and answer these questions.

- ○ Why does sodium float in water but sink in ethanol?
- ○ The general reaction in each case is the reduction of $H^+$ ions by sodium. Write a general ionic equation.
- ○ Judging from these reactions, which contains a higher concentration of $H^+$ ions, water or ethanol?
- ○ Which contains a higher concentration of $H^+$ ions, ethanol or a solution of phenol in ethanol?
- ○ Which is the stronger acid, water or ethanol?
- ○ Which is the stronger acid, phenol or ethanol?
- ○ Write a full equation for the reaction of ethanol with sodium.

The fact that ethanol reacts with sodium liberating hydrogen suggests that the alcohol contains some hydrogen ions and that these are being reduced by sodium:

$$2Na + 2H^+ \longrightarrow 2Na^+ + H_2$$

The fact that the reaction is slower than with water suggests that ethanol contains a lower concentration of $H^+$ than water, i.e. it is a weaker acid. In fact, the $K_a$ of ethanol is $10^{-18}$ whereas the value for water is $10^{-16}$. Remember, the higher the $K_a$, the stronger the acid (see section 22.9). We can see why ethanol is a weaker acid if we compare the equilibria:

$$CH_3CH_2OH + H_2O \rightleftharpoons \underset{\text{ethoxide ion}}{CH_3CH_2O^-} + H_3O^+$$

$$H_2O + H_2O \rightleftharpoons \underset{\text{hydroxide ion}}{HO^-} + H_3O^+$$

The ethoxide ion is more basic than the hydroxide ion, because of the tendency of alkyl groups to donate electrons. The $CH_3CH_2$ group donates electrons to the O, increasing the negative charge density on that atom and thus making it more ready to accept protons. Thus the equilibrium above is further to the left in the case of ethanol, so this is the weaker acid. Both water and ethanol are of course extremely weak acids compared with the substances we normally call acids. Correspondingly, the hydroxide and ethoxide ions are extremely strong bases.

In its reaction with sodium ethanol is converted to ethoxide ion, so the two products are hydrogen and sodium ethoxide:

$$2Na + 2CH_3CH_2OH \longrightarrow \underset{\text{sodium ethoxide}}{2CH_3CH_2O^-Na^+} + H_2$$

The reactions with sodium indicate that phenol is a stronger acid than ethanol. In fact it is also a stronger acid than water ($K_a$ for phenol $= 10^{-10}$), although sodium reacts more vigorously with water than with an ethanolic solution of phenol, because the latter only contains a fairly low concentration of phenol. If we consider the equilibrium:

$$H_2O + C_6H_5OH \rightleftharpoons \underset{\text{phenoxide ion}}{C_6H_5O^-} + H_3O^+$$

we can see why phenol is a stronger acid than aliphatic alcohols. In the phenoxide ion, the negative charge on the O can to some extent be delocalized round the ring. This reduces the tendency of the phenoxide ion to attract protons, i.e. reduces its strength as a base. Consequently, phenol is a stronger acid than aliphatic alcohols, though it is still weak.

As a consequence of its acidic nature, phenol is much more soluble in sodium hydroxide than in water. Hydroxide ions from sodium hydroxide remove hydrogen ions, displacing the above equilibrium to the right so that the phenol dissolves as sodium phenoxide.

Phenol used to be known as 'carbolic acid' and was used as one of the earliest antiseptics. The Edinburgh doctor Joseph Lister first used it in the 1860's to prevent wounds going septic after surgery. Unwittingly, it had been used as an antiseptic even before this, because the old-fashioned way of treating an amputation wound was to cover it with coal-tar, which contains phenol. Phenol is effective in killing bacteria, but it is also very corrosive to the skin, and has been largely replaced by other antiseptics. Many of these are derivatives of phenol and are better germicides. Two examples are shown in figure 29.4.

The antiseptic action of 'Coal Tar Soap' depended on the phenol and related compounds it contained.

2,4,6-trichlorophenol. 23 times as effective a germicide as phenol in a given concentration.

4-chloro-3,5-dimethylphenol. 280 times as effective as phenol.

**Figure 29.4** Derivatives of phenol used as antiseptics.

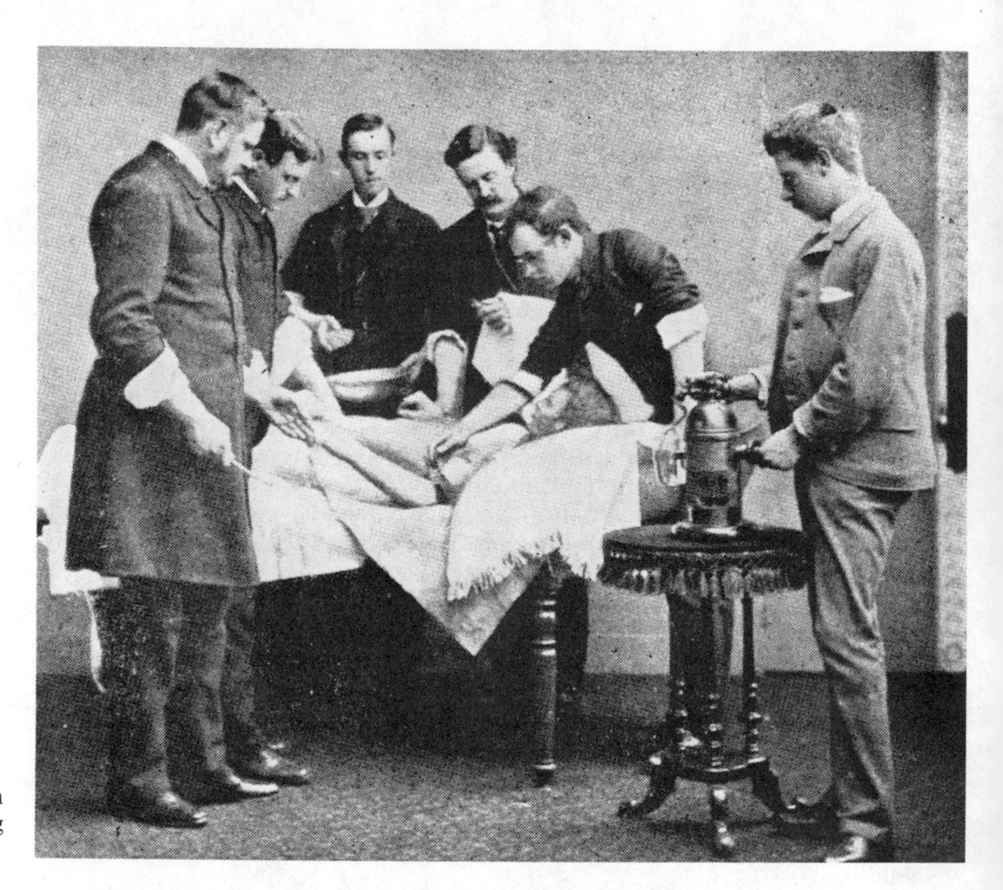

An early example of the use of antiseptics. In this operation a spray of phenol is being directed onto the wound.

ESTERIFICATION

In the presence of an acid catalyst, alcohols react with carboxylic acids to form esters. Water is eliminated, the alcohol cleaving at the O–H bond:

$$CH_3-\underset{\underset{O}{\|}}{C}-OH + CH_3CH_2-O-H \xrightarrow{H^+} CH_3-\underset{\underset{O}{\|}}{C}-O-CH_2CH_3 + H_2O$$

ethanoic acid ethanol ethyl ethanoate (an ester)

This reaction is considered more fully in chapter 31.

Phenol is again rather different from aliphatic alcohols: it does not react with acids directly. Phenyl esters can however be made by reacting phenol with an acyl halide (see section 28.7). For example:

$$CH_3-\underset{\underset{O}{\|}}{C}-Cl + C_6H_5OH \longrightarrow CH_3-\underset{\underset{O}{\|}}{C}-O-C_6H_5 + HCl$$

ethanoyl chloride phenyl ethanoate

Predict the formulae of the products of reactions between

- ○ propan-2-ol and sodium,
- ○ propan-1-ol and propanoic acid, in the presence of an acid catalyst.

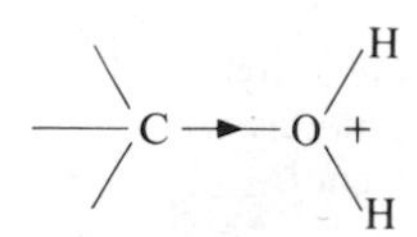


**Figure 29.5** Effect of protonation on polarization of the C—O bond in alcohols. The arrow indicates the direction of displacement of electrons in the bond.

## 29.6 Reactions involving cleavage of the C—O bond

REACTION WITH HALIDE IONS

In the presence of a strong acid, the OH group in an alcohol is protonated. The oxygen atom thus carries a positive charge, and tends to attract electrons very strongly from the adjacent carbon (figure 29.5). This induces a large positive charge on the carbon, making it susceptible to attack by nucleophiles. Thus, *in the presence of concentrated sulphuric*(VI) *acid*, ethanol reacts with $Br^-$ to form bromoethane:

$$CH_3CH_2{-}OH + H^+ \longrightarrow CH_3\overset{\delta+}{C}H_2{-}\overset{+}{O}H_2 \;(+\,Br^-) \longrightarrow CH_3CH_2Br + H_2O$$

(see section 28.6 for explanation of the significance of curved arrows).

Overall, this reaction amounts to the reaction of the alcohol with HBr:

$$CH_3CH_2OH + HBr \longrightarrow CH_3CH_2Br + H_2O$$

In some cases an intermediate carbonium ion may actually be formed before the $Br^-$ attacks. Cleavage of the C–O bond is greatly aided by protonation because it is much easier for the molecule to lose a neutral $H_2O$ molecule than a charged $^-OH$ ion. This is another example of a nucleophilic substitution reaction; compare it with the substitution reactions of halogenoalkanes (section 28.4). The reaction is normally carried out by heating ethanol under reflux with potassium bromide and concentrated sulphuric(VI) acid. Similar reactions occur between alcohols and $Cl^-$, $Br^-$ and $I^-$ in acid conditions, $Cl^-$ being the least reactive and $I^-$ the most.

OTHER HALOGENATION REACTIONS

There are several other reagents that are useful for replacing an —OH group with a halogen atom. Some of the more important are:

phosphorus pentachloride: $ROH + PCl_5 \rightarrow RCl + HCl + POCl_3$
phosphorus tribromide or triiodide (in practice a mixture of red phosphorus and bromine or iodine is used): $3ROH + PBr_3 \rightarrow 3RBr + H_3PO_3$

Phenol is again different in that it cannot be halogenated by normal reagents.

## 29.7 Reactions involving the carbon skeleton

So far we have considered reactions of alcohols in which part or all of the OH group is replaced. Alcohols also undergo several reactions which involve the carbon skeleton and the OH group simultaneously.

DEHYDRATION

*(i) Alkene formation.* Consider again the protonated form of ethanol that we discussed in the last section. This ion can readily lose water, forming a carbonium ion:

$$H{-}\underset{H}{\overset{H}{C}}{-}\underset{H}{\overset{H}{C}}{-}\underset{H}{\overset{+}{O}}{-}H \longrightarrow \left[H{-}\underset{H}{\overset{H}{C}}{-}\underset{H}{\overset{H}{C^+}}\right] + H_2O$$

In the reaction with HBr, this ion rapidly reacts with $Br^-$ to form bromoethane. In the absence of any nucleophile like $Br^-$, however, a different reaction may occur: the ion may lose $H^+$ and form ethene:

$$\left[H{-}\underset{H}{\overset{H}{C}}{-}\underset{H}{\overset{H}{C^+}}\right] \longrightarrow \underset{H}{\overset{H}{}}\!\!>C{=}C<\!\!\underset{H}{\overset{H}{}} + H^+$$

Thus in the presence of strong acids, ethanol forms ethene in what amounts to an overall dehydration reaction:

$$H-\underset{H}{\overset{H}{C}}-\underset{OH}{\overset{H}{C}}-H \longrightarrow \begin{matrix} H \\ H \end{matrix}\!\!>C{=}C<\!\!\begin{matrix} H \\ H \end{matrix} + H_2O$$

In practice the intermediate carbonium ion may never form: the $H^+$ and the $H_2O$ may leave simultaneously.

The reaction is normally carried out by heating ethanol at 170°C with excess concentrated sulphuric(VI) acid, when ethene is evolved and can be collected over water. The concentrated sulphuric(VI) acid can be regarded as behaving as a dehydrating agent, removing water from ethanol, though the actual mechanism as we have seen is more complex than this. Another way of preparing ethene from ethanol is by catalytic dehydration of ethanol vapour, using a heated catalyst of aluminium oxide or pumice stone in the apparatus shown in figure 29.6.

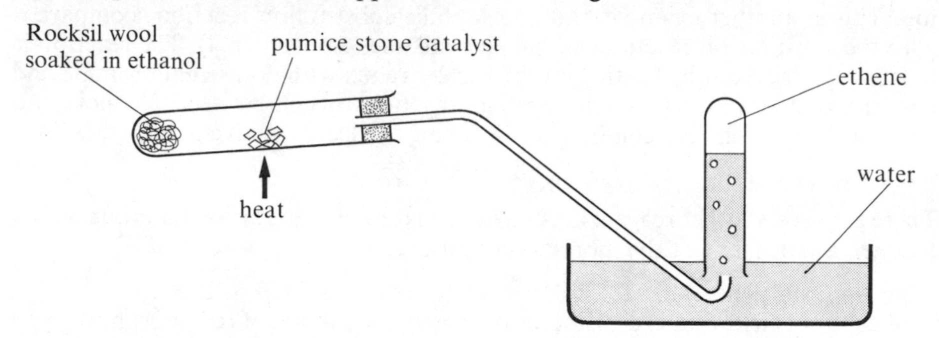


**Figure 29.6** Catalytic dehydration of ethanol.

*(ii) Ether formation.* The carbonium ion formed by ethanol in the presence of an acid is susceptible to attack by nucleophiles. Ethanol, with lone pairs on its oxygen atom, is itself a nucleophile, so we might expect it to attack the carbonium ion:

$$[CH_3CH_2^{\ +}] + H-\ddot{O}-CH_2CH_3 \longrightarrow [CH_3CH_2-\underset{H}{\overset{+}{O}}-CH_2CH_3]$$

This product can then lose an $H^+$ ion:

$$[CH_3CH_2\underset{H}{\overset{+}{O}}CH_2CH_3] \longrightarrow \underset{\text{ethoxyethane}}{CH_3CH_2OCH_2CH_3} + H^+$$

The product is ethoxyethane, an ether. Overall, this too is effectively a dehydration reaction:

$$\begin{matrix} CH_3CH_2OH \\ + \\ CH_3CH_2OH \end{matrix} \longrightarrow \begin{matrix} CH_3CH_2 \\ CH_3CH_2 \end{matrix}\!\!>O + H_2O$$

The dehydration of ethanol, then, can give two different products. Both this reaction and the one in which ethene is formed occur when ethanol is heated with concentrated sulphuric(VI) acid, but by adjusting the reaction conditions we can largely determine which product is formed. Since the formation of a molecule of the ether involves two molecules of ethanol and the formation of a molecule of the alkene involves only one, ether formation is favoured by having an excess of ethanol present. Compare this with the excess of sulphuric(VI) acid used in the preparation of the alkene.

Ethoxyethane is prepared by heating concentrated sulphuric(VI) acid with excess ethanol at 140°C. The ethoxyethane distils off. The properties of ethers are considered further in the next section.

The dehydration reactions considered here are applicable to alcohols in general, but not to phenols. Predict the formulae of the main products of the following reactions.

- ○ Butan-2-ol is heated with excess concentrated sulphuric(VI) acid.
- ○ Excess propan-1-ol is heated with concentrated sulphuric(VI) acid.
- ○ Propan-2-ol vapour is passed over heated aluminium oxide.

OXIDATION

Table 29.3 shows the effect of warming three different alcohols with acidified potassium dichromate(VI).

**Table 29.3** Effect of warming alcohols with acidified potassium dichromate(VI).

| Name | Formula | Observation |
|---|---|---|
| propan-1-ol | $CH_3CH_2CH_2OH$ | orange dichromate(VI) slowly turns green |
| propan-2-ol | $CH_3CHOHCH_3$ | orange dichromate(VI) slowly turns green |
| 2-methyl-propan-2-ol | $\begin{array}{c} CH_3 \\ \vert \\ CH_3{-}C{-}CH_3 \\ \vert \\ OH \end{array}$ | no change |

Look at the table and answer these questions.

- ○ Classify the three alcohols as primary, secondary and tertiary.
- ○ When potassium dichromate(VI) is reduced in acid solution, what is formed?
- ○ Which of the three alcohols is oxidized by acidified potassium dichromate(VI)?

Primary and secondary alcohols are readily oxidized by a variety of oxidants: acidified dichromate(VI), acidic or alkaline manganate(VII) or air in the presence of a catalyst. The initial product is a **carbonyl** compound (chapter 30) which in the case of a primary alcohol is an **aldehyde**:

$$\begin{array}{c} H\;\;H\;\;H \\ \vert\;\;\;\vert\;\;\;\vert \\ H{-}C{-}C{-}C{-}OH \\ \vert\;\;\;\vert\;\;\;\vert \\ H\;\;H\;\;H \\ \text{propan-1-ol} \end{array} \longrightarrow \begin{array}{c} H\;\;H \\ \vert\;\;\;\vert \\ H{-}C{-}C{-}C{\overset{\large O}{\diagup\!\!\diagup}} \\ \vert\;\;\;\vert\;\;\;\diagdown H \\ H\;\;H \\ \text{propanal, an} \\ \text{aldehyde} \end{array} + \begin{array}{c} 2H^+ + 2e^- \\ \text{electrons accepted} \\ \text{by oxidizer} \end{array}$$

Aldehydes are themselves readily oxidized to acids, thus:

$$\begin{array}{c} H\;\;H \\ \vert\;\;\;\vert \\ H{-}C{-}C{-}C{\overset{\large O}{\diagup\!\!\diagup}} \\ \vert\;\;\;\vert\;\;\;\diagdown H \\ H\;\;H \\ \text{propanal} \end{array} + H_2O \longrightarrow \begin{array}{c} H\;\;H \\ \vert\;\;\;\vert \\ H{-}C{-}C{-}C{\overset{\large O}{\diagup\!\!\diagup}} \\ \vert\;\;\;\vert\;\;\;\diagdown OH \\ H\;\;H \\ \text{propanoic acid} \end{array} + 2H^+ + 2e^-$$

The product of oxidizing a primary alcohol is therefore usually an acid, unless the aldehyde is distilled from the reaction mixture as it forms.

In the case of a secondary alcohol the product of oxidation is a **ketone**:

$$\begin{array}{c} H\;\;\;H\;\;\;H \\ \vert\;\;\;\;\vert\;\;\;\;\vert \\ H{-}C{-}C{-}C{-}H \\ \vert\;\;\;\;\vert\;\;\;\;\vert \\ H\;\;OH\;\;H \\ \text{propan-2-ol} \end{array} \longrightarrow \begin{array}{c} H\;\;\;\;\;\;\;\;H \\ \vert\;\;\;\;\;\;\;\;\;\vert \\ H{-}C{-}C{-}C{-}H \\ \vert\;\;\;\Vert\;\;\;\vert \\ H\;\;O\;\;H \\ \text{propanone-} \\ \text{a ketone} \end{array} + 2H^+ + 2e^-$$

Ketones are not readily oxidized, so the reaction stops at this point.

Tertiary alcohols cannot be readily oxidized since they have no hydrogen atom to be removed from the carbon atom carrying the OH group. Strong oxidizers cause their molecules to be broken up, giving a mixture of oxidation products.

Predict the products, if any, of oxidizing the following alcohols with acidified dichromate(VI):

- ethanol (product distilled off immediately),
- ethanol (reagents heated together under reflux for some time),
- 2-methylbutan-2-ol,
- butan-2-ol.

The oxidation of alcohols is important in industry and in living organisms. Methanal (commonly called formaldehyde and used in the production of plastics, see later) is manufactured by passing a mixture of methanol vapour and air over a silver catalyst at 500°C:

$$2\,H\!-\!\underset{\displaystyle H}{\overset{\displaystyle H}{\underset{|}{\overset{|}{C}}}}\!-\!OH + O_2 \xrightarrow[500^\circ C]{Ag} 2\,H\!-\!C\!\begin{matrix}\nearrow\!\!\!/ O\\ \searrow H\end{matrix} + 2\,H_2O$$

Propanone (acetone) is manufactured from propan-2-ol by a similar process, using air and a copper catalyst at 500°C. In these industrial processes, high temperature catalytic reactions are used in which air is the oxidizer, rather than aqueous oxidizers like dichromate(VI). This is of course for economic reasons: air is much cheaper than dichromate(VI) and although the catalyst is expensive it has a long life before it needs renewing.

The bacterial oxidation of ethanol to ethanoic acid (acetic acid) has long been a problem for wine-producers. The bacterium, *acetobacter*, uses air to oxidize ethanol in wine, producing a weak solution of ethanoic acid called vinegar. The bacterium uses this oxidative process as a source of energy. Once a bottle of wine has been opened it will turn to vinegar fairly quickly because of the considerable number of these bacteria in the air. One way of preventing this is to add extra alcohol to the wine so that the concentration of ethanol is too high for the bacteria to tolerate. Wine treated in this way is said to be *fortified*: sherry and port are examples.

Ethanol is disposed of by the body by oxidation: it is oxidized in the liver to ethanal. Very large concentrations of ethanol are too much for the liver to cope with and may cause damage to the organ.

The redox reaction between ethanol and dichromate(VI) is employed in 'breathalyzers' (see photograph). Ethanol, like all alcohols, undergoes complete oxidation

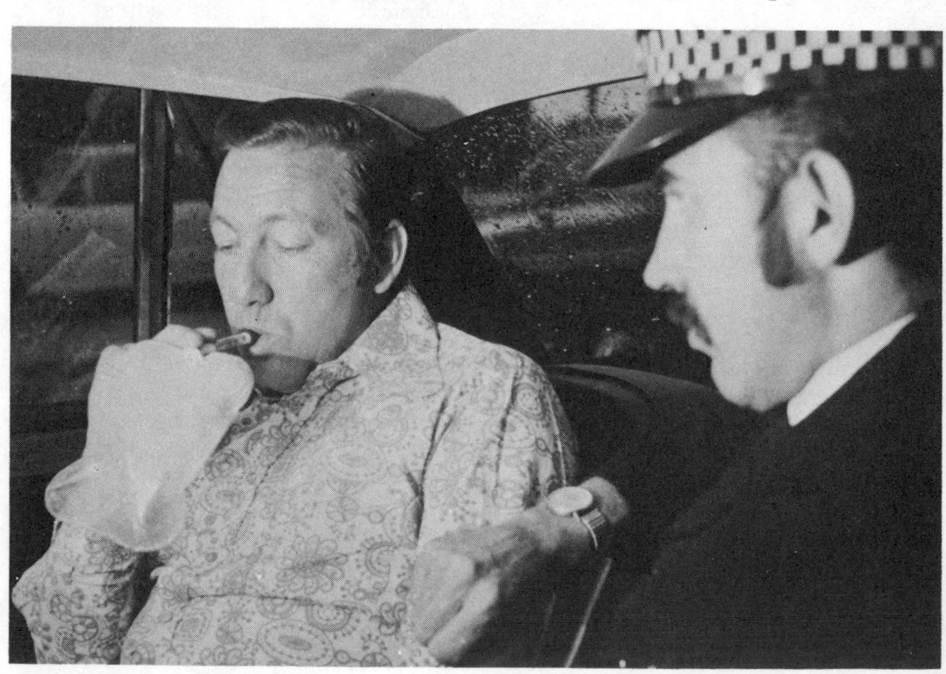

A 'breathalyzer' contains orange crystals of potassium dichromate(VI), which is reduced to green chromium(III) by ethanol. The colour change provides an indication of the level of ethanol vapour in a motorist's breath.

to $CO_2$ and $H_2O$ when heated in the presence of air, i.e. it undergoes combustion. It burns with a clean, smokeless flame and is occasionally used as a fuel (as in methylated spirit stoves). Its use as a fuel is limited though, because of its cost.

Phenol, like tertiary alcohols, cannot be oxidized to a carbonyl compound, though the aromatic ring with its high electron density is susceptible to attack by oxidizers.

A spirit burner. Ethanol burns with a very clean flame but its use is limited because it is a relatively expensive fuel.

### SUBSTITUTION REACTIONS OF THE AROMATIC RING IN PHENOL

If bromine water is added to a solution of phenol the bromine is immediately decolorized and a white precipitate is formed. This is a substitution reaction, the white precipitate being 2,4,6-tribromophenol:

$$C_6H_5OH + 3Br_2 \longrightarrow C_6H_2Br_3OH + 3HBr$$

2,4,6-tribromophenol

Benzene, it will be recalled, does not react with bromine except in the presence of a halogen-carrier catalyst. In the case of phenol, the lone pairs of electrons on the oxygen of the OH group become partially delocalized round the ring: this increases the electron density there and makes it more susceptible to attack by electrophiles. Phenol therefore undergoes electrophilic substitution much faster than benzene, the 2,4 and 6 positions being preferentially substituted (see section 27.10).

Phenol is nitrated far more readily than benzene. Dilute nitric(V) acid alone is sufficient, the products being 2- and 4-nitrophenol. A mixture of concentrated nitric(V) acid and concentrated sulphuric(VI) acid readily converts it to 2,4,6-trinitrophenol, commonly known as picric acid:

$$C_6H_5OH + 3HNO_3 \longrightarrow C_6H_2(NO_2)_3OH + 3H_2O$$

2,4,6-trinitrophenol

This substance is used as a high explosive: it decomposes spontaneously and exothermically, producing large volumes of CO, steam and $N_2$ which expand and produce an explosive shock.

*Plastics from phenol.* When phenol is heated with methanal in the presence of an acid or alkali catalyst, a hard, brittle plastic is formed. The initial stage of this reaction is substitution of the phenol in the 2 or 4 position:

$$C_6H_5OH + HCHO \longrightarrow HOC_6H_4CH_2OH$$

This product then undergoes a condensation reaction (see section 30.4) in which a molecule of water is eliminated between it and another molecule of phenol.

The product may then be substituted by another methanal molecule:

$$\text{HOC}_6\text{H}_4\text{CH}_2\text{C}_6\text{H}_4\text{OH} + \text{HCHO} \longrightarrow \text{HOC}_6\text{H}_4\text{CH}_2\text{C}_6\text{H}_3(\text{OH})\text{CH}_2\text{OH}$$

Condensation then occurs with a further molecule of phenol and the process continues until all the aromatic rings are substituted in the 2,4 and 6 positions and a molecular network like that shown in figure 29.7 has built up.

**Figure 29.7** Condensation polymer of phenol and methanal.

This polymer, because of its extensively cross-linked three-dimensional network, is very hard and rather brittle. It is a dark brown material, known as 'Bakelite'. It has the disadvantage that it sets hard on heating and cannot be remelted: this causes difficulties in moulding the plastic. Its hardness and cheapness are however a considerable advantage and it is widely used for such things as electric plugs and motor-car distributor caps. Plastics of this kind, which set hard and cannot be remelted, are called **thermosetting**. Plastics like polythene which can be moulded by melting are called **thermosoftening**.

A radio with 'Bakelite' case, manufactured just after the War. 'Bakelite' was the earliest synthetic plastic.

## 29.8 Ethers

Ethers are compounds with the general structure ROR′, where R and R′ are alkyl or aryl groups. They are named by regarding them as alkanes substituted by alkoxy groups. Thus,

$CH_3OCH_3$ methoxymethane,

$CH_3CH_2OCH_2CH_3$ ethoxyethane,

$CH_3OCH_2CH_3$ methoxyethane.

Ethoxyethane is an example of a **symmetrical** ether; methoxyethane is an **unsymmetrical** ether.

Since they have no capacity for hydrogen-bonding, ethers are far more volatile than alcohols, indeed they correspond closely in volatility to alkanes of comparable relative molecular mass. Thus the most commonly encountered ether, ethoxyethane (commonly called simply 'ether'), is a very volatile liquid of boiling point 35°C. This volatility makes ethoxyethane a dangerously flammable substance. Since its relative molecular mass is 74, ethoxyethane vapour is much denser than air, so it tends to diffuse slowly and stay at floor or bench level in a laboratory, where it is of course likely to be ignited by naked flames. Serious explosions and fires have been caused in this way.

The molecule of ethoxyethane is analogous in shape to that of water:

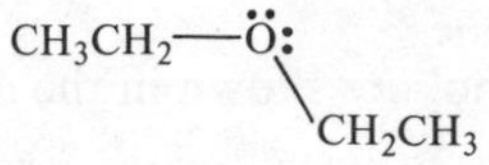

$$CH_3CH_2—\ddot{O}: \\ \quad\quad\quad\quad \diagdown CH_2CH_3$$

It is therefore slightly polar, but its polarity is not sufficient to have a significant effect on its properties. Ethoxyethane, like all ethers, is only slightly soluble in water and is a good solvent for non-polar substances. It is frequently used as a solvent in practical work because it is volatile and easily distilled off, but its use must be carefully controlled because of the fire risk. The small polarity in the molecule is not enough to make it significantly reactive toward nucleophiles or electrophiles: ethers are not much more reactive than alkanes.

Ethers do, however, show some reactivity in acid solution. The lone pairs on the oxygen atom enable them to act as bases: for example, ethoxyethane is protonated by concentrated sulphuric(VI) acid:

$$CH_3CH_2\ddot{\underset{..}{O}}CH_2CH_3 + H^+ \longrightarrow CH_3CH_2\overset{+}{\ddot{\underset{H}{O}}}CH_2CH_3$$

Ethoxyethane will therefore dissolve in concentrated sulphuric(VI) acid though it is almost insoluble in water. Protonation of the ether makes it more liable to cleavage since the carbon atoms now carry a considerable positive charge and are susceptible to attack by nucleophiles. Thus ethers are split by heating with concentrated hydriodic acid, HI:

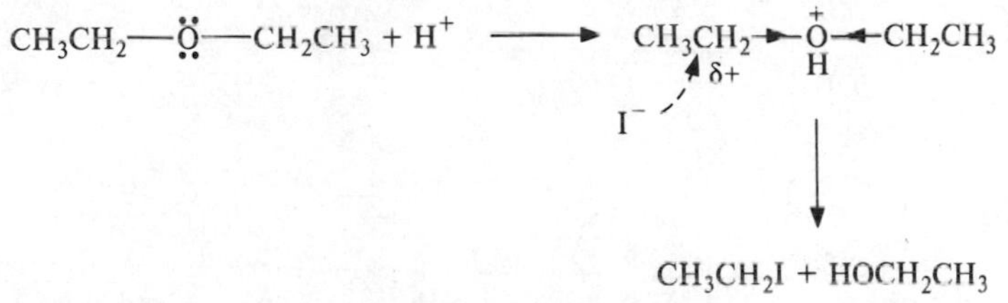

$$CH_3CH_2—\ddot{\underset{..}{O}}—CH_2CH_3 + H^+ \longrightarrow \underset{\delta+ \; \uparrow I^-}{CH_3CH_2} \rightarrow \overset{+}{\underset{H}{O}} \leftarrow CH_2CH_3$$

$$\downarrow$$

$$CH_3CH_2I + HOCH_2CH_3$$

$$\text{Overall: } CH_3CH_2OCH_2CH_3 + HI \longrightarrow CH_3CH_2I + CH_3CH_2OH$$

An interesting example of a **cyclic ether** is epoxyethane, $\underset{\diagdown O \diagup}{CH_2—CH_2}$

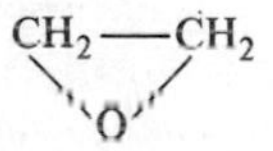

This compound has considerable ring-strain (section 25.7) so it is a good deal more reactive than straight-chain ethers. Its chemical reactivity makes it an important intermediate in industrial synthesis (see section 26.6).

## Summary

1 Alcohols are aliphatic compounds containing the OH group. They are named using the suffix *-ol*. Phenols are aromatic compounds with the —OH group attached directly to the ring.

2 Primary alcohols are of the form $RCH_2OH$; secondary alcohols are of the form $\begin{matrix} R_1 \diagdown \\ R_2 \diagup \end{matrix} CHOH$; tertiary alcohols are of the form $\begin{matrix} R_1 \diagdown \\ R_2 — \\ R_3 \diagup \end{matrix} C—OH$.

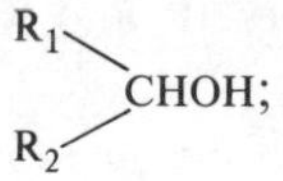

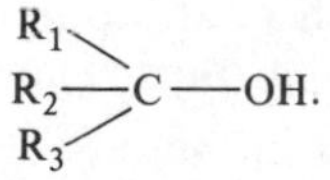

3 The physical properties of alcohols, especially the early members of the series, are strongly influenced by hydrogen bonding.

4 Alcohols can act as proton donors but are weaker acids than water. Phenols are considerably stronger acids.

5 Alcohols can act as bases, becoming protonated on the oxygen atom. This can lead to the formation of a carbonium ion and subsequent dehydration or substitution reactions.
6 Primary alcohols can be oxidized to aldehydes, then to acids. Secondary alcohols can be oxidized to ketones. Tertiary alcohols cannot be oxidized without fragmenting the molecule.
7 Some reactions of ethanol are shown in the diagram below.

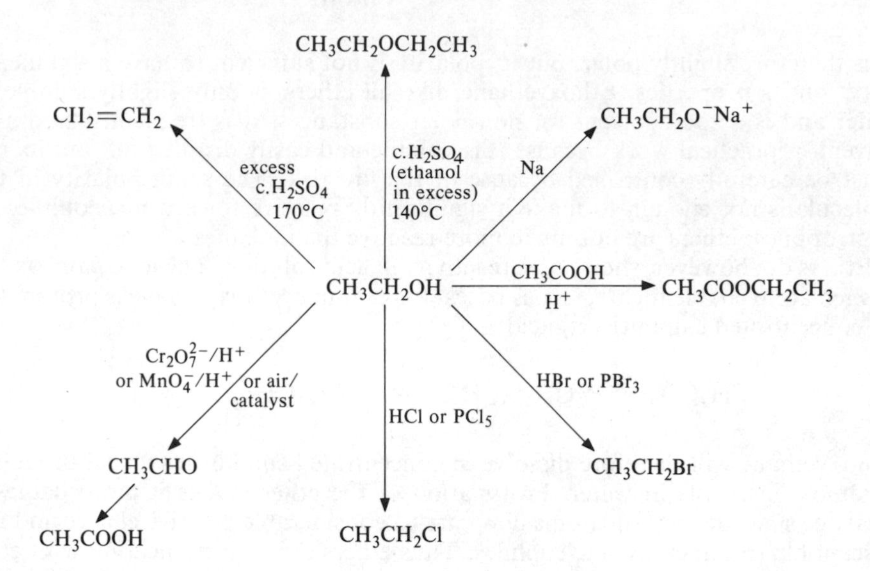


8 Some reactions of phenol are shown below.

OH
$NO_2$
+
OH
$NO_2$
dil $HNO_3$
OH
Na or NaOH
$O^-Na^+$
$c.HNO_3/$
$c.H_2SO_4$
$CH_3COCl$
OH
$O_2N$
$NO_2$
$NO_2$
$Br_2$
OOCCH$_3$
OH
Br
Br
Br

9 Ethers are compounds of the general form ROR′, where R and R′ are alkyl or aryl groups. They are rather unreactive.

## Study questions

1 Consider the following compounds.

A $CH_3CH_2-C(CH_3)(OH)-CH_3$

B $CH_3CH_2CHOHCH_3$

C $CH_3OCH_2CH_2CH_3$

D $CH_3CH_2CH_2CH_2OH$

E (benzene ring with OH and $CH_3$ in para positions)

(a) Name each compound.
(b) Which is a primary alcohol?
(c) Which is a tertiary alcohol?
(d) Which is a phenol?
(e) Which is an ether?

2 Refer again to the compounds in question 1.
(a) Which react with sodium metal?
(b) Which could be oxidized to an aldehyde?
(c) Which could be oxidized to a ketone?
(d) Which would be the strongest acid?
(e) Which would form an alkene when heated with excess concentrated sulphuric(VI) acid?
(f) Which has the lowest boiling point?

3 Table 29.4 gives some physical properties of water, ethanol and ethoxyethane.

**Table 29.4** Some physical properties of water, ethanol, and ethoxyethane.

| Name | Formula | $M_r$ | Boiling point/ °C | Density at 273 K/ $g\,cm^{-3}$ | Surface tension at 293 K/$N\,m^{-1}$ |
|---|---|---|---|---|---|
| water | $H_2O$ | 18 | 100 | 1.00 | 7.28 |
| ethanol | $CH_3CH_2OH$ | 46 | 78 | 0.79 | 2.23 |
| ethoxyethane | $CH_3CH_2OCH_2CH_3$ | 74 | 35 | 0.71 | 1.69 |

Explain in terms of intermolecular forces, why
(a) the boiling point of ethanol is greater than that of ethoxyethane,
(b) the density of water is greater than that of ethanol,
(c) the surface tension of water is greater than that of ethoxyethane.

4 An organic liquid **A** contains carbon, hydrogen and oxygen only. On combustion 0.463g of **A** gave 1.1g of carbon dioxide and 0.563g of water. When vaporized, 0.1g of **A** occupy 54.5cm³ at 208°C and 98.3 kPa (740mm Hg).
Standard pressure = 101 kPa (760mm Hg); 1 mole of a gas occupies 22.4dm³ at s.t.p.
(a) What is the percentage composition of **A**?
(b) Find the empirical formula of **A**.
(c) Calculate the relative molecular mass of **A**.
(d) Give the structures of possible non-cyclic isomers of **A**.
(e) Which isomers will react with sodium to give hydrogen?
(f) Which isomers will reduce acidified dichromate(VI) ion to green $Cr^{3+}$?

5 Suggest explanations for the following observations.
(a) Butan-1-ol is much more soluble in 5M hydrochloric acid than in water.
(b) 2,4,6-trinitrophenol (picric acid) is a much stronger acid than phenol ($K_a = 10^{-1}$ and $10^{-10}$ respectively).
(c) Phenol can be nitrated by dilute nitric(V) acid to give a mixture of 2-nitrophenol and 4-nitrophenol, whereas benzene is only nitrated by a mixture of concentrated nitric(V) and concentrated sulphuric(VI) acids.
(d) Ethanol is more acidic than 2-methylpropan-2-ol.
(e) Heating butan-2-ol with excess concentrated sulphuric(VI) acid produces a mixture of three isomeric alkenes.

6 Give the reagents and conditions you would use to carry out the following conversions.
(a) Ethanol to ethyl ethanoate (using ethanol as the only organic starting material).
(b) Ethanol to 1,2-dibromoethane.
(c) Ethene to ethanoic acid.
(d) Ethanol to propanonitrile.
(e) Bromoethane to ethoxyethane.

7 Suggest structures for the following compounds **X**, **Y** and **Z**. Explain the reactions involved in each case.
(a) **X**, $C_7H_8O$, burns with a smoky flame. **X** is only slightly soluble in water, but very soluble in sodium hydroxide solution.
(b) **Y**, $C_6H_{14}O$, is insoluble in water but soluble in concentrated sulphuric(VI) acid. **Y** has no reaction with sodium.
(c) **Z**, $C_5H_{12}O$, gives hydrogen when reacted with sodium metal but has no effect on acidified potassium dichromate(VI).

8 Predict the structure of the products of the following reactions.
(a) Butan-2-ol is warmed with acidified potassium manganate(VII).
(b) An aqueous solution of chlorine is added to phenol.
(c) Propan-2-ol is warmed with excess concentrated sulphuric(VI) acid.
(d) 2-Methylphenol is treated with propanoyl chloride.
(e) Methanol is treated with phosphorus pentachloride.

9 Primary alcohols can be thought of as compounds derived by replacing one hydrogen atom in a water molecule by an alkyl group. Ethers can be thought of as resulting from the replacement of both hydrogen atoms in water by alkyl groups. To what extent do the physical and chemical properties of water, alcohols and ethers fit in with this idea? Illustrate your answer by reference to ethanol and ethoxyethane.

# Carbonyl Compounds 30

## 30.1 The carbonyl group

In chapter 26 we looked at compounds containing C═C double bonds and saw that their typical reactions involved electrophilic addition. In the last chapter we looked at compounds containing the C—O single bond, which is polar and tends to bring about substitution reactions. We will now turn our attention to the carbonyl group, C═O, which is the functional group in aldehydes and ketones. We might expect this group to show similarity in its reactions to both $>C=C<$ and $-\overset{|}{\underset{|}{C}}-O-$

The double bond between C and O in the carbonyl group, like the double bond in alkenes, can be considered to consist of a $\sigma$-bond and a $\pi$-bond. Unlike the C═C group, however, the carbonyl group does not have an even electron distribution between the two atoms: there is a greater electron density over the more electronegative oxygen atom, as represented in figure 30.1.

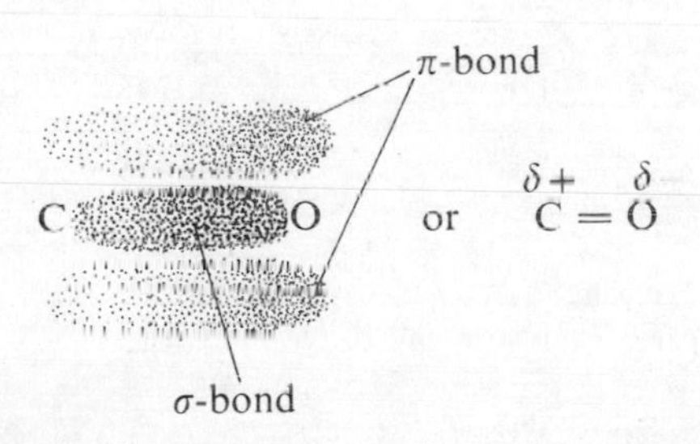


Figure 30.1 Electron distribution in the C═O bond.

This electron distribution makes the carbon atom attractive towards nucleophiles, which attack and bond to it, breaking the $\pi$-bond and resulting eventually in addition. With a general nucleophile $\overset{\delta-}{X}-\overset{\delta+}{Y}$:

$$>\overset{\delta+}{C}=\overset{\delta-}{O} \;+\; \overset{\delta-}{X}-\overset{\delta+}{Y} \longrightarrow \left[\, >C<^{O^-}_{X} \,\right] + Y^+ \xrightarrow{Y^+} >C<^{OY}_{X}$$

In most cases, Y is hydrogen. This sort of reaction, which is typical of the carbonyl group, is called **nucleophilic addition**: compare it with the *electrophilic* addition we have found to be typical of alkenes.

The reactions of the carbonyl group are important to the organic chemist as the group is common in biological molecules, particularly carbohydrates (section 30.7). Carbonyl compounds also have considerable industrial significance, for example as solvents and in the manufacture of thermosetting plastics (section 29.7).

## 30.2 Aldehydes and ketones: nature and nomenclature

Aldehydes and ketones both contain the carbonyl group, but differ in its position in the hydrocarbon skeleton. **Aldehydes** have the carbonyl group at the end of a chain: their general formula is therefore $R{-}\underset{\underset{O}{\|}}{C}{-}H$, usually written RCHO, where R is an alkyl or aryl group or hydrogen. **Ketones** have the carbonyl group in a non-terminal position in the chain: the general formula is $R{-}\underset{\underset{O}{\|}}{C}{-}R'$, usually written RCOR′ where R and R′ are alkyl or aryl groups. Thus aldehydes and ketones are structurally quite similar, but as we shall see their properties differ considerably and they are considered as different homologous series.

Aldehydes are named using the suffix **-al** after a stem indicating the number of carbon atoms (including the one in the carbonyl group). Thus $CH_3CHO$ is called ethanal. Ketones are named using the suffix **-one** after a stem indicating the number of carbon atoms, together with a number, if necessary, to indicate the position of the carbonyl group in the chain. Thus $CH_3COCH_2CH_2CH_3$ is called pentan-2-one. Table 30.1 gives the names and formulae of some important aldehydes and ketones.

**Table 30.1** Some important aldehydes and ketones.

| Formula | Systematic name | Other name | State at room temp. | Boiling point/K | Solubility in water |
|---|---|---|---|---|---|
| HCHO | methanal | formaldehyde | g | 254 | soluble |
| $CH_3CHO$ | ethanal | acetaldehyde | l | 294 | infinite |
| $CH_3CH_2CHO$ | propanal | propionaldehyde | l | 321 | soluble |
| $CH_3COCH_3$ | propanone | acetone | l | 329 | infinite |
| $CH_3COCH_2CH_3$ | butanone | methylethyl ketone | l | 353 | very soluble |
| $CH_3CH_2COCH_2CH_3$ | pentan-3-one | | l | 375 | very soluble |
| $C_6H_5$–CHO (benzene ring) | benzaldehyde | | l | 451 | slightly soluble |
| $C_6H_5$–$COCH_3$ (benzene ring) | phenylethanone | acetophenone | l | 475 | insoluble |

- Name the compounds:
  (a) $CH_3CH_2CH_2CHO$,
  (b) $CH_3CH_2COCH_2CH_2CH_3$.
- Give the formula of hexan-2-one.
- Why is there no such compound as ethanone?

PHYSICAL PROPERTIES

The polarity of the C=O group has considerable influence on the physical properties of aldehydes and ketones. The earlier members of both series are considerably less volatile than alkanes of corresponding relative molecular mass. Thus ethanal (acetaldehyde, $CH_3CHO$), with a boiling point of 20°C, is a liquid (though a very volatile one) at room temperature, while propane, $CH_3CH_2CH_3$, with the same relative molecular mass is a gas at room temperature (boiling point, −42°C). Note that the polar C=O group has less effect on inter-molecular forces than the —OH group, the latter being able to participate in hydrogen bonding. (Compare the boiling points of ethanal and ethanol—20°C and 78°C respectively.) The early members of the aldehydes and ketones are soluble in water, and will themselves dissolve both polar and non-polar solutes. Propanone (acetone), for example is a widely used industrial solvent. As expected, the polar C=O group has less and less influence on the physical properties of carbonyl compounds as the homologous series are ascended. Table 30.1 gives some physical properties of carbonyl compounds.

MAKING ALDEHYDES AND KETONES

In the laboratory these compounds can be made by the oxidation of alcohols, as described in section 29.7. Aldehydes are made by oxidizing primary alcohols, ketones by oxidizing secondary alcohols. This method is also used industrially for making methanal (from methanol) and propanone (from propan-2-ol). Most propanone, however, is now produced as a by-product from the manufacture of phenol (see study question 6, chapter 27). Some ethanal is still manufactured by the oxidation of ethanol, but most is made by direct oxidation of ethene (see study question 3 at the end of this chapter).

A 40% solution of methanal in water, called formalin, is used for preserving biological specimens.

## 30.3 Addition reactions of carbonyl compounds

The characteristic reaction of compounds containing the carbonyl group is nucleophilic addition:

$$\begin{matrix} R_1 \\ R_2 \end{matrix}\!\!>\overset{\delta+}{C}=\overset{\delta-}{O} + \overset{\delta-}{X}-\overset{\delta+}{Y} \longrightarrow \begin{matrix} R_1 \\ R_2 \end{matrix}\!\!>C<\!\!\begin{matrix} OY \\ X \end{matrix}$$

The readiness with which such an addition reaction occurs is largely determined by the size of the partial positive charge on the carbon atom of the carbonyl group.

- ○ Consider the cases when $R_1$ and $R_2$ are (a) both H, (b) $CH_3$ and H, (c) both $CH_3$. Place (a), (b) and (c) in order according to the magnitude of positive charge you would expect on the carbonyl carbon in each of the three compounds. (Remember that alkyl groups like $CH_3$ tend to donate electrons to groups to which they are attached.)
- ○ Place (a), (b) and (c) in order of readiness to undergo nucleophilic addition.
- ○ Name compounds (a), (b) and (c).
- ○ How would you expect benzaldehyde ($C_6H_5-\underset{\underset{O}{\|}}{C}-H$) to compare with ethanal in its readiness to undergo nucleophilic addition?

For the reasons suggested in the questions above, aldehydes tend to be more reactive than ketones in nucleophilic addition reactions. Methanal is the most reactive aldehyde. Benzaldehyde, because of the tendency of the benzene ring to delocalize the positive charge on the carbonyl carbon, is less reactive than aliphatic aldehydes. On the whole, though, the reactions of aromatic aldehydes and ketones (i.e. those with the carbonyl group attached directly to the benzene ring) are quite similar in nature to those of aliphatic ones.

The initial stage of addition to C=O is attack by a nucleophile. Compare this with C=C, where the initial stage of addition is attack by an electrophile. Carbonyl compounds thus tend to undergo addition reactions with rather different compounds to those which react with alkenes. Some examples are now given.

REACTION WITH HCN

In most addition reactions of carbonyl compounds, the molecule adding across the C=O double bond is of the form HX. A good example is HCN.

$$\underset{\text{ethanal}}{\begin{matrix} CH_3 \\ H \end{matrix}\!\!>C=O} + \overset{\delta+\,\delta-}{HCN} \longrightarrow \underset{\text{2-hydroxypropanonitrile}}{\begin{matrix} CH_3 \\ H \end{matrix}\!\!>C<\!\!\begin{matrix} OH \\ CN \end{matrix}}$$

$$(CH_3)_2C{=}O + HCN \longrightarrow (CH_3)_2C(OH)CN$$

propanone → 2-hydroxy-2-methylpropanonitrile

2-Hydroxy-2-methylpropanonitrile is an intermediate in the manufacture of methyl 2-methylpropenoate, the monomer for Perspex.

REACTION WITH SODIUM HYDROGENSULPHATE(IV)

Carbonyl compounds undergo addition reactions when shaken with saturated aqueous sodium hydrogensulphate(IV) ($NaHSO_3$). As expected, ethanal reacts more readily than propanone.

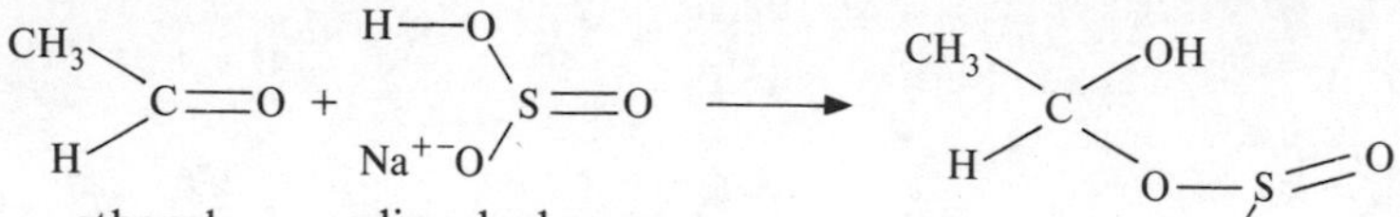

$$CH_3(H)C{=}O + H{-}O{-}S({=}O){-}O^-Na^+ \longrightarrow CH_3(H)C(OH){-}O{-}S({=}O){-}O^-Na^+$$

ethanal + sodium hydrogensulphate(IV)

Reaction 88% complete after 30 min at 25°C

$$(CH_3)_2C{=}O + H{-}O{-}S({=}O){-}O^-Na^+ \longrightarrow (CH_3)_2C(OH){-}O{-}S({=}O){-}O^-Na^+$$

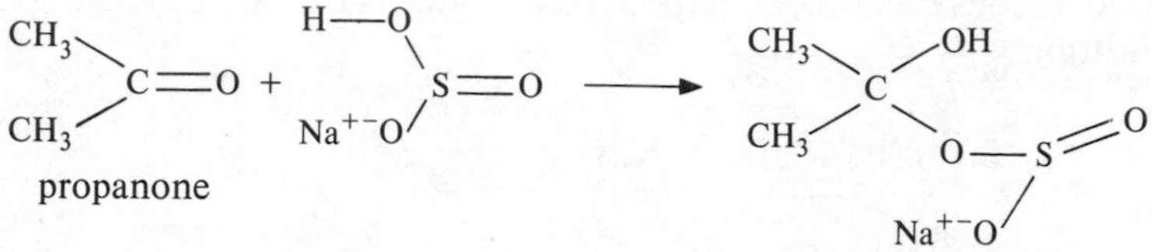

propanone

Reaction 47% complete after 30 min at 25°C

The products of these reactions are crystalline ionic compounds, soluble in water.

REDUCTION

Like the C=C bond, the C=O bond undergoes addition with hydrogen in the presence of a metal catalyst such as platinum or nickel. Aldehydes give primary alcohols, ketones give secondary alcohols.

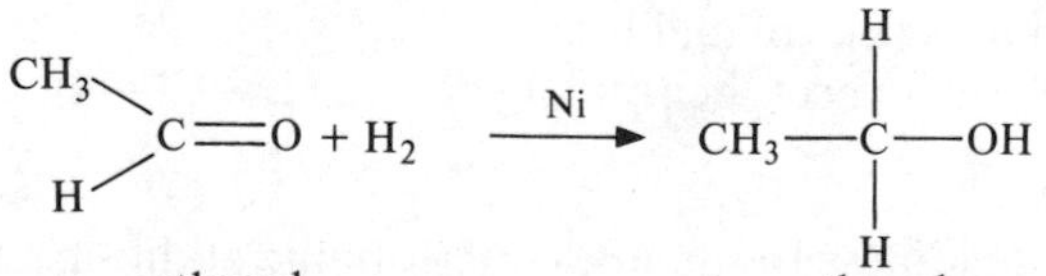

$$CH_3(H)C{=}O + H_2 \xrightarrow{Ni} CH_3{-}CH_2{-}OH$$

ethanal → ethanol

$$(CH_3)_2C{=}O + H_2 \xrightarrow{Ni} CH_3{-}CH(OH){-}CH_3$$

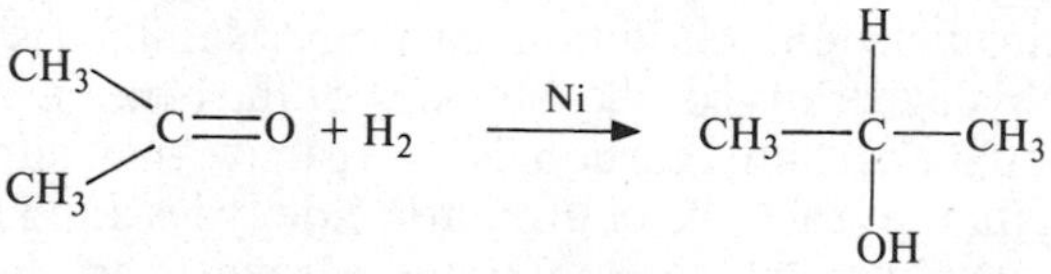

propanone → propan-2-ol

- ○ Write the formulae of the products you would expect when propanal reacts with (a) $H_2$ in the presence of a nickel catalyst, (b) HCN.
- ○ Write the formula of the product you would expect when (a) butanone, (b) benzaldehyde reacts with $NaHSO_3$.

'SELF ADDITION': POLYMERIZATION

Just as alkenes can be made to undergo addition polymerization, carbonyl compounds too can form polymers, by addition across the C=O double bond. Ketones are not reactive enough to polymerize easily, but aldehydes can readily be converted to a variety of addition polymers. As expected, methanal polymerizes most readily. Some of the polymers formed by methanal and ethanal are shown in table 30.2.

**Table 30.2** Polymers of methanal and ethanal.

| Method of preparation | Name | Structure | Uses |
|---|---|---|---|
| **1 From methanal** | | | |
| evaporate aqueous solution of methanal | poly (methanal) | $-O-CH_2-O-CH_2-O-CH_2-O-$ | high-strength plastic (making gear wheels, etc.), marketed as Delrin and Kematal |
| distil methanal from acidic solution | methanal trimer | $CH_2$<br>O O<br>$H_2C$ $CH_2$<br>O | |
| **2 From ethanal** | | | |
| add dilute acid to ethanal | ethanal trimer | $CH_3$<br>CH<br>O O<br>HC CH<br>$H_3C$ O $CH_3$ | hypnotic (sleep-inducing) drug |
| add acid to ethanal below 0°C | ethanal tetramer | $CH_3$<br>CH<br>O O<br>$CH_3-CH$ $HC-CH_3$<br>O O<br>CH<br>$CH_3$ | slug poison, solid fuel in portable stoves (commonly called metaldehyde) |

A plastic made from poly(methanal) is used to replace metal in machine parts such as gear wheels and clips.

## 30.4 Condensation reactions of carbonyl compounds

In some cases the addition reactions outlined above are followed by elimination of a molecule of water. Many of these elimination reactions involve derivatives of ammonia of the general form X-$NH_2$:

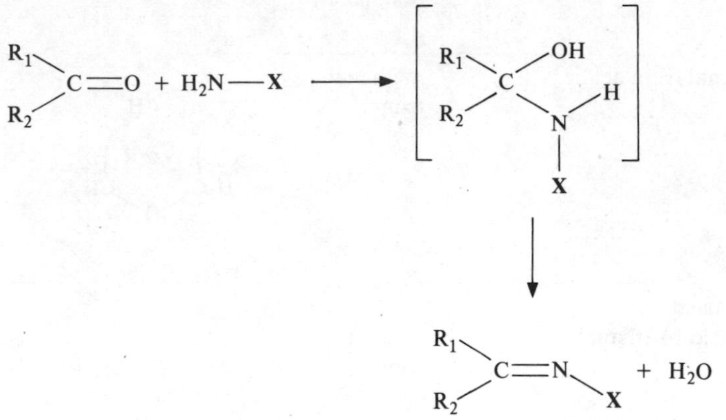

$$R_1R_2C{=}O + H_2N{-}X \longrightarrow \left[ R_1R_2C(OH){-}N(H){-}X \right] \longrightarrow R_1R_2C{=}N{-}X + H_2O$$

This sort of reaction, in which two molecules combine with the elimination of water, is called a **condensation** or **addition-elimination** reaction. Important examples are shown in table 30.3.

**Table 30.3** Condensation products of carbonyl compounds

| Reactant | Product |
|---|---|
| $NH_2OH$ hydroxylamine | 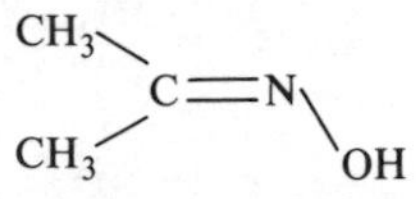 $(CH_3)_2C{=}N{-}OH$ (from propanone) |
| $NH_2NH_2$ hydrazine | 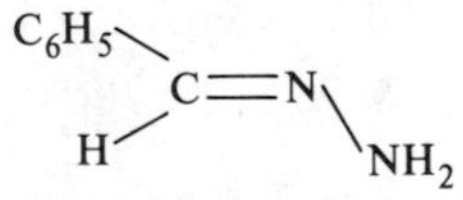 $C_6H_5(H)C{=}N{-}NH_2$ (from benzaldehyde) |
| 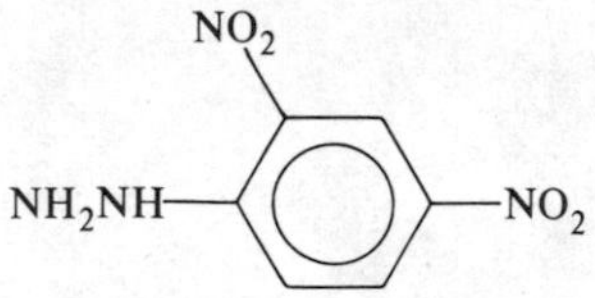 $NH_2NH{-}C_6H_3(NO_2)_2$ (2,4-dinitrophenylhydrazine) | 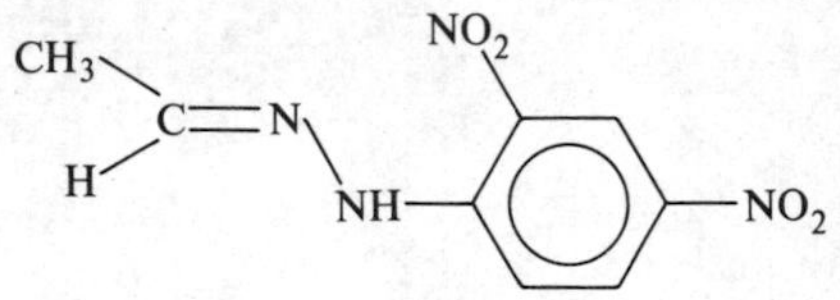 $CH_3(H)C{=}N{-}NH{-}C_6H_3(NO_2)_2$ (from ethanal) |

The products of condensation reactions between 2,4-dinitrophenylhydrazine and carbonyl compounds are all orange crystalline solids with well-defined melting points. They are therefore useful for identifying individual carbonyl compounds: the condensation product is prepared, its melting point measured accurately and the compound identified from tabulated known melting points.

## 30.5 Oxidation of carbonyl compounds

Table 30.4 shows the effect of some oxidizing agents on different carbonyl compounds.

**Table 30.4** Effect of warming different carbonyl compounds with oxidizing agents.

| | **Oxidizing agent** | |
|---|---|---|
| **Carbonyl compound** | **Complexed $Ag^+$ in alkaline solution (Tollen's reagent)** | **Complexed $Cu^{2+}$ in alkaline solution (Fehling's solution)** |
| **methanal** | silver mirror formed on walls of tube | copper metal formed together with copper(I) oxide |
| **ethanal** | silver mirror formed on walls of tube | red precipitate of copper(I) oxide formed |
| **propanone** | no reaction | no reaction |

Look at table 30.4 and answer these questions.

- ○ Write a half equation for the reduction of $Ag^+$ to Ag.
- ○ Which of the carbonyl compounds in the table are oxidized by $Ag^+$?
- ○ Suggest a compound to which ethanal might be oxidized.
- ○ To what oxidation state do methanal and ethanal respectively reduce $Cu^{2+}$?
- ○ Which is the more powerful reducer, methanal or ethanal?

Aldehydes carry a hydrogen atom next to their carbonyl group. This hydrogen is activated by the carbonyl group and readily oxidized to —OH. Aldehydes are therefore readily oxidized to carboxylic acids. For example, the oxidation of ethanal by $Ag^+$:

$$CH_3—\underset{\displaystyle O}{\underset{\|}{C}}—H + H_2O \longrightarrow CH_3\underset{\displaystyle O}{\underset{\|}{C}}—OH + 2H^+ + 2e^-$$

ethanoic acid

$$Ag^+ + e^- \longrightarrow Ag$$

Overall:

$$CH_3\underset{\displaystyle O}{\underset{\|}{C}}—H + H_2O + 2Ag^+ \longrightarrow CH_3\underset{\displaystyle O}{\underset{\|}{C}}—OH + 2H^+ + 2Ag$$

Methanal has two hydrogen atoms attached to the carbonyl group, both of which are oxidizable. Methanal is very readily oxidized and consequently a more powerful reducer than other aldehydes:

$$H—\underset{\displaystyle O}{\underset{\|}{C}}—H + H_2O \longrightarrow H—\underset{\displaystyle O}{\underset{\|}{C}}—OH + 2H^+ + \underset{\text{accepted by oxidizer}}{2e^-}$$

methanoic acid

Then:

$$H—\underset{\displaystyle O}{\underset{\|}{C}}—OH + H_2O \longrightarrow \left[HO—\underset{\displaystyle O}{\underset{\|}{C}}—OH\right] + 2H^+ + \underset{\text{accepted by oxidizer}}{2e^-}$$

carbonic acid

↓

$CO_2 + H_2O$

Thus methanal can reduce Cu(II) to Cu(O), while other aldehydes reduce it only to Cu(I).

Ketones have no hydrogen atom joined directly to the carbonyl group, so they are not readily oxidized. They therefore have no effect on mild oxidizing agents like those in table 30.4. Strong oxidizers like hot concentrated nitric(V) acid can oxidize ketones, but the effect is to rupture the molecule, forming at least two molecules of carboxylic acid.

The oxidizers in table 30.4 are commonly used to test for the aldehyde group.

*Fehling's solution* is made by mixing copper(II) sulphate(VI) solution with an alkaline solution containing 2,3-dihydroxybutanedioate ions (tartrate ions), which complex the $Cu^{2+}$ ions and prevent precipitation of copper(II) hydroxide from the alkaline solution.

*Tollen's reagent* is made by adding excess ammonia solution to a solution of silver(I) ions; it contains the $[Ag(NH_3)_2]^+$ ion in alkaline solution. Complexing of $Ag^+$ by $NH_3$ prevents precipitation of silver hydroxide.

Many other oxidizers can be used to convert aldehydes to carboxylic acids. In the laboratory, acidified dichromate(VI) is commonly used to prepare acids from aldehydes. Catalytic air oxidation of aldehydes was at one time used industrially to manufacture carboxylic acids, but nowadays they are mainly produced directly from alkanes (see section 31.1).

## 30.6 Effect of the carbonyl group on neighbouring atoms

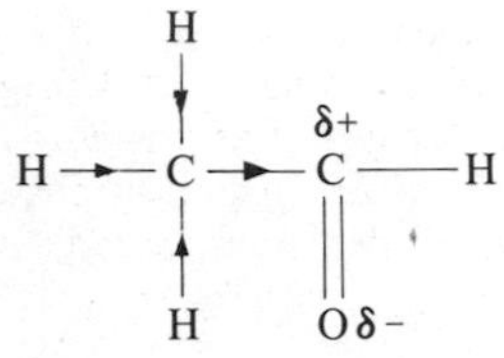

H, H → C → C(δ+) — H, C=O(δ−), H

**Figure 30.2** Polarizing effect of the carbonyl group in ethanal.

The carbonyl group is polar with a considerable partial positive charge on the carbon atom. This charge has the effect of withdrawing electrons from neighbouring carbon atoms, as shown in figure 30.2.

The result is to make the C–H bond of the neighbouring carbon more polar than normal, so that the hydrogen atoms are more readily replaced than those in alkanes. For example, ethanal reacts readily with chlorine even in the dark, forming trichloroethanal:

$$\underset{\text{ethanal}}{CH_3CHO} + 3Cl_2 \longrightarrow \underset{\text{trichloroethanal}}{CCl_3CHO} + 3HCl$$

Trichloroethanal is used in the manufacture of DDT (see figure 30.3).

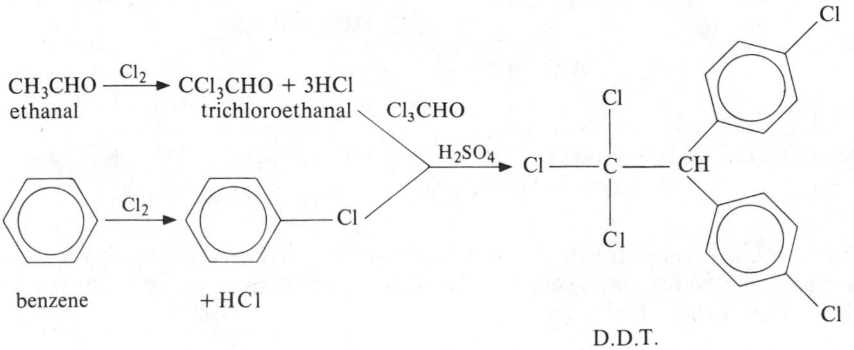

$$\underset{\text{ethanal}}{CH_3CHO} \xrightarrow{Cl_2} \underset{\text{trichloroethanal}}{CCl_3CHO} + 3HCl$$

benzene $\xrightarrow{Cl_2}$ chlorobenzene + HCl

$Cl_3CHO$ + chlorobenzene $\xrightarrow{H_2SO_4}$ D.D.T.

**Figure 30.3** Manufacture of D.D.T.

### THE IODOFORM REACTION

If chlorine and ethanal are warmed together in the presence of a base, the trichloroethanal initially formed reacts with the base to form trichloromethane, commonly known as chloroform:

$$\underset{\text{trichloroethanal}}{CCl_3CHO} + {}^-OH \longrightarrow \underset{\text{trichloromethane}}{CHCl_3} + \underset{\text{methanoate ion}}{HCOO^-}$$

A similar reaction occurs when ethanal is warmed with iodine and a base. The product in this case is triiodomethane (iodoform), $CHI_3$. This kind of reaction is not limited to ethanal; any compound of the general formula $CH_3COR$, where R is an alkyl group or hydrogen, will form $CHI_3$ when warmed with iodine and a base. Triiodomethane is a yellow solid with a characteristic smell. Being insoluble in water, it appears as a yellow crystalline precipitate. This reaction, known as the **iodoform reaction**, is a useful test for compounds of the form $CH_3COR$. The reaction is also given by compounds of the form $CH_3CHOHR$, since these are themselves oxidized by the iodine to $CH_3COR$.

Which of the following compounds would give a yellow precipitate of triiodomethane when heated with a solution of iodine in aqueous sodium carbonate?

- O $CH_3CH_2CHO$
- O $CH_3CH_2COCH_3$
- O $CH_3CHOHCH_3$
- O $HCHO$
- O $CH_3CH_2CH_2OH$

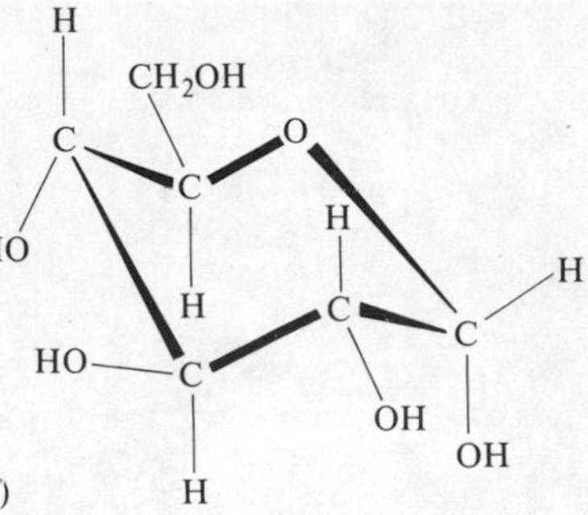

## 30.7 Sugars: naturally occurring carbonyl compounds

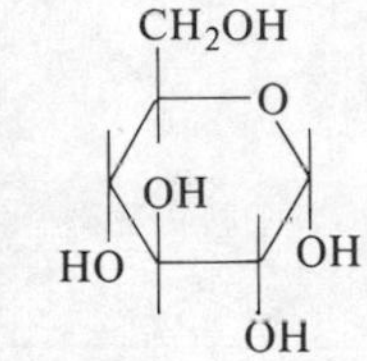

Sugars are sweet-tasting soluble carbohydrates. Carbohydrates derive their name from the fact that they are composed of carbon, hydrogen and oxygen with H and O in the ratio of 2:1 as in water. **Monosaccharides** like glucose (figure 30.4) are usually **pentoses** or **hexoses**, i.e. they contain 5 or 6 carbon atoms in their molecules. **Disaccharides** like sucrose consist of two monosaccharide molecules joined by the elimination of a molecule of water. **Polysaccharides** like starch are made up of many monosaccharide units joined together; their structure is considered further in section 8.8. Table 30.5 gives some examples of common carbohydrates. Notice that the monosaccharides all have asymmetric molecules: they therefore exhibit optical isomerism.

**Figure 30.4** Structure of glucose.
(i) Full structural formula, showing the puckered shape of the ring.
(ii) Skeletal formula. The hexagonal ring should be thought of as *perpendicular* to the paper, with the OH groups projecting above or below it. Hydrogen atoms attached to ring carbon atoms are not shown.

**Table 30.5** Some common carbohydrates.

| Name | Type | Structure | Occurrence |
|---|---|---|---|
| glucose | monosaccharide, aldose, hexose | 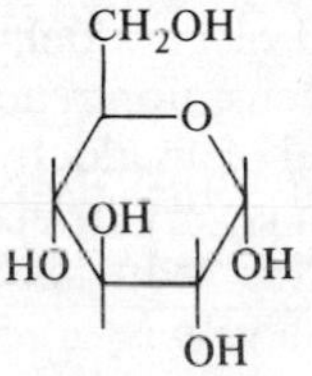 $CH_2OH$, O, OH, HO, OH, OH (ring structure) | occurs abundantly in plants and animals |
| fructose | monosaccharide, ketose, hexose | 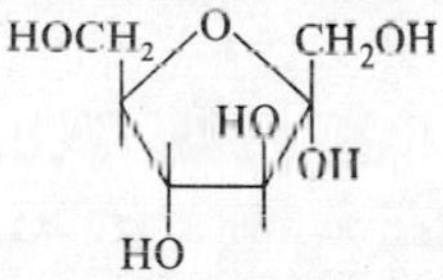 $HOCH_2$, O, $CH_2OH$, HO, OH, HO (ring structure) | in fruit and honey |
| ribose | monosaccharide, aldose, pentose | 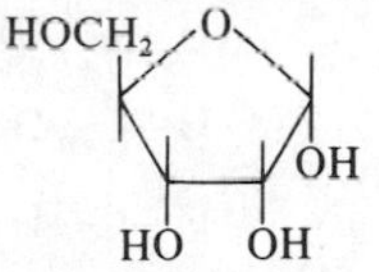 $HOCH_2$, O, OH, HO, OH (ring structure) | component of the molecules of ribonucleic acid (RNA) and vitamin B12 |
| sucrose | disaccharide | glucose – fructose | sugar-cane, sugar-beet (commonly simply called 'sugar') |
| maltose | disaccharide | glucose – glucose | malt |
| lactose | disaccharide | glucose – galactose | milk |
| starch | polysaccharide | chains of glucose units | plant storage organs, e.g. potato, wheat grain |
| cellulose | polysaccharide | chains of glucose units (linked differently to those in starch) | structural material of plants |

Carbohydrates—a monosaccharide, a disaccharide and a polysaccharide.

The most obvious feature of the structures of the mono- and disaccharides shown in table 30.5 is the presence of large numbers of OH groups. These give them a large capacity for hydrogen bonding, so they are involatile solids, soluble in water. The presence of OH groups on several adjacent carbon atoms in the molecule is thought to be responsible for the sweet taste of sugars.

As well as showing the properties of polyhydroxy compounds, sugars show many properties in solution that are typical of carbonyl compounds. For example, glucose gives a crystalline condensation compound with 2,4-dinitrophenylhydrazine (section 30.4). This is surprising since the structure of glucose shown in figure 30.4, contains no carbonyl group.

The carbonyl properties possessed by glucose arise from the fact that in addition to its normal ring form it can exist as an 'open-chain' form.

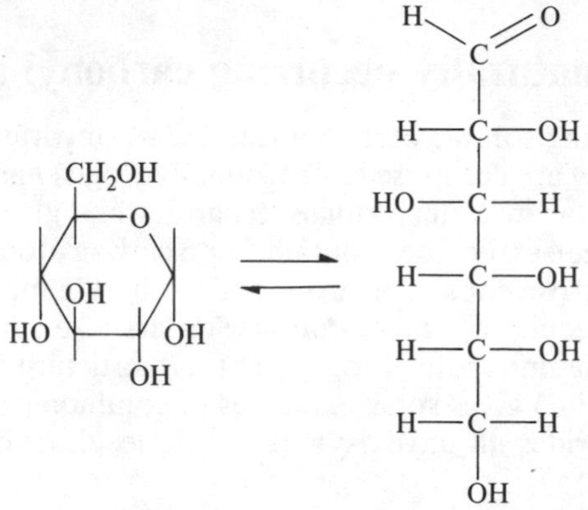

Ring form: six-membered ring with ring O, bearing $CH_2OH$, OH, HO, OH, OH $\rightleftharpoons$ Open-chain form:

H—C=O
H—C—OH
HO—C—H
H—C—OH
H—C—OH
H—C—H
OH

ring form    open-chain form

In aqueous solution about 1 % of glucose molecules exist in this open chain form, which can be clearly seen to carry an aldehyde group. Because of this group, glucose has several properties typical of an aldehyde and is sometimes called an **aldose**. Thus, in addition to the condensation reaction already mentioned, glucose shows the reducing properties typical of an aldehyde. The reduction of Fehling's solution (or Benedict's solution, which is very similar) is a standard test, frequently used by biologists, for glucose and other reducing sugars.

The open chain form of fructose is

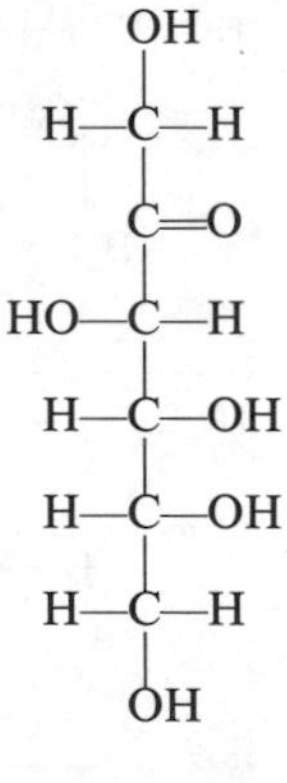

OH
H—C—H
C=O
HO—C—H
H—C—OH
H—C—OH
H—C—H
OH

Fructose is therefore a **ketose**.

The tendency of the open chain form of glucose and other sugars to revert to the ring form results from the tendency of the carbonyl group to undergo nucleophilic addition. The nucleophile involved is the oxygen atom of one of the OH groups of the molecule: an internal nucleophilic addition reaction occurs, forming a ring.

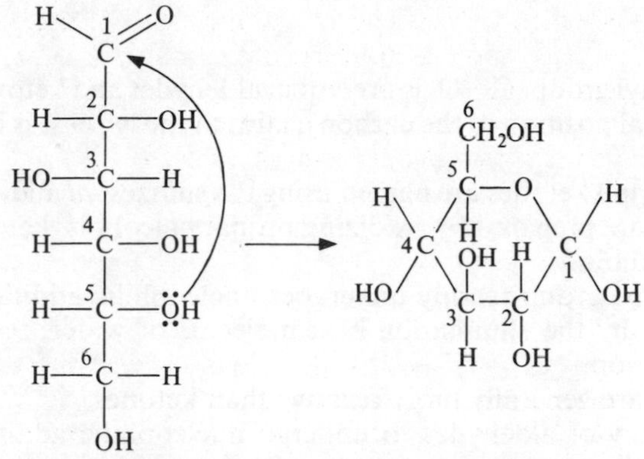

This reaction occurs spontaneously. Under normal conditions in aqueous solution the two forms exist in equilibrium, with the ring form predominating.

An understanding of the chemistry of carbohydrates is vital to an understanding of biology, since these molecules occur in all living organisms, as structural materials (e.g. cellulose), energy storage compounds (e.g. starch) and primary energy sources (e.g. glucose). The biological importance of carbohydrates is considered further in section 8.8.

Consider the following sugars. In each case the normal ring form is shown, with the open-chain form below it.

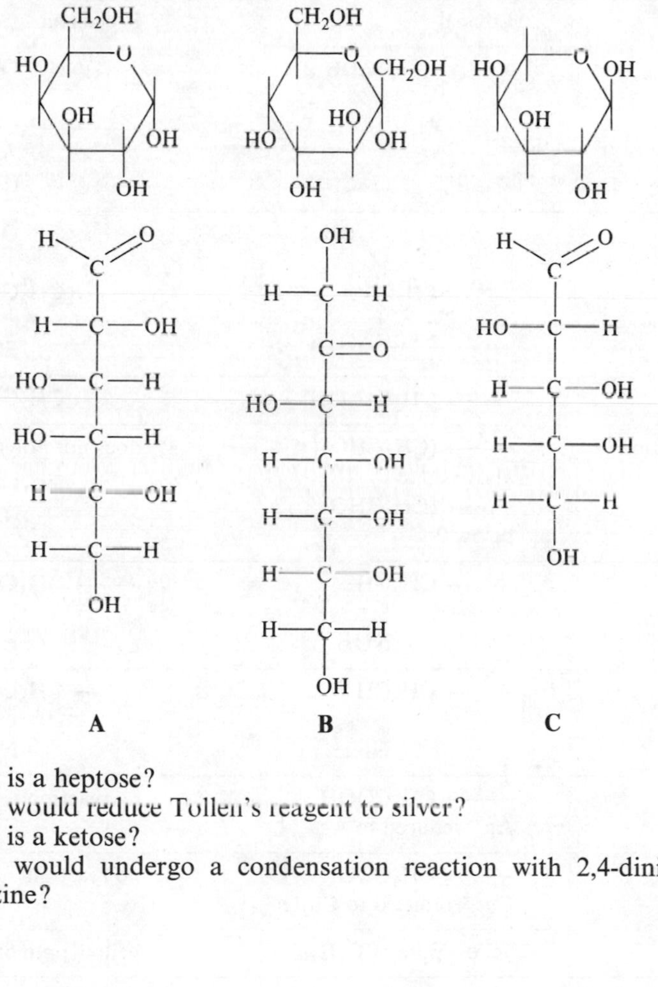

A B C

- ○ Which is a heptose?
- ○ Which would reduce Tollen's reagent to silver?
- ○ Which is a ketose?
- ○ Which would undergo a condensation reaction with 2,4-dinitrophenylhydrazine?

## Summary

1 The carbonyl group, C=O, is present in aldehydes and ketones. In the former it is in a terminal position in the carbon chain; in the latter it is in a non-terminal position.
2 Aldehydes and ketones are named using the suffixes *-al* and *-one* respectively.
3 Aldehydes are prepared by oxidizing primary alcohols, ketones by oxidizing secondary alcohols.
4 The carbonyl group readily undergoes nucleophilic addition; this is sometimes followed by the elimination of a molecule of water, resulting in a condensation reaction.
5 Aldehydes are generally more reactive than ketones.
6 The tendency of aldehydes to undergo nucleophilic addition makes them polymerize readily.
7 Aldehydes can be oxidized to carboxylic acids by a variety of reagents. Ketones are not readily oxidized.
8 The carbonyl group activates the hydrogen atoms on neighbouring carbon atoms, making them more readily substituted than those in alkanes.
9 Monosaccharides like glucose or fructose are polyhydroxy compounds usually containing five or six carbon atoms. They show many properties typical of carbonyl compounds.
10 Table 30.6 gives some of the important reactions of ethanal and propanone.

**Table 30.6** Important reactions of ethanal and propanone.

| Reagent | Reaction type | Ethanal | Propanone |
|---|---|---|---|
| HCN | addition | $\longrightarrow CH_3CH(OH)(CN)$ | $\longrightarrow CH_3C(OH)(CN)CH_3$ |
| $NaHSO_3$ | addition | $\longrightarrow CH_3CH(OH)(HSO_3^-Na^+)$ | $\longrightarrow CH_3C(OH)(SO_3^-Na^+)CH_3$ |
| $H_2/Ni$ | addition/reduction | $\longrightarrow CH_3CH_2OH$ | $\longrightarrow CH_3CHOHCH_3$ |
| dilute acid | addition/polymerization | $\longrightarrow (CH_3CHO)_3$ above 0°C<br>$\longrightarrow (CH_3CHO)_4$ below 0°C | does not polymerize |
| $NH_2OH$ | condensation | $\longrightarrow CH_3CH{=}NOH$ | $\longrightarrow CH_3C(=NOH)CH_3$ |
| $NH_2NH_2$ | condensation | $\longrightarrow CH_3CH{=}NNH_2$ | $\longrightarrow CH_3C(=NNH_2)CH_3$ |
| $Ag^+$ (complexed) (Tollen's reagent) | oxidation | $\longrightarrow CH_3COOH$<br>$Ag^+$ reduced to Ag | no reaction |
| $Cu^{2+}$ (complexed) (Fehling's solution) | oxidation | $\longrightarrow CH_3COOH$<br>$Cu^{2+}$ reduced to $Cu_2O$ | no reaction |
| $I_2$/base | iodoform reaction | yellow ppte of $CHI_3$ | yellow ppte of $CHI_3$ |

## Study questions

1 Consider the following compounds:

A $CH_3COCHCH_2CH_3$ (with $CH_3$ on the CH)

B $CH_3CH_2CH_2CH_2CHO$

C $CH_2OH$ … O … OH HO … HO … OH

D

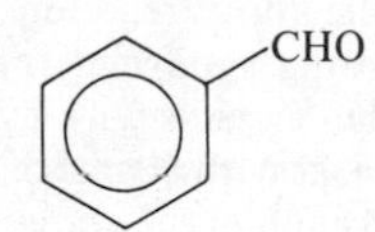


E

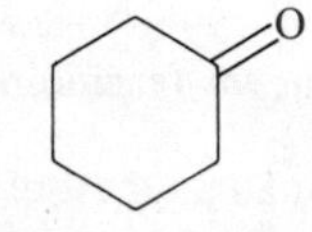


(a) Which are aldehydes?
(b) Which are ketones?
(c) Which is a hexose?
(d) Name compounds **A**, **B**, **D** and **E**.
(e) Which would produce a red precipitate of copper(I) oxide when boiled with Fehling's solution?
(f) Which would give a yellow crystalline precipitate when reacted with iodine in basic solution?
(g) Which would be reduced to a secondary alcohol by hydrogen in the presence of a nickel catalyst?

2 'The C=C and C=O groups might be expected to show chemical similarity, but in fact show very little.' Illustrate and discuss this statement, referring to a wide range of examples and to the underlying mechanistic principles.

3 This question is about the manufacture of ethanal. Read the passage below and then answer the questions on it.
Ethanal used to be made from ethyne, CH≡CH, which is readily manufactured from coke. This process is now being replaced by a method based on ethene. The overall process amounts to oxidation: ethene and oxygen are bubbled together through an aqueous solution containing $CuCl_2$ and $PdCl_2$ catalysts:

$$CH_2{=}CH_2 + \tfrac{1}{2}O_2 \xrightarrow[CuCl_2]{PdCl_2} CH_3CHO$$

The mechanism of the reaction involves initial attack on the ethene molecule by water. Since $H_2O$ is a nucleophile, it is necessary to render the ethene, which is normally subject only to electrophilic attack, attractive to nucleophiles. This is effected by the $Pd^{2+}$ catalyst, which forms a complex with the ethene, decreasing the electron density in the double bond and making the molecule prone to nucleophilic attack.

$$H_2\ddot{O}: \dashrightarrow \; CH_2{=}CH_2 \; \text{------} Pd^{2+}$$

The complex thus formed breaks down by a number of stages to form ethanal and palladium metal. The palladium metal is oxidized back to $Pd^{2+}$ by oxygen, the $CuCl_2$ being a catalyst for this oxidation.

(a) Why is ethanal an important industrial chemical?
(b) Why do you think ethene is replacing ethyne as the starting material for ethanal manufacture?
(c) The suggested mechanism involves the formation of a complex between ethene and $Pd^{2+}$. How do you think ethene is bonded to $Pd^{2+}$ in this complex?
(d) Suggest a reason why the formation of this complex renders ethene susceptible to nucleophilic rather than electrophilic attack.
(e) Could this method be used to manufacture *propanone*? If so, what starting material would be used instead of ethene?

4 Predict the formulae of the products of the following reactions.

(a) $CH_3COCH_2CH_3 + H_2 \xrightarrow{Ni}$
(b) $C_6H_5COCH_3 + NH_2OH \longrightarrow$
(c) $CH_3CH_2COCH_2CH_3 + HCN \longrightarrow$
(d) $C_6H_5CHO + KMnO_4 \longrightarrow$
(e) $CH_3CH_2CHO + Cl_2 \longrightarrow$

5 Write structural formulae for all compounds of molecular formula $C_4H_8O$ containing a carbonyl group. How would you distinguish between the different compounds, using simple chemical tests?

6 (a) A compound **A** has molecular formula $C_4H_6O_2$. **A** reacts with HCN to form compound **B**, $C_6H_8O_2N_2$. **A** is readily oxidized by acidified potassium dichromate(VI) to an acidic compound **C**, $C_4H_6O_4$. When 1.0g of **C** is dissolved in water and titrated with 1.0M sodium hydroxide, 16.9cm³ of sodium hydroxide are required for neutralization. Suggest structural formulae for **A**, **B** and **C** and explain the above reactions.

(b) A compound **X** contains 64.3% C, 7.1% H, and 28.6% O by mass. Its relative molecular mass is 56. **X** reduces Fehling's solution to copper(I) oxide. **X** reacts with hydrogen in the presence of a nickel catalyst: 0.1g of **X** was found to react with 80cm³ of hydrogen (measured at s.t.p.). Suggest a structural formula for **X** and explain the above reactions.

7 Suggest explanations for the following.

(a) Five different oxidation products of ethane-1,2-diol are known (excluding carbon dioxide and water).
(b) When ethanal is added to heavy water ($D_2O$) containing a small amount of base, tri-deuteroethanal ($CD_3CHO$) is formed.
(c) Trichloroethanal undergoes addition reactions far more readily than ethanal.

8 This question is about **Grignard reagents**, compounds that are of great use in organic synthesis for forming carbon-carbon bonds. Grignard reagents are compounds of general formula RMgX where X = Br or I. They are very reactive, giving rise to the highly nucleophilic ion $R^-$. When a solution of a Grignard reagent in dry ether is added to a carbonyl compound, the $R^-$ attacks the carbonyl group:

$$>C{=}O + R^- \longrightarrow >C(O^-)R$$

If water is now added to this product, an alcohol is formed:

$$>C(O^-)R + H_2O \longrightarrow >C(OH)R + {}^-OH$$

Overall:

$$>C{=}O + RMgBr + H_2O \longrightarrow >C(OH)R + MgBrOH$$

Predict the formulae of the compounds formed when the following are treated with the Grignard reagent methyl magnesium bromide, followed by water.

(a) Methanal
(b) Ethanal
(c) Propanone
(d) Carbon dioxide.

# Carboxylic Acids 31

## 31.1 Carboxylic acids

Why is a pickled onion like a stinging nettle?

They both owe their powerful effects to carboxylic acids: in the case of the pickled onion, to the ethanoic acid present in vinegar; in the case of stinging nettles to the methanoic acid injected by the fine hairs of the nettle when it stings you. Carboxylic acids and their derivatives occur widely in nature and are also present in many manufactured products, such as soaps and polyesters.

Carboxylic acids contain the carboxyl group, $-\underset{\underset{O}{\|}}{C}-OH$. Their general formula is usually written RCOOH, where R is an alkyl or aryl group or hydrogen. They are named using the suffix **-oic acid** after a stem indicating the number of carbon atoms (including the one in the carboxyl group). The first two members of the series, systematically named methanoic acid and ethanoic acid, are often called by their older names, formic acid and acetic acid.

Where two carboxyl groups are present, the suffix **-dioic acid** is used. Table 31.1 further illustrates the naming of carboxylic acids.

**Table 31.1** Some important carboxylic acids.

| Formula | Systematic name | Other name | Occurrence and uses |
|---|---|---|---|
| HCOOH | methanoic acid | formic acid (from Latin *formica*, an ant) | used in textile processing and as a grain preservative, ants use it as a poison, injecting it when they bite their victims, nettles do likewise |
| $CH_3COOH$ | ethanoic acid | acetic acid (from Latin *acetum*, vinegar) | in vinegar, used in making artificial textiles |
| $CH_3CH_2COOH$ | propanoic acid | propionic acid | calcium propanoate is used as an additive in bread manufacture |
| $C_6H_5COOH$ (COOH on benzene ring) | benzoic acid | | food preservative |
| COOH—COOH | ethanedioic acid | oxalic acid | in rhubarb leaves |

The C═O and O—H of the carboxyl group are in such close proximity that they modify one another's properties a great deal. Carboxylic acids therefore have many reactions that are different from those of both alcohols and carbonyl compounds.

The properties of the COOH group are modified only slightly when it is attached to a benzene ring. Aromatic carboxylic acids as a result have many properties in common with aliphatic ones.

### MAKING CARBOXYLIC ACIDS

In the laboratory, carboxylic acids are normally prepared by oxidizing primary alcohols (section 29.7) or aldehydes (section 30.5). They can also be prepared by hydrolysis of nitriles (section 32.2). The hydrolysis is carried out by heating with strong acid or base, e.g.

$$RCN + 2H_2O + HCl \longrightarrow RCOOH + NH_4Cl$$

Nitriles are themselves prepared from halogenoalkanes (section 28.5).

The oxidation of alcohols and aldehydes was at one time the major industrial method of preparing carboxylic acids, but nowadays the most important aliphatic acids are produced by direct oxidation of alkanes. A volatile petroleum fraction is oxidized in the liquid phase by passing air through it under pressure in the presence of a catalyst. Ethanoic acid is the major and most important product; by-products include methanoic and propanoic acids.

Benzoic acid is made industrially and in the laboratory by the oxidation of methylbenzene (chapter 27).

Consider the following acids:

A: $HOOC{-}CH_2{-}CH_2{-}COOH$

B: $CH_3CH_2CH_2COOH$

C: $CH_3{-}C_6H_4{-}COOH$ (methyl-substituted benzene ring with COOH)

D: $CH_3(CH_2)_{16}\,COOH$

- ○ Which is octadecanoic acid?
- ○ Which is a dibasic acid?
- ○ Which is butanoic acid?
- ○ Which could be prepared by oxidation of butane-1,4-diol?
- ○ Which is an aromatic acid?

### PHYSICAL PROPERTIES

Both these bottles contain glacial ethanoic acid. Pure ethanoic acid freezes at 17°C.

Ethanoic acid melts at 17°C and boils at 118°C. This means that although it is normally a liquid in British laboratories, it freezes in cold weather; in fact it is a fairly good rule that when the ethanoic acid freezes it is time to switch the heating on. Because of the readiness with which it freezes, and the similarity of solid ethanoic acid to ice, pure ethanoic acid is often described as 'glacial'. Dilute solutions of the acid, of course, freeze at about the same temperature as water.

The boiling point of ethanoic acid is higher than that of either ethanol (78°C), which has the same number of carbon atoms, or propan-1-ol (97°C), which has the same relative molecular mass. The relatively high boiling point of ethanoic acid and other carboxylic acids reflects a considerable degree of hydrogen bonding. Carboxylic acids form stronger hydrogen bonds than alcohols because their —OH group is more polarized due to the presence of the electron-withdrawing C=O group:

compare $R{-}C({=}O){-}O^{\delta-}{-}H^{\delta+}$ with $R{-}CH_2{-}O^{\delta-}{-}H^{\delta+}$

Furthermore, carboxylic acids have the possibility of forming doubly hydrogen-bonded dimers:

$$R{-}C{\overset{O\cdots H{-}O}{\underset{O{-}H\cdots O}{\rightleftharpoons}}}C{-}R$$

Carboxylic acids in the liquid and solid states exist mostly in this form (see chapter 8).

Their capacity for hydrogen-bonding makes the early members of the series miscible with water in all proportions, though as with all homologous series solubility decreases with increasing molecular size. Table 31.2 gives physical properties of some carboxylic acids.

**Table 31.2** Some properties of carboxylic acids.

| Acid | Formula | State at 25°C | Boiling point/°C | $K_a$ at 25°C/mole dm$^{-3}$ | Solubility in water |
|---|---|---|---|---|---|
| methanoic | $HCOOH$ | l | 101 | $1.6 \times 10^{-4}$ | infinite |
| ethanoic | $CH_3COOH$ | l | 118 | $1.7 \times 10^{-5}$ | infinite |
| propanoic | $CH_3CH_2COOH$ | l | 141 | $1.3 \times 10^{-5}$ | infinite |
| butanoic | $CH_3CH_2CH_2COOH$ | l | 164 | $1.5 \times 10^{-5}$ | infinite |
| octanoic | $CH_3(CH_2)_6COOH$ | l | 237 | $1.4 \times 10^{-5}$ | slightly soluble |
| chloroethanoic | $ClCH_2COOH$ | s | 189 | $1.3 \times 10^{-3}$ | very soluble |
| dichloroethanoic | $Cl_2CHCOOH$ | l | 194 | $5.0 \times 10^{-2}$ | infinite |
| trichloroethanoic | $Cl_3CCOOH$ | s | 196 | $2.3 \times 10^{-1}$ | very soluble |
| benzoic | $C_6H_5COOH$ | s | 249 | $6.4 \times 10^{-5}$ | slightly soluble |
| ethanedioic | $(COOH)_2$ | s | | $3.5 \times 10^{-2}$ (first dissociation) $4.0 \times 10^{-5}$ (second dissociation) | soluble |

Methanoic, ethanoic and propanoic acids have strong, sharp vinegary odours. The $C_4$ to $C_8$ acids (butanoic to octanoic) have very strong, unpleasant odours. Butanoic acid is responsible for the smell of rancid butter and is present in human sweat. Its smell can be detected at concentrations of $10^{-11}$ mole dm$^{-3}$ by man and at concentrations of $10^{-17}$ mole dm$^{-3}$ by dogs. The ability of a dog to track a man stems from its ability to detect carboxylic acids in the sweat from the man's feet. Each person's sweat glands produce a characteristic blend of carboxylic acids which can be detected and recognized by the sensitive nose of the dog.

A dog can be trained to track men by recognizing the characteristic blend of carboxylic acids in the sweat from a particular man's feet.

## 31.2 The carboxyl group

In chapter 29 we discussed the acidity of the OH group in alcohols, and found that alcohols were even less acidic than water. Carboxylic acids also contain this group, but they are much stronger acids than alcohols and water. Most of the acids we encounter in our food, such as ethanoic acid in vinegar and citric acid in lemons, are carboxylic acids. They appear to have a very strong acid taste; nevertheless, they are only present in low concentrations (vinegar is about 7% ethanoic acid) and they are weak acids compared to inorganic ones like hydrochloric acid.

When a carboxylic acid dissociates in water, it forms a carboxylate anion $RCOO^-$:

$$H_2O + R{-}C(=O){-}OH \longrightarrow R{-}C(=O){-}O^- + H_3O^+$$

X-ray diffraction studies of the anion show the two carbon-oxygen bonds to be of equal length. This implies that the two bonds are identical and that the negative charge and double bond character are distributed uniformly over the whole carboxylate group, delocalizing the negative charge:

$$R-C\overset{O}{\underset{O}{\lessgtr}}\ -$$

Thus each oxygen effectively carries half a negative charge. This delocalization of charge on the anion makes it less likely to become protonated again, so that the equilibrium

$$RCOOH + H_2O \rightleftharpoons RCOO^- + H_3O^+$$

is much further to the right than in the case of alcohols. Nevertheless, the majority of the acid is in the un-ionized form: 1.0M ethanoic acid is only 0.3% ionized.

The nature of the group to which COOH is attached in a carboxylic acid can have a considerable effect on the strength of the acid. In general, an electron-withdrawing group reduces the density of negative charge on the anion, and so increases the strength of the acid. An electron-donating group does the reverse.

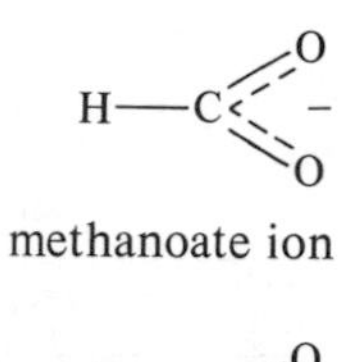


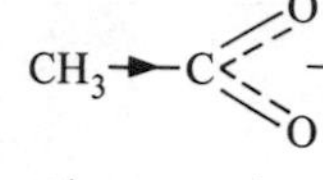


ethanoate ion

H
Cl—C—C
H
O
O

chloroethanoate ion

**Figure 31.1** Anions formed from methanoic, ethanoic, and chloroethanoic acids.

Consider the following acids:

$CH_3COOH$ $CH_2ClCOOH$ $CCl_3COOH$ $HCOOH$

Which would be the stronger acid:

- ○ $CH_3COOH$ or $CH_2ClCOOH$?
- ○ $CCl_3COOH$ or $CH_2ClCOOH$?
- ○ $CH_3COOH$ or $HCOOH$?

Table 31.2 gives the $K_a$ values for different carboxylic acids (remember, the higher the $K_a$, the stronger the acid). The electron-donating $CH_3$ group tends to increase the negative charge density on the carboxylate group in the ethanoate ion. Ethanoic acid is therefore weaker than methanoic, which carries no electron-donating alkyl group. On the other hand, the electron-withdrawing chlorine atom reduces the charge density on the chloroethanoate ion so chloroethanoic acid is stronger than ethanoic (figure 31.1).

The electron-donating effect of all alkyl groups is roughly equal, so unsubstituted aliphatic carboxylic acids with more than one carbon atom (ethanoic, propanoic etc.) are roughly equal in strength.

## 31.3 Salts of carboxylic acids

Like all acids, carboxylic acids react with bases to form salts. Table 31.3 gives some experimental data about benzoic acid.

**Table 31.3** Behaviour of benzoic acid in basic and acidic solution.

| Experiment | Result |
|---|---|
| 1 solid benzoic acid is shaken with water | most of the benzoic acid is undissolved |
| 2 sodium hydroxide is added to the aqueous suspension from experiment (1) | benzoic acid dissolves |
| 3 hydrochloric acid is added to the solution from experiment (2) | precipitate of benzoic acid appears |

Look at the table, then answer these questions.

- ○ Why is benzoic acid not very soluble in water?
- ○ Why does it dissolve in sodium hydroxide solution?
- ○ Why does hydrochloric acid precipitate benzoic acid from its solution in sodium hydroxide?

Carboxylic acids are weak acids and so only slightly dissociated in aqueous solution:

$$H_2O + RCOOH \rightleftharpoons RCOO^- + H_3O^+$$

Addition of a base removes $H^+$, shifting this equilibrium to the right, converting the acid to the carboxylate ion and forming the salt. Benzoic acid is much more soluble in basic solution than in water because in the former it is converted almost entirely to soluble benzoate ion. The addition of a strong acid to a solution of the carboxylate salt reverses the equilibrium, reforming free acid. Thus the anions of carboxylic acids are themselves basic: a molar solution of sodium ethanoate has a pH of about 9.5.

An interesting application of the formation of salts by carboxylic acids occurs with ethanedioic acid (oxalic acid). This acid is poisonous, because of the toxic nature of its anion, the ethanedioate ion. The normal way of treating people who have ingested an acid is to give them an alkali like sodium hydrogencarbonate to neutralize it. In the case of ethanedioic acid, though, this only makes the poisoning worse, because the addition of an alkali encourages the formation of the poisonous anion by removing $H^+$. The correct way to treat this kind of poisoning is to administer an alkali which can form an insoluble salt with ethanedioic acid, for example magnesium hydroxide (magnesium ethanedioate is insoluble).

Salts of carboxylic acids show marked differences in properties from the free acids, because they are ionic rather than molecular substances. Thus all such salts are solids, and they are soluble in water. Not surprisingly, as the hydrocarbon chain increases in length, the properties of these salts tend more and more to those of molecular hydrocarbons rather than ionic salts. The characteristics of both hydrocarbons and salts are possessed by salts of carboxylic acids with medium-length chains and this makes them valuable detergents.

## DETERGENTS

Detergents are substances that improve the cleaning properties of water. The most commonly encountered 'dirt' is that on human clothes and bodies: typically this consists of a mixture of natural fats, long-chain carboxylic acids, skin debris, silicon(IV) oxide and carbon. The main problem is caused by the greasy fats and carboxylic acids, which tend to bind the other materials to the skin or fabric surface and prevent them being washed away by water. A detergent works by enabling water to mix with and remove these greasy materials.

Soap was the first detergent, dating from Roman times, and it is still one of the commonest. Soap is a mixture of the salts of long-chain carboxylic acids: its manufacture is described in the next section. A typical salt present in soap is sodium octadecanoate (sodium stearate):

$$CH_3(CH_2)_{16}COO^-Na^+$$

or [zigzag chain] $COO^-Na^+$

- ○ Which end of this molecule would you expect to mix well with water?
- ○ Which end would you expect to mix well with grease?
- ○ Can you see how soap helps water to remove grease from a surface?
- ○ Why is hot water better than cold for washing in?

The charged $—COO^-$ group at the end of a soap chain enables it to dissolve in water. When soap dissolves, it does so in the form of spherical clusters called **micelles** (figure 31.2). These micelles are large enough to cause the solution to scatter light: this is why soapy water is cloudy. The non-polar hydrocarbon chain mixes well with greasy substances so that when soapy water comes into contact with grease, the non-polar end of the molecule mixes with the non-polar grease (figure 31.3).

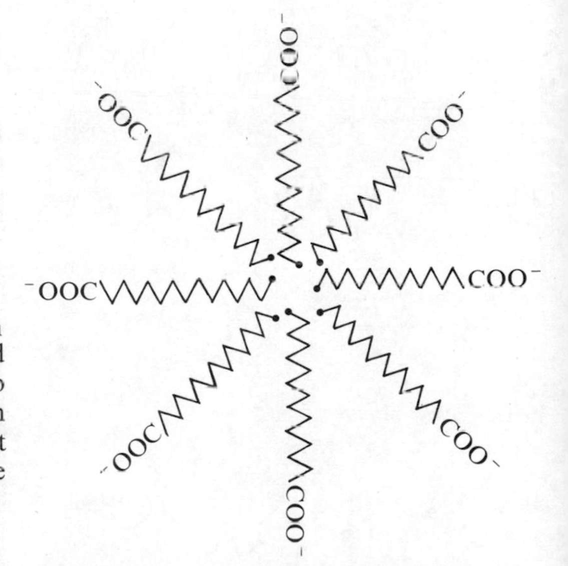


**Figure 31.2** Arrangement of soap chains in a micelle, shown in cross section. The charged ends are on the outside, since they are attracted by water. (Each micelle contains far more chains than shown here.)

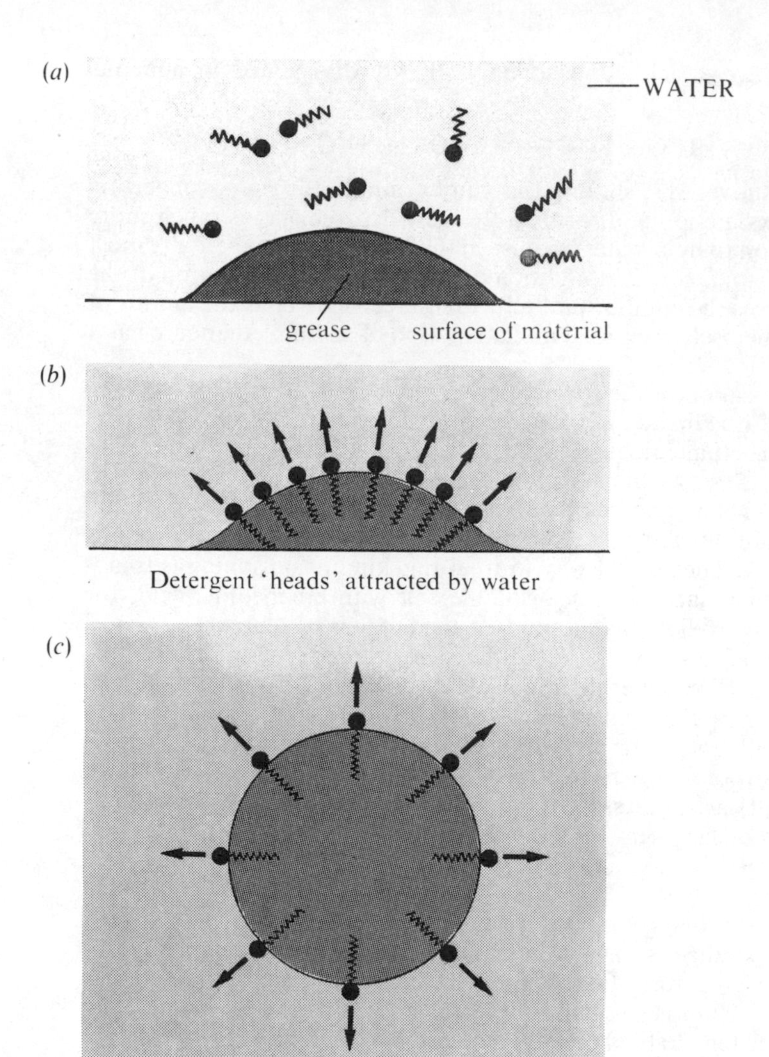


**Figure 31.3** Effect of detergent on grease.

This leaves the grease surrounded by an outer sheath of polar $COO^-$ groups which are attracted to water. The forces between water and grease are thus much increased, so that the latter is lifted off the surface in the form of small globules which can be washed away. The sheath of negatively charged groups causes the globules to repel one another, so they do not coagulate and redeposit on the surface.

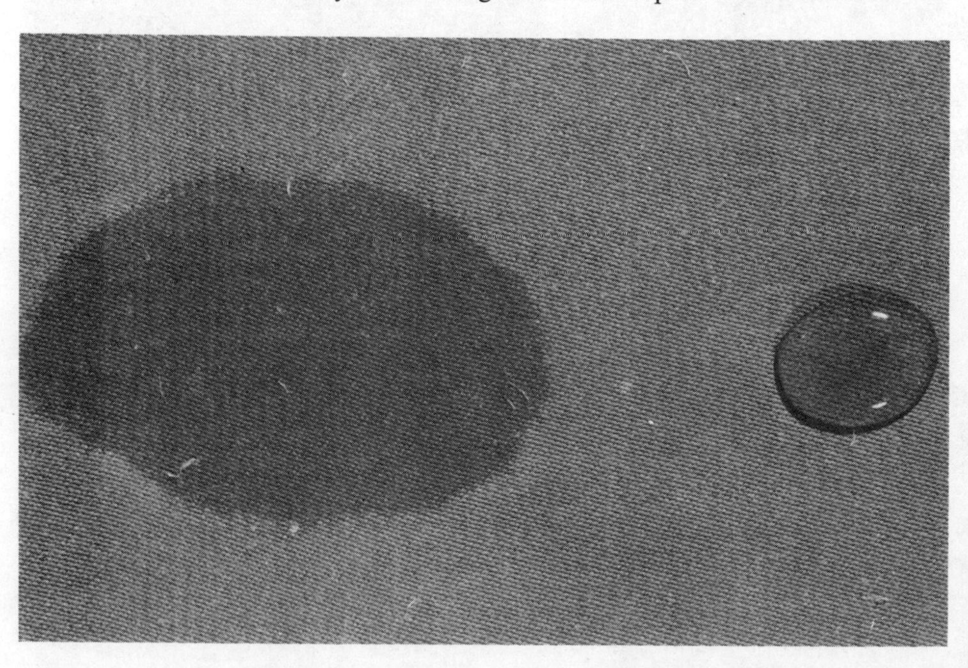

The effect of soap on the wetting ability of water. The droplet on the right is pure water; on the left is a soap and water solution.

SOAPLESS DETERGENTS

The property of detergency is not unique to soap: any molecule with a fairly large non-polar 'tail' and a charged or polar 'head' will have detergent properties. The head may carry a negative charge (**anionic** detergents are the most common), a positive charge (**cationic**), or may consist of a polar, uncharged group (**non-ionic**).

Soaps have the disadvantage that they form a 'scum' with hard water. Hard water contains calcium ions, which react with the soap to form the insoluble calcium salt of the carboxylic acid. This is the scum. For example, with sodium octadecanoate:

$$2CH_3(CH_2)_{16}COO^-(aq) + Ca^{2+}(aq) \longrightarrow (CH_3(CH_2)_{16}COO)_2Ca(s)$$

The scum is itself undesirable and its formation leads to the use of excessive quantities of soap.

Soapless detergents have the advantage that they form no scum with hard water. In addition, they are made from by-products of oil refining unlike soap. Soap is made from vegetable or animal fats which are expensive and could be better used for human consumption.

A typical and very common soapless detergent has the structure:

$CH_3(CH_2)_{11}$–C₆H₄–$SO_3^-Na^+$ i.e. [zig-zag chain]–C₆H₄–$SO_3^-Na^+$

Which are the water-attracting and water-repelling parts of this compound? This detergent is manufactured by the sequence shown in figure 31.4: look back to section 27.7 to remind yourself of the reactions involved. This kind of soapless detergent is present in well-known and well-advertised washing powders. Earlier synthetic detergents included branched-chain compounds with the structure:

[branched chain]–C₆H₄–$SO_3^-Na^+$

These are cheap to make, but because of their side chains cannot be degraded by bacteria, unlike the synthetic detergent shown earlier. They therefore tended to persist in water after sewage treatment and caused foaming in rivers and streams. They have now been superseded by detergents with unbranched side-chains.

$CH_3(CH_2)_9CH = CH_2$ + C₆H₆
dodecene

↓ $AlCl_3$, HCl (Friedel-Craft's reaction)

$CH_3(CH_2)_{11}$–C₆H₅
dodecylbenzene

↓ c. $H_2SO_4$

$CH_3(CH_2)_{11}$–C₆H₄–$SO_3H$

↓ NaOH

$CH_3(CH_2)_{11}$–C₆H₄–$SO_3^-Na^+$

**Figure 31.4** Manufacture of a soapless detergent.

Non-biodegradable detergents are responsible for the persistent foam to be seen in some rivers.

Synthetic detergents are gradually replacing soap. Many different compounds with detergent properties are now known and manufactured. They are used not only for cleaning purposes but also as wetting agents, foam-stabilizers and emulsifying agents, uses which all depend on their simultaneous compatability with water and water-repelling materials. Some examples are given in table 31.4.

**Table 31.4** Some synthetic detergents.

| Type | Non-polar portion | Polar portion | Uses |
|---|---|---|---|
| Non-ionic | $C_9H_{19}$—$C_6H_4$— | $O(CH_2CH_2O)_x\,CH_2CH_2OH$ ($x$ = 5 to 10) | liquid detergents<br>emulsifying agents<br>wetting agents |
| Cationic | $C_{15}H_{31}$— | $CH_2\overset{+}{N}(CH_3)_3Br^-$ | hair conditioner |
| Anionic | $C_{12}H_{23}$— | $COOCH_2CHOHCH_2OSO_3^-Na^+$ | toothpastes<br>shampoos |

## 31.4 Some important reactions of carboxylic acids

*Reaction with bases.* This has already been considered (section 31.3).

*Esterification.* This important reaction is considered in the next section.

*Reaction with phosphorus pentachloride.* You may recall from section 29.6 that phosphorus pentachloride reacts with the OH group in alcohols. What is the product of such a reaction? Given that carboxylic acids contain the OH group, what product would you expect from the reaction of say, ethanoic acid with phosphorus pentachloride?

Carboxylic acids react with phosphorus pentachloride to form acyl chlorides (section 28.7). For example

$$CH_3\underset{\substack{\|\\O}}{C}OH + PCl_5 \longrightarrow \underset{\text{ethanoyl chloride}}{CH_3\underset{\substack{\|\\O}}{C}Cl} + HCl + POCl_3$$

Other halogenating agents such as $SOCl_2$ and $PBr_3$ react similarly.

*Oxidation.* As a rule, carboxylic acids are not readily oxidized, being themselves the end products of the oxidation sequence

$$\text{primary alcohol} \longrightarrow \text{aldehyde} \longrightarrow \text{carboxylic acid.}$$

There are however two exceptions to this general rule.

(i) Methanoic acid and methanoates are readily oxidized, for example by potassium manganate(VII), to carbon dioxide:

$$\underset{\text{methanoic acid}}{HCOOH} \longrightarrow CO_2 + 2H^+ + \underset{\text{accepted by oxidizer}}{2e^-}$$

(ii) Ethanedioic acid and its salts are oxidized to carbon dioxide by warm potassium manganate(VII):

$$(COOH)_2 \longrightarrow 2CO_2 + 2H^+ + \underset{\text{accepted by oxidizer}}{2e^-}$$

*Reduction.* Carboxylic acids are rather hard to reduce, but powerful reducing agents like lithium tetrahydridoaluminate, $LiAlH_4$, will convert them to the corresponding primary alcohol:

$$R{-}\underset{\substack{\|\\O}}{C}{-}OH + 4H^+ + \underset{\text{from reducer}}{4e^-} \longrightarrow RCH_2OH + H_2O$$

Esters (section 31.5) are somewhat easier to reduce than acids: they can be converted to alcohols by the use of high-pressure hydrogen and a catalyst.

*Dehydration.* Methanoic and ethanedioic acids are exceptional as the only readily oxidized carboxylic acids. They are also the only acids that can be readily dehydrated.

- ○ What products would you expect to be formed when a molecule of water is eliminated from (a) a molecule of methanoic acid, HCOOH, (b) a molecule of ethanedioic acid, $(COOH)_2$?
- ○ What reagent would you employ as a dehydrating agent for these two compounds?

Methanoic acid is dehydrated to carbon monoxide when warmed with concentrated sulphuric(VI) acid:

$$HCOOH \xrightarrow{\text{conc.}H_2SO_4} CO + H_2O$$

Ethanedioic acid is dehydrated by the same method to carbon monoxide and carbon dioxide:

$$(COOH)_2 \xrightarrow{\text{conc.}H_2SO_4} CO + CO_2 + H_2O$$

Use your knowledge of the reactions of carboxylic acids to write the structural formulae of the products formed when

- ○ benzoic acid is treated with phosphorus pentachloride.
- ○ butanoic acid is treated with lithium tetrahydridoaluminate.
- ○ ethanedioic acid is treated with excess sodium hydroxide.

## 31.5 Esters

Figure 31.5 illustrates an experiment in which ethanol and ethanoic acid are heated in the presence of concentrated sulphuric(VI) acid. Study it, then answer the questions on the following page.

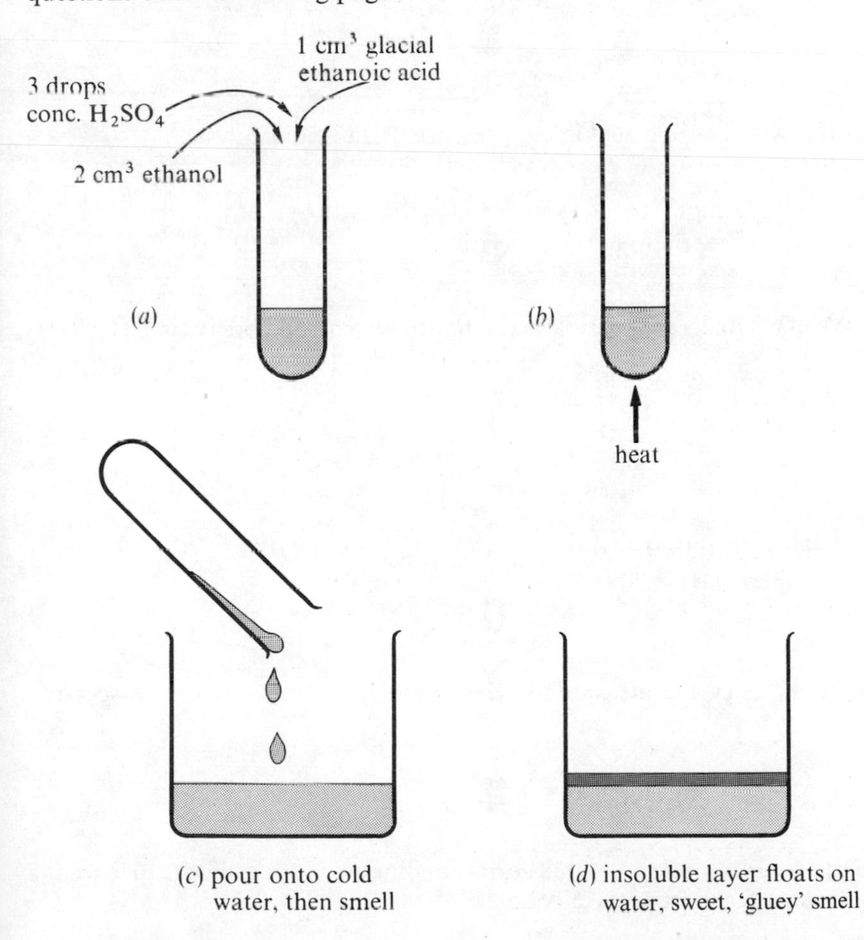


**Figure 31.5** Reaction between ethanol and ethanoic acid in the presence of concentrated sulphuric(VI) acid.

- ○ Does (a) ethanol, (b) ethanoic acid, (c) sulphuric(VI) acid, mix with water?
- ○ Do any of the substances above have a sweet, 'gluey' smell?
- ○ What can you say about the product of this reaction between ethanol and ethanoic acid?
- ○ Why do you think the reaction mixture is poured into cold water before smelling?

Ethanol and ethanoic acid react together to form an ester, ethyl ethanoate:

$$CH_3\underset{\underset{O}{\|}}{C}-OH + CH_3CH_2OH \longrightarrow CH_3\underset{\underset{O}{\|}}{C}-OCH_2CH_3 + H_2O$$

ethyl ethanoate

This reaction proceeds extremely slowly under ordinary conditions, but at an appreciable rate in the presence of a strong acid catalyst such as sulphuric or hydrochloric acid. This is a general method of preparing esters, and can be applied to any combination of acid and aliphatic alcohol:

$$R-\underset{\underset{O}{\|}}{C}-OH + R'OH \longrightarrow R-\underset{\underset{O}{\|}}{C}-OR' + H_2O$$

Notice that the bridging oxygen atom between the RCO and R′ groups in the ester could have come either from the alcohol or from the acid. As the equation stands, there is no way of telling from which molecule it originated. To put the problem another way, does the oxygen atom in the alcohol molecule end up in the ester or in the water?

The answer to this question was found by two American chemists, Roberts and Urey, using a technique known as **isotopic labelling**. The method is outlined in figure 31.6.

1 Prepare methanol 'labelled' with $^{18}O$.

$CH_3{}^{18}OH$

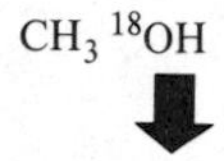

2 React this with benzoic acid in the presence of an acid catalyst.

$C_6H_5-COOH + CH_3{}^{18}OH$

3 Two sets of products are possible according to which bond breaks in $CH_3{}^{18}OH$.

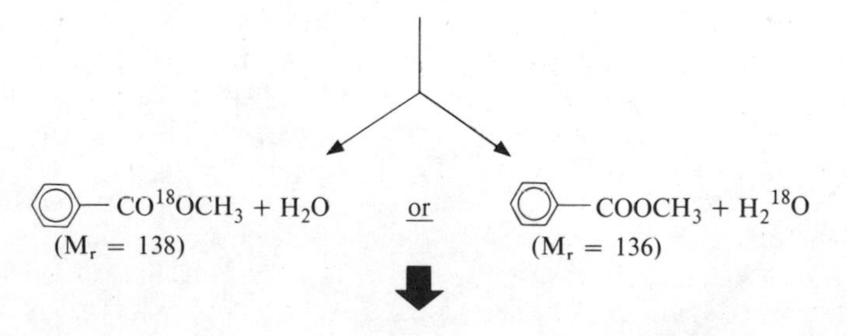


4 Separate the ester and measure its relative molecular mass in a mass spectrometer.

5 $M_r$ of the ester is found to be 138 corresponding to $C_6H_5-CO^{18}OCH_3$. Hence the bridging O must come from the alcohol.

**Figure 31.6** Investigation of the fate of oxygen atoms in an esterification reaction.

This experiment showed that the bridging O in fact comes from the alcohol; this information enabled chemists to propose a mechanism for the esterification reaction. Isotopic labelling is often used to investigate the mechanism of a chemical reaction: it has advanced our knowledge considerably, especially in the field of biochemistry.

### THE NATURE OF ESTERS

Esters are the products of condensation reactions between alcohols and carboxylic acids. They can also be prepared by reacting alcohols with acyl chlorides (section 28.7) or acid anhydrides (section 31.6). They are named by regarding them as alkyl (or aryl) derivatives of carboxylic acids. Thus the name is obtained from a stem indicating the alcohol from which the ester is derived, with the suffix **-yl**, followed by a stem indicating the acid, with the suffix **-oate**. So the ester derived from methanol and propanoic acid is called methyl propanoate. The name of an ester is most easily worked out from its formula by these stages:

(i) Divide the formula into two portions by mentally drawing a line after the bridging O of the COO group, e.g.

$$CH_3CH_2\underset{\underset{O}{\|}}{C}{-}O \,\vdots\, CH_2CH_2CH_2CH_3$$

(ii) Name the portion that does *not* carry the COO as an alkyl radical—*butyl* in this example.

(iii) Name the portion that carries the COO group as a carboxylate radical—*propanoate* in this example.

(iv) Combine (ii) and (iii) to give the name of the ester—*butyl propanoate*.

Try these questions to test your understanding of how esters are named.

- ○ Give the name of (a) $C_6H_5COOCH_3$,
  (b) $HCOOCH_2CH_3$.
- ○ Write the formula of (a) heptyl decanoate,
  (b) phenyl benzoate.
- ○ How would you prepare propyl propanoate?
- ○ Write the formula of the product formed when ethanoic acid and butan-2-ol are warmed together in the presence of an acid catalyst.

Unlike the acids and alcohols from which they are derived, esters have no free OH groups. They are therefore volatile compared with acids and alcohols of similar molecular mass and they are not very soluble in water. Ethyl ethanoate, for example, is a liquid, boiling point 77°C, and at 25°C its solubility is 8.5g per 100g of water. Volatile esters have characteristically pleasant, fruity smells. The flavour and fragrance of many fruits and flowers are due to mixtures of esters. Artificial fruit flavourings are made by mixing synthetic esters to give the approximate flavour (raspberry, pear, cherry, etc.) required. Organic acids are usually added to give the sharp taste characteristic of fruit. Artificial fruit flavours can only approximate to the real thing, because it would be too costly to include all the components of the complex mixture of compounds present in the real fruit.

The smell of some adhesives is very like that of ethyl ethanoate. Polystyrene cement, for example, consists of polystyrene dissolved in ethyl ethanoate; when the cement is applied the ethyl ethanoate evaporates, leaving behind the solid plastic which binds together the surfaces being joined.

Most yachts today have sails made of polyester. It is stronger than the natural fibres formerly used, and does not rot.

### POLYESTERS

Polyesters are synthetic fibres used as substitutes for cotton and wool, marketed under such trade names as 'Terylene' and 'Dacron'. As the name 'polyester' suggests, they are polymers joined by an ester linkage. By far the commonest polyester is 'Terylene' poly(ethane-1,2-diyl benzene-1,4-dicarboxylate), made by esterifying ethane-1,2-diol, $HOCH_2CH_2OH$, with benzene-1,4 dicarboxylic acid (terephthalic acid),

$$HOOC{-}C_6H_4{-}COOH$$

$$\ldots + HOCH_2CH_2OH + HOOC{-}C_6H_4{-}COOH + HOCH_2CH_2OH + HOOC{-}C_6H_4{-}COOH + {-}\ldots$$

$$\downarrow$$

$$\ldots{-}OCH_2CH_2OOC{-}C_6H_4{-}COOCH_2CH_2OOC{-}C_6H_4{-}CO{-}\ldots$$

$$+ H_2O \quad + H_2O \quad + H_2O \quad + H_2O$$

The ability of polyester to withstand higher temperatures than most plastics makes it a valuable material for the manufacture of food roasting bags.

It can be seen that having two OH groups on the alcohol and two COOH groups on the acid makes possible the formation of the polymer. This type of polymerization, in which the monomer units are joined by a condensation reaction, is called **condensation polymerization**.

Polyesters have great tensile strength. They are used as the bonding resin in glass fibre plastics as well as having wide application in the textile industry.

### ESTERIFICATION AS AN EQUILIBRIUM REACTION

The reaction of a carboxylic acid with an alcohol to form an ester proceeds fairly slowly, even in the presence of an acid catalyst. As well as being kinetically quite slow, esterification is also an equilibrium reaction that does not normally reach completion.

If known quantities of ethanol, ethanoic acid and hydrochloric acid catalyst are sealed together and left for two or three weeks an equilibrium is reached in which ethanol, ethanoic acid, ethyl ethanoate and water are all present, as well as unchanged acid catalyst.

$$CH_3CH_2OH + CH_3COOH \overset{H^+}{\rightleftharpoons} CH_3COOCH_2CH_3 + H_2O$$

If the reaction mixture is now titrated with standard sodium hydroxide, the amount of ethanoic acid present at equilibrium can be found, and from it the amounts of all the other components of the equilibrium mixture. From these results the equilibrium constant $K_c$ for the esterification reaction can easily be found: its value is about 4 at 25°C.

- O Write an expression for the equilibrium constant for the esterification of ethanol and ethanoic acid as in the last equation.
- O If 1 mole each of ethanoic acid and ethanol are allowed to reach equilibrium at 25°C, what will be the number of moles of ethyl ethanoate present at equilibrium? (Assume $K_c = 4$ at 25°C.)
- O What would you expect to happen if ethyl ethanoate were refluxed with an excess of water for a long time?

A mixture of 1 mole of ethanol and 1 mole of ethanoic acid brought to equilibrium at 25°C contains 2/3 mole each of ethyl ethanoate and water and 1/3 mole each of ethanol and ethanoic acid. The esterification reaction is thus far from complete and readily reversed. The reverse of esterification is **ester hydrolysis**, in which an ester reacts with water to form an alcohol and a carboxylic acid. Like esterification, ester hydrolysis is a slow reaction, but is speeded up by an acid catalyst. It is also catalysed by alkali, but in this case the carboxylic acid that is formed as a result of the hydrolysis reacts with excess alkali to form the carboxylate salt. This removes the carboxylic acid from the equilibrium mixture as it is formed, which means that ester hydrolysis can proceed to completion in the presence of alkali, though not in the presence of acid. For example:

$$CH_3COOCH_2CH_3 + {}^{-}OH \longrightarrow CH_3COO^- + CH_3CH_2OH$$

or

$$CH_3COOCH_2CH_3 + NaOH \longrightarrow CH_3COO^-Na^+ + CH_3CH_2OH$$

Esters are therefore hydrolysed more effectively in alkaline than in acidic solution, though in both cases the reaction is still quite slow and the mixture must be boiled under reflux. Strictly speaking the alkali is not acting catalytically in this reaction because, as the equation shows, the $^{-}OH$ is used up. The hydrolysis of an ester by alkali is sometimes called **saponification**, a word derived from Latin words meaning 'soap-making'. The reason for the use of this word will become clear shortly.

FATS

Fats and oils are naturally occurring esters used as energy-storing compounds by plants and animals.

They are derived from propane-1,2,3-triol, $CH_2OH—CHOH—CH_2OH$ (commonly known as glycerol or glycerine). This molecule has the capacity to combine with one, two or three molecules of carboxylic acid. In practice, most fats are triesters derived from propane-1,2,3-triol and a variety of long-chain carboxylic acids, sometimes called **fatty acids**. For example, a simple fat molecule is that derived from a molecule of propane-1,2,3-triol and three molecules of octadecanoic acid:

$$\begin{array}{l} CH_2OH \\ | \\ CHOH \\ | \\ CH_2OH \end{array} + \begin{array}{l} HOOC(CH_2)_{16}CH_3 \\ \\ HOOC(CH_2)_{16}CH_3 \\ \\ HOOC(CH_2)_{16}CH_3 \end{array} \longrightarrow \begin{array}{l} CH_2OOC(CH_2)_{16}CH_3 \\ | \\ CHOOC(CH_2)_{16}CH_3 \\ | \\ CH_2OOC(CH_2)_{16}CH_3 \end{array} + 3H_2O$$

1 mole of propane-1,2,3-triol; 3 moles of octadecanoic acid; 1 mole of propane-1,2,3-triyl trioctadecanoate

Fats are energy storage compounds. . . .

In naturally occurring fats each molecule is derived from two or three *different* fatty acids. Table 31.5 gives some common fatty acids; fifty or so are found in nature, the vast majority of them having an even number of carbon atoms in their molecule.

**Table 31.5** Some common fatty acids.

| Structure | Systematic name | Common name | Occurrence |
|---|---|---|---|
| $CH_3(CH_2)_{16}COOH$ | octadecanoic acid | stearic acid | mainly in animal fats |
| $CH_3(CH_2)_{10}COOH$ | dodecanoic acid | lauric acid | coconut oil, palm-kernel oil |
| $CH_3(CH_2)_{14}COOH$ | hexadecanoic acid | palmitic acid | most fats, especially palm oil |
| $CH_3(CH_2)_7CH{=}CH(CH_2)_7COOH$ | octadec-9-enoic acid | oleic acid | most fats, especially olive oil |

Fats containing a large proportion of unsaturated acids tend to have low melting points: many are liquid at room temperature and these are called **oils**. They can be converted to solid fats by hydrogenation (section 26.6).

Like all esters, fats can be hydrolysed. The products are propane-1,2,3-triol and a mixture of carboxylic acids. If the hydrolysis is carried out by boiling the fat with an alkali such as sodium hydroxide, the products are propane-1,2,3-triol and a mixture of the sodium salts of fatty acids. For example:

$$\begin{array}{l} CH_2OOC(CH_2)_{16}CH_3 \\ | \\ CHOOC(CH_2)_{16}CH_3 \\ | \\ CH_2OOC(CH_2)_{16}CH_3 \end{array} + 3NaOH \rightarrow 3CH_3(CH_2)_{16}COO^-Na^+ + \begin{array}{l} CH_2OH \\ | \\ CHOH \\ | \\ CH_2OH \end{array}$$

propane-1,2,3-triyl trioctadecanoate; sodium octadecanoate; propane-1,2,3-triol

Sodium octadecanoate is, of course, a soap.

This reaction is the basis of **soap-making**, one of the oldest chemical manufacturing processes. Soap is manufactured by boiling tallow (storage fat of cattle and sheep) or vegetable oil such as coconut oil with sodium or potassium hydroxide. Because fats contain a mixture of fatty acids, soaps are in practice a mixture of the salts of fatty acids. Potassium soaps are milder than sodium ones, so potassium hydroxide is the alkali used in the manufacture of toilet soaps. Propane-1,2,3-triol (glycerol) is a useful by-product of soap manufacture: it is used to make paints and to make the explosive 'nitroglycerine'.

## 31.6 Acid anhydrides

Acid anhydrides can be regarded as the products of a dehydration reaction between two molecules of carboxylic acid, though they are not readily formed by direct dehydration. For example, ethanoic anhydride:

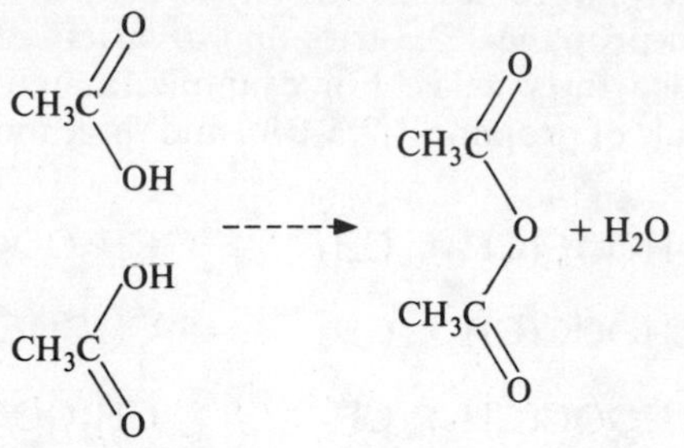

$CH_3C(=O)OH$ + $HO(O=)CCH_3$ - - - → $CH_3C(=O)-O-C(=O)CH_3 + H_2O$

ethanoic acid ethanoic anhydride

Ethanoic anhydride is normally made in the laboratory by the reaction of ethanoyl chloride with sodium ethanoate. You will recall from section 28.7 that ethanoyl chloride reacts readily with nucleophiles. The ethanoate ion is a nucleophile, so:

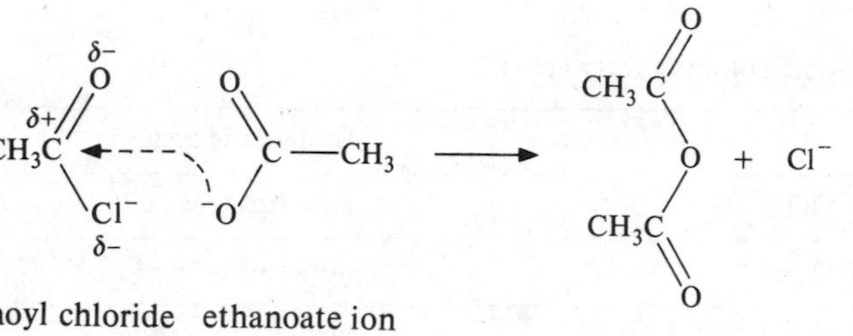

$CH_3\overset{\delta+}{C}(=\overset{\delta-}{O})\overset{\delta-}{Cl^-}$ ← - - - $^-O-C(=O)-CH_3$ ⟶ $CH_3C(=O)-O-C(=O)CH_3 + Cl^-$

ethanoyl chloride ethanoate ion

Industrially, ethanoic anhydride is made either from ethanal or from ethanoic acid (in the latter case, indirectly, not by direct dehydration).

Methanoic anhydride does not exist, since methanoic acid dehydrates to give carbon monoxide.

The oxygen atoms in ethanoic anhydride withdraw electrons from the carbons to which they are attached, giving them a large partial positive charge:

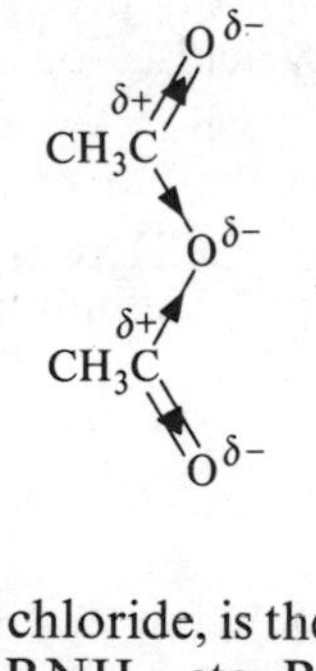

$CH_3\overset{\delta+}{C}(=O^{\delta-})-O^{\delta-}-\overset{\delta+}{C}(=O^{\delta-})CH_3$

Ethanoic anhydride, like ethanoyl chloride, is therefore readily attacked by nucleophiles such as $H_2O$, $NH_3$, ROH, $RNH_2$, etc. Representing a general nucleophile as $\overset{\delta+}{Y}-\overset{\delta-}{X}$, we can show the reaction as

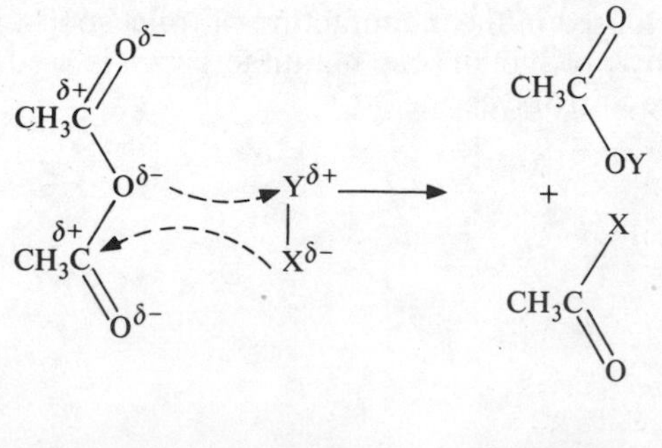

$CH_3\overset{\delta+}{C}(=O^{\delta-})-O^{\delta-}-\overset{\delta+}{C}(=O^{\delta-})CH_3$ + $Y^{\delta+}-X^{\delta-}$ ⟶ $CH_3C(=O)OY$ + $X-C(=O)CH_3$

In most cases, Y is hydrogen. Thus acid anhydrides, $(RCO)_2O$, act as **acylating agents**, joining the acyl group RCO— onto a nucleophile. In this respect they behave like acyl chlorides which are also acylating agents, though more vigorous ones. Compare the general acylating reactions of acid anhydrides and acyl chlorides:

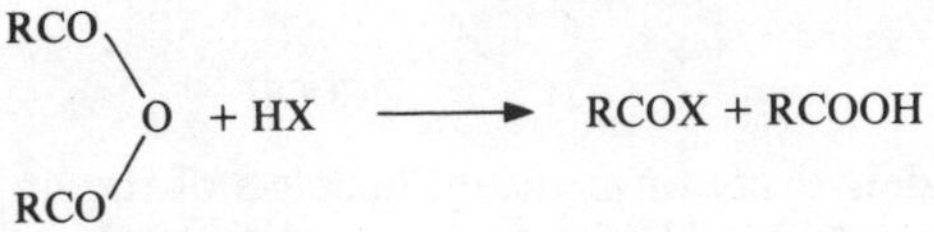

$$(RCO)_2O + HX \longrightarrow RCOX + RCOOH$$

acid anhydride

$$RCOCl + HX \longrightarrow RCOX + HCl$$

acyl chloride

Acid anhydrides react with all those nucleophiles which react with acyl chlorides. In the following reactions, ethanoic anhydride is used as an example, though the reactions apply quite generally.

With **water** to give the carboxylic acid:
e.g. $(CH_3CO)_2O + H_2O \longrightarrow 2CH_3COOH$

With **alcohols** to give esters:
e.g. $(CH_3CO)_2O + CH_3CH_2OH \longrightarrow CH_3COOCH_2CH_3 + CH_3COOH$
ethyl ethanoate

With **ammonia** to give amides:
e.g. $(CH_3CO)_2O + NH_3 \longrightarrow CH_3CONH_2 + CH_3COOH$
ethanamide

With **amines** to give substituted amides:
e.g. $(CH_3CO)_2O + CH_3CH_2NH_2 \longrightarrow CH_3CONHCH_2CH_3 + CH_3COOH$

Predict the formulae of the products of reactions between the following pairs of substances:

- ethanoic anhydride and phenol,
- propanoic anhydride and methylamine,
- butanoic anhydride and ammonia.

Ethanoic anhydride is used industrially in preference to ethanoyl chloride where an ethanoylating agent is required. This is because it is cheaper to make and less vigorous in its reactions. Aspirin, for example, is manufactured by ethanoylating 2-hydroxybenzoic acid:

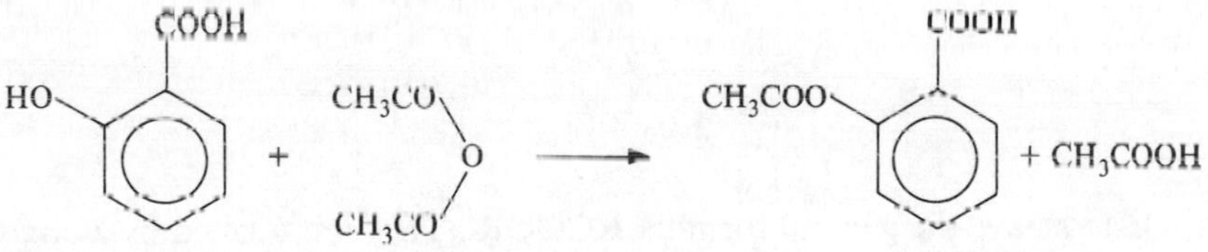

$$HOC_6H_4COOH + (CH_3CO)_2O \longrightarrow CH_3COOC_6H_4COOH + CH_3COOH$$

2-hydroxybenzoic acid → aspirin

The man-made fibre cellulose triethanoate (sold under the trade name 'Tricel') is made by ethanoylating cellulose with ethanoic anhydride. Three of the OH groups on each glucose molecule of the cellulose chain are ethanoylated, giving a fibre with good 'wash and wear' properties.

Certain dicarboxylic acids can form internal anhydrides by eliminating a molecule of water from one rather than two molecules of the acid. For example, *cis*-butenedioic acid (maleic acid) readily eliminates water when heated, forming its anhydride:

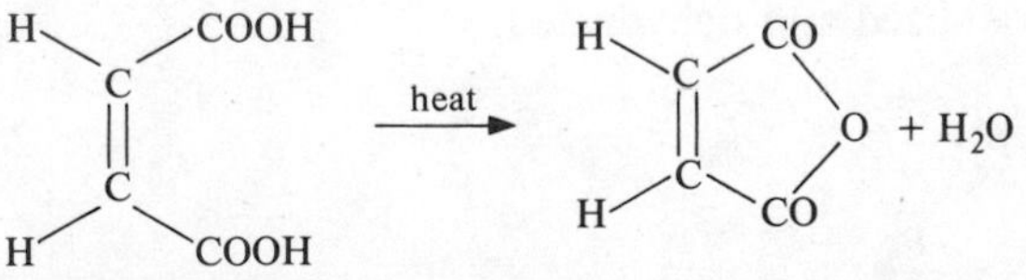

$$HC(COOH){=}C(COOH)H \xrightarrow{heat} HC(CO){=}C(CO)H\text{ (ring via O)} + H_2O$$

*cis*-butenedioic acid
(maleic acid)

However, the *trans* isomer of this compound, commonly called fumaric acid,

HOOC H
C
||
C
H COOH

does not readily dehydrate on heating, because of the impossibility of forming a cyclic internal anhydride. This is an unusual example of a pair of *cis* and *trans* isomers being chemically distinguishable.

## Summary

1 Carboxylic acids possess the carboxyl group, —COOH. They are named using the suffix *-oic acid.*
2 Carboxylic acids are prepared by oxidizing primary alcohols or aldehydes or by the hydrolysis of nitriles.
3 The —COOH group is strongly acidic compared to the —OH group, but nevertheless carboxylic acids are weak compared with mineral acids such as hydrochloric acid. The strength of a carboxylic acid is increased by the presence of electron-withdrawing substituents.
4 Some important reactions of ethanoic acid are shown in the following diagram:

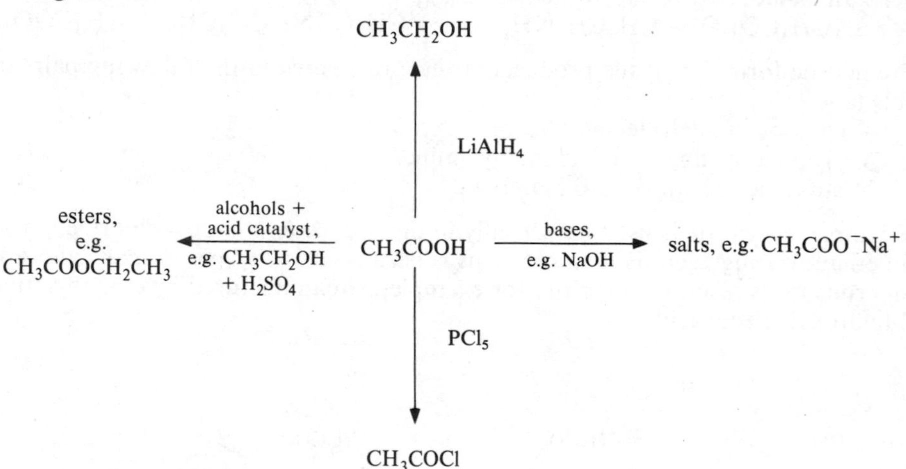


5 Esters have the general formula RCOOR′. They are formed by a condensation reaction between an alcohol and a carboxylic acid, usually in the presence of an acid catalyst.
6 Esterification is a reversible reaction. Esters can be hydrolyzed by boiling with dilute acid or alkali.
7 Acid anhydrides have the general formula $(RCO)_2O$. They are acylating agents, with the ability to join the RCO— group to nucleophiles.
8 Detergents are substances that improve the cleaning power of water. They have both a polar and a non-polar portion in their molecules. Soap is a detergent; soaps are the sodium or potassium salts of long-chain carboxylic acids.
9 Fats are esters derived from propane-1,2,3-triol and long-chain carboxylic acids. Alkaline hydrolysis of fats produces soaps.
10 Methanoic acid and ethanedioic acid differ from other carboxylic acids in being readily oxidized and dehydrated.

## Study questions

1 Consider the following compounds:

**A** $CH_3CHCOOH$

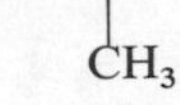


**B**

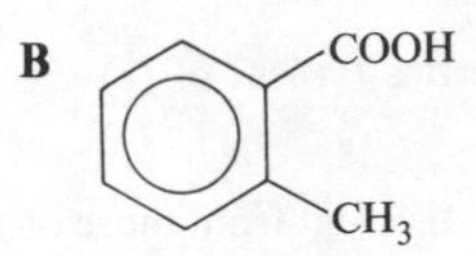


**C** $CH_3CH_2COOCH_3$

**D** $HOOCCH_2CH_2COOH$

**E** $CH_3CH_2CO$
O
$CH_3CH_2CO$

(a) Which is an ester?
(b) Which is a dibasic acid?
(c) Which is an acid anhydride?
(d) Name each compound.
(e) Which would be almost insoluble in water, but would slowly dissolve when boiled with sodium hydroxide solution?
(f) Which would be almost insoluble in water, but soluble in cold sodium hydroxide solution?
(g) Which would form a pleasant-smelling liquid when warmed with ethanol and concentrated sulphuric(VI) acid?
(h) Which would react with ammonia to form a mixture of propanamide and propanoic acid?

2 Predict the formulae of the products of the following reactions.
(a) $CH_3CH_2COOH + PBr_3 \longrightarrow$
(b) $C_6H_5COOH + LiAlH_4 \longrightarrow$
(c) $CH_3COO(CH_2)_4CH_3 + NaOH(aq) \xrightarrow{\text{boil}}$
(d) $CH_3COOH + Ca(OH)_2(aq) \longrightarrow$
(e) $CH_3CO$ O $CH_3CO$ + $C_6H_5NH_2$ $\longrightarrow$

3 (a) When a carboxylate salt (e.g. $RCOO^-Na^+$) is heated with 'soda lime' (a mixture of sodium and calcium hydroxides) a *decarboxylation* reaction occurs, in which $CO_2$ is effectively eliminated from the molecule, forming an alkane:

$$RCOO^-Na^+ + NaOH \longrightarrow \underset{\text{an alkane}}{RH} + Na_2CO_3$$

What do you think would be formed when *soap* is heated with soda lime?
(b) Suggest explanations for the following observations.
(i) Butane, propan-1-ol, propanal and ethanoic acid all have approximately the same relative molecular mass, but their boiling points are 273 K, 370 K, 321 K and 391 K respectively.
(ii) When dilute hydrochloric acid is added to an aqueous solution of soap, a white insoluble substance is formed.

4 Three isomeric acids **A**, **B** and **C** have molecular formula $C_8H_6O_4$ and all contain a benzene ring. In each case, one mole of the acid will react with 2 moles of sodium hydroxide. Suggest structures for the acids.
When the three acids are separately heated, **A** and **B** melt without decomposing, but **C** loses a molecule of water at about 250°C to form **D**, $C_8H_4O_3$. Suggest structures for **C** and **D**.

5 How would you carry out the following conversions in the laboratory?
(a) $CH_3CHO$ to $CH_3COOCH_2CH_3$
(b) $CH_2{=}CH_2$ to $CH_3COOH$
(c) $CH_3CH_2COOH$ to $(CH_3CH_2CO)_2O$
(d)

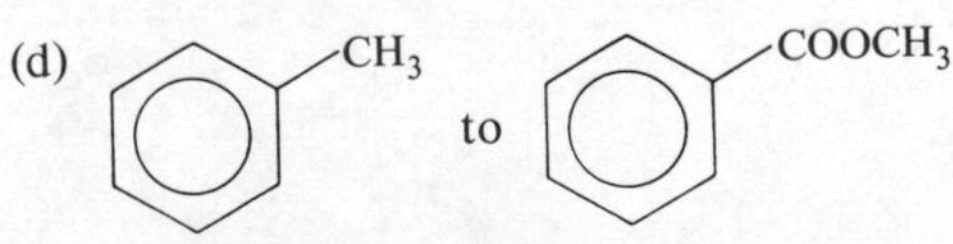


(e) $CH_3COOH$ to $CH_3CONH_2$

6 What simple chemical tests would you use to distinguish one compound from the other in the following pairs:
(a) $CH_3CH_2CHO$ and $CH_3CH_2COOH$,
(b) $HCOOH$ and $CH_3COOH$,
(c) $CH_3COCH_2CH_3$ and $CH_3COOCH_2CH_3$,
(d) $CH_3COOCH_3$ and $HCOOCH_2CH_3$,
(e) $CH_2FCOO^-Na^+$ and $CH_3COO^-Na^+$.

7 To what extent does the carboxyl group, $—\underset{\underset{O}{\|}}{C}—OH$, show properties typical of (a) the $—\underset{\underset{O}{\|}}{C}—$ group, (b) the —OH group? To what extent are its properties different from those of either of these groups?

8 Write structural formulae for all acids and esters of molecular formula $C_4H_8O_2$. What simple chemical tests would you use to distinguish between the different esters?

9 Consider the following compounds:

**A** $CH_2OOC(CH_2)_{14}CH_3$
$|$
$CHOOC(CH_2)_{16}CH_3$
$|$
$CH_2OOC(CH_2)_{14}CH_3$

**B** $HOCH_2CH_2OH$

**C** $CH_3(CH_2)_{15}COOH$

**D** $CH_3(CH_2)_{15}SO_3^-Na^+$

**E** $HOOC(CH_2)_4COOH$

(a) Which is a fatty acid?
(b) Which is an anionic detergent?
(c) Which would form soap on boiling with sodium hydroxide solution?
(d) Which pair of compounds could together be used to make a polyester?
(e) Which is a fat?
(f) Which would be oxidized to carbon dioxide and water on boiling with acidified potassium manganate(VII)?
(g) Which might form a cyclic anhydride?

# Organic Nitrogen Compounds 32

## 32.1 Dyestuffs and the development of the organic chemical industry

The structure of the organic chemical industry is very different from its inorganic counterpart. The latter is far more static, tending to stick for long periods to established processes producing a small number of essential chemicals in very large quantities—for example, sulphuric acid, sodium hydroxide, and steel. In organic chemistry, on the other hand, the number of possible compounds and the number of routes by which they can be made is far greater, so the industry is more dynamic, constantly discovering new compounds to do a particular job, and new ways of making them. Pharmaceuticals and polymers are just two areas of the organic chemical industry that show this rapid evolution. With new compounds and new processes to be investigated, the research chemist is more closely involved.

The evolution of the organic chemicals industry started with the discovery of synthetic dyestuffs. Until the 1850's dyes were all produced from living materials, usually plants. Consequently, they were expensive, limited in colour range and not always technically efficient. In 1856 in London, W. H. Perkin, an 18-year-old student of the German chemist Hofmann, was attempting to prepare the drug quinine from aniline. Although he failed, he managed, quite by accident, to produce a brilliant purple dye that was named mauve. This was the first synthetic dyestuff, the forerunner of many other dyes produced from aniline, or phenylamine as it is now called. The presence of the $NH_2$ group attached to a benzene ring makes phenylamine peculiarly suited to the preparation of coloured materials, particularly **azo dyes**, as we shall see later.

The synthetic dyestuffs industry was dominated by Britain, its country of origin, until the 1870's. However, Germany, many of whose chemists had gained experience of synthetic dyestuffs in Britain, slowly began to draw ahead and to develop an extraordinarily dynamic dyestuffs industry. This was partly due to the German education system, which placed great importance on the training of scientists in universities. Germany thus had the large reserve of trained scientists needed to carry out the research fundamental to the development of a new industry. Then, as now, the key to the development of science and technology was education.

By the First World War Britain was importing most of its dyes from Germany, and producing only 20 % of its own needs. Indeed, even after the outbreak of war, dyes had to be imported covertly from Germany to dye British soldiers' uniforms.

The expertise in applied research acquired by the German chemical industry in the development of dyestuffs stood it in good stead as it diversified into pharmaceuticals (aspirin was first used in Germany in 1898), synthetic polymers, photographic chemicals and many other fields. By the eve of the First World War the German chemical industry was superior to that of any other country.

The organic chemical industry has evolved a long way since its start with dyestuffs, and these materials now account for only a very small proportion of the total tonnage of organic chemicals produced. Dyestuffs are manufactured on a fairly small scale by relatively lengthy and complex processes: their production tends to resemble scaled-up laboratory methods rather than the giant processes used in the manufacture of petrochemicals. Nevertheless, dyestuffs are essential materials and their high price makes them commercially important. Section 32.7 gives further details of the production of azo dyes.

Sir William Perkin, discoverer of the dye mauve.

The Antwerp plant of the giant German chemical company BASF. The initials stand for Badische Anilin und Soda Fabrik, indicating the origin of the company in the dye industry.

## 32.2 Important organic nitrogen compounds

In this chapter we shall be mainly concerned with compounds containing the $NH_2$ group, but we will meet briefly other nitrogen compounds. The range of nitrogen compounds covered and the methods used to name them are dealt with in this section.

### AMINES

Amines can be considered as compounds formed by substituting hydrogen atoms of ammonia with alkyl or aryl groups. If one of the hydrogen atoms in $NH_3$ is substituted, we get a compound of the form $RNH_2$, called a **primary amine**. Such compounds are named using the suffix **-ylamine** after a stem indicating the number of carbon atoms in the molecule: thus $CH_3NH_2$ is methylamine. Substitution of two of the hydrogens of $NH_3$ gives compounds of the form $R_1R_2NH$, called **secondary amines**, named in a similar way to primary amines: thus $(CH_3)_2NH$ is dimethylamine. You should be able to work out the general structure and nomenclature of **tertiary amines**. Table 32.1 gives some examples of amines and other nitrogen compounds together with their systematic and trivial names.

The prefix **amino-** can be used to indicate the presence of an $NH_2$ group in molecules containing more than one functional group, e.g. aminoethanoic acid.

**Table 32.1** Some important organic nitrogen compounds.

| Formula | Type of compound | Name | Other name |
|---|---|---|---|
| $CH_3CH_2NH_2$ | primary amine | ethylamine | aminoethane |
| $C_6H_5NH_2$ (benzene ring with $NH_2$) | primary amine | phenylamine | aniline |
| $(CH_3)_3N$ | tertiary amine | trimethylamine | — |
| $CH_3CONH_2$ | amide | ethanamide | acetamide |
| $C_6H_5NO_2$ (benzene ring with $NO_2$) | nitro-compound | nitrobenzene | — |
| $CH_3CN$ | nitrile | ethanonitrile | acetonitrile |
| $CH_3CH(NH_2)COOH$ | amino acid | 2-aminopropanoic acid | alanine |

### AMIDES

Not to be confused with amines, **amides** have the general structure $RCNH_2$ (with =O on the C). The presence of the C=O group 'next door' profoundly affects the properties of the $—NH_2$ group, so amides behave very differently from amines. Amides are named using the suffix **-amide** after a stem indicating the number of carbon atoms in the molecule, including that in the C=O group. Thus $CH_3CONH_2$ is ethanamide.

### NITRO COMPOUNDS

Compounds containing the $—NO_2$ group are called **nitro-compounds**. Only aromatic ones need concern us here. They are named using the prefix **nitro-**. Thus $C_6H_5NO_2$ is called nitrobenzene.

### NITRILES (CYANO-COMPOUNDS)

Compounds containing the cyano group $—C{\equiv}N$ are called **nitriles** or **cyano-compounds**. They are named using the suffix **-onitrile** after the stem indicating the number of carbon atoms, including that in the CN group. Thus $CH_3CN$ is ethanonitrile.

### AMINO ACIDS

As their name implies, **amino acids** contain both the $—NH_2$ and $—COOH$ groups. By far the most important are those in which the $—NH_2$ and the $—COOH$ are both attached to the same carbon, thus:

$$R—\underset{\displaystyle NH_2}{\underset{|}{CH}}—COOH$$

Most of the amino acids found in nature are of this form. Amino acids are named as amino derivatives of carboxylic acids, rather than vice-versa; thus

$$CH_3\underset{\displaystyle NH_2}{\underset{|}{C}}HCOOH$$

is called 2-aminopropanoic acid. Since many natural amino acids have quite complex structures, they are often referred to by trivial names.

Name the following compounds:
- ○ $CH_3CH_2CH_2CONH_2$
- ○ $CH_3CH_2CN$
- ○ $(CH_3CH_2)_2NH$

Give the formulae of the following compounds:
- ○ 2-nitrophenylamine
- ○ aminoethanoic acid
- ○ propylamine
- ○ propanamide

## 32.3 The nature and occurrence of amines

The $—NH_2$ group is very widespread in biological molecules, especially proteins. Normally the group is associated with other functional groups: free amines are relatively rare in nature. They do occur in decomposing protein material, however, formed by the action of bacteria on amino acids. For example, $NH_2(CH_2)_4NH_2$ and $NH_2(CH_2)_5NH_2$ are found in decaying animal flesh, as their common names, putrescine and cadaverine, evocatively suggest. Di- and trimethylamine are found in rotting fish and are partly responsible for its peculiar smell.

The physical and chemical properties of the early members of the amine series resemble those of ammonia. Their smell is very similar to that of ammonia, though with a slightly fishy character. Like ammonia, the early amines are gaseous and very soluble in water, their high solubility being attributable to hydrogen bonding between the $—NH_2$ group and water molecules.

The peculiar smell of fish is mainly due [to] amines.

Table 32.2 gives some physical properties of amines. Can you suggest a reason why trimethylamine has a lower boiling point than dimethylamine?

**Table 32.2** Physical properties of ammonia and some amines.

| Name | Formula | State at 25°C | Boiling point/°C |
|---|---|---|---|
| ammonia | $NH_3$ | g | −33 |
| methylamine | $CH_3NH_2$ | g | −6 |
| ethylamine | $CH_3CH_2NH_2$ | g | 17 |
| propylamine | $CH_3CH_2CH_2NH_2$ | l | 49 |
| butylamine | $CH_3CH_2CH_2CH_2NH_2$ | l | 78 |
| phenylamine | $C_6H_5NH_2$ | l | 184 |
| dimethylamine | $(CH_3)_2NH$ | g | 7 |
| trimethylamine | $(CH_3)_3N$ | g | 3 |
| triethylamine | $(CH_3CH_2)_3N$ | l | 90 |

Phenylamine, with its large hydrocarbon portion is only sparingly soluble in water and has appreciable solubility in organic solvents. Its ability to dissolve in fats means that it is readily absorbed through the skin. Since it is toxic, great care must be exercised when using phenylamine.

## 32.4 Making amines

There are many ways of making amines; four of the most important methods are listed here.

FROM AMMONIA AND HALOGENOALKANES

We have already seen (section 28.5) that halogenoalkanes undergo substitution reactions with ammonia to form a mixture of primary, secondary and tertiary amines:

$$RCl + NH_3 \longrightarrow RNH_2 + HCl$$

$$RCl + RNH_2 \longrightarrow R_2NH + HCl$$

$$RCl + R_2NH \longrightarrow R_3N + HCl$$

In practice the salt of the amine is formed by combination of the HCl with the amine (section 32.5). The shortcoming of this method is the need to separate any required product from the mixture; this is usually done by distillation. Nevertheless, the method is widely used both in the laboratory and industrially, because of the ready availability of the starting materials.

BY REDUCTION OF NITRO-COMPOUNDS

This method is normally used only for aromatic amines.

Aromatic nitro-compounds are readily prepared by the nitration of aromatic hydrocarbons (see section 27.6). Reduction of nitro-compounds in acid solution produces amines:

$$C_6H_5NO_2 + 6H^+ + 6e^- \longrightarrow C_6H_5NH_2 + 2H_2O$$

nitrobenzene → phenylamine

In the laboratory the reducing agent used is normally tin in concentrated hydrochloric acid:

$$Sn \longrightarrow Sn^{4+} + 4e^-$$

Overall:

$$2\ C_6H_5NO_2 + 12H^+ + 3Sn \longrightarrow 2\ C_6H_5NH_2 + 4H_2O + 3Sn^{4+}$$

This method is used industrially to manufacture phenylamine, the reducing agent in this case being iron in concentrated hydrochloric acid.

BY REDUCTION OF NITRILES

Primary amines can be produced by the reduction of nitriles with powerful reducing agents such as lithium tetrahydridoaluminate ($LiAlH_4$):

$$R{-}C{\equiv}N + 4H^+ + \underset{\text{from reducer}}{4e^-} \longrightarrow RCH_2NH_2$$

This method is useful in the laboratory but because of the relatively high cost of the reagents it is not used industrially.

BY REDUCTION OF AMIDES

Amides, like nitriles, are reduced by $LiAlH_4$ to primary amines:

$$RCONH_2 + 4H^+ + 4e^- \longrightarrow RCH_2NH_2 + H_2O$$

How would you prepare:
- ○ propylamine from chloroethane (2 stages),
- ○ ethylamine from chloroethane,
- ○ 2-methylphenylamine from methylbenzene (2 stages)?

## 32.5 Amines as bases

Figure 32.1 illustrates an experiment in which acid is added to a solution of ethylamine.

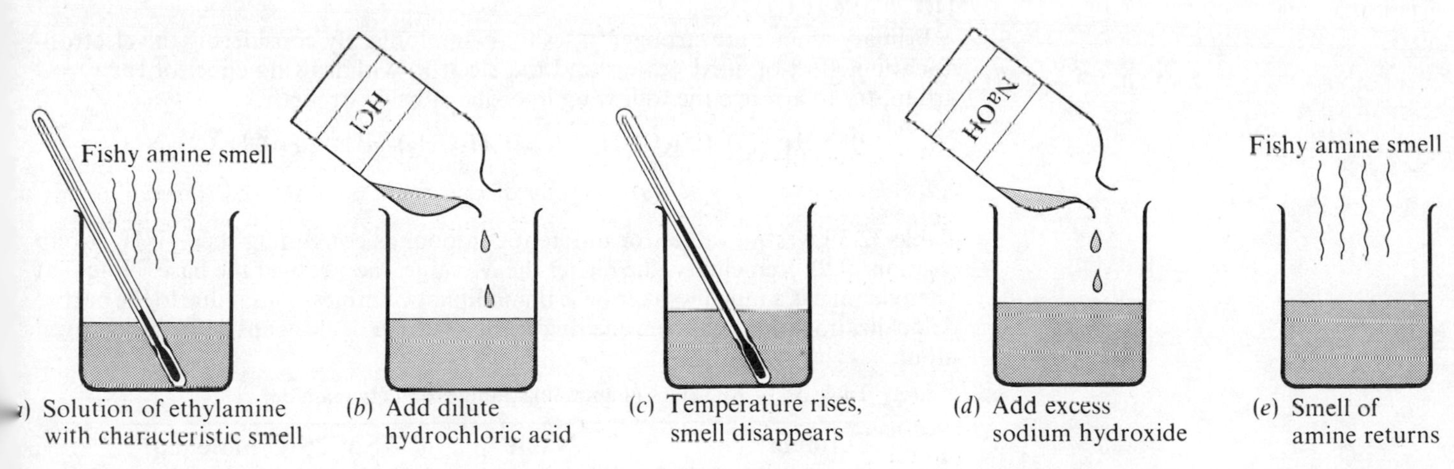


**Figure 32.1** Effect of acid on aqueous ethylamine.

Answer these questions.
- ○ What evidence is there for a chemical reaction between ethylamine and hydrochloric acid?
- ○ Why does the smell of ethylamine disappear when hydrochloric acid is added?
- ○ Why does the smell reappear when sodium hydroxide is added?

Like ammonia and all primary amines, ethylamine carries a lone pair of electrons on its nitrogen atom. This enables it to form a dative bond to a hydrogen ion:

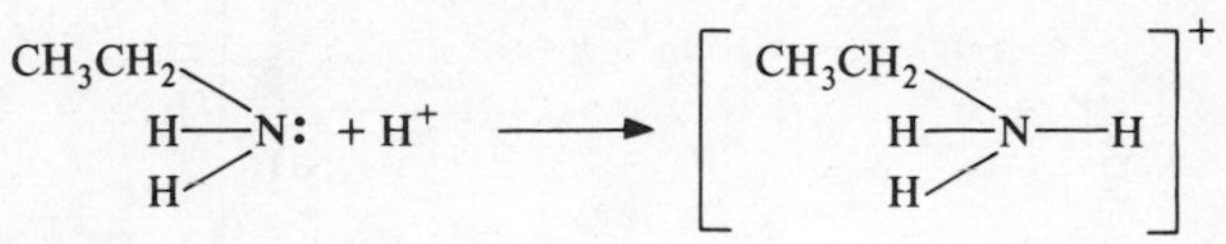

$$CH_3CH_2-\ddot{N}H_2 + H^+ \longrightarrow [CH_3CH_2-NH_3]^+$$

Ethylamine is therefore a base like ammonia.

When an acid is added to a solution of ethylamine, a salt is formed—in this case **ethylammonium chloride**, $CH_3CH_2NH_3{}^+Cl^-$. Like all salts it is involatile and therefore has no smell. When a strong base such as sodium hydroxide is added to this salt, protons are removed from it, reforming the free amine:

$$CH_3CH_2NH_3{}^+ + {}^-OH \longrightarrow CH_3CH_2NH_2 + H_2O$$

Compare this series of reactions with the familiar corresponding reactions of ammonia:

$$NH_3 + H^+ \longrightarrow NH_4{}^+$$
$$NH_4{}^+ + {}^-OH \longrightarrow NH_3 + H_2O$$

Amine salts are soluble white crystalline solids, similar to ammonium compounds.

Figure 32.2 shows another experiment that can be carried out with ethylamine and HCl. What would you expect to happen? Why?

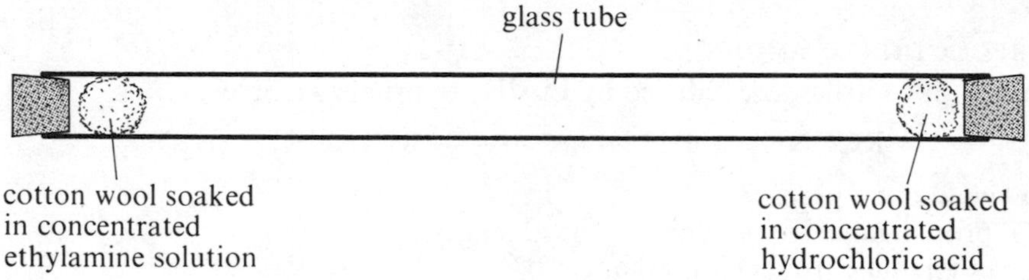

**Figure 32.2** An experiment with ethylamine and HCl. The apparatus is set up as shown. What happens?

IS ETHYLAMINE A STRONGER OR WEAKER BASE THAN AMMONIA?

To answer this question, refer to the following equilibria:

$$CH_3CH_2NH_2 + H^+ \rightleftharpoons CH_3CH_2NH_3{}^+$$
$$NH_3 + H^+ \rightleftharpoons NH_4{}^+$$

The stronger the base, the further the equilibrium is to the right. The position of equilibrium depends on the stability of the cation on the right hand side. Recalling the electron-donating effect of alkyl groups, which will be the more stable cation, $NH_4{}^+$ or $CH_3CH_2NH_3{}^+$?

Primary amines are stronger bases than ammonia. By considering the electron-donating effect of alkyl groups and the electron-withdrawing effect of the C=O group, try to arrange the following in order of basic strength:

(*a*) $NH_3$ (*b*) $CH_3C(=O)NH_2$ (*c*) $(CH_3CH_2)_2NH$ (*d*) $CH_3CH_2NH_2$

Table 32.3 gives $K_b$ values for different compounds containing the $-NH_2$ group (section 22.9). Remember, the higher the $K_b$ value, the stronger the base. Note that phenylamine is a much weaker base than aliphatic amines. This is due to the partial delocalization round the benzene ring of the lone pair of electrons from the nitrogen atom.

**Table 32.3** $K_b$ values of ammonia, amines and ethanamide.

| Formula | Name | $K_b$ at 25°C/mole dm$^{-3}$ |
|---|---|---|
| $NH_3$ | ammonia | $1.8 \times 10^{-5}$ |
| $CH_3CH_2NH_2$ | ethylamine | $5.4 \times 10^{-4}$ |
| $(CH_3CH_2)_2NH$ | diethylamine | $1.3 \times 10^{-3}$ |
| $CH_3CONH_2$ | ethanamide | $10^{-15}$ |
| $C_6H_5NH_2$ | phenylamine | $5 \times 10^{-10}$ |

## 32.6 Other reactions of amines

WITH ACYL CHLORIDES

The lone pair of electrons on the nitrogen atom gives amines nucleophilic character. Thus they tend to react readily with electrophiles like ethanoyl chloride. For example:

$$C_6H_5NH_2 + CH_3COCl \longrightarrow C_6H_5NHCOCH_3 + HCl$$

N-phenylethanamide (m.p. 114°C)

The products of this type of reaction are (unlike the amines from which they are made) crystalline solids with sharp melting points. They are therefore useful for identifying (*characterizing*) unknown amines: the acyl derivative is prepared, its melting point is taken and this is then checked against the accepted melting points of known amides in published tables.

WITH NITRIC(III) ACID

Nitric(III) acid ($HNO_2$) is rather unstable, so when it is used as a chemical reagent it is usually generated *in situ* from sodium nitrate(III) and hydrochloric acid.

When ethylamine is reacted with nitric(III) acid, nitrogen is rapidly evolved and a mixture of ethanol, ethene and other products is formed. However, when phenylamine is reacted with nitric(III) acid at low temperatures (below 10°C), a clear solution is formed, but no nitrogen appears. If, however, this solution is warmed, nitrogen is evolved and phenol is formed.

These observations are explained as follows. When amines react with nitric(III) acid, **diazonium compounds** are formed:

$$RNH_2 + HNO_2 \longrightarrow \underset{\text{diazonium ion}}{R{-}\overset{+}{N}{\equiv}N} + {}^{-}OH + H_2O$$

The $-\overset{+}{N}{\equiv}N$ group is rather unstable and in the case of ethyldiazonium ion, decomposition occurs instantaneously:

$$CH_3CH_2{-}\overset{+}{N}{\equiv}N \longrightarrow N_2 + [CH_3\overset{+}{C}H_2] \xrightarrow{H_2O} CH_3CH_2OH + H^+$$

$$[CH_3\overset{+}{C}H_2] \longrightarrow CH_2{=}CH_2 + H^+$$

When the $-\overset{+}{N}{\equiv}N$ group is attached to a benzene ring, though, the ion is stabilized to some extent by the delocalized electrons of the ring. Benzenediazonium ion is therefore much more stable than its aliphatic counterparts, though it nevertheless decomposes readily above 10°C. Aromatic diazonium compounds are considered further in the next section.

REACTIONS OF THE AROMATIC RING IN PHENYLAMINE

The lone pair of electrons on the nitrogen atom in phenylamine tends to get partly delocalized round the ring; we have already seen how this reduces the basic strength of phenylamine relative to aliphatic amines. Another result is that the electron density round the ring in phenylamine is considerably enhanced: this makes it undergo electrophilic substitution much more readily than benzene. For example, phenylamine reacts with bromine water even in the absence of a halogen carrier:

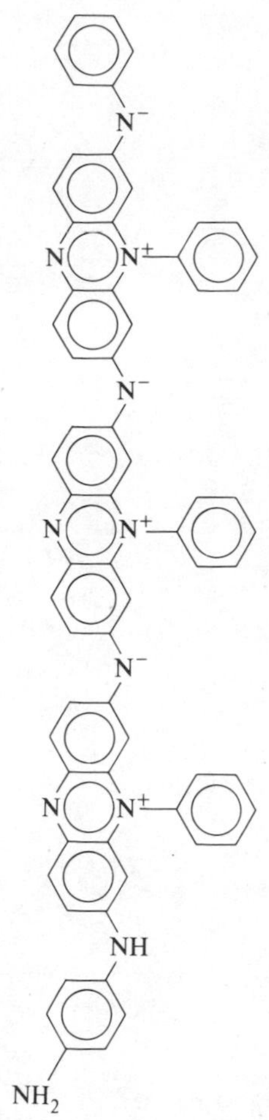

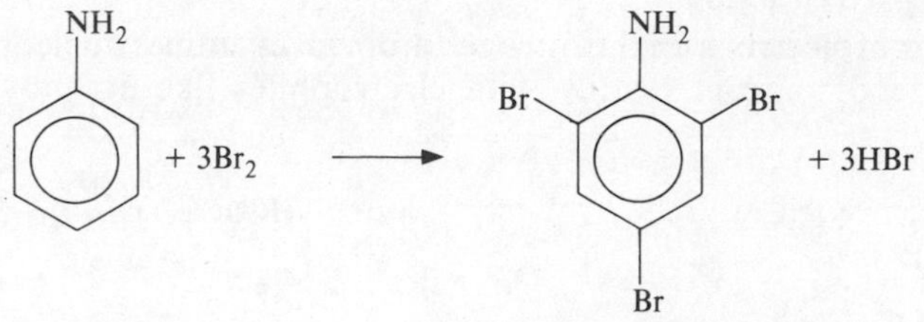

$$C_6H_5NH_2 + 3Br_2 \longrightarrow C_6H_2Br_3NH_2 + 3HBr$$

2,4,6-tribromophenylamine

The high electron density in the phenylamine molecule renders it very susceptible to oxidation. Pure phenylamine is colourless, but it quickly darkens due to atmospheric oxidation. Chemical oxidizers can convert it to a host of different products, including the important pigment Aniline Black (figure 32.3).

An important ring-substituted derivative of phenylamine is 4-aminobenzenesulphonamide:

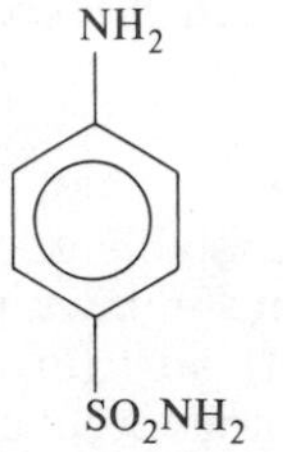

$NH_2$ / $SO_2NH_2$

This compound and other compounds derived from it are the active agents in the **sulphonamides**, one of the earliest groups of anti-bacterial drugs. First used in 1935, the sulphonamide drugs played an important part in the decline of puerperal or child-bed fever, once a major cause of death in childbirth (see figure 32.4). They also helped to achieve the vast reduction in deaths from pneumonia and tuberculosis over the last 40 years. They are still being used today though they have been largely superseded by antibiotics such as penicillin.

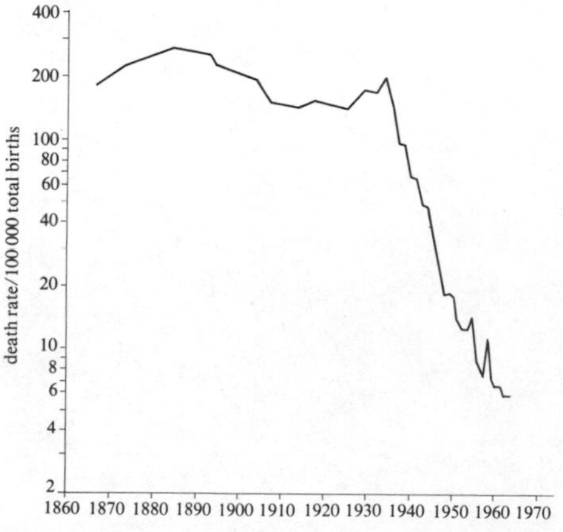

$N^-$ $N^+$ $NH$ $NH_2$

**Figure 32.3** The pigment Aniline Black is a mixture of several substances. The structure of one of them is believed to be that shown here (the + and − charges are effectively delocalized over the molecule, which is electrically neutral overall).

death rate/100 000 total births

400 200 100 80 60 40 20 10 8 6 4 2

1860 1870 1880 1890 1900 1910 1920 1930 1940 1950 1960 1970

**Figure 32.4** Maternal mortality death rate from childbed fever. Deaths per 100 000 total births, 1850–1964, in England and Wales (logarithmic scale). Sulphonamides were introduced in 1935.

## 32.7 Diazonium salts

The only stable diazonium salts are aromatic ones (section 32.6), and even these are not particularly stable. In aqueous solution benzenediazonium chloride decomposes above about 5–10°C, and the compound is explosive when solid. Its reactive nature makes it useful in synthesis, however.

Benzenediazonium chloride is prepared by adding a cold solution of sodium nitrate(III) to a solution of phenylamine in concentrated hydrochloric acid below 5°C:

$$C_6H_5NH_2 + HNO_2 + HCl \longrightarrow C_6H_5N_2^+Cl^- + 2H_2O$$

Owing to the explosive nature of the solid, the compound is always used in solution.

Consider the structure of the benzenediazonium ion, $C_6H_5—\overset{+}{N}{\equiv}N$.

- What stable molecule can be eliminated from the ion?
- What would be the structure of the product left when this molecule is eliminated?
- What kind of reaction would this product tend to undergo?

The benzenediazonium ion reacts readily with nucleophiles:

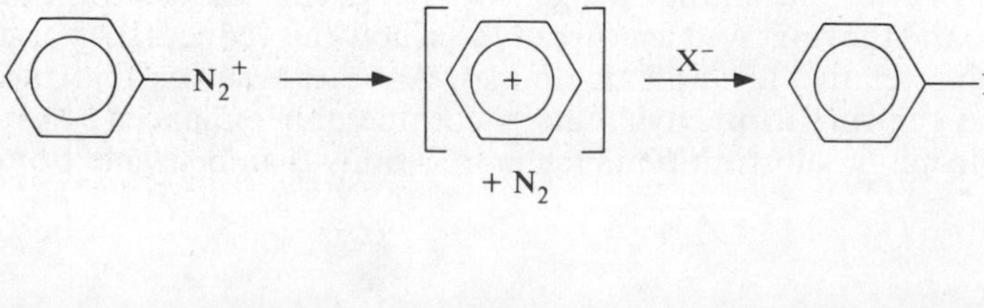

$$C_6H_5N_2^+ \longrightarrow [C_6H_5^+] + N_2 \xrightarrow{X^-} C_6H_5X$$

The introduction of the diazonium group is therefore a means of rendering the aromatic ring susceptible to *nucleophilic* substitution (the characteristic mode of reaction of an aromatic ring is normally *electrophilic* substitution). This makes diazonium compounds useful in synthesis. For example:

$$C_6H_5N_2^+ + I^- \longrightarrow C_6H_5I + N_2$$ (warm benzenediazonium chloride with KI solution)

$$C_6H_5N_2^+ + H_2O \longrightarrow C_6H_5OH + N_2 + H^+$$ (warm the aqueous solution)

$$C_6H_5N_2^+ + Cl^- \longrightarrow C_6H_5Cl + N_2$$ (warm benzenediazonium chloride with CuCl catalyst)

### COUPLING REACTIONS

The positive charge on the $—N_2^+$ group of the benzenediazonium ion means that this group is itself a strong electrophile. Thus we might expect it to attack another benzene ring, particularly one that has an electron-donating group such as —OH attached to it:

$$C_6H_5—\overset{+}{N}\equiv N + C_6H_5—OH \longrightarrow C_6H_5—N=N—C_6H_4—OH + H^+$$

(4-hydroxyphenyl)azobenzene

This is an example of a **coupling reaction**. If a cold solution of benzenediazonium chloride is added to a cold solution of phenol in sodium hydroxide, a bright orange precipitate is immediately formed. This is (4-hydroxyphenyl)azobenzene. It is an example of an **azo compound**: many different azo compounds can be formed by coupling reactions between diazonium compounds and activated aromatic rings. They are all brightly coloured, their colour being associated with the extensive delocalized electron systems they possess, delocalization extending from one ring through the —N=N— group to the next ring.

Manufacture of a dye, *c.* 1925.

Coupling reactions are important in the dyestuffs industry in the production of azo dyes. Many different colours can be obtained by adjusting the structure of the azo compound, which can be quite complex as figure 32.5 shows.

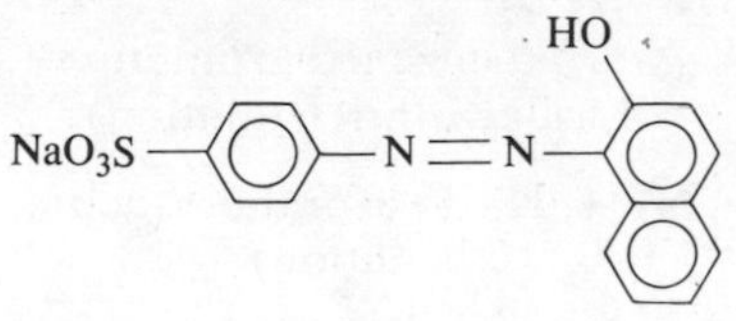

'Acid Orange 7' (bright reddish-orange)

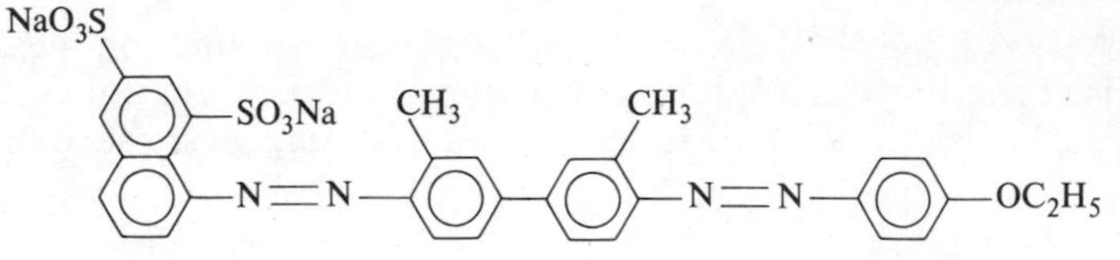

'Direct Red 39' (bluish-red)

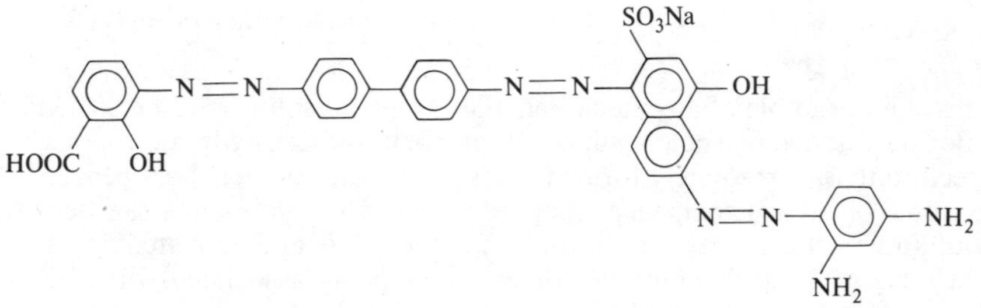

'Direct Brown 57' (reddish-brown)

**Figure 32.5** Structures of some azo dyes.

Unlike diazonium compounds, azo compounds are quite stable, so they do not fade or lose their colour.

Write the structures of the products you would expect to be formed at each stage when:

- O phenylamine is dissolved in excess concentrated hydrochloric acid,
- O sodium nitrate(III) solution is added to the cooled solution from the first stage,
- O the product from the second stage is added to a fresh solution of phenylamine.

## 32.8 Amino acids and proteins

All proteins are chemically similar, but they perform many different functions in living things. They act as structural materials (e.g. skin and fingernails), they make up the muscle fibres that enable animals to move, and they are the basic material of the enzymes that catalyse the many chemical reactions that are the driving force of all organisms. Protein molecules are chemically very similar, yet their diverse functions mean there must be many different ways of putting such molecules together.

Proteins are long-chain molecules made by linking together relatively small molecules called amino acids (see p. 513). There are 22 different amino acids found in nature and many proteins contain several thousand amino acid units. For a protein containing 5 000 such units the number of possible arrangements using 22 different amino acids is $22^{5000}$ (i.e. about $10^{6700}$), so it is not surprising that proteins come in many different forms. Indeed, the surprising thing is that a given organism can control precisely the form of protein it produces.

Amino acids were introduced in section 32.2. Table 32.4 gives some examples of naturally-occurring amino acids. All amino acids except aminoethanoic acid have

**Table 32.4** Some naturally-occurring amino acids. In this table X stands for the —CH($NH_2$)COOH group common to all these amino acids.

| Formula | Common name |
|---|---|
| X—H | glycine |
| X—$CH_3$ | alanine |
| X—$CH_2$COOH | aspartic acid |
| X—$CH_2$ 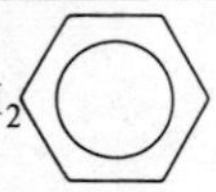 | phenylalanine |
| X—$CH_2$OH | serine |
| X—$(CH_2)_4NH_2$ | lysine |
| X—$CH_2$SH | cysteine |

four different groups attached to the central carbon atom (see figure 32.6); they therefore exhibit optical isomerism. All naturally occurring amino acids exist as the L isomer.

Amino acids carry at least two functional groups, $—NH_2$ and —COOH. They therefore show the properties of both amines and carboxylic acids. Consider the simplest amino acid, aminoethanoic acid (glycine), $H_2NCH_2COOH$.

- ○ Write the structure of the ion formed by aminoethanoic acid when it reacts with a base.
- ○ Write the structure of the ion formed when it reacts with an acid.
- ○ Could aminoethanoic acid form an ion under neutral conditions?

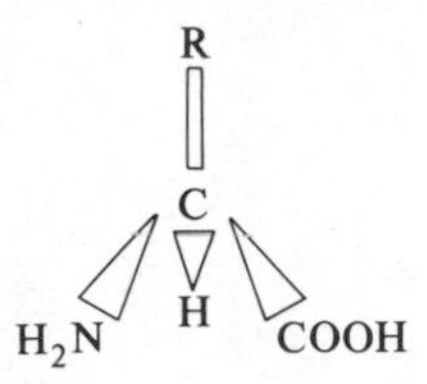


**Figure 32.6** The asymmetry of amino acid molecules.

In neutral solution and in the solid state, aminoethanoic acid exists as a dipolar ion:

$$H_3\overset{+}{N}—CH_2—COO^-$$

The ion is formed as a result of an internal acid-base reaction: the COOH group donates a proton to the $NH_2$ group. This kind of ion is called a **zwitterion**, from the German word 'zwei' meaning two. The ionic character of aminoethanoic acid accounts for its high solubility and high melting point (232°C).

Aminoethanoic acid can thus exist in three forms, depending on the pH:

| $H_3\overset{+}{N}—CH_2—COOH$ | $H_3\overset{+}{N}—CH_2—COO^-$ | $H_2N—CH_2—COO^-$ |
|---|---|---|
| acid conditions | neutral conditions | basic conditions |

Other amino acids behave similarly, though the pH ranges in which the three forms exist differ for different amino acids, the situation being complicated by the presence in some amino acids of $NH_2$ and COOH groups elsewhere in the molecule.

### THE PEPTIDE LINK

The amino acids in a protein chain are linked by the elimination of a molecule of water between the $—NH_2$ of one amino acid and the —COOH of the next. Such a link is called a **peptide link**:

$$H_2N—\underset{H}{\overset{R_1}{C}}—C(=O)OH \; + \; H_2N—\underset{H}{\overset{R_2}{C}}—C(=O)OH$$

$$\downarrow$$

$$H_2N—\underset{H}{\overset{R_1}{C}}—\underbrace{\overset{O}{\overset{\|}{C}}—\overset{H}{\overset{|}{N}}}_{\text{peptide link}}—\underset{H}{\overset{R_2}{C}}—C(=O)OH$$

$$+ \; H_2O$$

This kind of linkage is not easily formed under laboratory conditions, but it forms readily in living systems under the influence of enzymes. Long chains of amino acids, called **polypeptides**, can be formed in this way. Elaboration of the chain leads to the formation of proteins, the structure of which is considered further in section 8.8.

The formation of a peptide link between amino acids is reversible: proteins can be hydrolyzed to reform amino acids. Hydrolysis can be achieved by boiling with aqueous acid, but is carried out more effectively by digestive enzymes like pepsin and trypsin.

Hydrolysis is used to identify the amino acids present in a protein. The hydrolyzed protein is subjected to paper chromatography, whereupon the individual amino acids separate and can be identified.

Buttons made from casein, a protein obtained from milk.

## 32.9 Nylon

Naturally occurring fibres like wool, cotton and silk are gradually being replaced by synthetic fibres which are cheaper and often have better wearing and washing characteristics. Consumers tend to be conservative, though, and the demand is for synthetic fibres that feel and look like natural ones. It is not therefore surprising that several synthetic fibres are structurally similar to natural ones.

Wool and silk are protein fibres, cotton is cellulose. The commonest man-made fibres in Britain are in fact only semi-synthetic, being based on natural cellulose. These are the fibres known as rayon (viscose) and cellulose triacetate ('Tricel'). They are made by chemically regenerating cellulose (wood pulp, straw etc.) in a form suitable for use as a textile. (See section 8.8. for the structure of cellulose.)

Nylon, one of the major wholly synthetic fibres, has some structural similarity to protein fibres like wool and silk. Like proteins, nylon is a condensation polymer. The monomer units are joined by a link similar to the peptide link. There are several forms of nylon, one of the commonest being nylon 66, which is made by a condensation reaction between 1,6-diaminohexane and hexanedioic acid. The product is named nylon 66 because both monomers contain 6 carbon atoms.

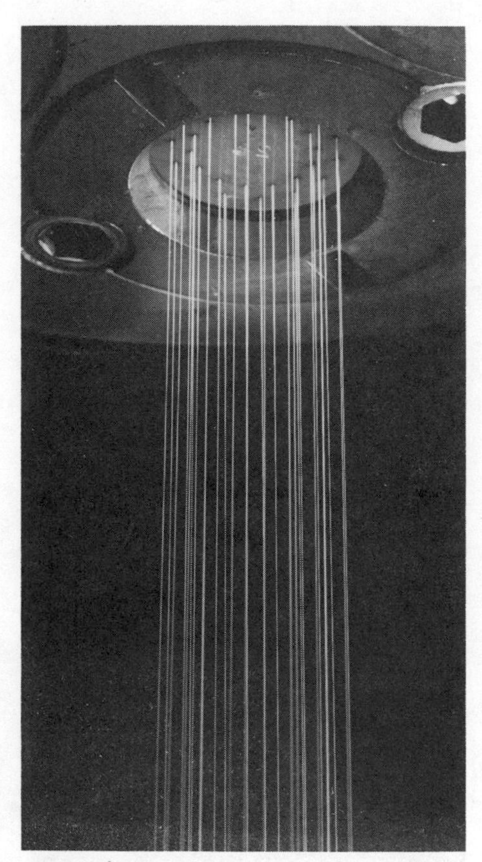

Extruding nylon through spinnarets.

$$\underset{\text{1,6-diaminohexane}}{H_2N{-}(CH_2)_6{-}NH_2} \quad \underset{\text{hexanedioic acid}}{HOOC{-}(CH_2)_4{-}COOH} \quad \underset{\text{1,6-diaminohexane}}{H_2N{-}(CH_2)_6{-}NH_2}$$

$$\downarrow$$

$$-HN{-}(CH_2)_6{-}NH{-}\underset{\|\atop O}{C}{-}(CH_2)_4{-}\underset{\|\atop O}{C}{-}NH{-}(CH_2)_6{-}NH-$$

$$+\ H_2O \qquad\qquad + H_2O$$

The structural similarity of this polymer to a polypeptide is evident. The —CONH— grouping is also present in amides (see section 32.2), so nylons are sometimes given the general name **polyamides**. The 1,6-diaminohexane and hexanedioic acid used in this process are obtained from phenol or cyclohexane by the route outlined in figure 32.7.

phenol $\xrightarrow{H_2}$ cyclohexanol + cyclohexanone

cyclohexane $\xrightarrow{O_2}$ cyclohexanol + cyclohexanone

cyclohexanol + cyclohexanone $\xrightarrow[\text{oxidize}]{HNO_3}$ hexanedioic acid ($C_4H_8(COOH)_2$: COOH, COOH) $\xrightarrow{NH_3}$ ammonium hexanedioate ($COO^-NH_4^+$, $COO^-NH_4^+$)

ammonium hexanedioate $\xrightarrow{-4H_2O}$ hexanedionitrile (CN, CN) $\xrightarrow{H_2}$ 1,6-diamino-hexane ($CH_2NH_2$, $CH_2NH_2$)

1,6-diaminohexane + hexanedioic acid $\xrightarrow{\text{heat}}$ $\left[\ CONH(CH_2)_6NHCO(CH_2)_4\ \right]_n$ nylon 66

**Figure 32.7** Outline of the manufacture of 1,6-diaminohexane, hexanedioic acid, and hence nylon 66.

The polymer produced in this way is melted and extruded through small holes called spinnarets: as the jets emerge they are cooled and solidify. The fibres so produced contain molecular chains that are tangled and disarrayed: the fibres are now stretched or **drawn**, with the result that the molecules become more aligned and the fibre is strengthened (see figure 32.8).

Nylon 66 can be more readily produced on a laboratory scale by activating the hexanedioic acid by converting it to hexanedioyl chloride. You will recall from chapter 29 that acyl chlorides are more reactive than the carboxylic acids from which they are produced. Hexanedioyl chloride reacts readily with 1,6-diaminohexane at room temperature to form nylon 66, with the elimination of HCl.

$$\underset{\text{1,6-diaminohexane}}{H_2N{-}(CH_2)_6{-}NH_2} \quad \underset{\text{hexanedioyl chloride}}{ClOC{-}(CH_2)_4{-}COCl} \quad \underset{\text{1,6-diaminohexane}}{H_2N{-}(CH_2)_6{-}NH_2}$$

$$\downarrow$$

$$-HN-(CH_2)_6-NH-\underset{\overset{\|}{O}}{C}-(CH_2)_4-\underset{\overset{\|}{O}}{C}-NH-(CH_2)_6-NH- \quad + HCl \qquad + HCl$$

This reaction forms the basis of the well-known 'nylon rope trick' (see figure 32.9). The reaction is not used industrially because of the high cost of hexanedioyl chloride.

Despite its structural similarity to wool, nylon lacks the softness and moisture-absorbing properties of the natural fibre. It is, however, harder wearing and has good 'wash and wear' characteristics. One of the earliest uses of nylon was as a substitute for silk in the manufacture of stockings: the elasticity and strength of the fibre make it ideal for this purpose. About 75% of UK nylon consumption goes into clothing. Other uses include tufted carpets, tyre cords and machine gear wheels and bearings.

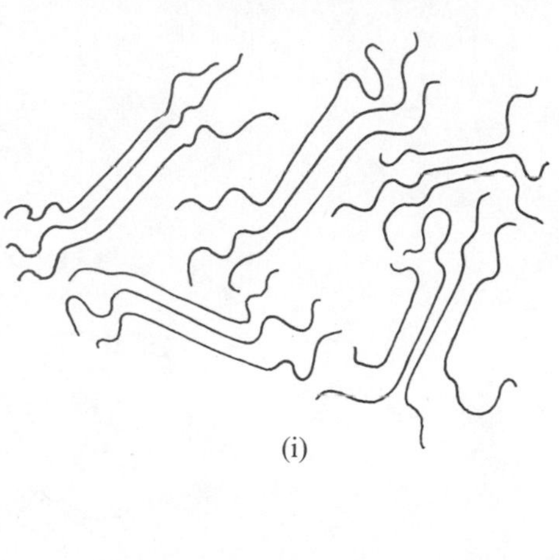


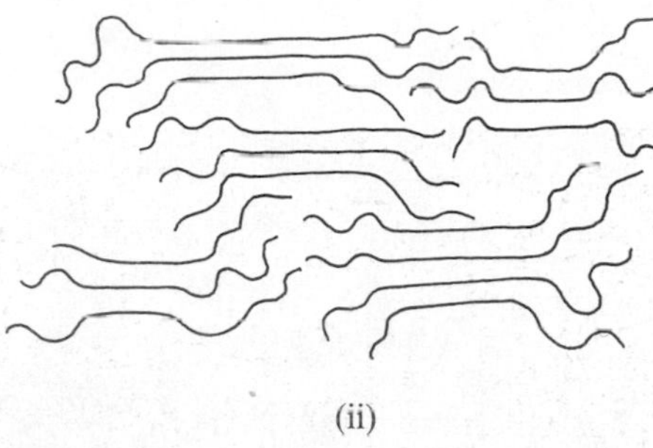


**Figure 32.8** Arrangement of polymer chains, (i) before, and (ii) after drawing.

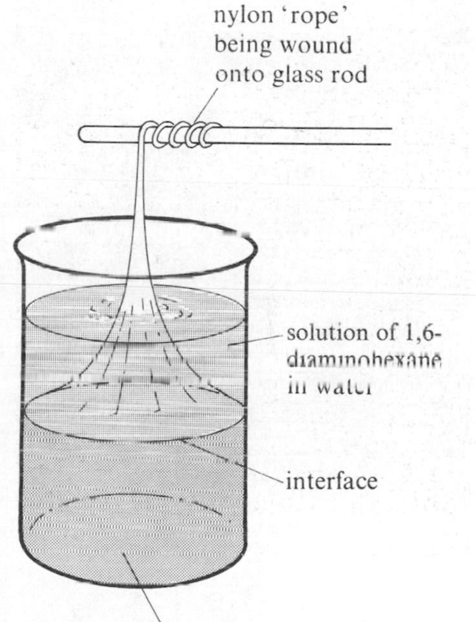


**Figure 32.9** The nylon rope trick. The hexanedioyl chloride and 1,6-diaminohexane react at the interface between the aqueous and organic layers, and a nylon 'rope' can be continuously pulled out.

In the 1940's nylon was newly discovered and nylon stockings were a luxurious novelty.

## Summary

1 Primary amines have the structure $RNH_2$. They are named using the suffix *-ylamine*.

2 Four ways of making amines are summarized below.

(a) $RCl \xrightarrow{NH_3} RNH_2$

(b) $RNO_2 \xrightarrow{\text{reduce}} RNH_2$ (R = aryl)

(c) $RCN \xrightarrow{\text{reduce}} RCH_2NH_2$

(d) $RCONH_2 \xrightarrow{\text{reduce}} RCH_2NH_2$

3 Amines, particularly the early members of the series, show considerable physical and chemical similarity to ammonia.

4 Amines are basic. Alkyl amines are stronger bases than ammonia, aryl amines are weaker.

5 This diagram shows some important reactions of amines.

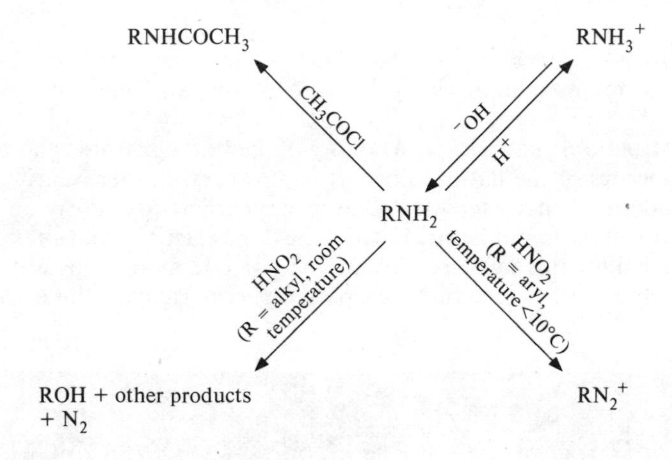


6 This diagram shows some important reactions of benzenediazonium chloride.

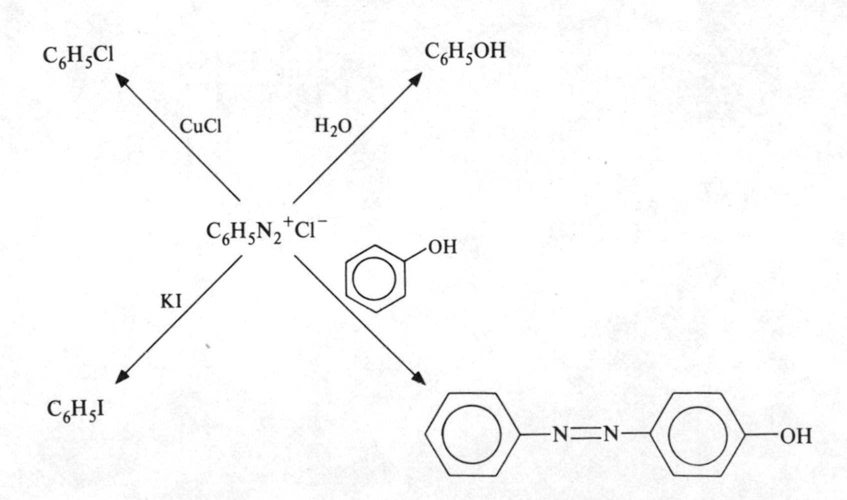


7 Amino acids contain both —$NH_2$ and —COOH groups and can behave as bases and as acids.

8 Proteins contain chains of amino acids joined by a condensation linkage between the —COOH of one amino acid and the —$NH_2$ of the next. This is called a peptide link.

9 Nylons are condensation polymers containing the —CONH— group. Nylon 66 is made by a condensation reaction between $H_2N(CH_2)_6NH_2$ and $HOOC(CH_2)_4COOH$.

## Study questions

1 Consider the following compounds:

**A** $CH_3CHCH_3$ with $CN$ on the central carbon

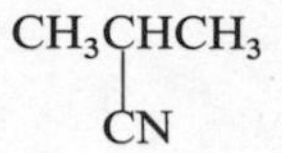

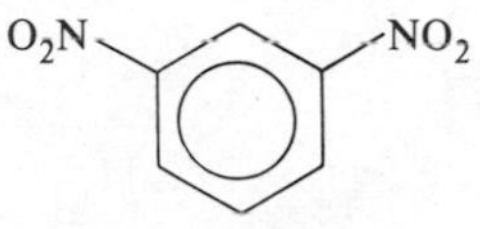

**B** $CH_3CH_2CH_2NH_2$

**C** $O_2N$ $NO_2$ (benzene ring)

**D** $CH_3$ $NH$ $CH_3CH_2$

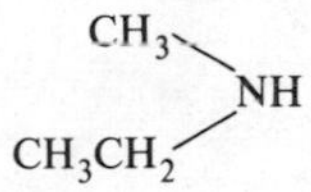

**E** $CH_3CH_2CONH_2$

**F** $(CH_3CH_2CH_2)_3N$

(a) Which is a primary amine?
(b) Which is a nitrile?
(c) Which is an amide?
(d) Which is a tertiary amine?
(e) Name each compound.

2 (a) Place the following in expected order of basic strength. Give reasons for your answer.

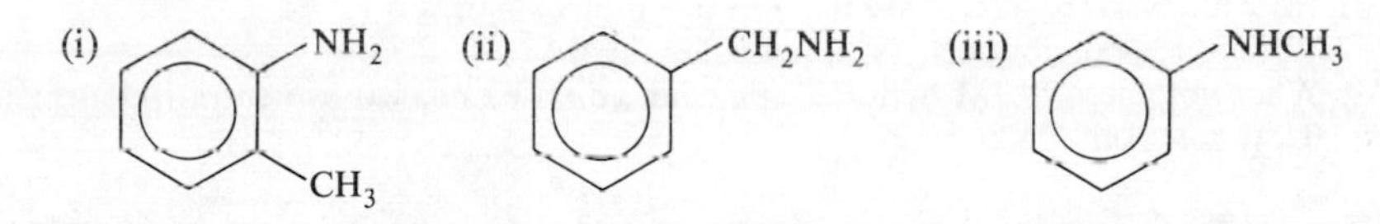


(b) Place the following in expected order of boiling point. Give reasons for your answer.

(i) $CH_3CH_2CH_2NH_2$ (ii) $CH_3CH_2CH_2CH_3$ (iii) $CH_3CH_2CH_2OH$

3 Draw up a table comparing the principle physical and chemical properties of methylamine and ammonia. Do the two compounds in general behave similarly? Do they show more or less similarity than methanol and water show?

4 Consider the following compounds:

**A** $CH_3$ $NH_2$ (benzene ring)

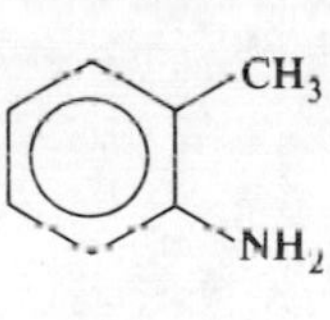

**B** $CH_3CH_2NH_2$

**C** $(CH_3CH_2)_2NH$

**D** $H_2N(CH_2)_5NH_2$

(a) Which would be among the products of the reaction of chloroethane with ammonia?
(b) Which would be converted to a diazonium compound by the action of nitric(III) acid below 10°C?
(c) Which could be made by reduction of ethanonitrile?
(d) Which would react with hydrochloric acid in the ratio one mole of the compound to two moles of hydrochloric acid?
(e) Which is the weakest base?
(f) Which could be one of the reagents in the manufacture of a nylon (polyamide)?

5 Suggest explanations for the following.
(a) Ethylamine can be readily made from chloroethane and ammonia, but it is difficult to make phenylamine from chlorobenzene and ammonia.
(b) If electrodes are placed in a solution of aminoethanoic acid at pH 0 the aminoethanoic acid migrates to the cathode. If the same experiment is repeated at pH 14 the aminoethanoic acid migrates to the anode.
(c) Phenylamine is much more soluble in dilute hydrochloric acid than in water.

6 In the following, explain how you would convert the first compound to the second. Each conversion may involve one or more steps.

(a) $CH_2{=}CH_2$ to $H_2NCH_2CH_2NH_2$

(b)

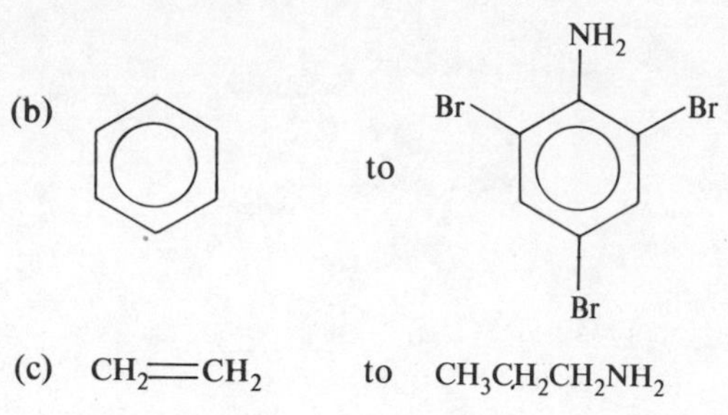


(c) $CH_2{=}CH_2$ to $CH_3CH_2CH_2NH_2$

(d) $CH_3CH_2NH_2$ to $CH_3CH_2NH{-}C({=}O){-}C_6H_5$

(e) $C_6H_5NO_2$ to $C_6H_5OH$

7 A compound **X** containing carbon, hydrogen and nitrogen only has relative molecular mass 88. When reacted with nitric(III) acid, 0.1g of **X** released 50.9cm$^3$ of nitrogen gas, measured at s.t.p.

(a) How many moles of nitrogen gas are produced by 1 mole of **X** when it reacts with nitric(III) acid?
(b) How many $NH_2$ groups are there in one molecule of **X**?
(c) Write three possible structural formulae for **X**.
(d) What volume of 0.1M hydrochloric acid would be needed to neutralize 50cm$^3$ of a 0.2M solution of **X**?

8 Copy out the following reaction sequences, inserting the formulae of the products formed in the blank spaces.

(a) $CH_3COCl \xrightarrow{NH_3} \quad \xrightarrow{LiAlH_4} \quad \xrightarrow{HCl}$

(b) $CH_3CH_2CH_2CH_2Br \xrightarrow{NH_3}$ (three products), one of which is $CH_3CH_2CH_2CH_2NH_2 \xrightarrow{CH_3COCl}$

(c) 2-nitrophenylamine ($C_6H_4(NH_2)NO_2$) $\xrightarrow[HCl\ 5°C]{NaNO_2,} \quad \xrightarrow[warm]{KI(aq)}$

9 Suppose a molecule of each of the amino acids serine, lysine and alanine are joined together by peptide links to form a tripeptide.

(a) How many different tripeptides are possible?
(b) Write the structure of each different tripeptide. (Refer to table 32.4 for the structures of these amino acids.)

10 A compound **A** of molecular formula $C_7H_8$ was treated with a mixture of concentrated nitric(V) acid and concentrated sulphuric(VI) acid to form two isomeric compounds, one of which was **B**, $C_7H_7O_2N$. **B** could be converted by reduction to **C**, $C_7H_9N$. **C** reacted with dilute hydrochloric acid to form **D**, $C_7H_{10}NCl$. When cold sodium nitrate(III) solution was added to a cold solution of **D** in hydrochloric acid, a solution of **E**, $C_7H_7N_2Cl$, was produced. When a cold solution of **E** was added to a cold solution of phenol, a brightly-coloured substance **F** was produced; when the solution **E** was warmed alone an unreactive gas was evolved and **G**, $C_7H_8O$, was formed.
Give possible structural formulae for **A** to **G**.

# Index

# Acknowledgements

Acknowledgement is due to the following for permission to use photographs:

The Science Museum (pp.1 *centre right*, 2, 9 *below right*, 14, 41, 70 *below and top*, 80, 83 *right*, 109, 228, 277, 278 *below*, 279 *top right*, 282 *top*, 293 *both*, 386, 387, 424 *bottom*, 425, 427, 453 *bottom*, 511)
Mansell Collection (pp. 1 *centre left*, 9 *top right and centre*, 131, 218, 237, 442)
Dr. D. W. Thompson (pp.1, 83 *left*)
National Westminster Bank (p.4)
Keystone Press (pp. 5, 40, 65, 72, 148, 161, 196, 249 *centre*, 250 *top left*, 282 *bottom*, 351, 399 *bottom*, 400, 428 *left*, 433, 495, 505)
John Freeman (pp.10, 20, 21 (by courtesy of the Royal Institution))
Ron Chapman (pp.12, 53, 219)
Imperial Chemical Industries Limited (pp.13 *top*, 222, 328 *bottom left*, 354, 378, 416 *bottom*, 417 *left and centre*, 444)
Bryn Campbell (pp.17, 246)
Novosti Press Agency (p.24)
Professor F. G. A. Stone (p.26)
Dr. J. H. Holloway (pp.27, 61)
Atomic Energy Research Establishment Harwell (p.43)
Aerofilms Limited (p.44)
Cavendish Laboratory, University of Cambridge (pp.39, 46)
Camera Press (pp.62 *right and left*, 72, 128, 219 *bottom right*, 436, 487, 497)
Mullard Limited (p.33)
Ullstein (p.65)
U.K. Atomic Energy Authority (pp.71, 74, 75, 76, 142, 217)
British Museum (Natural History) (p.77)
Heather Angel (pp.99, 336, 337)
John Hillelson Agency (pp.100, 215 *bottom right and left*, 234, 235 *bottom*)
Professor Watson Fuller (p.104)
North East London Polytechnic (p.110)
Louis R. Wolberg (courtesy of Dover Publications Incorporated) (p.115)
Mayflower Studio (p.116)
Mary Evans Picture Library (pp.126, 160, 236, 323, 329, 459)
Shell (pp.13 *bottom*, 378 *left*, 390, 401 *top*, 428 *right*)
NASA (p.148)
Associated Press Limited (pp.185, 197, 396)
Dr. J. D. Dodge (p.204)
Department of the Environment (p.215)
Alcan Limited (pp.223, 228)
British Aircraft Corporation (p.225)
John Antill (p.226)
The British Aluminium Company Limited (p.229)
The British Tourist Authority (pp.231, 294)
The Chilean Nitrate Industry (p.232)
Hodder and Stoughton Limited (p.240)
Bruce Coleman Limited (pp.247, 300, 335)
John Walmsley (pp.250 *bottom right and bottom left*, 369, 378 *right*, 392, 420)
Elisabeth Photo Library (p.251)
Chemical Defence Establishment Portondown (p.250)
Australian News Information (p.273)
Homer Sykes (pp.279 *middle*, 298)
British Steel Corporation (pp.278, 279 *top*, 281, 285)
Fotofirst (pp. 286, 287 *both*)
Rio Tinto Zinc Photographic Library (p.278 *top left*)
Mark Edwards (pp. 299, 443)
BASF (pp.324 *right*, *centre*, 325 *right and left*, 504)
Nicholas Horne (p.328 *bottom right*)
Tony Othen (pp.379, 393, 399, 411, 457, 465, 480, 486, 496)
Punch (p.395)
BP Chemicals Limited (pp.416 *top*, 417 *right*, 446 *right*, 513)
London Fire Brigade (p.413)
Glaxo Limited (p.446 *left*)
Metropolitan Police (p.464)
Radio Times Hulton Picture Library (pp.424, 466)
Fox Photo's Limited (p.453 *top*)
Claridge Communications Limited (p.475)
Gareth J. Jones (pp.419, 473)
Unilever (p.490)
Leyland cars (p.400)
Time-Life Inc. (p.416 *centre*)
Royal College of Surgeons (p.460)
Will Green (p.491)
National Portrait Gallery (p.503)
Courtaulds Limited (p.514)
CERN (p.71)
De Beers (pp.116, 249 *bottom right*)
Associated Octel Co Ltd (p.185)
United Press International (p.235 *top*)
ICI Agricultural Division (p.257)
Stewart Keating (p.277 *top*)
Industrial Diamond Information Bureau (p.116)
David Alexander Simson (p.401)
Director of the Institute of Geological Sciences (p.215 *top right*)
ARC Weed Research Organisation (p.370)
Blue Circle Cement (p.302)
Brunel University (p.382)
*Cover photograph*; John G. Ross, courtesy of Robert Harding Associates